Beginning AutoCAD® 2026

EXERCISE WORKBOOK

Beginning AutoCAD® 2026

EXERCISE WORKBOOK

by
Cheryl R. Shrock
Professor, retired
Drafting Technology
Orange Coast College, California
Autodesk Authorized Author

Updated for AutoCAD® 2026
and Revised by
Steve Heather
Former Lecturer of
Machine Shop Engineering and
Computer Aided Design.
AutoCAD Beta Tester

INDUSTRIAL PRESS, INC.

Industrial Press, Inc.
1 Chestnut Street
South Norwalk, Connecticut 06854 U.S.A.
Phone: 203-956-5593
Toll-Free in USA: 888-528-7852
Email: info@industrialpress.com

Title: *Beginning AutoCAD© 2026 Exercise Workbook*
Authors: Cheryl Shrock and Steve Heather
Library of Congress Control Number: On file with the Library of Congress

ISBN (print): 978-0-8311-3700-7
ISBN (eBook): 978-0-8311-9673-8
ISBN (ePub): 978-0-8311-9674-5
ISBN (eMobi): 978-0-8311-9675-2

Publisher: Judy Bass
Editor: Laura Brengelman
Cover Designer: Jeff Weeks

Printed and bound in the United States of America

Beginning AutoCAD® 2026
First Printing

books.industrialpress.com
ebooks.industrialpress.com

As always, many thanks are due to Cheryl Shrock for allowing me to continue on with her Exercise Workbook series. I also thank all at Industrial Press and Autodesk who support my efforts in updating and revising these invaluable teaching and learning texts.

Steve Heather

AutoCAD® Books and eBooks from Industrial Press

Beginning AutoCAD **2018** .... ISBN 978-0-8311-3615-4
Beginning AutoCAD **2018** eBook............ ISBN 978-0-8311-9437-6
Advanced AutoCAD **2018** .... ISBN 978-0-8311-3616-1
Advanced AutoCAD **2018** eBook............ ISBN 978-0-8311-9440-6

Beginning AutoCAD **2019** .... ISBN 978-0-8311-3626-0
Beginning AutoCAD **2019** eBook............ ISBN 978-0-8311-9473-4

Beginning AutoCAD **2020** .... ISBN 978-0-8311-3639-0
Beginning AutoCAD **2020** eBook............ ISBN 978-0-8311-9515-1

Beginning AutoCAD **2021** .... ISBN 978-0-8311-3659-8
Beginning AutoCAD **2021** eBook............ ISBN 978-0-8311-9579-3

Beginning AutoCAD **2022** .... ISBN 978-0-8311-3666-6
Beginning AutoCAD **2022** eBook............ ISBN 978-0-8311-9591-5
Advanced AutoCAD **2022** .... ISBN 978-0-8311-3667-3
Advanced AutoCAD **2022** eBook............ ISBN 978-0-8311-9594-6

Beginning AutoCAD **2023** .... ISBN 978-0-8311-3679-6
Beginning AutoCAD **2023** eBook............ ISBN 978-0-8311-9609-7

Beginning AutoCAD **2024** .... ISBN 978-0-8311-3686-4
Beginning AutoCAD **2024** eBook............ ISBN 978-0-8311-9627-1

Beginning AutoCAD **2025** .... ISBN 978-0-8311-3693-2
Beginning AutoCAD **2025** eBook............ ISBN 978-0-8311-9647-9

Beginning AutoCAD **2026** .... ISBN 978-0-8311-3700-7
Beginning AutoCAD **2026** eBook............ ISBN 978-0-8311-9673-8

AutoCAD Pocket Reference
8th Edition, Releases 2018–2019 .... ISBN 978-0-8311-3627-7
AutoCAD Pocket Reference
8th Edition, eBook .... ISBN 978-0-8311-9476-5

AutoCAD 3D Modeling............................ ISBN 978-0-8311-3613-0
AutoCAD 3D Modeling eBook .... ISBN 978-0-8311-9428-4

AutoCAD 2D Resource Kit
Digital Edition eBook.............................. ISBN 978-0-8311-9503-8

For information about these and other bestselling titles, visit:
books.industrialpress.com
ebooks.industrialpress.com

Table of Contents

Lesson 6

Lesson 7

Lesson 8

Lesson 9

Lesson 10

Lesson 11

Lesson 12

Lesson 13

Lesson 14

Lesson 15

Lesson 16

Lesson 17

Lesson 18

Lesson 19

Lesson 20

Lesson 21

Lesson 22

Lesson 23

Lesson 24

Lesson 25

Lesson 26

Lesson 27

Lesson 28

Lesson 29

Lesson 30

Appendixes

Index

Final Notes about AutoCAD®

INTRODUCTION

SETTING UP AUTOCAD

After reading and completing the instructions in this introductory chapter, you will have:

- Downloaded and saved the provided preset drawing files, so you can get started right away practicing drawing commands
- Configured your system for using AutoCAD 2026
- Customized your wheel mouse for efficiency in navigating and using the many AutoCAD commands and features

Introduction

About this Workbook

The *Beginning AutoCAD® 2026 Exercise Workbook* is designed for classroom instruction or self-study. It is suitable for both inch and metric users. There are 30 lessons. Each lesson starts with step-by-step instructions followed by exercises designed for practicing the commands you learned within that lesson.

You may find the order of instruction in this Workbook somewhat different from most textbooks. The approach we take is to first familiarize you with the drawing commands. After you are comfortable with the drawing commands, you will then be taught to create your own setup drawings.

To accomplish this method and enable you to get started using AutoCAD right away, we supply you with the preset drawings, **"inch-helper.dwg"** and **"metric-helper.dwg."** These drawings are ready for you to open and use as templates for your initial practice exercises. For the first eight lessons, you do not need to worry about settings, **you just learn how to use the commands you need to draw**.

How to get the supplied drawings?

The file **"workbook-helper.zip"** should be downloaded from our website:

https://books.industrialpress.com/autocad/workbook-helper.zip

Enter the address into your web browser and the download will start automatically. Once the file has been downloaded, you can unzip it to extract both drawing files.

Printing

The exercises in this Workbook have been designed to be printed on any **8-1/2" X 11" (letter size) or 210mm X 297mm (A4 size) paper**. You will be able to print your exercises and view your results as you progress through the lessons. If you would like to configure a large format printer, refer to **Appendix-A** to **"Add a Printer / Plotter."**

About the Authors

Cheryl R. Shrock is a retired Professor and Chairperson of Computer Aided Design at Orange Coast College in Costa Mesa, California, where she had taught since 1990. Previous to teaching, Cheryl owned and operated a commercial product and machine design company, where designs were created and documented using CAD. She also is an Autodesk® registered author. This Workbook was created with ground-breaking instructive materials based on a combination of her extensive teaching skills and industry experience.

Steve Heather is a former Lecturer of Machine Shop Engineering and Computer Aided Design in England, UK. Previous to teaching and for more than 30 years, he worked as a Precision Engineer in the aerospace and defense industries. For more than a dozen years, Steve has been a Beta Tester for Autodesk®, testing the latest AutoCAD® software. He also is a member of the AutoCAD Customer Council. Steve can be contacted for questions or comments about this book at info@industrialpress.com.

Configuring Your System

AutoCAD allows you to customize its configuration. While you are using this Workbook, it is necessary for you to make some simple changes to your configuration so our configurations are the same. This will ensure that the commands and exercises work as expected. The following instructions will guide you through these changes.

1. Start AutoCAD
   A. To start AutoCAD, follow the instructions on pages 1-2 through 1-4. Then return to this page and go to **Step 2**.

2.
   A. Type ***options*** (not case sensitive) anywhere within the main drawing area then press the ***<Enter>*** key.
   B. Type ***options*** on the Command Line then press the ***<Enter>*** key.

The text that you type will appear in the **Dynamic Input box** or on the **Command Line**, as shown below:

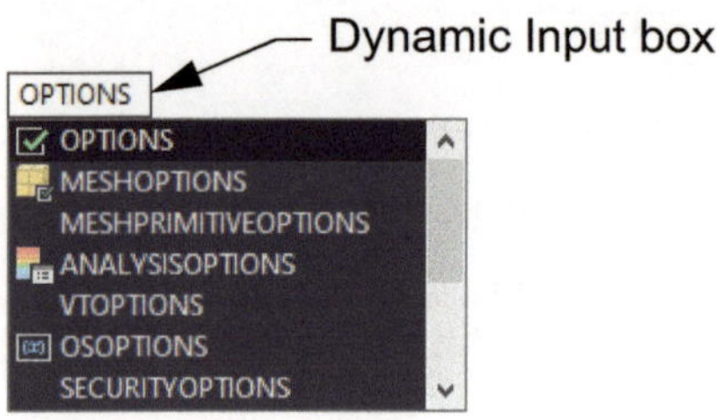

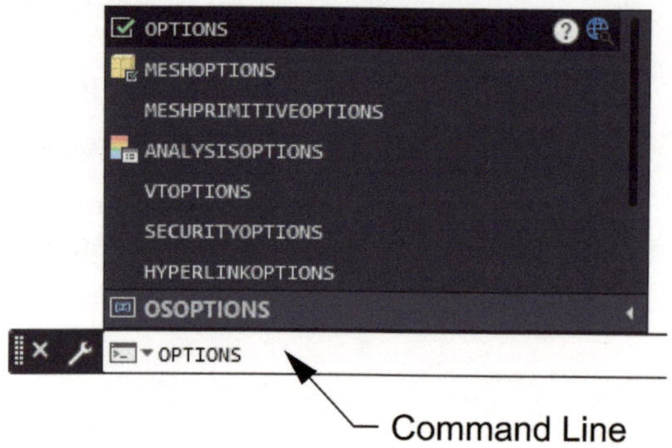

Note: AutoCAD LT users may find that some of the settings appear slightly different. But the settings in both AutoCAD and AutoCAD LT are mostly the same.

3. Select the **Display** Tab and change the settings on your screen to match the dialog box below.

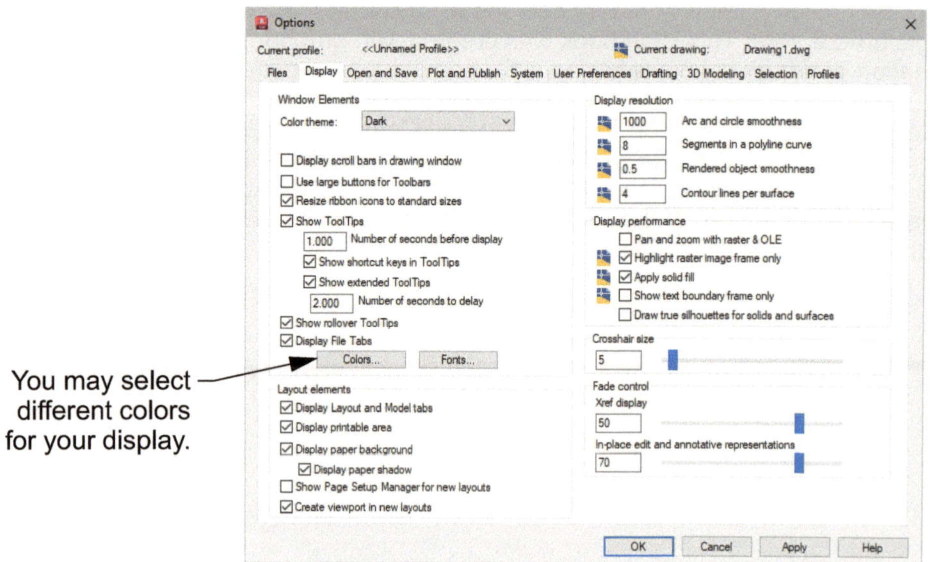

4. Select the **Open and Save** Tab and change the settings on your screen to match the dialog box below.

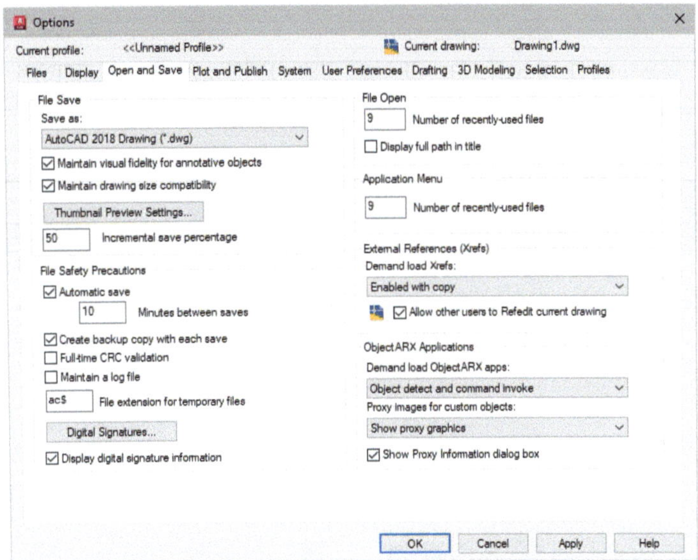

5. Select the **Plot and Publish** Tab and change the settings on your screen to match the dialog box below.

You may use this setting or select the actual name of your printer.

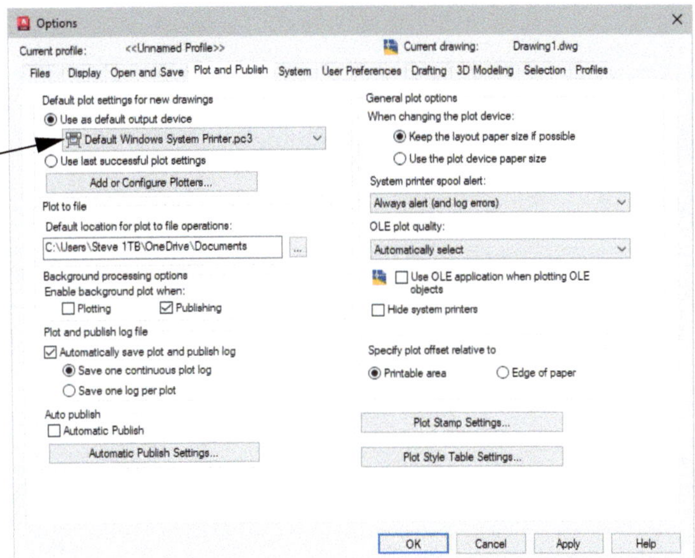

6. Select the **System** Tab and change the settings on your screen to match the dialog box below.

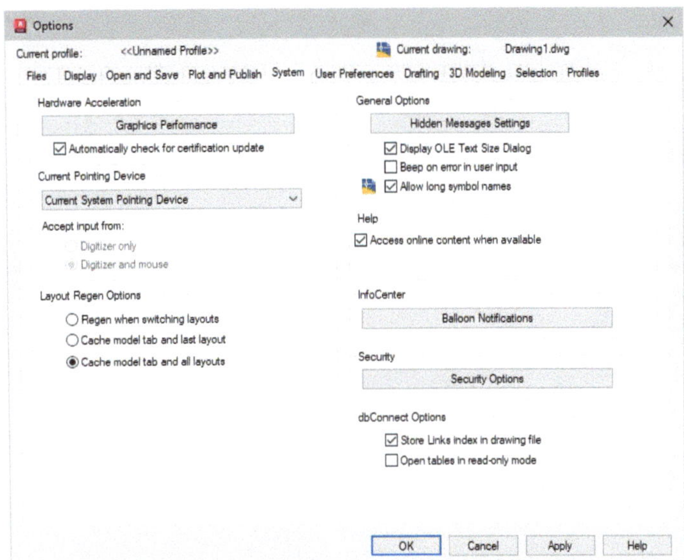

7. Select the **User Preferences** Tab and change the settings on your screen to match the dialog box below.

Right-click Customization...
Select this button and change the settings shown below.

Select **Inches** (Inch users).
or
Select **Millimeters** (Metric users).

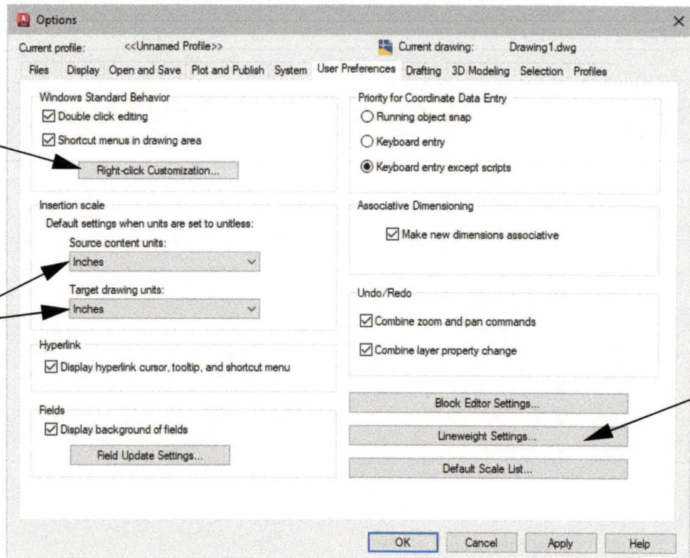

Lineweight Settings...
Select this button and change the settings shown below.

8. After making the setting changes shown below, select the **Apply & Close** button.

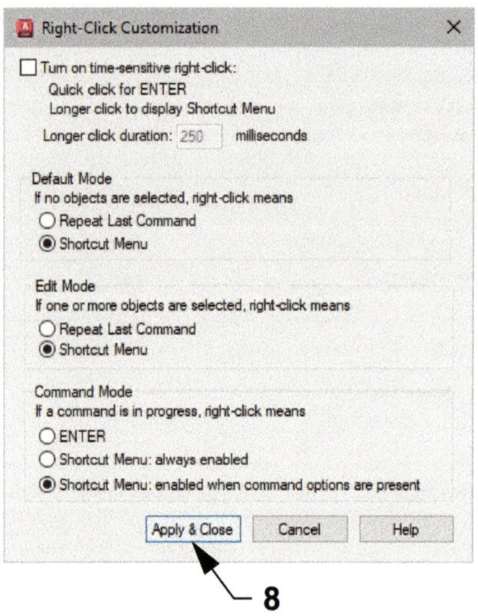

Select **Millimeters** (Metric users).

Select **Inches** (Inch users).

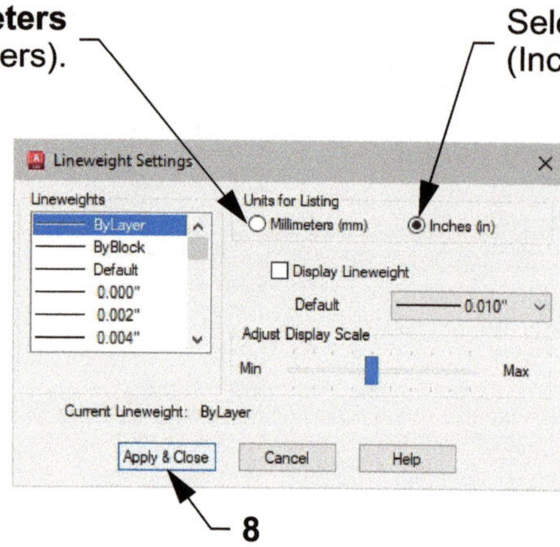

9. Select the **Drafting** Tab and change the settings on your screen to match the dialog box below.

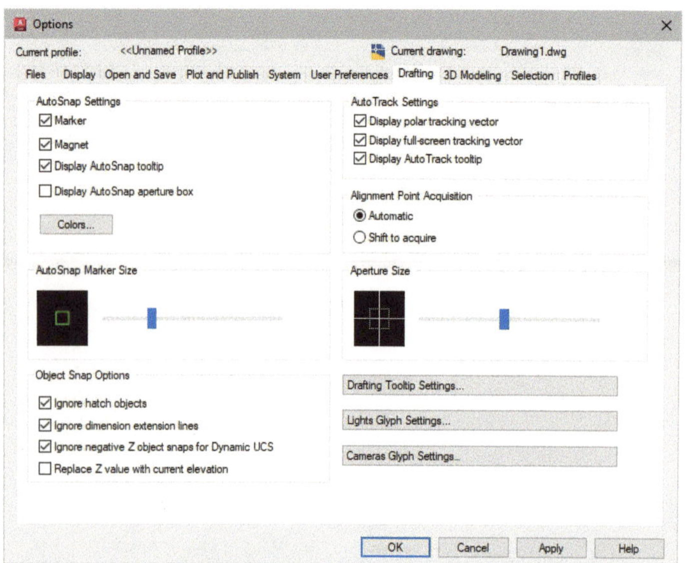

10. Select the **Selection** Tab and change the settings on your screen to match the dialog box below.

Note: The 3D Modeling Tab was skipped. For more on 3-dimensional modeling, see the *Advanced AutoCAD® Exercise Workbook* and *AutoCAD® 3D Modeling Exercise Workbook*.

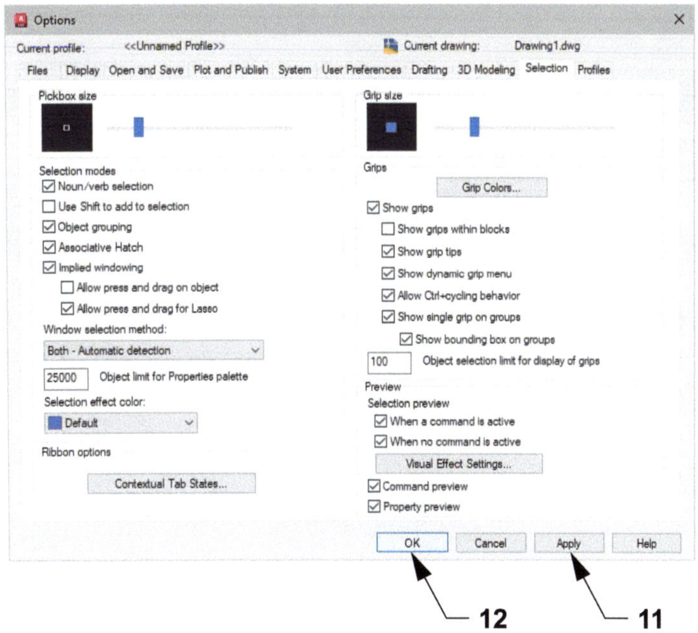

11. Select the **Apply** button.

12. Select the **OK** button.

13. Now you should be back to the main AutoCAD screen.

If you select the "**X**" in the top right corner of the **Options** dialog box before you select the **Apply** button and then the **OK** button, the **Options - Changes Not Saved** dialog box will appear.

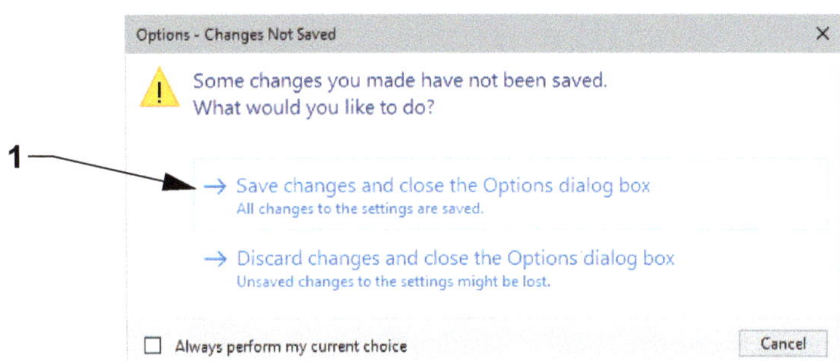

If you selected the "**X**" instead of the **Apply** and then the **OK** buttons and see this box, do the following:

1. Select **Save changes and close the Options dialog box**.

2. The changes you made in the Options dialog box will be saved and the Options dialog box will close. If you want to recheck the changes have been saved, reopen the Options dialog box.

Customizing Your Wheel Mouse

Using a wheel mouse is by far the easiest and most efficient method of navigating your way around AutoCAD and for creating 2-dimensional (2D) drawings.

The standard wheel mouse has left and right buttons and a central wheel that also doubles as a button. The default functions for the buttons and wheel are as follows.

The **right-hand** button is used to **enter** a command or for the shortcut **Menu**.

The **wheel** is used either for **Zooming** and **Panning**, or it can be used for **Zooming** and displaying the **Object Snap Menu**. The default is Zooming and Panning.

The **left-hand** button is used for **input** during commands and cannot be reprogrammed.

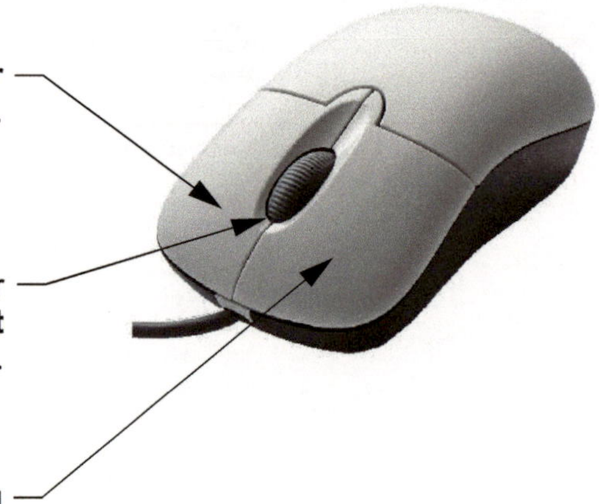

Note: The following describes how to select the wheel functions. After you understand the functions, you may choose to change the setting.

To change the setting you must use the **MBUTTONPAN** variable.

MBUTTONPAN setting 1:
(This is the factory setting.)

Zoom: Rotate the wheel forward to zoom in.
Rotate the wheel backward to zoom out.

Zoom Extents: Double click the wheel to view entire drawing.

Pan: Press the wheel and drag the mouse to move the drawing on the screen.

MBUTTONPAN setting 0:

Zoom: Rotate the wheel forward to zoom in.
Rotate the wheel backward to zoom out.

Object Snap: The Object Snap Menu will appear when you press the wheel.

To change the setting:

1. Type ***mbuttonpan*** and then press ***<Enter>***.
2. Enter ***0*** or ***1*** and then press ***<Enter>***.

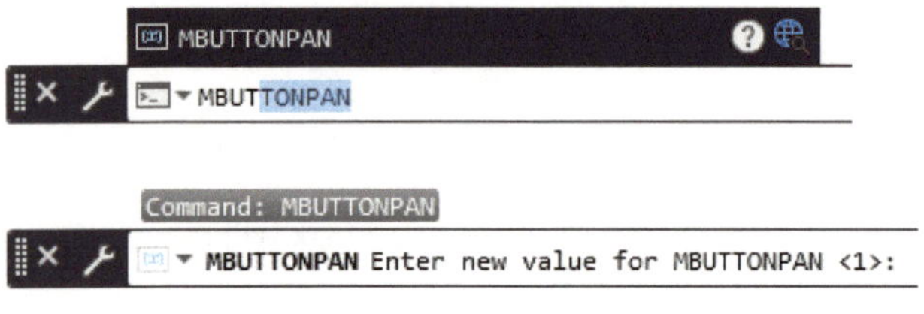

Command Line

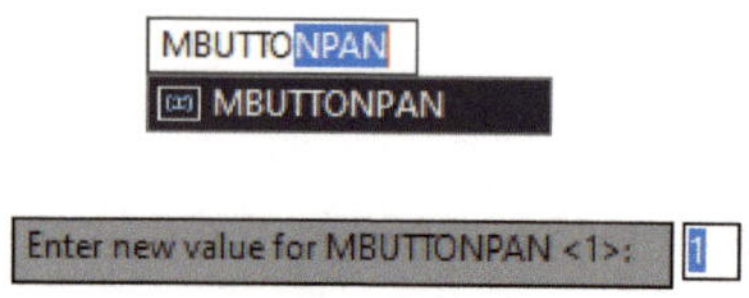

Dynamic Input

LESSON 1

LEARNING OBJECTIVES

After completing this lesson, you will be able to:

- Start the AutoCAD program
- Understand AutoCAD workspaces
- Recognize all of the features in the AutoCAD Window
- Understand what the keyboard function keys are used for
- Use the Artificial Intelligence Autodesk Assistant help system

Starting AutoCAD

To Start AutoCAD use one of the two methods below. *(Be patient, it may take a few minutes to load. AutoCAD is a large program.)*

1. Select **Start / AutoCAD 2026** (folder) **/ AutoCAD 2026**
   or
   Double click on the **AutoCAD 2026** desktop shortcut icon.

2. The **"Start"** page should appear.

Note: Your computer should be connected to the internet.

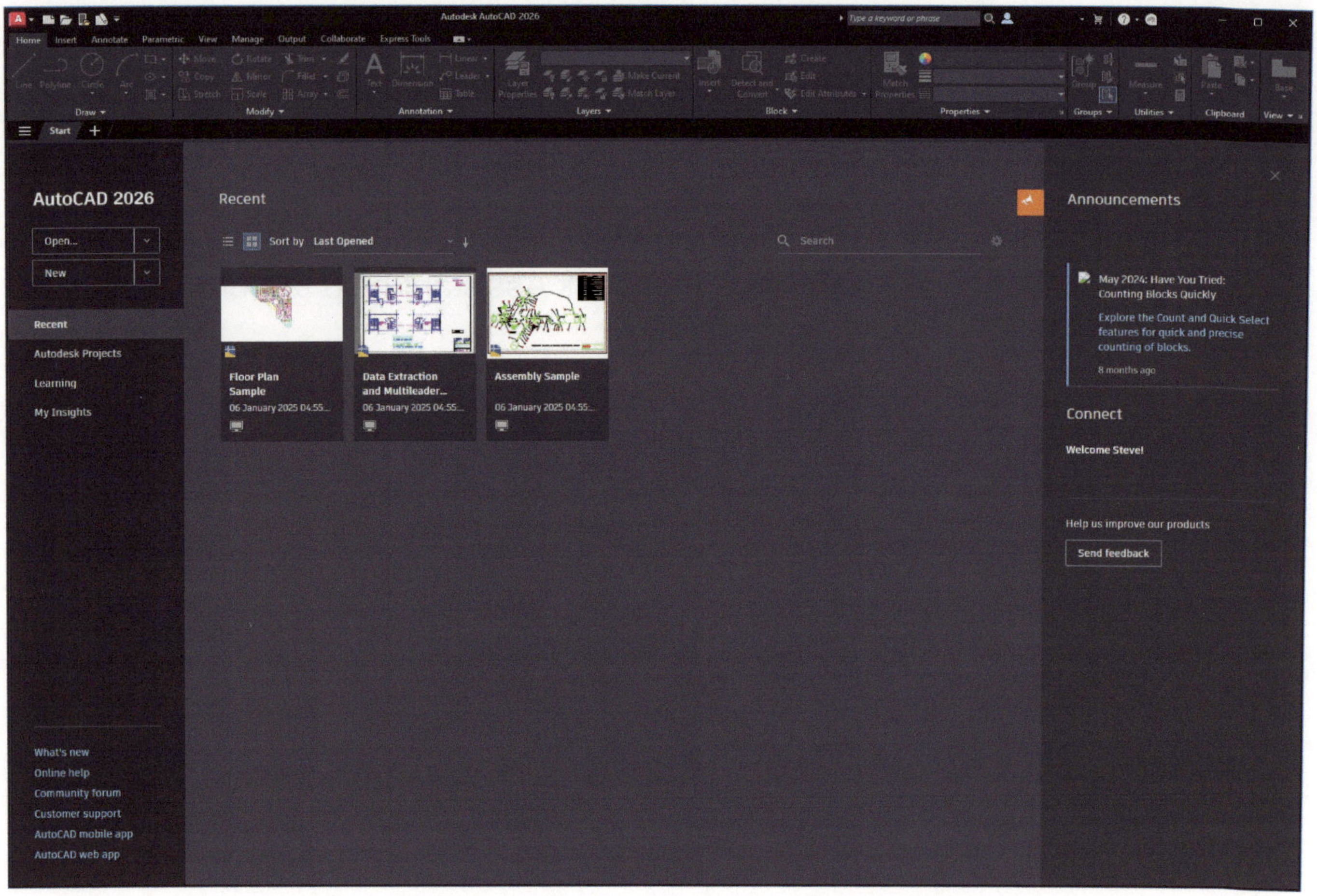

The left-hand panel contains the **Open** drop-down list where you can open an existing drawing file, sheet set, or explore sample drawing files that come pre-loaded. It also contains the **New** drop-down list where you can start a new drawing file using a recent drawing template, browse for existing templates, or get more templates online.

The center panel contains thumbnails of recently opened drawing files. There are three examples of recently opened drawing files shown in the **Recent** panel above.

The right-hand panel contains announcements for product updates, surveys you can take part in, and helpful tips and suggestions on using AutoCAD.

You can start a new drawing by selecting the **New** drop-down list and then selecting **Browse templates...**

1. Select the **New** drop-down list.
2. Select **Browse templates...**

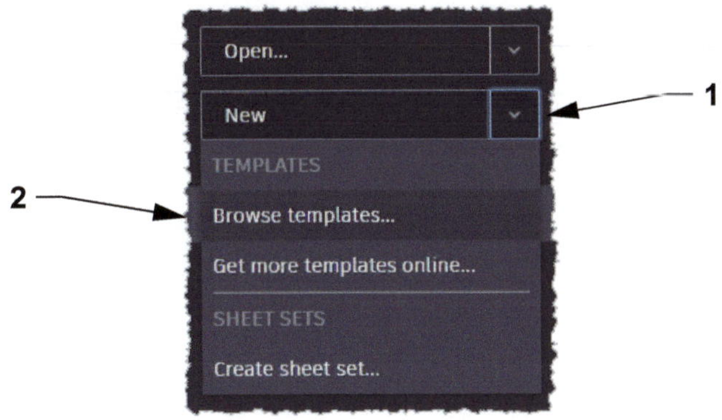

3. In the **Select Template** dialog box, select a drawing template file.
4. Select **Open**.

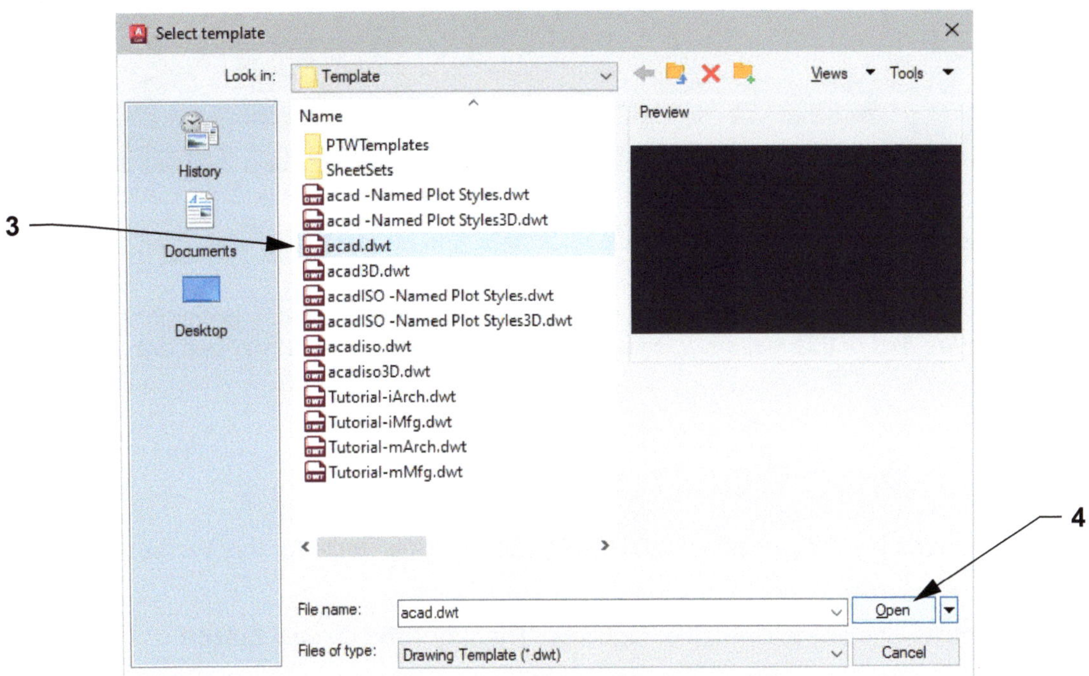

Note: Be sure to select the correct template in the dialog box:
The drawing template **acad.dwt** is for inch measurements.
The drawing template **acadiso.dwt** is for metric measurements.

Continued on the next page...

Note: If your screen does not appear as shown below, go to the **Intro** section of this Workbook and follow the steps for configuring AutoCAD to match the Workbook configuration.

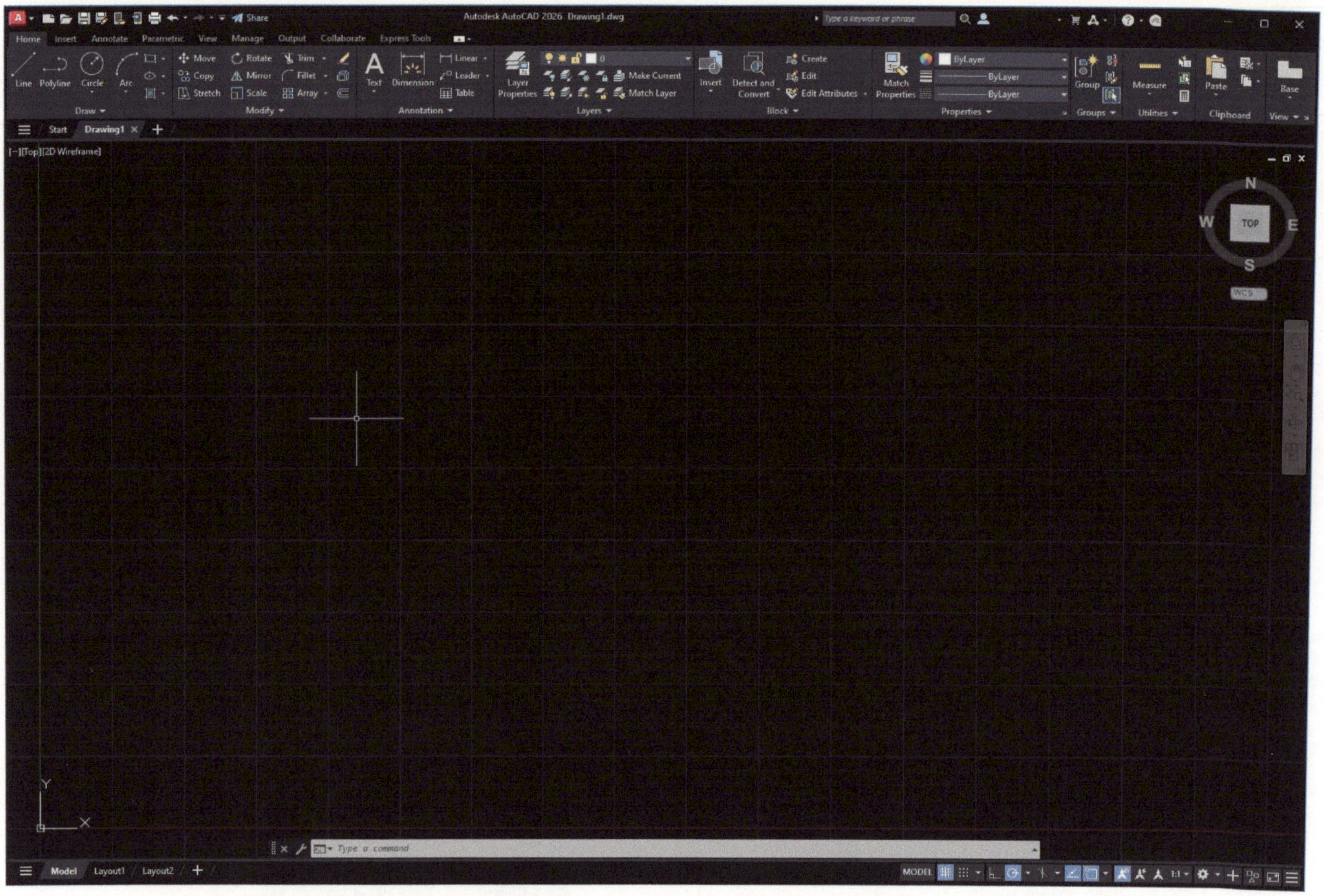

Note: The following section will describe each area and element of the AutoCAD interface.

While you probably would like to jump ahead to start drawing, please be patient and go through the rest of this lesson. It is very important that you understand and are familiar with AutoCAD's interface.

AutoCAD Application Window

Important: I have changed my 2-dimensional (2D) background color to white for this Workbook. Yours may be another color.

You may change the color of many areas using: **Options / Display Tab / Colors button**

AutoCAD comes with Dark and Light Color themes. The screen shot above shows the Dark theme. I have changed to the Light theme; this is just for clarity. You may choose either Color theme.

You may change the Color theme by using: **Options / Display Tab / Color theme**

If the remainder of your screen does not appear as shown on the next page, go to the **Intro** section of this Workbook and follow the steps for configuring AutoCAD to match the Workbook configuration.

The following pages discuss the elements of the AutoCAD interface and application window, providing an introduction to workspaces, documents, display options, and various features, terms, options, and ways to find and use different commands.

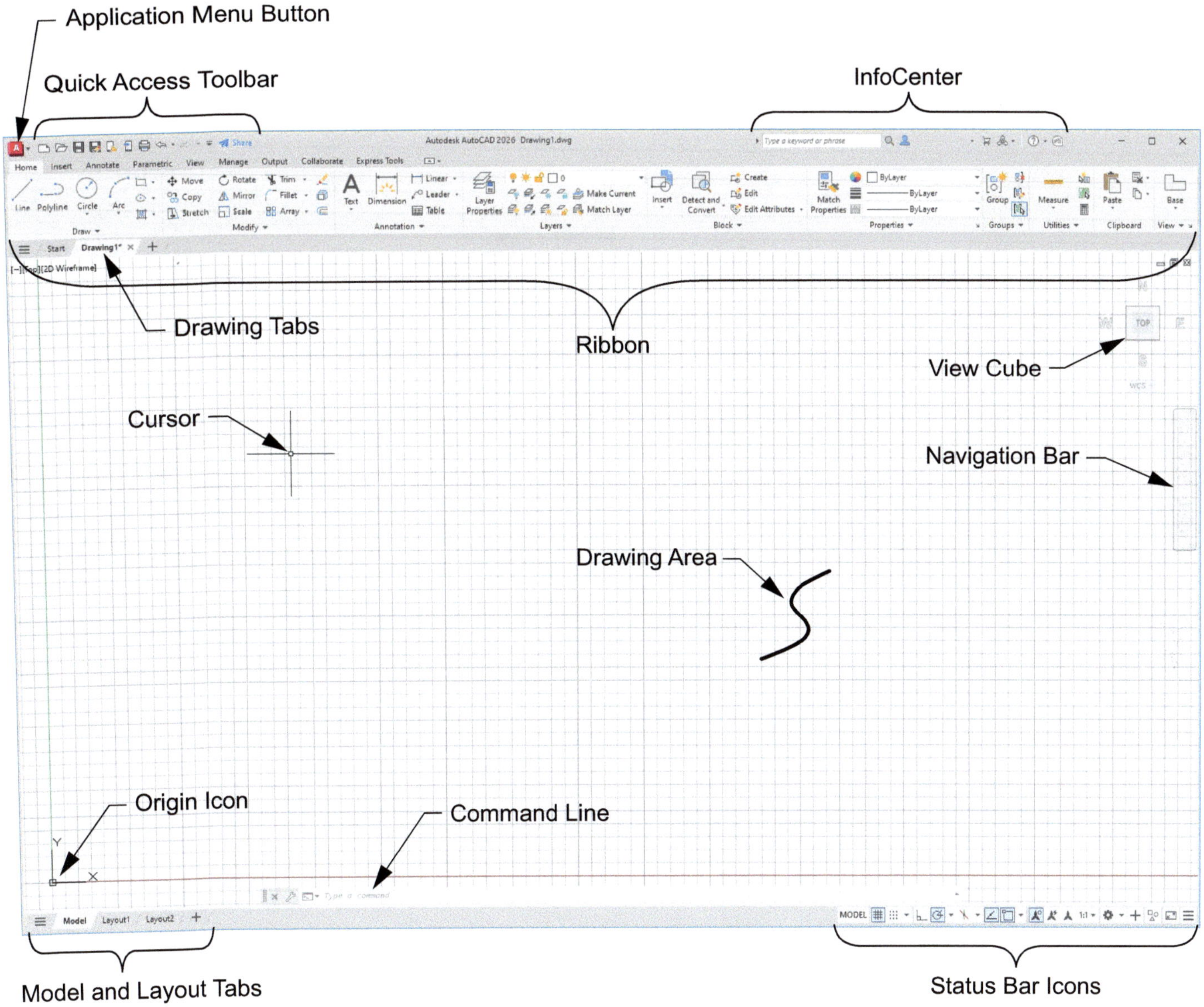

Workspace

Workspaces control the display of Ribbons, Tabs, Menus, Toolbars, and Palettes. AutoCAD gives you the option of deciding how you would like them displayed. When you use a workspace, only the Menus, Toolbars, and Palettes that are relevant to a task are displayed. For example, if you selected the 3D Modeling workspace, only 3D Menus, Toolbars, and Palettes would be displayed.

There are 3 preset workspaces.

Drafting & Annotation

This workspace (shown above) is the default display. It displays the necessary Ribbons, Tabs, Menus, Toolbars, and Palettes for 2D drafting. We will be using this workspace for all lessons within this Workbook.

3D Basics

This 3D Basics workspace provides a simple workspace with the most basic tools for creating and visualizing 3D models.

3D Modeling

The 3D Modeling workspace provides access to the vast array of 3D tools in AutoCAD. (For more on 3D modeling, see the *Advanced AutoCAD® and AutoCAD® 3D Modeling Exercise Workbooks*.)

How to select a Workspace

1. Selecting a workspace is easy. Select the ▾ on the **Workspace Switching** icon located on the **Status Bar** at the bottom right corner of the screen.

2. Select one of the workspace names displayed.

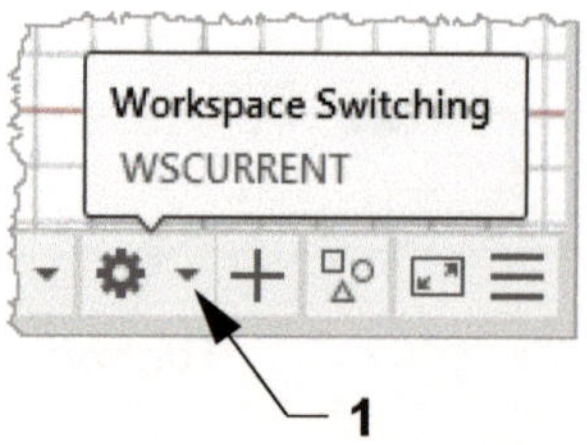

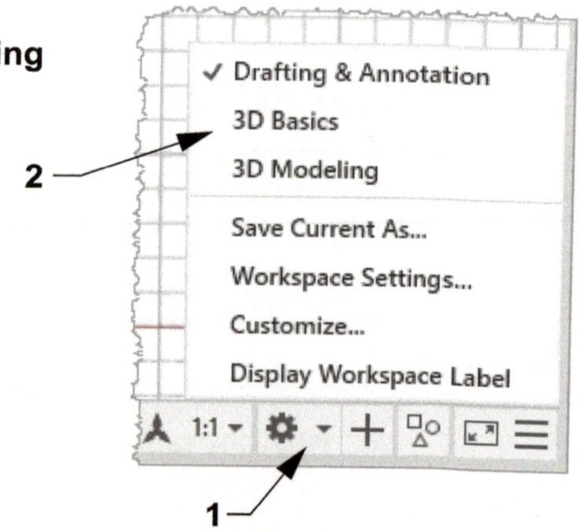

Note: The **"Drafting & Annotation"** workspace will be used in this Workbook.

Application Menu

The Application Menu provides easy access to common tools. Each of the tools will be discussed later in the Workbook.

1. Click on the **Application Menu** button in the upper left corner of the AutoCAD display screen (**"A"**).

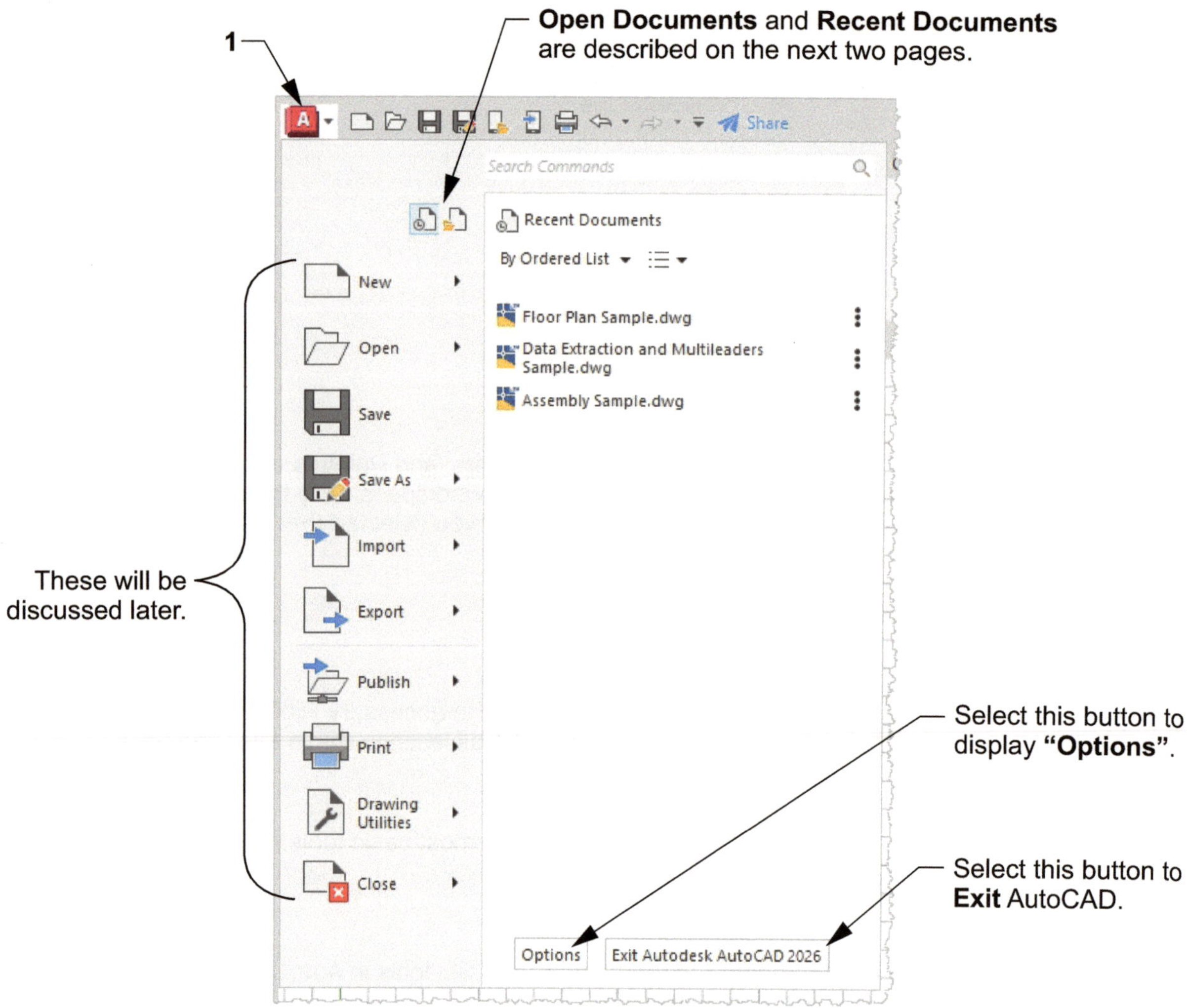

Open Documents

First let me emphasize, this is **not** a method to **"open"** a drawing file.

Open Documents is a list of all documents that are already open within AutoCAD.

Display choices: The list of documents may be displayed as icons or images. If you hover the cursor over a document name, a preview image will appear.

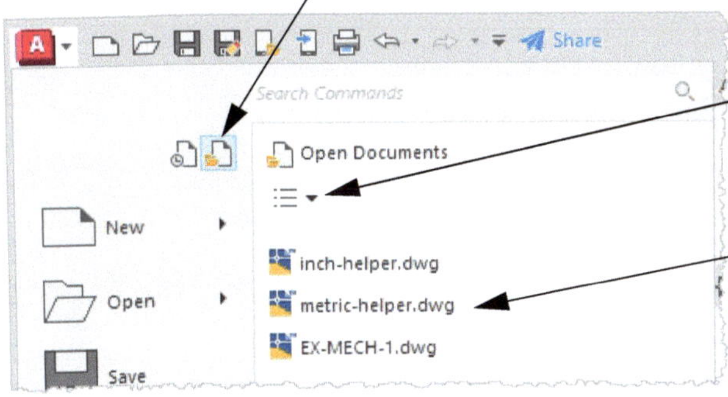

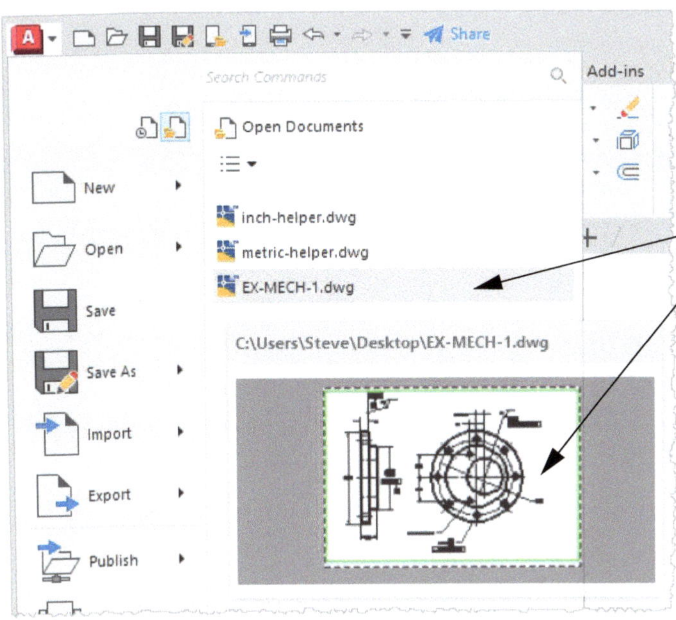

Recent Documents List

When you select **Recent Documents** button, a list of the recently viewed documents will appear.

Display choices: This list may be displayed as icons or images and may be sorted in an ordered list or grouped by date or file type. If you hover the cursor over a document name, a preview image will appear.

Pinned Files: You can keep a file listed, regardless of files that you save later, by using the push pin button to the right (shown on the next page). The file is displayed at the bottom of the list until you turn off the push pin button.

Continued on the next page...

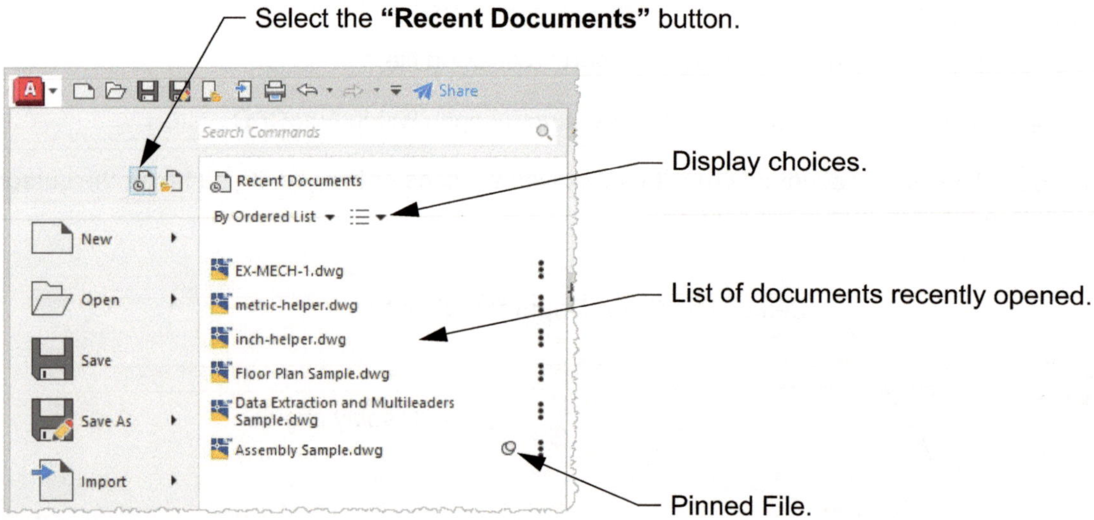

How to Pin a File to the Recent Documents List

1. Select the **Recent Documents** button.
2. Move the Cursor to the right-hand side of the file you want to pin. The push pin will appear.
3. Left click on the push pin button.
4. The file will now be pinned to the Recent Documents List.

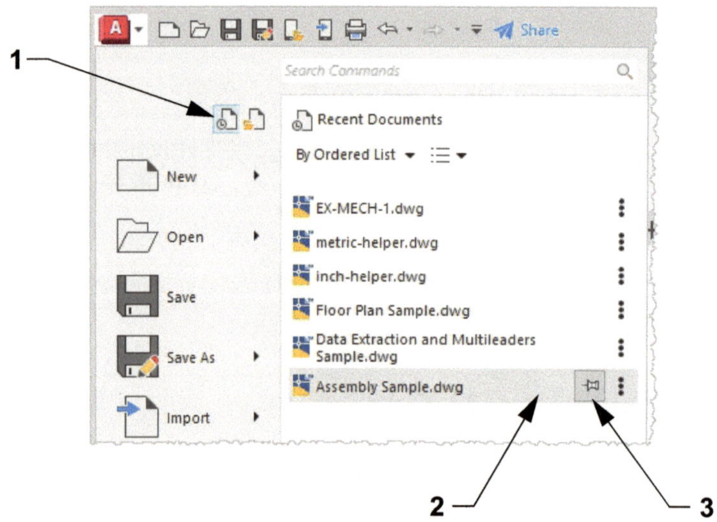

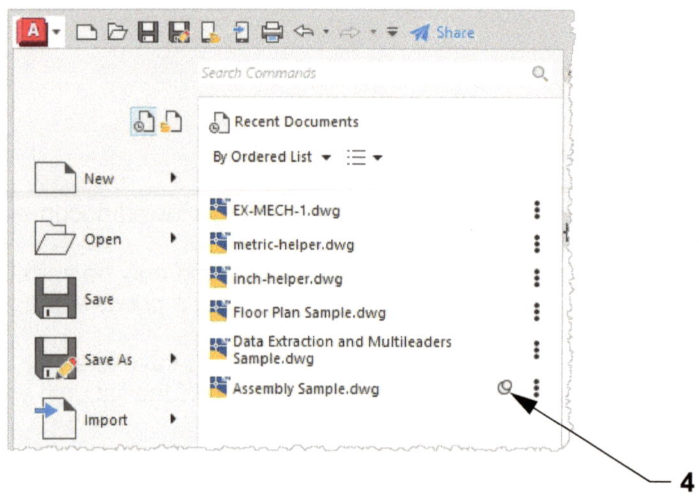

You also can remove a file from the Recent Documents List if you no longer need it.

How to Remove a File from the Recent Documents List

1. Select the **Recent Documents** button.
2. Move the Cursor to the right-hand side of the file you want to remove.
3. Left click on the button with the three vertical filled circles. A menu will appear
4. Left click on **Remove from List**.

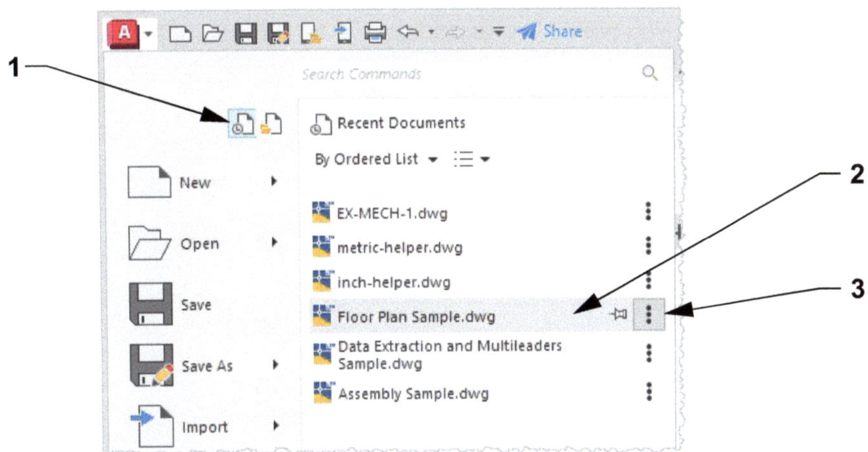

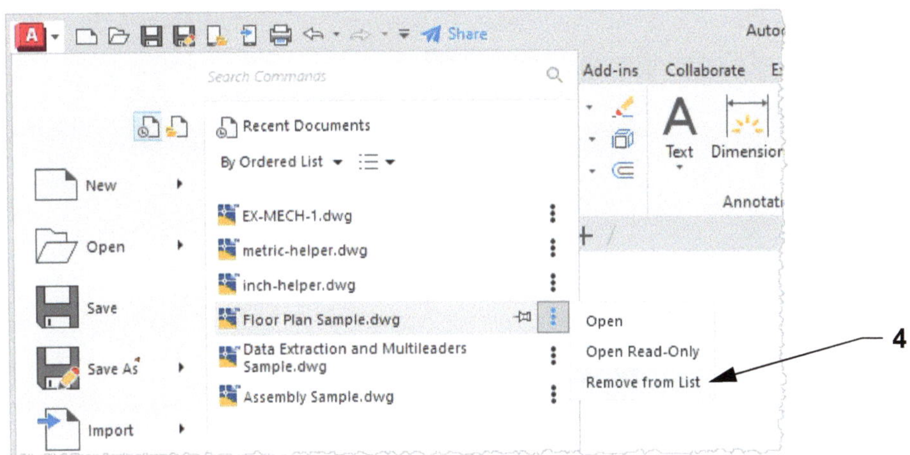

5. The selected file has been removed from the Recent Documents List.

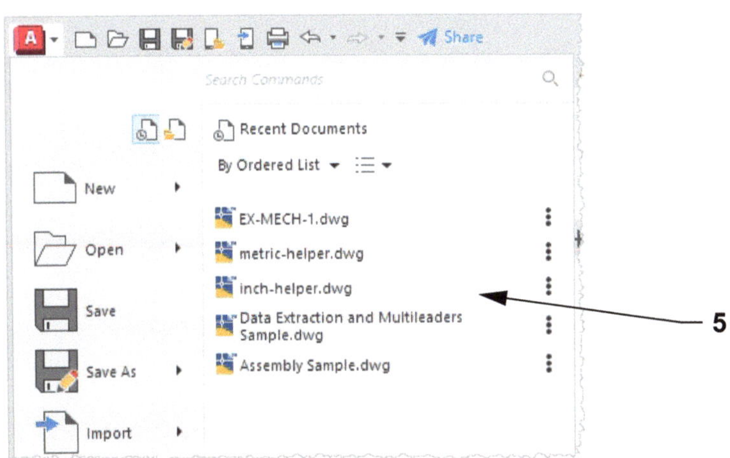

Note: You also can remove a pinned file from the Recent Documents list. You **do not** have to unpin it first.

Quick Access Toolbar

The **Quick Access Toolbar** is located in the top left-hand corner of the AutoCAD Window. It includes the most commonly used tools, such as New, Open, Save, Save as, Print, Undo, and Redo.

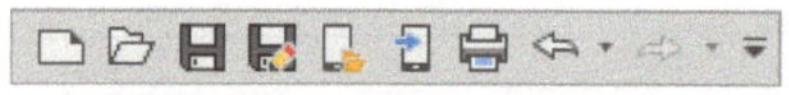

How to Customize the Quick Access Toolbar

You can add tools with the Customize User Interface dialog box.

Example:

After you have completed Lesson 4, you will find that you will be using **"Zoom All"** often. So I added the Zoom All tool to the Quick Access Toolbar. If you would like to add the Zoom All tool, or any other tool, to your Quick Access Toolbar, follow the steps below.

1. Place the Cursor on the Quick Access Toolbar and press the right mouse button.

2. Select **"Customize Quick Access Toolbar"** from the menu.

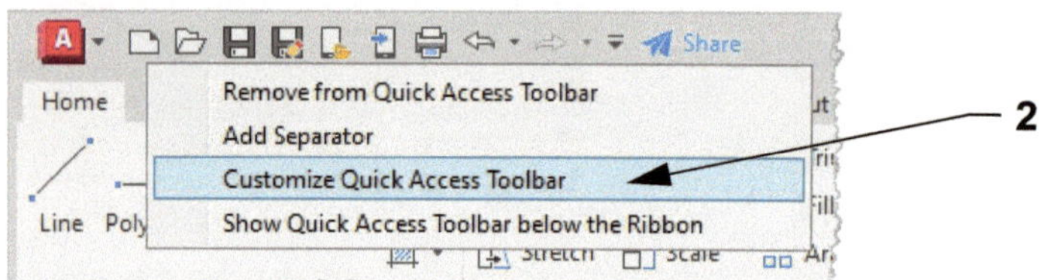

3. Scroll through the list of Commands to **"Zoom, All"**.

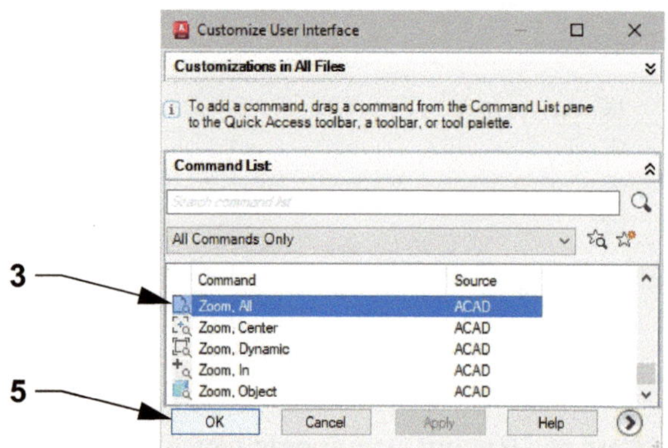

4. Press the Left mouse button on **"Zoom, All"** and drag it to a location on the Quick Access Toolbar and drop it by releasing the left mouse button.

5. Select the **OK** button at the bottom of the Customize User Interface dialog box.

The Customize User Interface dialog box will disappear and the new Quick Access Toolbar is saved to the current workspace.

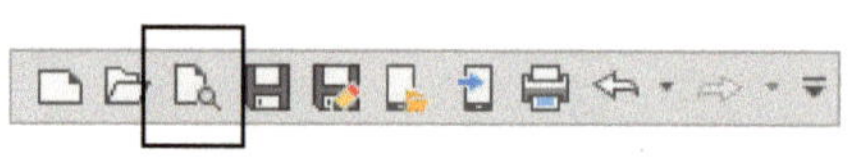

How to Remove a Tool

Place the cursor on the tool to remove and press the right mouse button. Select **"Remove from Quick Access Toolbar".**

Ribbon

The **Ribbon** provides access to the AutoCAD tools. The **Tabs** contain multiple **Panels**. Each **Panel** contains multiple tools. When you select a **Tab**, a new set of **Panels** will appear.

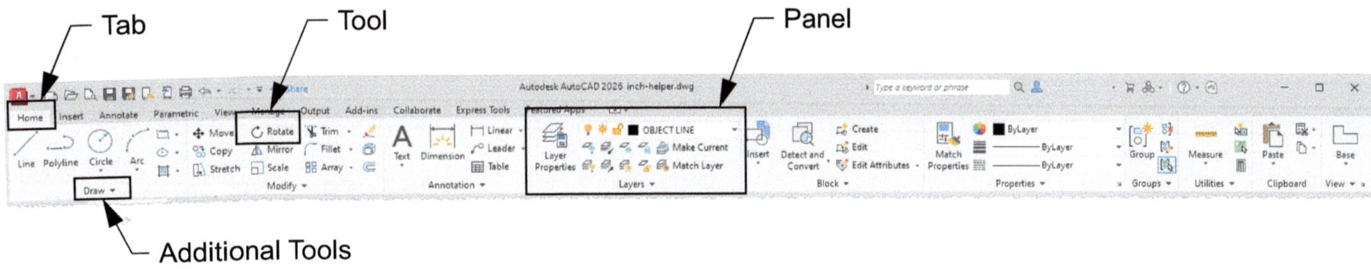

Additional Tools: If you select the ▾ symbol at the bottom of the Panel, the Panel will expand to access "Additional Tools".

Control the display of Tabs and Panels

Right click on the Ribbon and select which Tabs or Panels you choose to display. The check mark confirms the Tab or Panel is already displayed.

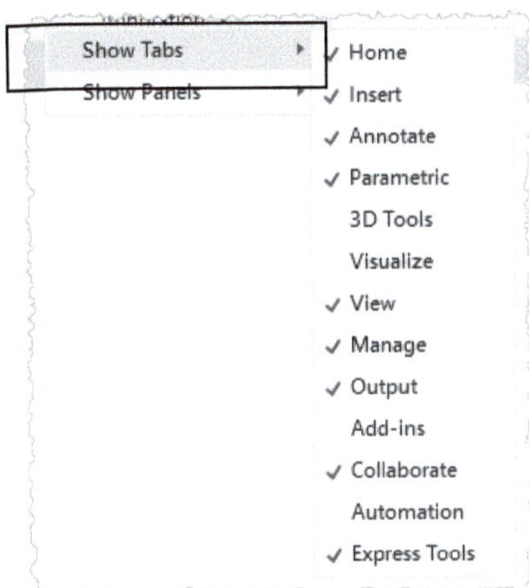

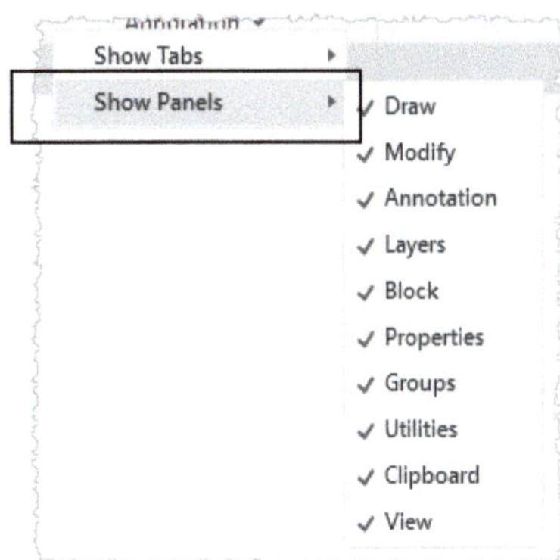

Control the Tab order

If you would like to change the order of the Tabs, click and drag the Tab horizontally to the new location.

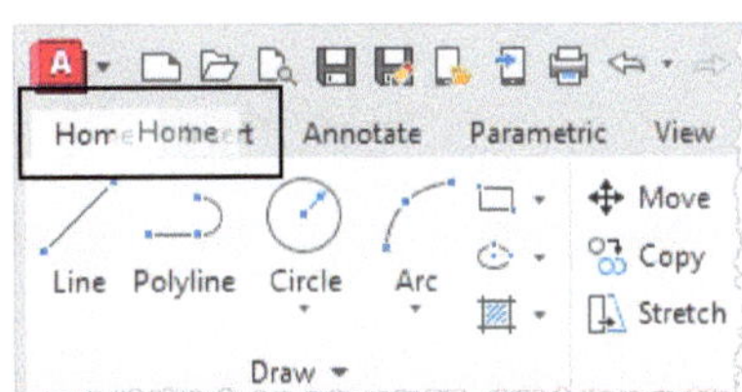

Floating Panels

If you prefer to separate a Panel from the Ribbon, you may drag the Panel off the Ribbon to a new location on the screen.

Status Bar

The **Status Bar** is located on the bottom of the screen. It displays the current settings. These settings can be turned **on** or **off** by clicking on one of the buttons or by pressing a corresponding function key. For example, ***<F2>***, ***<F3>***, etc.

When an icon is turned on, it will be displayed in blue.

Model and Layout Tabs
This will be discussed fully in Lesson 26.

Status Bar Icons

Status Bar Icons

The Status Bar provides you with a set of commonly used drawing tools like Grid Display, Snap, Object Snap, and Isometric Drafting. You can choose to remove some or all of them, or you can choose to add more tools.

How to add or remove Tools from the Status Bar

1. Select and click on the **Customization** button.

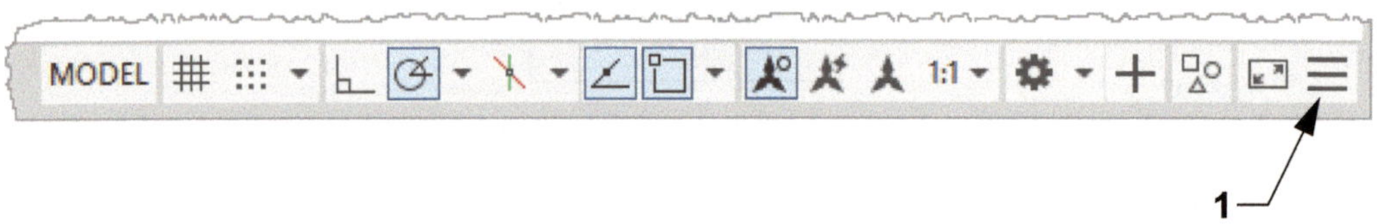

2. Check or uncheck the tools you want to add or remove from the menu.

3. Left click in the main drawing area to close the menu.

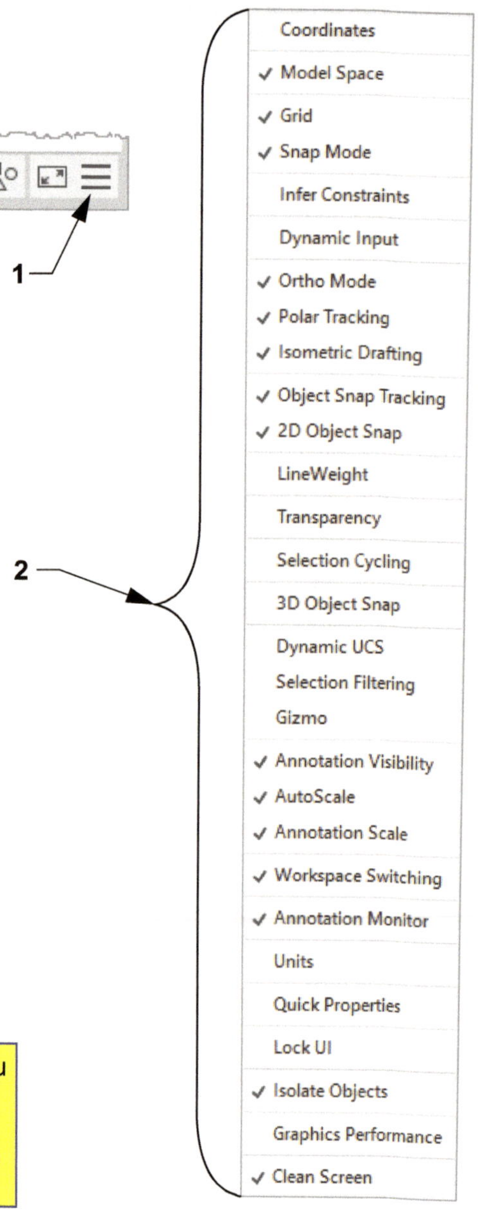

Note: The tools on the Status Bar may change, depending on whether you are in Model Space or Paper Space.

A brief description of each tool will be shown on the next four pages. Each tool will be discussed in more detail in future lessons.

Status Bar Tool Button Descriptions

I have enabled all the tool buttons and broken them down into two sections starting from the left-hand side.

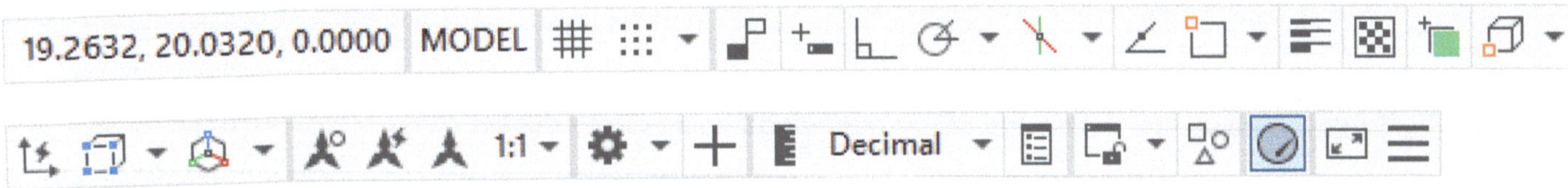

X Y Z

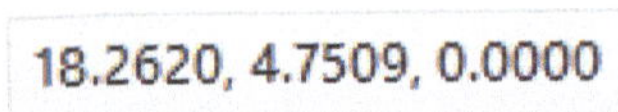

Coordinates

The coordinates display the location of the cursor in reference to the Origin. The Origin is currently in the lower left corner of the drawing area. These numbers will change as you move the cursor.

First set of numbers represents the horizontal movement of the cursor (**X axis**).
Second set of numbers represents the vertical movement of the cursor (**Y axis**).
Third set of numbers represents the **Z axis**, which is used for 3D and not discussed in this book.

Model 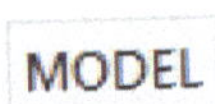

The **Model** / **Paper** button allows you to work in either Model Space or Paper Space without leaving the Layout Tab. When you switch to a Layout Tab, this button automatically switches to **Paper**.

Grid 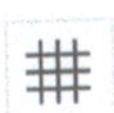 (You may also use **<F7>** to toggle **on** or **off**.)

The criss-cross lines in the Drawing Area are called the Grid. It is only a drawing aid and will not print. The default spacing is 1 unit of measurement. You may change the Grid spacing at any time by typing ***DS***, then press ***<Enter>***, and then select the **Snap and Grid** Tab from the **Drafting Settings** dialog box.

Snap Mode 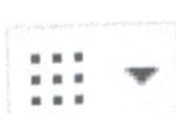 (You may also use **<F9>** to toggle **on** or **off**.)

Increment Snap controls the incremental movement of the cursor. If it is **on**, the cursor will "snap" in an incremental movement. If it is **off**, the cursor will move smoothly. You may set the increments by clicking on the down ▾ arrow and selecting **Snap Settings...**

You can also choose whether to use **Grid Snap** or **Polar Snap** on the same menu.

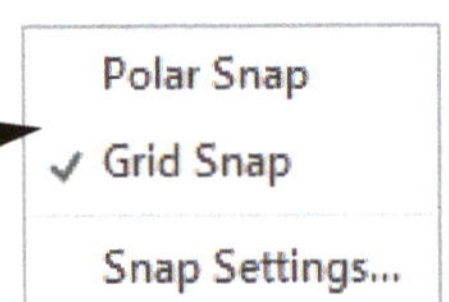

Infer Constraints 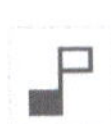 (**Note:** Not used in this Workbook.)

Inferred Geometric Constraints automatically applies coincident constraints for Endpoint, Midpoint, Center, Node, and Insertion Object Snaps.

Dynamic Input  (You may also use **<F12>** to toggle **on** or **off**.)

When Dynamic Input is **on**, you can enter coordinate values in tooltips near the cursor. More on this in Lesson 11.

Ortho Mode 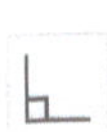 (You may also use **<F8>** to toggle **on** or **off**.)

Ortho restricts the movement of the cursor to horizontal or vertical. When Ortho is **on**, the cursor moves only horizontally or vertically. When Ortho is **off**, the cursor moves freely in any direction.

Continued on the next page...

Polar Tracking 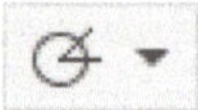 (You may also use ***<F10>*** to toggle **on** or **off**.)
Polar Tracking restricts cursor movement to specified increments along a Polar angle. More on this in Lesson 11.

You may set the increments by clicking on the down ▾ arrow and selecting **Tracking Settings...**, or you can select one of the predefined angles from the list.

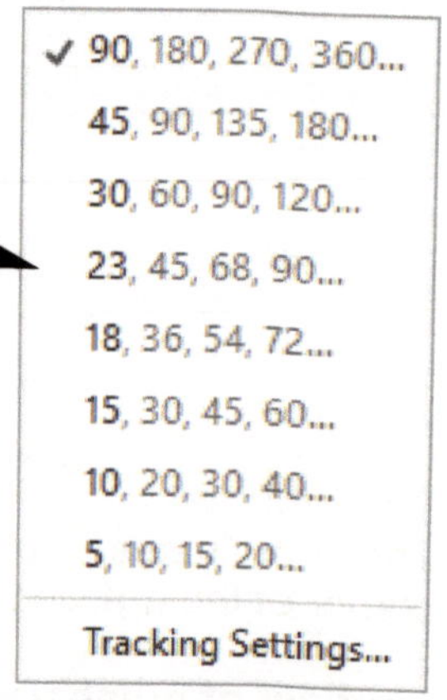

Isometric Drafting 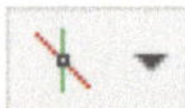
Isometric drawing allows you to simulate a 3D object by aligning along 3 axes; these are **Top**, **Right**, and **Left**, and are called **Isoplanes**. When the button is enabled, you can toggle through the Isoplanes by pressing the ***<F5>*** key. Isometric Drafting is discussed in the *Advanced AutoCAD® Exercise Workbook*.

You may also change Isoplanes by clicking the down ▾ arrow and selecting from the menu.

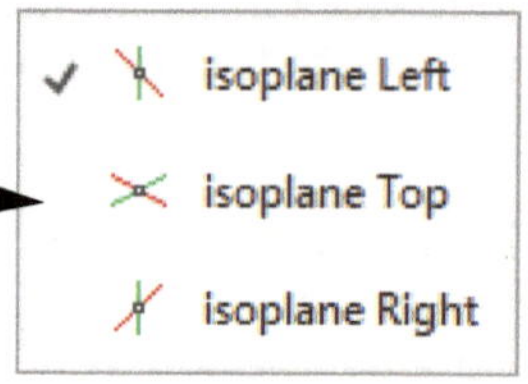

Object Snap Tracking  (You may also use ***<F11>*** to toggle **on** or **off**.)
Object Snap Tracking controls the display of Object Snap reference lines, AutoSnap marker, tooltip, and magnet.

2D Object Snap 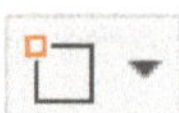 (You may also use ***<F3>*** to toggle **on** or **off**.)
When 2D Object Snap is **on**, the cursor will "snap" to preset locations on 2D objects. More on this in Lesson 4.

You may also add or remove Object Snaps by clicking the down ▾ arrow and selecting from the menu.

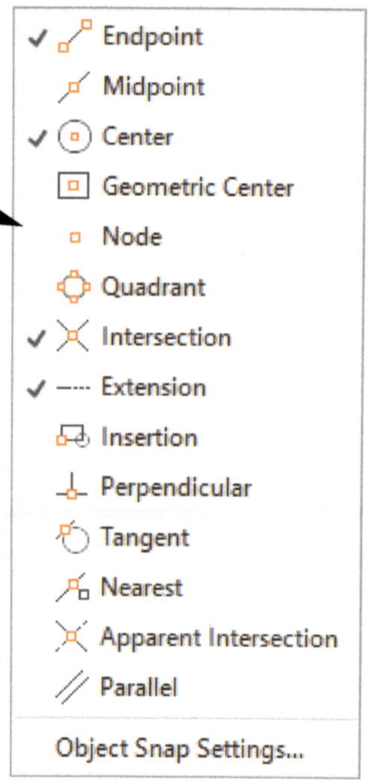

Continued on the next page...

Lineweight

Lineweight displays the width assigned to each object. When it is **on**, the lineweights are visible. You can change the lineweight settings by right clicking on the button and selecting **Lineweight Settings**. More on this in Lesson 3.

Transparency

When Transparency Show/Hide is **on**, all transparent layers will be displayed. If it is **off**, no transparent layers will be displayed. More on this in Lesson 3.

Selection Cycling

Selection cycling allows you to select objects that are overlapping. This is most useful when creating 3D models as discussed in the *Advanced AutoCAD® Exercise Workbook*.

3D Object Snap

(You may also use ***<F4>*** to toggle **on** or **off**.)

When 3DOsnap is **on**, the cursor will "snap" to preset locations on 3D objects. This option is discussed in the *Advanced AutoCAD® Exercise Workbook*.

Dynamic UCS

(You may also use ***<F6>*** to toggle **on** or **off**.)

Dynamic User Coordinate System changes the grid plane to follow the XY plane of the dynamic UCS. It is used for 3D; refer to the *Advanced AutoCAD® Exercise Workbook* and *AutoCAD® 3D Modeling Exercise Workbook*.

Selection Filtering

Selection filtering allows you to filter whether certain faces, edges, vertices, or solid history subobjects are highlighted when you roll over them, very useful in complex 3D.

Gizmo

Gizmo tools help you move, rotate, or scale an object or set of objects along a 3D Plane, and are discussed in the *Advanced AutoCAD® Exercise Workbook*.

Annotation Visibility

When switched **on**, the Annotation Visibility tool displays or hides the visibility of annotative objects at the current scale.

AutoScale

When switched **on**, the AutoScale tool automatically updates annotative objects to support the annotation scale when the annotation scale is changed.

Annotation Scale

1:1 ▾

The Annotation Scale tool displays the current annotation scale. You can change the scale by clicking on the down ▾ arrow and selecting from the list of predefined scales or you may create a custom scale. You can also display the scale in percentages by selecting **Percentages** from the list.

Workspace Switching

Workspace Switching allows you to change the workspace environment: you can choose between Drafting & Annotation, 3D Basics, and 3D Modeling. You can change the workspace by clicking on the down ▾ arrow and selecting from the list.

Annotation Monitor

Provides feedback regarding the state of Associative annotations when using parametric dimensioning; it is discussed in the *Advanced AutoCAD® Exercise Workbook*.

Continued on the next page...

Units

The Units tool allows you to change the display style of the Drawing Units. You can choose between Decimal, Architectural, Engineering, Fractional, and Scientific. You can change the drawing unit display by clicking on the down ▾ arrow and selecting from the list.

Quick Properties

If **on**, Quick Properties displays the properties of the object selected. If **off**, the Quick Properties box will not appear. More on this in Lesson 12.

Lock User Interface

The Lock User Interface tool allows you to lock or unlock Toolbars, Panels, Windows, and Floating Toolbars, Panels, and Windows. Click on the down arrow to select from the 4 options.

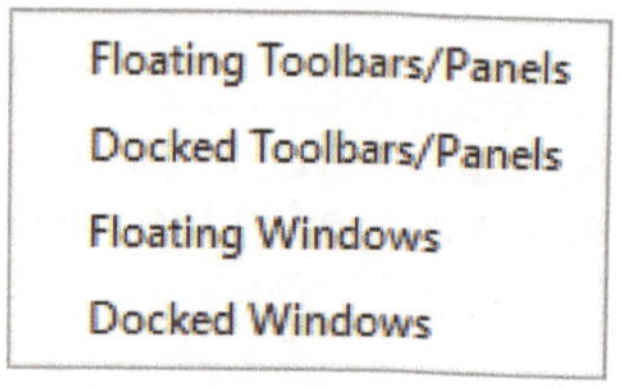

Isolate Objects

You can choose to isolate objects by keeping them visible on the screen, all other objects will be hidden. Or you can choose to hide objects. To isolate or hide objects, left click on the **Isolate Objects** tool button and select either **Isolate Objects** or **Hide Objects** from the list. To restore all hidden objects, left click on the **Isolate Objects** tool button and select **End Object Isolation** from the list.

Graphics Performance

The Graphics Performance tool examines your graphics card and 3D display driver and determines whether to use software acceleration or hardware acceleration. You can change the performance settings by right clicking on the tool button and selecting **Graphics Performance**, then changing any settings required in the dialog box.

Clean Screen

(You may also use ***<Ctrl+0>*** to toggle **on** or **off**.)

When Clean Screen is selected, it will hide all tool Palettes, Windows, and Ribbons from the screen, leaving you with a larger drawing area to work with. You can restore all the Palettes, Windows, and Ribbons by selecting the Clean Screen tool button again.

Keyboard Function Keys and What They Are Used For

F1 Opens the Help Window.

F2 Displays an Extended Command History list.

F3 Turns the 2D Object Snaps on or off.

F4 Turns the 3D Object Snaps on or off.

F5 Toggles Isoplanes between Top, Right, or Left.

F6 Turns the Dynamic UCS on or off.

F7 Turns the Grid on or off.

F8 Turns Orthographic Mode on or off.

F9 Turns Snap Mode on or off.

F10 Turns Polar Tracking on or off.

F11 Turns Object Snap Tracking on or off.

F12 Turns Dynamic Input on or off.

Floating Command Line

When you first start AutoCAD, if the software has not been modified, the **Command Line** will be displayed at the bottom of the screen, as shown below.

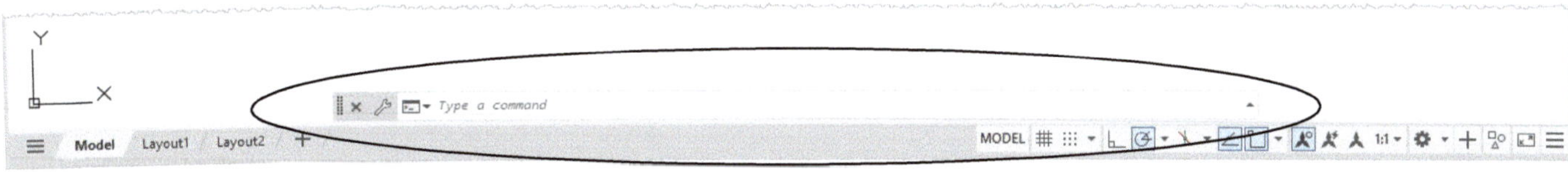

This is where AutoCAD will prompt you for information and you will enter commands, values, and select options. Basically, this is how you communicate with AutoCAD.

You may "dock" the Command Line at the top or bottom of the AutoCAD Window or let it float in the drawing area.

To move the Command Line, place the cursor on the left end grip, press the left mouse button, and drag the Command Line to a desired location.

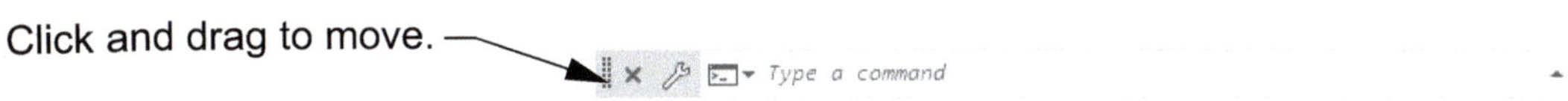

To "dock" the Command Line, drag it to the top or bottom of the drawing area. It will snap to the edge. You can't dock the Command Line to the sides. You may also drag it below the drawing area as shown below.

CAD TIP! You may toggle the Command Line **on** and **off** using ***<Ctrl+9>*** on your keyboard.

Command Line

How to enter a command on the Command Line

1. Type the first letter of a command, such as ***c*** for **circle**.

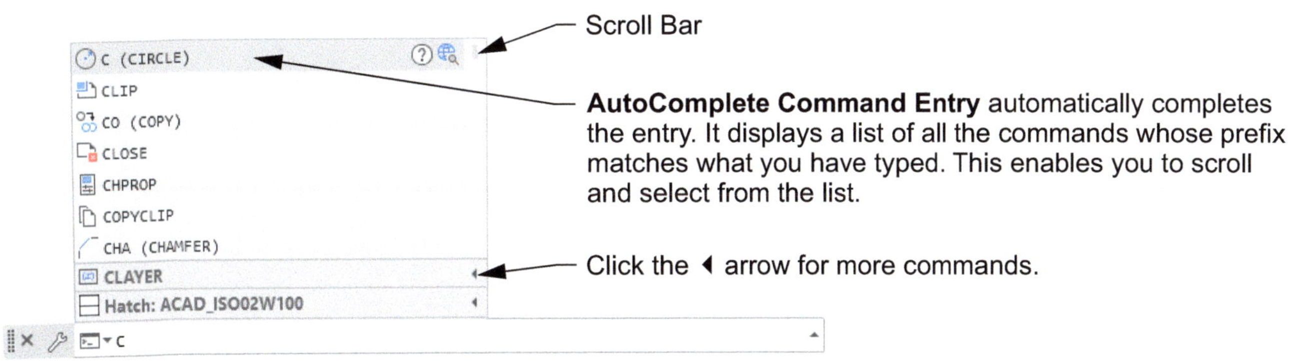

2. A list of commands that begin with the letter **C** will appear. Select the desired command from the list.

Continued on the next page...

3. When you enter a command such as Circle, the **prompt** and **options** will be displayed on the Command Line.

4. The **prompt** for the Circle command asks you to:

   "Specify center point for circle" or [3P 2P Ttr (tan tan radius)]:

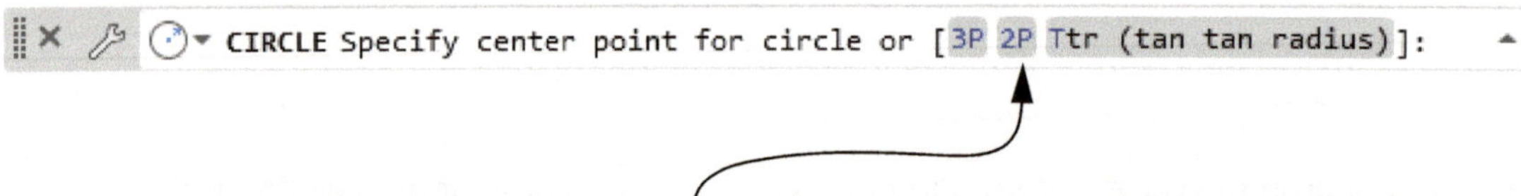

The information within the **[ ]** brackets are options that you may select.

Note: Clickable options are displayed in blue. Options displayed in black must be typed or selected from the option menu. This will be discussed more in Lesson 2.

Command and Prompt History

As you enter commands, AutoCAD records them as "history". You may display this history by pressing ***<F2>*** or by selecting the up arrow at the right-hand end of the Command Line.

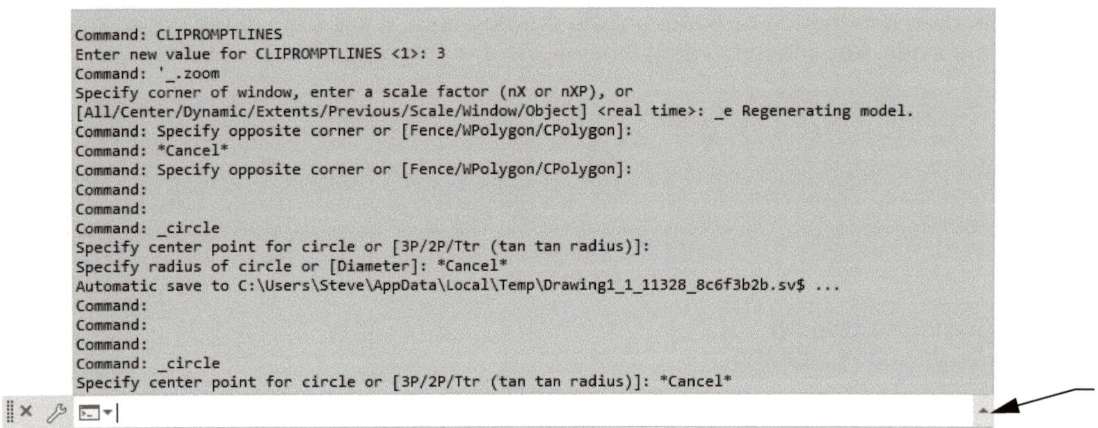

Select the up arrow to display the history.

Command Line Tools

The **Recent Commands** tool displays recently selected commands.

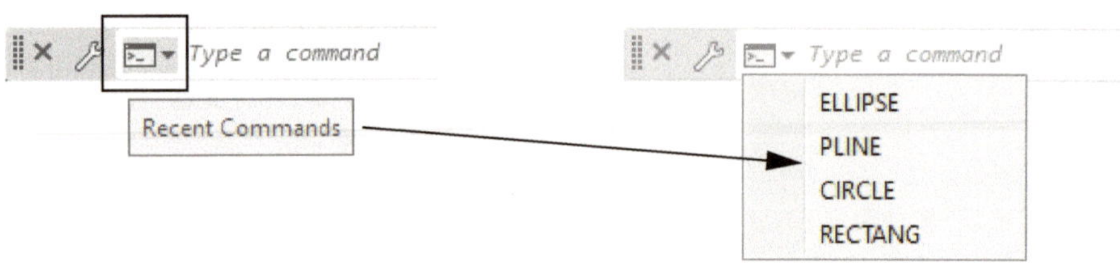

Continued on the next page...

The **Customize** tool allows you to select options for the AutoComplete by selecting **"Input Settings"**. You can also control how many lines of history are displayed and the degree of transparency for the Command Line.

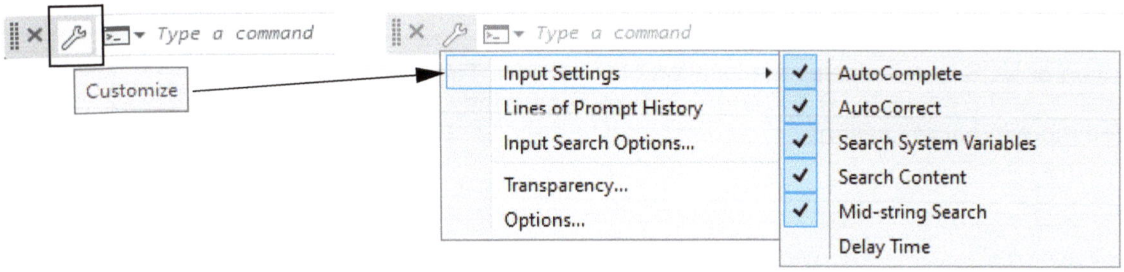

Dynamic Input

Dynamic Input is another method of inputting commands and values and selecting options. To use Dynamic Input, you must turn on the **Dynamic Input** button in the Status Bar, shown on page 1-12.

If you choose to use Dynamic Input, the command will be entered in the tooltip box beside the cursor.

How to enter a command using Dynamic Input

1. Place the cursor in the Drawing Area.
2. Type the first letter of a command, such as *c* for circle.
3. A list of commands that begin with the letter **C** will appear. Select the command from the list.

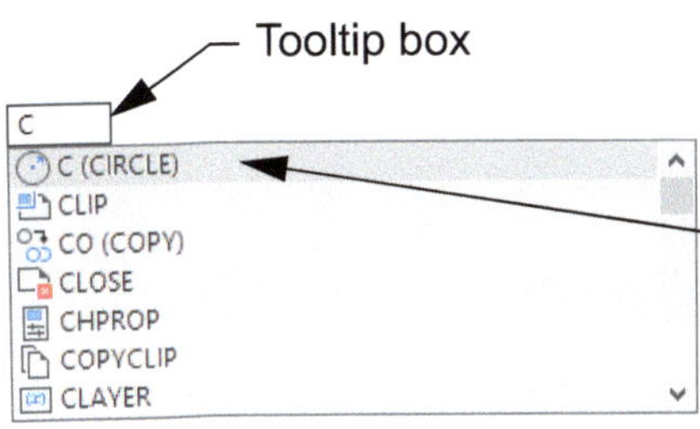

AutoComplete Command Entry automatically completes the entry. It displays a list of all the commands whose prefix matches what you have typed. This enables you to scroll and select from the list.

4. If you press the ↓ arrow, the options will appear below the prompt.

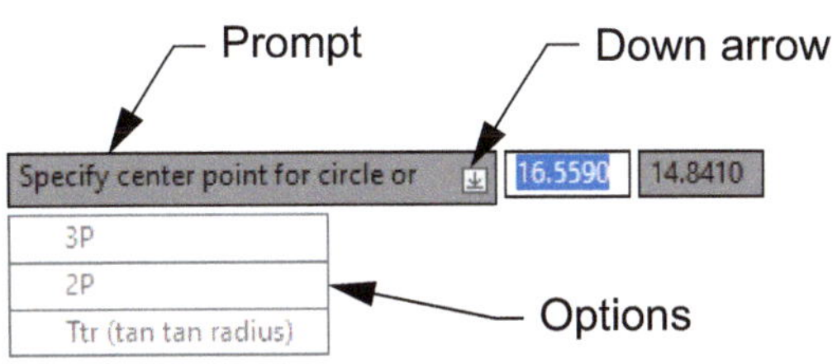

Note: The command entry also is being displayed on the Command Line. Whether to use the Command Line or Dynamic Input is your choice. This will be discussed more in Lesson 2.

Drawing Area

The Drawing Area is the large open area of the screen. This is where you will draw. Consider this your paper. The color of this area can be changed using **Options / Display Tab / Color**.

Origin Icon

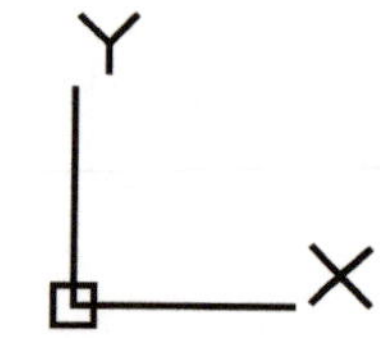

The Origin icon or UCS icon indicates the location of the Origin. The Origin is where the coordinates X, Y, and Z originate. The X and Y coordinates for the Origin is 0,0. This will be discussed more in future Lessons.

Cursor

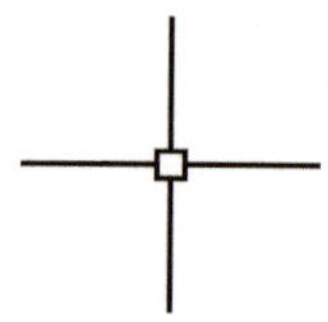

The Cursor is located within the Drawing Area. The movement of the pointing device, such as a mouse, controls the movement of the cursor. You will use the cursor to locate points, make selections and draw objects. The size can be changed using **Options / Display Tab / Crosshair Size**.

InfoCenter

The InfoCenter is a tool to search for information. It is located in the upper right corner of the screen.

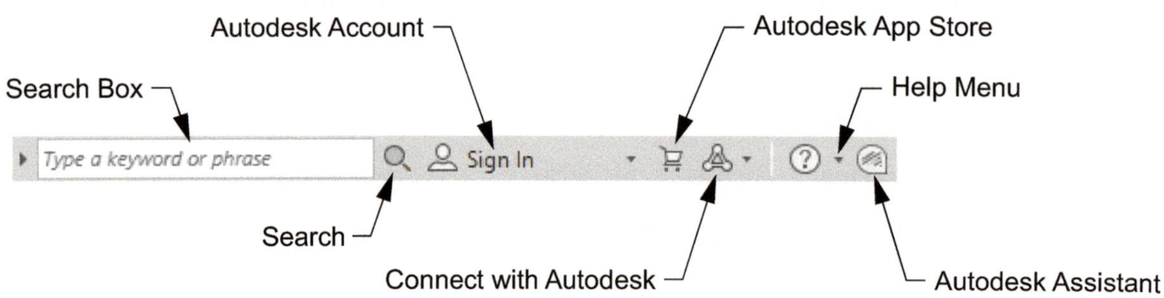

Search Box

The InfoCenter allows you to search for information by typing key words or a question in the **"Help Box"**. After typing, press ***<Enter>***.

Search

Displays multiple search options.

Autodesk Account

Sign in to your Autodesk online account to access services that integrate with your desktop software.

Autodesk App Store

Launches the Autodesk App Store website.

Connect with Autodesk

Allows you to connect to the Autodesk online community.

Help Menu

Displays the Help Window.

Autodesk Assistant

Opens the Autodesk Assistant Palette.

ViewCube and Navigation Bar

The ViewCube and the Navigation Bar are used primarily in the 3D mode. They enable you to view and rotate the 3D model. We will not be using these tools in this Workbook. Refer to the *Advanced AutoCAD® Exercise Workbook* and *AutoCAD® 3D Modeling Exercise Workbook*.

Since we are not using these tools, you may choose to turn their display off. Follow the easy instructions below to turn the display **off** or **on**.

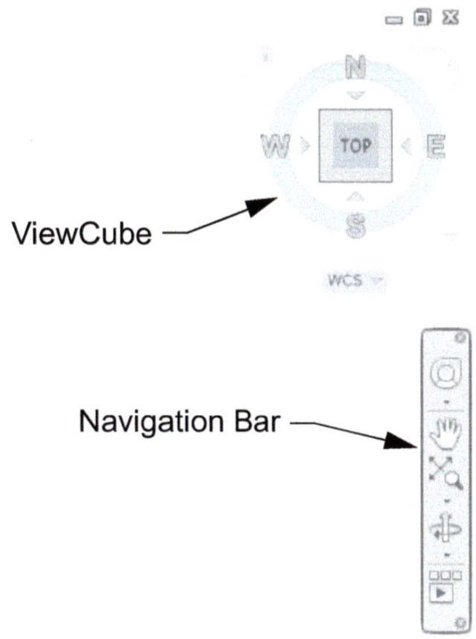

How to turn off the ViewCube and Navigation Bar

There are 2 methods.

Method 1.

1. Type ***navbardisplay*** and then press ***<Enter>***.
2. Type ***0*** and then press ***<Enter>***. [0 = off 1 = on]
3. Type ***navvcubedisplay*** and then press ***<Enter>***.
4. Type ***1*** and then press ***<Enter>***.

The **navvcubedisplay** has 4 settings:

0 = Not displayed in 2D or 3D.
1 = Displayed in 3D only (Select this one for the Workbook).
2 = Displayed in 2D only.
3 = Displayed in both 2D and 3D.

Method 2.

1. Select the **View** Tab.
2. Left click on the **ViewCube** and **Navigation Bar** buttons.

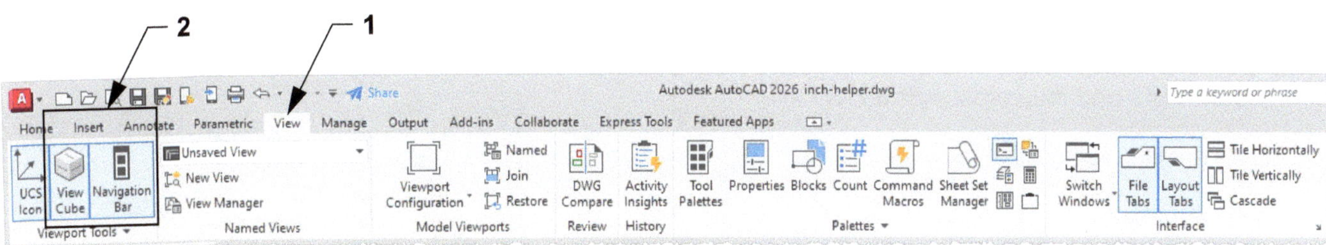

Note: The buttons will be displayed in blue when switched on.

ToolTip Help

Basic ToolTip

When you hover your cursor over a tool, an initial ToolTip will appear telling you the name of the tool with a brief description.

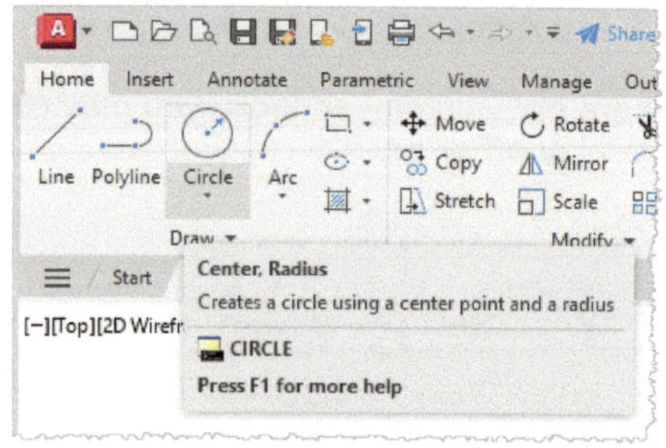

Extended ToolTip

If you hover just a little longer, a graphic display, directly from the Help system, will appear.

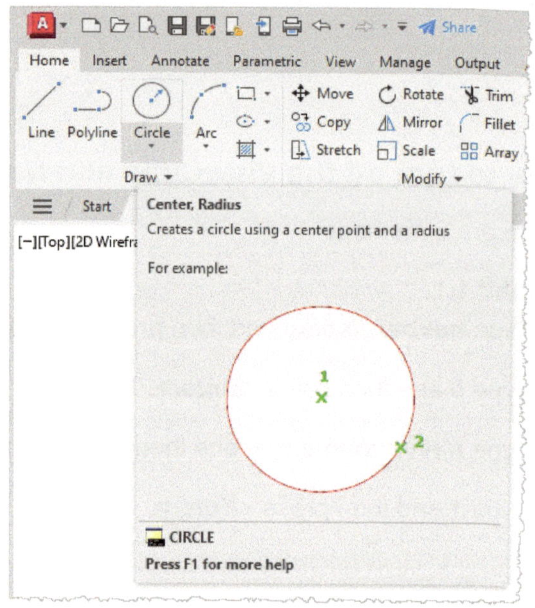

How to turn off ToolTips

After you become familiar with AutoCAD, you will want to turn these off. Or you may just want to delay the extended ToolTips.

1. Type ***options*** and then press ***<Enter>***.
2. Select the **Display** Tab.
3. Uncheck boxes.

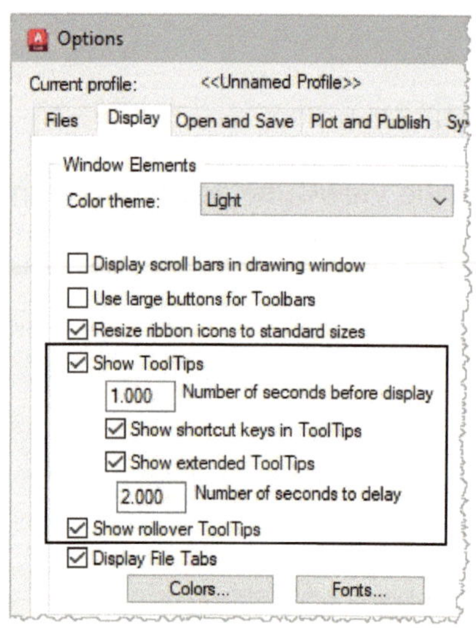

Autodesk Assistant

Autodesk® has added to AutoCAD® a smarter interactive help system called **Autodesk Assistant**. This artificial intelligence (AI) driven feature can be used to answer AutoCAD-related questions that you might have, such as "*How to create a Construction Line?*" If the Autodesk Assistant cannot or does not sufficiently answer your question, you can choose to connect online with an Autodesk agent for further assistance.

When you start AutoCAD, a **Meet the Autodesk Assistant** balloon notification will appear at the top of the Application Window for a few seconds. You can choose to open Autodesk Assistant at this time by clicking on the **Launch the Assistant now** text in this notification. You also can open the Autodesk Assistant at any other time while using AutoCAD.

Left click on the text to open Autodesk Assistant.

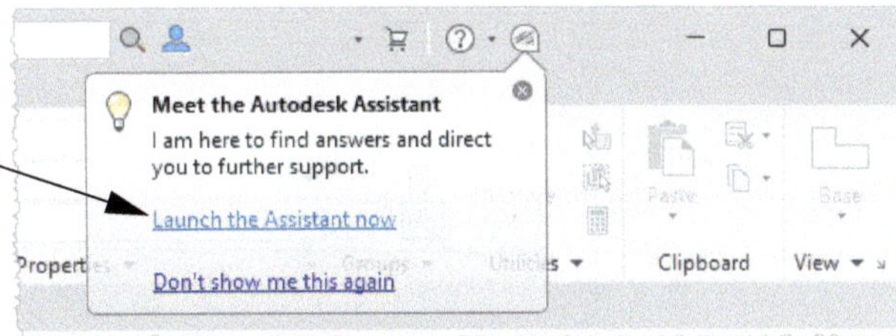

How to Open and Use Autodesk Assistant

1. Left click on the **Autodesk Assistant** button on the **InfoCenter** Toolbar.

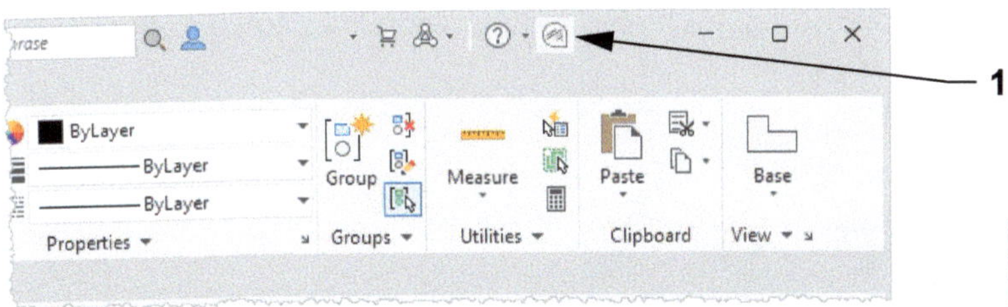

2. The **Autodesk Assistant** Palette will open.

3. If you want to see examples of how best to ask a question, left click on **Show examples**.

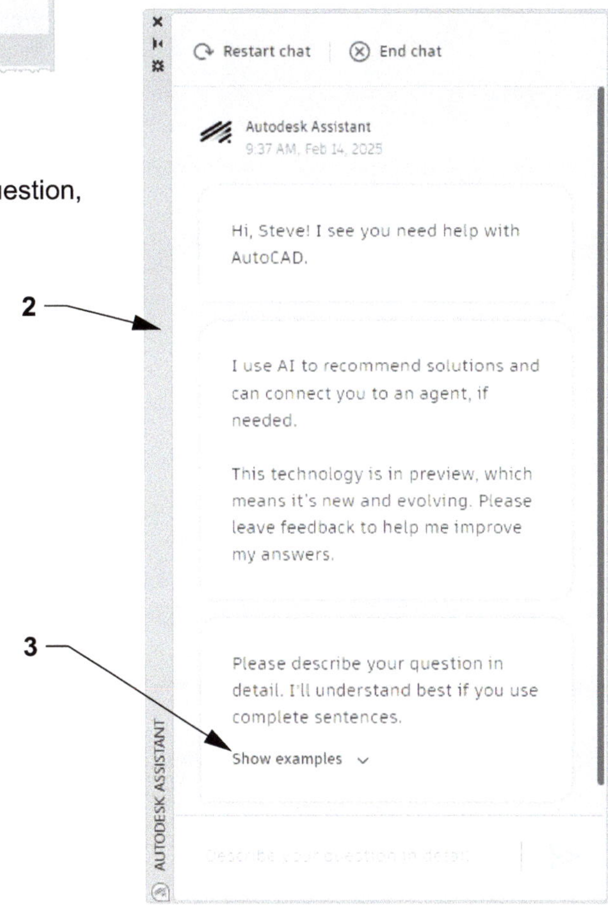

Continued on the next page...

4. Some examples of how to ask a question will be displayed in the window.
5. You can hide the question examples by left clicking on **Hide examples**.
6. When you are ready, type a question in the bottom window, for example, "*How to create a Construction Line?*"
7. Left click on the chevron or press ***<Enter>***.

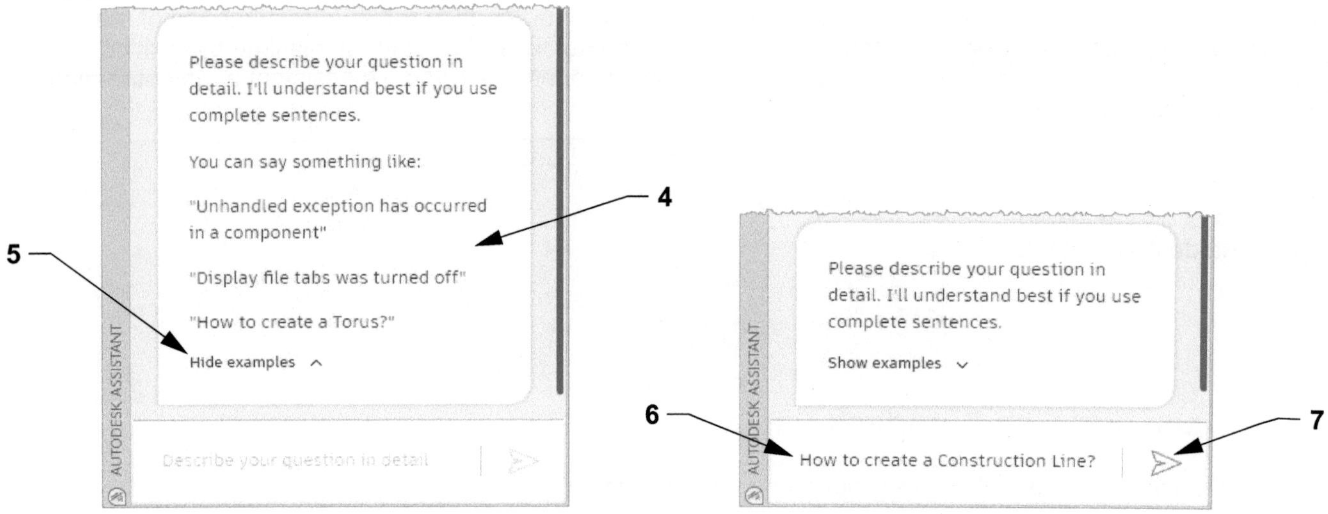

8. The Autodesk Assistant will display an answer to your question. It also may provide links to find other information related to your question.

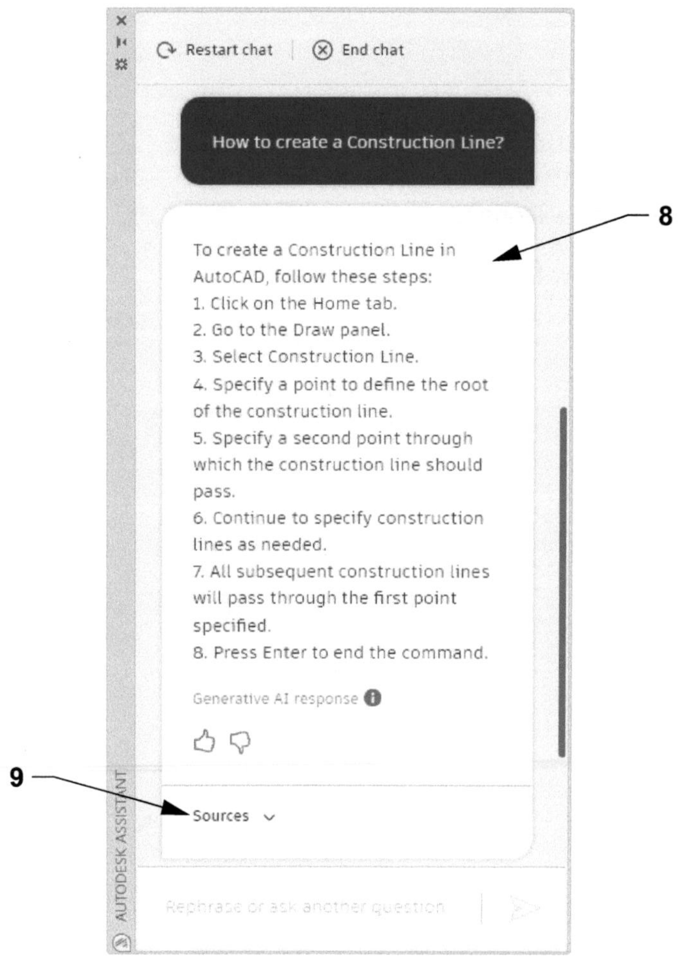

9. If you require more information based on the question you asked, left click on **Sources**.

Continued on the next page...

10. A list of other sources are displayed in the window.

11. You can also connect with an Autodesk Agent by left clicking on **Connect with an agent**.

Note: Your ability to connect with and receive additional information from an Autodesk agent may depend on the type and level of your AutoCAD subscription.

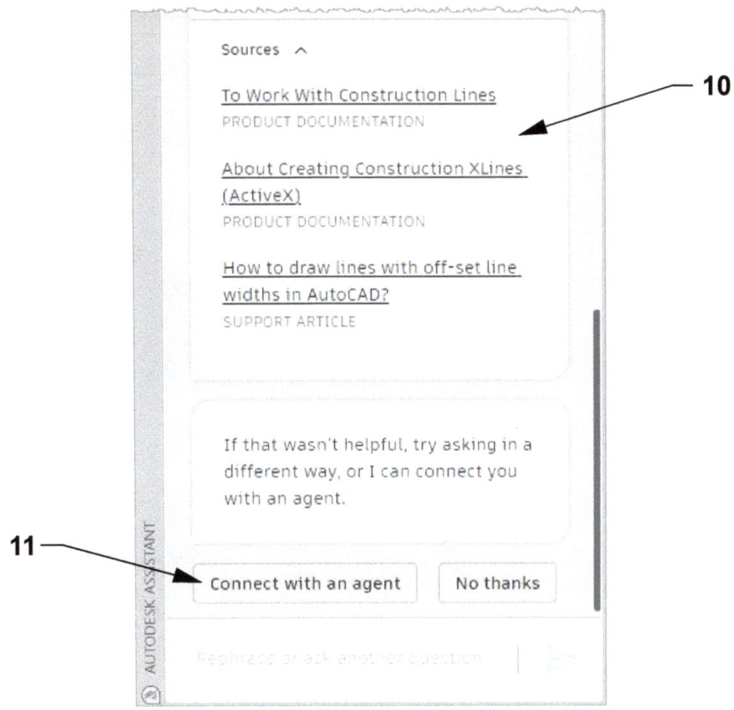

12. If you are satisfied with the answer(s) to the question you asked, select **End chat**.

13. You also can ask the AutoCAD Assistant another question by selecting **Restart chat**.

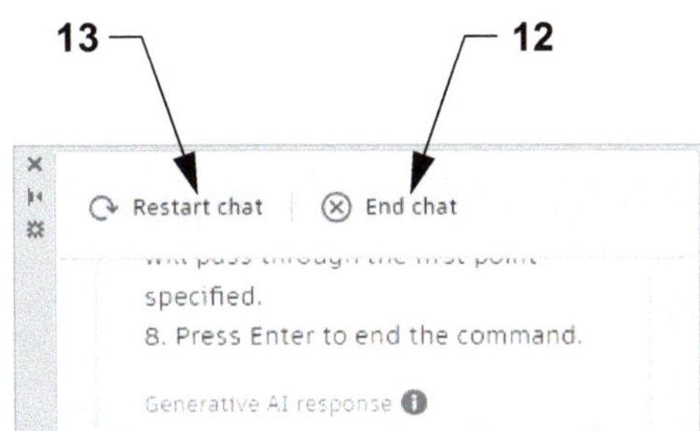

Continued on the next page...

14. After selecting **End chat**, the **End chat** window will appear. Left click on **Yes** to end the current chat session.

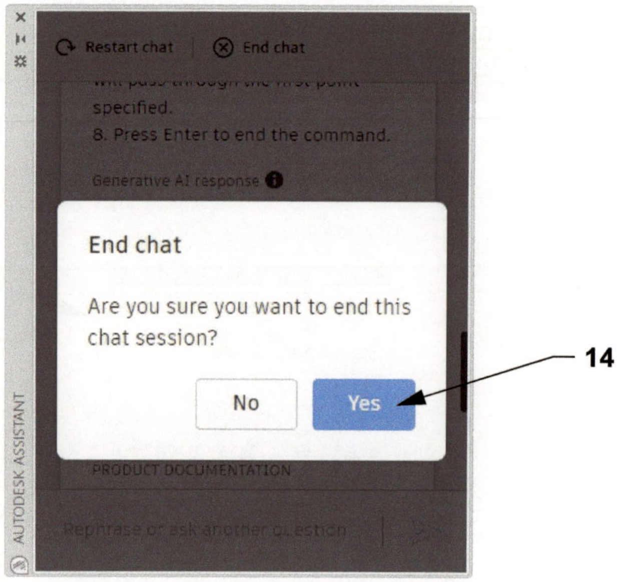

15. Select the **Close Window** text to close the Autodesk Assistant Palette.

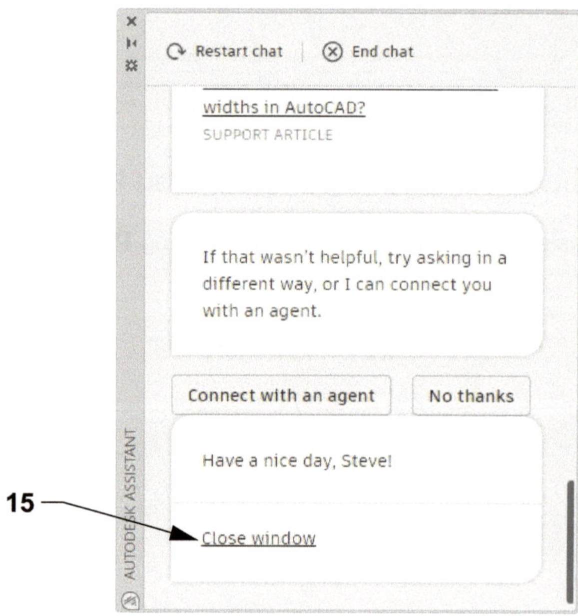

LESSON 2

LEARNING OBJECTIVES

After completing this lesson, you will be able to:

- Create and use a Template
- Select a Command
- Draw, Select, and Erase Objects
- Start a New Drawing
- Open an Existing Drawing
- Use the Activity Insights Palette
- Open Multiple Drawings
- Use Floating File Tabs
- Save, Backup, and Recover a Drawing
- Exit AutoCAD

Creating a Template

The first item on the learning agenda is **how to create a template file** from a drawing file. **This is important:** You will need this template to complete Lessons 2 through 8.

First you need to download a drawing file.

A. Type the website address shown below into your web browser, then press ***<Enter>***.

https://books.industrialpress.com/autocad/workbook-helper.zip

B. The workbook-helper file will download automatically.

C. Save the downloaded file to your desktop, then unzip the file to extract both drawings.

Now you will create a template. (This will be a very easy task.)

1. Start AutoCAD, if you haven't already. (Refer to page 1-2.)
2. Select the **Open** tool from the **Quick Access Toolbar**. (Refer to page 1-10.)

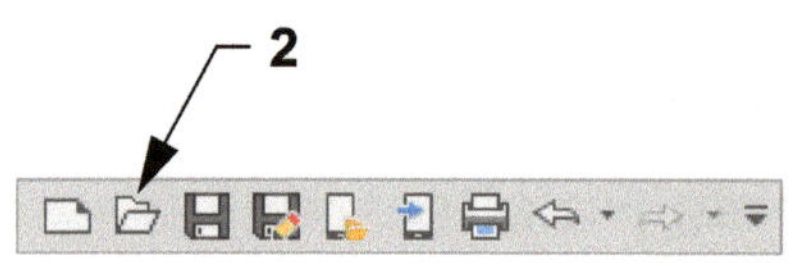

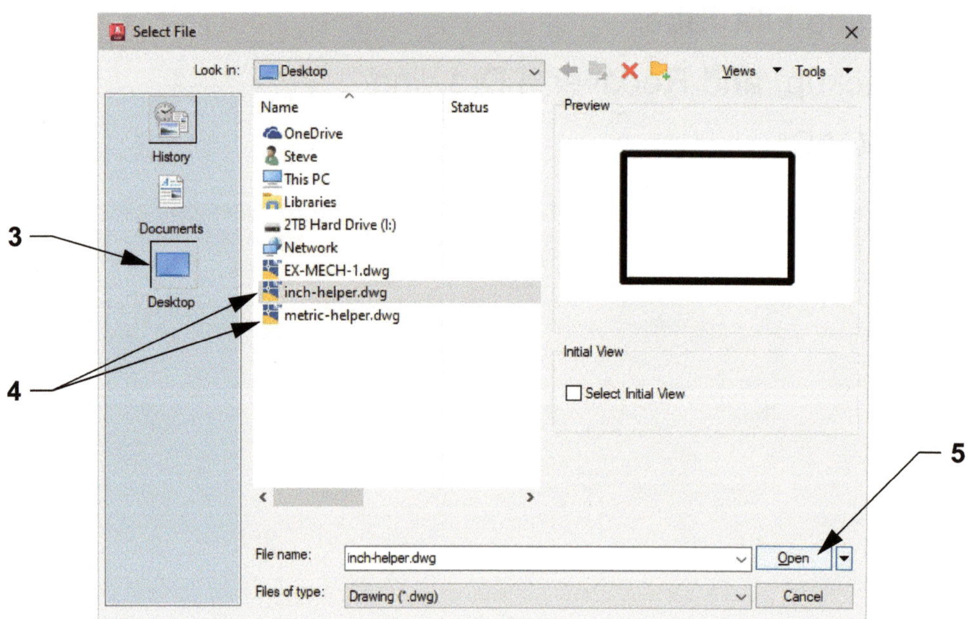

3. Select the **Desktop** directory
4. Select either the **inch-helper** or **metric-helper** drawing.
5. Select the **Open** button located in the lower right corner.

Continued on the next page...

Your screen should appear as shown below.

I created the **Rectangular shape** that appears in the drawing area. I have designed the exercises that follow to fit on a 11" X 8-1/2" sheet of paper for inch users, or to fit on a 297mm x 210mm sheet of paper for metric users. This will enable you to easily print them on any letter or A4 size printer. The rectangle represents an 11" X 8-1/2" or 297mm x 210mm sheet of paper, depending on which file you use. While completing the exercises within this Workbook, please try to draw all objects within this rectangle.

The criss-cross lines are **Grids**. I have set them to display every 1 inch vertically and horizontally for inch users, and every 25mm for metric users. You will learn more about Grids in Lesson 3. For now, notice that the grids are 11 horizontally and 8.5 vertically for inch users, and 11.88 horizontally and 8.4 vertically for metric users. Grids are merely a visual aid and will not print. The size may be changed at any time and they may be turned **on** or **off** easily by selecting the **"Grid"** button on the Status Bar, or by pressing the ***<F7>*** key. (Refer to page 1-13.)

Notice the **drawing file name** is displayed here.

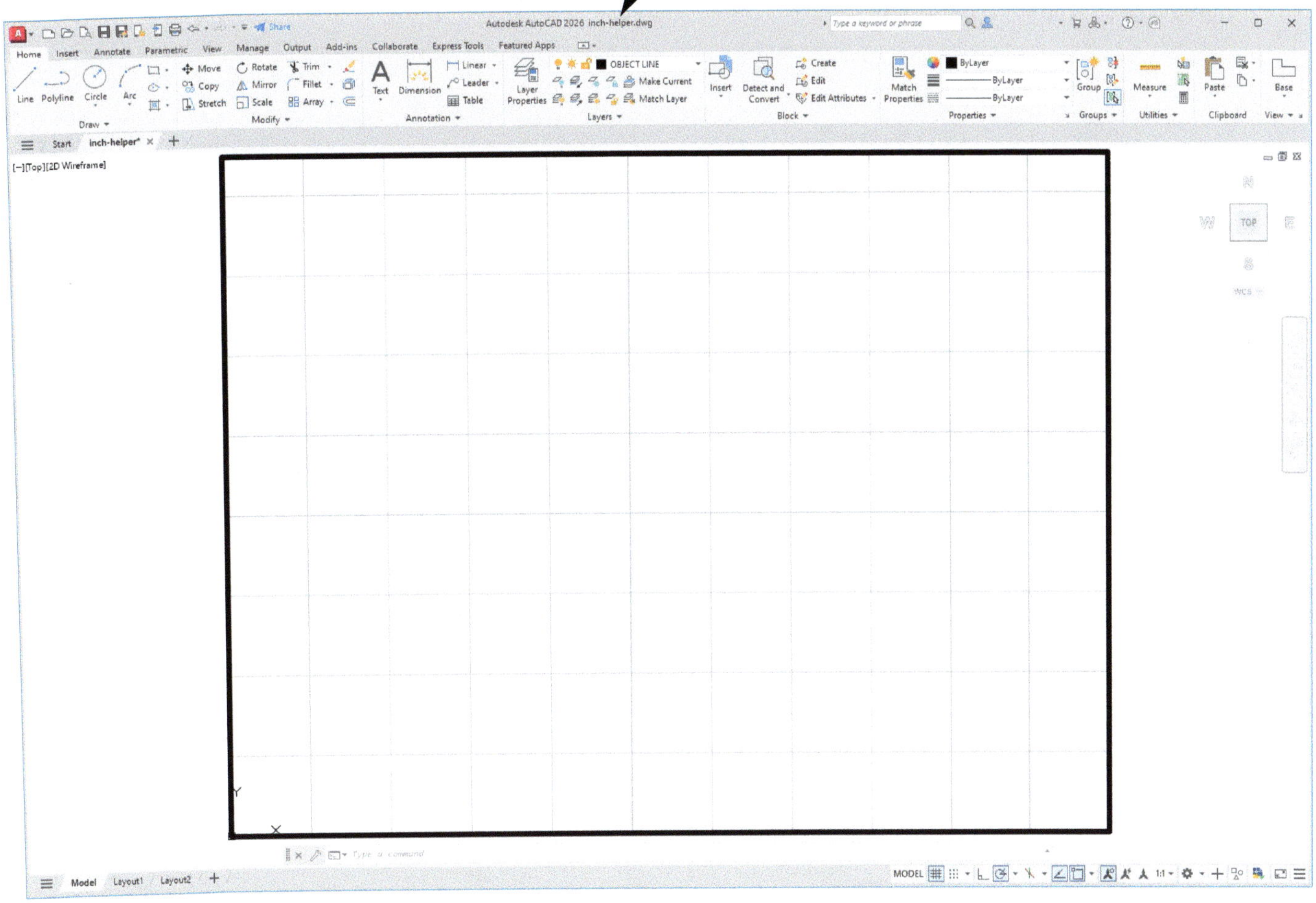

The next step is to create a template from this drawing.

6. Select the **"Application Menu"** button. 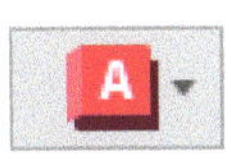

7. Select **Save As** ▸ (Click on the arrow not the words "Save As", as shown on the next page.)

8. Select **"Drawing Template"**.

Continued on the next page...

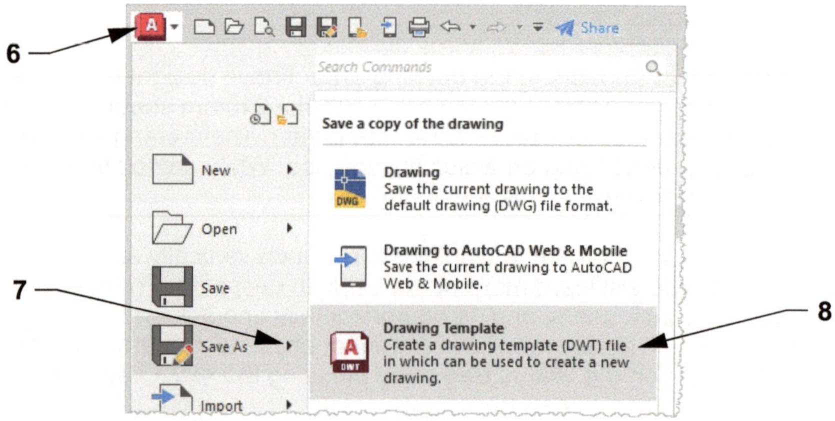

9. The name of the new file should already be highlighted in the **"File name"** box; if it's not, just type in **inch-helper** or **metric-helper** in the **File name** box. Do not type the extension **.dwt**. AutoCAD will add it automatically.

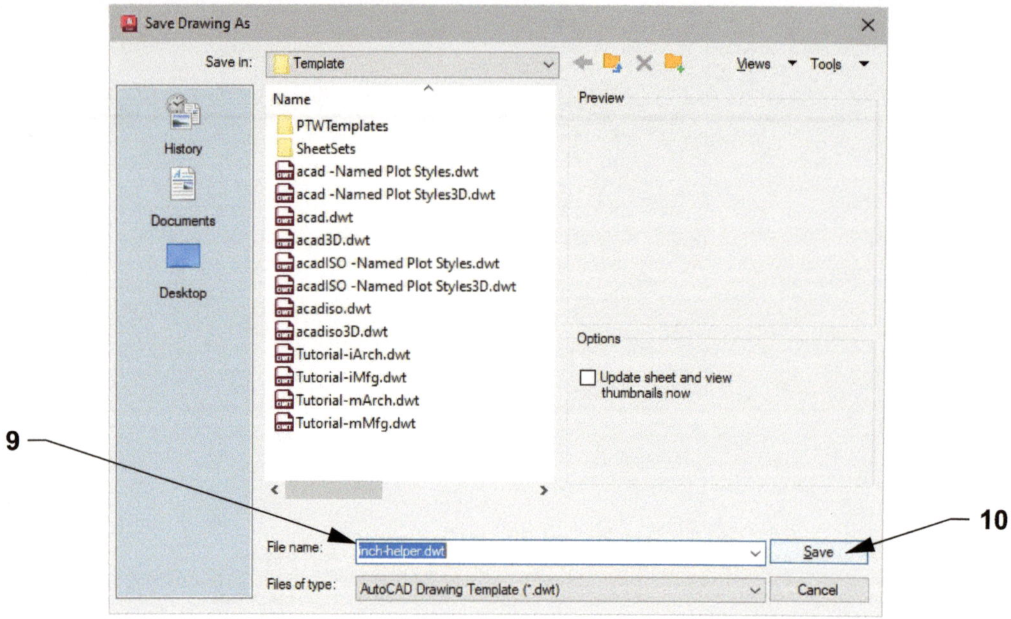

10. Select the **Save** button.

11. Type the description and make the selections as shown below.

12. Select the **OK** button.

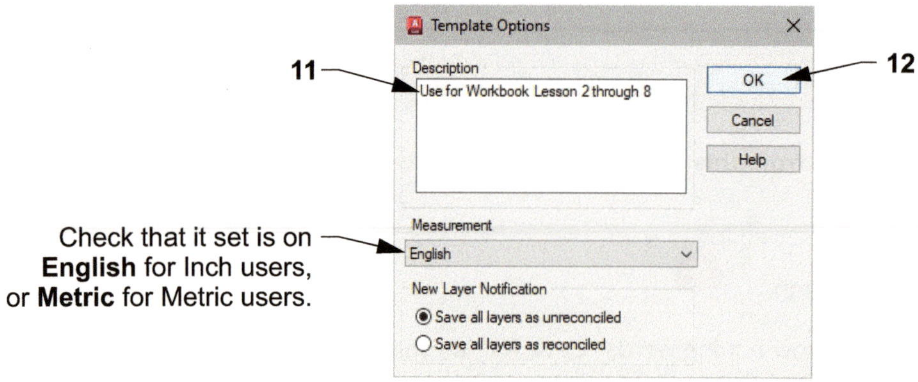

Note: Now you have a template to use for Lessons 2 through 8. At the beginning of each exercise, you will be instructed to start a **new** drawing using either the **inch-helper.dwt** or **metric-helper.dwt**. Using a template as a master setup drawing is very good CAD management.

Using a Template

The template that you created following the instructions on the previous pages will be used for Lessons 2 through 8. Many variables have been preset in this template. This will allow you to start drawing immediately. You will learn how to set those variables before you complete this Workbook, but for now you will concentrate on learning the AutoCAD commands and hopefully have some fun.

How to use a Template

1. Select the **New** tool from the **Quick Access Toolbar.**

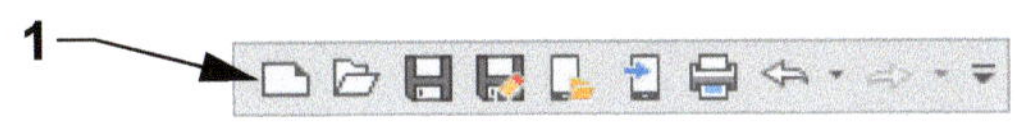

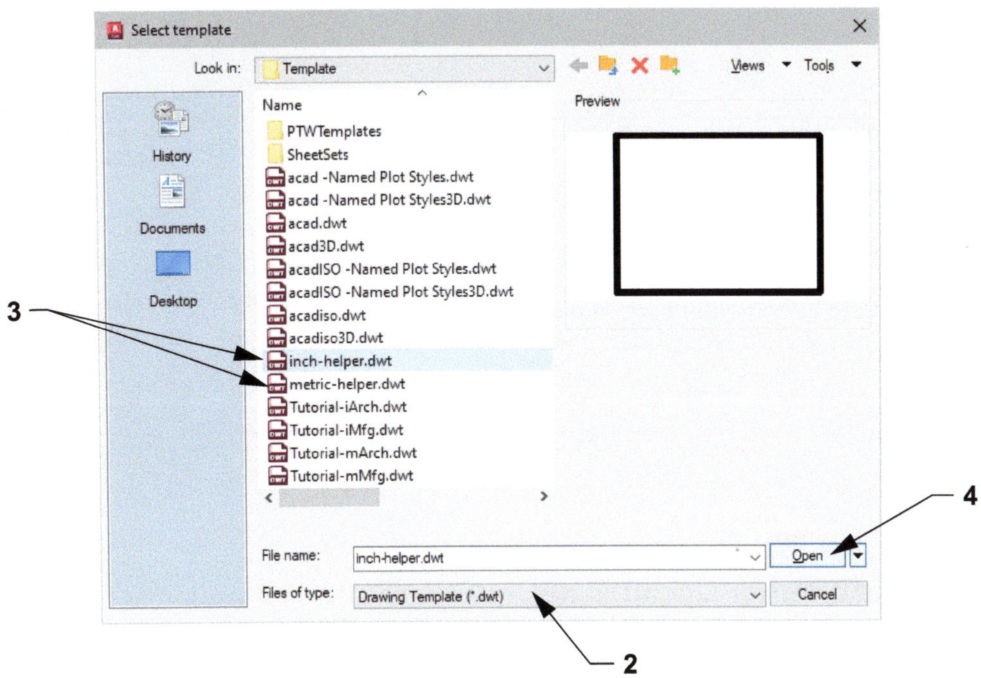

2. Select **Drawing Template (*.dwt)** from the **"Files of type"** if not already selected.

3. Select either the **inch-helper.dwt** or **metric-helper.dwt** from the list of templates.

   Note: If you do not have these templates, refer to page 2-2 for instructions.

4. Select the **Open** button.

Selecting a Command

AutoCAD provides you with two different methods for selecting commands. One is **selecting a tool from the Ribbon**, the other is **typing the command**. Both methods will accomplish the same end result. You decide which method you prefer.

Instructions for Method 1 are shown on the next page, followed by Method 2.

Continued on the next page...

Method 1. Selecting a tool from the Ribbon

1. First select a **Tab** such as **Home**.
2. Locate the correct **Panel** such as **Draw**.
3. Select a **tool** such as **Circle**.

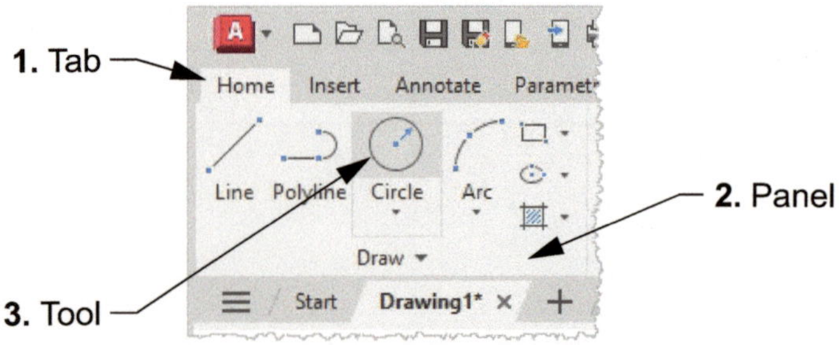

If the tool includes multiple types, it will have a down arrow ▾

If you select the down arrow, a sub-menu will appear.

Select the desired type such as **2-Point**.

The latest selection will then become the current displayed tool, because AutoCAD assumes that you may need that tool again.

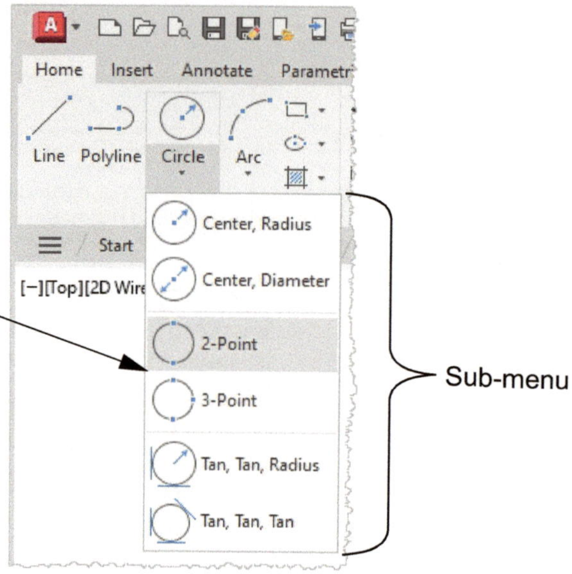

Method 2. Keyboard entry, Command Line

You may type commands on the **Command Line** (shown below) or in the **Dynamic Input Tooltip** box (shown on the next page). It depends on whether you have Dynamic Input turned **on** or **off**.

How to enter a command on the Command Line

1. **Important:** Place the cursor in the Command Line area.
2. Type the first letter of a command, such as ***c*** for **circle**.

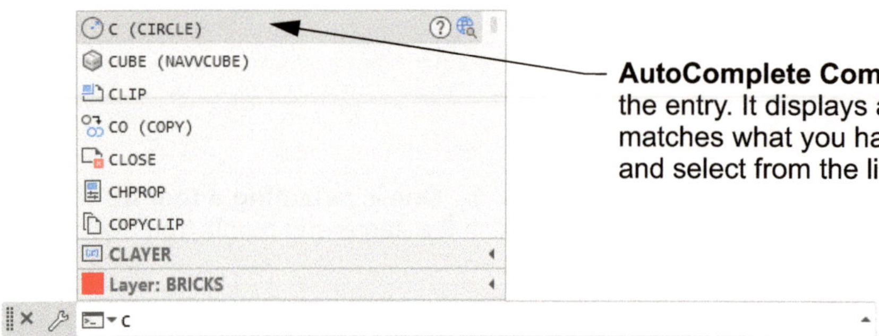

AutoComplete Command Entry automatically completes the entry. It displays a list of all the commands whose prefix matches what you have typed. This enables you to scroll and select from the list.

Continued on the next page...

3. A list of commands that begin with the letter **C** will appear. Scroll down and select a command from the list.

4. When you select a command such as **Circle**, the **prompt** and **options** will be displayed on the Command Line.

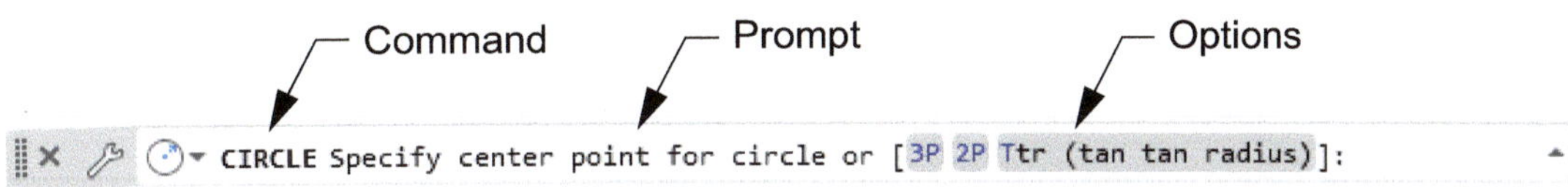

5. The **prompt** for the Circle command asks you to:

   "Specify center point for circle" or [3P 2P Ttr (tan tan radius)]:

The information within the **[ ]** brackets are options that you may select.

Method 2. Keyboard entry, Dynamic Input

Dynamic Input is another method of inputting commands, values, and select options. To use Dynamic Input, you must turn **on** the **Dynamic Input** button in the Status Bar, shown on page 1-13. 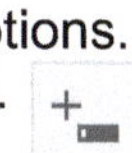

If you choose to use Dynamic Input, the command will be entered in the Tooltip box beside the cursor.

How to enter a command using Dynamic Input

1. **Important:** Place the cursor in the Drawing Area.

2. Type the first letter of a command, such as ***c*** for **circle**.

3. A list of commands that begin with the letter **C** will appear. Select the command from the list.

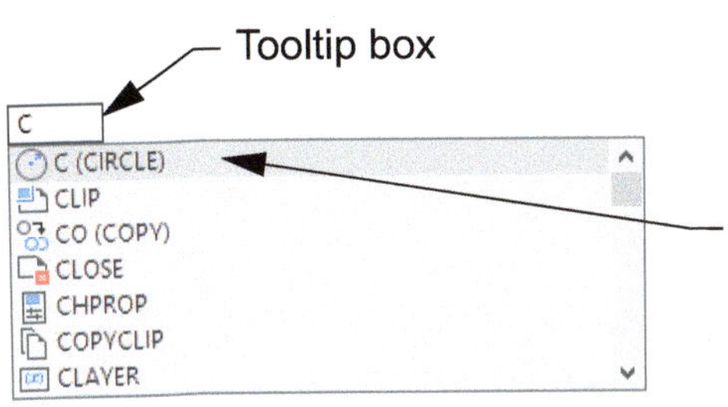

AutoComplete Command Entry automatically completes the entry. It displays a list of all the commands whose prefix matches what you have typed. This enables you to scroll and select from the list.

4. If you press the ↓ arrow, the options will appear below the prompt.

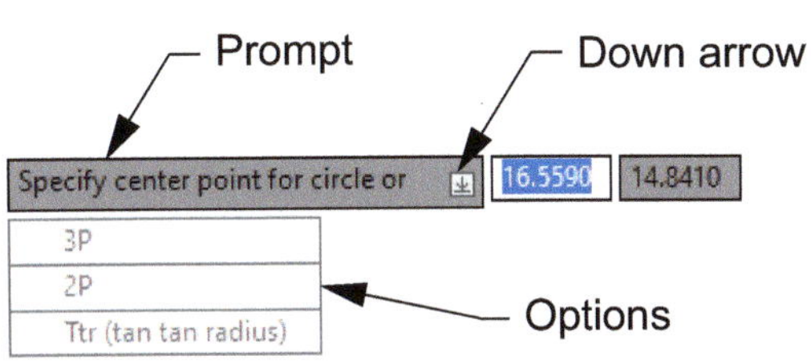

Notice that the command entry also is being displayed on the Command Line. Whether you use the Command Line or Dynamic Input is your choice.

Drawing Lines

A **Line** can be **one segment** or a **series of connected segments**. But each segment is an individual object.

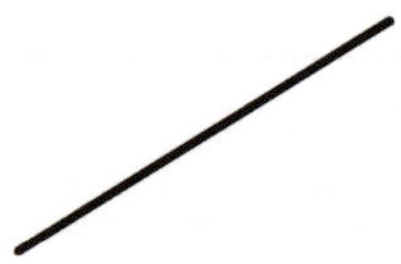

One segment
One object

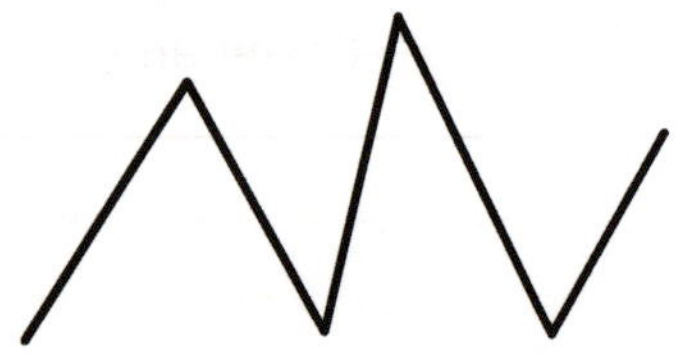

Series of connected segments
5 objects

Start the **Line command** using one of the following methods:

Ribbon = Home Tab / Draw Panel /
or
Keyboard = *L* *<Enter>*

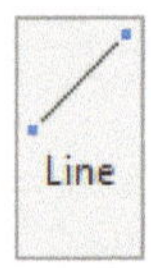

Lines are drawn by specifying the locations for each endpoint.
Move the cursor to the location of the **"first"** endpoint (**1**), then press the left mouse button and release. (Click and release, do not Click and Drag.) Move the cursor again to the **"next"** endpoint (**2**) and press the left mouse button. Continue locating **"next"** endpoints until you want to stop drawing lines.

There are 2 ways to stop drawing a line: press the ***<Enter>*** key **or** press the ***<Spacebar>***.

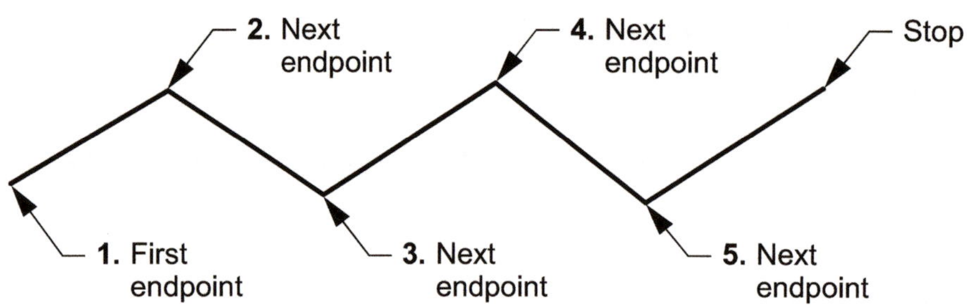

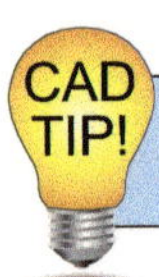

To **quickly** repeat the Line command, press the ***<Spacebar>***.

Horizontal and Vertical Lines

To draw a Line perfectly horizontal or vertical, select the **Ortho** mode by selecting the **Ortho** button on the Status Bar or pressing the ***<F8>*** key.

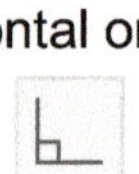

Try the following example:

1. Select the **Line** command.
2. Place the **first endpoint** anywhere in the drawing area.
3. Turn **Ortho on** by selecting the **Ortho** button or ***<F8>***. (The **"Ortho"** button will change to a neon blue when **on**.)

Continued on the next page...

4. Move the cursor to the right and press the left mouse button to place the **next endpoint**. (The line should appear perfectly horizontal.)

5. Move the cursor down and press the left mouse button to place the **next endpoint**. (The line should appear perfectly vertical.)

6. Now turn **Ortho off** by selecting the **Ortho** button. (The **"Ortho"** button will change to gray when **off**.)

7. Now move the cursor up and to the right on an angle (the line should move freely now) and press the left mouse button to place the **next endpoint**.

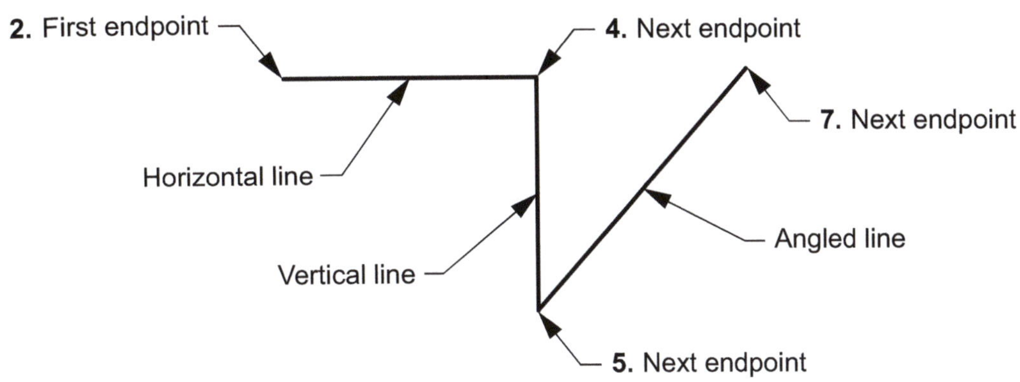

CAD TIP! Ortho can be turned **on** or **off** at any time while you are drawing. It can also be turned **on** or **off** temporarily by holding down the ***<Shift>*** key.

Closing Lines

If you have drawn two or more line segments, **the endpoint of the last line segment** can be automatically connected to the **first endpoint**, with the space between them closed up, by using the **Close** option.

Try the following example:

1. Select the **Line** command.
2. Place the **first endpoint**.
3. Place the **next endpoint**.
4. Place the **next endpoint**.
5. Type ***C <Enter>***.

   Or

5. Press the **right** mouse button and select **Close** from the **Shortcut Menu**.

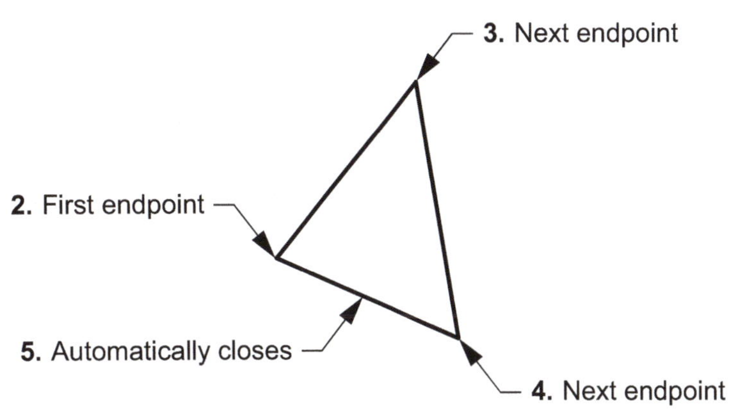

Continued on the next page...

What is the Shortcut Menu?

The **Shortcut Menu** gives you quick access to command options.

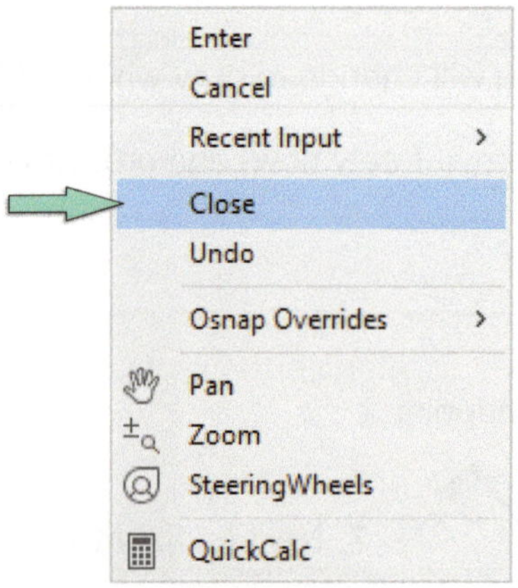

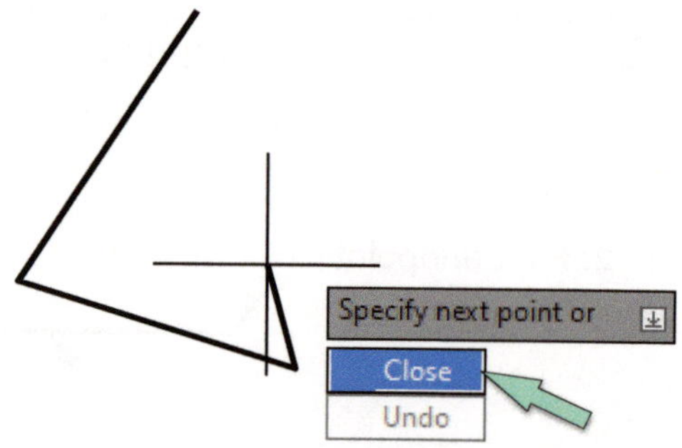

Using the Shortcut Menu:

Press the right mouse button. The Shortcut Menu will appear. Select an option.

Using the Dynamic Input down arrow:

You may use the right mouse button or press the down arrow ↓ and the options will appear below the Dynamic Input prompt.

Methods for Selecting Objects

Many AutoCAD commands prompt you to **"select objects"**. This means selecting the objects that you want the command to affect. There are 3 methods:

Method 1. Pick, is very easy and should be used if you have only 1 or 2 objects to select.

Method 2. Window selection, is a little more difficult, but once mastered it is extremely helpful and time saving.

Method 3. Lasso selection, is a little more difficult than Window selection, but again, once mastered it is very useful and will save you time. Practice the following examples.

Method 1. Pick

Place the cursor on the object, but do not press the mouse button yet. The object will highlight. This appearance change is called **"Rollover Highlighting"**. It gives you a preview of which object AutoCAD is recognizing. Press the left mouse button to actually select the highlighted object.

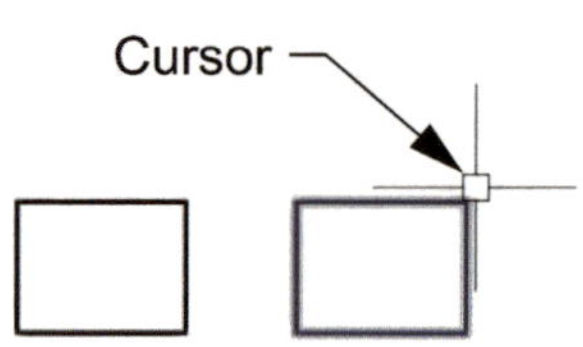

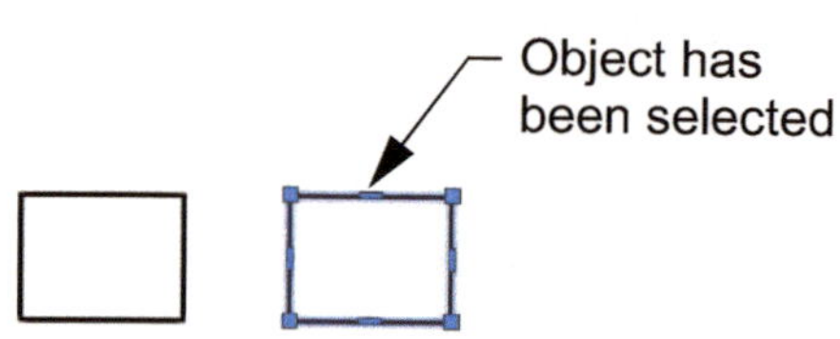

Method 2. Window selection, Crossing and Window

Crossing:

Place your cursor in the area **up** and to the **right** of the objects that you want to select (**P1**) and press the left mouse button. (**Do not** hold the mouse button down. Just press and release.) Then move the cursor **down** and to the **left** (**P2**) and press the left mouse button again. (**Note:** The Window will be ***green*** and the outer line will be ***dashed***.) **Only** the objects that this Window **crosses** will be selected.

Continued on the next page...

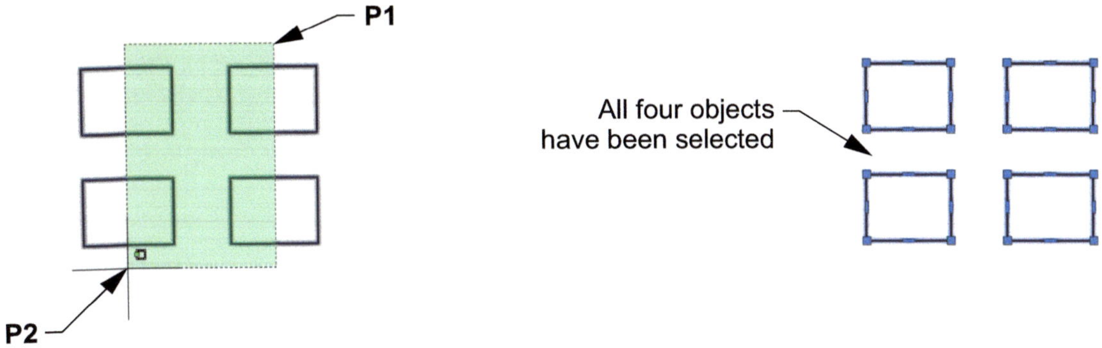

In the example above, all four rectangles have been selected because the Crossing Window **crosses** a portion of each.

Window:

Place your cursor in the area **up** and to the **left** of the objects that you want to select (**P1**) and press the left mouse button. (**Do not** hold the mouse button down. Just press and release.) Then move the cursor **down** and to the **right** of the objects (**P2**) and press the left mouse button again. (**Note:** The Window will be ***blue*** and the outer line will be ***solid***.) **Only** the objects that this Window **completely encloses** will be selected.

In the example below, only two rectangles have been selected. (The other two rectangles are **not** completely enclosed in the **Window**.)

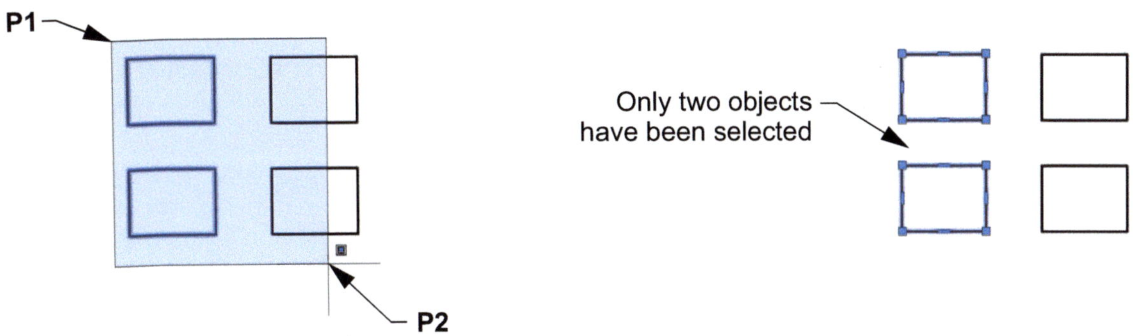

Method 3. Lasso selection, Crossing, Window, and Fence

Crossing:

Place your cursor in the area **up** and to the **right** of the objects that you want to select (**P1**), then press and hold the left mouse button. (**Do not** release the mouse button.) Then move the cursor in an counter-clockwise direction until you have crossed the objects you want to select (**P2**) and then release the left mouse button. (**Note:** The Lasso Window will be ***green*** and the outer line will be ***dashed***.) **Only** the objects that the Lasso Window **crosses** will be selected.

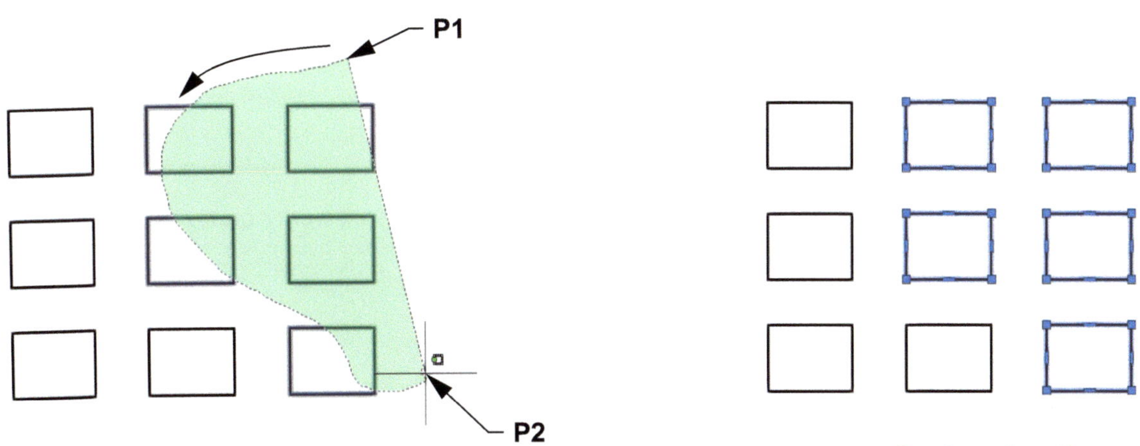

Continued on the next page...

Window:

Place your cursor in the area **up** and to the **left** of the objects that you want to select (**P1**), then press and **hold** the left mouse button. (**Do not** release the mouse button.) Then move the cursor in a clockwise direction until you have completely enclosed the objects you want to select (**P2**), and then release the left mouse button. (**Note:** The Window will be ***blue*** and outer line will be ***solid***.) Only the objects that this Window **completely encloses** will be selected.

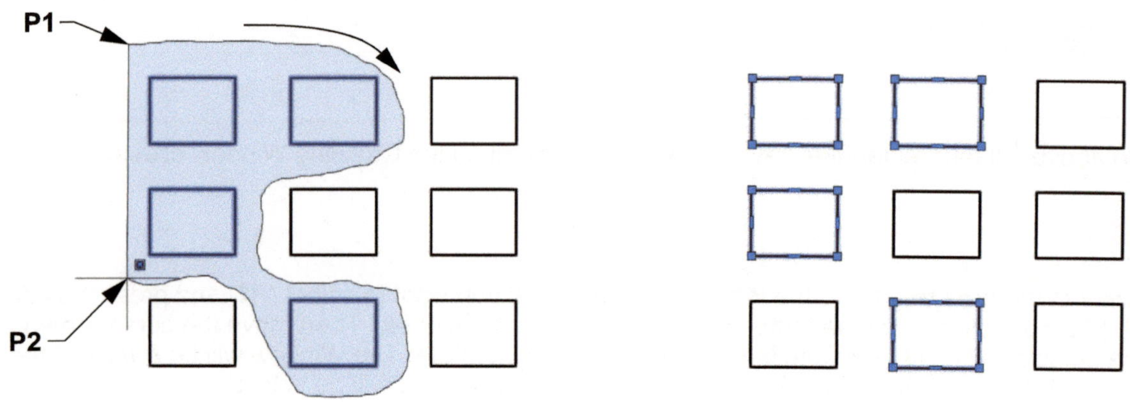

Fence:

With the **Fence** option of the **Lasso** selection you can place the mouse cursor in any position you choose. For this example, place your cursor at (**P1**), then press and **hold** the left mouse button. (**Do not** release the mouse button.) Move the mouse until you see either the green or blue lasso, then press the ***<Spacebar>*** until you see just a **Dashed Fence Line**. Move the mouse over the objects you want to select (**P2**), and then release the left mouse button. Only the objects that the Fence line crosses will be selected.

You may have to press the ***<Spacebar>*** twice to activate the **Fence Line** option.

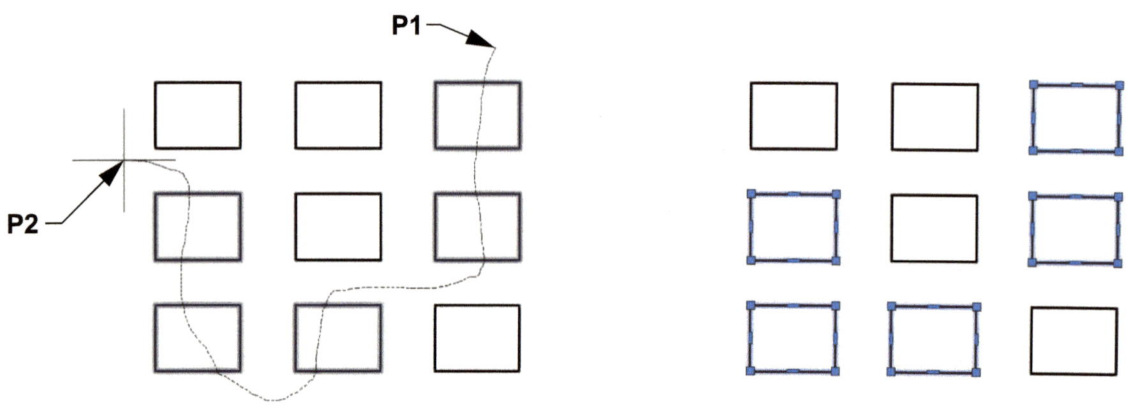

Erase

There are three methods to erase (delete) objects from the drawing. They all work equally well. You decide which one you prefer to use.

Method 1.

Select the **Erase** command first and then select the objects.

Continued on the next page...

Example:

1. Start the **Erase** command using one of the following:

   Ribbon = Home Tab / Modify Panel /
   or
   Keyboard = *E <Enter>*

2. The following will appear on the Command Line:

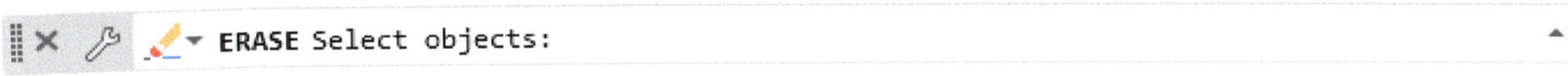

3. Select one or more objects and then press ***<Enter>*** to erase the object(s).

Method 2.

Select the objects first and then press the ***<Delete>*** key.

Example:

1. Select the object(s) to be erased.
2. Press the ***<Delete>*** key.

Method 3.

Select the objects first and then select **Erase** from the Shortcut Menu.

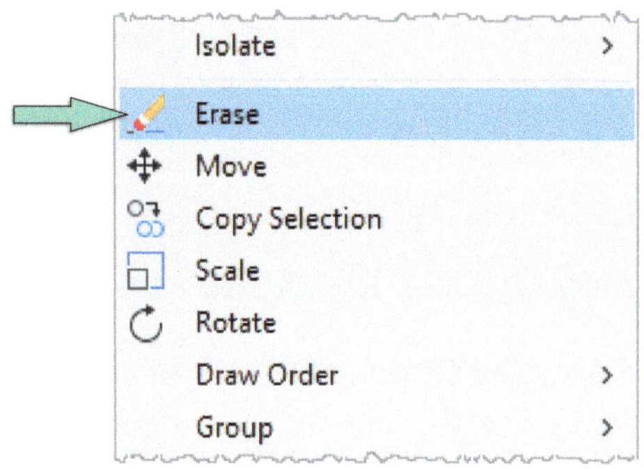

Example:

1. Select the object(s) to be erased.
2. Press the right mouse button.
3. Select **Erase** from the Shortcut Menu using the left mouse button.

Very Important:
If you want the erased objects to return, select the **Undo tool** from the **Quick Access Toolbar**. This will **Undo** the last command.

Undo and Redo

The **Undo** and **Redo** tools allow you to undo or redo **previous commands**. For example, if you erase an object by mistake, you can **undo** the previous "erase" command and the object will reappear. So don't panic if you do something wrong. Just use the **Undo** command to remove the previous commands.

The **Undo** and **Redo** tools are located in the **Quick Access Toolbar**.

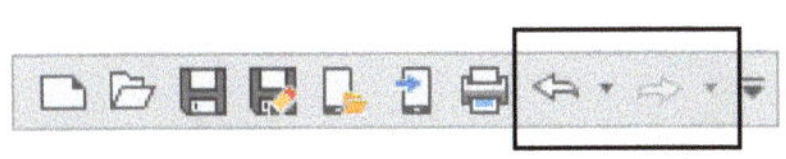

CAD TIP! You may **undo** any commands used during a work session until you close the drawing.

Continued on the next page...

How to use the Undo tool

1. Draw a line, circle, and a rectangle.

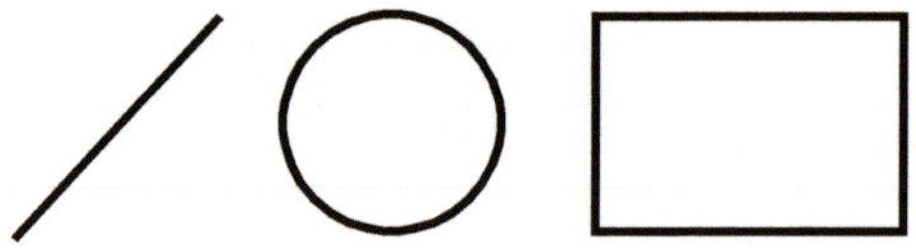

Your drawing should look approximately like this.

2. Next erase the circle and the rectangle.

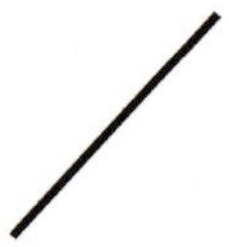

(The circle and the rectangle disappear.)

3. Select the **Undo** arrow. 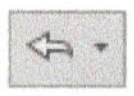

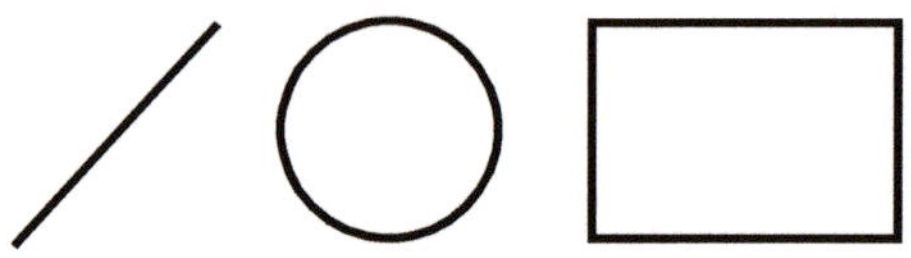

You have now deleted the **Erase** command operation. As a result, the erased objects reappear.

How to use the Redo tool

Select the **Redo** arrow 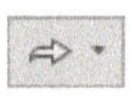 and the circle and rectangle will disappear again.

Starting a New Drawing

Starting a new drawing means that you want to start with a previously created Template file. That is why I taught you *"How to create a template"* at the beginning of this lesson. You will use either the **inch-helper.dwt** or the **metric-helper.dwt** template each time you are instructed to start a new drawing.

Note: Do not use the **New** tool if you want to **open** an **existing drawing**. Refer to the next page to open an existing drawing file.

How to start a new drawing

1. Select the **New** tool from the **Quick Access Toolbar**.

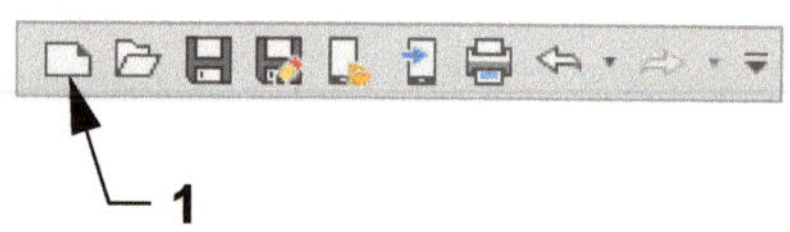

2. Select either the **inch-helper.dwt** or the **metric-helper.dwt** from the list of templates.

Note: If you do not have these templates, refer to page 2-2 for instructions.

Continued on the next page...

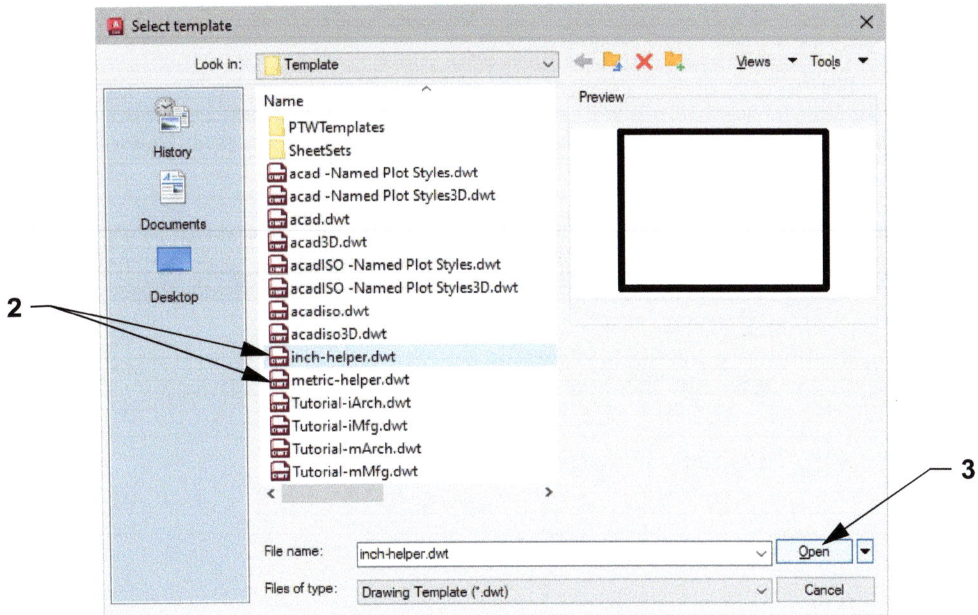

3. Select the **Open** button.

Opening an Existing Drawing File

Opening an **Existing Drawing File** means that you would like to open, on the screen, a drawing that has been previously created and saved. Usually you are opening it to continue working on it or to make some changes.

1. Select the **Open** tool from the **Quick Access Toolbar**.

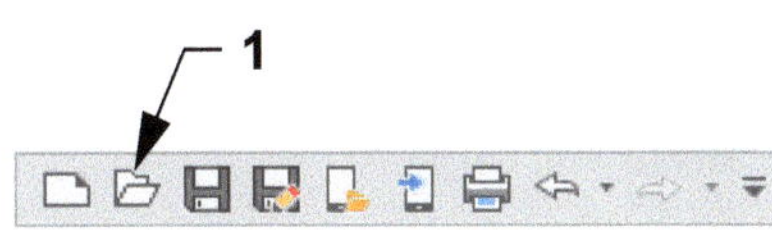

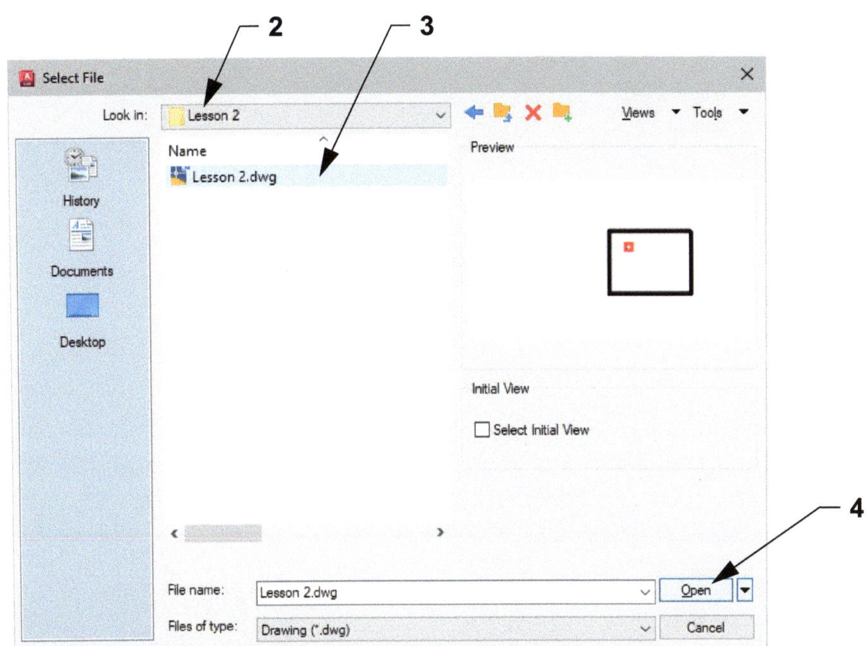

2. Locate the Directory and Folder in which the file had previously been saved.

3. Select the file that you want to open.

4. Select the **Open** button.

Activity Insights Palette

The **Activity Insights Palette** provides users with access to a What's Changed summary of what has been done in each creation and edit session of a drawing. Activity Insights is available on any drawing stored locally, on a local area network (LAN), or in Autodesk Docs.

The Activity Insights What's Changed summary is connected to the drawing and not to individual users. This means that collaborators working on the drawing can view the same insights, and their actions will be added to the updated summary. You must be connected to the internet to receive the What's Changed summary.

Note: The What's Changed summary will not show in the Activity Insights Palette if you stored your drawing on a cloud storage provider such as Microsoft OneDrive, Dropbox, Box, or Google Drive.

How to open the Activity Insights Palette

1. Start a new drawing.
2. Select the **View** Tab.
3. Select the **Activity Insights** tool on the **History** Panel.

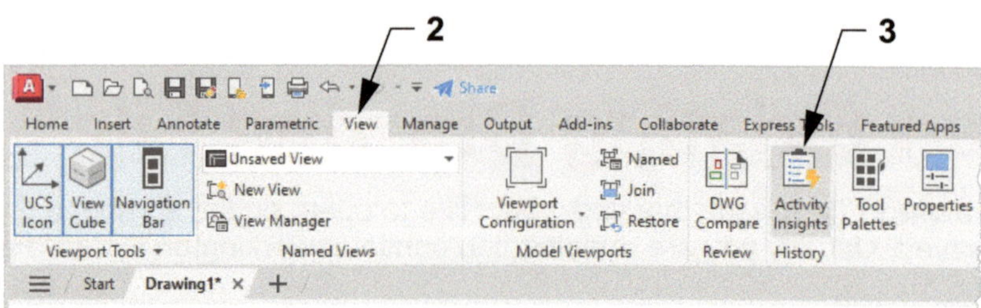

The Activity Insights Palette will appear.

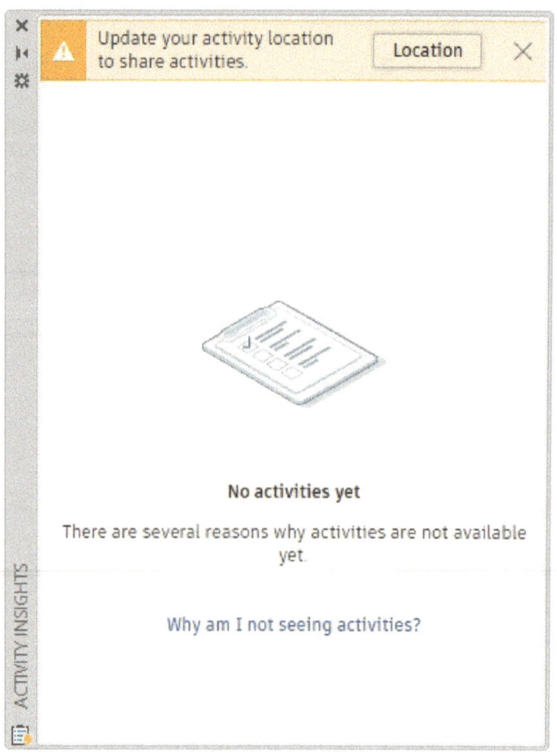

Because you have created a new drawing, no activities will show in the Palette. Once you save the drawing, activities will then start to show in the Palette.

Continued on the next page...

By default, the location for the Activity Insights data is stored locally in a folder. The Activity Insights data location can be changed to any location you choose.

If you are a single user and do not want to share your drawing's Activity Insights data with other users, you can leave the data in the default location that only you have access to. If you do want to share your drawing's Activity Insights data with other users, you can move the Activity Insights data to any location you choose, provided that location is on a LAN or in Autodesk Docs.

How to change the Activity Insights data location

1. Select the **Location** button on the Activity Insights Pallete.
2. In the **Browse for folder** dialog box, select the **Look In** drop-down list.
3. Choose the location where you want to store the Activity Insights data.

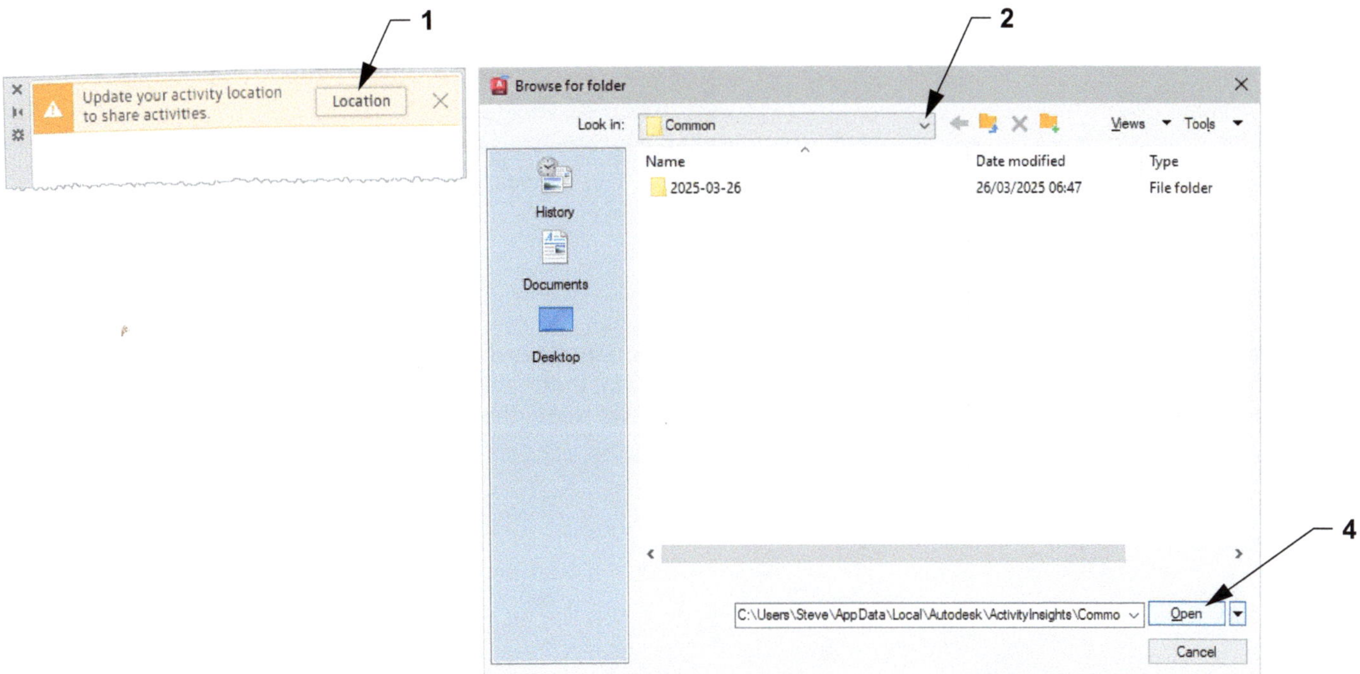

4. Select the **Open** button to close the **Browse for folder** dialog box.

The Activity Insights data is now stored in the new location you chose, and the Location button will disappear from the Palette.

Note: Remember, if you want to share your drawings Activity Insights data with other users, make sure you choose a location on a LAN or in Autodesk Docs.

After starting a new drawing and then saving it, the activities will start to show in the Activity Insights Palette. Depending on the speed of the Autodesk servers, these activities may take a few seconds to appear. The example below shows the **New drawing created** activity in the Palette.

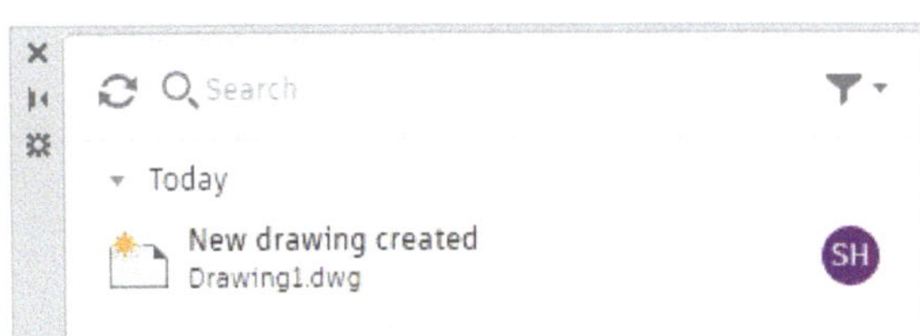

Continued on the next page...

If you left click on an activity, the **Activity Properties** information will appear.

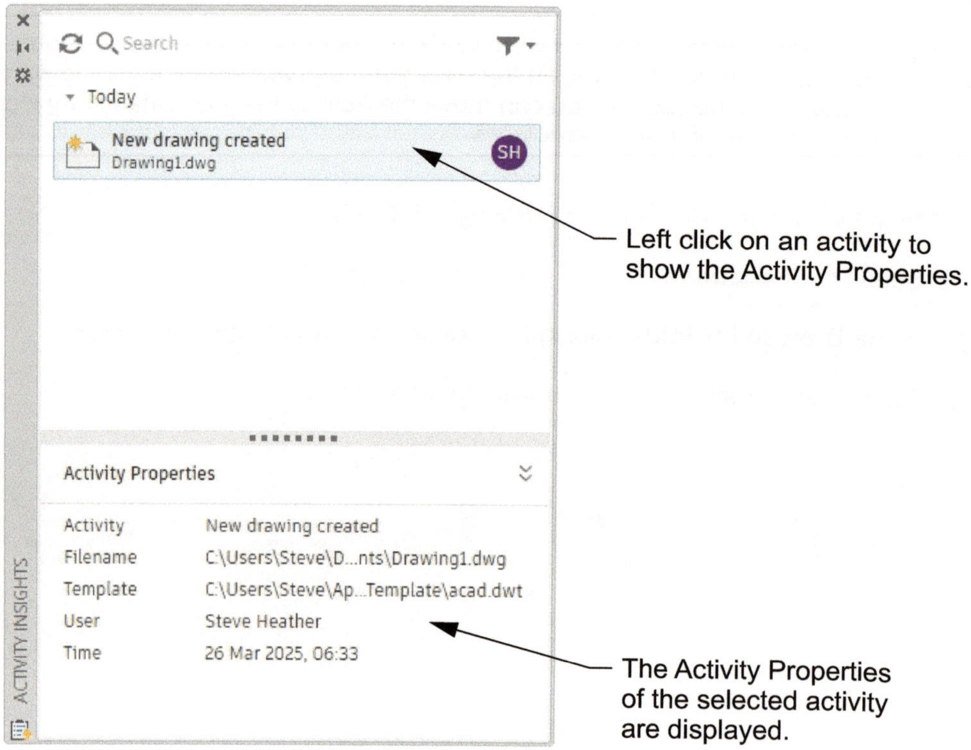

If you open an existing drawing and then left click on the previously **Saved** activity, the Activity Properties will show all the changes to the drawing since the last save of that drawing. In the example below, the Activity Properties shows what has changed since the previous save, the Workflows used, the Editing time, and the File size.

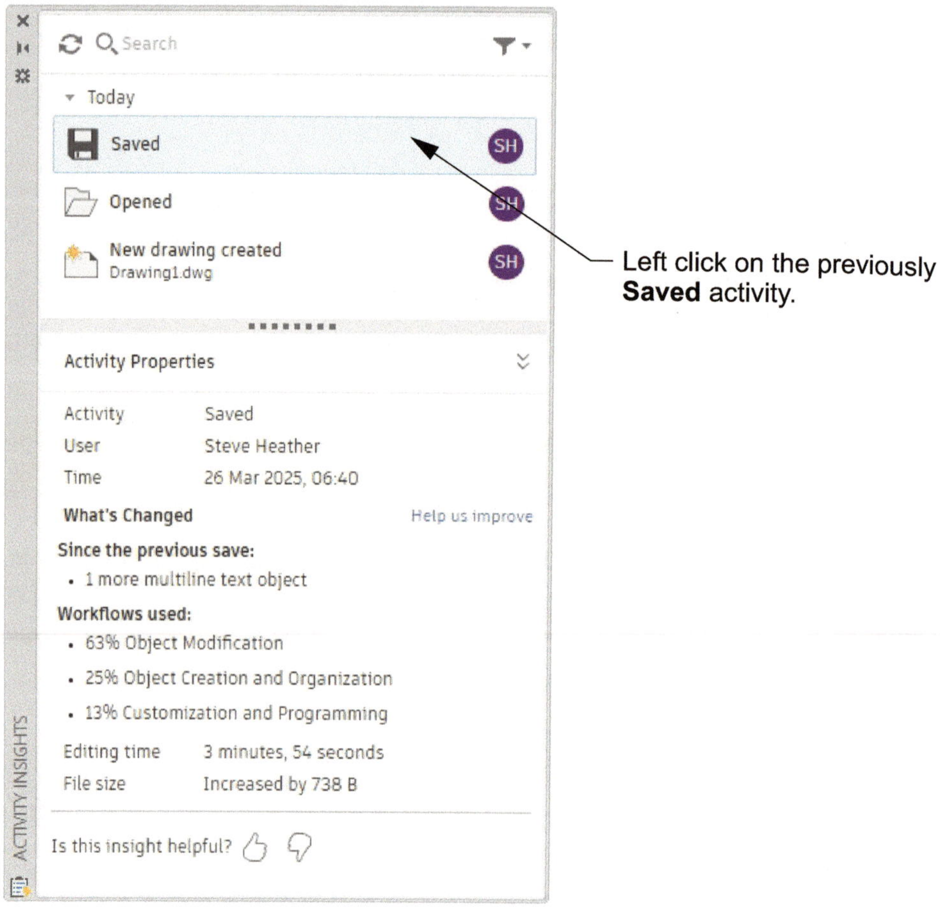

Continued on the next page...

You also can choose to **Filter** the activities in the Activity Insights Palette. You can choose to filter by **Time**, **Users**, and **Activities**.

How to Filter the Activities

1. Select the **Filter** button on the Activity Insights Pallete.
2. In the **Filter** drop-down list, left click to select any of the filters. (You can choose more than one filter.)
3. Select the **Filter** button again to close the Filter drop-down list.

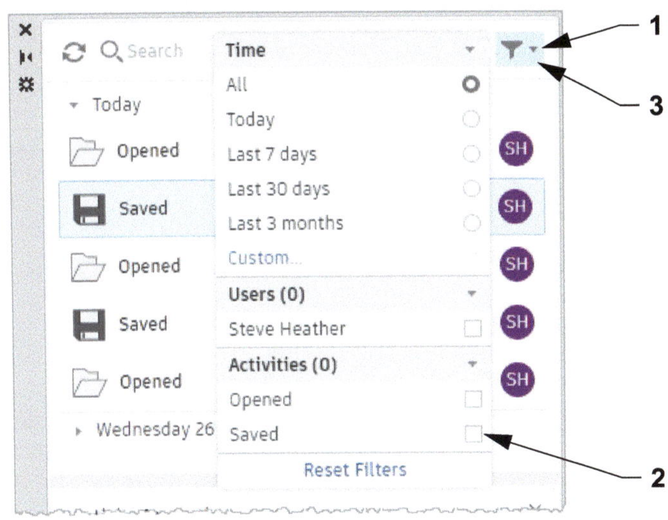

In the example below, the **Saved** Activities Filter has been selected. This means that all other activities will be hidden from the Activity Insights Palette.

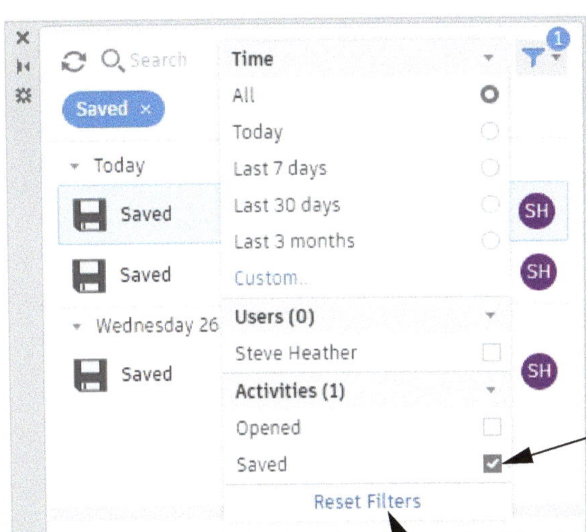

When the Saved activities filter has been selected, only the Saved activities will show in the Palette.

To reset the Filters back to default, simply left click on **Reset Filters** in the Filter drop-down list.

Opening Multiple Files

The **File Tabs** tool allows you to have multiple drawings open at the same time. If the File Tabs tool is switched **on** (on by default), you can open existing saved drawings or create new ones.

The **File Tabs** tool is located on the **Interface** Panel of the **View** Tab, and is a neon blue color when switched on.

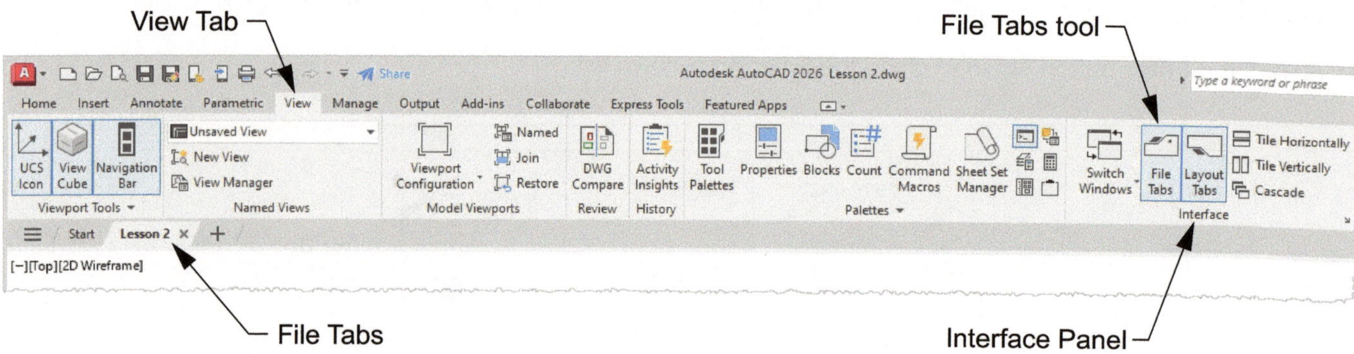

How to open an existing saved drawing from the File Tabs

1. Right mouse click on the **+** icon.
2. Select **Open** from the menu.

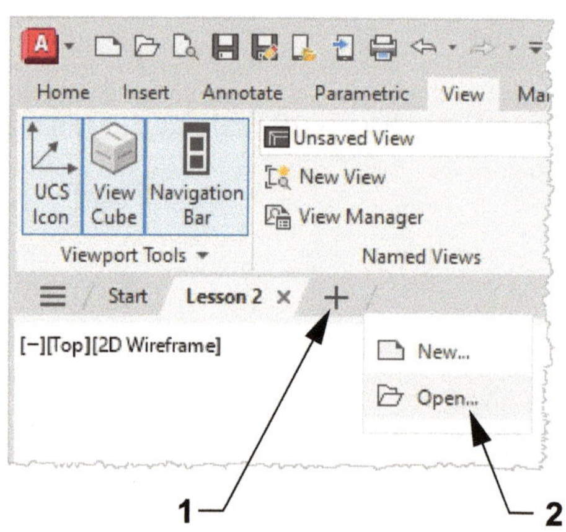

3. Locate the Directory and Folder for the previously saved file. (Refer to page 2-15.)
4. Select the file that you want to open.
5. Select the **Open** button.

How to open a new drawing from the File Tabs

1. Right mouse click on any **File Tab** and select **New** from the menu.
2. Select **Drawing Template (*.dwt)** from the **Files of type** drop-down list (as shown on the next page).
3. Select the **Template** you require.
4. Select the **Open** button.

Continued on the next page...

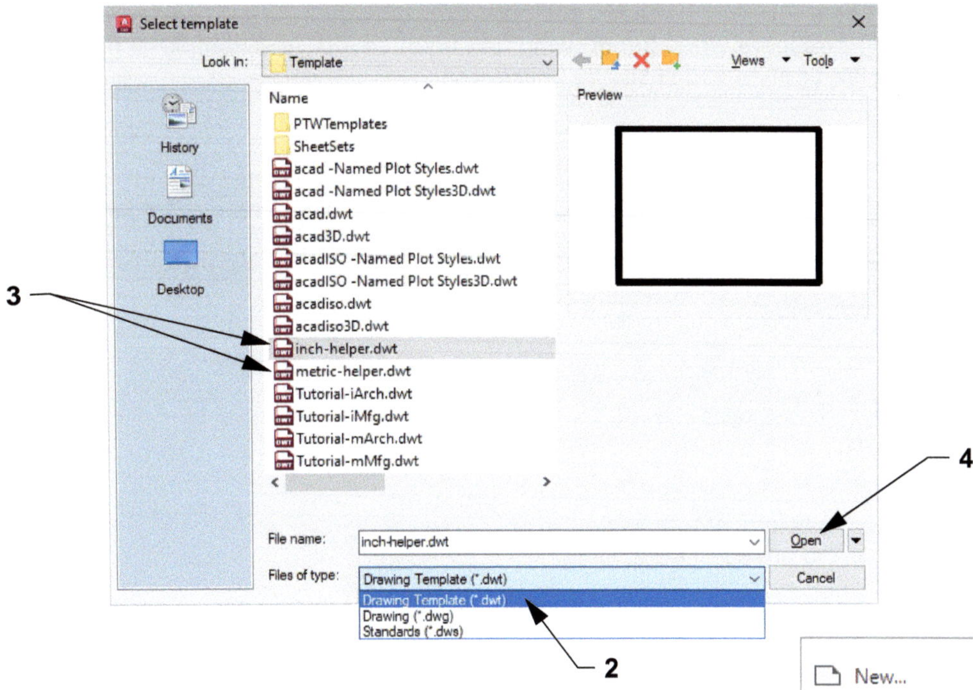

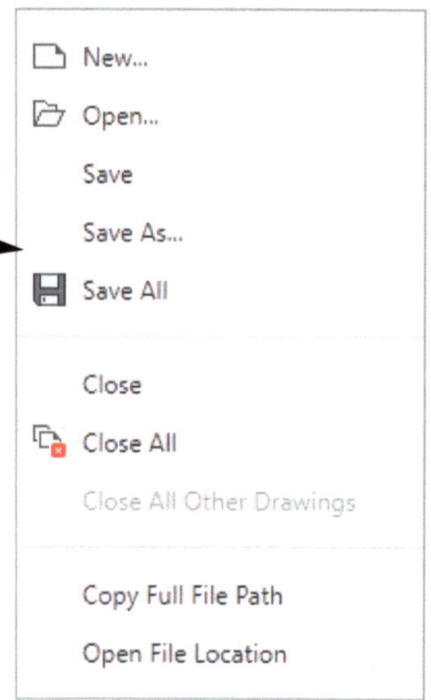

If you right mouse click on any **File Tab**, a menu appears with various options, including closing all open drawing Tabs except the one you just clicked on.

You can also start a new drawing using the default template by left clicking on the **+** icon.

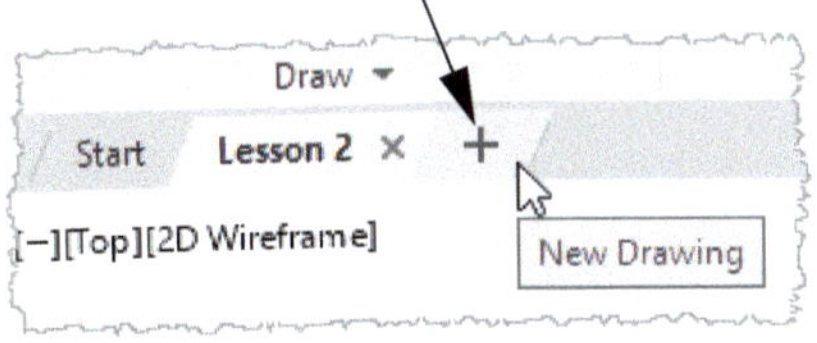

The **File Tabs** drawing previews allow you to quickly change between open drawings. If you hover your mouse over any open File Tab, a preview of the Model and the Layout Tabs are displayed. You can click on any of the previews to take you to that particular open drawing or view, as shown in the example below.

Click on any of the previews to instantly switch to that Model or Layout Tab.

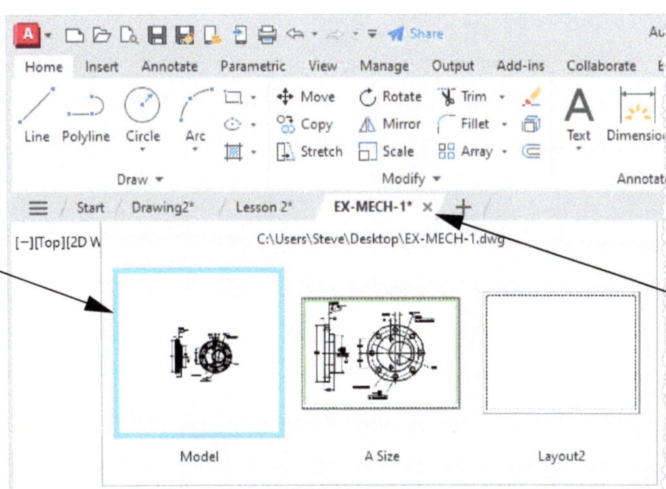

Hover the mouse over any File Tab to see a preview of the Model and Layout Tabs.

Continued on the next page...

If an asterisk is displayed on a File Tab, it means that particular drawing has not been saved since it was last modified. The asterisk will disappear when the drawing has been saved. An example of having two separate drawings open that have not been saved is shown below.

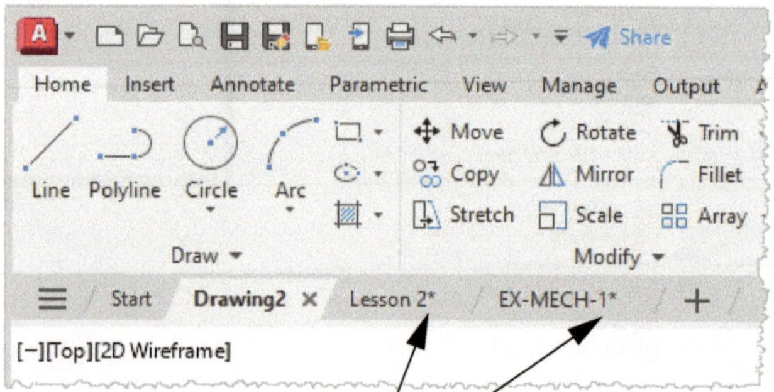

These two drawings are showing the asterisk and have not been saved.

Using Floating File Tabs

You can drag a File Tab off the main AutoCAD Application Window to make it a separate drawing window. This means that you could have several drawing files open at the same time with each file in its own drawing window, without having to switch between File Tabs. The floating File Tabs are particularly useful if you have two or more monitors, because you can have a different drawing file open on each monitor.

You can switch between the main AutoCAD Application Window and any floating File Tab by left-clicking inside the drawing area of that separate file's drawing window. A floating File Tab window also can be resized to make it larger or smaller.

How to create a floating File Tab

1. Left click and hold down the left mouse button (do not release the button) on the File Tab you want as a floating File Tab.

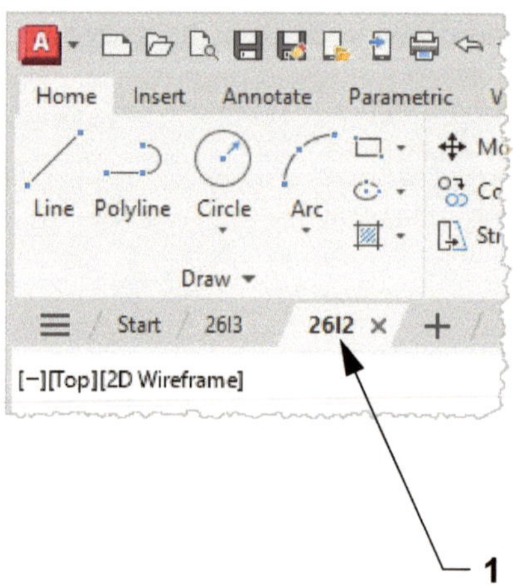

Continued on the next page...

2. Drag the File Tab away from the main AutoCAD Application Window.

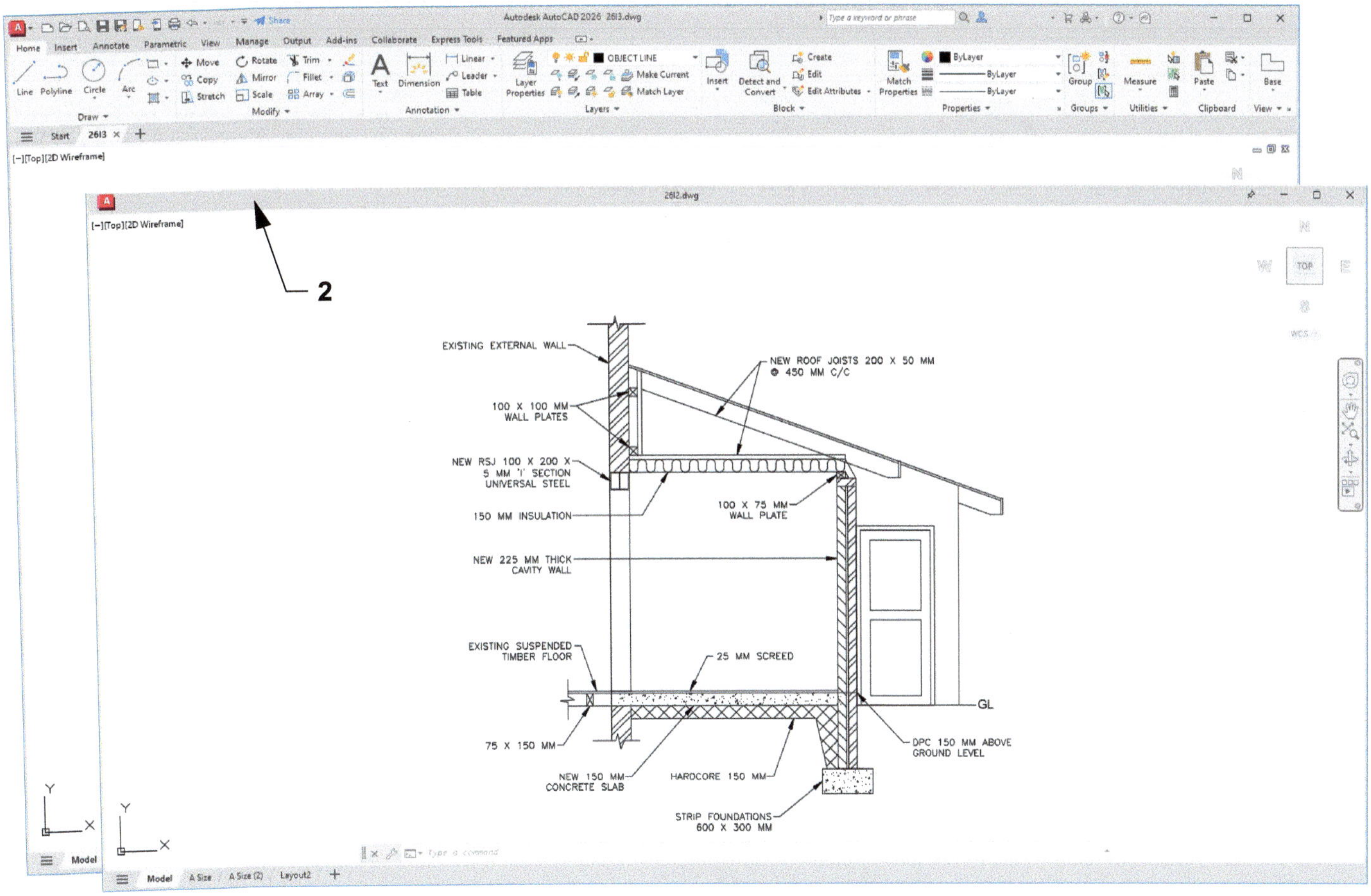

3. Drag the File Tab to the location you want for the floating File Tab. This could be on the same screen or on a second monitor.

4. When you are happy with the location of the floating File Tab, release the left mouse button.

Main AutoCAD Application Window.

Floating File Tab.

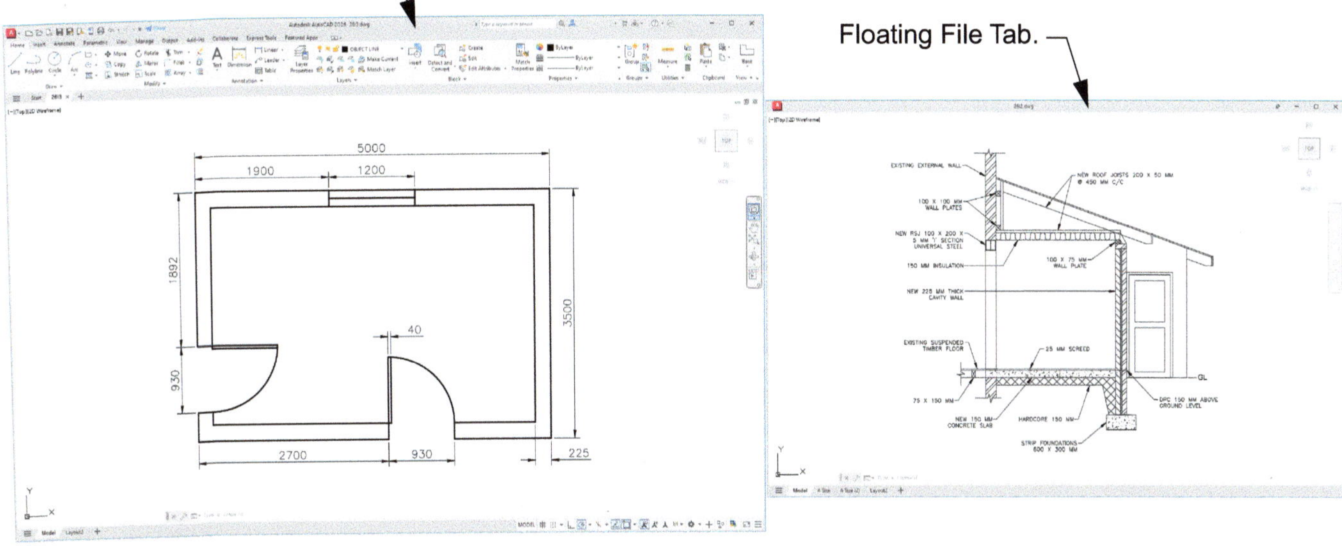

How to move a floating File Tab back to the main AutoCAD Application Window

1. Left click and hold down the left mouse button (do not release the button) on the Title Bar of the floating File Tab you want to move back to the main AutoCAD Application Window.

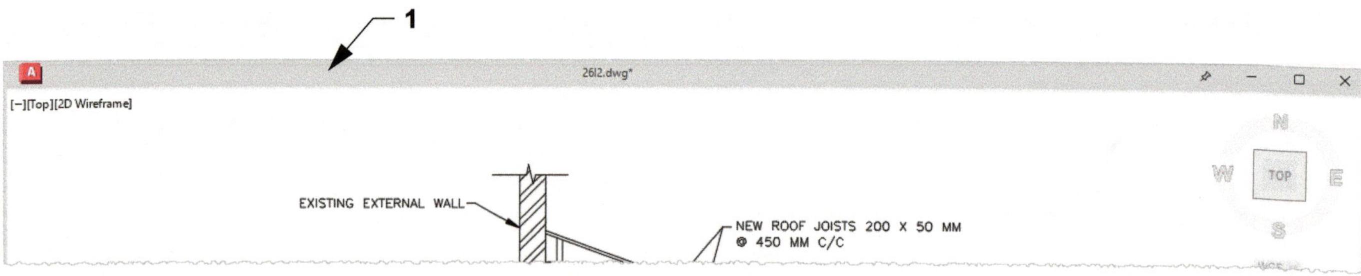

2. Drag the File Tab back to the main AutoCAD Application Window, until you see the entire drawing area turn a translucent blue in color.

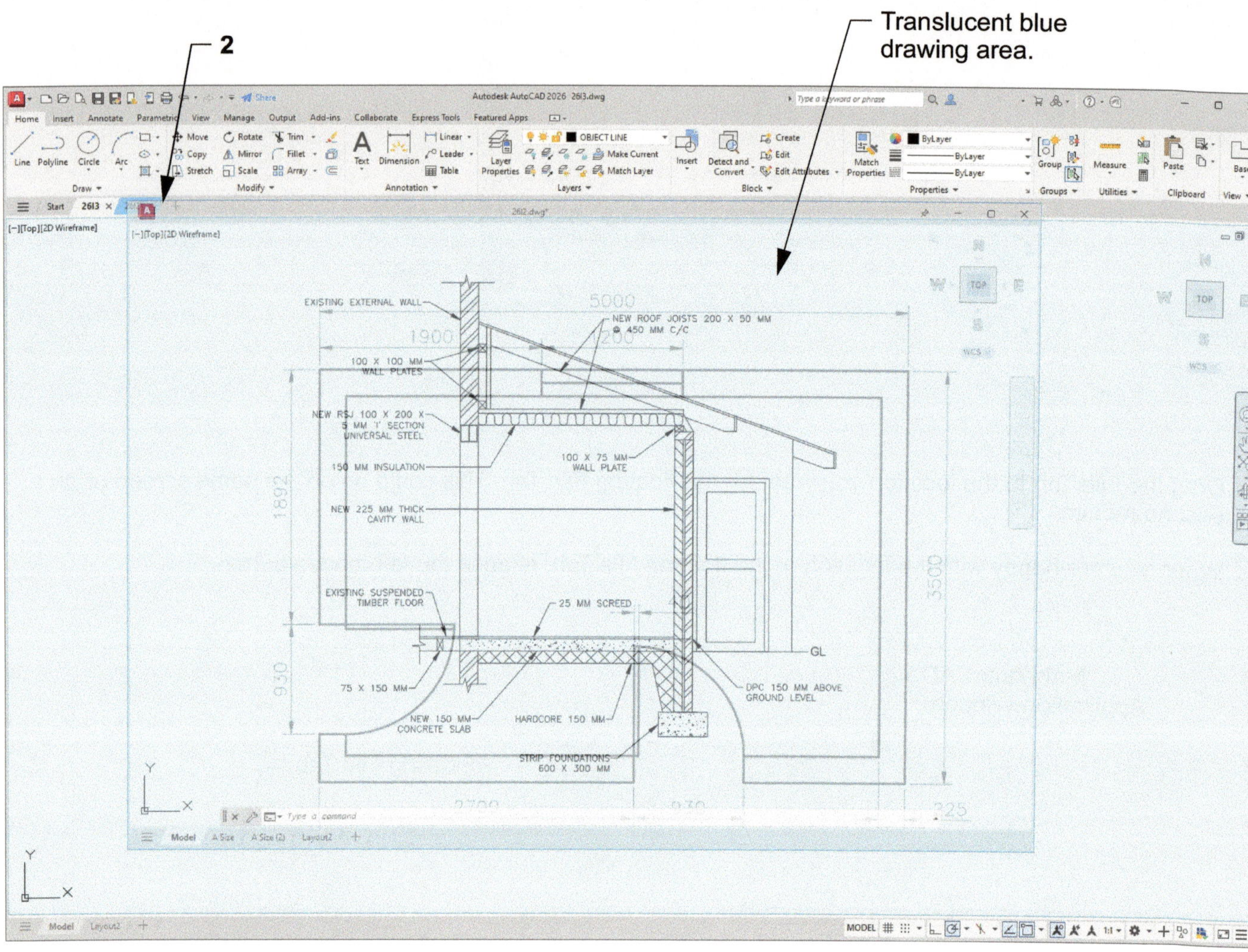

3. Release the left mouse button.

4. The floating File Tab will now be back on the main AutoCAD Application Window.

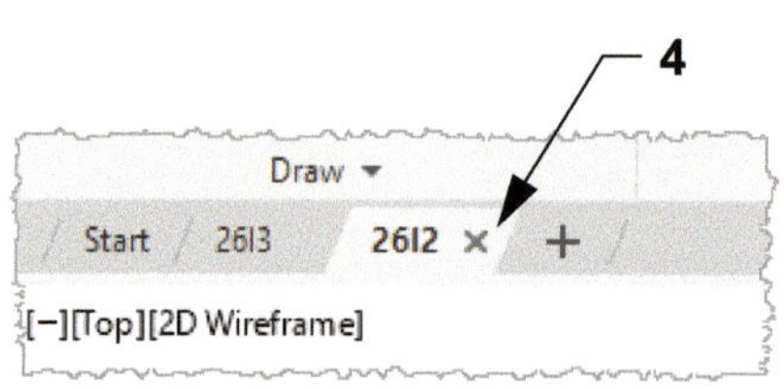

You also can move a floating File Tab back to the main AutoCAD Application Window by right clicking on the Title Bar of the floating File Tab, and then selecting **Move to File Tab** from the menu.

Right click on the Title Bar, then select **Move to File Tab**.

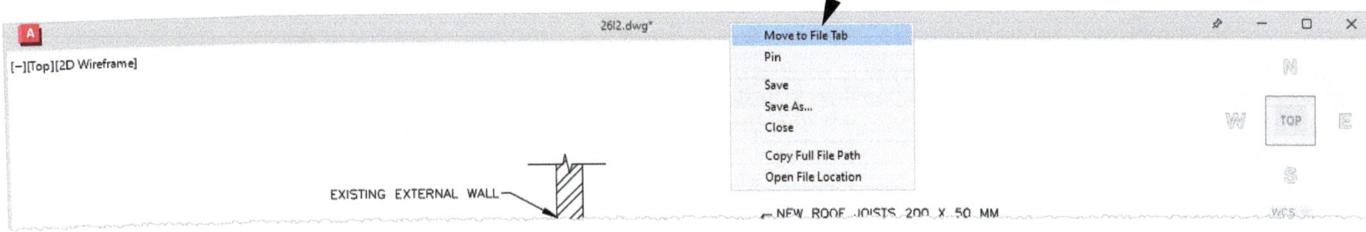

You can Pin a floating File Tab so that it remains on top of any other open or overlapping drawing file windows.

How to Pin a floating File Tab

1. Right click on the Title Bar of the floating File Tab you want to Pin.
2. Select **Pin** from the menu.

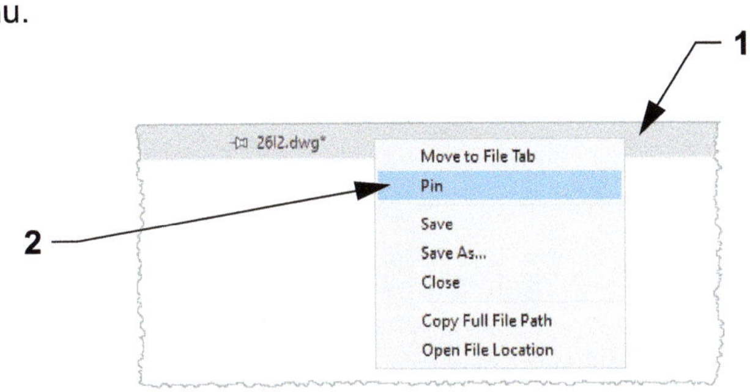

3. The floating File Tab is now pinned so that it remains on top of any other open or overlapping drawing file windows.

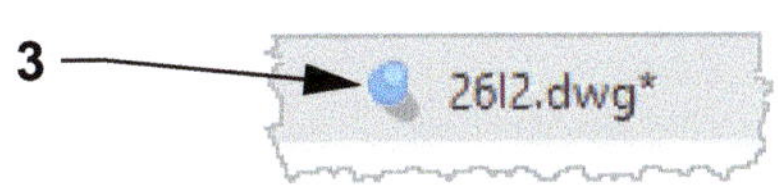

How to Unpin a floating File Tab

1. Right click on the Title Bar of the floating File Tab you want to Unpin.
2. Select **Unpin** from the menu.

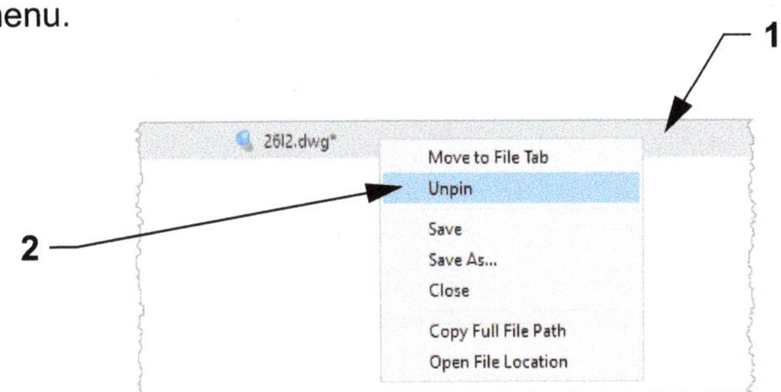

3. The floating File Tab is now unpinned. It will only remain on top of any other open or overlapping drawing file windows if you left click inside the file's drawing area.

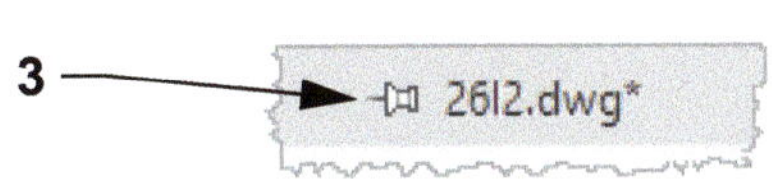

Saving a Drawing File

After starting a new drawing, it is best practice to save it immediately. Learning how to save a drawing correctly is almost more important than making the drawing. If you can't save correctly, you will lose the drawing and hours of work.

There are two commands for saving a drawing: **Save** and **Save As**. I prefer to use **Save As**.

The **Save As** command always pauses to allow you to choose where you want to store the file and what name to assign to the file. This may seem like a small thing, but it has saved me many times from saving a drawing on top of another drawing by mistake.

The **Save** command will automatically save the file either back to where you retrieved it or where you last saved a previous drawing. Neither may be the correct destination. And the file you are saving may replace a file with the same name. So play it safe, use **Save As** for now.

1. Select the **Save As** command using one of the following:

   Quick Access Toolbar = 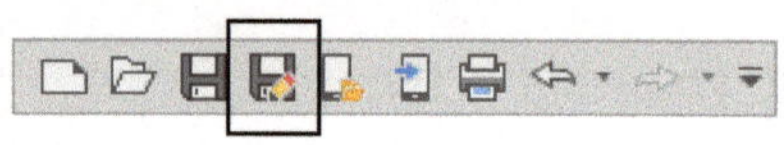
   or
   Application Menu = Save As / Drawing
   or
   Keyboard = *SA <Enter>*

2. Select the **"Save in:"** location.
(This is where the file will be saved.)

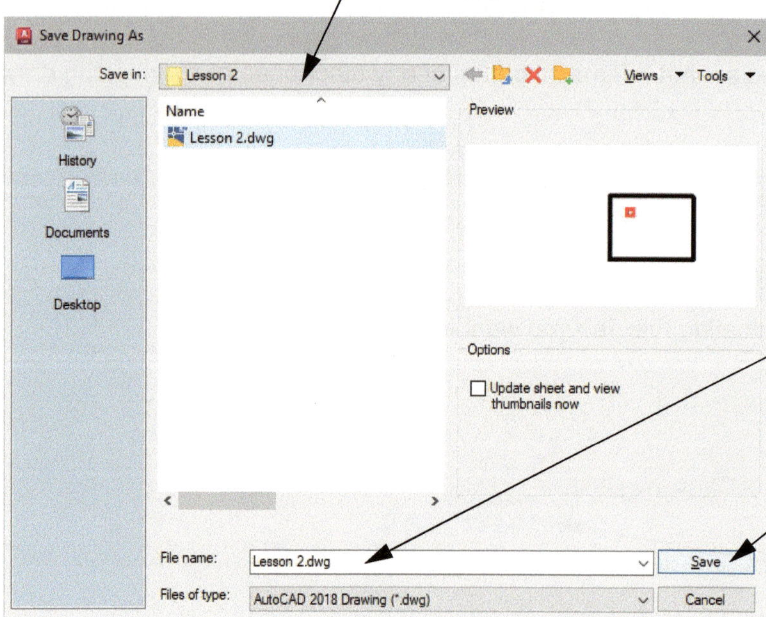

3. Type the new file name here.

4. Select the **"Save"** button.

Automatic Save

If you turn the **Automatic save** option **on**, your drawing is saved at specified time intervals. These temporary files are automatically deleted when a drawing closes normally. The default save time is every 10 minutes. You may change the save time intervals and where you would prefer the Automatic Save files to be saved.

How to set the Automatic Save intervals

1. Type ***options*** and then press ***<Enter>***.
2. Select the **Open and Save** Tab.
3. Enter the desired **Minutes between saves**.
4. Select the **OK** button.

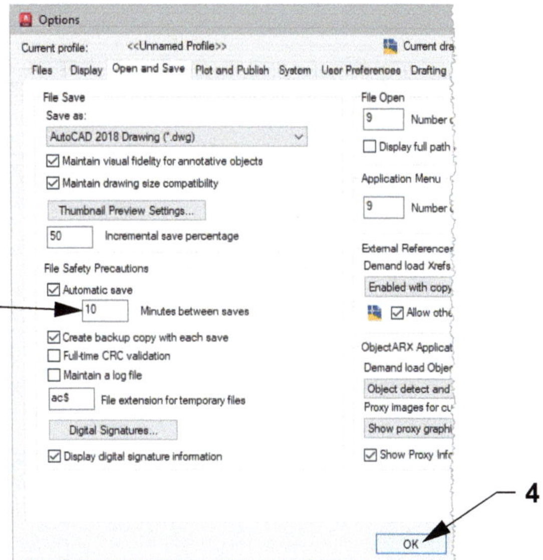

How to change the Automatic Save location

1. Type ***options*** and then press ***<Enter>***.
2. Select the **Files** Tab.
3. Locate the **Automatic Save File Location** and click on the **+** to display the path.
4. Double click on the path.
5. Browse to locate the Automatic Save Location desired and highlight it.
6. Select **OK**.

(The browse box will disappear and the new location path should be displayed under the Automatic Save File Location heading.)

7. Select **OK** to accept the change.

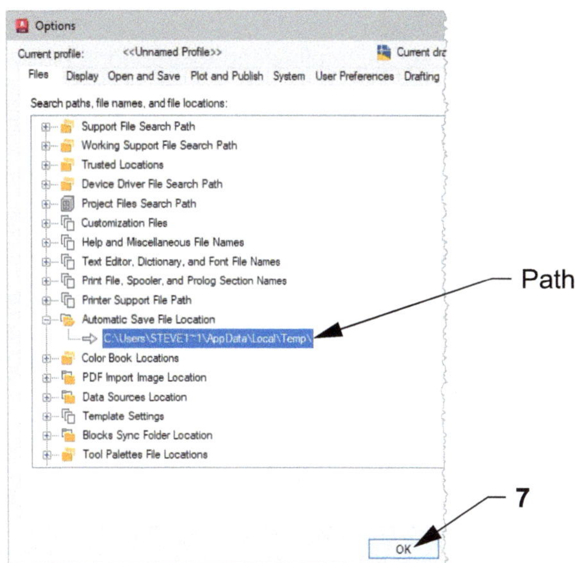

Back Up Files and Recover

When you save a drawing file, AutoCAD creates a file with a **.dwg** extension. For example, if you save a drawing as **12b**, AutoCAD saves it as **12b.dwg**.

The next time you save that same drawing, AutoCAD replaces the old version of the file with the new version and renames the old version **12b.bak**. The old version is now a backup file. (Only 1 backup file for each drawing file is stored.)

Continued on the next page...

How to open a backup file

You can't open a **".bak"** file. It must first be renamed with a **".dwg"** file extension.

How to view the list of backup files

The backup files will be saved in the same location as the drawing file. You must use Windows Explorer to locate the **.bak** files.

How to rename a backup file

1. Right click on the file name.

2. Select **"Rename"**.

3. Change the **.bak** extension to **.dwg** and then press ***<Enter>***.

How to recover a Drawing

In the event of a program failure or a power failure, any open files should be saved automatically. (Refer to page 2-27.)

When you attempt to re-open the drawing, the **Drawing Recovery Manager** will display a list of all drawing files that were open at the time of a program or system failure. You can preview and open each **.dwg** or **.bak** file to choose which one should be saved as the primary file.

Exiting AutoCAD

To safely exit AutoCAD, follow the instructions below.

1. Save all open drawings using **Save** or **Save As**.

2. Start the **Exit** procedure using one of the following:

Ribbon = None
or
Application Menu = Exit Autodesk AutoCAD 2026
or
Keyboard = *Exit <Enter>*

If any changes have been made to the drawing since the last **Save** or **Save As**, the warning box shown below will appear asking if you want to Save the changes?

Select **Yes**, **No**, or **Cancel**.

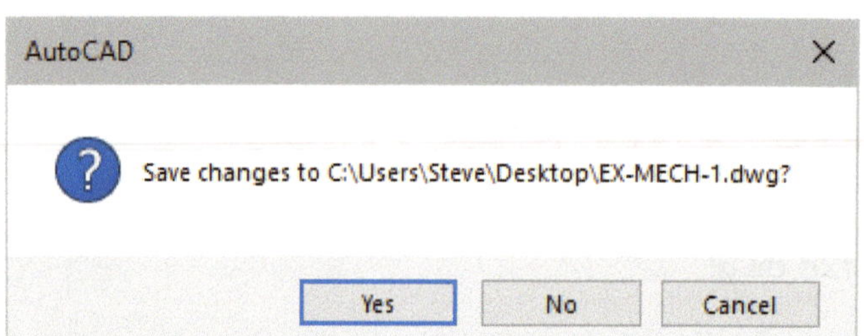

Exercise 2A

Instructions:

1. Start a **New** file using either the **inch-helper.dwt** or the **metric-helper.dwt**
2. **Draw** the objects at right using:
   A. **Line** command.
   B. **Ortho <*F8*> on** when drawing horizontal and vertical lines.
   C. **Ortho <*F8*> off** when drawing lines on an angle.
   D. Turn **Increment Snap <*F9*> on**.
   E. Turn **Osnap <*F3*> off**.
   F. Turn **Grid <*F7*> on**.
   G. Use the **Close** option.
3. **Save** the drawing as: **Ex-2A**

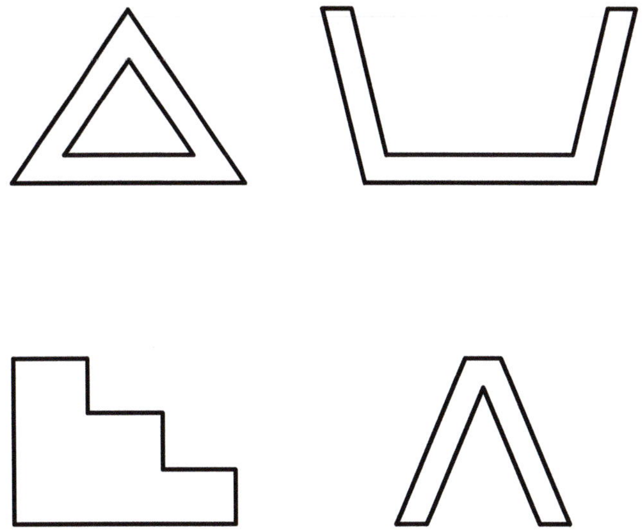

Exercise 2B

Instructions:

1. **Open Ex-2A**, if not already open.
2. **Erase** the missing lines as shown.
   A. Turn **Osnap <*F3*> off** (it will be easier to move the cursor around accurately).
3. Save the drawing as: **Ex-2B**

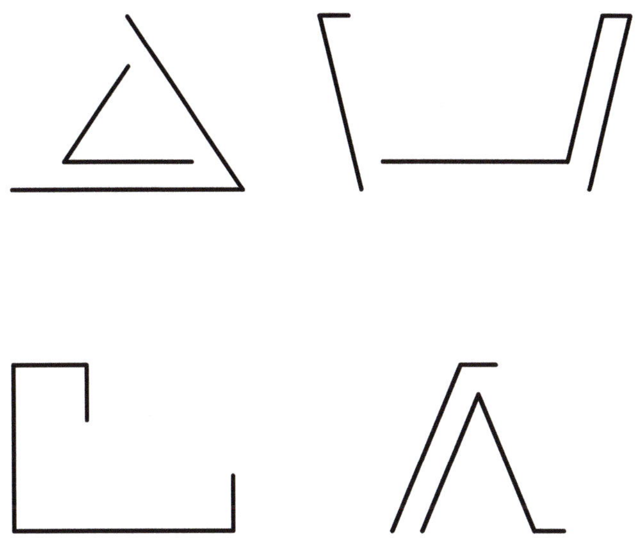

Exercise 2C

Instructions:

1. Start a **New** file using either the **inch-helper.dwt** or the **metric-helper.dwt**
2. **Draw** the objects at right using:
   A. **Line** command.
   B. **Ortho** ***<F8>*** **on** when drawing horizontal and vertical lines.
   C. **Ortho** ***<F8>*** **off** when drawing lines on an angle.
   D. Turn **Increment Snap** ***<F9>*** **on**.
   E. Turn **Osnap** ***<F3>*** **off**.
   F. Turn **Grid** ***<F7>*** **on**.
   G. Use the ***<Shift>*** key to toggle Ortho **on** and **off**.
3. **Save** the drawing as: **Ex-2C**

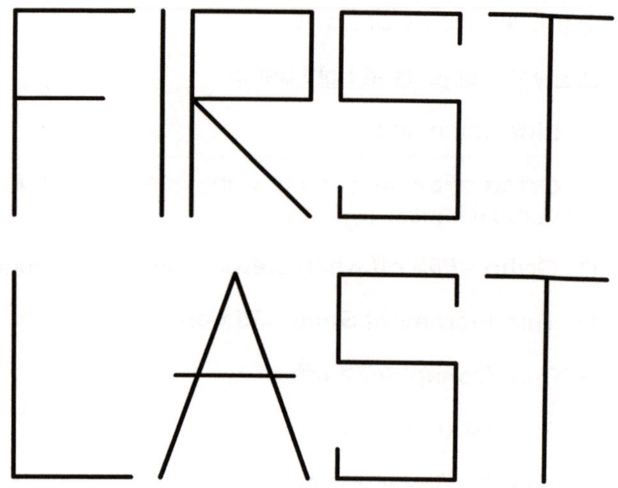

Exercise 2D

Instructions:

1. Start a **New** file using either the **inch-helper.dwt** or the **metric-helper.dwt**
2. **Draw** the objects at right using:
   A. **Line** command.
   B. **Ortho** ***<F8>*** **on** when drawing horizontal and vertical lines.
   C. **Ortho** ***<F8>*** **off** when drawing lines on an angle.
   D. Turn **Increment Snap** ***<F9>*** **on**.
   E. Turn **Osnap** ***<F3>*** **off**.
   F. Turn **Grid** ***<F7>*** **on**.
   G. Use the ***<Shift>*** key to toggle Ortho **on** and **off**.
3. **Save** the drawing as: **Ex-2D**

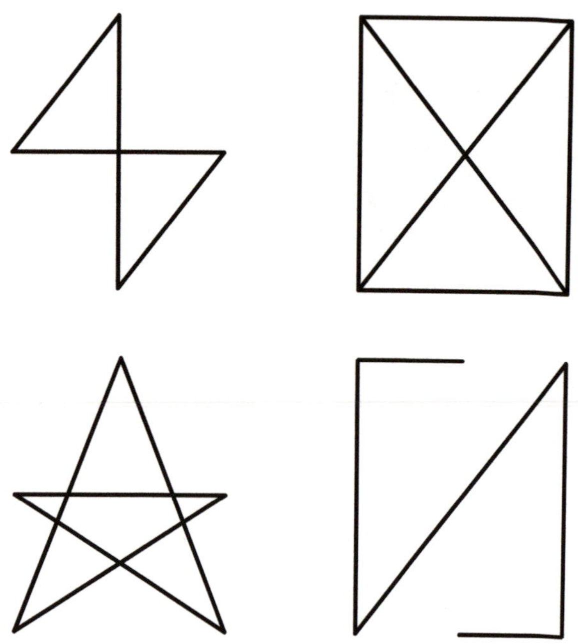

LESSON 3

LEARNING OBJECTIVES

After completing this lesson, you will be able to:

- Create a Circle using 6 different methods
- Create Rectangles with Chamfers, Fillets, Width, and Rotation
- Set Grids and Increment Snap
- Draw using Layers
- Control Layers
- Create Layers

Circle

There are six options to create a circle.

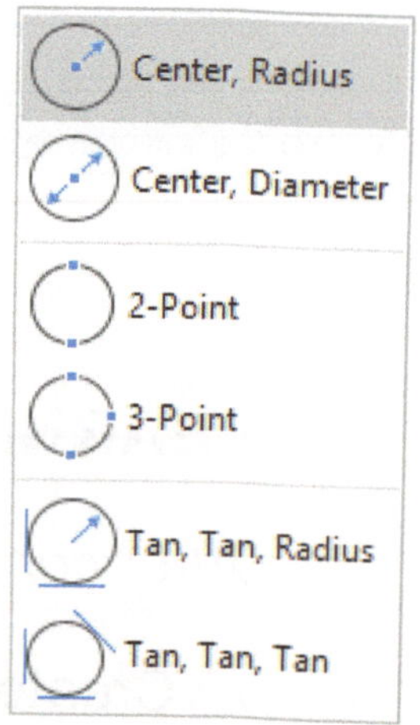

The default option is **"Center, Radius"**. (Probably because that is the most common method of creating a circle.)

We will try the **"Center, Radius"** option first.

1. Start the **Circle** command by using one of the following:

   Ribbon = Home Tab / Draw Panel /
   or
   Keyboard = *C <Enter>*

   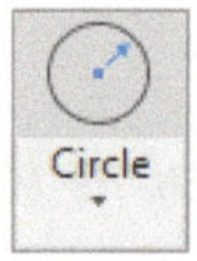

2. The following will appear on the Command Line:

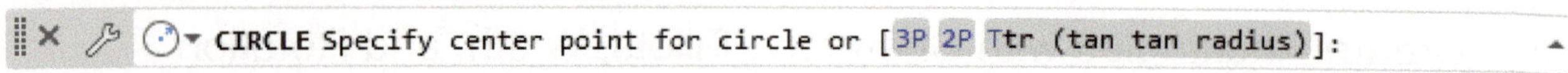

3. Locate the center point for the circle by moving the cursor to the desired location in the drawing area (**P1**) and then press the left mouse button.
4. Now move the cursor away from the center point and you should see a circle forming.
5. When the circle is the size desired (**P2**), press the left mouse button, or type the radius and then press ***<Enter>***.

Note: To use one of the other methods described below, first select the Circle command, then select one of the other Circle options.

Center, Radius: (Default option)

1. Specify the center location (**P1**).
2. Specify the radius (**P2**). (Define the radius by moving the cursor or by typing the radius.)

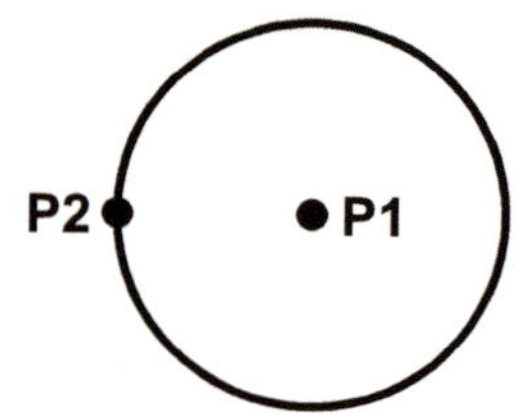

Center, Diameter:

1. Specify the center location (**P1**).
2. Specify the diameter (**P2**). (Define the diameter by moving the cursor or by typing the diameter.)

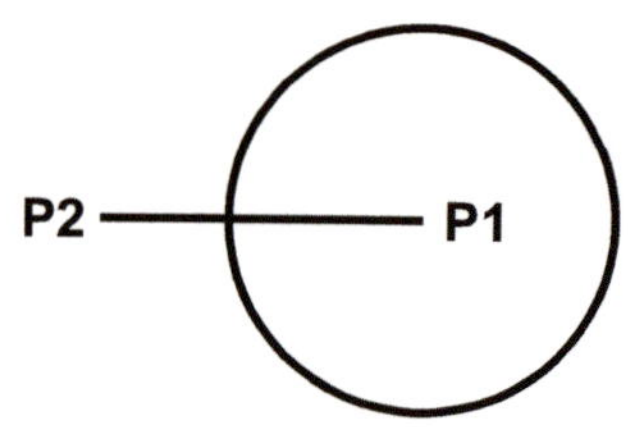

2-Point:

1. Select the 2-Point option.
2. Specify the 2 points (**P1** and **P2**) that will determine the diameter.

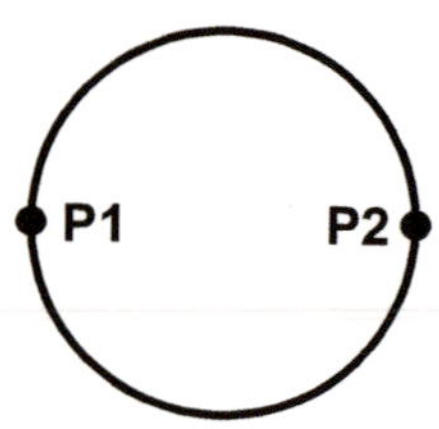

3-Point:

1. Select the 3-Point option.
2. Specify the 3 points (**P1**, **P2**, and **P3**) on the circumference. The circle will pass through all three points.

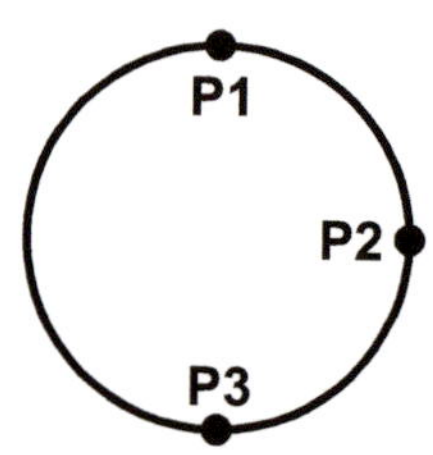

Tan, Tan, Radius:

1. Select the Tan, Tan, Radius option.

2. Select two objects (**P1** and **P2**) for the circle to be tangent to by placing the cursor on each of the objects and pressing the left mouse button.

3. Specify the radius.

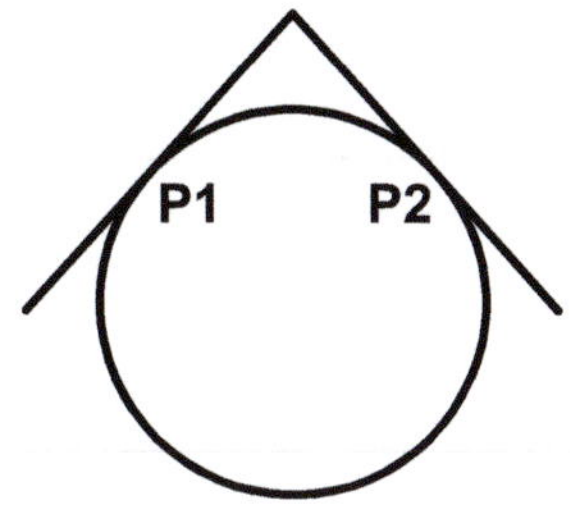

Tan, Tan, Tan:

1. Select the Tan, Tan, Tan option.

2. Select three objects (**P1**, **P2**, and **P3**) for the circle to be tangent to by placing the cursor on each of the objects and pressing the left mouse button.

   (AutoCAD will calculate the diameter automatically.)

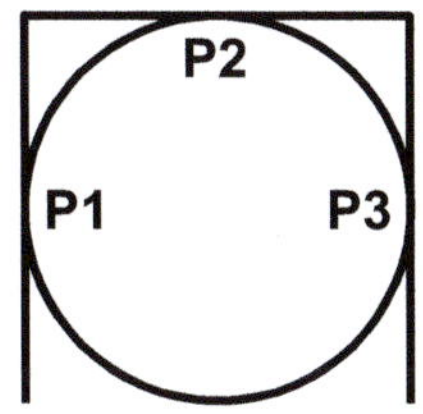

Rectangle

A rectangle is a closed rectangular shape. It is one object not four lines.
You can specify the length, width, area, and rotation options.
You can also control the type of corners on the rectangle—fillet, chamfer, or square—and the width of the line.

First, let's start with a simple rectangle using the cursor to select the corners.

1. Start the **Rectangle** command by using one of the following:

   Ribbon = Home Tab / Draw Panel /
   or
   Keyboard = *REC <Enter>*

2. The following will appear on the Command Line:

```
RECTANG Specify first corner point or [Chamfer Elevation Fillet Thickness Width]:
```

3. Specify the location of the first corner by moving the cursor to a location (**P1**) and then press the left mouse button.

4. The following will appear on the Command Line:

```
RECTANG Specify other corner point or [Area Dimensions Rotation]:
```

5. Specify the location of the diagonal corner (**P2**) by moving the cursor diagonally away from the first corner (**P1**) and then pressing the left mouse button.

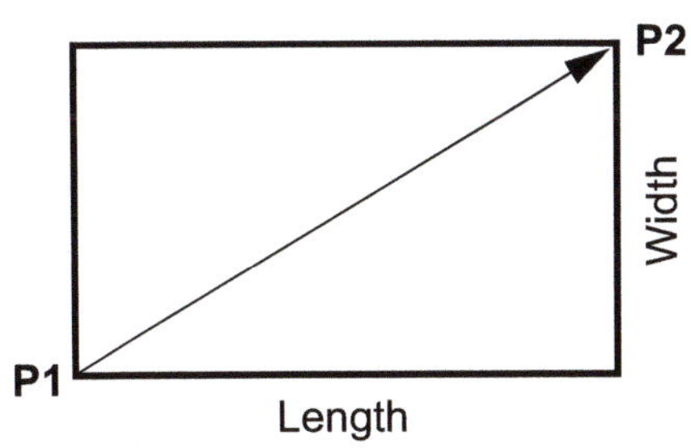

If you know the dimensions for the length and width of the rectangle, refer to the next page for an alternative to Step 5.

5. Type ***D <Enter>*** (or click on the blue letter **"D"**).

   Specify length for rectangles <0.000>: ***Type the desired length and then press <Enter>***.

   Specify width for rectangles <0.000>: ***Type the desired width and then press <Enter>.***

   Specify other corner point or [Area Dimensions Rotation]: ***Move the cursor up, down, right or left to specify where you want the second corner relative to the first corner and then press <Enter> or press the left mouse button.***

Options: Chamfer, Fillet, and Width

The following options are only available before you place the first corner of the rectangle.

> **Note:** Sizes shown in brackets [...] are for metric users. Enter the numbers without the brackets. **Example:** [12.7] just enter 12.7

Chamfer

A chamfer is an angled corner. The Chamfer option automatically draws all 4 corners with chamfers simultaneously and all the same size. You must specify the distance for each side of the corner as distance 1 and distance 2.

Example: Rectangle with Dist1 = 0.50" [12.7 mm] and Dist2 = 0.25" [6.35 mm].

1. Select the Rectangle command.
2. Type ***C <Enter>*** (or click on the blue letter **"C"**).
3. Enter ***0.50 [12.7]*** for the first distance.
4. Enter ***0.25 [6.35]*** for the second distance.
5. Place the first corner (**P1**).
6. Place the diagonal corner (**P2**).

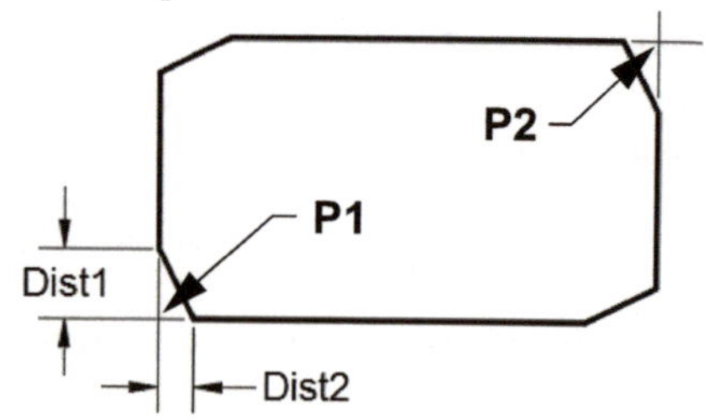

Fillet

A fillet is a rounded corner. The Fillet option automatically draws all 4 corners with fillets (all the same size). You must specify the radius for the rounded corners.

Example: Rectangle with 0.50" [12.7 mm] radius corners.

1. Select the Rectangle command.
2. Type ***F <Enter>*** (or click on the blue letter **"F"**).
3. Enter ***0.50 [12.7]*** for the radius.
4. Place the first corner (**P1**).
5. Place the diagonal corner (**P2**).

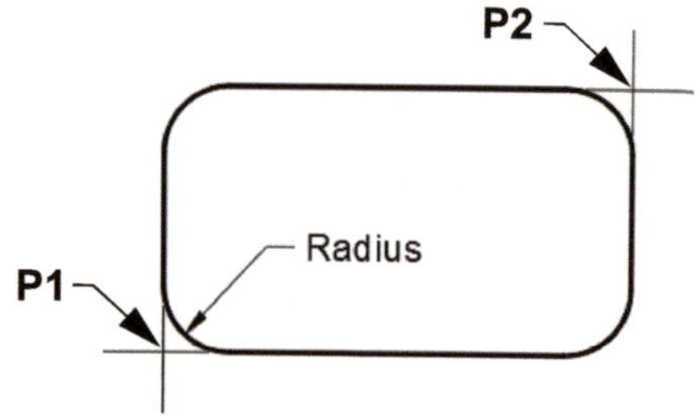

> **Note:** You must set Chamfer and Fillet back to "0" before defining the width. Unless you want fat lines and Chamfered or Filleted corners.

Width

The Width option defines the width of the rectangle lines. (Do not confuse this with the "Dimensions" option, where you specify the Length and Width of a rectangle. This Width option makes the lines appear fatter.)

Example: Rectangle with a line width of 0.50" [12.7 mm].

1. Select the Rectangle command.
2. Type ***W <Enter>*** (or click on the blue letter **"W"**).
3. Enter ***0.50 [12.7]*** for the width.
4. Place the first corner (**P1**).
5. Place the diagonal corner (**P2**).

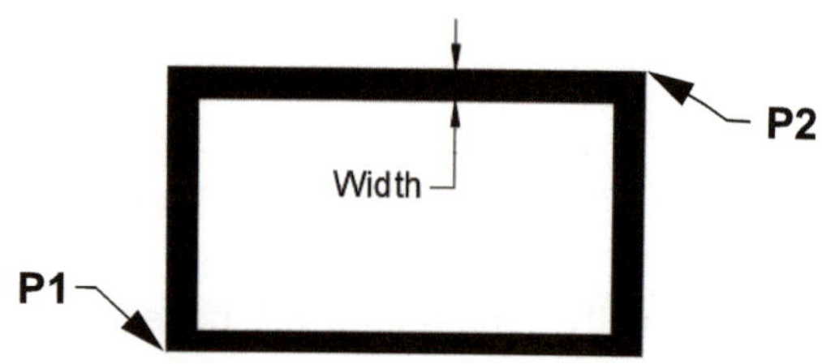

Options: Area and Rotation

The following options are available after you place the first corner of the rectangle.

Area

Area creates a rectangle using the area and either a length or a width. If the Chamfer or Fillet option is active, the area includes the effect of the chamfers or fillets on the corners of the rectangle.

Example: Rectangle with an area of 6" [152.4 mm] and a length of 2" [50.8 mm].

1. Select the Rectangle command.
2. Place the first corner (**P1**).
3. Type ***A <Enter>*** (or click on the blue letter **"A"**).
4. Enter ***6 [152.4] <Enter>*** for the area.
5. Select ***L <Enter>*** for the length (or click the blue letter **"L"**).
6. Enter ***2 [50.8] <Enter>*** for the length.

   (The width will be calculated automatically.)

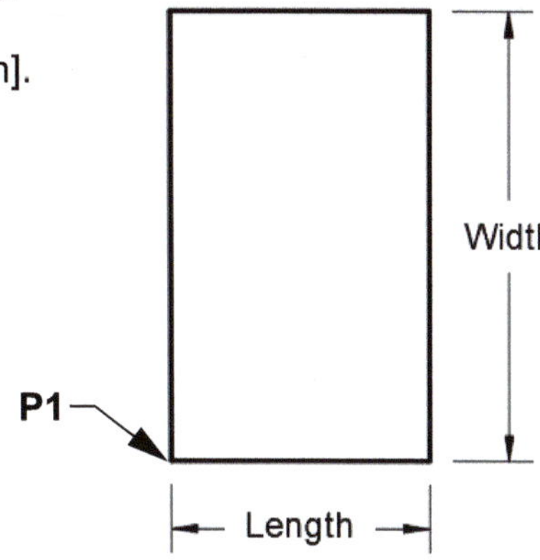

Rotation

You may select the desired rotation angle after you place the first corner and before you place the second corner. The basepoint (pivot point) is the first corner.

Note: This option will not affect rectangles already in the drawing. But all new rectangles within the drawing will also be rotated until you reset the rotation to **0**.

Example: Rectangle with a rotation angle of 45 degrees.

1. Select the Rectangle command.
2. Place the first corner (**P1**).
3. Type ***R <Enter>*** (or click on the blue letter **"R"**).
4. Enter ***45 <Enter>***.
5. Place the diagonal corner (**P2**).

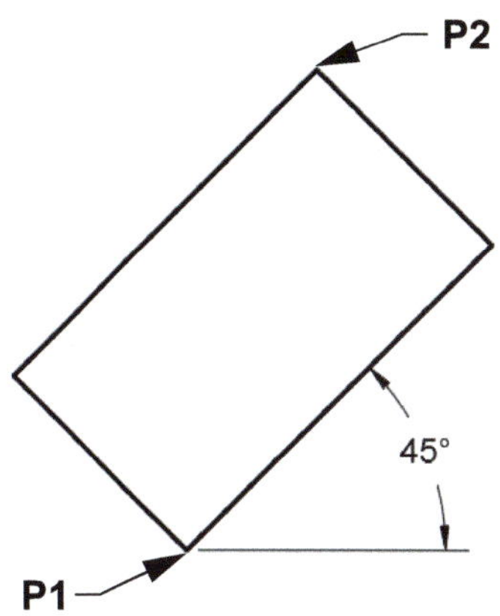

Grid and Increment Snap

Grid is the criss-cross lines in the drawing area. The grid is only a drawing aid to assist you in aligning objects and visualizing the distances between them. **The Grid will not plot**. (Refer to page 1-13.)

Increment Snap controls the movement of the cursor. If it is **off**, the cursor will move smoothly. If it is **on**, the cursor will jump in an **incremental** movement. (Refer to page 1-13.)

The **Drafting Settings** dialog box allows you to set **Increment Snap** and **Grid** spacing. You may change the Grid spacing and Increment Snap at anytime while creating a drawing. The settings are only drawing aids to help you visualize the size of the drawing and control the movement of the cursor.

1. Select **Drafting Settings** by using one of the following:

   Keyboard = *DS <Enter>*
   or
   Status Bar = Right click on the Snap or Grid button and select Settings.

Continued on the next page...

2. The dialog box shown below will appear showing inch or metric units depending on whether you selected an **inch** or **metric template**.

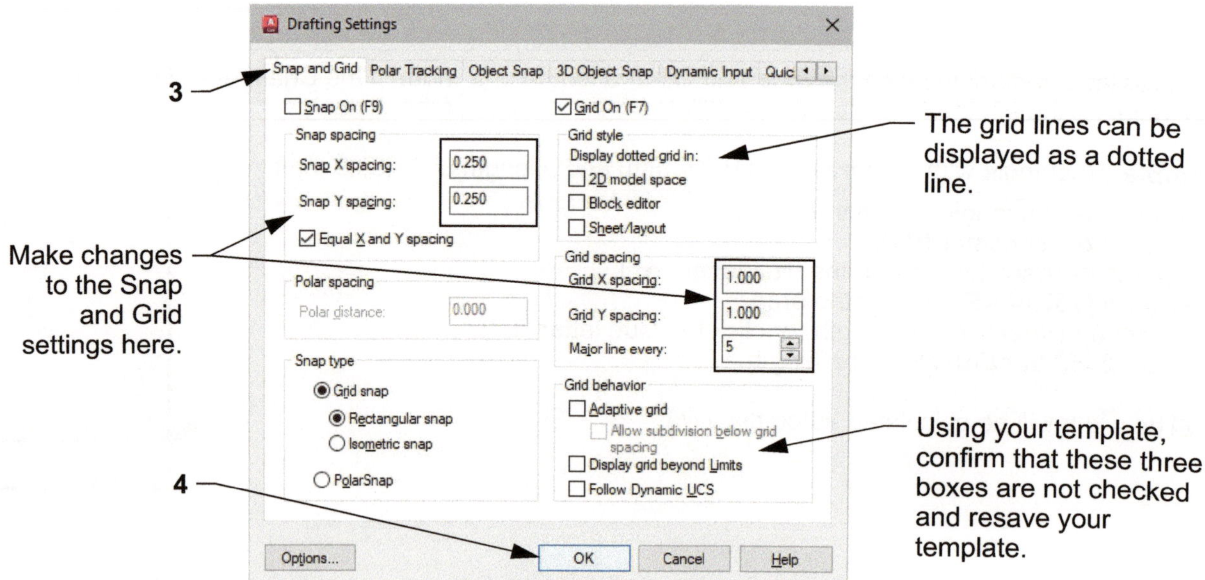

3. Select the **"Snap and Grid"** Tab.

4. Make your changes and select the **OK** button to save them. If you select the **Cancel** button, your changes will not be saved.

Layers

A **Layer** is like a transparency. Have you ever used an overhead light projector? Remember those transparencies that are laid on top of the light projector? You could stack multiple sheets, but the projected image would have the appearance of one document. Layers are basically the same. Multiple layers can be used within one drawing.

The example, on the right, shows three layers. One for annotations (text), one for dimensions, and one for objects.

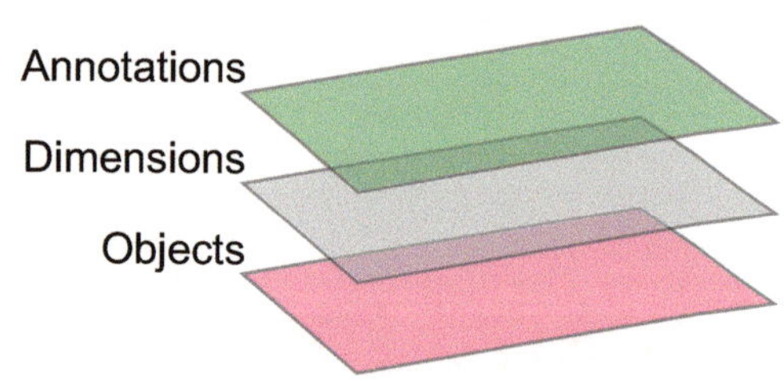

How to use Layers

First you select the layer and then you draw the objects.

Always select the layer first and then draw the objects.

It is good "drawing management" to draw related objects on the same layer. For example, in an architectural drawing, you would select the "Walls" layer and then draw the floor plan.

Then you would select the "Electrical" layer and draw the electrical objects.

Then you would select the "Plumbing" layer and draw the plumbing objects.

Each layer can then be controlled independently. If a layer is **Frozen**, it is **not visible**. When you **Thaw** the layer, it becomes **visible** again. (Refer to the following pages for detailed instructions on controlling layers.)

How to select a Layer

1. Go to **Ribbon = Home Tab / Layers Panel**.

2. Select the drop-down arrow ▼.

3. Highlight the desired layer and press the left mouse button.

Continued on the next page...

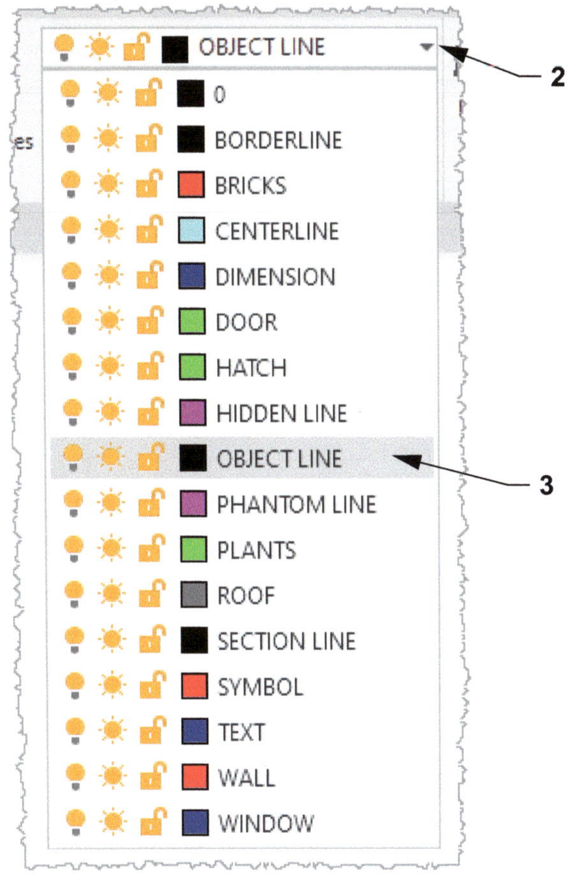

Note: The selected layer becomes the **current** layer. All objects will be located on this layer until you select a different layer.

Controlling Layers

The following controls can be accessed using the Layer drop-down arrow ▼.

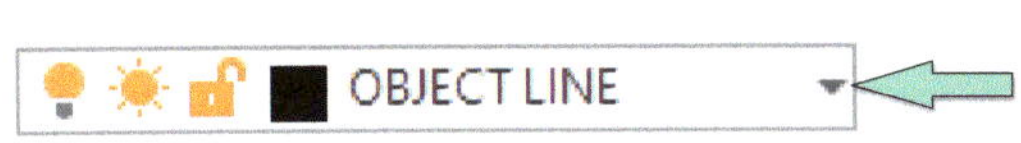

Layers On or Off

If a layer is **on**, it is **visible**. If a layer is **off**, it is **not visible**. **Only** layers that are **on** can be **edited** or **plotted**.

Freeze or Thaw Layers

Freeze and Thaw are very similar to **on** and **off**. A Frozen layer is **not visible** and a Thawed layer **is visible**. Only Thawed layers can be edited or plotted.

Additionally:

A. Objects on a Frozen layer **cannot** be accidentally erased.

B. When working with large and complex drawings, freezing saves time because Frozen layers are not **regenerated** when you zoom in and out.

Lock or Unlock Layers

Locked layers are visible but **cannot be edited**. They are visible so they **will** be plotted.

Continued on the next page...

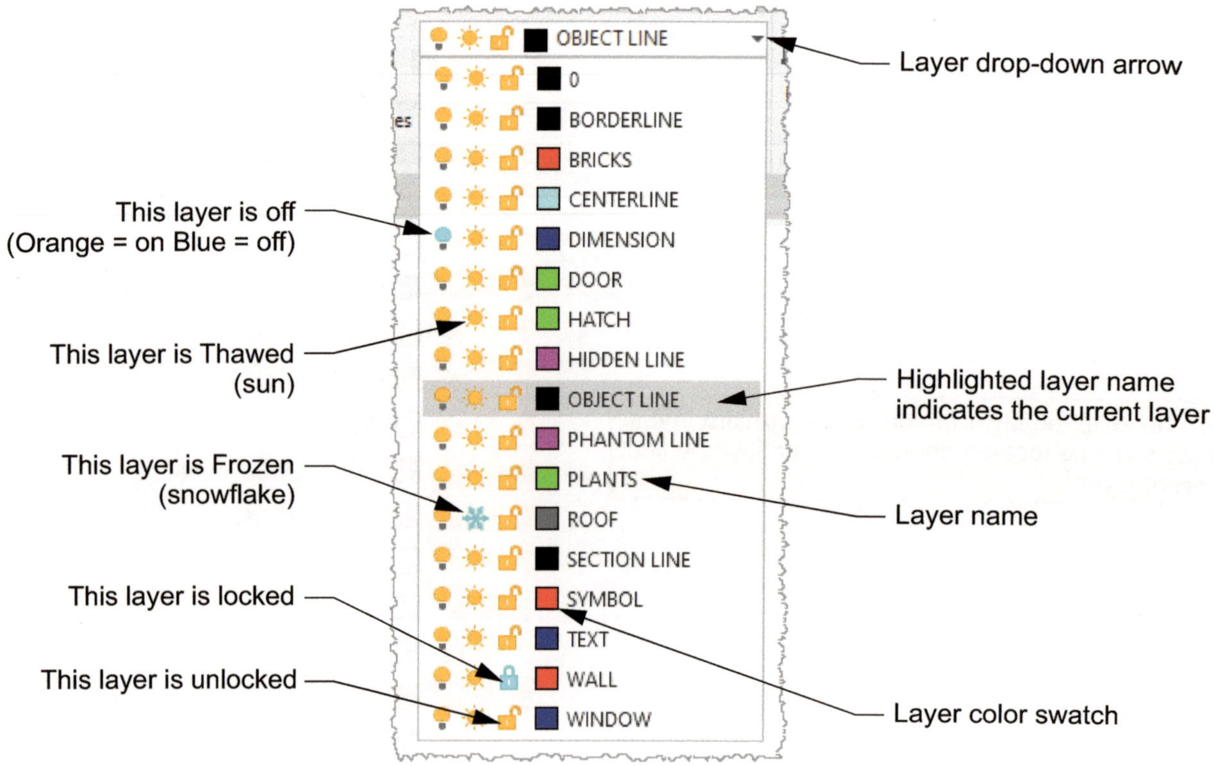

Layer Properties Manager

To access the following options you must use the **Layer Properties Manager**. You may also access the options listed on the previous page within this dialog box.

To open the **Layer Properties Manager** use one of the following.

Ribbon = Home Tab / Layers Panel /
or
Keyboard = *LA <Enter>*

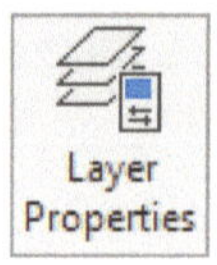

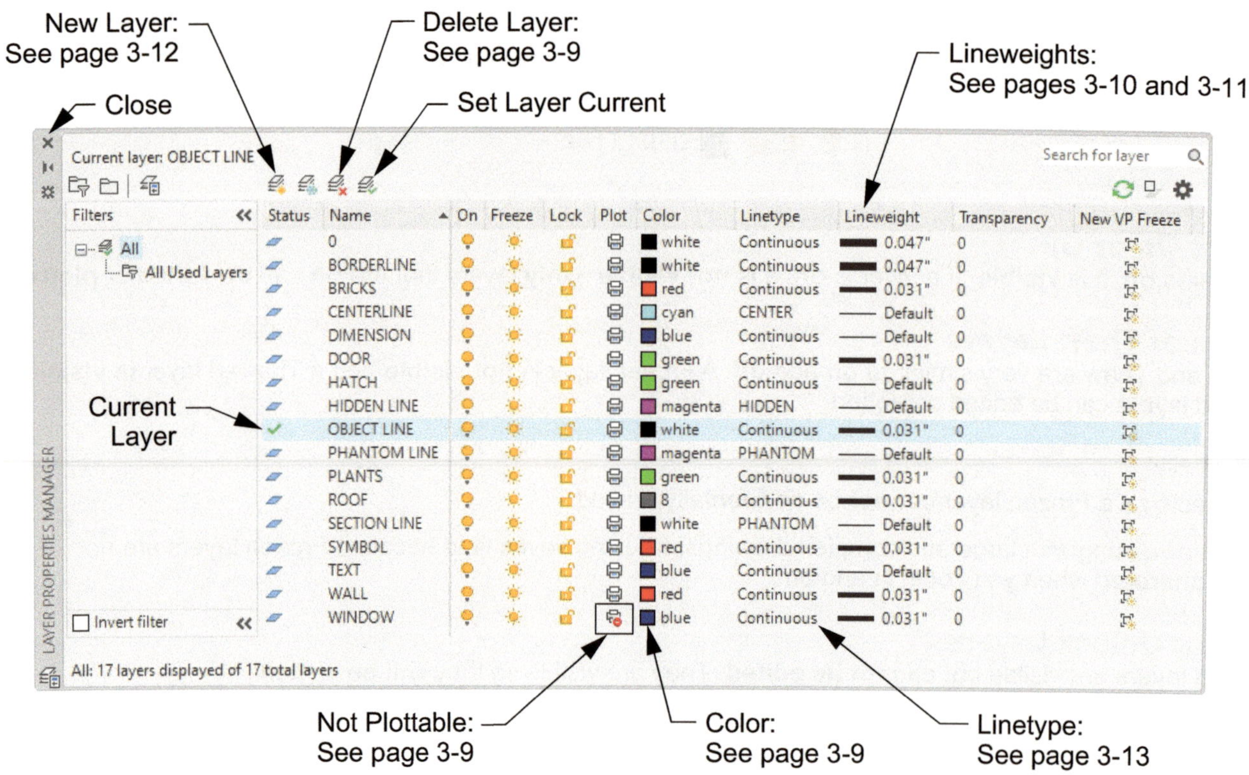

How to delete an existing layer

1. Highlight the layer name to be deleted.
2. Select the **Delete Layer** tool.

   or

1. Highlight the layer name to be deleted.
2. Right click and select **Delete Layer**.

Plot or Not Plottable layer

This tool prevents a layer from plotting even though it is visible within the Drawing Area. A **Not Plottable** layer will not be displayed when using **Plot Preview**. If the Plot tool has a slash, the layer will not plot.

Layer Color

Color is not merely to make a pretty display on the screen. Layer colors can help define objects. For example, you may assign green for all doors. Then, at a glance, you could identify the door and the layer by its color.

Here are some additional things to consider when selecting the colors for your layers.

Consider how the colors will appear on the paper.
(Pastels do not display well on white paper.)

Consider how the colors will appear on the screen.
(Yellow appears well on a black background but not on white.)

How to change the color of a layer

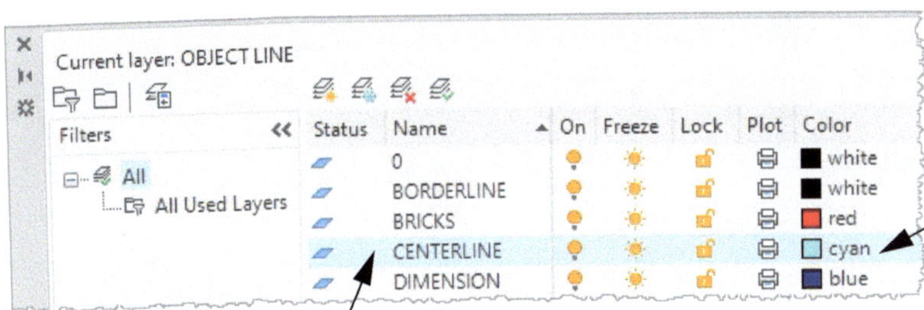

2. Select the color swatch or name.

1. Select the layer that you want to change.

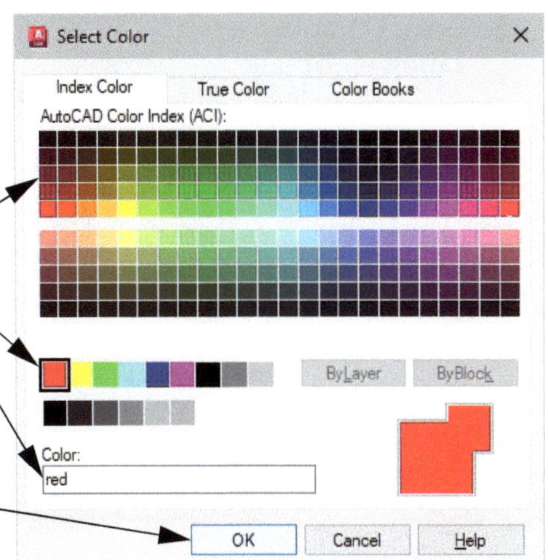

3. Select the color from the Index or primary color strip. (The color name or number will appear in the color box.)

4. Select the **OK** button.

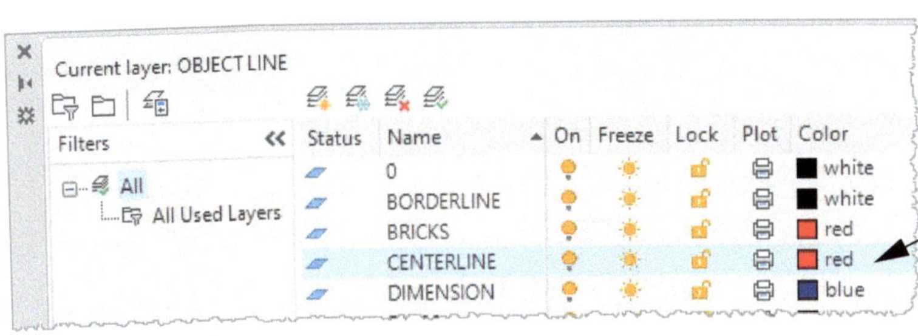

5. The color selected will appear on the layer line.

Lineweights

Lineweight refers to **"how heavy or thin is the object line"**.

It is "good drawing management" to establish a contrast in the lineweights of entities.

In the example below, the rectangle has a heavier lineweight than the dimensions. The contrast in lineweights makes it easier to distinguish between entities.

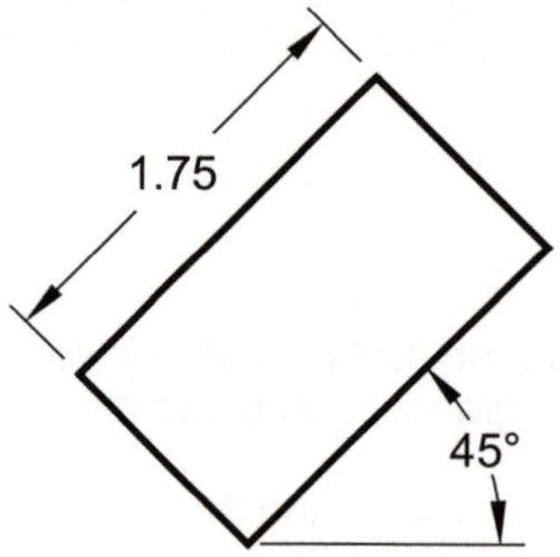

Lineweight Settings

Lineweights are plotted with the exact width of the lineweight assigned. But you may adjust how they are **displayed on the screen**. (Refer to Step 4 below.)

Important: Before assigning lineweights you should first select the **Units for Listing** and **Adjust Display Scale** as shown below.

1. Select the **Lineweight Settings** box using one of the following:

   Keyboard = *LW <Enter>*
   or
   Status Bar = Right click on the Lineweight button and then click on Lineweight Settings.

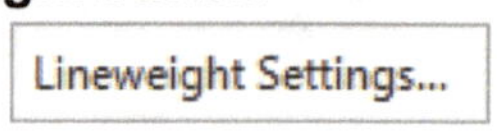

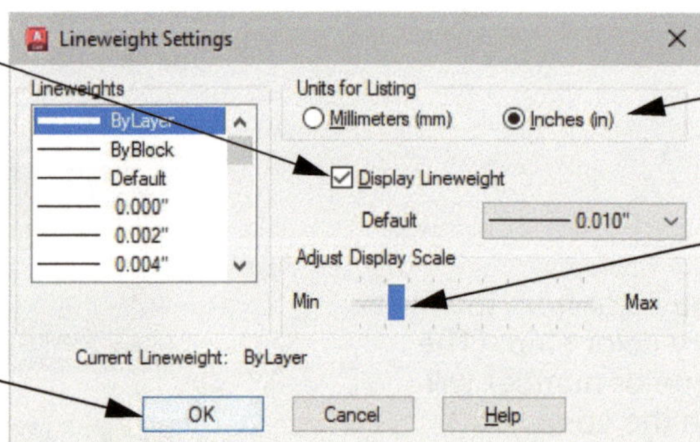

2. Select **Inches** or **Millimeters**.

3. Select **Display Lineweight** box.

4. Click and drag to adjust the **"Displayed Scale"** of the lineweight. (**Note:** This will not affect the width when plotted, it is just for on-screen display.)

5. Select **OK** Button.

Note: These settings will be saved to the computer not the drawing, and will remain until you change them. You may have to shut down AutoCAD and then restart the program for the changes to take effect.

How to assign Lineweights to Layers

Note: Before assigning **Lineweights** to layers, make sure your **Lineweight settings** (**Units for Listing** and **Adjust Display Scale**) are correct. (See above.)

1. Select the **Layer Properties Manager** using one of the following:

Ribbon = Home Tab / Layers Panel /
or
Keyboard = *LA <Enter>*

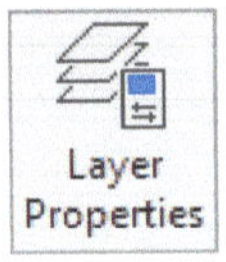

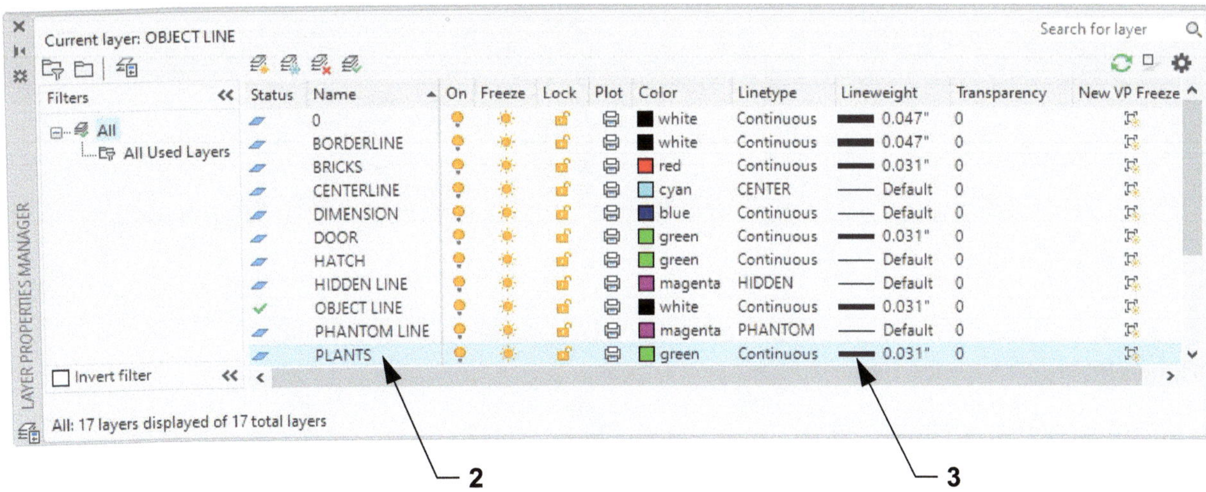

2. Highlight a layer (click on the name).
3. Click on the Lineweight for that layer.
4. Scroll and select a Lineweight from the list.
5. Select the **OK** button.

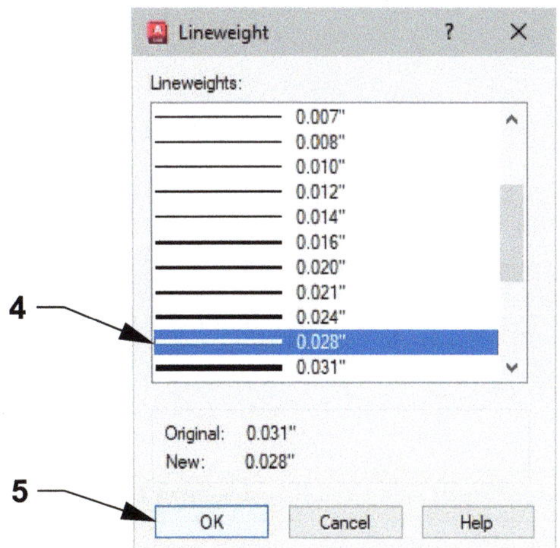

Note: Lineweight selections will be saved within the **current** drawing and will not affect any other drawing. The images above show lineweight sizes for the **inch-helper.dwt**. Metric sizes will appear if you are using the **metric-helper.dwt**.

Transparency

Each layer may be assigned a transparency percentage from **0** to **90** percent. **0** would not be transparent at all and **90** would be **90** percent transparent.

How to assign Transparency to Layers

1. Select the **Layer Properties Manager** using one of the following:

Ribbon = Home Tab / Layers Panel /
or
Keyboard = *LA <Enter>*

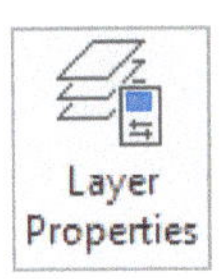

Continued on the next page...

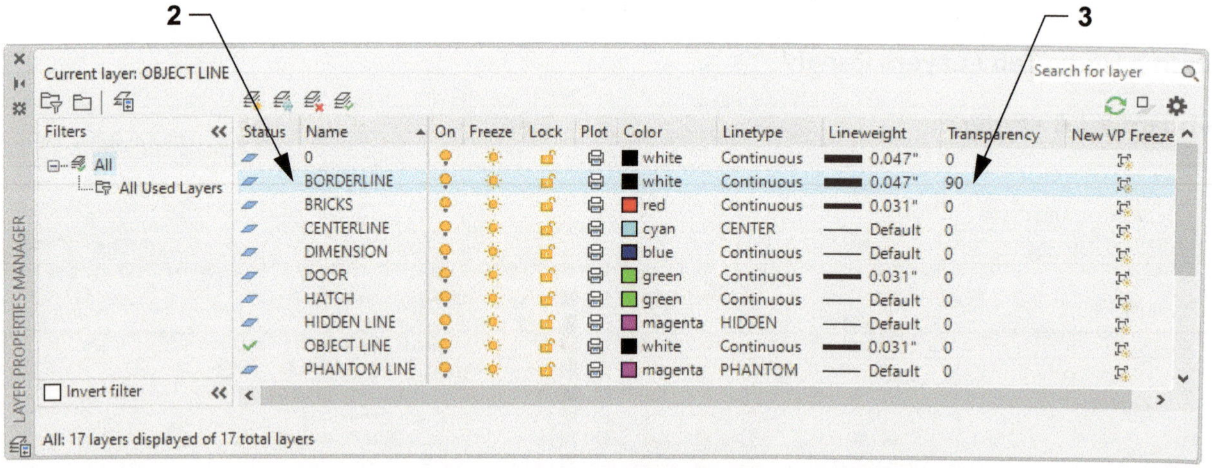

2. Highlight a layer (click on the name).
3. Click on Transparency for that layer.
4. Select a Transparency value from the list.
5. Select the **OK** button.

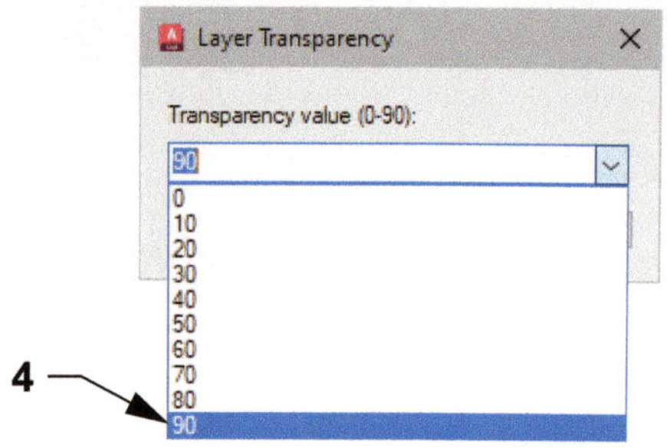

Controlling Transparent display

You may toggle the display of Transparent objects on or off by selecting the Transparency button on the Status Bar.

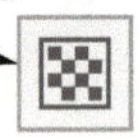

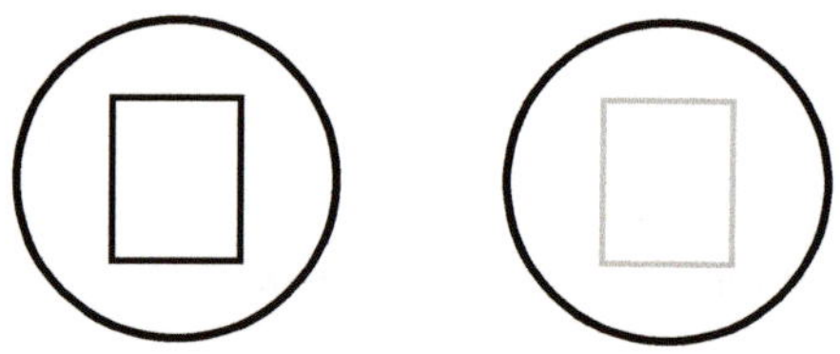

Transparency off   Transparency on

Note: Transparency selections will be saved within the current drawing and will not affect any other drawing.

Plotting Transparent Objects

Plotting transparency is disabled by default. To plot transparent objects, check the Plot transparency option in either the Plot dialog box or the Page Setup dialog box. This will be discussed in Lesson 26.

Creating New Layers

Using layers is an important part of managing and controlling your drawing. It is better to have too many layers than too few. You should draw like objects on the same layer. For example, place all doors on the “door” layer or Centerlines on the “Centerline” layer.

When you create a new layer, you will assign a **name**, **color**, **linetype**, **lineweight**, **transparency** and whether or not it should **plot**.

1. Select the **Layer Properties Manager** using one of the following:

Ribbon = Home Tab / Layers Panel /
or
Keyboard = ***LA <Enter>***

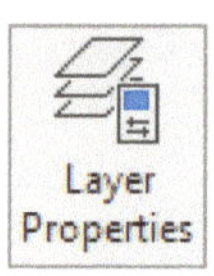

Continued on the next page...

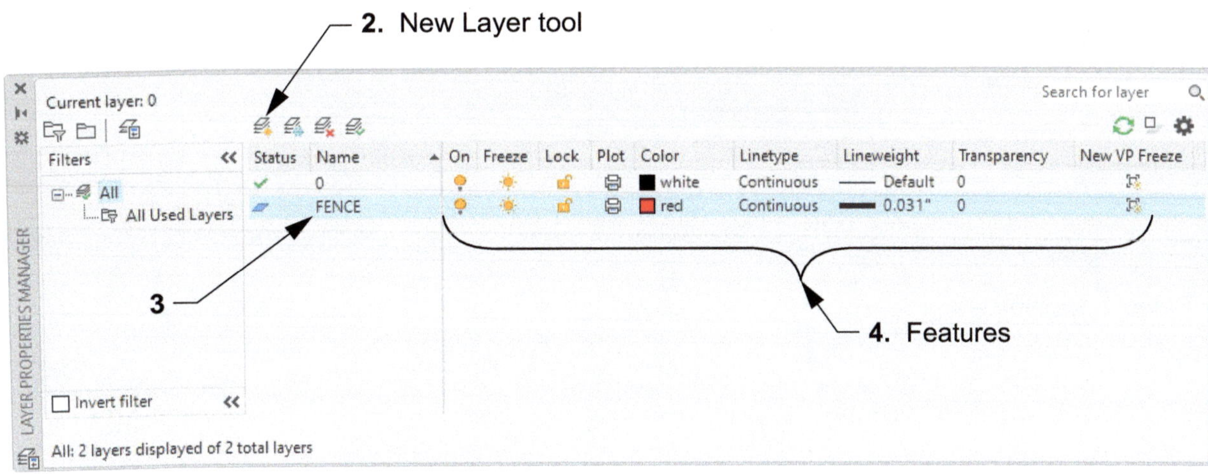

2. Select the **New Layer** tool and a new layer will appear.
3. Type the new layer name and press ***<Enter>***.
4. Select any of the **features** and a dialog box will appear.

Note: Refer to the previous pages for controlling and selecting layer colors, lineweights, and transparency.

Loading and Selecting Layer Linetypes

In an effort to conserve data within a drawing file, AutoCAD automatically loads only one linetype called **"continuous"**. If you would like to use other **linetypes**, such as "DASHED" or "FENCELINE1", you must **Load** them into the drawing as follows:

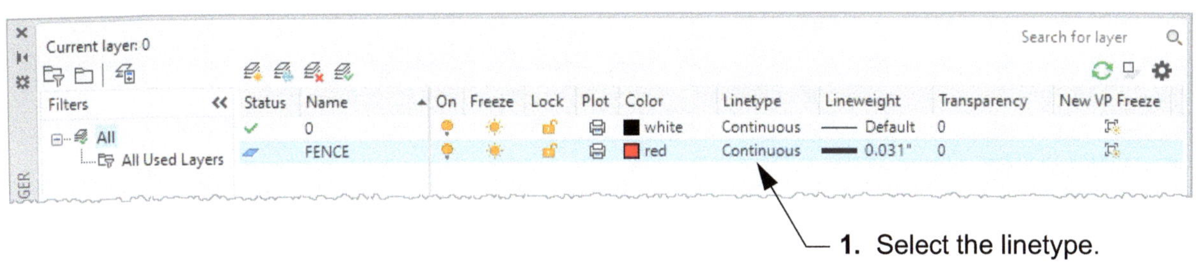

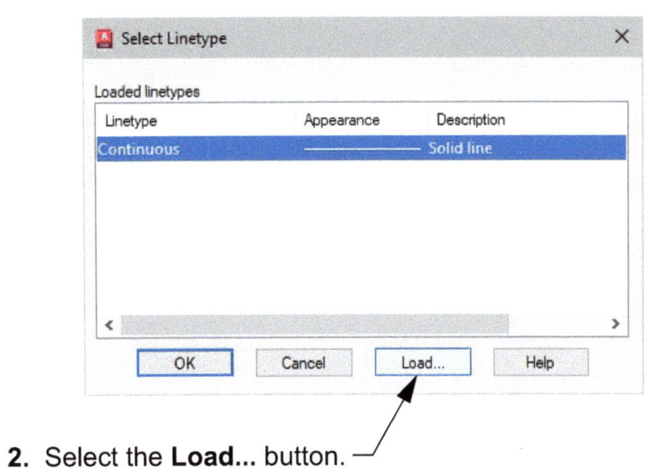

Continued on the next page...

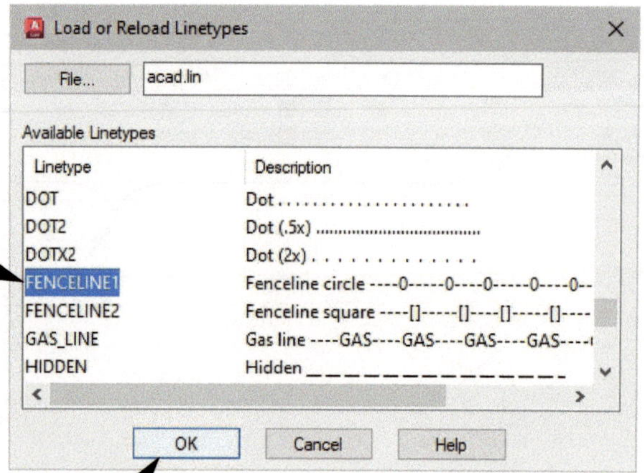

3. Select a linetype.

4. Select the **OK** button.

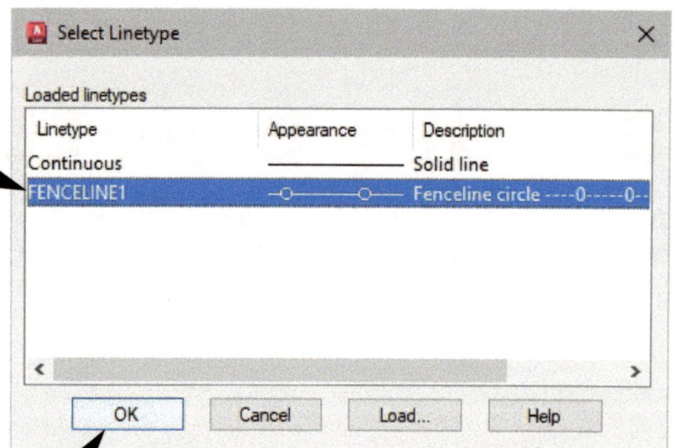

5. Select a linetype to assign to the layer.

6. Select the **OK** button.

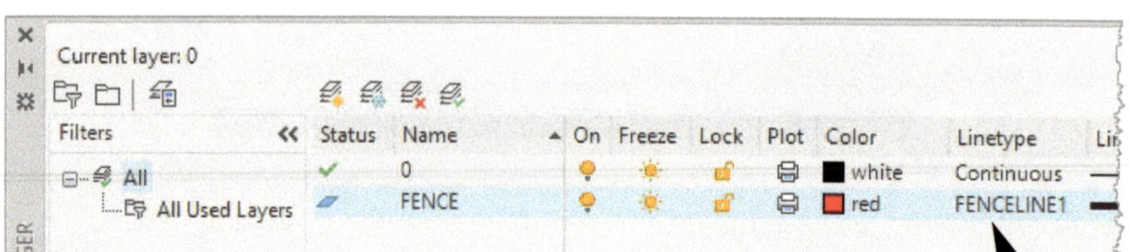

7. Linetype appears on the layer line.

Exercise 3A

Instructions:

1. Start a **New** file using either the **inch-helper.dwt** or the **metric-helper.dwt**
2. **Draw** the **Lines** using:

   A. **Line** command.

   B. **Ortho <*F8*> on**.

   C. **Turn Increment Snap <*F9*> on**.

   (Do not enter the text. You will learn text soon.)
3. Select the appropriate layer, then select the Line command, then draw a line.
4. Select the appropriate layer, then select the Line command, then draw a line.
5. Select the appropriate layer, then select the Line command, then draw a line.

   Are you noticing a pattern here?
6. **Save** the drawing as: **Ex-3A**

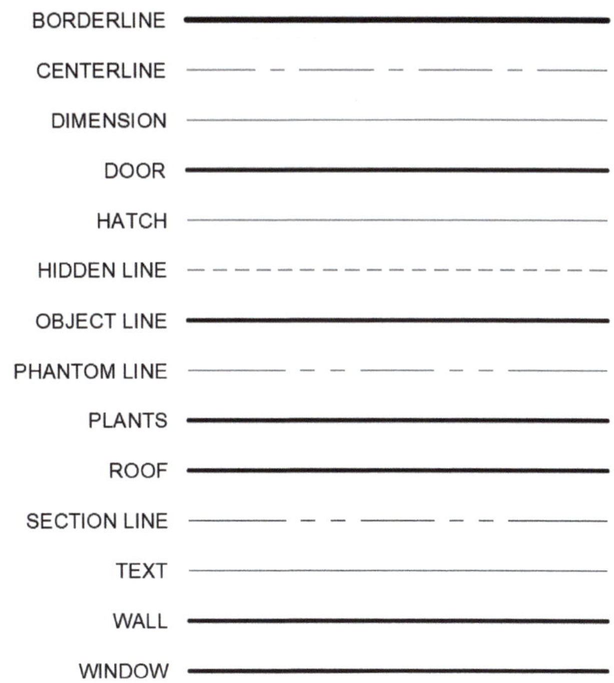

Exercise 3B

Instructions:

1. Start a **New** file using either the **inch-helper.dwt** or the **metric-helper.dwt**
2. Change the **Increment Snap** to **0.20" [5.08 mm]** and **Grid spacing** to **0.40" [10.16 mm]**. If you have the Snap and Grids set correctly, it will be easy.
3. Make sure **Snap** and **Grid** Status Bar buttons are on.
4. Draw the objects using Layer Object Line.

   Review pages 3-2 through 3-5 for Circle and Rectangle options.
5. **Do not dimension** (you will learn dimensioning soon).
6. Save the drawing as: **Ex-3B**

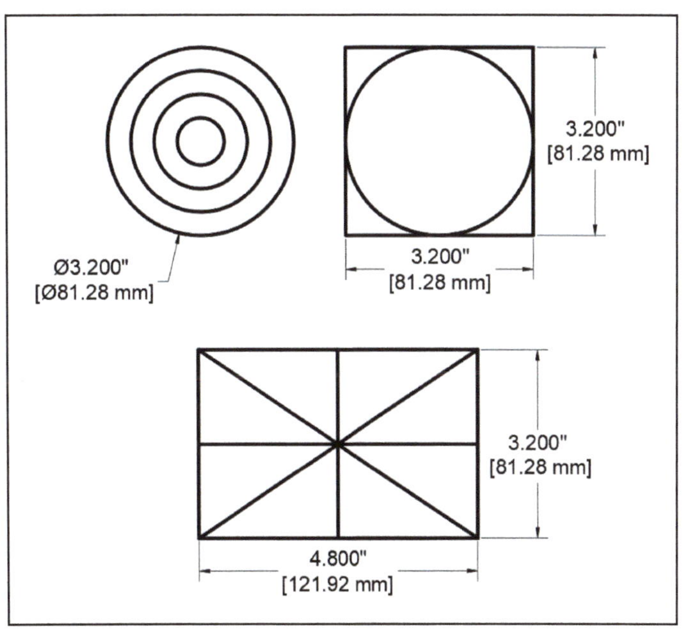

Exercise 3C

Instructions:

1. Start a **New** file using either the **inch-helper.dwt** or the **metric-helper.dwt**
2. **Draw** the **Rectangles** below using the options:

   Dimension, **Chamfer**, **Fillet**, **Width**, and **Rotation**
3. Use Layer Object Line.
4. Save the drawing as: **Ex-3C**

Rectangle Dimensions:

Length = **3" [76.2 mm]**
Width = **2" [50.8 mm]**
Chamfer = Dist1 = **0.50" [12.7 mm]**
Dist2 = **0.50" [12.7 mm]**

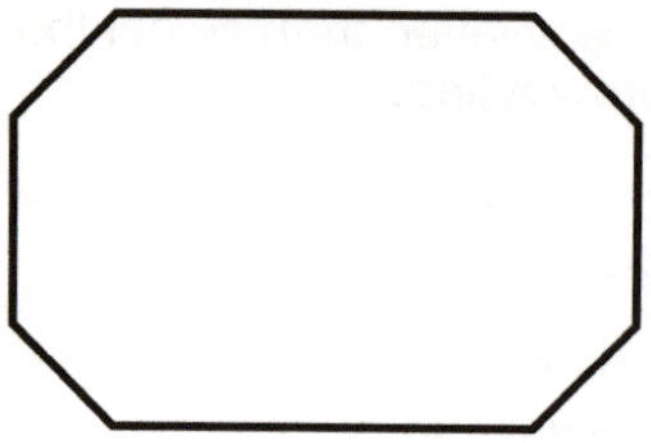

Rectangle Dimensions:

Length = **3" [76.2 mm]**
Width = **2" [50.8 mm]**
Fillet = **0.50" [12.7 mm]**

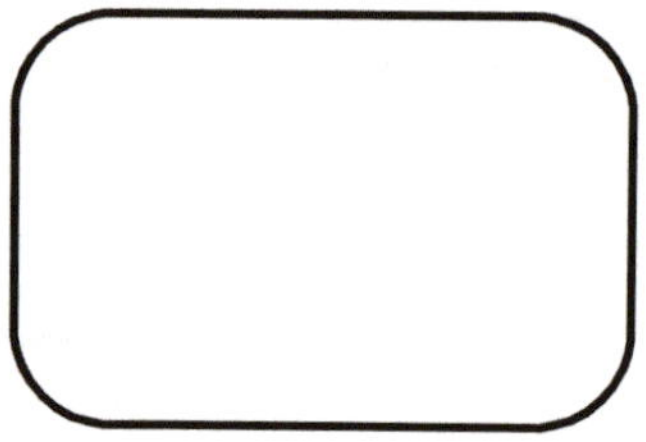

Rectangle Dimensions:

Length = **3" [76.2 mm]**
Width = **2" [50.8 mm]**
Rotation = **45°**
Chamfer = **0**
Fillet = **0**

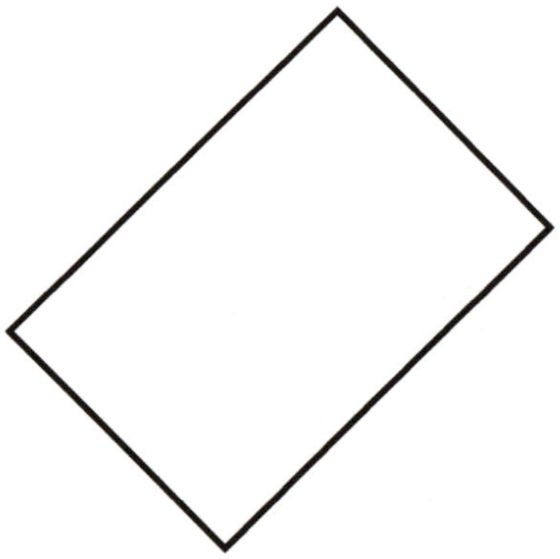

Rectangle Dimensions:

Length = **3" [76.2 mm]**
Width = **2" [50.8 mm]**
Line Width = **0.200" [5.08 mm]**
Rotation = **0°**
Chamfer = **0**
Fillet = **0**

Exercise 3D

Instructions:

1. Start a **New** file using either the **inch-helper.dwt** or the **metric-helper.dwt**
2. **Important: Draw** the objects **using the following layers**:

   Roof, Wall, Window, Door, and Plants.
3. You can change Snap and Grid settings to whatever you like.
4. You decide when to turn Ortho and Snap **on** or **off**.
5. All objects must be placed accurately. All lines must intersect exactly.
6. Save the drawing as: **Ex-3D**

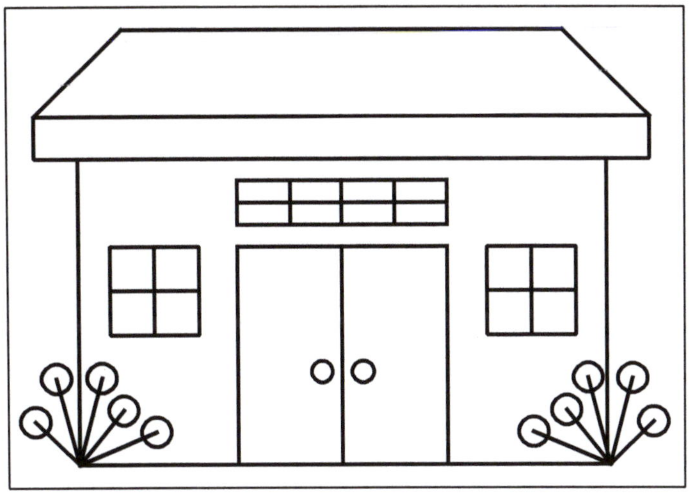

Exercise 3E

Instructions:

1. Open **Ex-3D**. (Use **Open** not **New** to open an existing drawing.)
2. Make Layer Object Line current.

   (I am having you do that because you can't Freeze or Lock a layer that is current and I want to make sure that none of the layers that you used in the drawing are current.)
3. **Freeze** the following layers: **Window** and **Plants** (Do not use erase).
4. **Lock** Layer Roof.
5. Try to erase any of the roof lines.

   You can't because the Roof layer is **Locked**.
6. Save the drawing as: **Ex-3E**

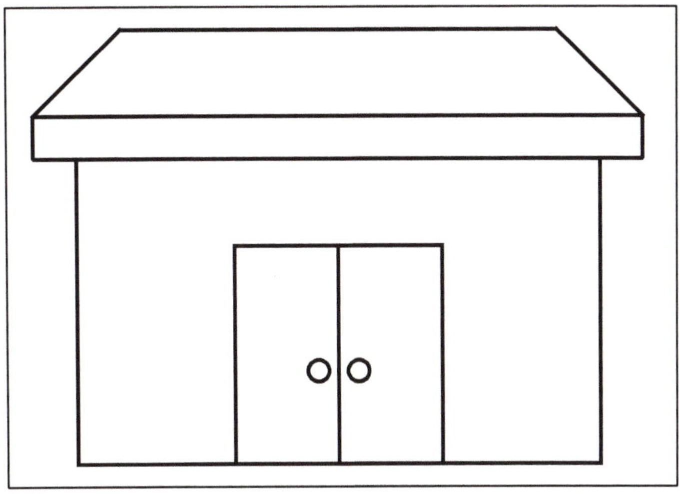

Exercise 3F

Instructions:

1. Start a new file using template: **acad.dwt** (inch) or **acadiso.dwt** (metric).
2. Load linetypes **ZIGZAG**, **GAS_LINE**, and **DASHED**. (Refer to page 3-13.)

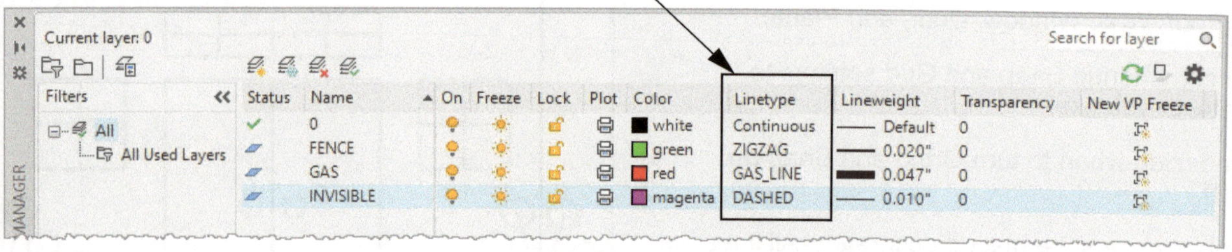

3. Create 3 layers named Fence, Gas, and Invisible. (Refer to page 3-12.)

   Assign the name, color, linetype, and lineweight as shown above.

Note: Lineweights for metric users are: **ZIGZAG = 0.5 mm, GAS_LINE = 1.2 mm, DASHED = 0.25 mm**.

4. **Draw** the **Lines** below using the new layers:
   A. Select the layer.
   B. Select the **Line** command.
   C. Draw the lines approximately **5 inches [127 mm]** long.
   D. **Ortho** ***<F8>*** **on**.

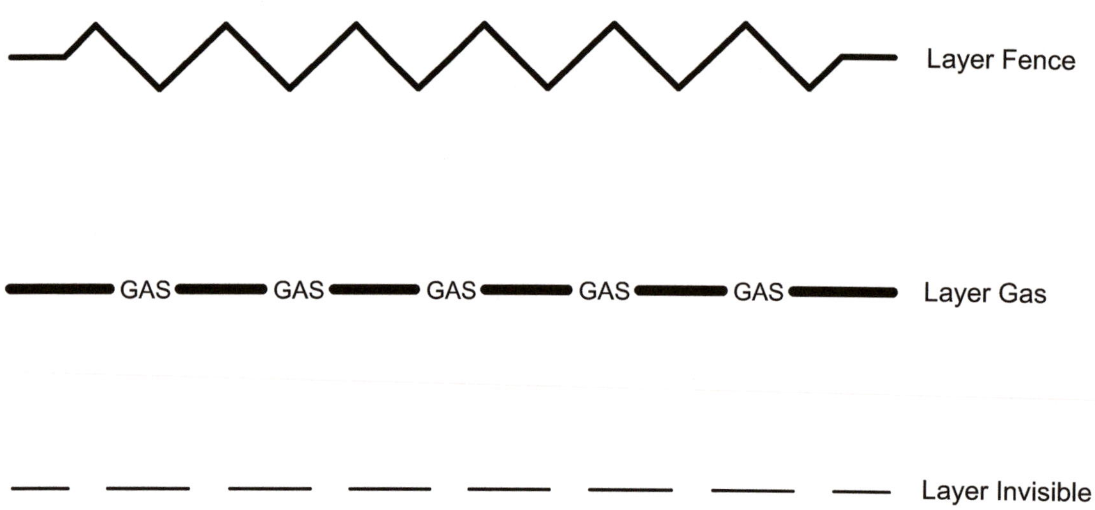

5. Save the drawing as: **Ex-3F**

LESSON 4

LEARNING OBJECTIVES

After completing this lesson, you will be able to:

- Understand Object Snap
- Use Running Object Snap
- Use the Zoom options to view the Drawing
- Change the Drawing Limits
- Select the Units of Measurement and Precision

Object Snap

In Lesson 3, you learned about **Increment Snap**. Increment Snap enables the cursor to move in an incremental movement. So you could say your cursor is "snapping to increments" preset by you.

Now you will learn about **Object Snap**. If Increment Snap snaps to increments, what do you think Object Snap snaps to? That's right: "objects". Object Snap enables you to snap to objects in very specific and accurate locations on the objects. For example, the endpoint of a line or the center of a circle.

How to select from the Object Snap Menu

1. You must select a command, such as **Line**, before you can select Object Snap.

2. While holding down the ***<Shift>*** key, press the right mouse button. The menu shown at right should appear.

3. Highlight and press the left mouse button to select an Object Snap.

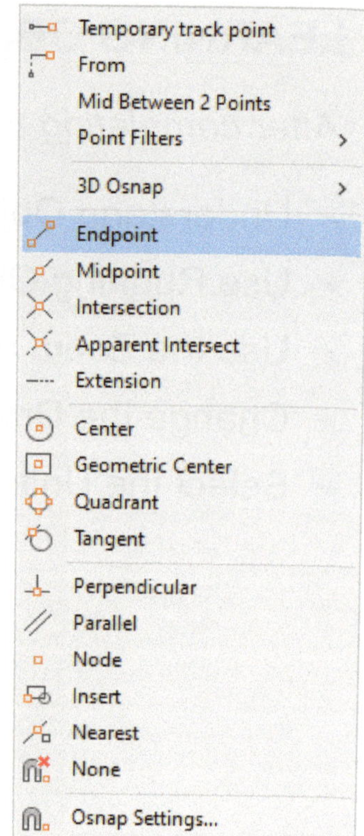

Note: The Object Snaps discussed in this lesson include Endpoint, Midpoint, Intersection, Center, Geometric Center, Quadrant, Tangent, and Perpendicular. The remaining Object Snaps will be discussed in future lessons.

Object Snap Definitions

Object Snap is used when AutoCAD prompts you to place an object. Object Snap allows you to place objects very accurately. A step-by-step example of **"How to use Object Snap"** is shown on the next page.

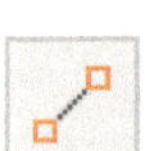 **END**point

Snaps to the closest endpoint of a line, arc, or polygon segment. Place the cursor on the object close to the end and the cursor will snap like a magnet to the end of the line.

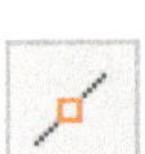 **MID**point

Snaps to the middle of a line, arc, or polygon segment. Place the cursor anywhere on the object and the cursor will snap like a magnet to the midpoint of the line.

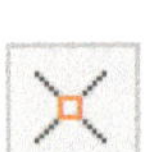 **INT**ersection

Snaps to the intersections of any two objects. Place the cursor directly on top of the intersection or select one object and then the other and AutoCAD will locate the intersection.

CENter — Snaps to the center of an arc, circle, or donut. Place the cursor on the object, or at the approximate center location, and the cursor will snap like a magnet to the center.

Geometric CEnter — Snaps to the centroid of any closed polylines and splines. Place the cursor anywhere on the object and the cursor will snap to the centroid of any closed polyline or spline.

QUAdrant — Snaps to a 12:00, 3:00, 6:00, or 9:00 o'clock location on a circle or ellipse. Place the cursor on the circle near the desired quadrant location and the cursor will snap to the closest quadrant.

TANgent — Calculates the tangent point of an arc or circle. Place the cursor on the object as near as possible to the expected tangent point.

PERpendicular — Snaps to a point perpendicular to the object selected. Place the cursor anywhere on the object, then pull the cursor away from the object and press the left mouse button.

CAD TIP! You may also type the **3 bold letters** shown rather than selecting from the menu. For example, type in ***CEN*** and then press ***<Enter>*** for the Center Snap.

How to Use Object Snap

The following is an example of attaching a line segment to previously drawn vertical lines. The new line will go from the upper endpoint, to the midpoint, to the lower endpoint.

1. Turn off **SnapMode**, **OrthoMode**, and **Object Snap** on the **Status Bar**. (Gray is **off**.)

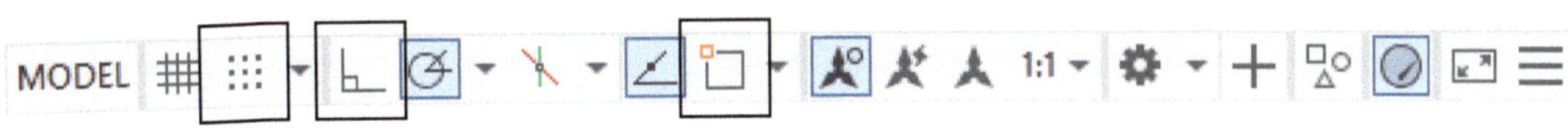

2. Select the **Line** command.
3. Draw two vertical lines as shown below (they don't have to be perfectly straight).

4. Select the **Line** command again.
5. Hold the ***<Shift>*** key down and press the right mouse button.

Continued on the next page...

6. Select the Object Snap **Endpoint** from the Object Snap Menu.

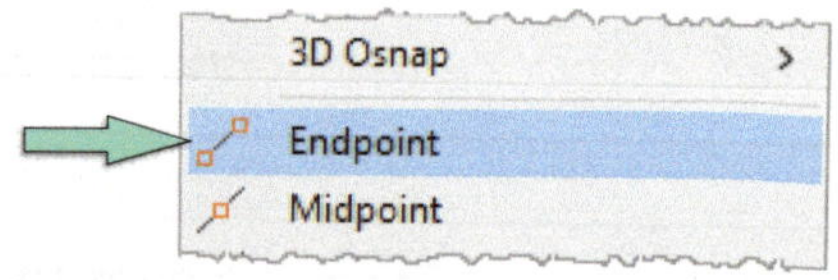

7. Place the cursor close to the upper endpoint of the left-hand line.

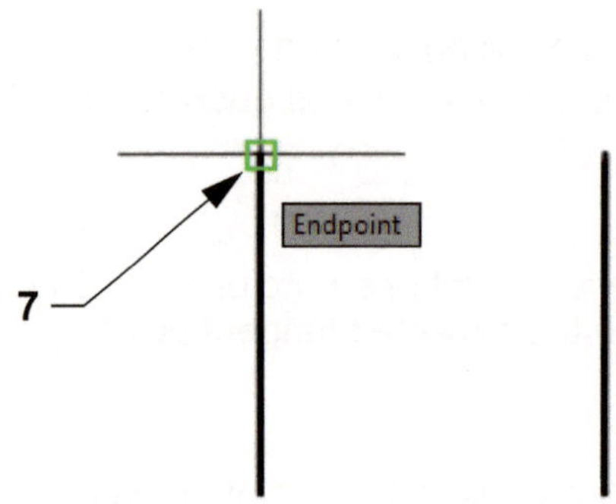

The cursor should snap to the end of the line like a magnet. A little square and an **"Endpoint"** tooltip will be displayed.

8. Press the left mouse button to attach the new line to the upper endpoint of the previously drawn vertical line. (Do not end the Line command yet.)

9. Now hold the ***<Shift>*** key down and press the right mouse button and select the **Midpoint** Object Snap option.

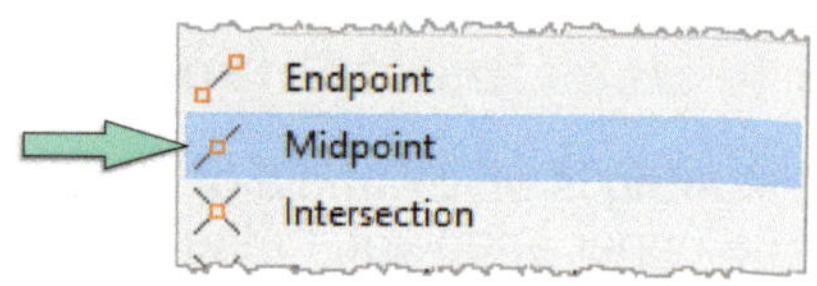

10. Move the cursor to approximately the middle of the right-hand vertical line.

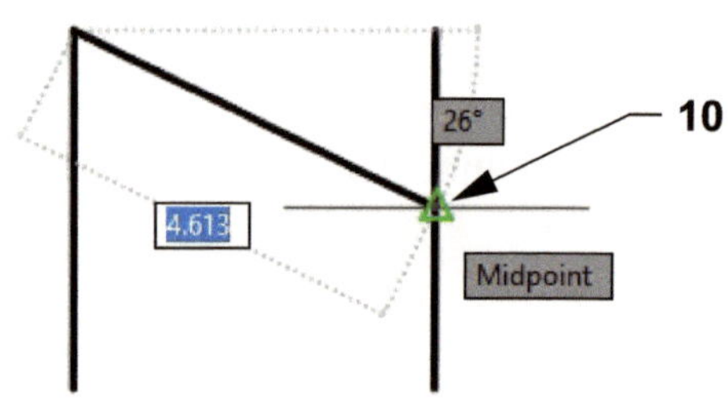

The cursor should snap to the midpoint of the line like a magnet. A little triangle with a **"Midpoint"** tooltip will be displayed.

11. Press the left mouse button to **attach** the new line to the midpoint of the previously drawn vertical line. (Do not end the Line command yet.)

12. Now hold the ***<Shift>*** key down and press the right mouse button and select the Object Snap **Endpoint** again.

13. Move the cursor close to the lower endpoint of the left-hand vertical line.

Continued on the next page...

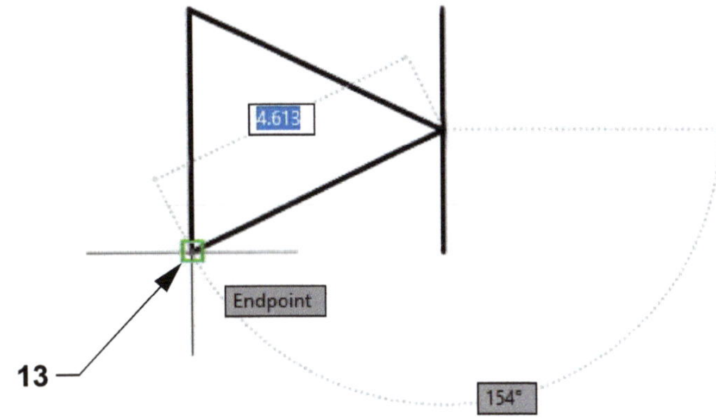

The cursor should snap to the end of the line like a magnet. A little square and an **"Endpoint"** tooltip will be displayed.

14. Press the left mouse button to **attach** the new line to the endpoint of the previously drawn vertical line.

15. Stop the **Line** command and disconnect by pressing ***<Enter>***.

Running Object Snap

Selecting Object Snap is not difficult, but AutoCAD has provided you with an additional method to increase your efficiency by allowing you to preset frequently used Object Snap options. This method is called **Running Object Snap**.

When **Running Object Snap** is active, the cursor will automatically snap to any preset Object Snap locations thus eliminating the necessity of invoking the Object Snap Menu for each locations.

First, you must **set the Running Object Snaps**, and **second**, you must **turn on the Running Object Snap option**.

How to set the Running Object Snap

1. Select the **Running Object Snap** dialog box using one of the following:

   Keyboard = ***DS <Enter>***
   or
   Status Bar = Left click on the Object Snap button down arrow ▾ and select Object Snap Settings...

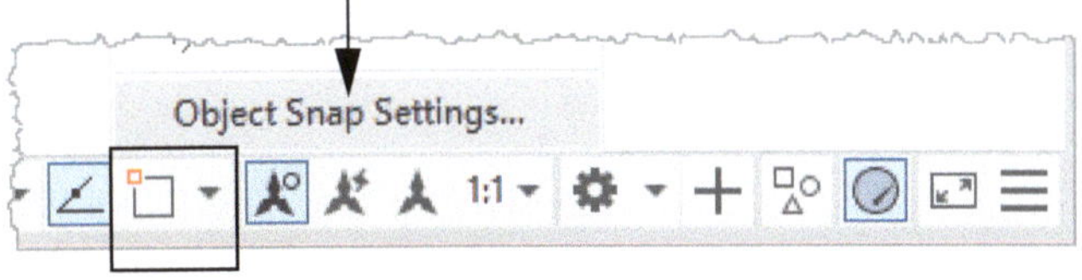

2. Select the **Object Snap** Tab.

3. Select the Object Snaps desired.

   (In the example on the next page, Object Snap **Endpoint**, **Midpoint**, and **Intersection** have been selected.)

Continued on the next page...

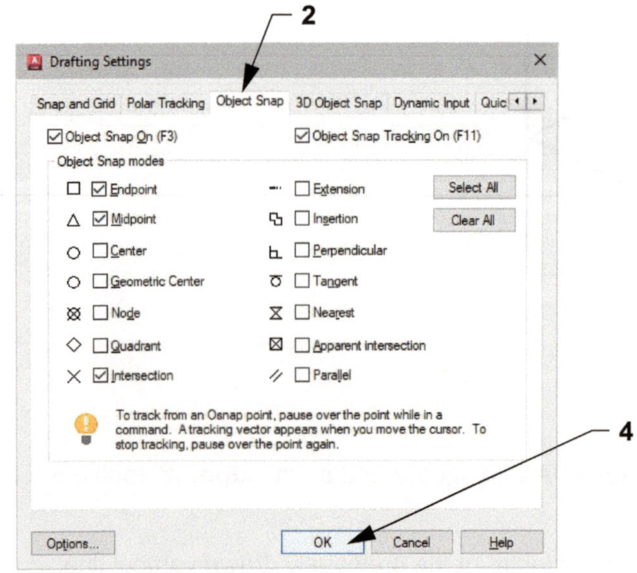

Note: Try not to select more than three or four Object Snaps at one time.

If you select too many, the cursor will flit around trying to snap to multiple snap locations. And possibly snap to the wrong location.

You will lose control, and it will confuse you.

4. Select the **OK** button.

How to turn on the Running Object Snap

1. Turn on the **Object Snap** button on the **Status Bar**. (Blue is **on**.)

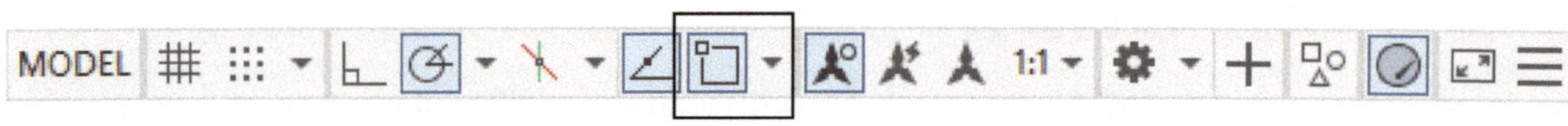

2. Now try drawing the line from the endpoint to the midpoint again, **but this time do not select the "Object Snap Menu"**. Just move the cursor close to the endpoint and the cursor will automatically snap to the end of the line.

3. Move the cursor to approximately the middle of the right-hand vertical line and the cursor will automatically snap to the midpoint of the line.

4. Move the cursor close to the lower endpoint of the left-hand vertical line and the cursor will automatically snap to the lower endpoint of the left line.

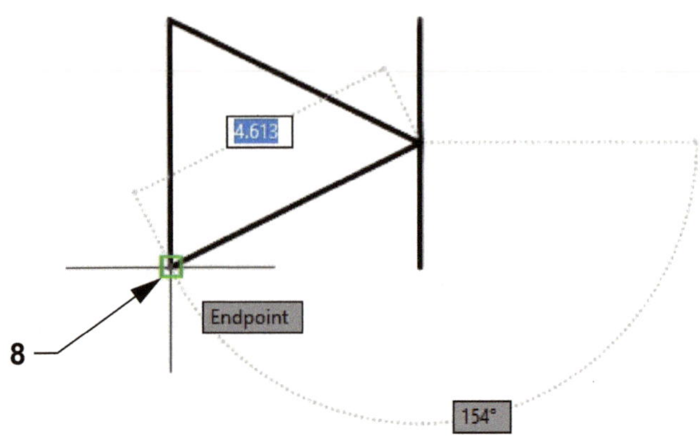

Note: Running Object Snap is very handy. But remember not to select more than three or four at a time. The selections will fight each other, and you may end up snapping to a location that you did not want.

If you want to snap to a location that is not preset, just select the Object Snap Menu, as shown on page 4-2, and select the one you want. Running Object Snap and Object Snap work together very well, but it may take a little practice.

Zoom

The **Zoom** command is used to move closer to or farther away from an object. This is called Zooming In and Out.

The Zoom commands are located on the **Navigate** Panel of the **View** Tab and are **off** by default. Select the **View** Tab, then right click on any Panel, select **Show Panels**, and activate the **Navigate** Panel. (To Show Panels, refer to page 1-11.)

1. Select the **Zoom** command by using the following:

   Ribbon = View Tab / Navigate Panel

2. Select the ▾ down arrow to display all of the selections.

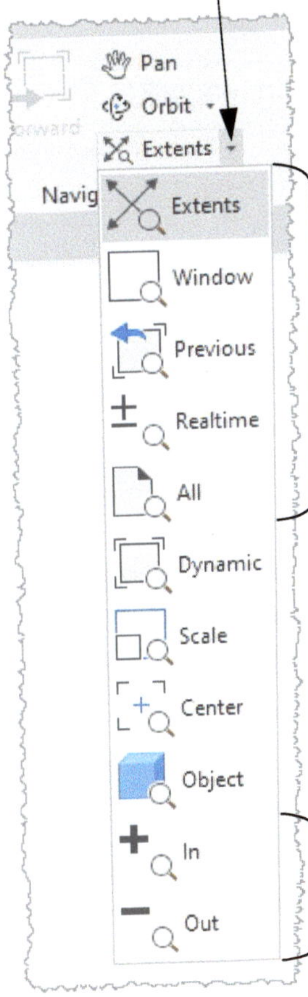

These are the most commonly used zoom tools:

Extents Displays all objects in the drawing file, even objects outside of the drawing limits.

Window Zoom in on an area by specifying a Window around the area.

Previous Returns the screen to the previous display. (Limited to 10.)

Realtime Interactive Zoom: You can zoom in or out by moving the cursor vertically up or down while pressing the left mouse button. To stop, press the ***<Esc>*** key.

All Zooms to drawing limits or Extents, whichever is greater.

In or **Out** Moves in two times or out two times.

You may also select the **Zoom** commands using one of the following:

Right click and select Zoom from the Shortcut Menu. (Refer to page Intro-4 for right click settings.)

Keyboard = *Z <Enter>*. Select from the options listed.

How to use Zoom / Window

1. Select **Zoom / Window**.
2. Create a Window around the objects you want to enlarge. (Creating a Window is a similar process to drawing a rectangle. It requires a first corner and then a diagonal corner.)

Continued on the next page...

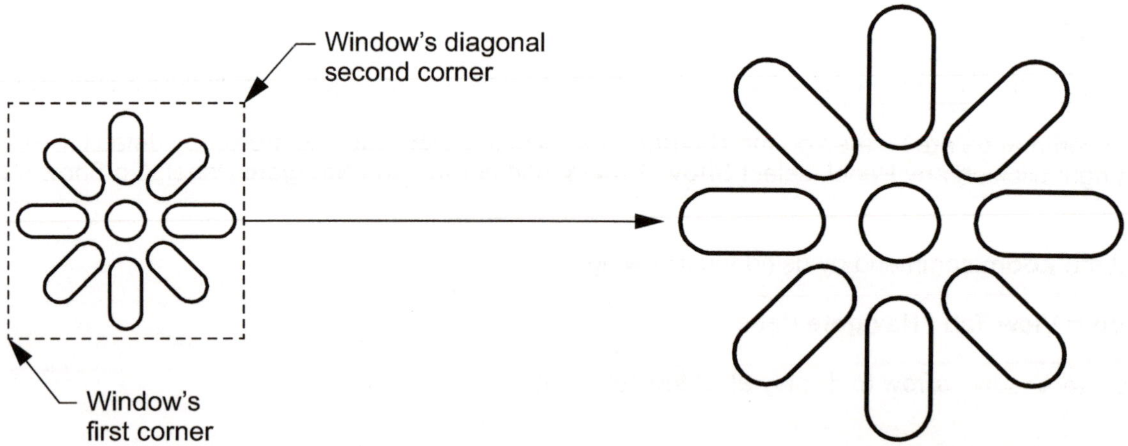

Magnified to this view

Note: The objects have been magnified, but the actual size has not changed.

How to return to Original View

1. Type ***Z <Enter> A <Enter>*** (this is a shortcut for **Zoom / All**).

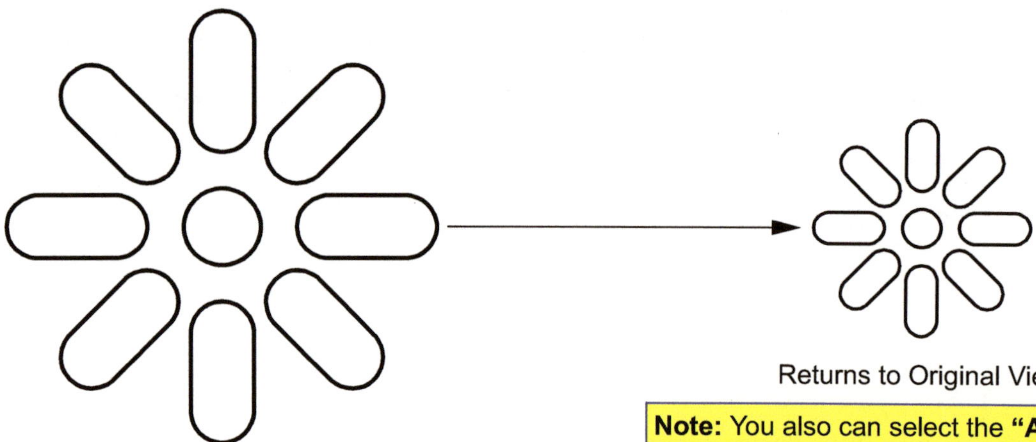

Returns to Original View

Note: You also can select the **"All"** tool from the Panel shown on the previous page.

Drawing Setup

When drawing with a computer, you must **"set up your drawing area"** just as you would on your drawing board if you were drawing with pencil and paper. You must decide what **size** paper you will need, what **Units of measurement** you will use (feet and inches or decimals, etc.) and how **precise** you need to be. In AutoCAD these decisions are called **"Setting the Drawing Limits, Units, and Precision"**.

Drawing Limits

Consider the drawing limits as the **size of the paper** you will be drawing on. You will first be asked to define where the lower left corner should be placed, then the upper right corner, similar to drawing a rectangle. An 11" x 8.5" or 297 mm x 210 mm piece of paper would have a **lower left corner** of **0,0** and an **upper right corner** of **11, 8.5**, or 297, 210. *(11 or 297 is the horizontal measurement X axis, and 8.5 or 210 is the vertical measurement Y axis.)*

How to Set the Drawing Limits

1. Start a **New** drawing using either the **inch-helper.dwt** or the **metric-helper.dwt**. (Refer to page 2-5.)

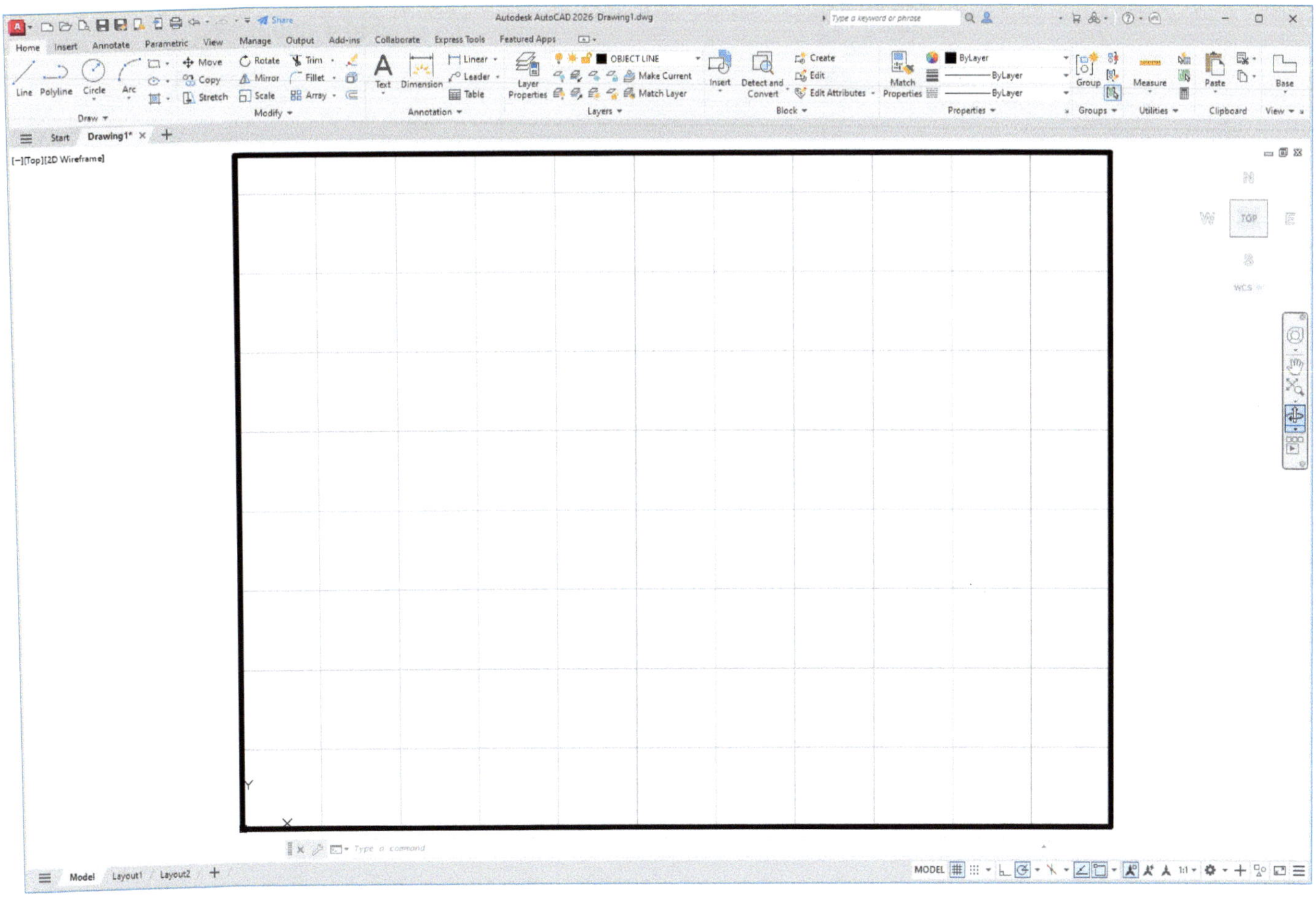

2. Select the **Drawing Limits** command by typing ***Limits*** and then press ***<Enter>***.
3. The following prompt will appear:

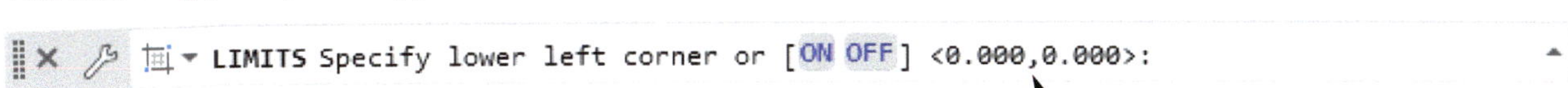

Displays the current lower left corner coordinates before change.

4. Type the X,Y coordinates ***0,0 <Enter>*** for the new lower left corner location of your piece of paper. Make sure you separate the two figures with a comma.
5. The following prompt will appear:

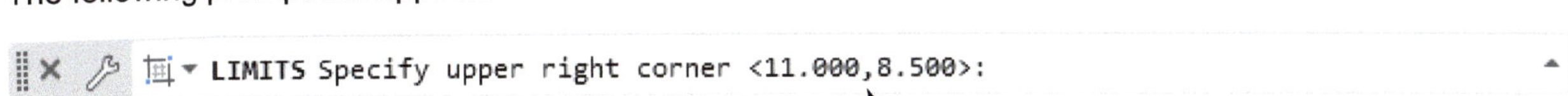

Displays the current upper right corner coordinates before change.

Continued on the next page...

6. Type the X,Y coordinates ***36,24 <Enter>*** if you are using the **inch template**, or ***914.4,609.6 <Enter>*** if you are using the **metric template**, for the new upper right corner of your piece of paper. Make sure you separate the two figures with a comma.

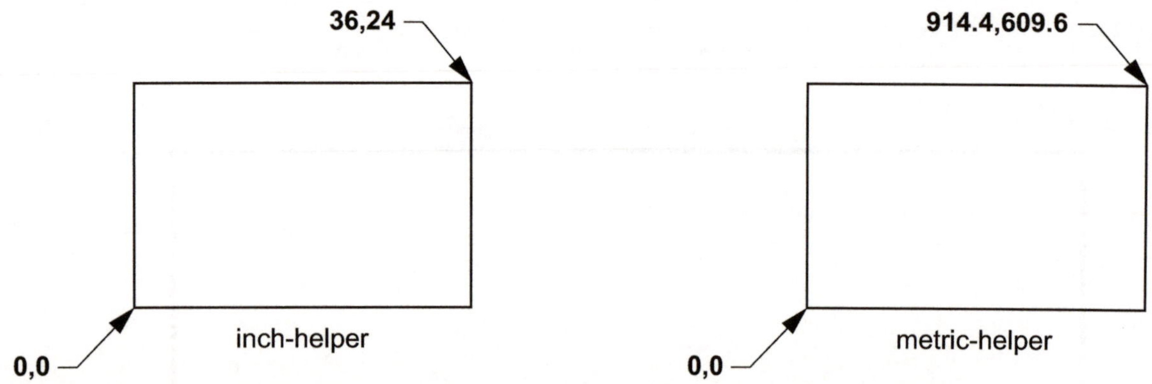

Note: Visually, the screen has not changed. After you complete the next step, it will.

7. This next step is very important:

   Type ***Z <Enter> A <Enter>*** to make the screen display the new drawing limits. (This is the shortcut for **Zoom / All**. Refer to page 4-8.)

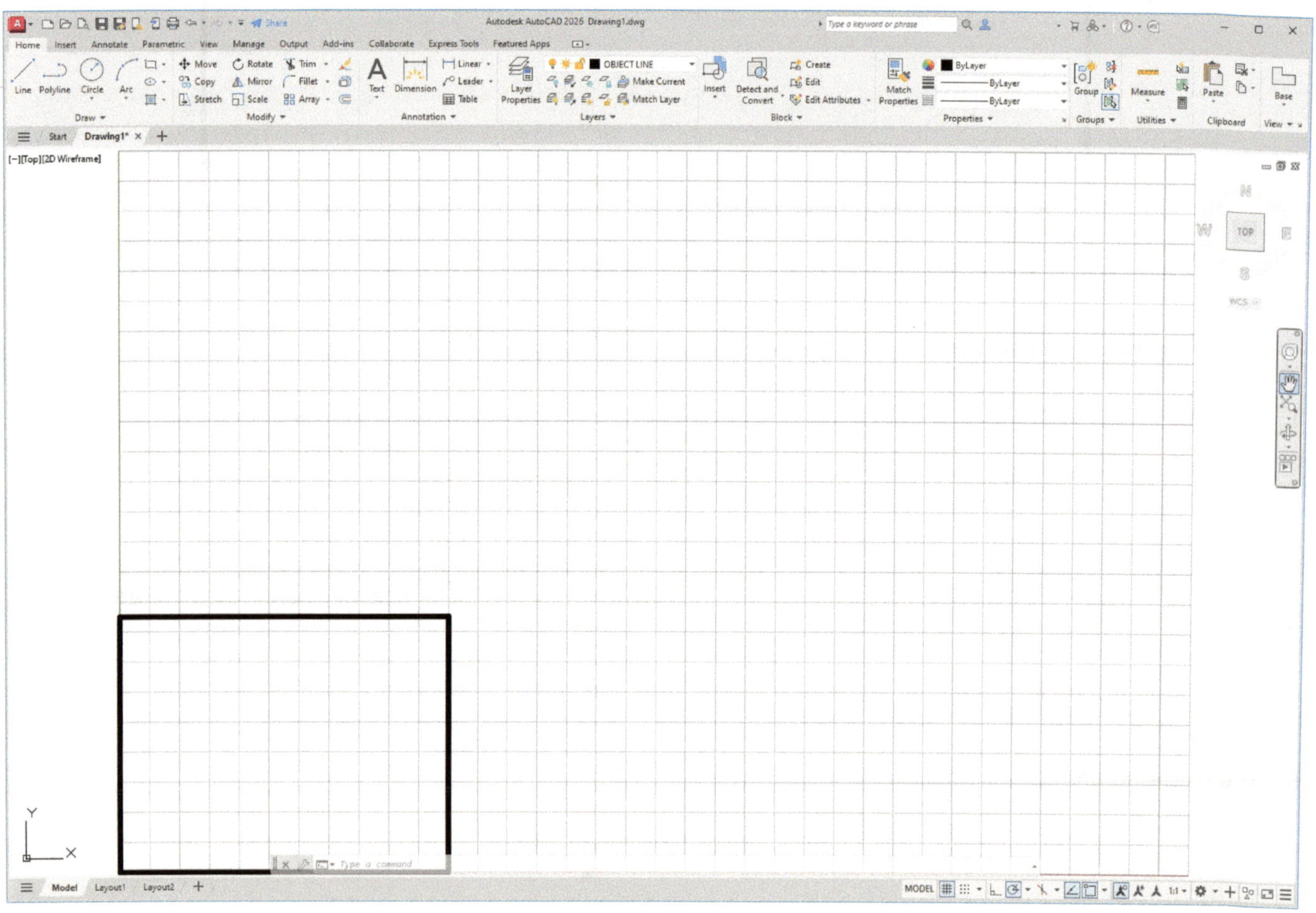

Note: The drawing limits (area) are now 36" or 914.4 mm **wide** X 24" or 609.6 mm **high**. The rectangle did not change size or location, but the drawing area around it got larger so the rectangle appears to have moved and gotten smaller. Think about it.

Display Grid within Limits

If the Grid behavior setting is ☐ **Display grid beyond Limits** (no check mark), it is turned **off**, and the grid will only be displayed within the Limits that you set.

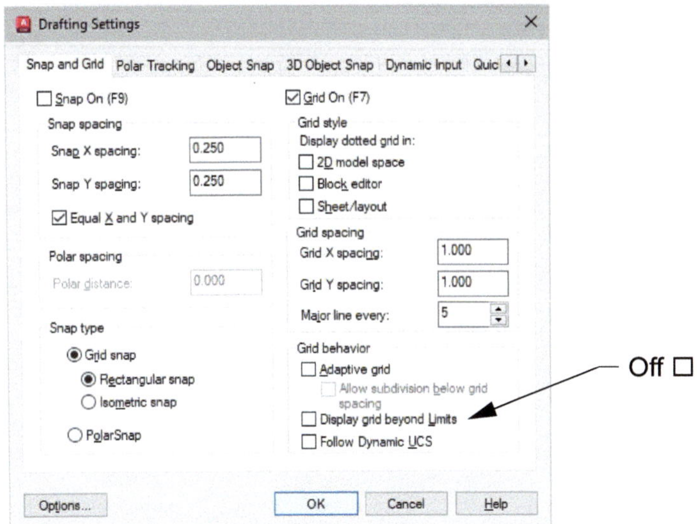

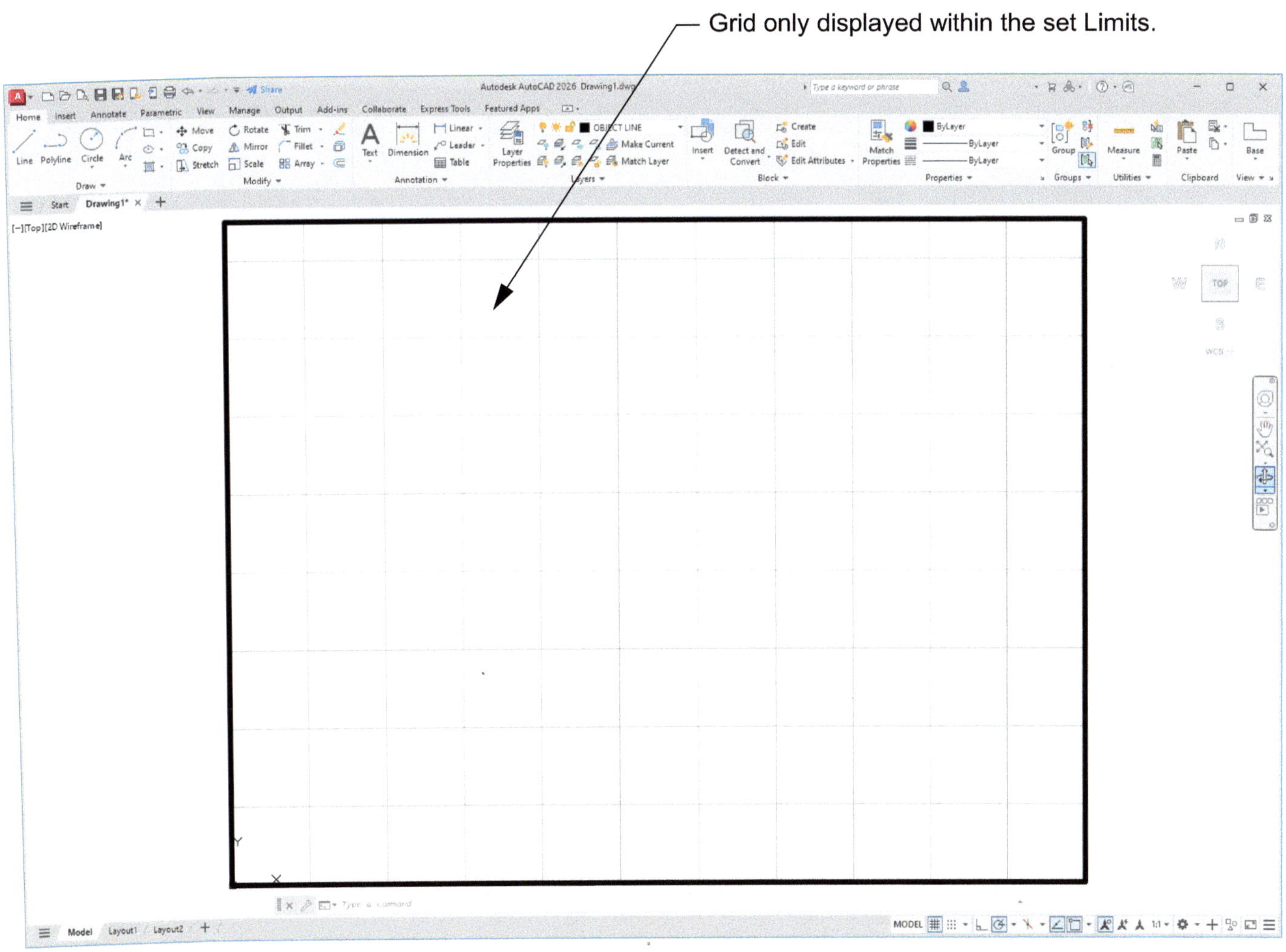

Display Grid beyond Limits

If the Grid behavior setting is ☑ **Display grid beyond Limits** (check mark), it is turned **on**, and the grid will be displayed beyond the Limits that you set.

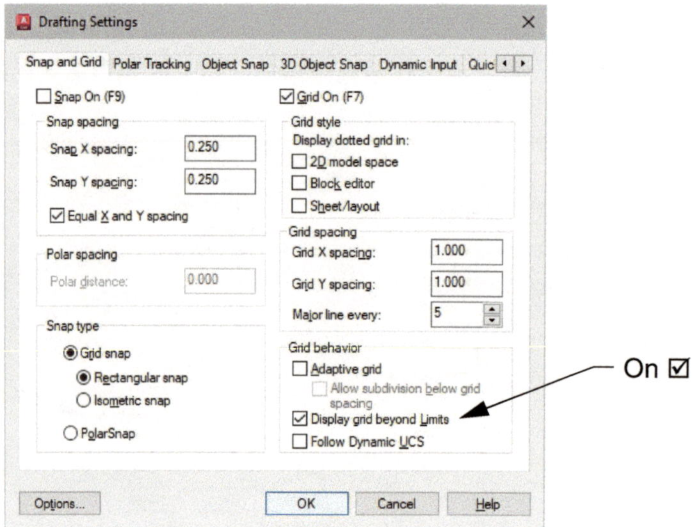

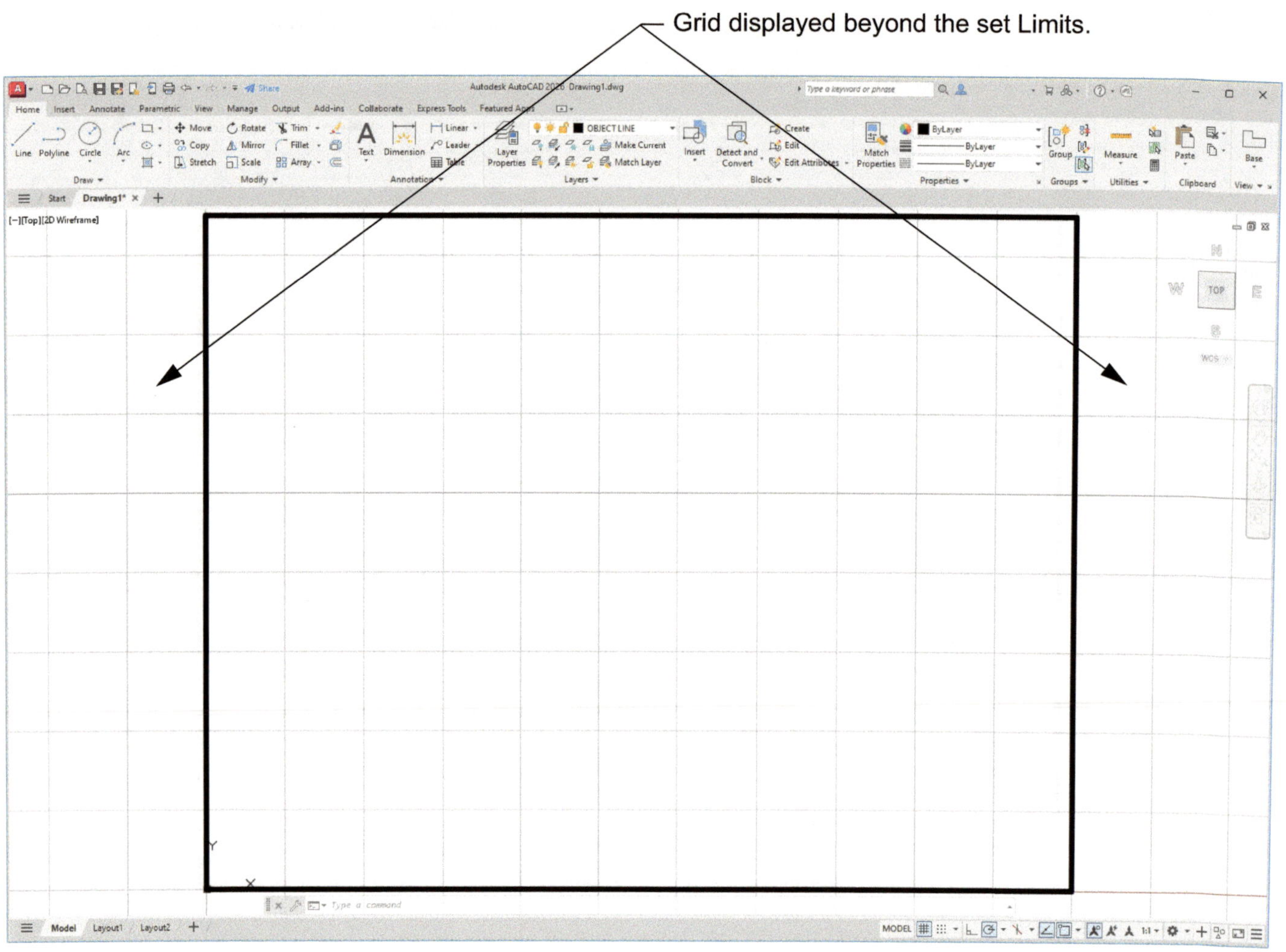

Units and Precision

You now need to select what unit of measurement with which you want to work. You might select Decimal (0.000) or Architectural (0'-0"). Next you should select how precise you want the measurements. For example, do you want the measurement limited to a 3-place decimal or the nearest 1/8".

How to Set the Units and Precision

1. Select the **Units** command using one of the following:

Application Menu = Drawing Utilities / Units
or
Keyboard = *Units <Enter>*

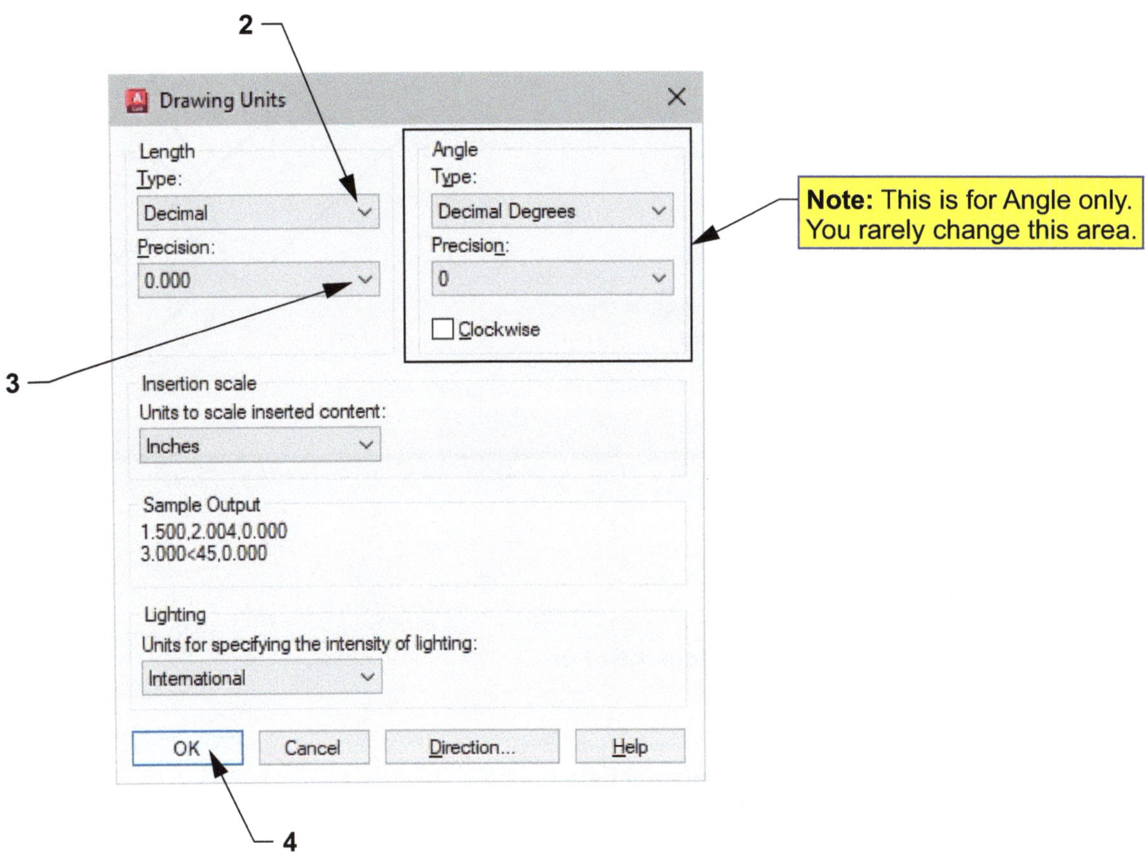

2. **Type:** Select the down arrow and select: **Decimal** or **Architectural**.

3. **Precision:** Select the down arrow and select the appropriate **Precision** associated with the "type". Examples: 0.000 for Decimals or 1/16" for Architectural.

4. Select the **OK** button to save your selections.

Easy, yes?

Exercises 4A & 4B

Exercise 4A

Instructions:

1. Start a **New** file using either the **inch-helper.dwt** or the **metric-helper.dwt**
2. Set **Units** and **Precision**:
   Units = **Fractional** for inch users or **Decimal** for metric users.
   Precision = **1/2"** for inch users or **0** for metric users.
3. Set **Drawing Limits**:
   Lower Left corner = **0,0** Upper Right Corner = **20,15** [**508,381** metric users].
4. Make sure you use **Zoom / All** after setting Drawing Limits.
5. Erase the rectangle that appears with the template. It will appear too small.
6. Turn **off** the **Snap** and **Ortho**. (Your cursor should move freely.)
7. Draw the objects shown using:
   Circle, Center Radius, and Line (Use Layer = Object Line) Object Snap = Center and Tangent
8. **Save** the drawing as: **Ex-4A**

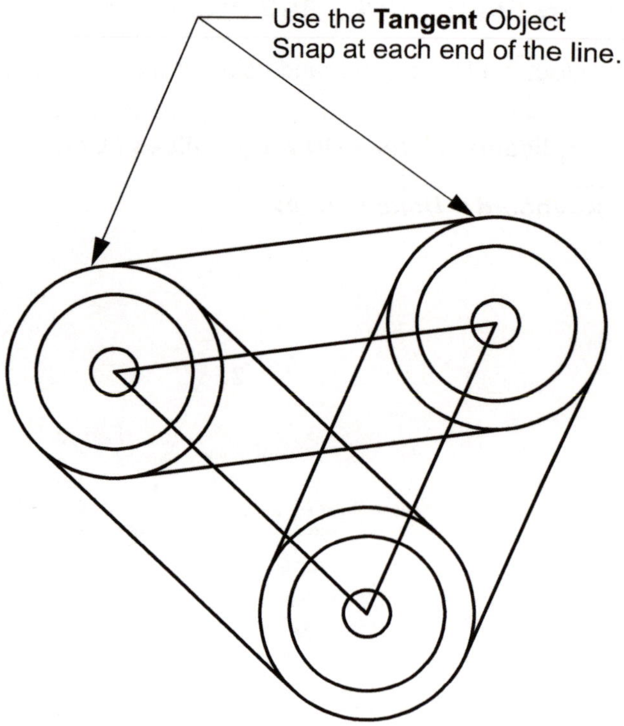

Exercise 4B

Instructions:

1. Start a **New** file using either the **inch-helper.dwt** or the **metric-helper.dwt**
2. Set Units and Precision:
   Units = **Fractional** for inch users or **Decimal** for metric users.
   Precision = **1/4"** for inch users or **0** for metric users.
3. Set **Drawing Limits**:
   Lower Left corner = **0,0** Upper Right Corner = **12,9** [**304.8,228.6** metric users].
4. Make sure you use **Zoom / All** after setting Drawing Limits
5. Turn **off** the **Snap** and **Ortho**. (Your cursor should move freely.)
6. Draw the objects shown using:
   Circle, Center Radius (Use Layer = Object Line)
   Line (Use Layer = Hidden Line)
   Object Snap = Quadrant
7. **Save** the drawing as: **Ex-4B**

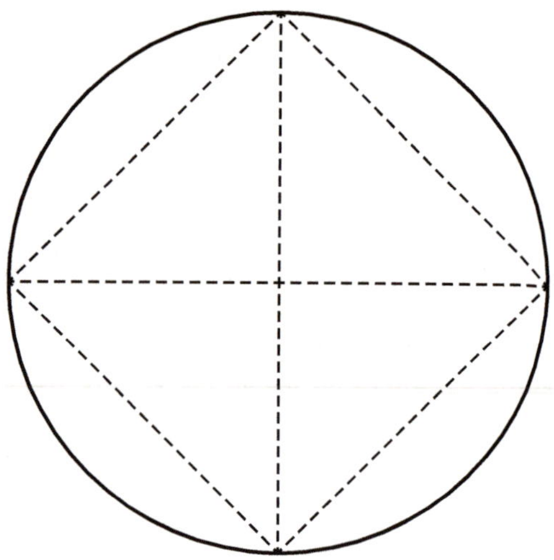

Exercise 4C

Instructions:

1. Start a **New** file using either the **inch-helper.dwt** or the **metric-helper.dwt**
2. Set **Units** and **Precision**:
   Units = **Architectural** for inch users or **Decimal** for metric users.
   Precision = 1/2" for inch users or **0** for metric users.
3. Set **Drawing Limits**:
   Lower Left corner = **0,0** Upper Right Corner = **25,20** [**635,508** metric users].
4. Make sure you use **Zoom / All** after setting Drawing Limits.
5. Erase the rectangle that appears with the template.
6. Turn **off** the **Snap** and **Ortho**.
7. Draw the objects showne using:
   Line (Use Layer = Object Line)
   Object Snap = Perpendicular
8. **Save** the drawing as: **Ex-4C**

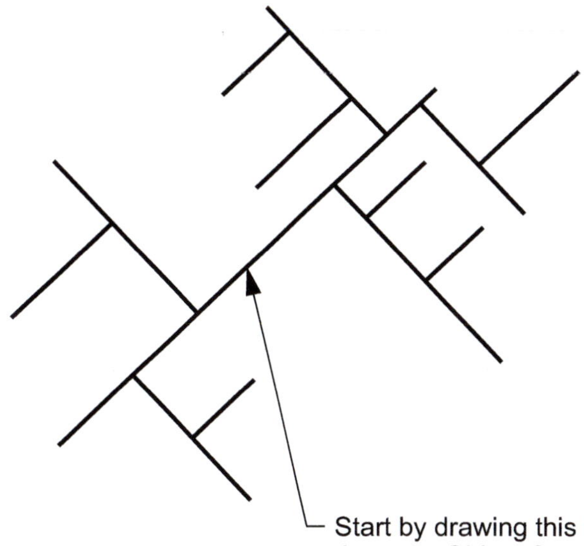

Exercise 4D

Instructions:

1. Start a **New** file using either the **inch-helper.dwt** or the **metric-helper.dwt**
2. Set **Units** and **Precision**:
   Units = **Decimal**   **Precision = 0.00**
3. Set **Drawing Limits**:
   Lower Left corner = **0,0** Upper Right Corner = **12,9** [**304.8,228.6** metric users].
4. Make sure you use **Zoom / All** after setting Drawing Limits.
5. Turn **off** the **Snap** and **Ortho**.
6. Draw the objects showne using:
   Line (Use Layer = Object Line)
   Object Snap = Midpoint
7. **Save** the drawing as: **Ex-4D**

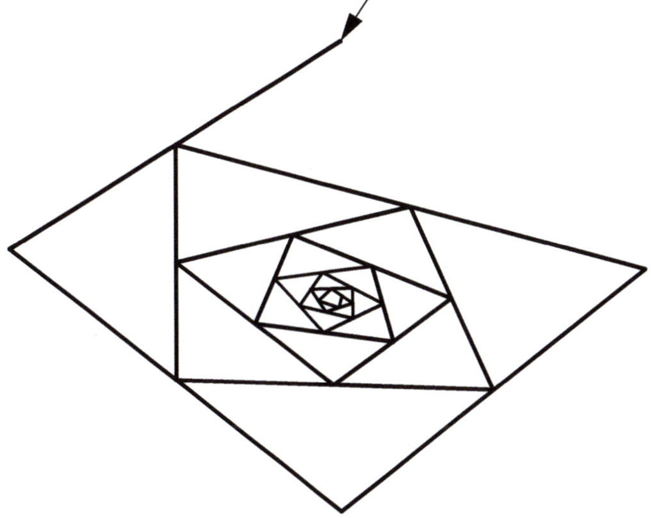

Exercises 4E, 4F, & 4G

Exercise 4E

Instructions:

1. Start a **New** file using either the **inch-helper.dwt** or the **metric-helper.dwt**
2. Turn **off** the **Snap** and **Ortho**.
3. Draw the objects shown using:
   Line (Use Layer = Object Line)
   Object Snap = Intersection
4. **Save** the drawing as: **Ex-4E**

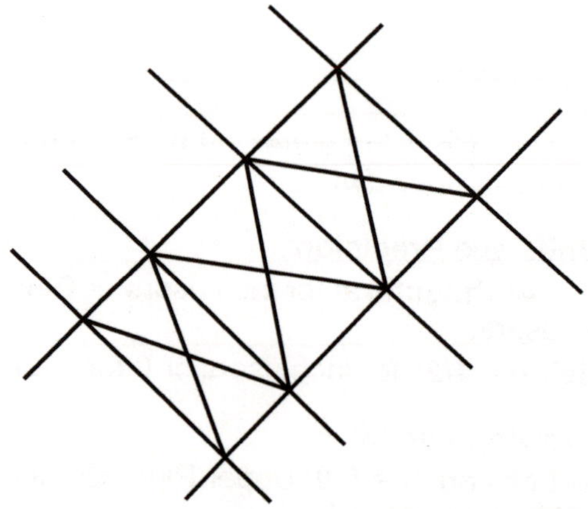

Exercise 4F

Instructions:

1. Start a **New** file using either the **inch-helper.dwt** or the **metric-helper.dwt**
2. Turn **off** the **Snap** and **Ortho**.
3. Draw the 2 circles on Layer Object Line with the following radii: **1.5"** [**38.1 mm**] and **3.5"** [**88.9 mm**]
   (Use Object Snap Center so both circles will have the same center.)
4. Draw the lines using Layers Object Line and Centerline
5. Use Object Snaps Quadrant and Tangent
6. **Save** the drawing as: **Ex-4F**

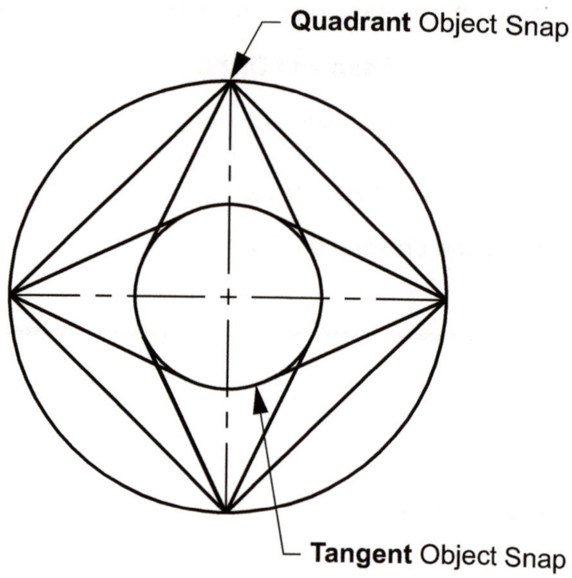

Exercise 4G

Instructions:

1. Start a **New** file using either the **inch-helper.dwt** or the **metric-helper.dwt**
2. Turn **off** the **Snap** and **Ortho**.
3. Draw the rectangle on Layer Object Line with Length **6.5"** [**165.1 mm**] and Width **4.5"** [**114.3 mm**]
4. Draw the circle using Layer Object Line with a radius of **1.5"** [**38.1 mm**]
5. Draw the lines using Layer Centerline
6. **Save** the drawing as: **Ex-4G**

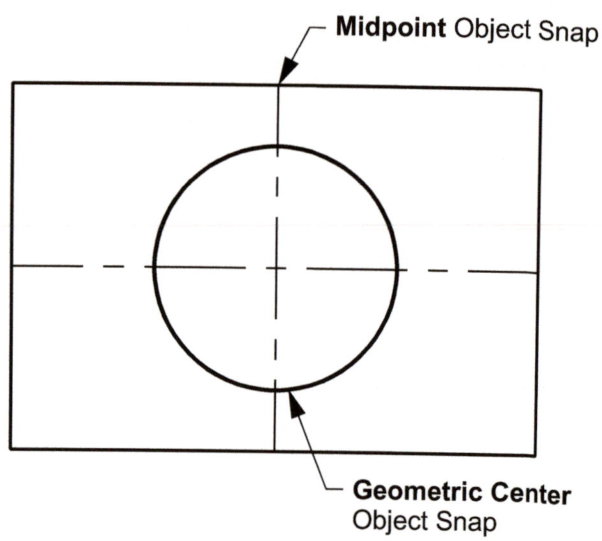

LESSON 5

LEARNING OBJECTIVES

After completing this lesson, you will be able to:

- Draw an Inscribed or Circumscribed Polygon
- Create an Ellipse using two different methods
- Define an Elliptical Arc
- Create Donuts
- Define a Location with a Point
- Select various Point Styles
- Use 3 new Object Snap modes

Polygon

A polygon is an object with multiple edges (flat sides) of equal length. You may specify from 3 to 1024 sides. A polygon appears to be multiple lines, but in fact it is one object. You can specify the **center** and a **radius** or the **edge length**. The **radius** size can be specified **Inscribed** or **Circumscribed**.

Method 1. Center, Radius

1. Select the **Polygon** command using one of the following:

   Ribbon = Home Tab / Draw Panel /
   or
   Keyboard = *POL <Enter>*

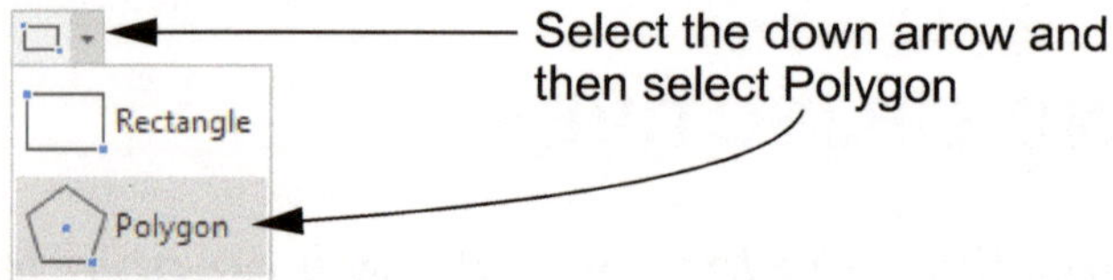

   The following prompt will appear on the Command Line:

```
POLYGON POLYGON Enter number of sides <4>:
```

2. Type the number of sides and then press ***<Enter>***.

   The following prompt will appear on the Command Line:

```
POLYGON Specify center of polygon or [Edge]:
```

3. Specify the center location (**P1**).

   The following prompt will appear on the Command Line:

```
POLYGON Enter an option [Inscribed in circle Circumscribed about circle] <I>:
```

4. Type ***I*** or ***C*** and then press ***<Enter>***.

   The following prompt will appear on the Command Line:

```
POLYGON Specify radius of circle:
```

5. Type the radius or locate with the cursor (**P2**).

Note: The dashed circle is shown only as a reference to help you visualize the difference between Inscribed and Circumscribed. Notice that the radius is the same (**2"**) [**50.8 mm**], but the polygons are different sizes. Selecting **Inscribed** or **Circumscribed** is important.

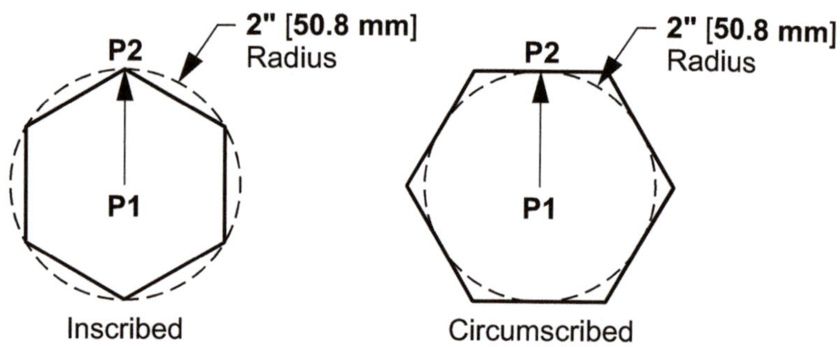

Method 2. Edge

1. Select the **Polygon** command using one of the options shown above:

   The following prompt will appear on the Command Line:

```
POLYGON POLYGON Enter number of sides <4>:
```

2. Type the number of sides and then press ***<Enter>***.

Continued on the next page...

The following prompt will appear on the Command Line:

```
POLYGON Specify center of polygon or [Edge]:
```

3. Type ***E*** and then press ***<Enter>***.

   The following prompt will appear on the Command Line:

```
POLYGON Specify first endpoint of edge:
```

4. Place the first endpoint of edge (**P1**).

   The following prompt will appear on the Command Line:

```
POLYGON Specify first endpoint of edge: Specify second endpoint of edge:
```

5. Place the second endpoint of edge (**P2**).

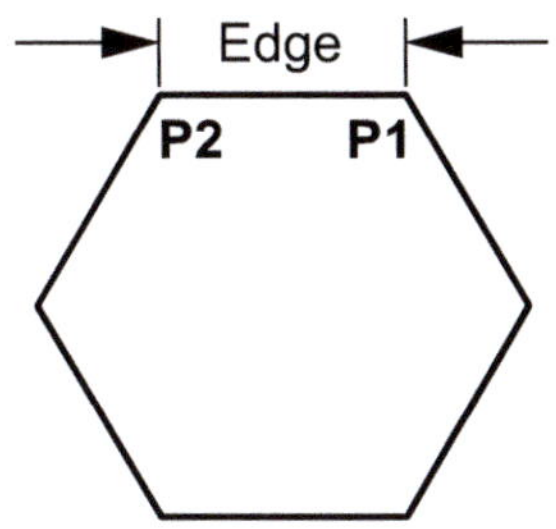

Ellipse

There are three methods to draw an ellipse. You may (1) specify 3 points of the axes, (2) define the center point and the axis points or (3) define an elliptical arc. The following 3 examples illustrate each of the methods.

Method 1. Axis, End

1. Select the **Ellipse** command using one of the following:

   Ribbon = Home Tab / Draw Panel /
   or
   Keyboard = *EL <Enter>*

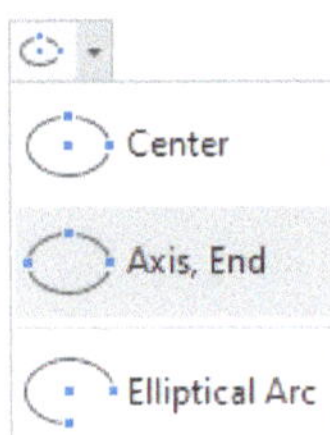

   The following prompt will appear on the Command Line:

```
ELLIPSE Specify axis endpoint of ellipse or [Arc Center]:
```

2. Place the first endpoint of either the **major** or **minor** axis (**P1**).

   The following prompt will appear on the Command Line:

```
ELLIPSE Specify other endpoint of axis:
```

3. Place the other endpoint of the first axis (**P2**).

   The following prompt will appear on the Command Line:

```
ELLIPSE Specify distance to other axis or [Rotation]:
```

4. Place the point perpendicular to the first axis (**P3**).

Continued on the next page...

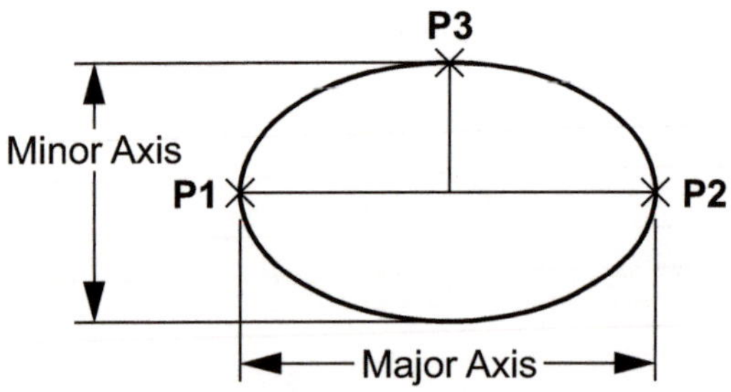

Specifying the Major Axis first (**P1**/**P2**), and then the Minor Axis (**P3**).

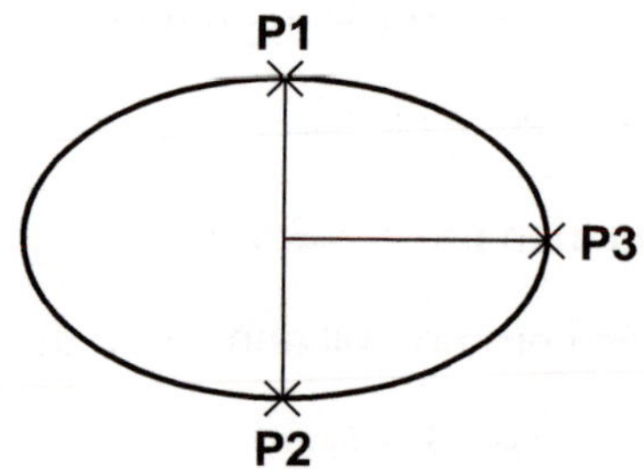

Specifying the Minor Axis first (**P1**/**P2**), and then the Major Axis (**P3**).

Method 2. Center

1. Select the **Ellipse** command using one of the following:

Ribbon = Home Tab / Draw Panel /
or
Keyboard = *EL <Enter>*

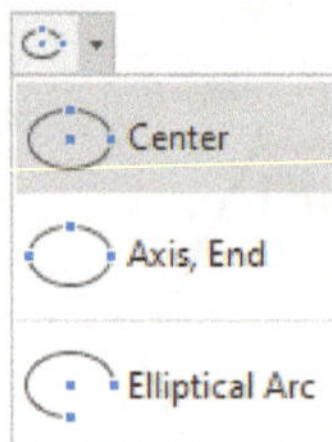

The following prompt will appear on the Command Line:

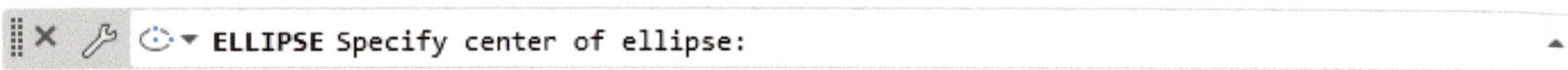

2. Place the center of the ellipse (**P1**).

The following prompt will appear on the Command Line:

3. Place the first axis endpoint (either axis) (**P2**).

The following prompt will appear on the Command Line:

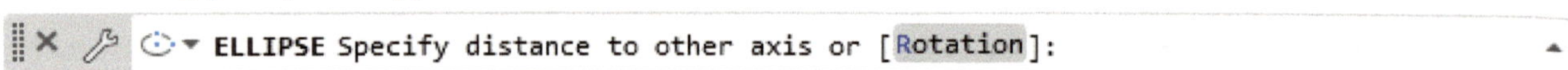

4. Place the point perpendicular to the first axis (**P3**).

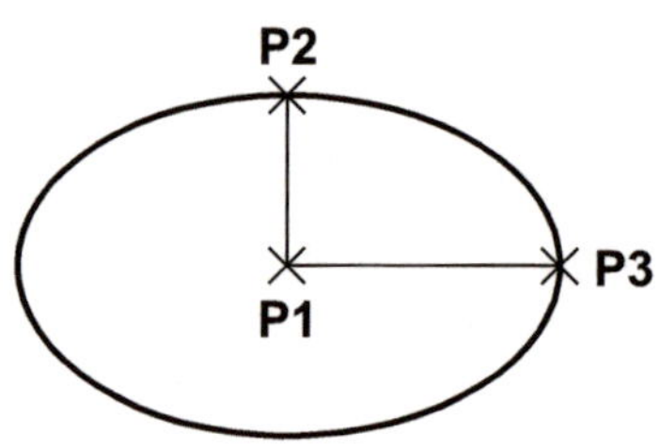

Method 3. Elliptical Arc

1. Select the **Ellipse** command using one of the following:

Ribbon = Home Tab / Draw Panel /
or
Keyboard = *EL <Enter>*

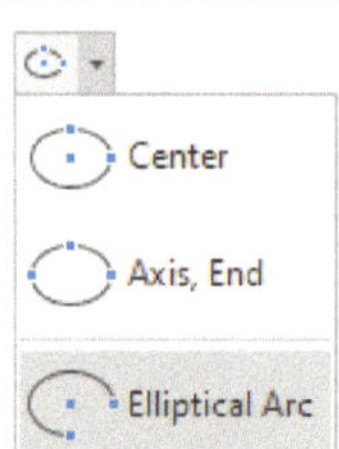

Continued on the next page...

The following prompt will appear on the Command Line:

```
ELLIPSE Specify axis endpoint of elliptical arc or [Center]:
```

2. Type in ***C*** and then press ***<Enter>***.

The following prompt will appear on the Command Line:

```
ELLIPSE Specify center of elliptical arc:
```

3. Place the center of the elliptical arc (**P1**).

The following prompt will appear on the Command Line:

```
ELLIPSE Specify endpoint of axis:
```

4. Place endpoint of the first axis (**P2**).

The following prompt will appear on the Command Line:

```
ELLIPSE Specify distance to other axis or [Rotation]:
```

5. Place the endpoint perpendicular to the first axis (**P3**).

The following prompt will appear on the Command Line:

```
ELLIPSE Specify start angle or [Parameter]:
```

6. Place the **start** angle (**P4**).

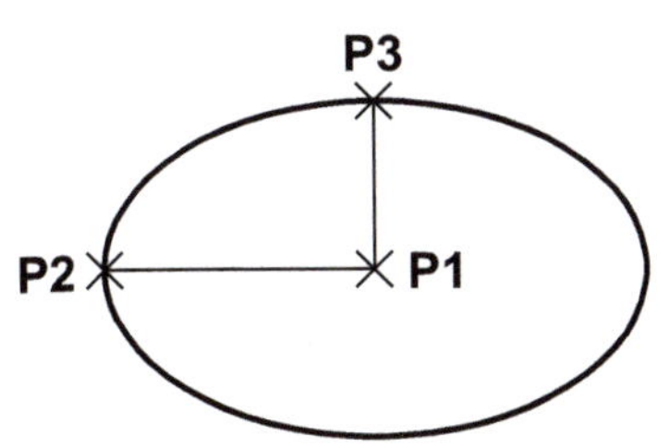

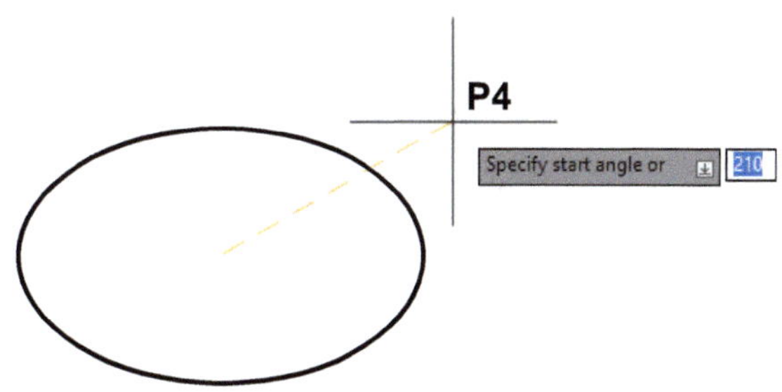

The following prompt will appear on the Command Line:

```
ELLIPSE Specify end angle or [Parameter Included angle]:
```

7. Place the **end** angle (**P5**).

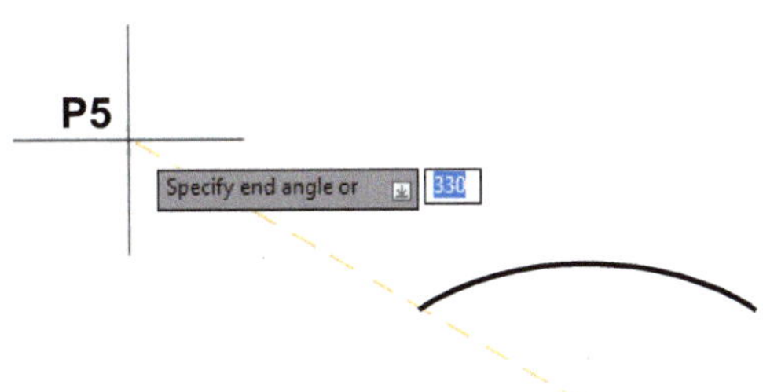

Donut

A Donut is a circle with **width**. You will define the **Inside** and **Outside** diameters.

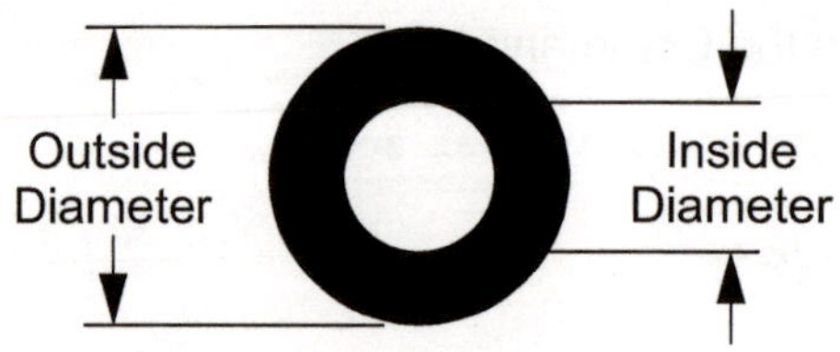

1. Select the **Donut** command using one of the following:

   Ribbon = Home Tab / Draw Panel / 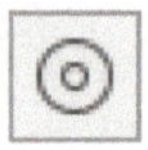
   or
   Keyboard = *DO <Enter>*

   The following prompt will appear on the Command Line:

   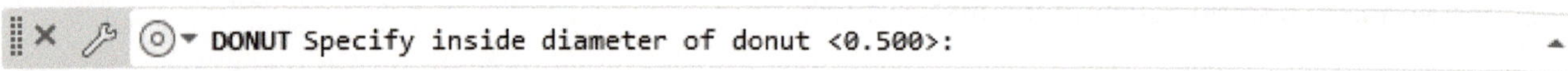

2. Type the inside diameter and then press ***<Enter>***.

   The following prompt will appear on the Command Line:

   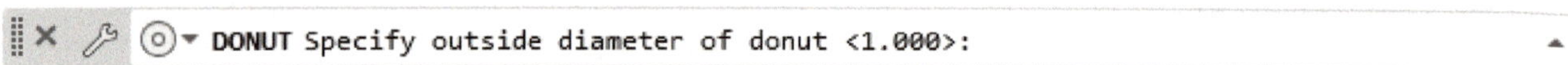

3. Type the outside diameter and then press ***<Enter>***.

   The following prompt will appear on the Command Line:

   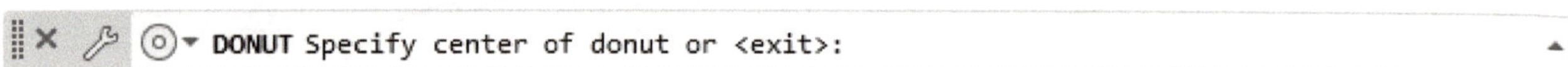

4. Place the center of the first donut.

   The following prompt will appear on the Command Line:

   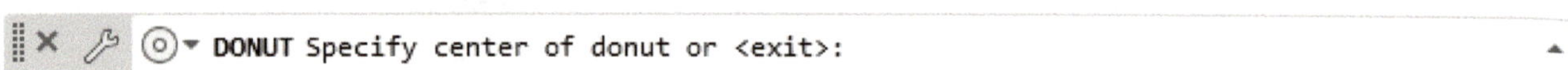

5. Place the center of the second donut or press ***<Enter>*** to stop.

Note: AutoCAD will continue to create more donuts until you press ***<Enter>*** to stop the command.

Controlling the "Fill Mode"

1. Command: Type ***Fill*** and then press ***<Enter>***.
2. **FILL** Enter mode [ON OFF] <on>: Type ***on*** or ***off*** and then press ***<Enter>***.
3. Type ***Regen*** and then press ***<Enter>*** to regenerate the drawing to show the latest setting of the **Fill** mode.

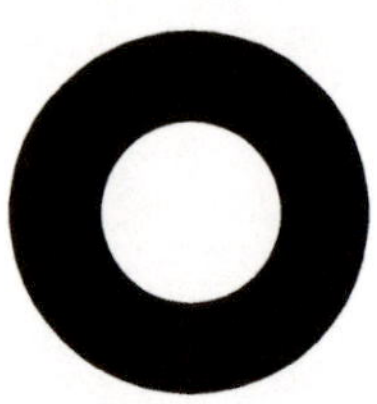
Fill = **On**

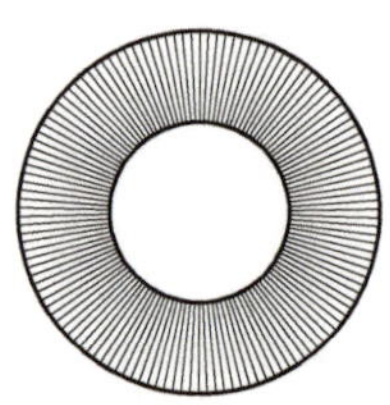
Fill = **Off**

Point

Points are used to locate a point of reference or location. A point may be represented by one of many Point Styles shown below in the Point Style box.

Note: The only Object Snap option that can be used with Point is Node. Refer to the next page for more information on Node Object Snap.

How to Use the Point Command

1. Select the **Point** command using one of the following:

   Ribbon = Home Tab / Draw Panel ▾ / 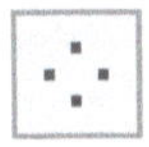
   or
   Keyboard = *PO <Enter>*

2. The following prompt will appear on the Command Line:

3. Place the point location.

   The following prompt will appear on the Command Line:

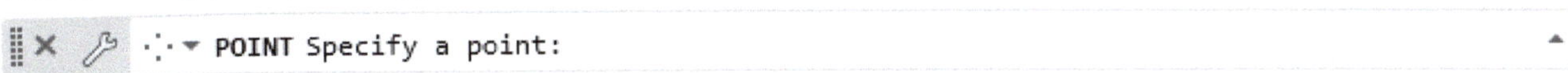

4. Place another point or press the ***<Esc>*** key to stop placing points.

How to Select a "Point Style"

1. Open the **Point Style** dialog box:

   Ribbon = Home Tab / Utilities Panel ▾ / Point Style...
   or
   Keyboard = *ddptype <Enter>*

2. The Point Style dialog box will appear.
3. Select a Point Style tile.
4. Select the **OK** button.

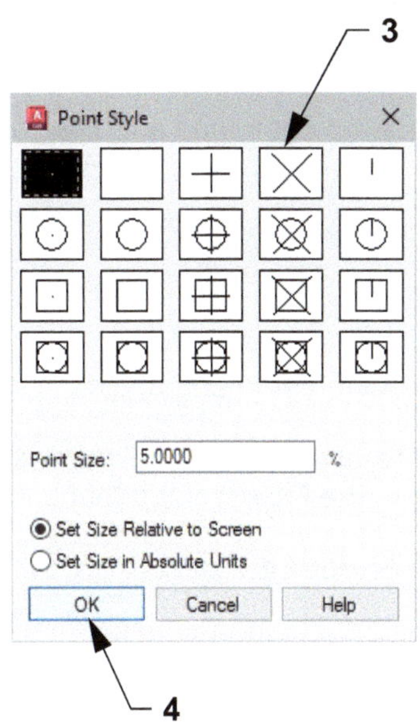

Point Size

Set Size Relative to Screen

Sets the point display size as a percentage of the screen size. The point display does not change when you zoom in or out.

Set Size in Absolute Units

Sets the point display size as the actual units you specify under Point Size. Points are displayed larger or smaller when you zoom in or out.

More Object Snaps

NODe

This option snaps to the object **"Point"** described on the previous page. Select **Node** Object Snap and place the cursor on the **Point**. The cursor will snap to the **Point**.

Note: This is the only Object Snap that you can use with the object **Point**.

NEArest

Snaps to the nearest location on an object. For example, if you want to attach a line somewhere on a circle between quadrants:

Select the **Line** command and then select **Nearest** Object Snap. Place the cursor anywhere on the circumference of the circle and press the left mouse button. The line will now be accurately attached to the circle at the location you selected.

NO TOOL **M2P**

Mid Between 2 Points locates a midpoint between two points.

CAD TIP! You may select the **M2P** option from the Object Snap Menu using ***<Shift>* + right click**, or you may type ***M2P <Enter>*** when prompted for an endpoint.

How to use M2P

1. Select the **Line** command and draw 2 parallel lines.

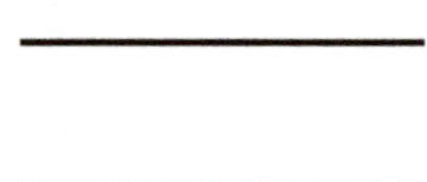

2. Select the **Line** command again.
3. Type ***M2P*** and then press ***<Enter>***. Or hold down the ***<Shift>*** key and right click, and select **Mid Between 2 Points** from the Object Snap Menu.
4. Using the **Endpoint** Object Snap, select each of the 2 endpoints (**P1**) and (**P2**).

P1
P2

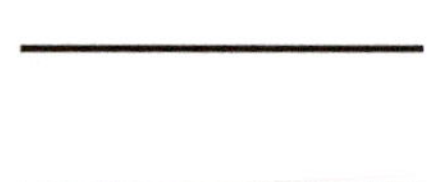

5. The new line's first endpoint should start **exactly** midpoint between those 2 endpoints.

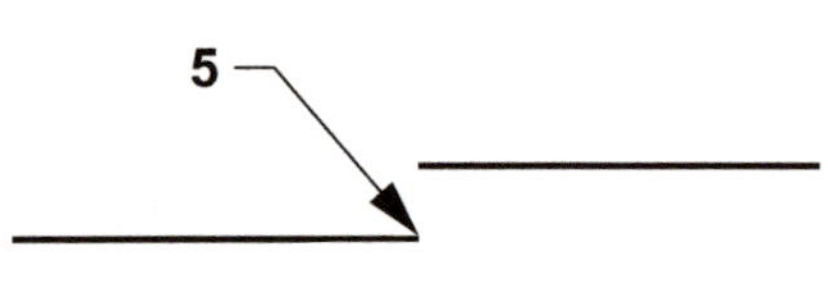

Exercise 5A

Instructions:

1. Start a **New** file using either the **inch-helper.dwt** or the **metric-helper.dwt**
2. Draw the circle first using the **Center, Radius** option.
3. Draw the **Circumscribed** polygon next using Object Snap **Center** to locate the center of the polygon at the center of the circle, and **Quadrant** Object Snap to locate the radius of the polygon on the circle.
4. Draw the lines last using Object Snaps **Midpoint** and **Endpoint**.
5. **Ortho *<F8>* on**.
6. **Increment Snap *<F9>* off** (it will get in your way).
7. Use Layer Object Line.
8. **Save** the drawing as: **Ex-5A**

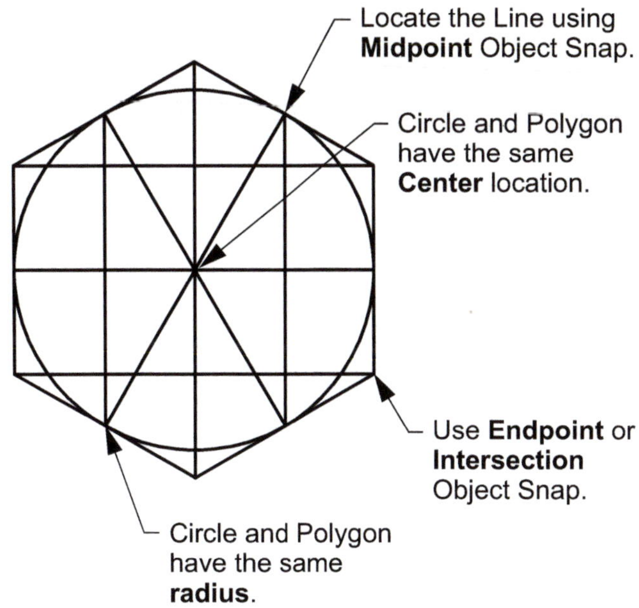

Exercise 5B

Instructions:

1. Start a **New** file using either the **inch-helper.dwt** or the **metric-helper.dwt**
2. Select the **Point Style** shown.
3. Draw the **Point** first.
4. Draw the **Inscribed** polygons.
   A. Locate the center of each polygon using Object Snap **Node**.
5. **Ortho *<F8>* on**.
6. **Increment Snap *<F9>* off**.
7. Use Layer Object Line.
8. **Save** the drawing as: **Ex-5B**

Note: Place the polygon points with the cursor. If you type a radius, the polygon will automatically rotate and locate the flat at the bottom.

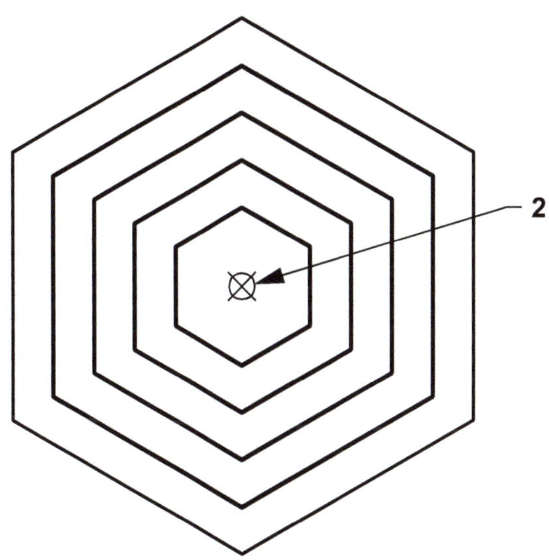

Exercise 5C

Instructions:

1. Start a **New** file using either the **inch-helper.dwt** or the **metric-helper.dwt**
2. Draw the objects shown using **Ellipse**, **Line**, and **Donut**.
3. Use Object Snaps **Quadrant**, **Center**, and **Nearest**.
4. **Ortho *<F8>* on**.
5. **Increment Snap *<F9>* off**.
6. Use any layer you like.
7. **Save** the drawing as: **Ex-5C**

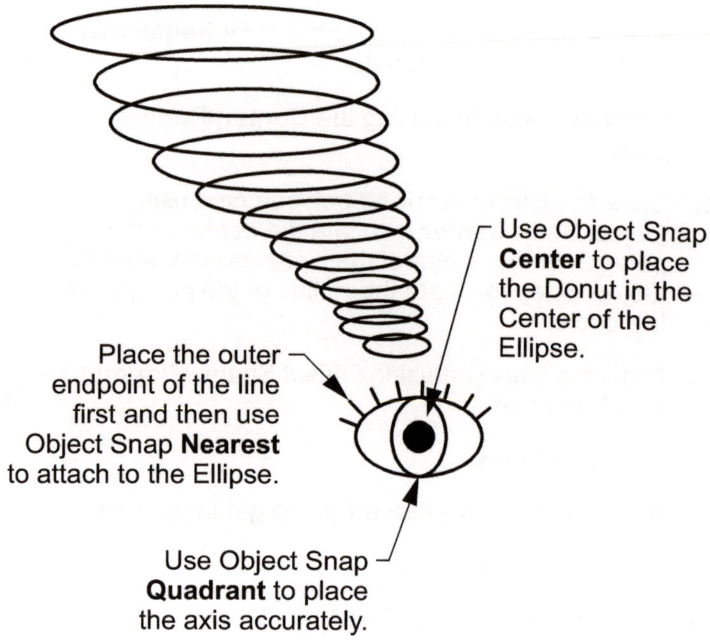

Note: This is the **"eye"** of the tornado. (Just a little fun.)

Exercise 5D

Instructions:

1. Start a **New** file using either the **inch-helper.dwt** or the **metric-helper.dwt**
2. Select the **Point Style** shown.
3. Draw the **Point** first.
4. Draw the **Ellipses**.
   A. Locate the center of each ellipse using Object Snap **Node**.
   B. Locate the axis using Object Snap **Quadrant**.
5. **Ortho *<F8>* on**.
6. **Increment Snap *<F9>* off**.
7. Use Layer Object Line.
8. **Save** the drawing as: **Ex-5D**

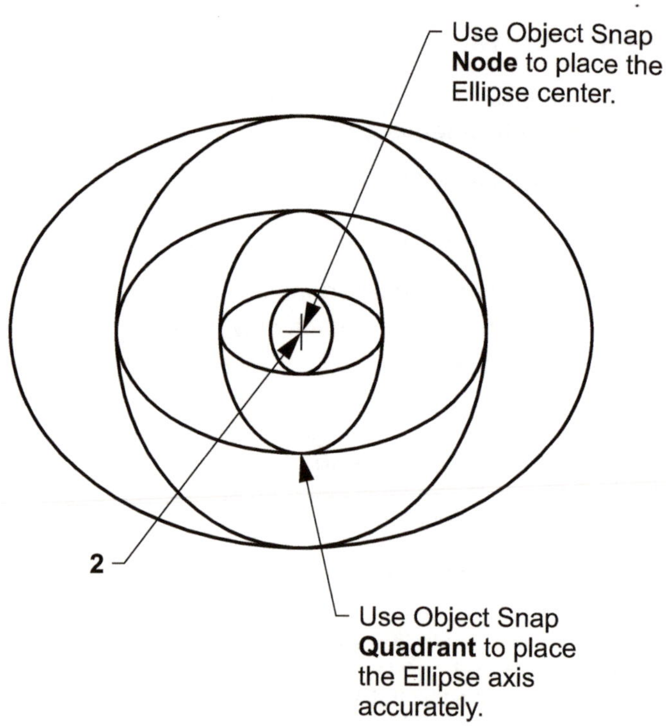

Exercises 5E & 5F

Exercise 5E

Instructions:

Step 1.

1. Start a **New** file using either the **inch-helper.dwt** or the **metric-helper.dwt**
2. Draw the 3 **Donuts** shown.
3. Use Object Snap **Center** to place the donut centers accurately.
4. **Increment Snap <*F9*> off**.
5. Use Layer Object Line.

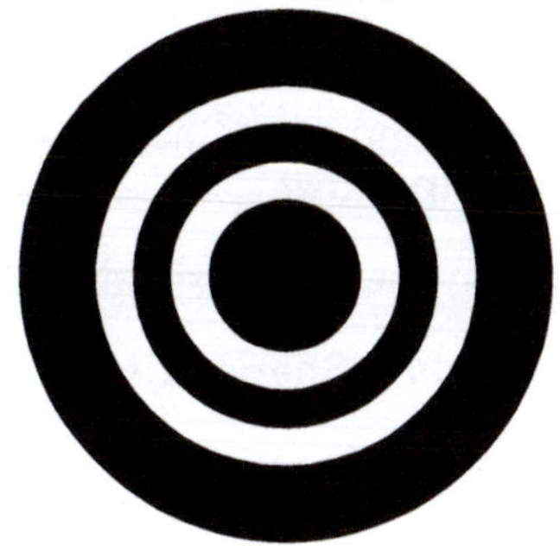

Donut sizes

A. ID = **0**	OD = **1" [25.4 mm]**
B. ID = **1.5" [38.1 mm]**	OD = **2" [50.8 mm]**
C. ID = **2.5" [63.5 mm]**	OD = **3.5" [88.9 mm]**

Step 2.

6. Turn the **Fill mode off**.
7. Type ***Regen <Enter>***.
8. **Save** the drawing as: **Ex-5E**

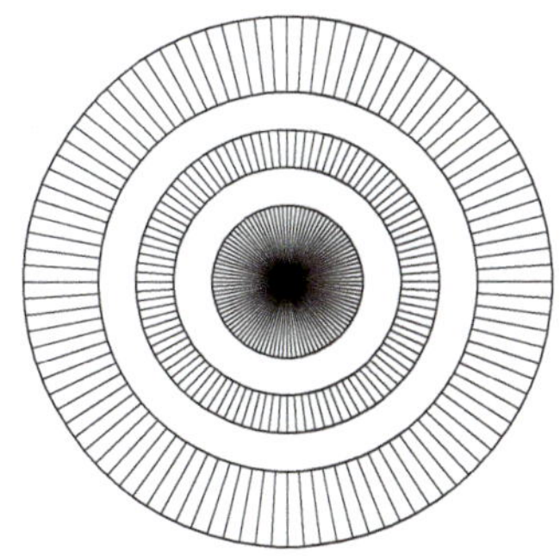

Exercise 5F

Instructions:

1. Start a **New** file using either the **inch-helper.dwt** or the **metric-helper.dwt**
2. Draw the objects using **Point**, **Polygon**, **Ellipse**, and **Donut**.
5. **Ortho <*F8*> on**.
6. **Increment Snap <*F9*> off**.
7. Use whatever layers you like.
8. **Save** the drawing as: **Ex-5F**

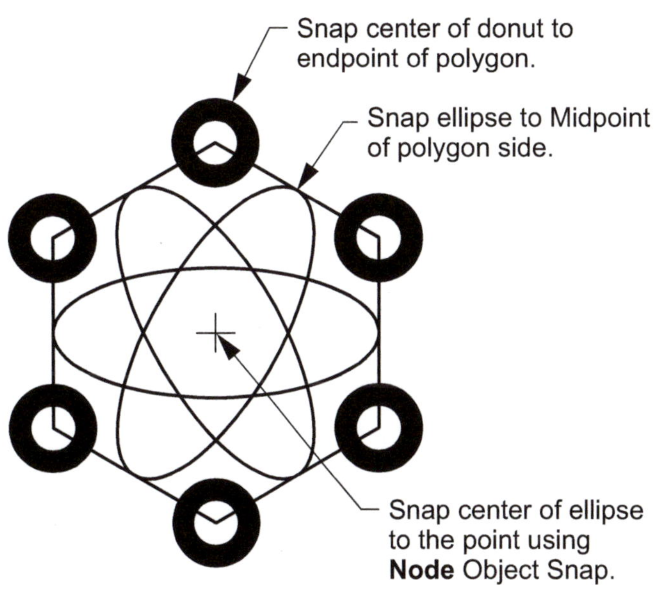

Exercise 5G

Instructions:

1. Start a **New** file using either the **inch-helper.dwt** or the **metric-helper.dwt**
2. Draw the zig zag lines approximately as shown.
3. Set **Running Object Snap** to **Endpoint**.
4. **Ortho** ***<F8>*** **on**.
5. **Increment Snap** ***<F9>*** **off**.
6. Draw the circles at the **"Midpoint Between 2 Points"** (**P1** and **P2**).
7. Use Layer Object Line.
8. **Save** the drawing as: **Ex-5G**

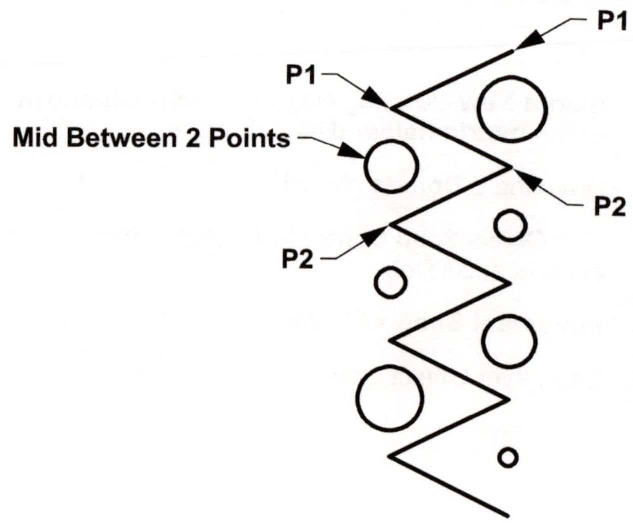

Exercise 5H

Instructions:

1. Start a **New** file using either the **inch-helper.dwt** or the **metric-helper.dwt**
2. Draw the **Rectangle** on the left first. (Any size.)
3. Draw the **Polygon** next using the **"Edge"** option.
4. Draw the **Ellipses**.
   A. Use Object Snap **Endpoint** to place the **Edge** points accurately.
5. **Ortho** ***<F8>*** **on**.
6. Use Layer Object Line.
7. **Save** the drawing as: **Ex-5H**

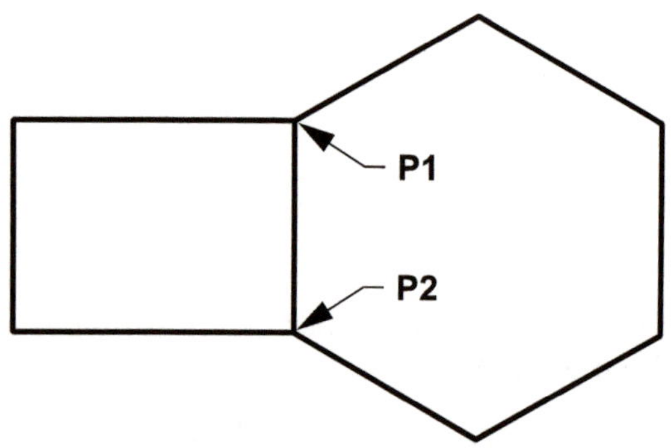

LESSON 6

LEARNING OBJECTIVES

After completing this lesson, you will be able to:

- Use the Break Command
- Trim objects to other objects
- Extend objects to other objects
- Move an object to a new location
- Explode objects to their primitive entities

Break

The **Break** command allows you to break an object at a single point (Break at Point) or between two points (Break). I think of it as breaking a single line segment into two segments or taking a bite out of an object.

Method 1. Break at a Single Point

How to break one line into two seperate objects with no visible space in between.

1. Select the **Break at Point** command by using:

   Ribbon = Home Tab / Modify Panel ▾ /

   The following prompt will appear on the Command Line:

   ```
   BREAKATPOINT Select object:
   ```

2. Select the object to break (**P1**).

   The following prompt will appear on the Command Line:

   ```
   BREAKATPOINT Specify break point:
   ```

3. Select the break location (**P2**) accurately.

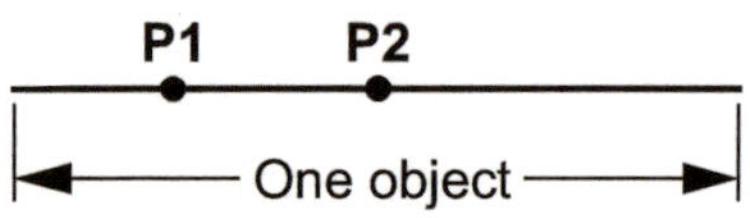

Before break
One Line segment

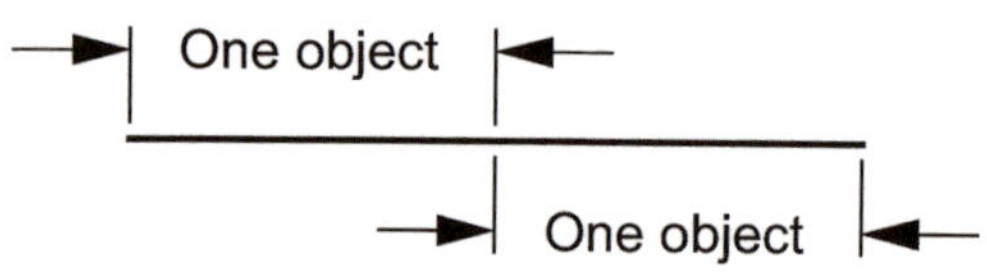

After Break
Two Line segments

Note: The single line is now two lines, but there is no gap in between the two lines. For example, if you break a 2 inch long line at its center, it would become two 1 inch lines butted together.

Method 2. Break between two Points

Take a bite out of an object. Use this method if the **location** of the break is **not important**.

1. Select the **Break** command using one of the following:

   Ribbon = Home Tab / Modify Panel ▾ /
   or
   Keyboard = *BR <Enter>*

   The following prompt will appear on the Command Line:

   ```
   BREAK Select object:
   ```

2. Pick the first break location (**P1**).

   The following prompt will appear on the Command Line:

   ```
   BREAK Specify second break point or [First point]:
   ```

3. Pick the second break location (**P2**).

Continued on the next page...

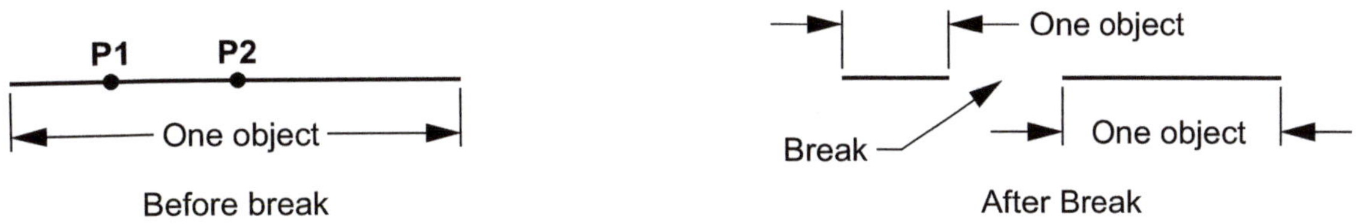

Note:
A. Circles break counter-clockwise (ccw).
B. Circles cannot be broken with one point. You must use two points.

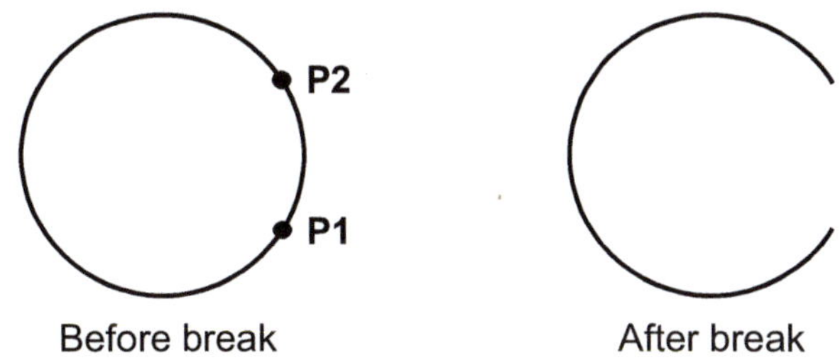

Method 3. Break between two Points

The following method is the same as Method 2 above; however, use this method if the location of the break is **very specific**.

1. Select the **Break** command.

   The following prompt will appear on the Command Line:

   ```
   BREAK Select object:
   ```

2. Select the object to break (**P1**) anywhere on the object.

   The following prompt will appear on the Command Line:

   ```
   BREAK Specify second break point or [First point]:
   ```

3. Type ***F*** and then press ***<Enter>***.

   The following prompt will appear on the Command Line:

   ```
   BREAK Specify first break point:
   ```

4. Select the first break location (**P2**) accurately.

   The following prompt will appear on the Command Line:

   ```
   BREAK Specify second break point:
   ```

5. Select the second break location (**P3**) accurately.

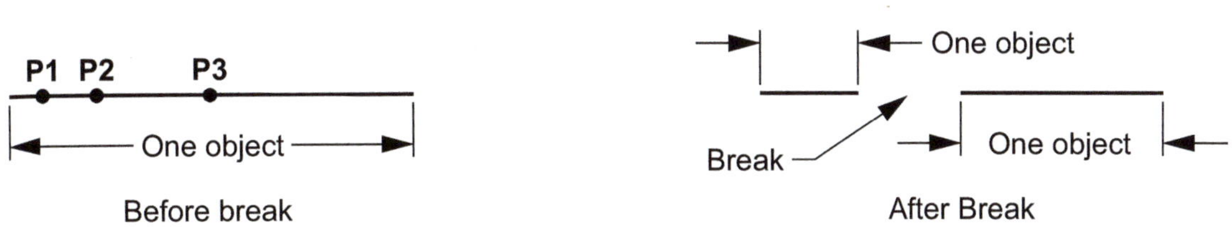

Trim

The **Trim** command is used to trim objects to other objects. You can **select objects individually**, or by using a **crossing fence** by picking two locations, or you can use **press and drag** to start a **freehand selection path**.

Method 1. Selecting objects to trim individually

1. Select the **Trim** command using one of the following:

   Ribbon = Home Tab / Modify Panel /
   or
   Keyboard = ***TR <Enter>***

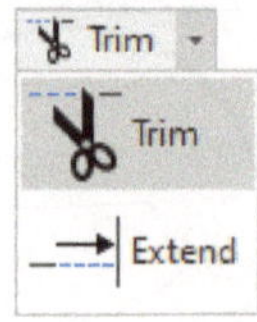

   The following prompt will appear on the Command Line:

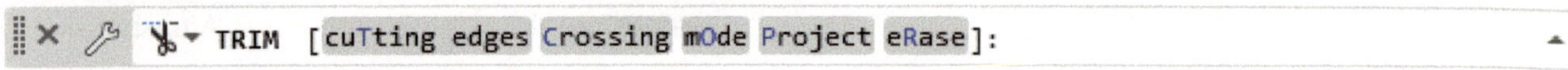

2. Select the object that you want to trim (**P1**).

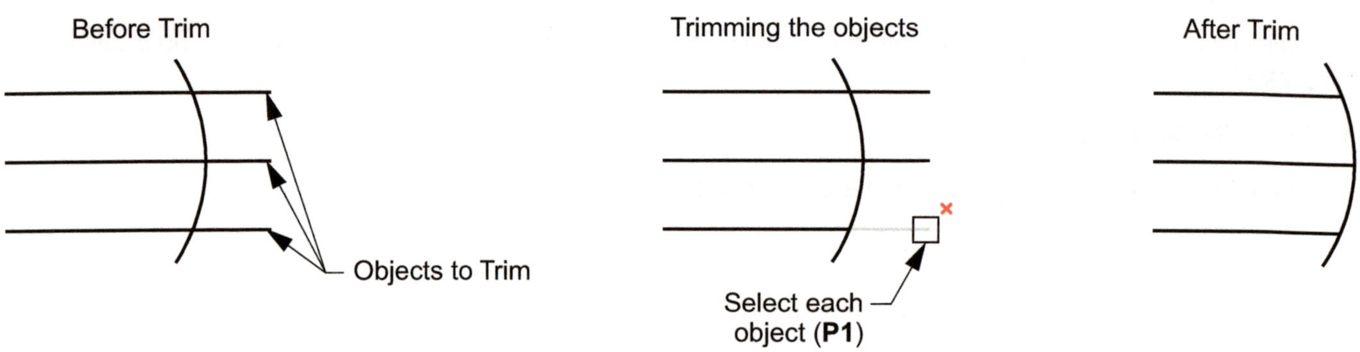

After you select your first object to Trim, a new Command Line will appear.

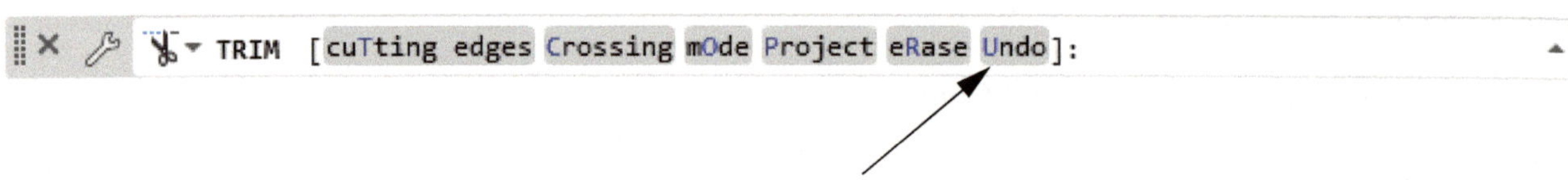

If you have trimmed an object by mistake, you can use the **Undo** command to "undo" the trimmed object and place it back to its original length before it was trimmed.

How to use the Undo command during Trim

1. Type ***U*** and then press ***<Enter>***.
2. The last object you trimmed will reappear.

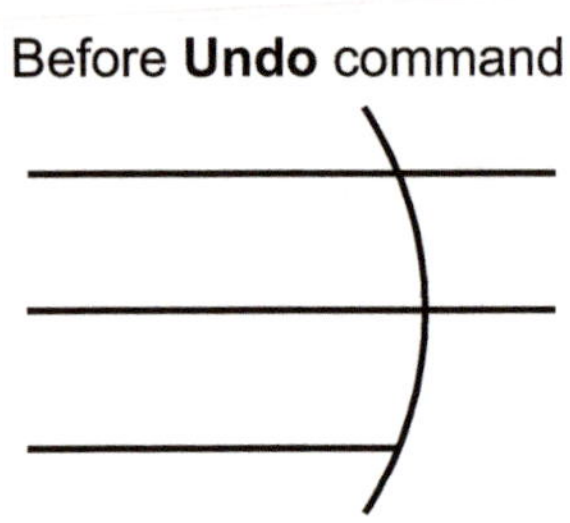

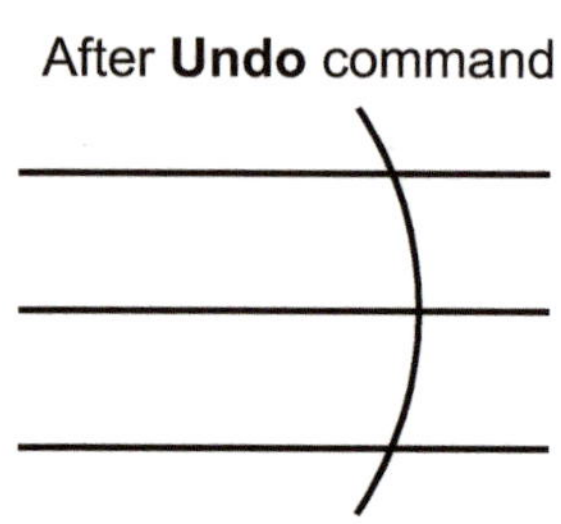

Method 2. Selecting objects to trim using a crossing fence

1. Select the **Trim** command.
2. Select the objects that you want to trim by selecting the first point at (**P1**), by left clicking and then releasing the mouse button.
3. Select a second point at (**P2**). Again, left click and release.

Note: You will see a dashed fence line when moving the cursor to (**P2**).

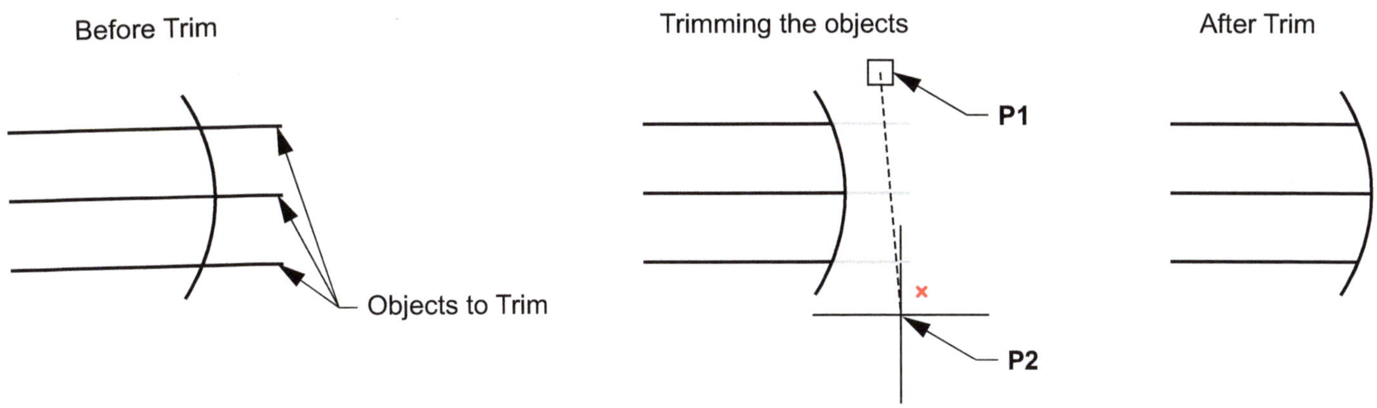

Method 3. Selecting objects to trim using a press and drag freehand path

1. Select the **Trim** command.
2. Select the objects that you want to trim by selecting the first point at (**P1**), by left clicking and **holding down** the mouse button, but **do not** release it yet.
3. Drag the cursor around the area of the objects you wish to trim, to (**P2**), then release the left mouse button.

Note: You will see a dashed fence line when dragging the cursor around the objects. In the example below, you could drag the cursor clockwise or counter-clockwise, whichever way you prefer.

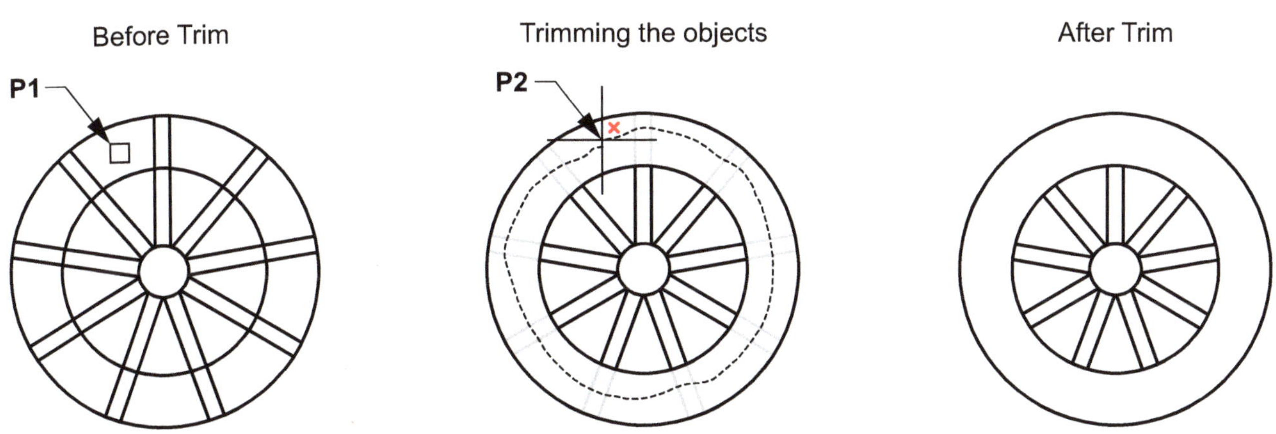

The lines in between the two outer circles need to be trimmed.

Extend

The **Extend** command is used to extend objects to other objects. You can **select objects individually**, or by using a **crossing fence** by picking two locations, or you can use **press and drag** to start a **freehand selection path**.

Method 1. Selecting objects to extend individually

1. Select the **Extend** command using one of the following:

   Ribbon = Home Tab / Modify Panel /
   or
   Keyboard = *EX <Enter>*

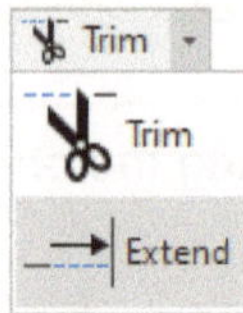

   The following prompt will appear on the Command Line:

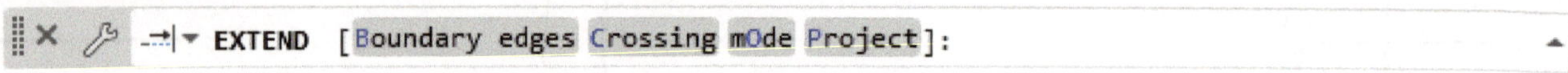

2. Select near the end of the object that you want to extend (**P1**).

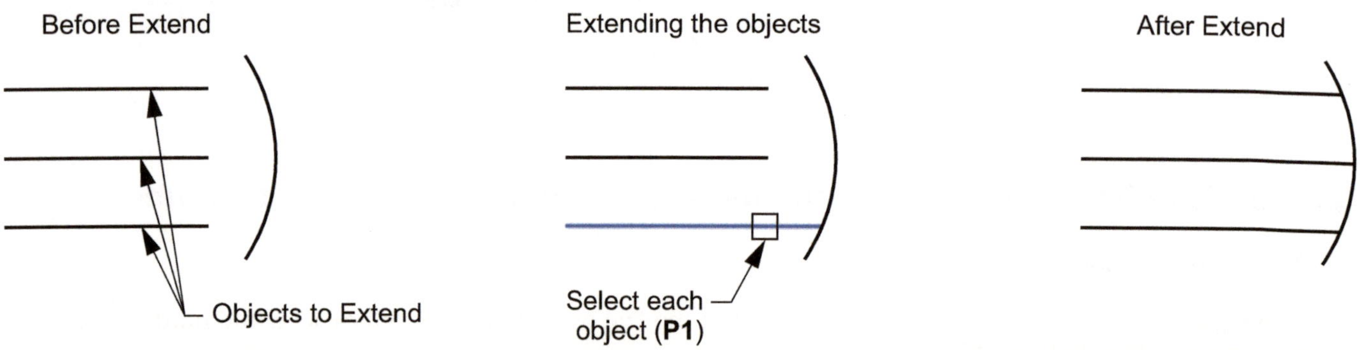

Note: Make sure you select the object near the end you want to extend. If you select the other end, AutoCAD will try to find another object to extend to.

Method 2. Selecting objects to extend using a crossing fence

1. Select the **Extend** command.
2. Select the objects that you want to extend by selecting the first point at (**P1**). by left clicking and then releasing the mouse button.
3. Select a second point at (**P2**). Again, left click and release.

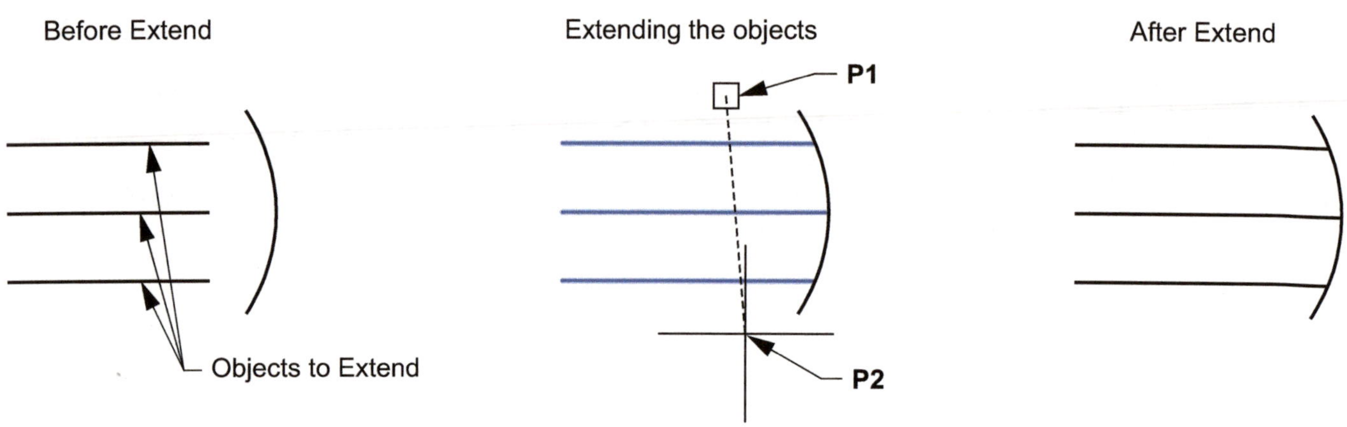

Method 3. Selecting objects to extend using a press and drag freehand path

1. Select the **Extend** command.

2. Select the objects that you want to extend by selecting the first point at (**P1**), by left clicking and **holding down** the mouse button, but **do not** release it yet.

3. Drag the cursor around the area of the objects you wish to extend, to (**P2**), then release the left mouse button.

Note: You will see a dashed fence line when dragging the cursor around the objects. In the example below, you could drag the cursor clockwise or counter-clockwise, whichever way you prefer. Either way, the freehand circle should be close to where you want the objects extended. In the example below, if your freehand circle was closer to the smaller inner circle, AutoCAD would extend the lines inside that small circle instead of extending them to the outer circle.

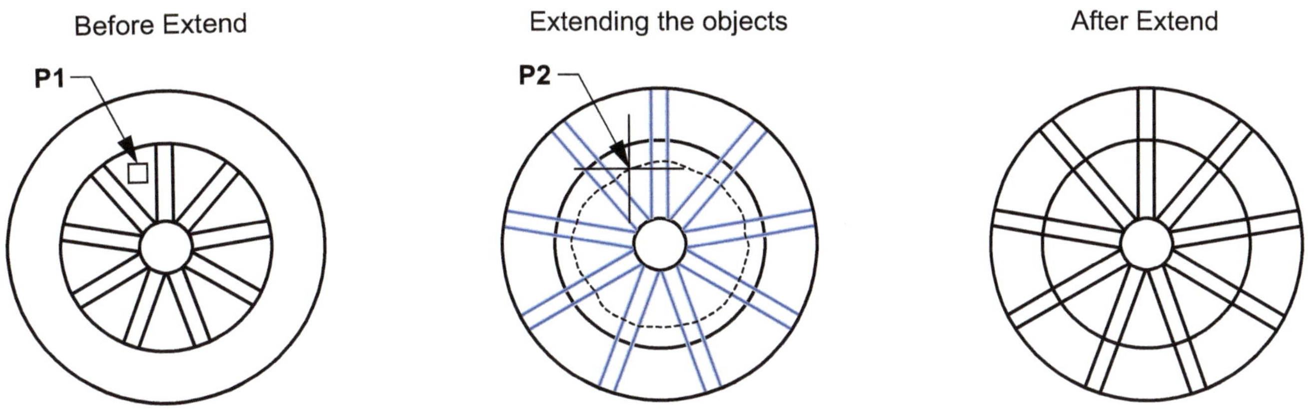

All the lines need to be extended to the outer circle.

Note: If you extend an object by mistake, you can use the **Undo** command to "undo" the extended object and place it back to its original length before it was extended. (Refer to page 6-4.)

CAD TIP! You may toggle between **Extend** and **Trim** (page 6-4). Hold down the ***<Shift>*** key and the **Trim** command will activate. Release the ***<Shift>*** key and you return to **Extend**.

Move

The **Move** command is used to move object(s) from their current location (basepoint) to a new location (second displacement point).

1. Select the **Move** command using one of the following:

Ribbon = Home Tab / Modify Panel / 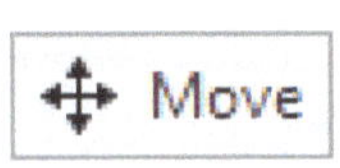

or
Keyboard = *M <Enter>*

The following prompt will appear on the Command Line:

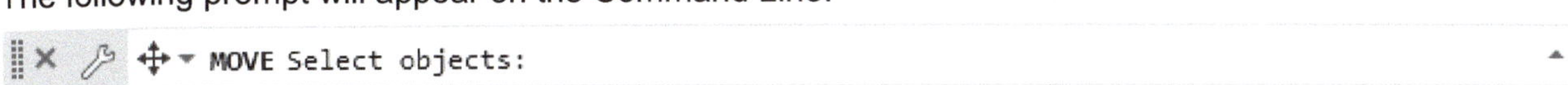

Continued on the next page...

2. Select the object(s) that you want to move (**P1**).

3. Select more objects or to stop selecting objects, select ***<Enter>***.

   The following prompt will appear on the Command Line:

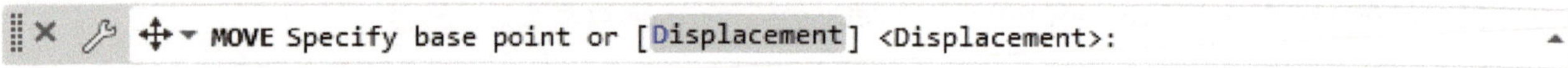

4. Select a location (**P2**) (usually a point on the object).

   The following prompt will appear on the Command Line:

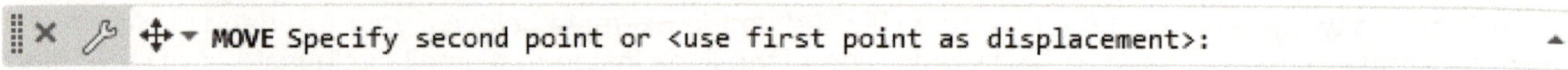

5. Move the object to its new location (**P3**) and press the left mouse button.

Warning: If you press ***<Enter>*** instead of actually picking a new location (**P3**) with the mouse, AutoCAD will send it into **Outer Space**. If this happens, just select the **Undo** tool and try again.

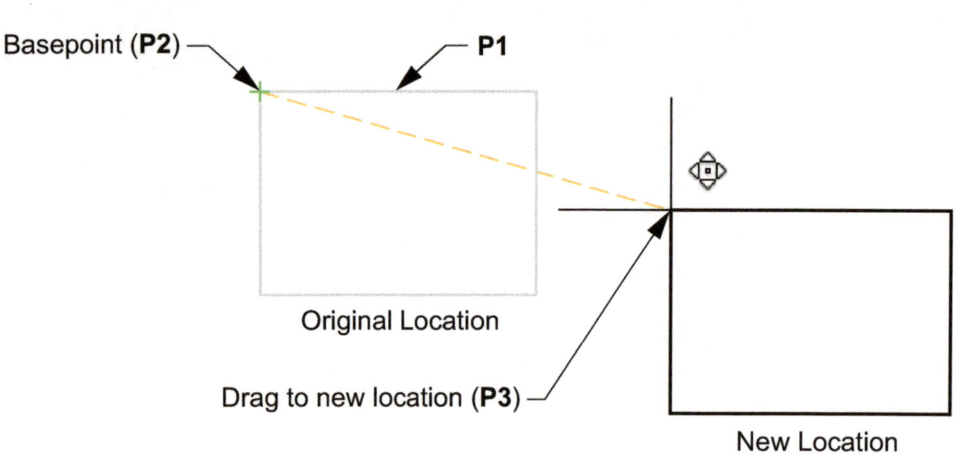

Drag

The **Drag** option allows you to quickly **move** or **copy** one or more objects.

Example:

1. Draw a circle.

2. Select the circle.

   Five little boxes appear. These are called grips and allow you to edit the object. Grips will be discussed more in future lessons.

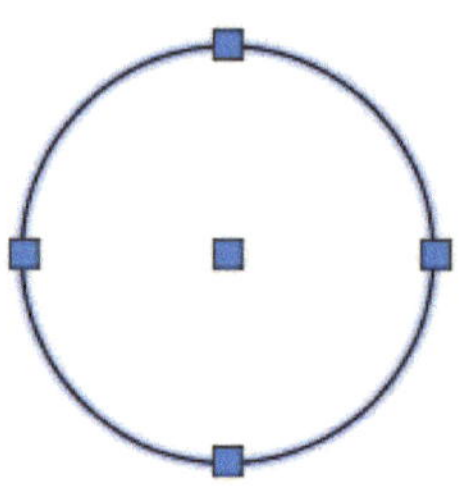

Continued on the next page...

3. Click on the circle and hold the right-hand mouse button down as you drag the circle to the right.

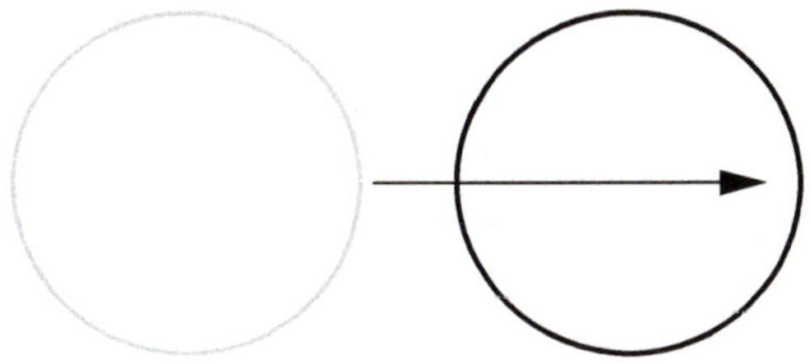

4. When the dragged circle is in the desired location, release the right mouse button.

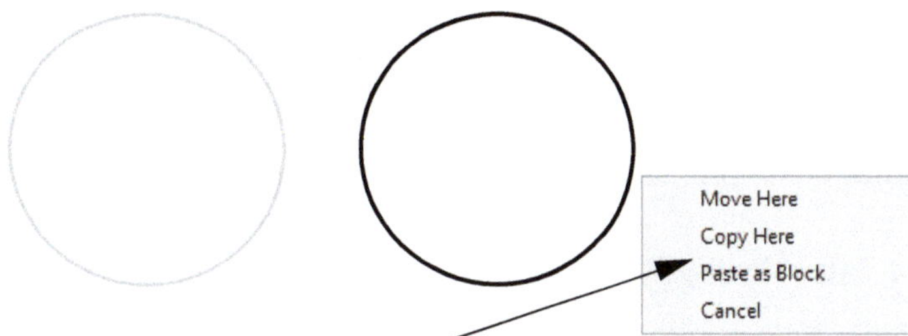

5. Select from any of the four options in the menu.

Nudge

The **Nudge** option allows you to nudge objects in orthogonal increments.

Snap mode affects the distance and direction in which the objects are nudged:

Nudge objects with **Snap** mode turned **off**: Objects move two pixels at a time.

Nudge objects with **Snap** mode turned **on**: Objects are moved in increments specified by the current **Snap** spacing.

Note: To set the Increment Snap spacing refer to page 3-5.

Example:

1. Draw a circle.

2. Select the circle.

   Five little boxes appear. These are called grips and allow you to edit the object. Grips will be discussed more in future lessons.

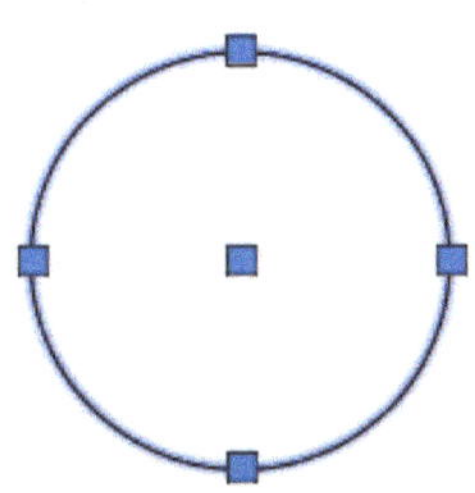

3. Hold down the ***<Ctrl>*** key and press one of the Arrow keys ← → ↑ ↓

Continued on the next page...

Remember: The distance the object moves depends on whether you have the **Snap** mode **on** or **off**. (Refer to the note on the previous page.)

CAD TIP! If you want to nudge an object precise distances, set the distance in the **Snap** and **Grid** Tab of the **Drafting Settings** dialog box and turn on **Snap** mode ***<F9>*.**

Explode

The **Explode** command changes (explodes) an object into its primitive objects. For example, a rectangle is originally one object. If you explode it, it changes into four lines. Visually, you will not be able to see the change unless you select one of the lines.

1. Select the **Explode** command using one of the following:

   Ribbon = Home Tab / Modify Panel /
   or
   Keyboard = *X <Enter>*

   The following prompt will appear on the Command Line:

   ```
   EXPLODE Select objects:
   ```

2. Select the object(s) you want to explode and then press ***<Enter>***.

Before Explode

One Object
(Rectangle)

After Explode

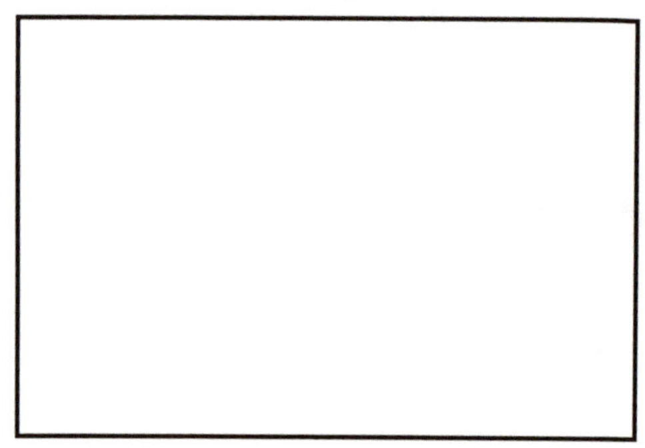

Four Objects
(Four Lines)

(Notice there is no visible difference. But now you have four lines instead of one rectangle).

Try this: Draw a rectangle and then click on it. The entire object highlights. Now explode the rectangle, then click on it again. Only the line you clicked on should be highlighted. Each line that forms the rectangular shape is now an individual object.

Exercises 6A & 6B

Exercise 6A

Instructions:

1. Start a **New** file using either the **inch-helper.dwt** or the **metric-helper.dwt**
2. Draw the objects.
   A. Use Object Snap **Midpoint** to locate the center for the circles.
   B. **Ortho <F8> on** when drawing horizontal and vertical lines.
   C. Turn **Increment Snap <F9> on**.
   D. Use Layer Object Line.

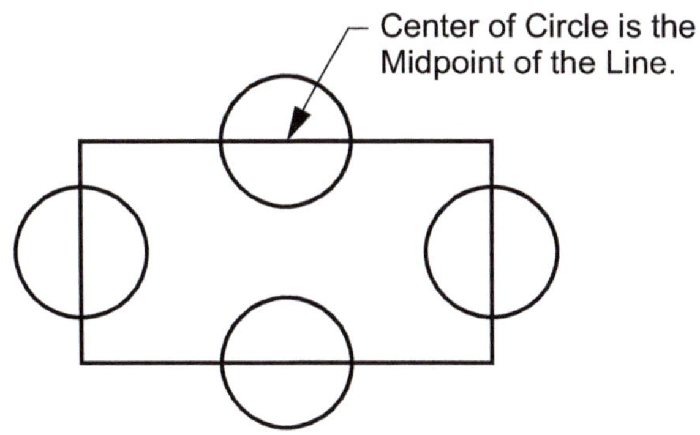

3. **Trim** the circles and rectangle to match the illustration.
4. **Save** the drawing as: **Ex-6A**

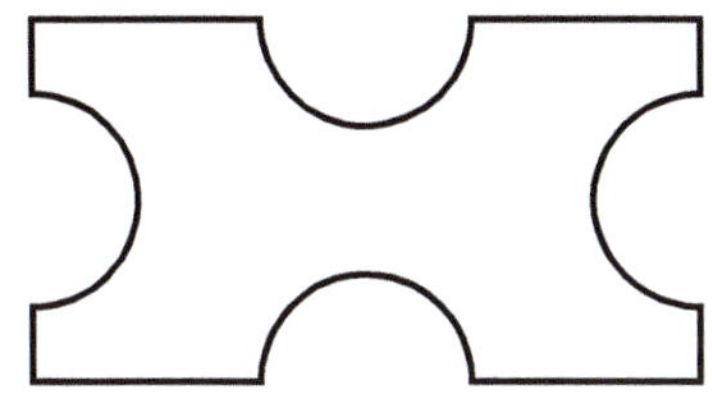

Exercise 6B

Instructions:

1. Start a **New** file using either the **inch-helper.dwt** or the **metric-helper.dwt**
2. Draw the objects.
   A. Circles should have the same center.
   B. **Ortho <F8> on** when drawing horizontal and vertical lines.
   C. **Ortho <F8> off** when drawing lines on an angle.
   D. Turn **Increment Snap <F9> off**.
   E. Use Layer Object Line.

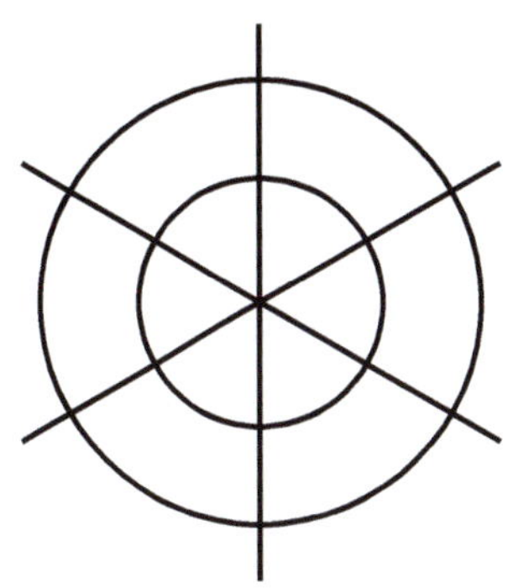

3. **Trim** the lines to match the illustration at right.
4. **Save** the drawing as: **Ex-6B**

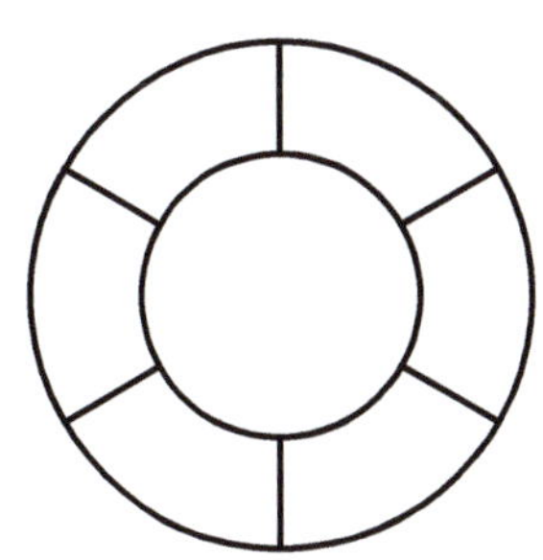

Note: You can trim the lines **individually** or by using the **press and drag freehand path** method.

Exercise 6C

Instructions:

1. Start a **New** file using either the **inch-helper.dwt** or the **metric-helper.dwt**
2. Draw the lines exactly as shown.
   A. **Ortho *<F8>* on** when drawing horizontal and vertical lines.
   B. Turn **Increment Snap *<F9>* on.**
   C. Turn **Osnap *<F3>* off.**
   D. Use Layer Object Line.

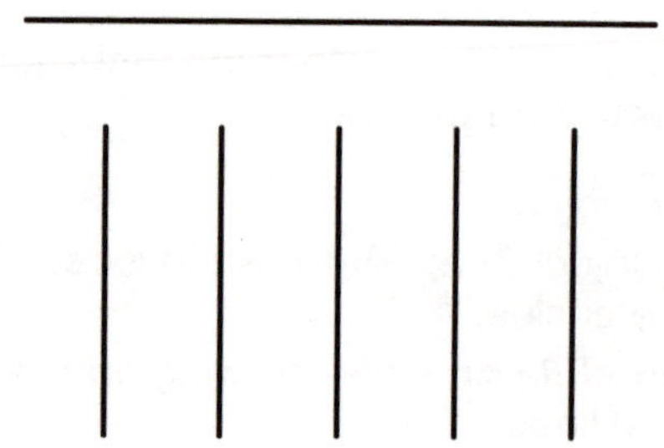

3. **Extend** the vertical lines **individually** to intersect with the horizontal line as shown.
4. **Save** the drawing as: **Ex-6C**

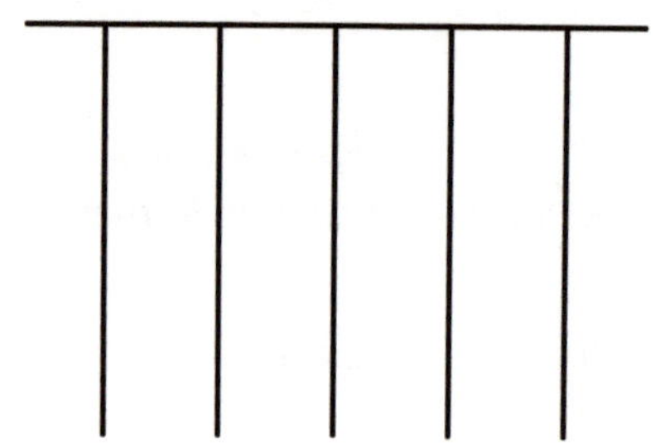

Exercise 6D

Instructions:

1. Start a **New** file using either the **inch-helper.dwt** or the **metric-helper.dwt**
2. Draw the lines exactly as shown.

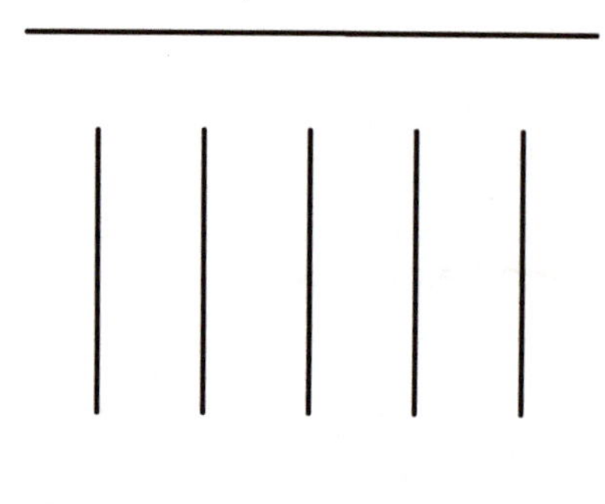

3. **Extend** the vertical lines to intersect with both horizontal lines using the **crossing fence** method by selecting the points at **P1** and **P2**.
4. **Save** the drawing as: **Ex-6D**

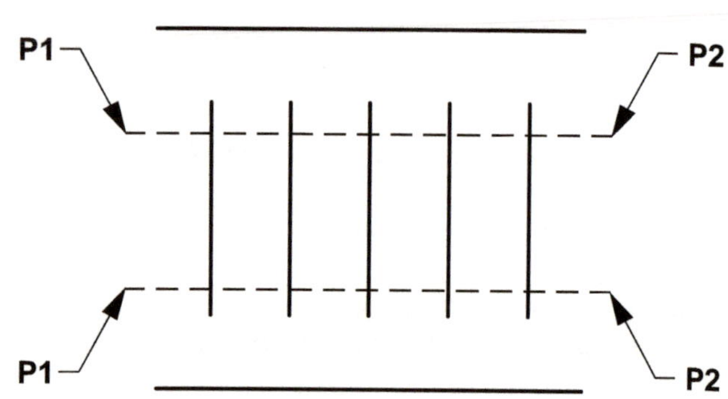

Exercises 6E & 6F

Exercise 6E

Instructions:

1. Start a **New** file using either the **inch-helper.dwt** or the **metric-helper.dwt**
2. Draw the lines using rectangles, circles, and lines (for the X's). Use Layer Object Line.

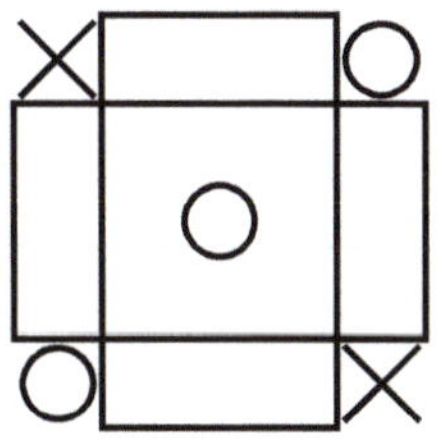

3. Modify the drawing to appear as shown using: **Explode** and **Erase**.
   Note: You must **Explode** a rectangle before you can **Erase** an individual line.
4. **Save** the drawing as: **Ex-6E**

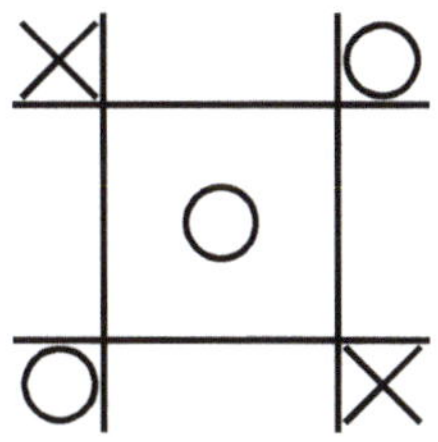

Exercise 6F

Instructions:

1. Start a **New** file using either the **inch-helper.dwt** or the **metric-helper.dwt**
2. Draw the objects below using all of the commands you have learned in the previous lessons. Have some fun with this, creating whatever layers you want.
3. **Save** the drawing as: **Ex-6F**

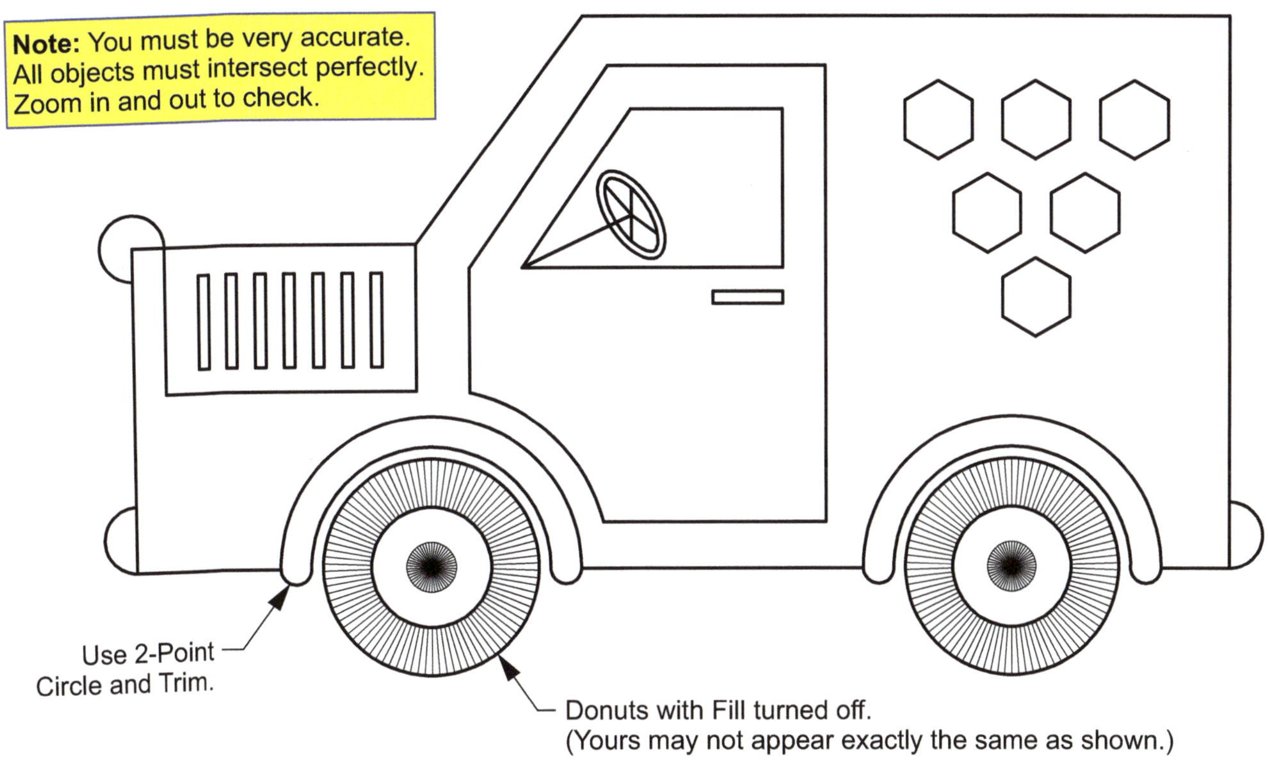

Notes:

LESSON 7

LEARNING OBJECTIVES

After completing this lesson, you will be able to:

- Copy objects.
- Create a Mirrored Image.
- Add Rounded Corners.
- Add Chamfered Corners.

Copy Multiple Copies

The **Copy** command creates a duplicate set of the objects selected. The **Copy** command is similar to the **Move** command.

The difference between Copy and Move commands:

The **Move** command **merely moves** the objects to a new location.
The **Copy** command **makes a copy**, and you select the location for the new copy.

The steps required are:

1. Select the objects to be copied.
2. Select a basepoint.
3. Select a new location for the new copy.

Note: For an optional copy method using Drag and Copy Here, please refer to page 6-9.

How to use the Copy command to make multiple copies

1. Select the **Copy** command using one of the following:

   Ribbon = Home Tab / Modify Panel / Copy
   or
   Keyboard = *CO <Enter>*

   The following prompt will appear on the Command Line:

   ```
   COPY Select objects:
   ```

2. Select the objects you want to copy and then press ***<Enter>***.

   The following prompt will appear on the Command Line:

   ```
   COPY Specify base point or [Displacement mOde] <Displacement>:
   ```

3. Select a basepoint (**P1**).

   The following prompt will appear on the Command Line:

   ```
   COPY Specify second point or [Array] <use first point as displacement>:
   ```

4. Select the new location (**P2**).

   The following prompt will appear on the Command Line:

   ```
   COPY Specify second point or [Array Exit Undo] <Exit>:
   ```

5. Select the new location (**P2**) for the next copy.

   The following prompt will appear on the Command Line:

   ```
   COPY Specify second point or [Array Exit Undo] <Exit>:
   ```

6. Select the new location (**P2**) for the next copy, or press ***<Enter>*** to stop specifying the location of more copies.

Refer to the example on the next page.

Continued on the next page...

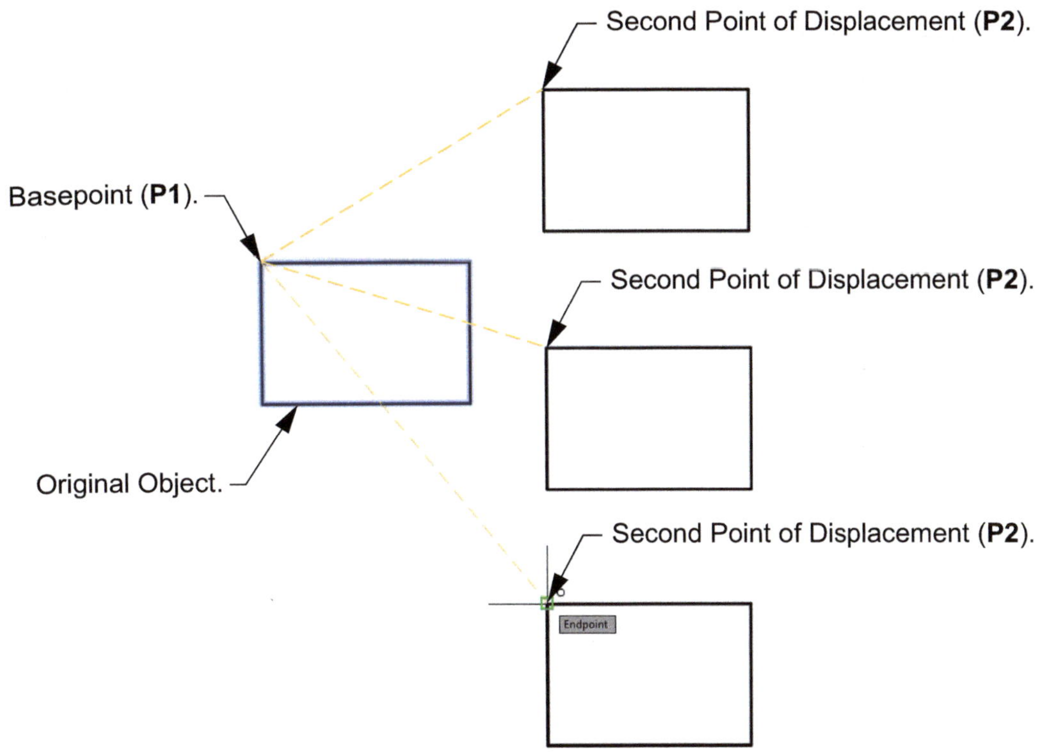

Note: The Copy command continues to make copies until you press ***<Enter>*** to exit.

How to select single or multiple copy mode

If you prefer to have AutoCAD stop the Copy command automatically after a single copy, you may change the copy mode to **Single**.

After you have selected the object(s) to copy, the following prompt will appear on the Command Line:

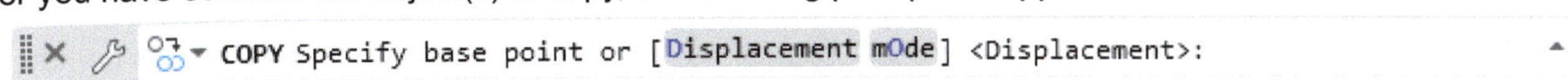

If you select the option **mOde**, you may select **Single** or **Multiple** copy mode.

Copy "Array" Option

The **Copy** command allows you to make an **Array** of copies.

After you have selected the basepoint (**P1**), the following prompt appears:

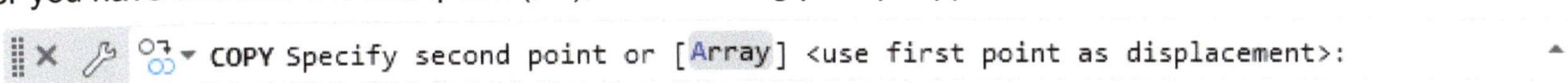

If you select the option **Array**, the following prompt will appear on the Command Line:

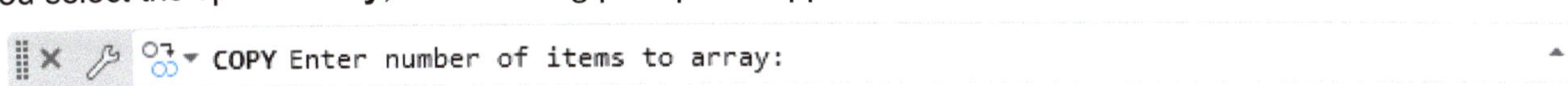

1. Enter the number of items to Array and then press ***<Enter>***.

   The following prompt will appear on the Command Line:

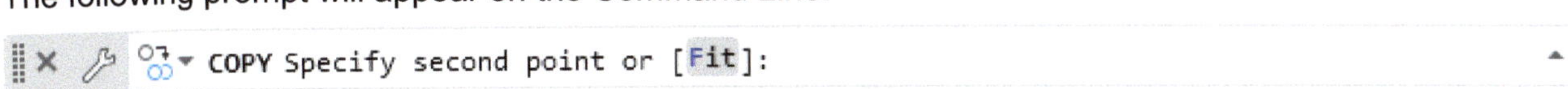

2. Place the second point (**P2**) or select ***F*** and then press ***<Enter>***.

Continued on the next page...

Place Second Point

The space between each copy is determined by the spacing between the original basepoint and the second point.

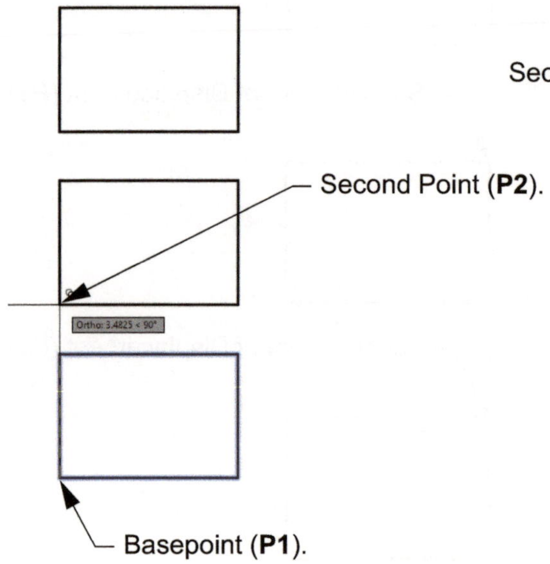

Use Fit option

The copies are evenly spaced between the basepoint and the second point.

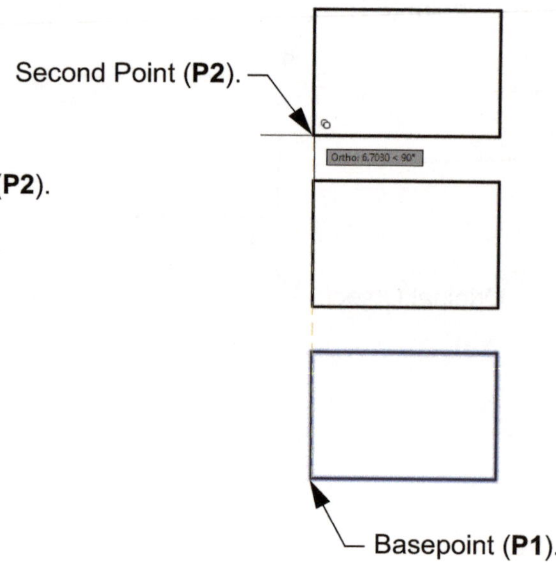

Note: The Array option within the Copy command is a quick method to create multiple copies. But AutoCAD has a more powerful and accurate Array command, which is described in Lesson 13.

Mirror

The **Mirror** command allows you to make a mirrored image of any objects you select. You can use this command for creating right- or left-hand parts, or to draw half of a symmetrical object and mirror it to save drawing time.

1. Select the **Mirror** command using one of the following:

   Ribbon = Home Tab / Modify Panel / Mirror
   or
   Keyboard = *MI* *<Enter>*

   The following prompt will appear on the Command Line:

   ```
   MIRROR Select objects:
   ```

2. Select the objects to be mirrored and then press ***<Enter>***.

   The following prompt will appear on the Command Line:

   ```
   MIRROR Specify first point of mirror line:
   ```

3. Select the first end of the mirror line (**P1**).

   The following prompt will appear on the Command Line:

   ```
   MIRROR Specify second point of mirror line:
   ```

Continued on the next page...

4. Select the second end of the mirror line (**P2**).

   The following prompt will appear on the Command Line:

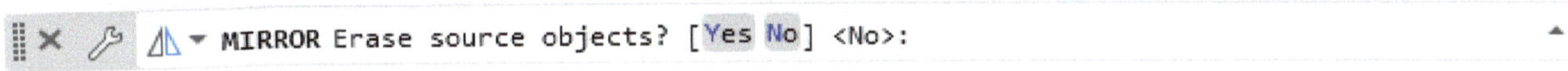

5. Select ***Y*** if you want to erase the original object, or ***N*** if you want to keep the original object.

Note: The placement of the "Mirror Line" is important. You may make a mirrored copy **vertically**, **horizontally**, or on an **angle**.

Mirror Line "Vertical"

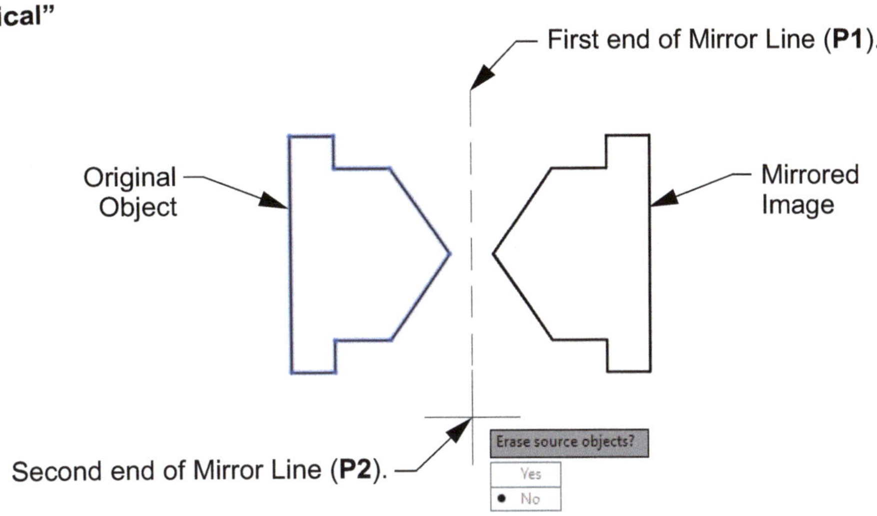

Mirror Line "Horizontal"

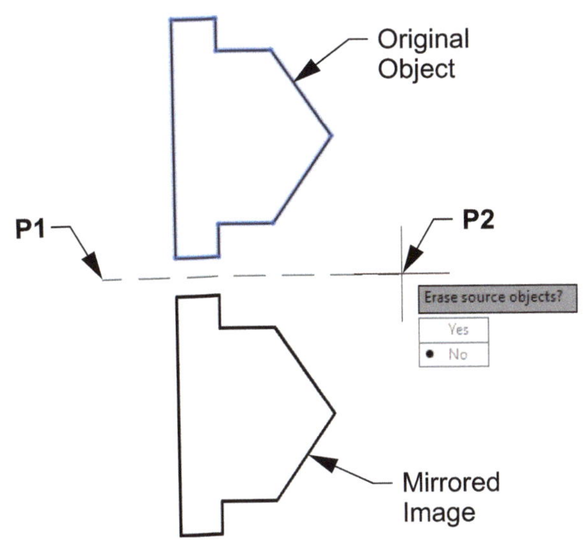

Mirror Line "Angled"

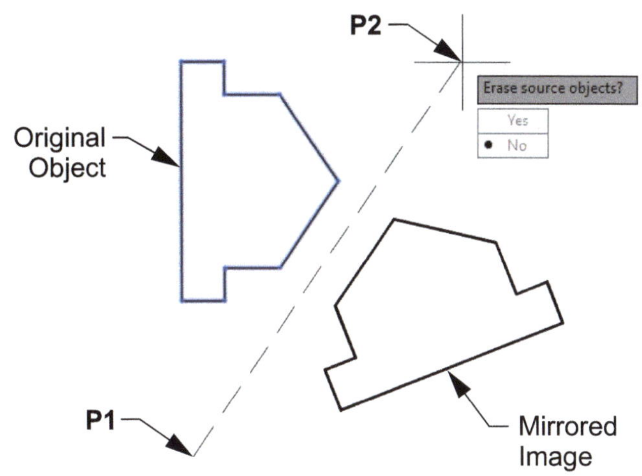

How to control text when using the Mirror command

(Do the following **before** you use the Mirror command.)

1. At the Command Line type: ***mirrtext*** and then press ***<Enter>***.

2. If you **want** the text to mirror (reverse reading), type: ***1*** and then press ***<Enter>***.
   If you **do not want** the text to mirror, type: ***0*** and then press ***<Enter>***.

MIRRTEXT SETTING = 1 MIRRTEXT SETTING = 0

Fillet

The **Fillet** command will create a radius between two objects. The objects do not have to be touching. If two parallel lines are selected, it will construct a full radius.

Radius a corner

1. Select the **Fillet** command using one of the following:

   Ribbon = Home Tab / Modify Panel / Fillet
   or
   Keyboard = *F <Enter>*

   The following prompt will appear on the Command Line:

   ```
   FILLET Select first object or [Undo Polyline Radius Trim Multiple]:
   ```

2. Type ***R*** and then press ***<Enter>***.

   The following prompt will appear on the Command Line:

   ```
   FILLET Specify fillet radius <0.0000>:
   ```

3. Type the radius you require and then press ***<Enter>***.

   The following prompt will appear on the Command Line:

   ```
   FILLET Select first object or [Undo Polyline Radius Trim Multiple]:
   ```

4. Select the first object to be filleted (left click).

   The following prompt will appear on the Command Line:

   ```
   FILLET Select second object or shift-select to apply corner or [Radius]:
   ```

5. Select the second object to be filleted (left click).

> **Note:** When you place the Cursor on the second object, AutoCAD displays the Fillet and allows you to change the radius before it is actually drawn. If you choose to change the radius, select the **Radius** option, enter a new radius value, and then select the second object.

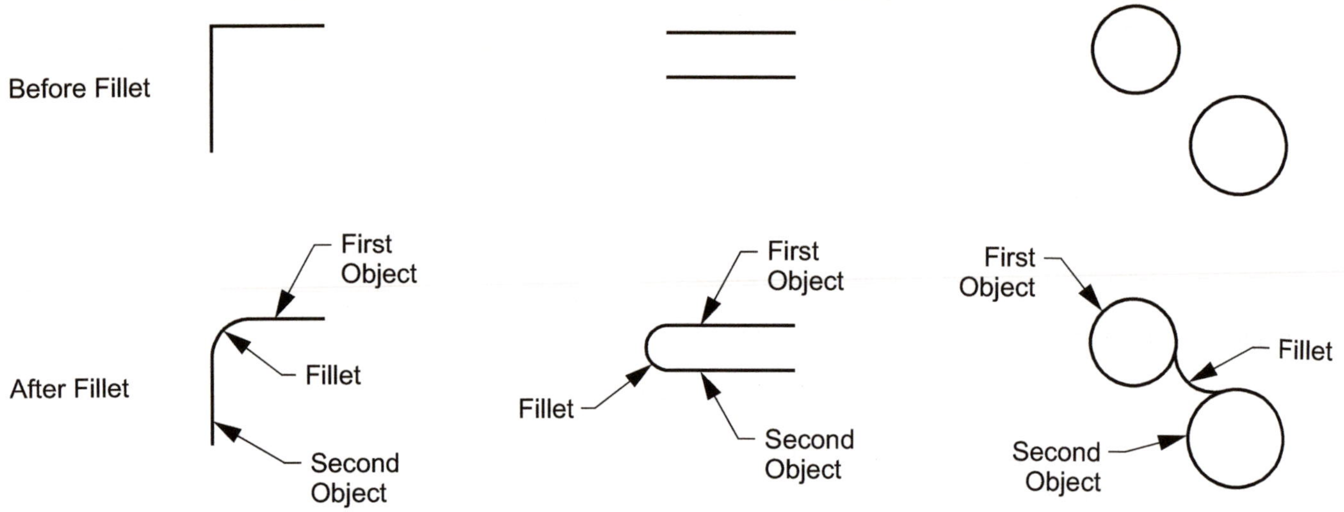

Continued on the next page...

The **Fillet** command may also be used to create a square corner.

Square corner

1. Select the **Fillet** command.

   The following prompt will appear on the Command Line:

   ```
   FILLET Select first object or [Undo Polyline Radius Trim Multiple]:
   ```

2. Select the first object (**P1**).

   The following prompt will appear on the Command Line:

   ```
   FILLET Select second object or shift-select to apply corner or [Radius]:
   ```

3. Hold down the ***<Shift>*** key while selecting the second object (**P2**).

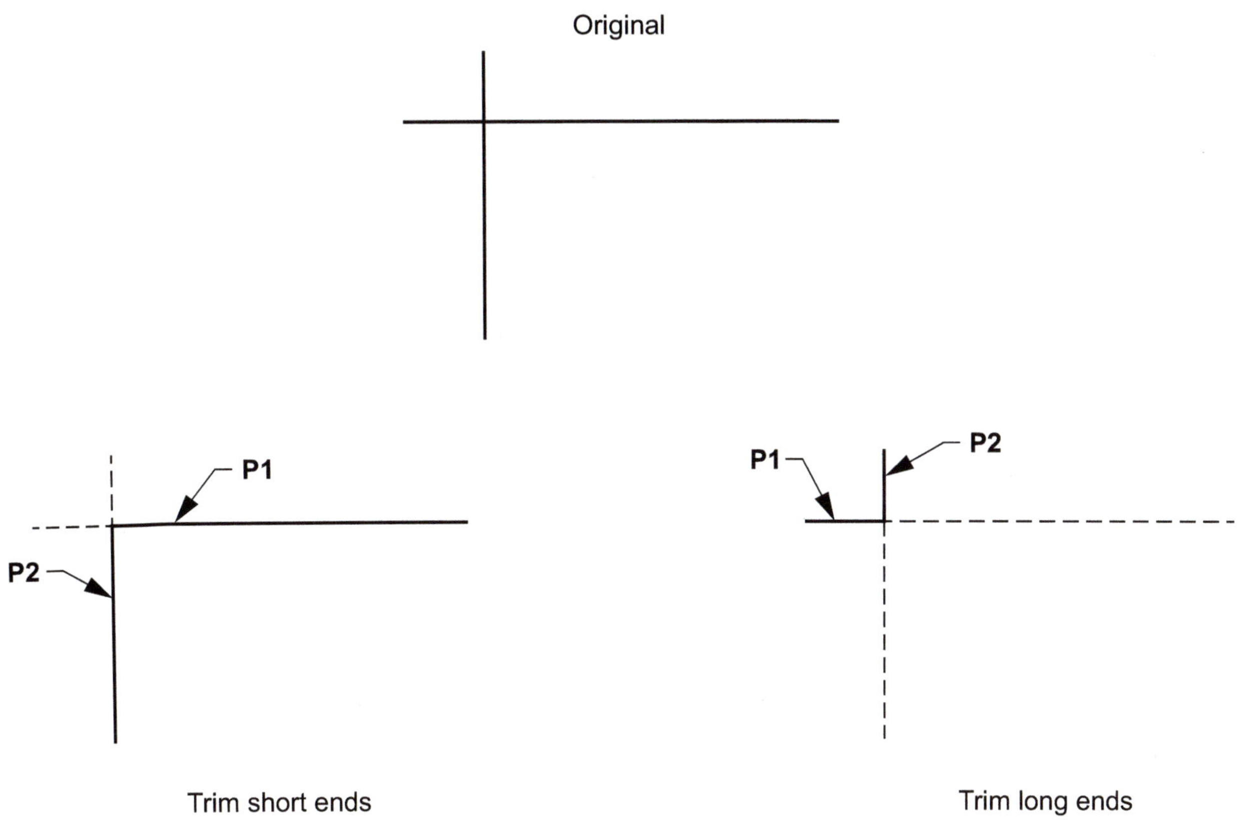

Options:

Polyline: This option allows you to fillet all intersections of a polyline in one operation, such as all four corners of a rectangle.

Trim: This option controls whether the original lines are trimmed to the end of the arc or remain the original length. (Set to Trim or No trim.)

Multiple: Repeats the fillet command until you press ***<Enter>*** or the ***<Esc>*** key.

Chamfer

The **Chamfer** command allows you to create a chamfered corner on two lines. There are two methods: **Distance** and **Angle**.

Method 1. Distance

Distance method requires input of a distance for each side of the corner.

1. Select the **Chamfer** command using one of the following:

   Ribbon = Home Tab / Modify Panel /
   or
   Keyboard = *CHA <Enter>*

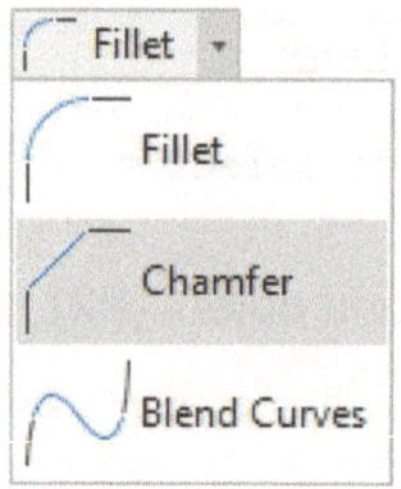

   The following prompt will appear on the Command Line:

```
CHAMFER Select first line or [Undo Polyline Distance Angle Trim mEthod Multiple]:
```

2. Select ***D*** and then press ***<Enter>***.

   The following prompt will appear on the Command Line:

```
CHAMFER Specify first chamfer distance <0.000>:
```

3. Type the distance for the first side and then press ***<Enter>***.

   The following prompt will appear on the Command Line:

```
CHAMFER Specify second chamfer distance <0.000>:
```

4. Type the distance for the second side and then press ***<Enter>***.

   The following prompt will appear on the Command Line:

```
CHAMFER Select first line or [Undo Polyline Distance Angle Trim mEthod Multiple]:
```

5. Select the first line to be chamfered (Distance 1).

   The following prompt will appear on the Command Line:

```
CHAMFER Select second line or shift-select to apply corner or [Distance Angle Method]:
```

6. Select the second line to be chamfered (Distance 2).

Note: When you place the cursor on the second line, AutoCAD displays the chamfer and allows you to change the distances before it is actually drawn.

If you choose to change the distance, select the Distance option, enter new distance values, and then select the second line.

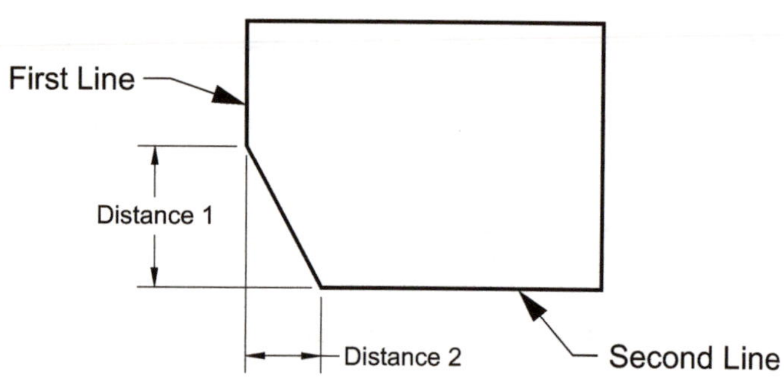

Method 2. Angle

Angle method requires input for the length of the line and an angle.

1. Select the **Chamfer** command.

   The following prompt will appear on the Command Line:

   ```
   CHAMFER Select first line or [Undo Polyline Distance Angle Trim mEthod Multiple]:
   ```

2. Select ***A*** and then press ***<Enter>***.

   The following prompt will appear on the Command Line:

   ```
   CHAMFER Specify chamfer length on the first line <0.000>:
   ```

3. Type the chamfer length and then press ***<Enter>***.

   The following prompt will appear on the Command Line:

   ```
   CHAMFER Specify chamfer angle from the first line <0>:
   ```

4. Type the angle and then press ***<Enter>***.

   The following prompt will appear on the Command Line:

   ```
   CHAMFER Select first line or [Undo Polyline Distance Angle Trim mEthod Multiple]:
   ```

5. Select the first line to be chamfered (the length side).

   The following prompt will appear on the Command Line:

   ```
   CHAMFER Select second line or shift-select to apply corner or [Distance Angle Method]:
   ```

6. Select the second line to be chamfered (the angle side).

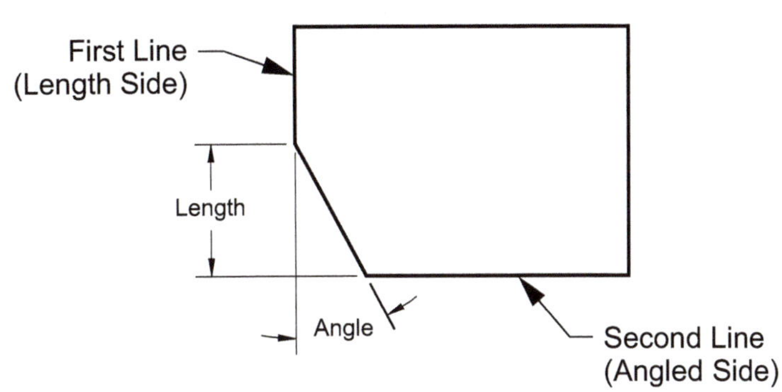

Options:

Polyline: This option allows you to chamfer all intersections of a polyline in one operation, such as all four corners of a rectangle.

Trim: This option controls whether the original lines are trimmed or remain after the corners are chamfered. (Set to Trim or No trim.)

mEthod: Allows you to switch between **Distance** and **Angle** method. The distance or angle must have been set previously.

Multiple: Repeats the Chamfer command until you press ***<Enter>*** or the ***<Esc>*** key.

Exercise 7A

Instructions:

1. Start a **New** file using either the **inch-helper.dwt** or the **metric-helper.dwt**
2. Draw the lines and one circle shown below.

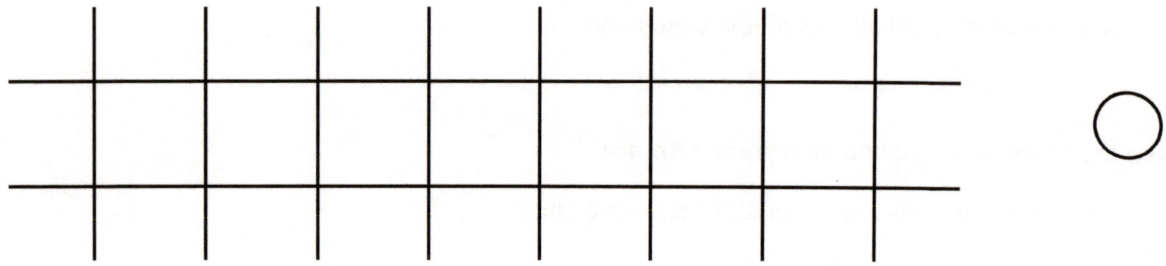

Before Copy

3. Copy the circle to all the other locations shown below.
   A. Select the **Copy** command.
   B. Select the circle.
   C. Select the basepoint on the original circle.

      Note: The basepoints are different.

   D. Select the new location (second point of displacement).

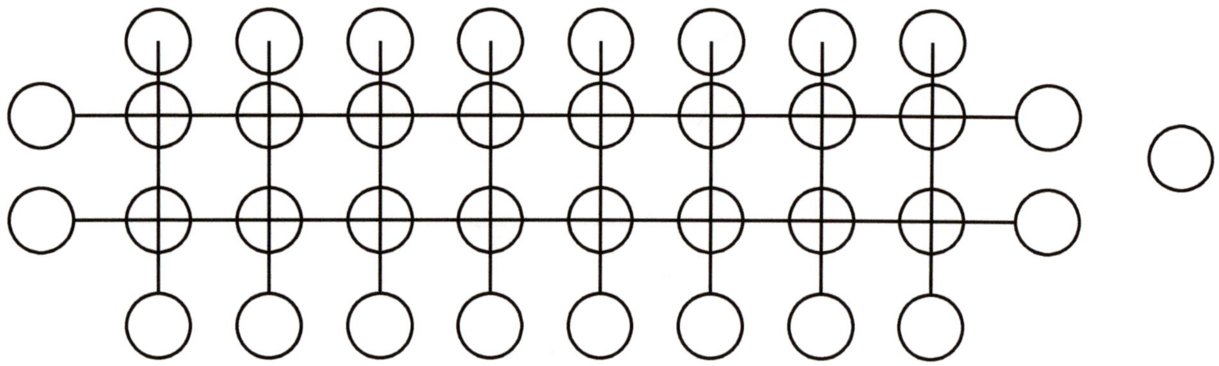

After Copy

4. **Save** the drawing as: **Ex-7A**

Note: Step-by-step instructions are shown on the next page.

Exercise 7A Helper

1. Select the **Copy** command and then select the circle to copy.
2. Select the basepoint: the center of the circle (use Object Snap Center).
3. Set Running Object Snap to **Endpoint** and **Intersection** and place all 24 circles.

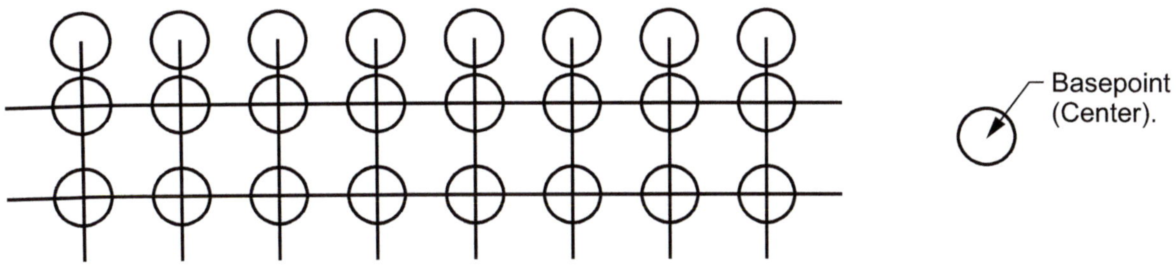

1. Select the **Copy** command and then select the circle to copy.
2. Select the basepoint: the quadrant (12 o'clock) this time.
3. Place the eight bottom circles accurately using the **Endpoint** Object Snap.

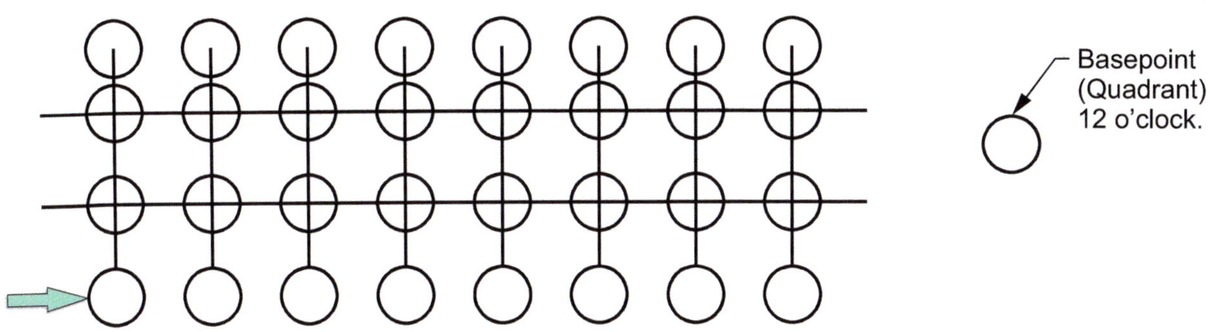

1. Select the **Copy** command and then select the circle to copy.
2. Select the basepoint: the quadrant (3 o'clock) this time.
3. Place the two circles on the left accurately using the **Endpoint** Object Snap.

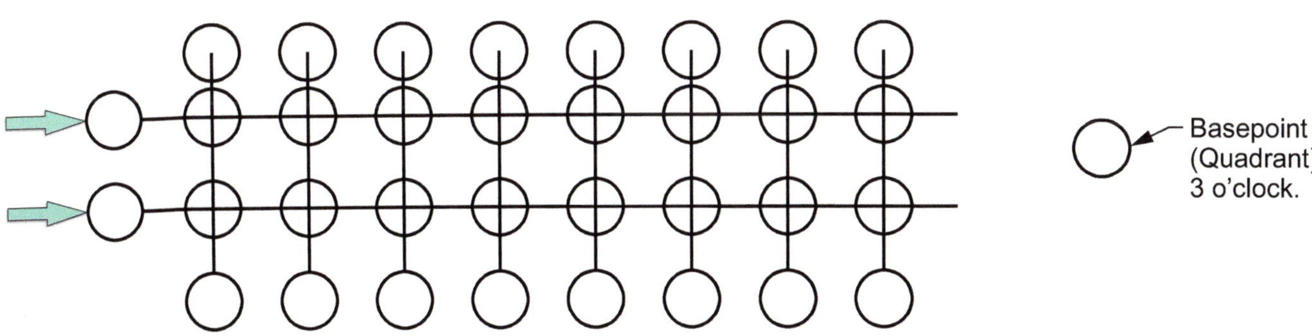

Exercises 7B & 7C

Exercise 7B

Instructions:

1. Start a **New** file using either the **inch-helper.dwt** or the **metric-helper.dwt**
2. Draw the rectangle shown.

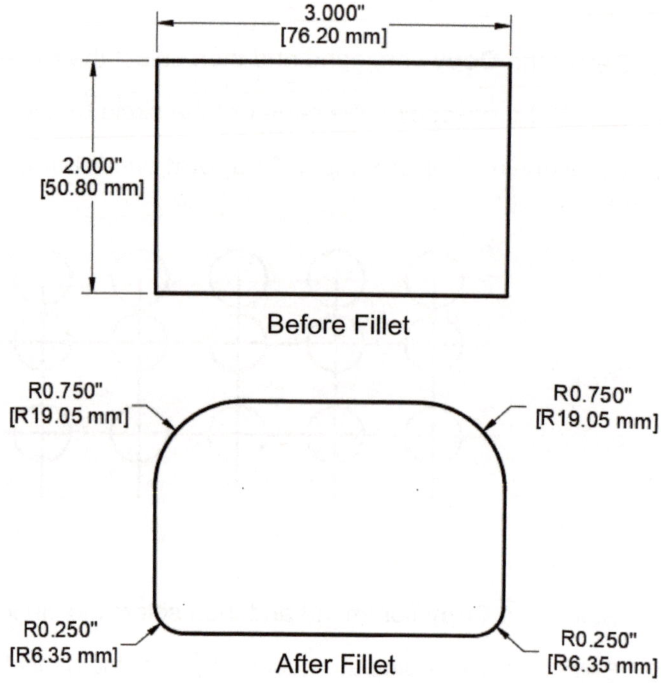

3. Radius the corners using the **Fillet** command.
4. **Save** the drawing as: **Ex-7B**

Exercise 7C

Instructions:

1. Start a **New** file using either the **inch-helper.dwt** or the **metric-helper.dwt**
2. Draw the three rectangles shown below.
3. Chamfer the corners using the **Chamfer** command.
4. **Save** the drawing as: **Ex-7C**

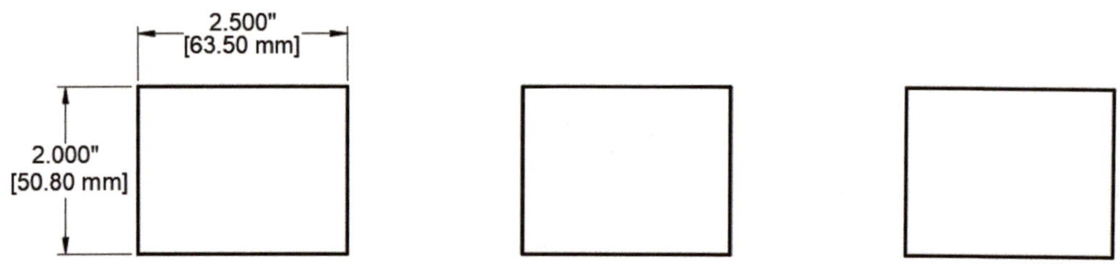

Before Chamfer

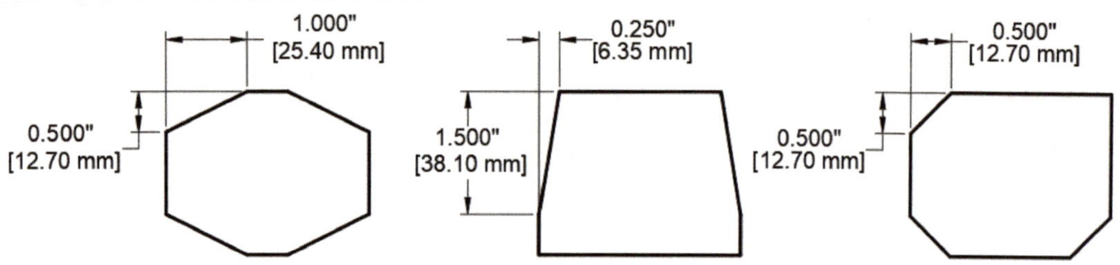

After Chamfer

Exercise 7D

Instructions:

1. Start a **New** file using either the **inch-helper.dwt** or the **metric-helper.dwt**
2. Draw the rectangle shown below.
3. Chamfer the corners using the **Chamfer** command.
4. **Save** the drawing as: **Ex-7D**

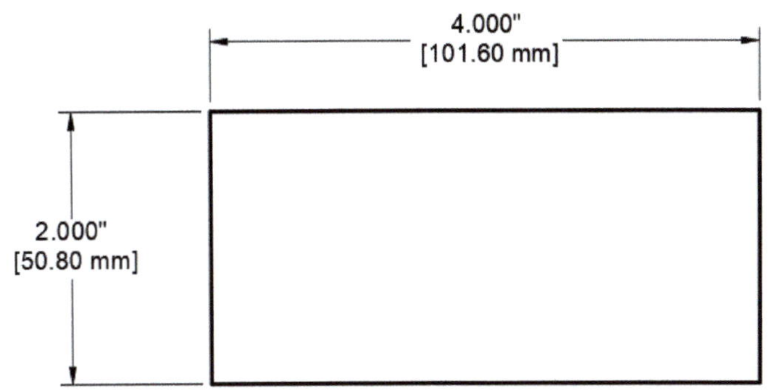

Before Chamfer

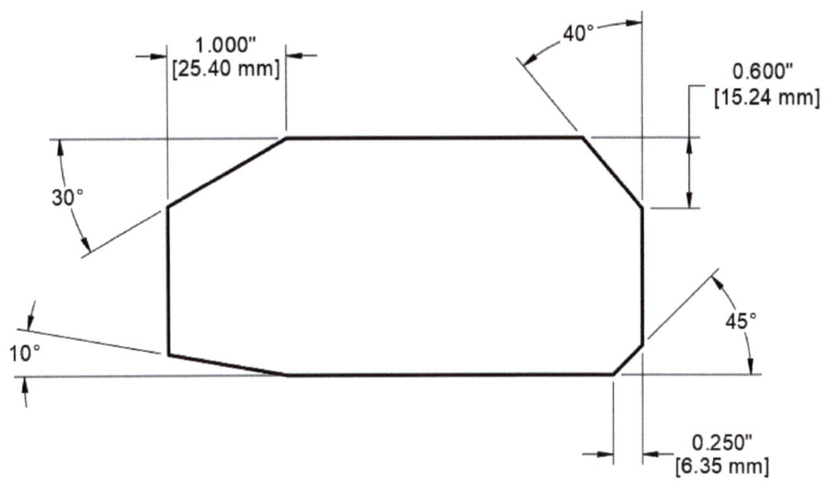

After Chamfer

Exercise 7E

Instructions:

1. Start a **New** file using either the **inch-helper.dwt** or the **metric-helper.dwt**
2. Draw the half the house shown below. Have fun using all of the commands you have learned: Trim, Extend, Copy, Rectangle, 2-Point Circle, Running Object Snap: Nearest, Perpendicular, Quadrant, and Endpoint.

   Use Layers Roof, Wall, Window, Door, Plants, and Bricks.

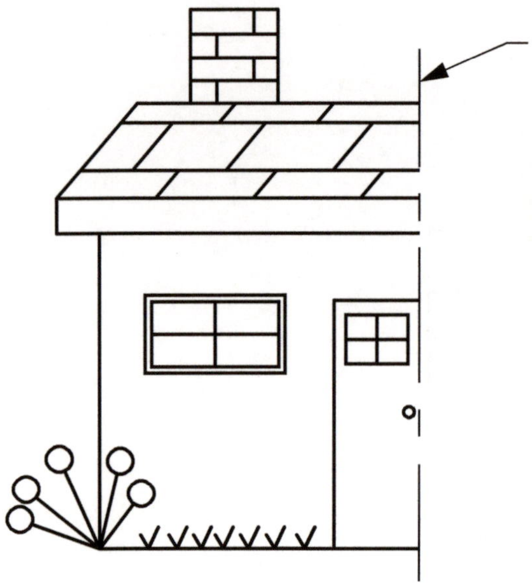

Hint: Draw this line first. Also snap to it for the Mirror line. Your halves should match perfectly.

Note: You must be very accurate. All objects must intersect perfectly. Zoom in and out to check.

3. Now create a Mirrored image of the half house:

   A. Select the Mirror command.
   B. Select the objects to be Mirrored (you could use Window selection).
   C. Select the first endpoint of the Mirror line (Ortho **ON**).
   D. Select the second endpoint of the Mirror line.
   E. Answer **NO** to "Erase Source Objects?"

4. **Save** the drawing as: **Ex-7E**

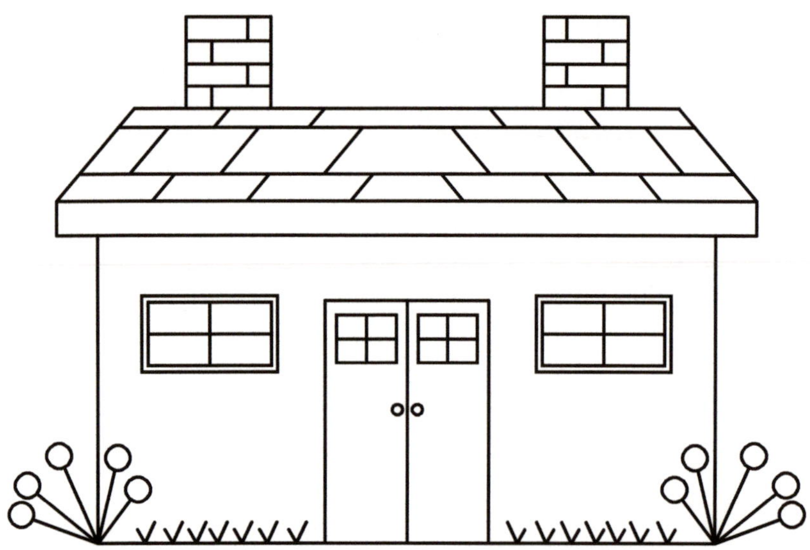

Can you imagine how this might save you valuable time?

LESSON 8

LEARNING OBJECTIVES

After completing this lesson, you will be able to:

- Add a Single Line of text to the drawing
- Add a paragraph using Multiline Text
- Control Tabs, Indents, and use the Spell Checker
- Add Columns
- Edit existing Single Line Text and Multiline Text

Single Line Text

Single Line Text allows you to draw one or more lines of text. The text is visible as you type. To place the text in the drawing, you may use the default **start point** (the lower left corner of the text), or use one of the many styles of justification described on the next page. Each line of text is treated as a separate object.

Using the default start point

1. Select the **Single Line Text** command using one of the following:

Ribbon = Annotate Tab / Text Panel /
or
Keyboard = *DT <Enter>*

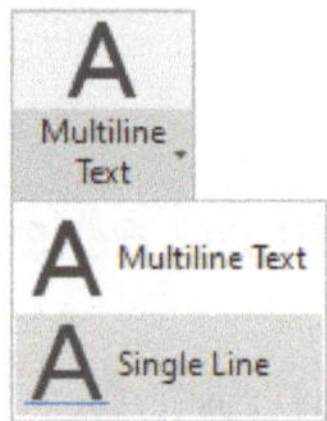

The following prompt will appear on the Command Line:

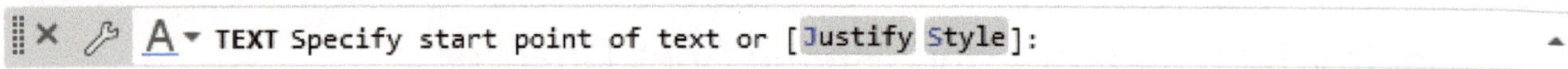

2. Place the cursor where the text should start and then **left click**.

The following prompt will appear on the Command Line:

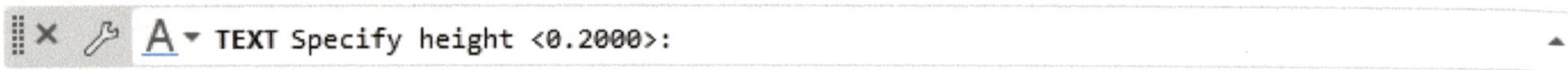

3. Type the height of your text and then press ***<Enter>***.

The following prompt will appear on the Command Line:

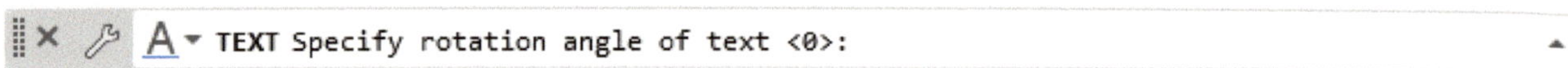

4. Type the rotation angle of your text and then press ***<Enter>***.

5. Type the text you require and then press ***<Enter>*** at the end of the sentence.

6. Type the next sentence you require or press ***<Enter>*** to stop the Single Line Text command.

Using justification

If you need to be very specific, where your text is located, you must use the justification option. For example, if you want your text in the middle of a rectangular box, you would use the justification option **"Middle"**.

The following is an example of Middle justification. (Use the inch or metric template.)

1. Draw a rectangle 6" [152.4 mm] wide and 3" [76.2 mm] high.

2. Draw a diagonal line from one corner to the other corner.

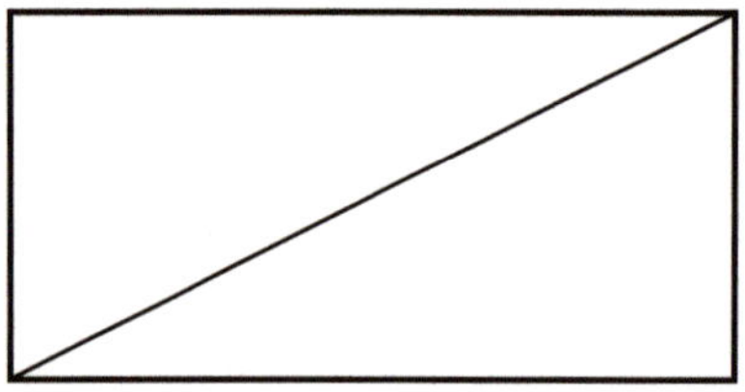

Continued on the next page...

3. Select the Single Line Text command.

   The following prompt will appear on the Command Line:

   ```
   TEXT Specify start point of text or [Justify Style]:
   ```

4. Type ***J*** and then press ***<Enter>***.

   The following prompt will appear on the Command Line:

   ```
   TEXT Enter an option [Left Center Right Align Middle Fit TL TC TR ML MC MR BL BC BR]:
   ```

5. Type ***M*** and then press ***<Enter>***.

   The following prompt will appear on the Command Line:

   ```
   TEXT Specify middle point of text:
   ```

6. Snap to the midpoint of the diagonal line (left click).

   The following prompt will appear on the Command Line:

   ```
   TEXT Specify height <0.2000>:
   ```

7. Type in ***1 [25.4]*** and then press ***<Enter>***.

   The following prompt will appear on the Command Line:

   ```
   TEXT Specify rotation angle of text <0>:
   ```

8. Type in ***0*** and then press ***<Enter>***.

9. Type the text **HHHH** and press ***<Enter>*** and then press ***<Enter>*** again to stop the command.

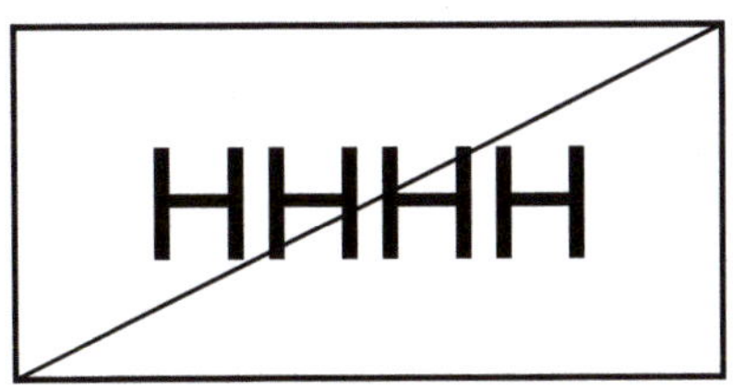

Note: Also refer to Exercise 8C for **"Mid Between 2 Points"** method.

Other justification options

Align

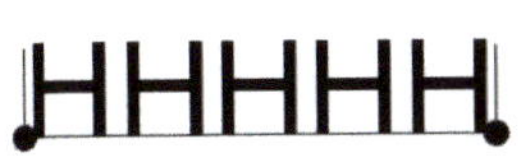

Aligns the line of text between two points specified. The height is adjusted automatically.

Fit

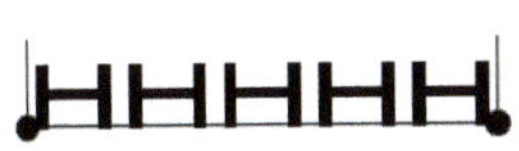

Fits the text between two points specified. The height is specified by you and does not change.

Center

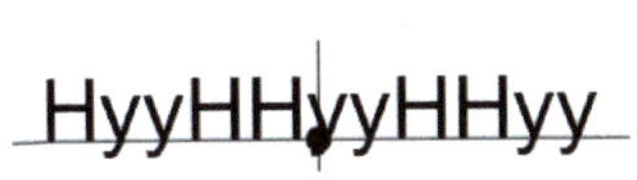

This is a tricky one. Center is located at the bottom center of uppercase letters.

Continued on the next page...

Middle

If only uppercase letters are used, **Middle** is located in the middle, horizontally, and vertically. If both uppercase and lowercase letters are used, **Middle** is located in the middle, horizontally, and vertically, but considers the lowercase letters as part of the height.

Right

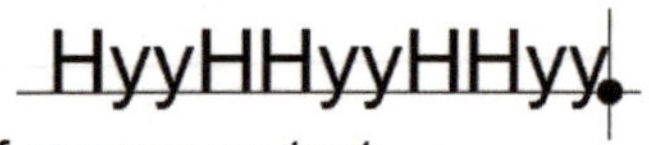

Bottom right of uppercase text.

TL, TC, TR

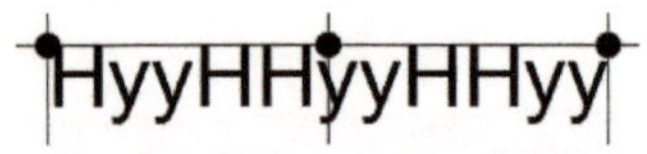

Top left, Top center, and Top right of uppercase and lowercase text.

ML, MC, MR

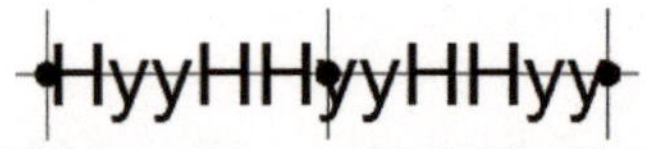

Middle left, Middle center, and Middle right of uppercase text. (Notice the difference between "Middle" and "MC".)

BL, BC, BR

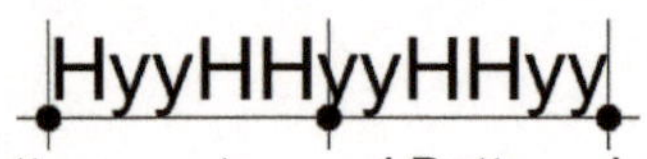

Bottom left, Bottom center, and Bottom right of lowercase text. Notice the different location for **BR** and **Right** shown above. **BR** considers the lowercase letters with tails as part of the height.

Multiline Text

Multiline Text command allows you to easily add a sentence, paragraph, or tables. The Mtext editor has most of the text editing features of a word processing program. You can underline, bold, italic, add tabs for indenting, change the font, line spacing, and adjust the length and width of the paragraph.

When using Mtext, you must first define a text boundary box. The text boundary box is defined by entering where you want to start the text (first corner) and approximately where you want to end the text (opposite corner). It is very similar to drawing a rectangle. The paragraph text is considered one object rather than several individual sentences.

Using Multiline Text

1. Select the **Multiline Text** command using one of the following:

Ribbon = Annotate Tab / Text Panel /
or
Keyboard = *MT <Enter>*

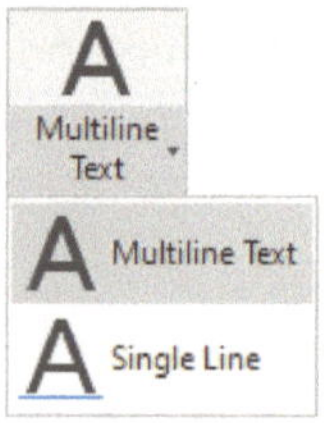

The Command Line will list the current style, text height, and annotative setting.

Current text style: "Standard" Text height: 0.200 Annotative: No

Note: You will not see these settings as they move up into the Command Prompt History. **Annotative** will be discussed in Lessons 26 and 27.

The cursor will then appear as crosshairs with the letters **"abc"** attached. These letters indicate how the text will appear using the current font and text height.

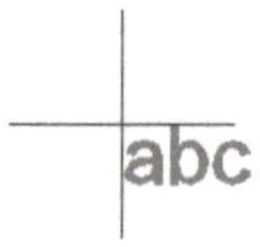

Continued on the next page...

The following prompt will appear on the Command Line:

```
MTEXT Specify first corner:
```

2. Place the cursor at the upper left corner of the area where you want to start the new text boundary box and press the left mouse button (**P1**).

   The following prompt will appear on the Command Line:

```
MTEXT Specify opposite corner or [Height Justify Line spacing Rotation Style Width Columns]:
```

3. Move the cursor to the right and down (**P2**) and press left mouse button.

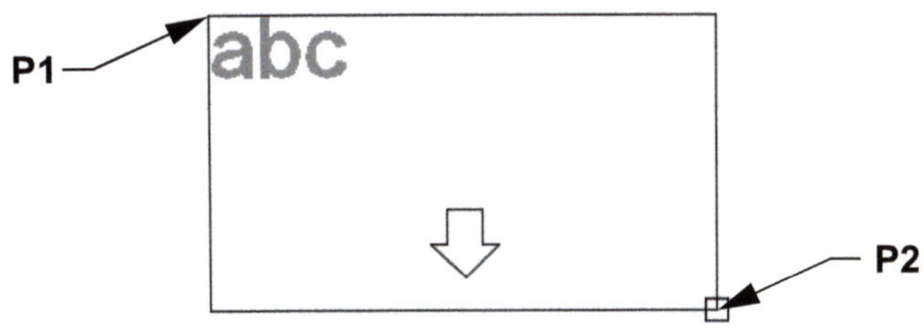

The Text Editor Ribbon will appear.

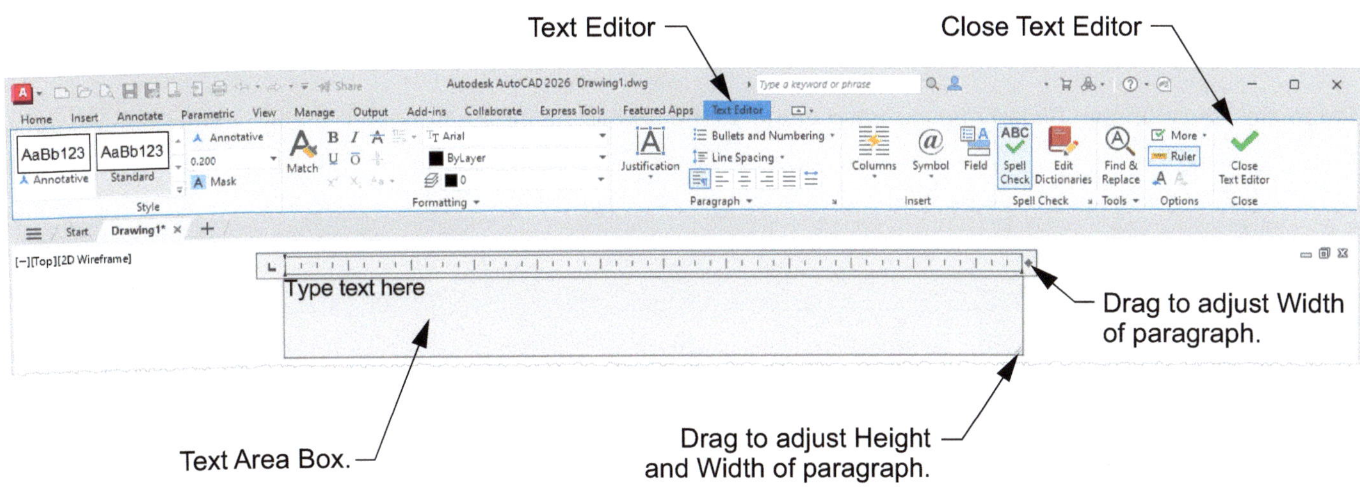

The **Text Editor** allows you to select the Text Style, Font, Height, etc. You can add features such as bold, italics, underline, and color.

The **Text Area Box** allows you to enter the text, add tabs, indent, adjust left-hand margins, and change the width and height of the paragraph.

4. After you have entered the text in the Text Area box, select the **Close Text Editor** tool.

How to change the "abc", on the crosshairs, to other letters

You can personalize the letters that appear attached to the crosshairs using the **"MTJIGSTRING"** system variable (10 characters max). The letters will simulate the appearance of the font and height selected but will disappear after you place the lower right corner (**P2**).

1. Type ***mtjigstring*** and then press ***<Enter>*** on the Command Line.
2. Type the new letters and then press ***<Enter>***.

Note: The letters will be saved to the computer, not the drawing. They will appear anytime you use Multiline Text and will remain until you change them again.

Tabs, Indents, and Spell Check

Tabs

Setting and removing Tabs is very easy.

The increments are determined by the text height. (Example: If the text height is 1", you may quickly place a Tab at any 1" mark on the ruler.) To be more specific refer to page 8-8.

Set or change the stop positions at anytime, using one of the following methods:

Place the cursor on the "Ruler" where you want the Tab and left click. A little dark **"L"** will appear. The Tab is set. If you would like to remove a Tab, just click and drag it off the ruler and it will disappear.

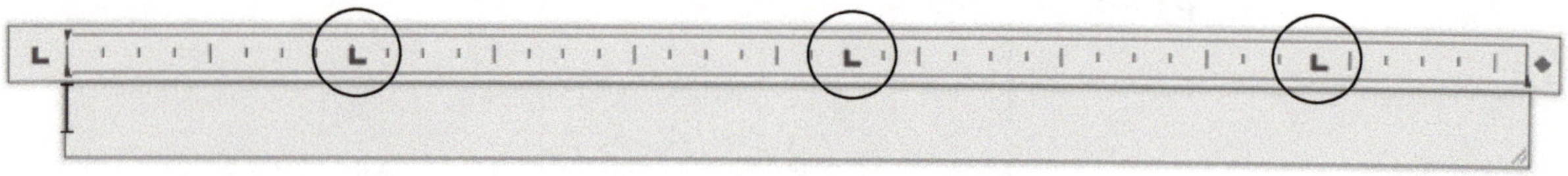

Indents

Sliders on the ruler show indention relative to the left side of the text boundary box. The top slider indents the first line of the paragraph, and the bottom slider indents the other lines of the paragraph. (Also refer to page 8-8.)

You may change their positions at anytime, using the following method:

Place the cursor on the "Slider" and click and drag it to the new location.

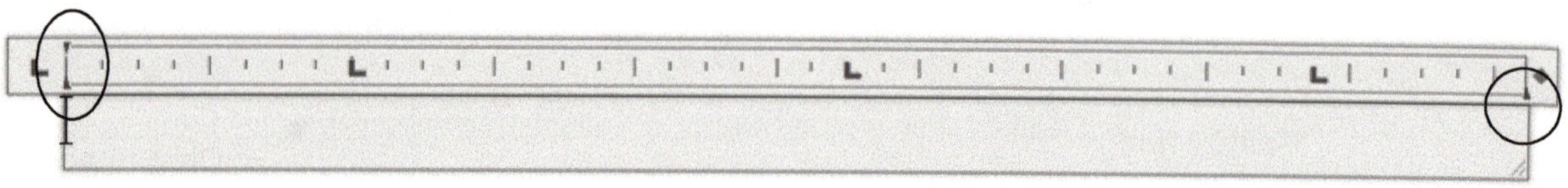

Spell Check

If you have Spell Check **on**, you will be alerted as you enter text with a red line under the misspelled word. Right click on the word and AutoCAD will give you some choices.

1. Select the text that you want to Spell Check. (Click once on the sentence.)

2. Select the **Annotate Tab / Text Panel /** 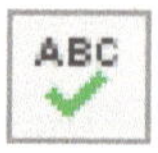

   The **Check Spelling** dialog box will appear.

3. Select **Start**.

   If AutoCAD finds any words misspelled, it will suggest a change. You may select **Change** or **Ignore**. When finished, a message will appear stating **"Spelling check complete"**.

4. Select **OK**.

5. Select **Close** to close the **Check Spelling** dialog box.

Columns

Static Columns

1. Right click in the **Text Box Area** and select **Columns**.
2. Select **Column Settings...**

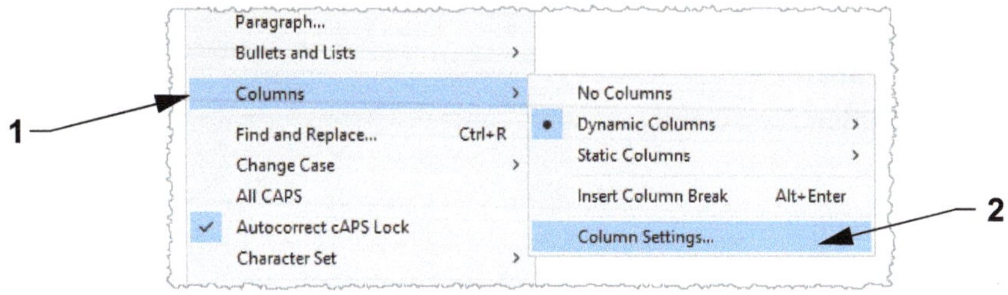

The **Column Settings** dialog box will appear.

3. Select **Static Columns**.
4. Select:
   A. Column Number.
   B. Height.
   C. Width of Column.
   D. Width of Gutter.

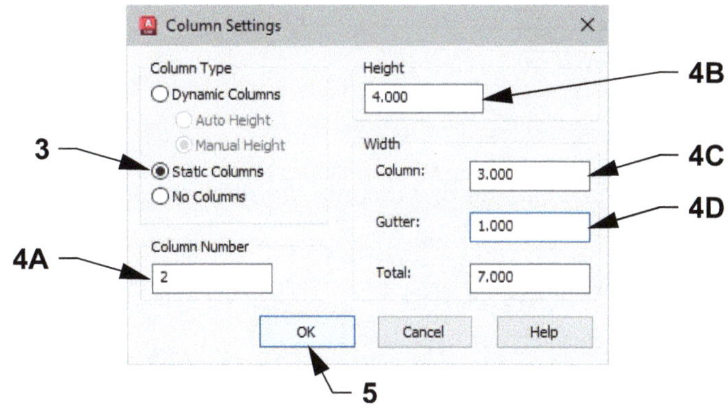

5. Select the **OK** button.
6. The Text Area should appear as shown below with two columns divided with a gutter.
7. Start typing in the left-hand box. When you fill the left-hand box, the text will start to spill over into the right-hand box.

You may also adjust and make changes to the width and height using the drag tools.

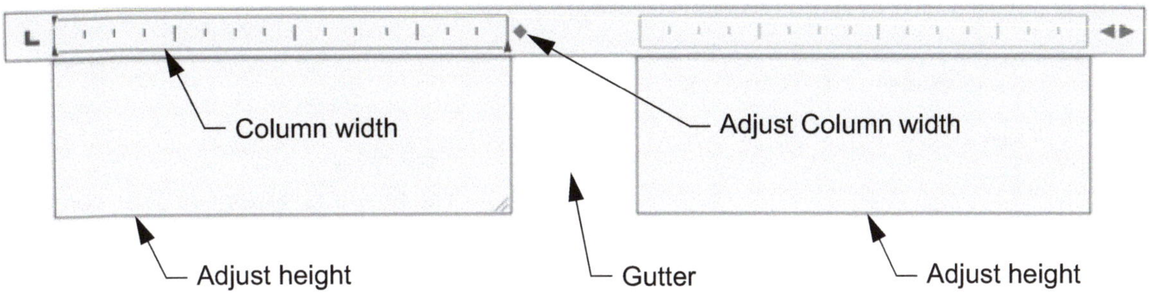

Dynamic Columns

1. Right click in the **Text Box Area** and select **Columns**.
2. Select **Column Settings...** (Refer to Static Columns above.)

Continued on the next page...

The **Column Settings** dialog box will appear.

3. Select **Dynamic Columns**.

4. Select **Auto Height**.

5. Select:
   A. Height.
   B. Width of Column.
   C. Width of Gutter.

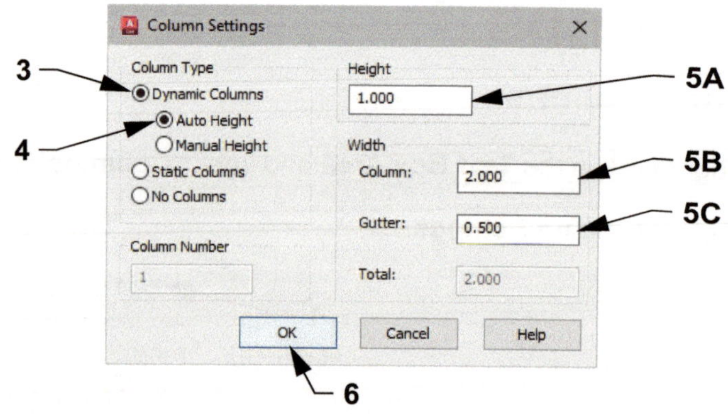

6. Select the **OK** button.

7. The Text Area will first appear with one column with the width and height you set.

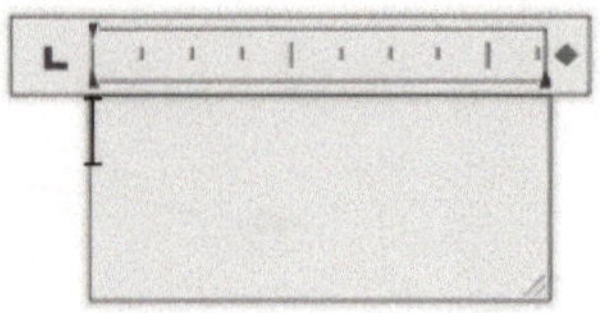

8. When the text fills the first column, another column will appear.

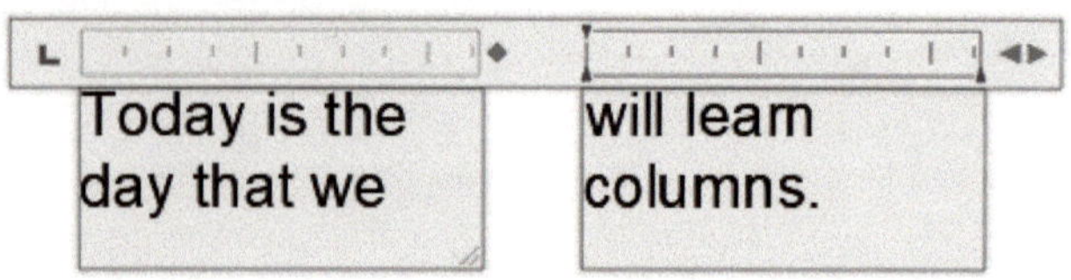

When the second column fills, another column will appear. You may also adjust and make changes to the width and height using the drag tools.

Paragraph and Line Spacing

You may set the Tabs, Indent, and Line Spacing for individual paragraphs.

1. Right click in the **Text Box Area** and select **Paragraph...**

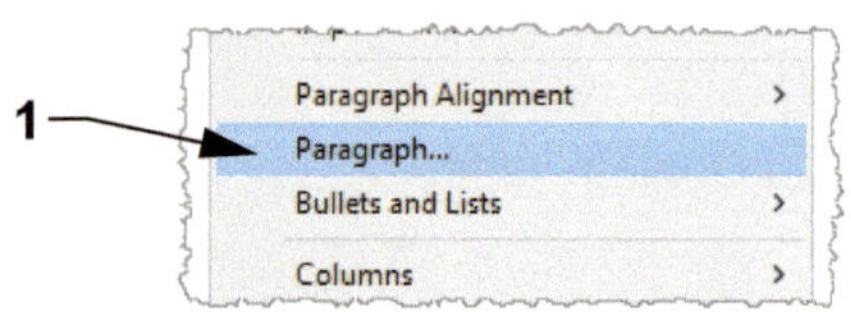

The **Paragraph** dialog box will appear.

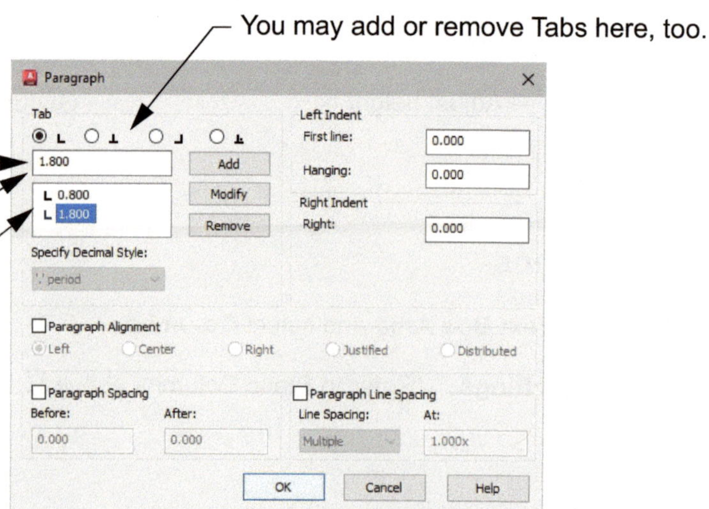

To Add:
1. Type spacing.
2. Select **Add** button.

To Modify:
1. Select the current spacing.
2. Modify the spacing.
3. Select **Modify** button.

To Remove:
1. Select from the list.
2. Select **Remove** button.

Editing Text

Single Line Text

Editing **Single Line Text** is somewhat limited compared to Multiline Text. In the example below, you will learn how to edit the text within a Single Line Text sentence.

Note: In Lesson 12, you will learn additional options for editing Single Line text by using the **Properties** command.

1. Double click on the **Single Line Text** you want to edit. The text will highlight.

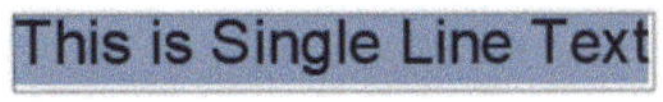

2. Make the changes in place and then press ***<Enter> <Enter>***.

Multiline Text

Multiline Text is as easy to edit as it is to input originally. You may use any of the text options shown on the Text Editor Tab.

1. Double click on the Multiline text you want to edit.
2. Highlight the text that you want to change, using click and drag.

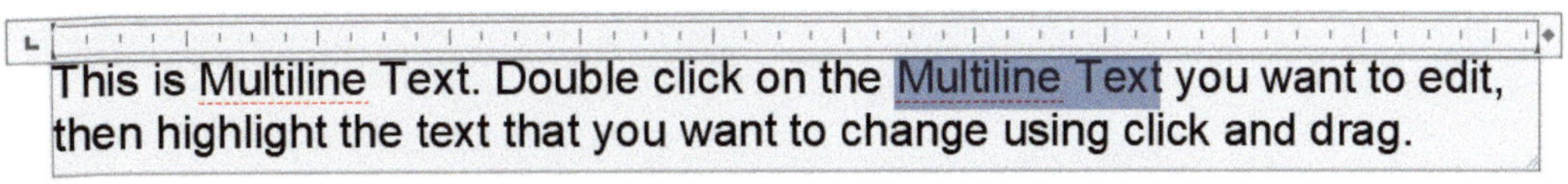

3. Make the changes, then select the **Close Text Editor** tool.

How to edit other Multiline Text features

1. Double click on the Multiline Text that you want to edit.
2. Right click in the **Text Box Area**.

   The menu shown below will appear.

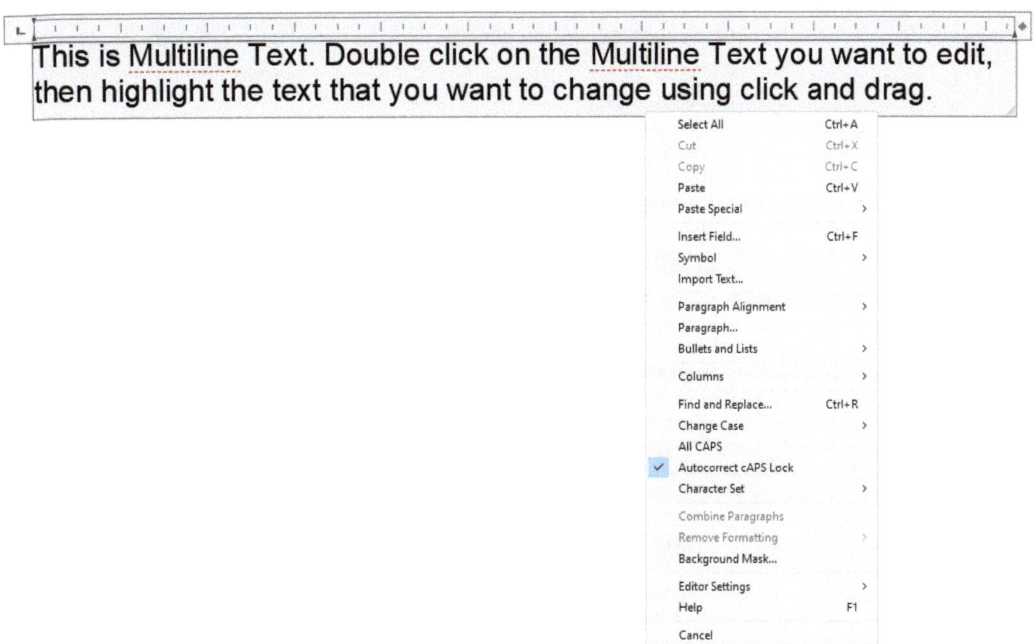

How to add a border around Multiline Text

You also may add a **border** around Multiline Text. (Not available for Single Line Text.)

1. Left click on the Multiline Text that you want to edit.

2. Right click and select **Properties** from the menu.

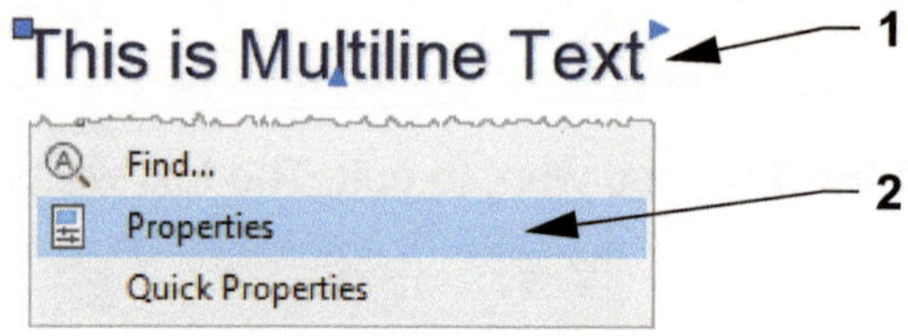

3. In the Properties dialog box, select **Yes** in the **Text Frame** drop-down list. You can also change the **Color**, **Linetype**, and **Lineweight** of the text border.

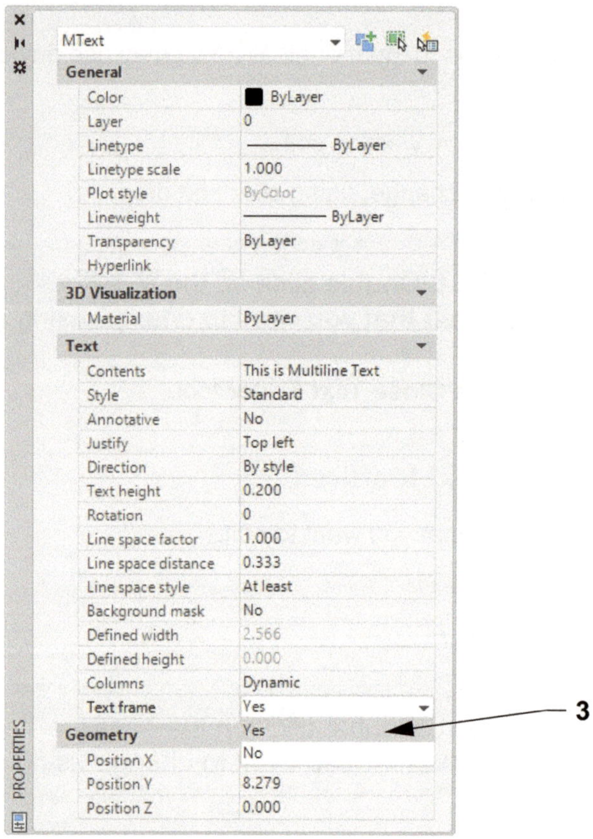

4. A border will be placed around the selected Multiline Text.

5. You can resize the border by selecting either of the two arrow grips, or you can move the Multiline Text and border by selecting the square grip.

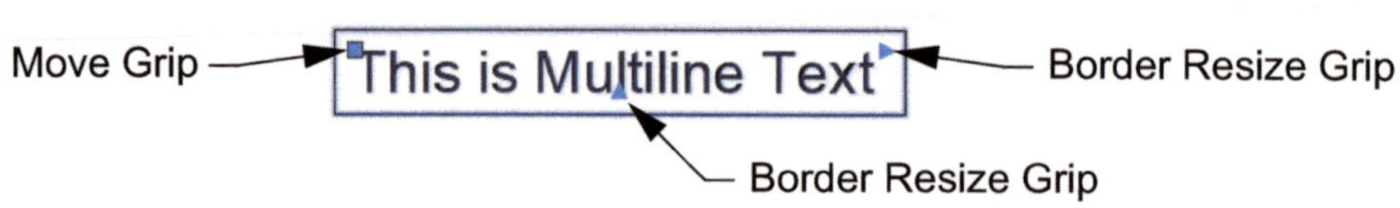

6. Close the Properties dialog box and then press the ***<Esc>*** key to cancel the grips.

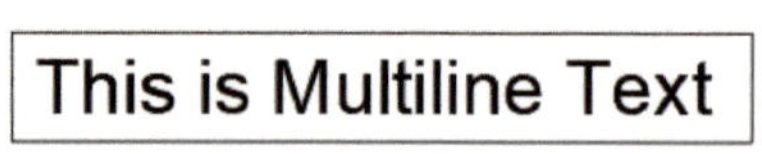

Exercise 8A-Inch

Instructions:

1. Start a **New** file using either the **inch-helper.dwt**
2. Duplicate the text shown below using **Single Line Text**.
3. Use Layer Text.
4. Follow the instructions in each block of text. To start the text in the correct location that is stated in each example, move the cursor while watching the coordinate display.
5. **Save** the drawing as: **Ex-8A-I**

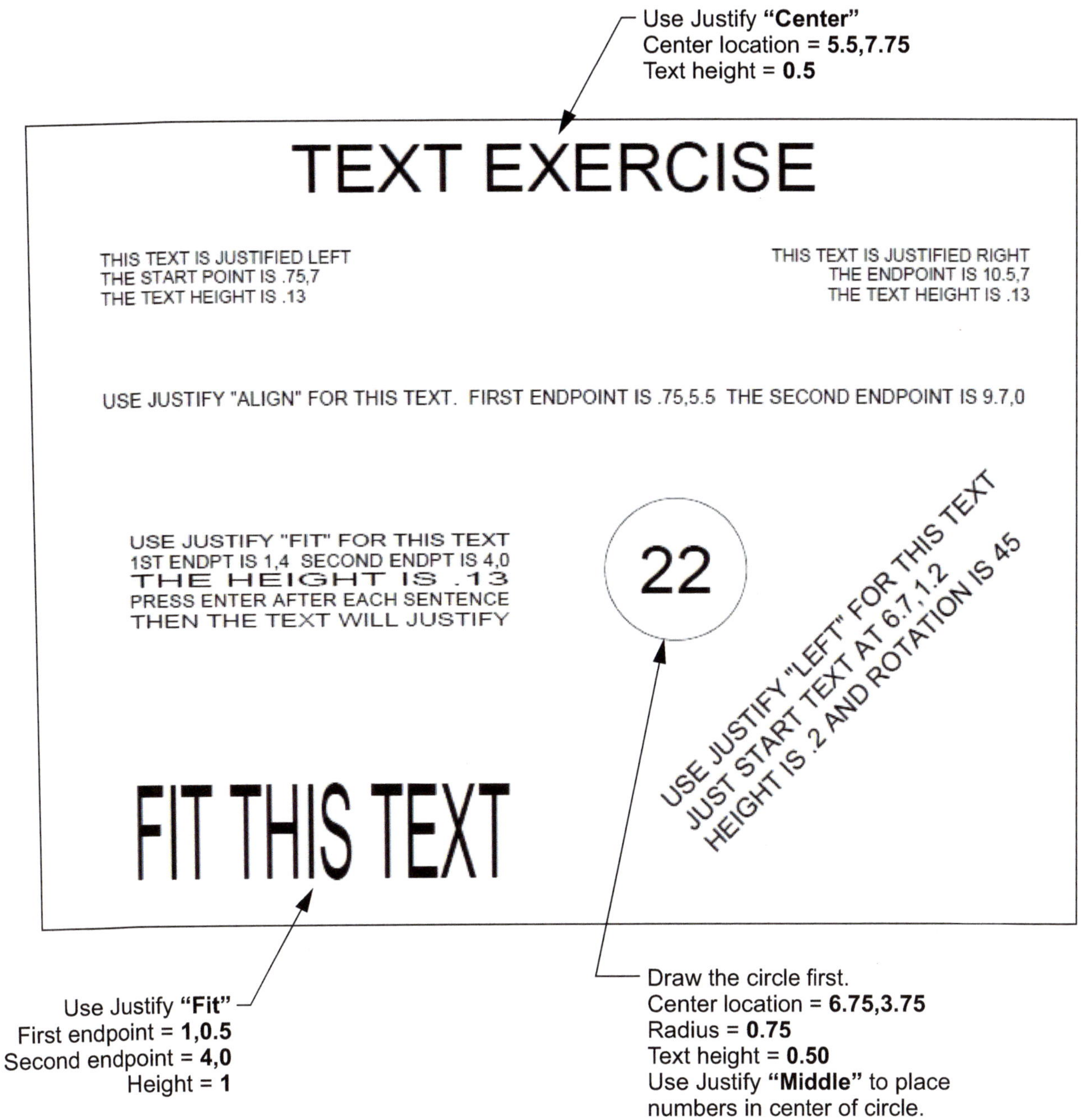

Exercise 8A-Metric

Instructions:

1. Start a **New** file using either the **metric-helper.dwt**
2. Duplicate the text shown below using **Single Line Text**.
3. Use Layer Text.
4. Follow the instructions in each block of text. To start the text in the correct location that is stated in each example, move the cursor while watching the coordinate display.
5. **Save** the drawing as: **Ex-8A-M**

Use Justify **"Center"**
Center location = **148.5,191**
Text height = **12.7**

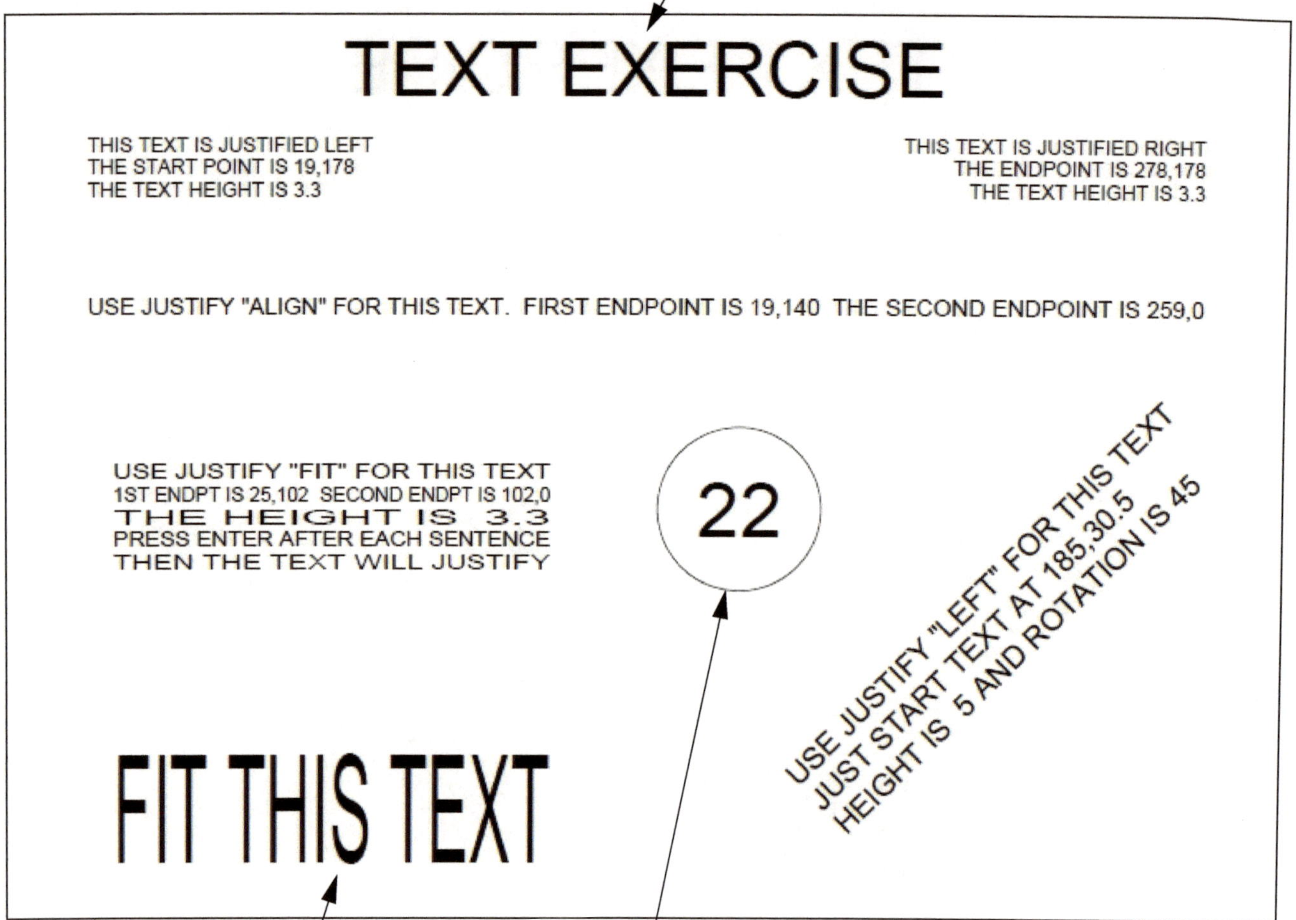

Use Justify **"Fit"**
First endpoint = **25,12.7**
Second endpoint = **102,0**
Height = **25**

Draw the circle first.
Center location = **171,95**
Radius = **19**
Text height = **12.7**
Use Justify **"Middle"** to place numbers in center of circle.

Exercise 8B

Instructions:

1. Start a **New** file using either the **inch-helper.dwt** or the **metric-helper.dwt**
2. Duplicate the text shown below using **Multiline Text**.
3. Use Layer Text.
4. Select **Multiline Text**.
5. Use Text Style: **Standard**.
6. Use font: **SansSerif**.
7. Text Height: **0.250" [6.35 mm]**
8. Follow the directions below. You may make changes to the settings as you type or you may enter all of the text and then go back and edit it. Your choice.
9. Enter all text shown below.
10. **Save** the drawing as: **Ex-8B**

The following is an exercise for Tabs, Indent, Bold and Underline.

1. This sentence should be indented 1" [25.4 mm].

    a. This sentence should be indented 1.5" [38.1 mm].

And now back to the left margin.

Isn't this fun?

STUDENT NAME	STUDENT ID	GRADE
Susie Que	1234567	A
John Smith	8910116	B

Set Tabs to **4.5" [114.3 mm] and 8" [203.2 mm]**
Clear all previous Tabs.

Exercise 8C

Instructions:

1. Start a **New** file using either the **inch-helper.dwt** or the **metric-helper.dwt**

   The following exercise is designed to teach you how to insert text into the exact middle of a rectangular area using **Single Line Text**.

2. Draw a **6" [152.4 mm]** wide by **3" [76.2 mm]** high rectangle.
3. Select **"Single Line Text"**.
4. Use Justify: **Middle**.
5. Use **"M2P"** Object Snap to locate the middle of the rectangle. (Refer to page 5-8.)
   A. Type ***m2p <Enter>*** on the Command Line.
   B. Using Object Snap **"Endpoint"**, snap to (**P1**) corner and the diagonal corner (**P2**).
6. Use Text Height: **1" [25.4 mm]**
7. Rotation: **0**
8. Type the word **Middle** and then press ***<Enter> <Enter>***.
9. **Save** the drawing as: **Ex-8C**

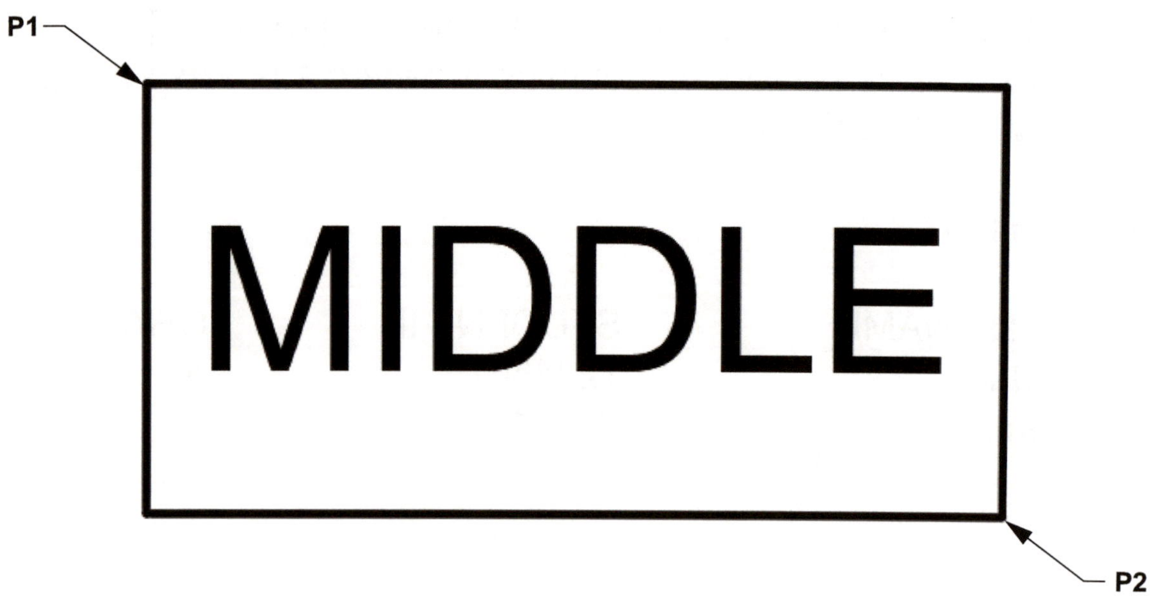

Exercise 8D

Instructions:

1. Start a **New** file using either the **inch-helper.dwt** or the **metric-helper.dwt**

   The following exercise is designed to teach you how to insert text into the exact middle of a rectangular area using **Multiline Text**.

2. Draw a **6" [152.4 mm]** wide by **3" [76.2 mm]** high rectangle.
3. Select **"Multiline Text"**.
4. Start the text boundary box at the upper left corner (**P1**) of the rectangle. Place the opposite corner at the lower right corner (**P2**) of the rectangle. (Use **"Endpoint"** Object Snap to be accurate.)
5. Select Text Style Standard: **1" [25.4 mm], Sans Serif** and **Middle Center**.
6. Type the word **Middle** and then select **Close Text Editor**.
7. **Save** the drawing as: **Ex-8D**

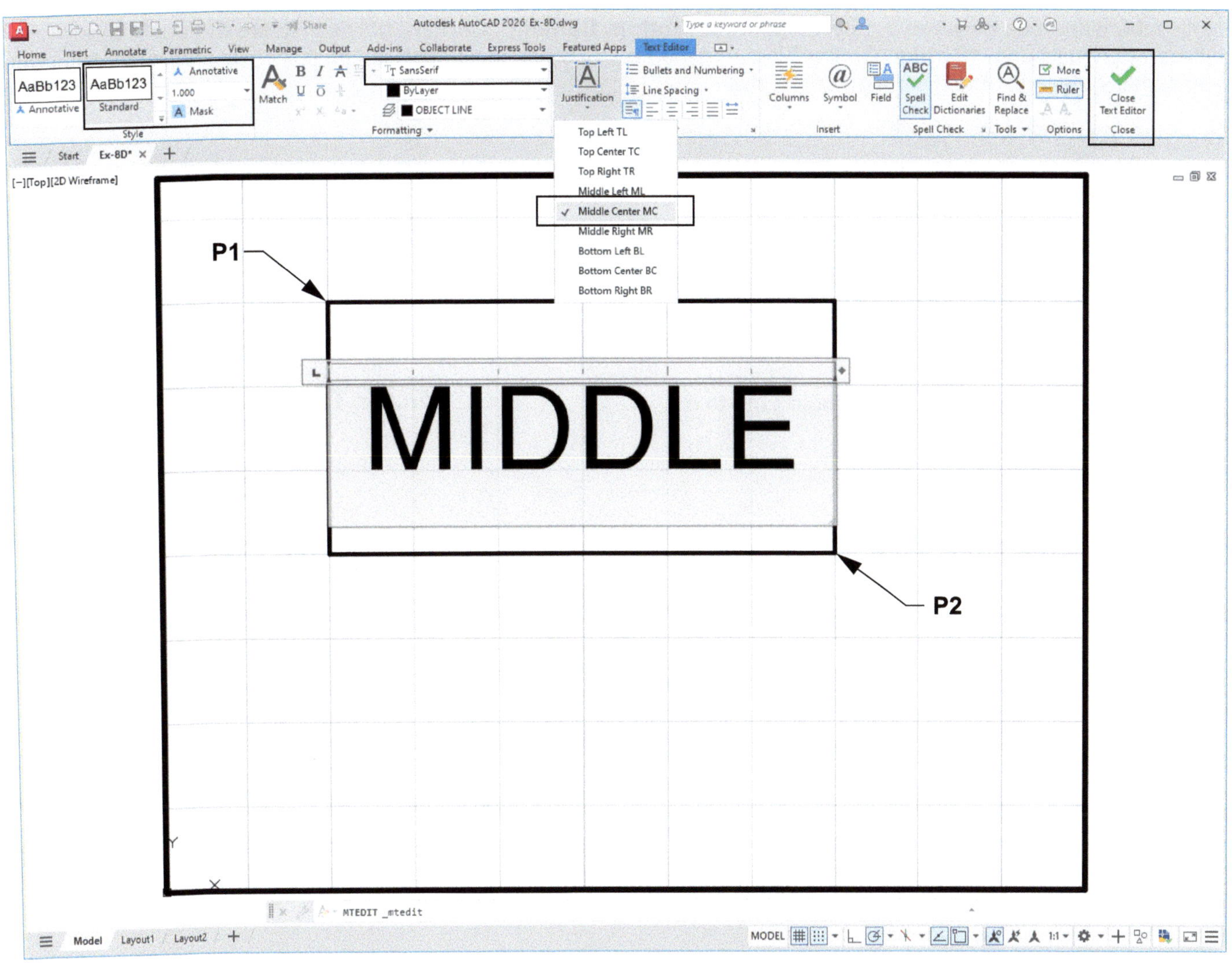

Exercise 8E

Instructions:

1. Start a **New** file using either the **inch-helper.dwt** or the **metric-helper.dwt**
2. Draw two **6" [152.4 mm]** long lines as shown.
3. Select **Single Line Text**.
   A. Select **Justify - Center**.
   B. Use "Midpoint" Snap to place the justification point at the midpoint of the line.
   C. Use text height **1" [25.4 mm]** and rotation angle **0**.
   D. Type the word **Happy** (use uppercase and lowercase) and then press ***<Enter> <Enter>***.

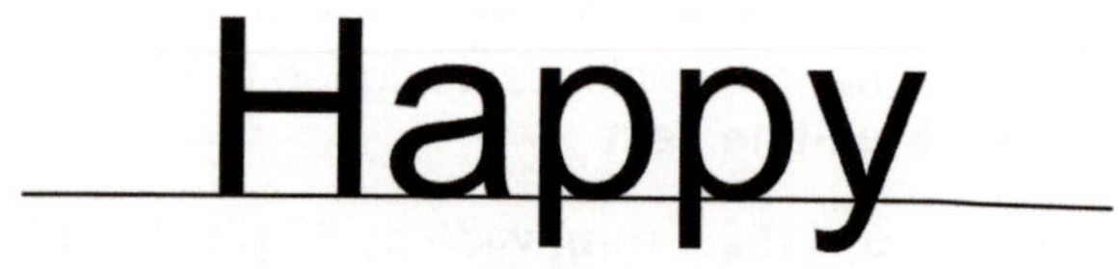

4. Select **Single Line Text** again and type the same word on the second line but this time use **Justify - BC** (Bottom Center).

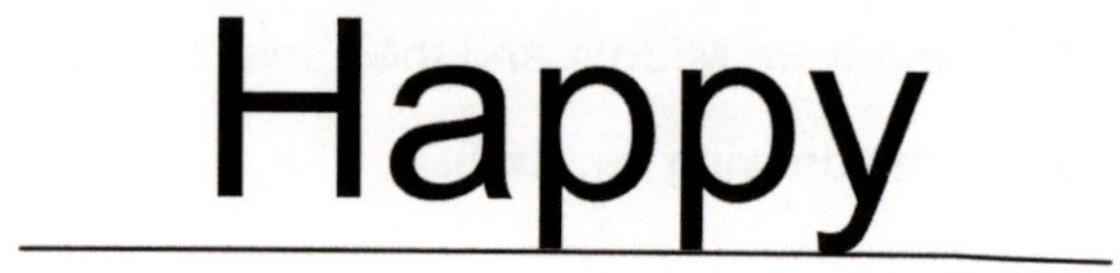

Notice the difference between Justify: **Center** and **Bottom Center**? **"Center"** only considers the uppercase letters when justifying. **"Bottom Center"** is concerned about those lowercase letters. Can you see how you could accidentally place your text too high or too low?

5. **Save** the drawing as: **Ex-8E**

Exercise 8F

Instructions:

1. Start a **New** file using either the **inch-helper.dwt** or the **metric-helper.dwt**
2. Draw the two sentences using **Single Line text**. (Use Layer Text.)

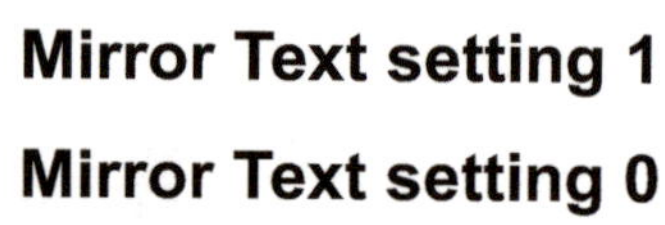

3. Change the **Mirrtext** setting:
   A. Type ***mirrtext <Enter>***
   B. Type ***1 <Enter>***
4. Using the **Mirror** command, mirror the first sentence using a vertical mirror line (**P1** and **P2**).

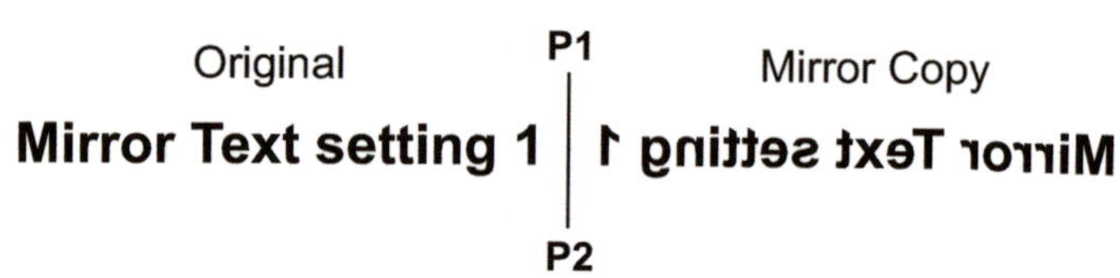

5. Now change the **Mirrtext** command setting to **0**
6. Mirror the **second** sentence.

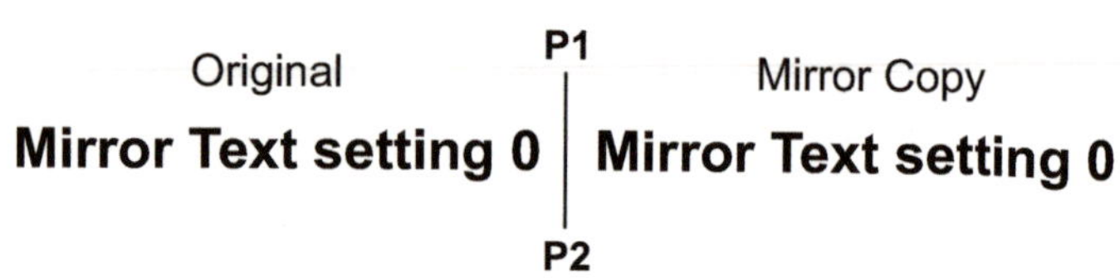

Notice the difference in the mirrored copy. Sometimes you will want the mirrored text to be reversed and sometimes you will not. Now you know how to control it.

7. **Save** the drawing as **Ex-8F**

LESSON 9

LEARNING OBJECTIVES

After completing this lesson, you will be able to:

- Understand the Origin
- Draw objects accurately using coordinate input
- Input Absolute and Relative coordinates
- Input using Direct Distance Entry
- Measure the Distance, Angles, and Areas of objects
- Identify a Location within the drawing
- Create your own Master Border Template
- Print from Model Space

Coordinate Input

In the previous lessons, you have been using the cursor to place objects. In this lesson, you will learn how to place objects in **specific locations** by entering coordinates. This process is called **coordinate input**. This is not difficult, so do not start to worry.

AutoCAD uses the **Cartesian Coordinate System**. The Cartesian Coordinate System has three axes, **X**, **Y**, and **Z**.

The **X** is the horizontal axis. (Right and Left.)
The **Y** is the vertical axis. (Up and Down.)
The **Z** is perpendicular to the X and Y plane.
(The **Z** axis is discussed in the *Advanced AutoCAD® Exercise Workbook* and *AutoCAD® 3D Modeling Exercise Workbook*.)

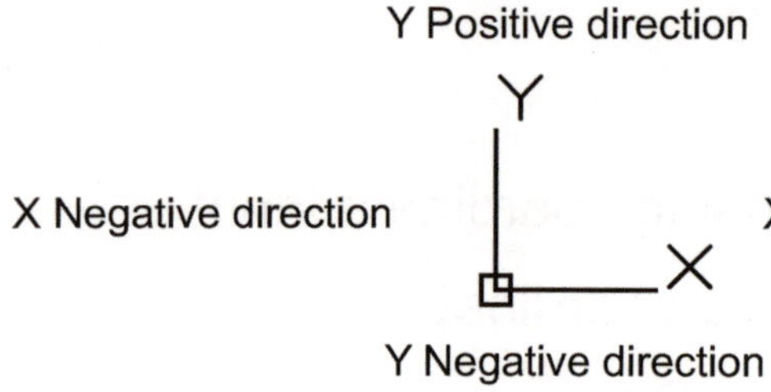

Look at the User Coordinate System (**UCS**) icon in the lower left corner of your screen. The X and Y are pointing in the positive direction.

The location where the X, Y, and Z axes intersect is called the **Origin**. The Origin always has a coordinate value of **X = 0**, **Y = 0**, and **Z = 0 (0,0,0)**.

When you move the **cursor** away from the Origin, in the positive direction, the X and Y coordinates are positive. When you move the **cursor** in the opposite direction, the X and Y coordinates are negative.

Using this system, every point on the screen can be specified using positive or negative X and Y coordinates.

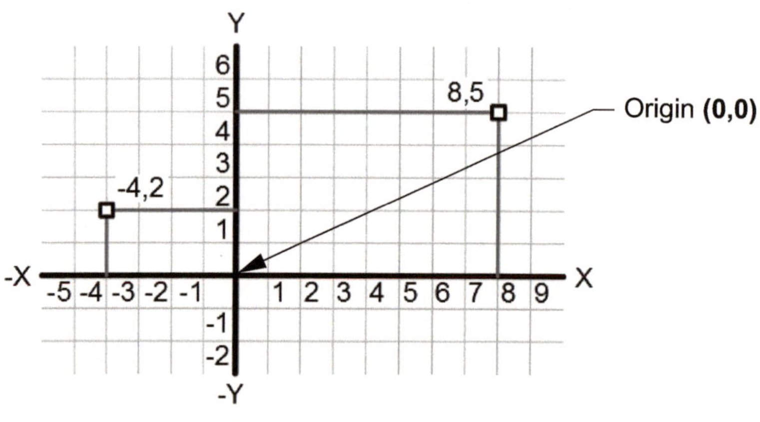

Very important: While working on Lessons 9 and 10 it is best to **turn off** Dynamic Input.

The Dynamic Input button is not displayed by default. To display this button in the Status Bar and to turn it off, refer to the instructions on pages 1-12 and 1-13.

Note that when Dynamic Input is switched off, the button is gray:

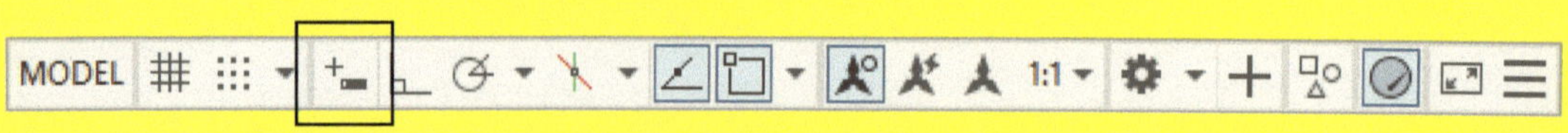

For more information on Dynamic Input, refer to Lesson 11.

There are three types of Coordinate Input: **absolute**, **relative**, and **Polar**. (Absolute and relative coordinates are discussed in this lesson, and Polar coordinates will be discussed in Lesson 11.)

Absolute Coordinates

When inputting absolute coordinates, the input format is: **X,Y** (that is: X **comma** Y). Absolute coordinates come **from the Origin** and are typed as follows: **8,5**

The first number (8) represents the **X axis** (horizontal) distance **from the Origin**, and the second number (5) represents the **Y axis** (vertical) distance **from the Origin**. The two numbers must be separated by a **comma**.

An absolute coordinate of **4,2** will be **4** units to the **right** (horizontal) and **2** units **up** (vertical) **from the current location of the Origin**.

An absolute coordinate of **-4, -2** will be **4** units to the **left** (horizontal) and **2** units **down** (vertical) **from the current location of the Origin**.

The following are examples of absolute coordinate input. Notice where the **Origin** is located in each example.

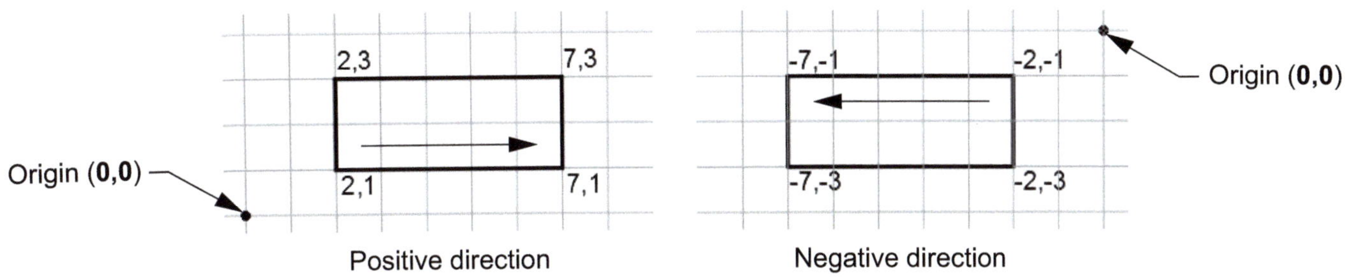

Relative Coordinates

Relative coordinates come **from the last point entered**. (Not from the Origin.)

The first number represents the **X axis** (horizontal) and the second number represents the **Y axis** (vertical) just like the absolute coordinates.

To distinguish the relative coordinates from absolute coordinates, the two numbers must be preceded by an **@** symbol in addition to being separated by a **comma**.

A relative coordinate of **@5,2** will go to the **right 5** units and **up 2** units from the last point entered.

A relative coordinate of **@-5, -2** will go to the **left 5** units and **down 2** units from the last point entered.

The following is an example of relative coordinate input.

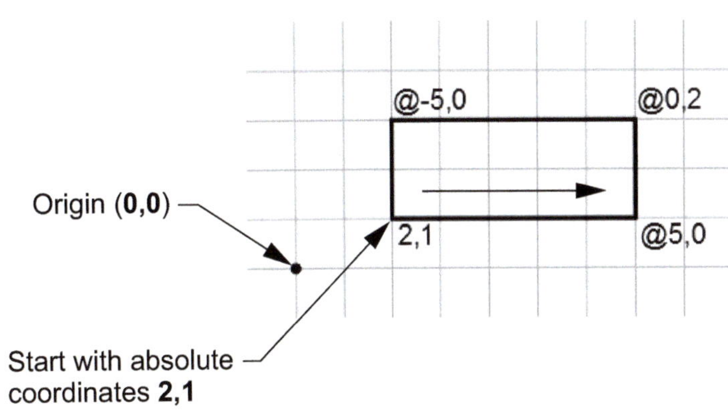

Examples of Coordinate Input

Scenario 1.

If you want to draw a line with the first endpoint "at the Origin" and the second endpoint 3 units in the positive X direction:

1. Select the **Line** command.
2. You are prompted for the first endpoint: Type ***0,0 <Enter>***.
3. You are then prompted for the second endpoint: Type ***3,0 <Enter>***.

What did you do?

The first endpoint coordinate input, **0,0** means that you do not want to move away from the Origin. You want to start **"on"** the Origin.

The second endpoint coordinate input, **3,0** means that you want to move **3** units in the positive X axis. The **"0"** means you do not want to move in the Y axis. So the line will be exactly horizontal.

Scenario 2.

If you want to start a line **1** unit to the right of the origin and **1** unit above and the line will be **4** units in length, perfectly vertical:

1. Select the **Line** command.
2. You are prompted for the first endpoint: Type ***1,1 <Enter>***.
3. You are prompted for the second endpoint: Type ***@0,4 <Enter>***.

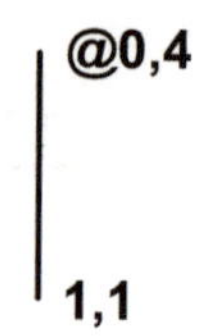

What did you do?

The first endpoint coordinate input, **1,1** means you want to move **1** unit in the X axis direction and **1** unit in the Y axis direction.

The second endpoint coordinate input **@0,4** means you do not want to move in the X axis "from the last point entered". But you do want to move in the Y axis "from the last point entered". (Remember the @ symbol is only necessary if you are not using **Dynamic Input**.)

Scenario 3.

Now try drawing five connecting line segments. (Watch for the negatives.)

1. Select the **Line** command.
2. First endpoint: ***2,4 <Enter>***.
3. Second endpoint: ***@ 2, –3 <Enter>***.
4. Second endpoint: ***@ 0, –1 <Enter>***.
5. Second endpoint: ***@ –1,0 <Enter>***.
6. Second endpoint: ***@ –2,2 <Enter>***.
7. Second endpoint: ***@ 0,2 <Enter> <Enter>***.

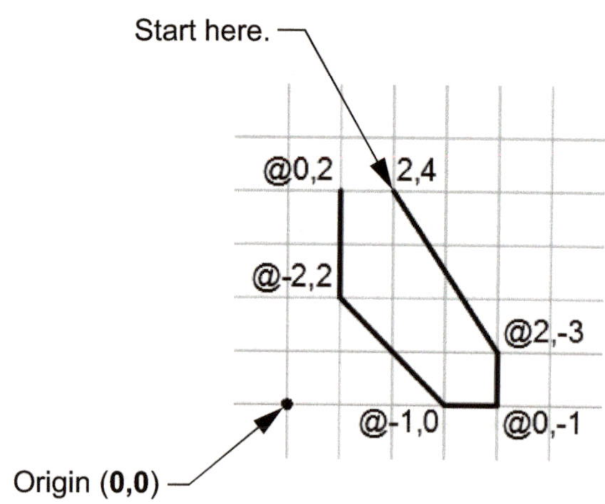

CAD TIP! If you enter an incorrect coordinate, just hold down the ***<Shift>*** key and press ***U*** then ***<Enter>***. The last segment will disappear, and you will have another chance at entering the correct coordinate.

Direct Distance Entry (DDE)

Direct Distance Entry is a combination of keyboard entry and cursor movement. **DDE** is used to specify distances in the horizontal or vertical axes from the **last point entered**. DDE is a **Relative Input**. Since it is used for horizontal and vertical movements, **Orthomode** must be **on**.

Note: To specify distances on an angle, refer to **Polar Coordinate Input** in Lesson 11.

Using DDE is simple. Just move the cursor and type the distance. Negative and positive is understood automatically by moving the cursor up (positive), down (negative), right (positive) or left (negative) from the last point entered. No **minus** or **@** sign necessary.

Moving the cursor to the right and typing ***5 <Enter>*** tells AutoCAD that the **5** is positive and horizontal.

Moving the cursor to the left and typing ***5 <Enter>*** tells AutoCAD that the **5** is negative and horizontal.

Moving the cursor up and typing ***5 <Enter>*** tells AutoCAD that the **5** is positive and vertical.

Moving the cursor down and typing ***5 <Enter>*** tells AutoCAD that the **5** is negative and vertical.

Example:

1. **Orthomode** must be **on**. Grid **off**.
2. Select the **Line** command.
3. Type: ***1,2 <Enter>*** to enter the first endpoint using absolute coordinates.
4. Now move your cursor to the right and type: ***5 <Enter>***.
5. Now move your cursor up and type: ***4 <Enter>***.
6. Now move your cursor to the left and type: ***5 <Enter> <Enter>*** (the second ***<Enter>*** ends the Line command).

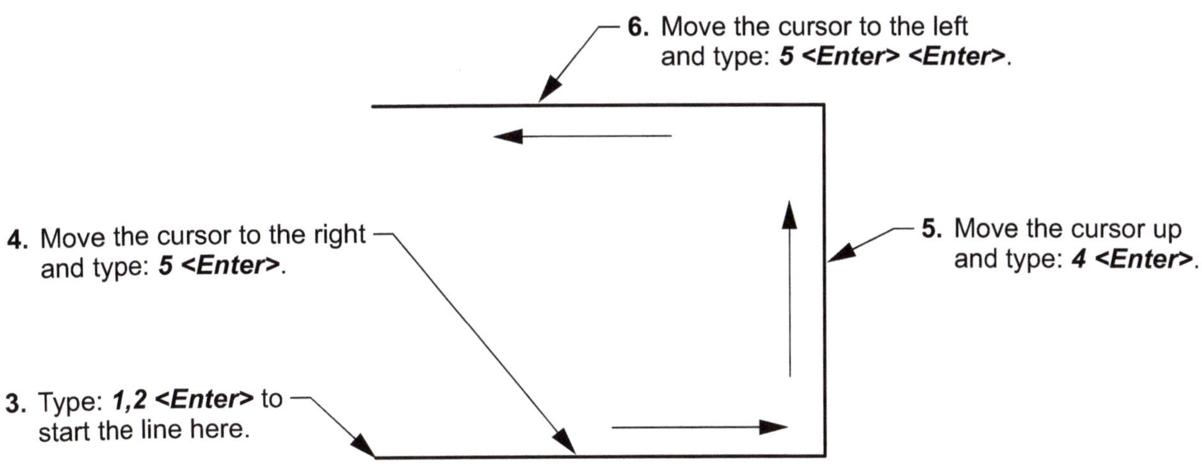

Measure Tools

The following tools are very useful to confirm the location or size of objects.

The **Measure** tools enables you to measure the **Distance**, **Radius**, **Angle**, **Area**, or **Volume** of a selected object. You may wish to use the individual measuring tools or the **Quick Measure** tool, which allows you to measure the dimensions, distances, angles, radii, and areas of objects in a 2-dimensional drawing. The default option is **Quick**.

Continued on the next page...

1. Select the **Measure** tools using one of the following:

Ribbon = Home Tab / Utilities Panel / Measure ▾
or
Keyboard = *MEA <Enter>*

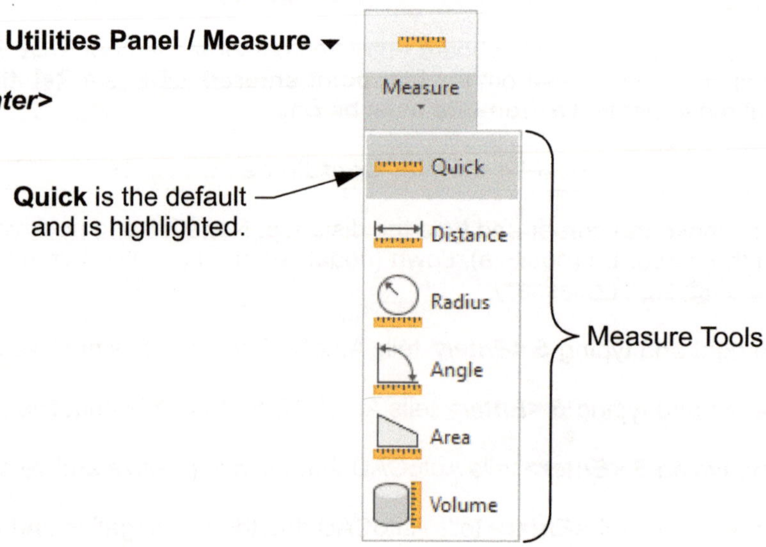

2. Select one of the tools and follow the instructions on the Command Line.

Example using the Quick Measure tool:

1. Create a right-angled triangle as shown.

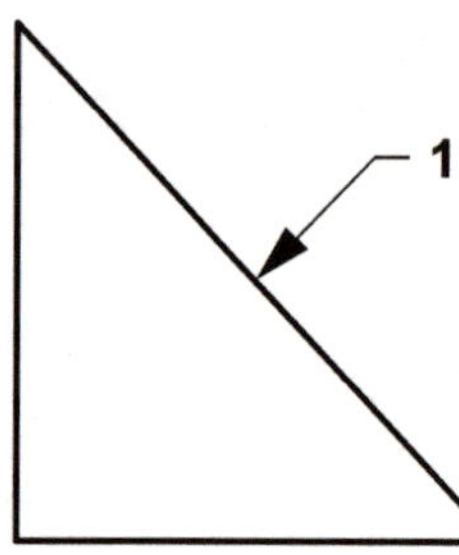

2. Select the **Quick** option of the Measure tools.
3. Place the yellow crosshair cursor inside the triangle.
4. The dimensions are displayed for the three sides of the triangle along with two angles. The yellow square in the bottom left corner represents a 90 degree angle (right angle).

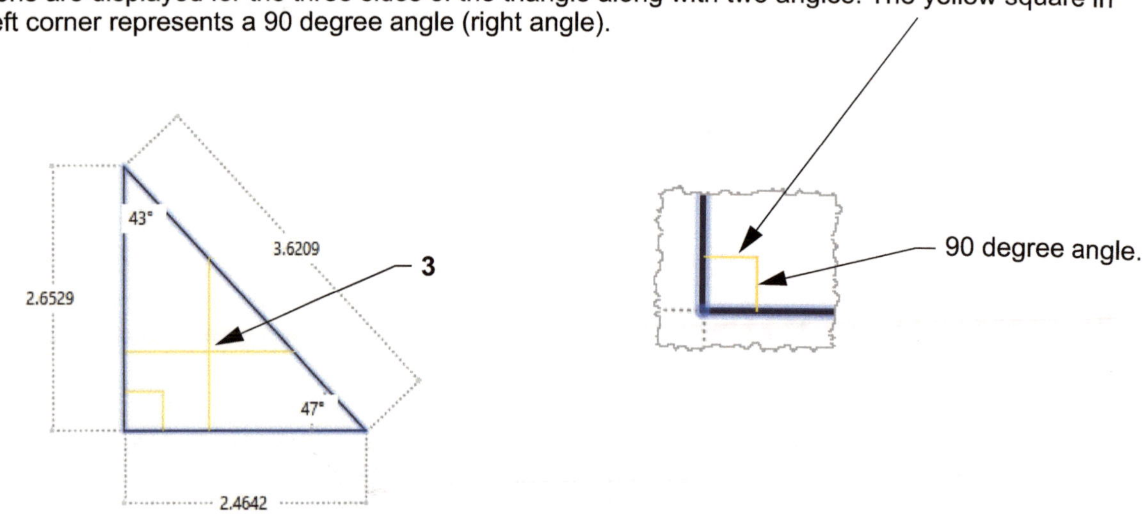

5. To close the **Quick Measure** option of the Measure tools, press the ***<Esc>*** key.

The **Quick Measure** option of the Measure tools can also measure the distance between two objects or the radius of a corner.

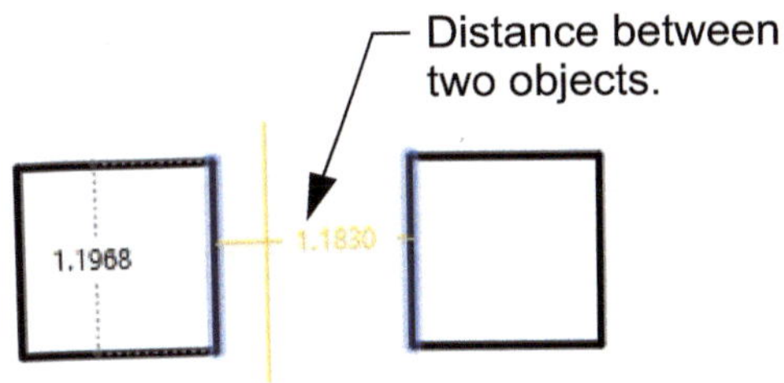

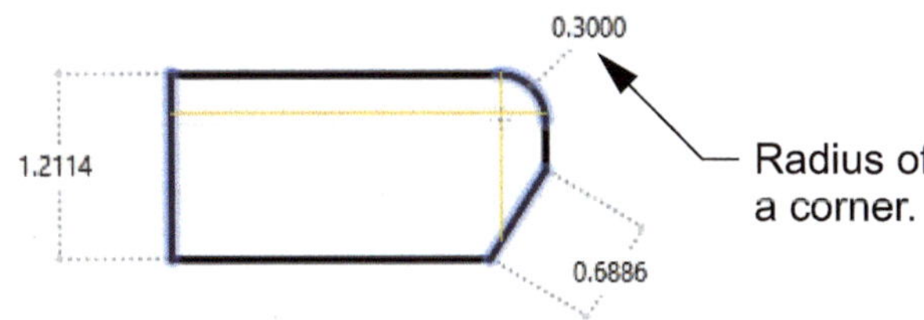

The **Quick Measure** option of the Measure tools can also measure the **Area** and **Perimeter** of a closed object.

How to measure the Area and Perimeter of a closed object using Quick Measure

1. Select the **Quick** option of the Measure tools.
2. Place the yellow crosshair cursor inside the closed object and then left click.
3. The closed object will highlight green and the Area and Perimeter dimensions will be displayed in a rectangular box.

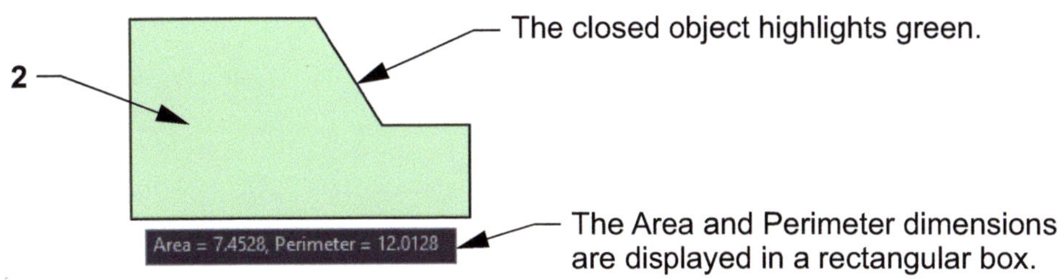

You may also select multiple closed objects to obtain the **combined** Area and Perimeter of those objects.

How to measure the Area and Perimeter of multiple closed objects

1. Select the **Quick** option of the Measure tools.
2. Place the yellow crosshair cursor inside the first closed object and then left click.
3. Hold down the ***<Shift>*** key and then left click inside the next closed object.
4. The closed objects will highlight green and the **combined** Area and Perimeter dimensions will be displayed in a rectangular box.

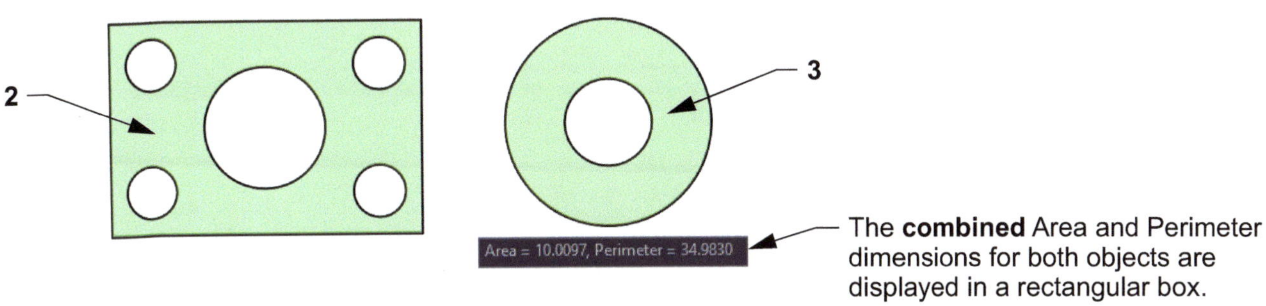

CAD TIP! To improve performance or to prevent any confusion when using the **Quick** option of the Measure tools, zoom into complex areas or objects.

ID Point

The **ID Point** command will list the **X**, **Y**, and **Z** coordinates of the point that you select. The coordinates listed will be from the **Origin**.

1. Select the **ID Point** command using one of the following:

   Ribbon = Home Tab / Utilities Panel / ▾
   or
   Keyboard = *ID <Enter>*

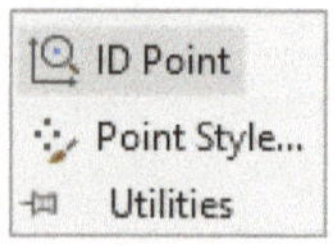

   The following will appear on the Command Line:

2. Select a location point, such as the center of a circle.

3. The **X**, **Y**, and **Z** location coordinates for the center of the circle will be displayed above the Command Line and in the Dynamic Input box.

Example using the ID Point tool:

1. Select the **ID Point** tool.

2. Snap to the center of the circle.

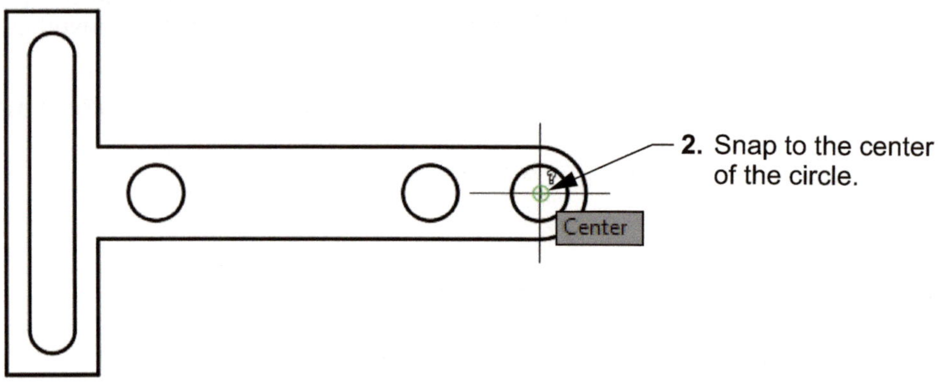

3. Coordinates, from the **Origin**, are displayed above the Command Line and in the Dynamic Input box.

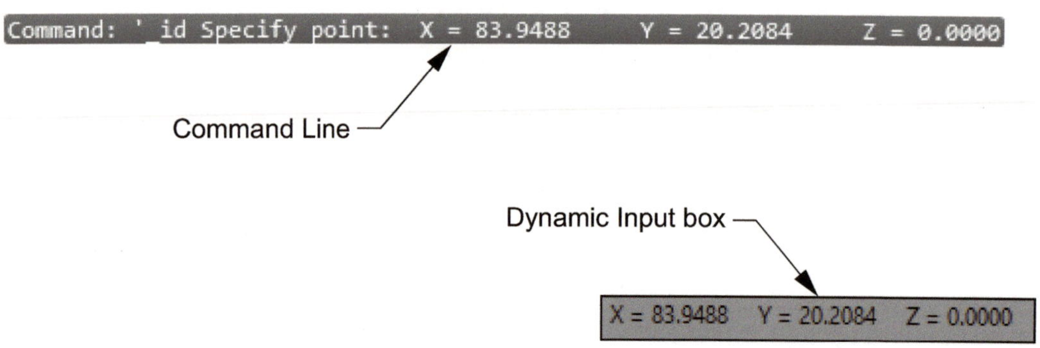

Exercise 9A: Creating a Border

In this exercise, you will create a master template named **"Border A-Inch"** for inch users, or **"Border A-Metric"** for metric users. This template will be used for the exercises in Lessons 9 through 25.

Instructions:

Follow the easy steps below. **Do not skip any.**

Step 1. Select the Settings

1. Start a **New** file using either the **inch-helper.dwt** or the **metric-helper.dwt**
2. Set **Units** (Refer to page 4-13.)
   Units = Decimal
   Precision = **0.000**
3. Set Drawing Limits (Refer to page 4-9.)
   Lower Left Corner = **0,0**
   Upper Right Corner = **11,8.5** for **inch** users. **297,210** for **metric** users.
4. Show the new limits using: (Refer to page 4-7.)
   Zoom / All
5. **Important:** Change Lineweight settings to **Inches** for inch users and to **Millimeters** for metric users. Then **Adjust Display Scale**. (Refer to the bottom of page 3-10.)
6. Set **Grids** and **Snap** (Refer to page 3-5, 4-11, and 4-12.)
   Snap = **0.125" [5 mm]** (This is important, it will make it easier to draw the lines.)
   Grids = **1.000" [25 mm]**

Step 2. Draw the Border Lines

1. Select Layer Borderline.
2. Draw the border below using the dimensions shown.

Note: When complete, your **Border A-Inch** or **Border A-Metric** template will have an all around border of **3/8"** for inch users, and a border of **10 mm** for metric users.

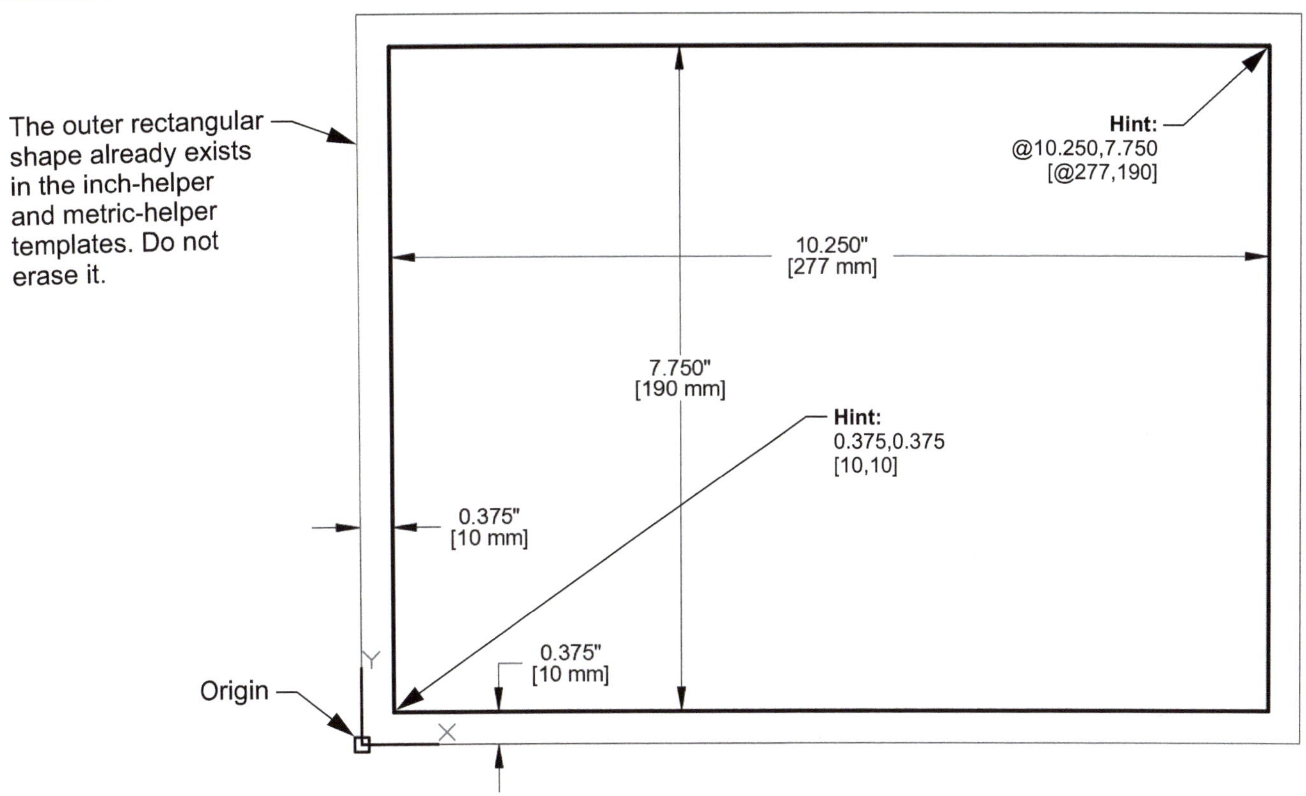

Continued on the next page...

Step 3. Draw the Title Block Lines

1. Select Layer Borderline.
2. Draw the three **Title Block** lines as shown below using the dimensions shown.

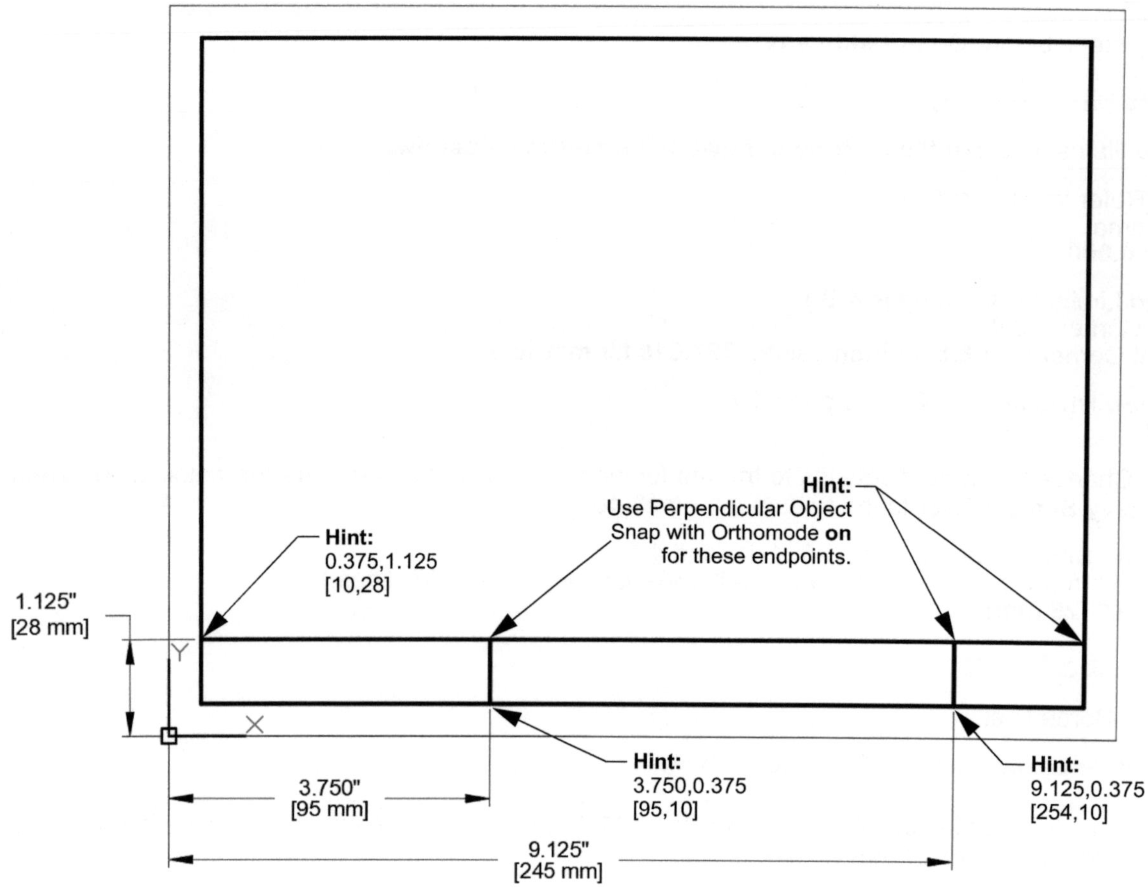

Step 4. Enter the Text

1. Select Layer Text.
2. Select the **Multiline Text** command.
3. Select the upper left corner and then the opposite corner.

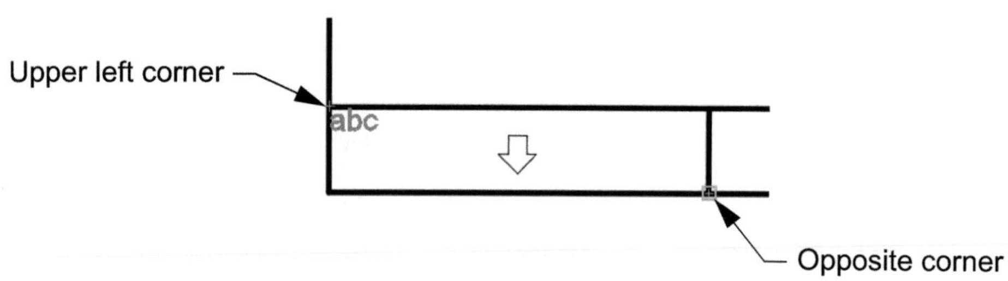

4. Change the text size to **0.250" [6.5 mm]**
5. Select **Bold**.
6. Select justification **Middle Center**.
7. Type the text shown on the next page.
8. Repeat the above for the remaining two title boxes.

Continued on the next page...

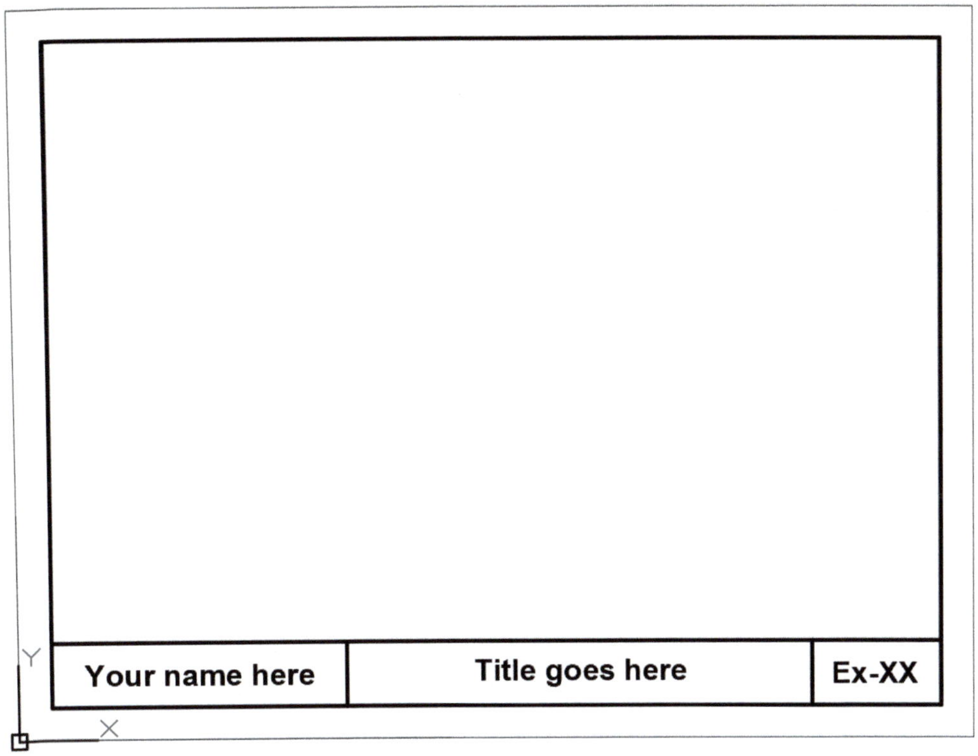

Step 5. Save the border as a Template

1. Select the **Application Menu**. (Refer to page 2-3.)
2. Select **Save As** ▸
3. Select **Drawing Template**.
4. Enter the new template name: **Border A-Inch** for inch users, or **Border A-Metric** for metric users.
5. Select **Save** button.
6. Enter the description: **Use for Lessons 9 through 25**
7. Select the **OK** button.

You now have a template to use for Lessons 9 through 25. At the beginning of each exercise, you will be instructed to start a **New** drawing using **Border A-Inch.dwt** for inch users, or **Border A-Metric.dwt** for metric users.

You will **edit** the **Title** and the **Ex-XX** to match the exercise. Editing makes it much easier. You will not need to change the location of the text, just edit it.

If you would like to print this border, follow the steps on the following pages.

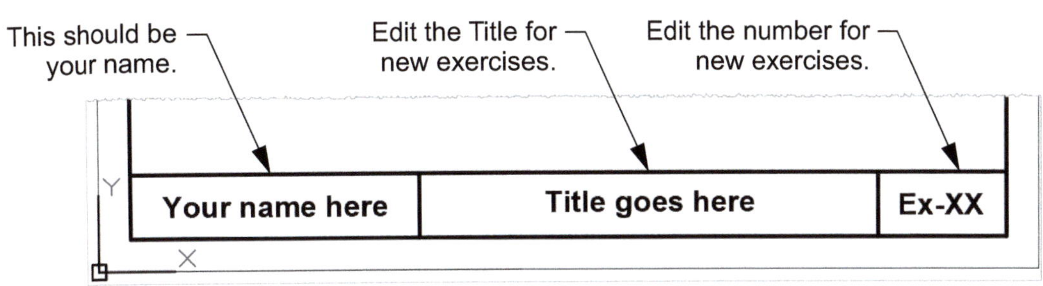

Basic Plotting from Model Space

Note: More Advance plotting methods will be explained in Lessons 26 and 27.

1. **Important:** Open the drawing you want to plot, if it is not already open.
2. Select: **Zoom / All** to center the drawing within the plot area.
3. Select the **Plot** command using one of the following:

Quick Access Toolbar = 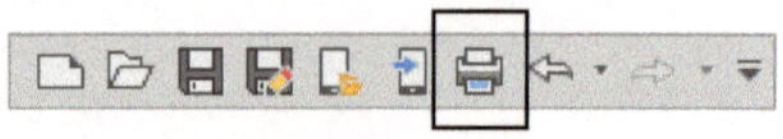
or
Ribbon = Output Tab / Plot Panel / 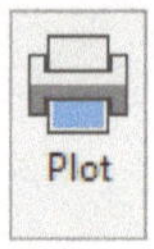

or
Keyboard = *PLOT <Enter>*

The **Plot - Model** dialog box will appear.

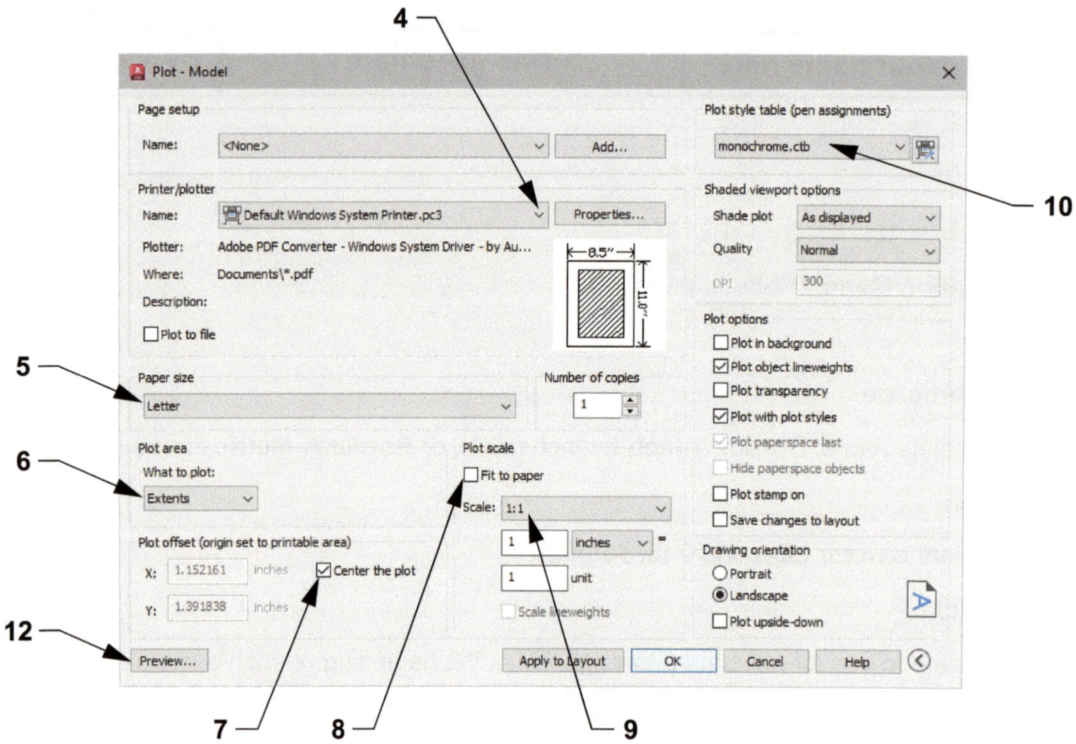

4. Select your printer from the drop-down list or use the settings for "Default Windows System Printer.pc3". **Note:** If your printer is not shown in the list, you should configure your printer within AutoCAD. This is not difficult. (Refer to Appendix-A for step-by-step instructions.)
5. Select the Paper size: **Letter** for inch users or **A4** for metric users.
6. Select the Plot Area: **Extents**
7. Select the Plot Offset: **Center the plot**
8. Uncheck the **Fit to paper** box
9. Select Plot Scale **1:1**
10. Select the Plot Style table: **monochrome.ctb** (for all black)
    acad.ctb (for color)

Continued on the next page...

11. If the following dialog box appears, select **Yes**

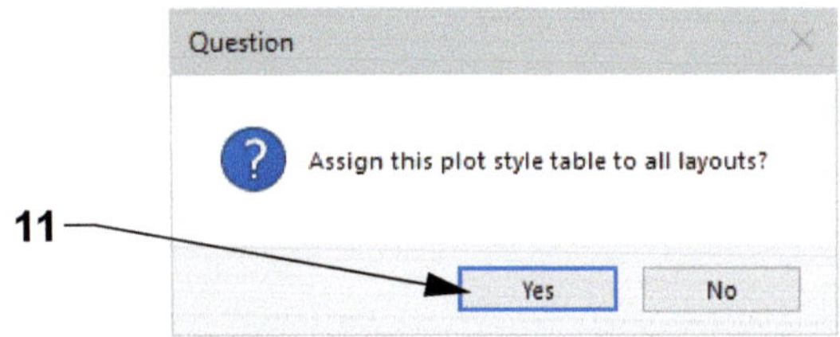

12. Select the **Preview...** button.

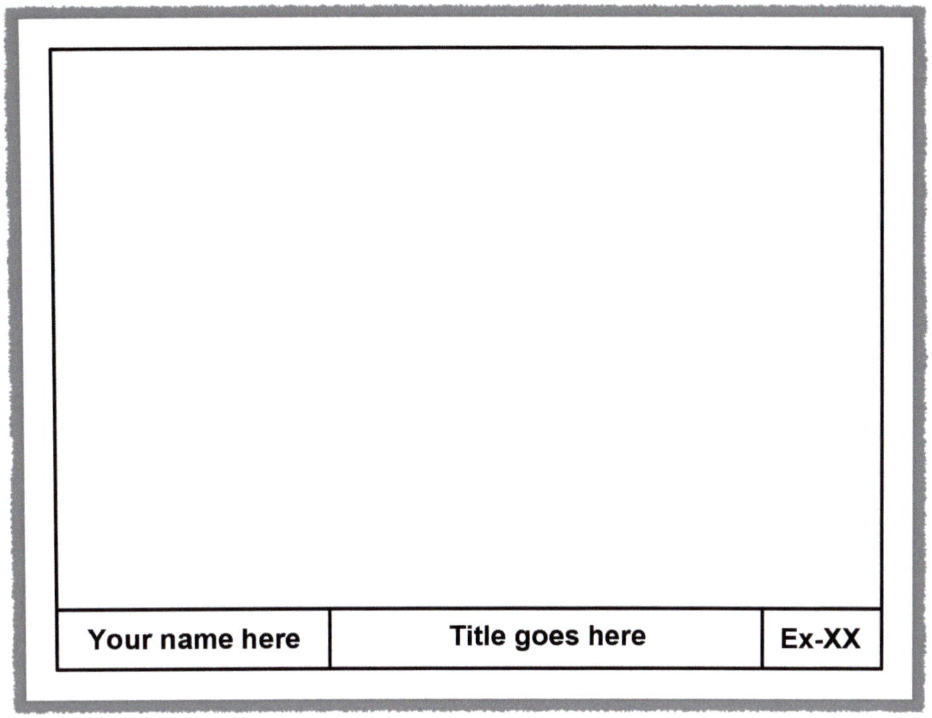

Does your display appear as shown above? If yes, press ***<Enter>*** and proceed to Step 13. If not, go back and review Steps 1 through 11.

You have just created a **Page Setup**. All of the settings you have selected can now be saved. You will be able to recall these settings for future plots using this Page Setup. To save the Page Setup you need to **Add** it to the Model Tab within this drawing.

13. Select the **Add...** button.

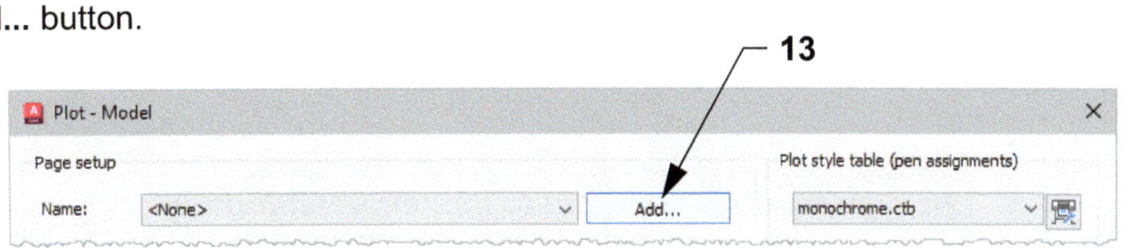

14. Type the new Page Setup name **Class Model A**

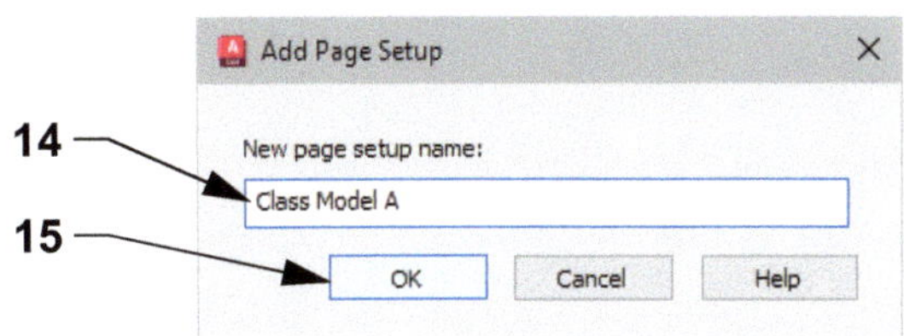

15. Select the **OK** button.

Continued on the next page...

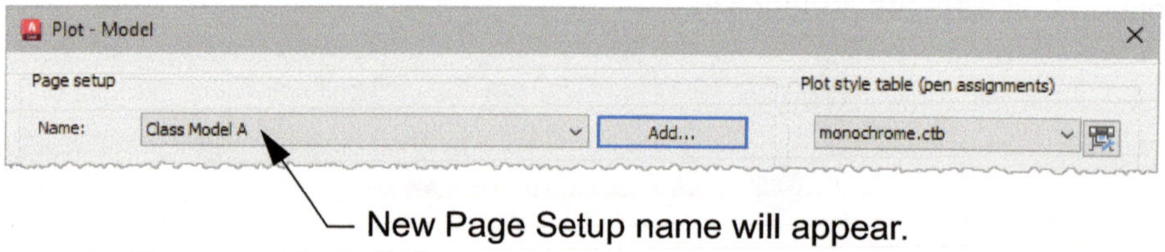

16. Select the **Apply to Layout** button.

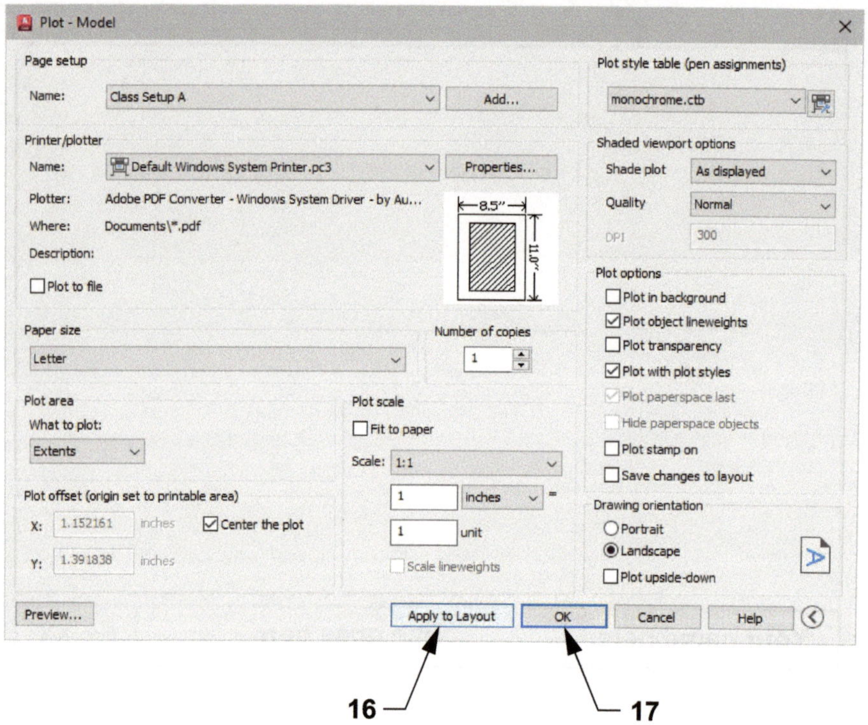

17. Select the **OK** button to send the drawing to the printer or select **Cancel** if you do not want to print the drawing at this time. The Page Setup will still be saved.

18. Save the entire drawing again as a template using **Save as – Drawing Template**: **Border A-Inch** for inch users, or **Border A-Metric** for metric users. (Refer to page 9-11.)

    The Page Setup will be saved to the drawing and available to select in the future. You will not have to select all the individual settings again unless you choose to change them.

Now go tape your printout to the refrigerator door. This is quite an accomplishment.

Exercise 9B

Instructions:

1. Start a **New** file using the **Border A-Inch.dwt** or the **Border A-Metric.dwt**
2. Draw the objects below using **absolute** and **relative** coordinates.
3. Use Layer Object Line.
4. Edit the **Title** and **Ex-XX** by double clicking on the text. Do not erase and replace.
5. Do not dimension (we will get to that soon).
6. **Save** as **Ex-9B**
7. Plot using Page Setup **Class Model A**. (Refer to pages 9-12 through 9-14.)

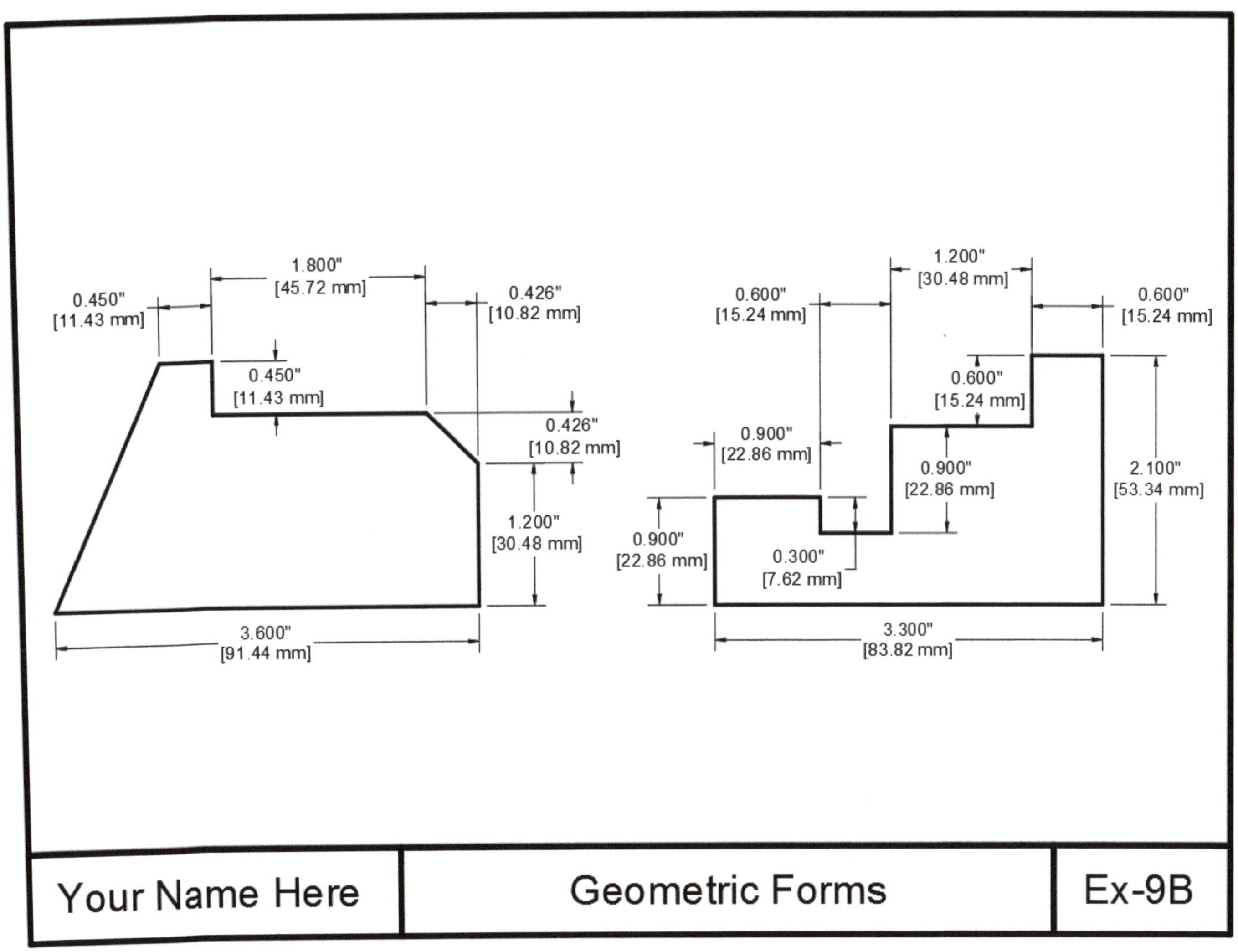

Exercise 9C

Instructions:

1. Start a **New** file using the **Border A-Inch.dwt** or the **Border A-Metric.dwt**
2. Draw the objects below using **Direct Distance Entry** (DDE).
3. Use Layer Object Line.
4. Edit the **Title** and **Ex-XX** by double clicking on the text. Do not erase and replace.
5. Do not dimension (we will get to that soon).
6. **Save** as **Ex-9C**
7. Plot using Page Setup **Class Model A**. (Refer to pages 9-12 through 9-14.)

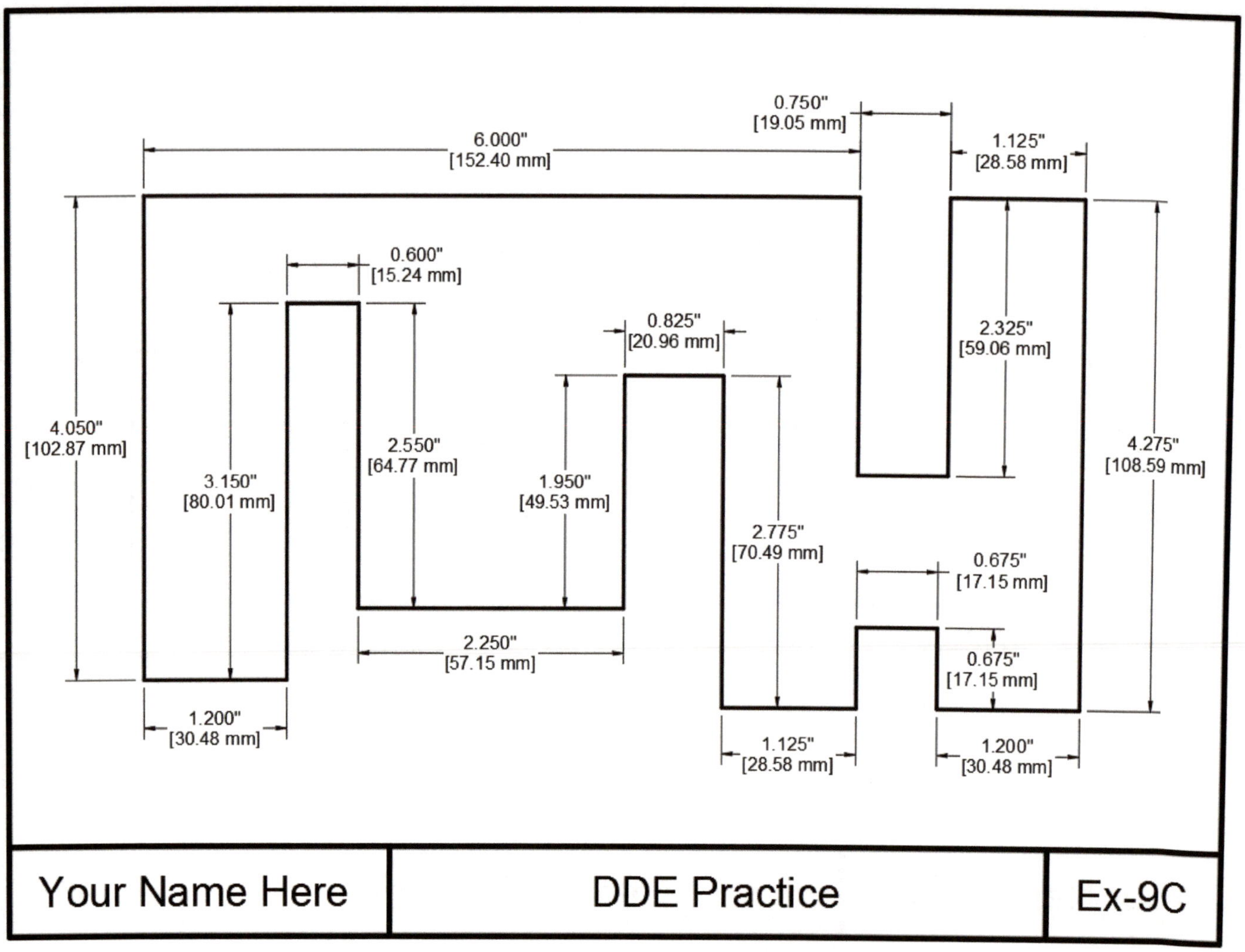

LESSON 10

LEARNING OBJECTIVES

After completing this lesson, you will be able to:

- Move the Origin
- Control the display of the UCS icon

Moving the Origin

As previously stated in Lesson 9, the **Origin** is where the X, Y, and Z-axes intersect. The Origin's (0,0,0) default location is in the lower left corner of the drawing area. But you can move the Origin anywhere on the screen using the UCS command. (The default location is designated as the **"World"** option or **WCS**, World Coordinate System. When it is moved, it is UCS, User Coordinate System.)

You may move the Origin many times while creating a drawing. This is not difficult and will make it much easier to draw objects in specific locations.

Refer to the examples on page 10-3.

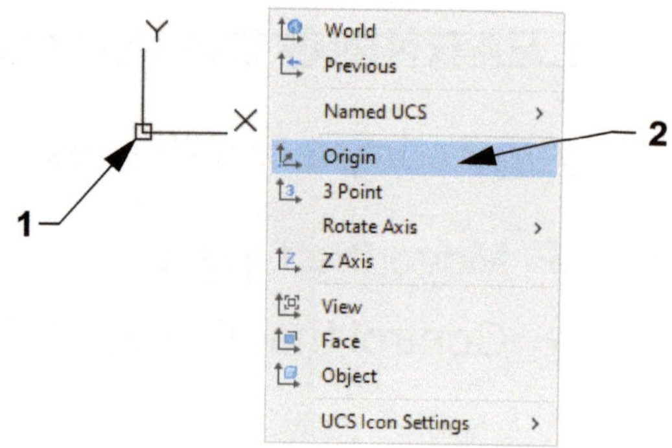

How to move the Origin

1. Right click on the **Origin** icon.
2. Select **Origin** from the Shortcut Menu.
3. Place the new Origin location by entering coordinates or pressing the left mouse button.

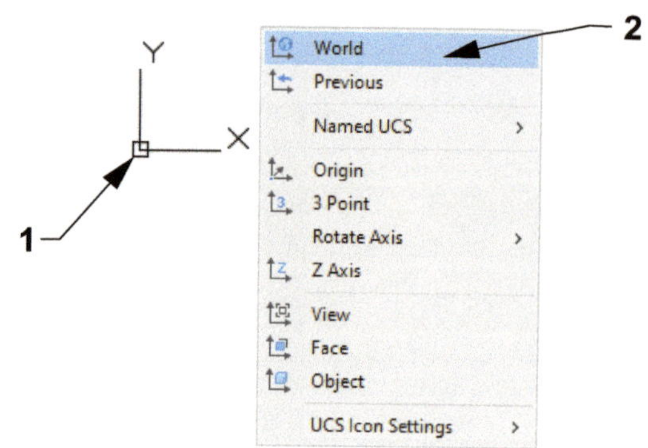

How to return the Origin

To the default "World" location (the lower left corner)

1. Right click on the **Origin** icon.
2. Select **World** from the Shortcut Menu.

CAD TIP! You can quickly return to the default "World" or WCS location by typing in ***UCS <Enter>*** and then ***W <Enter>***.

There is also another method of moving the UCS Origin to a new location.

Another way to move the Origin

1. Left click on the **Origin** icon.
2. Hover your mouse over the blue square base grip until it turns red. A Shortcut Menu will appear.

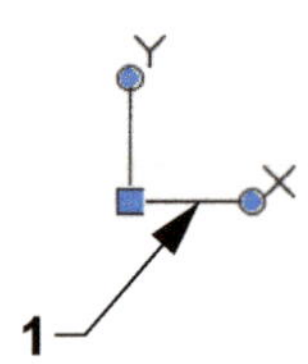

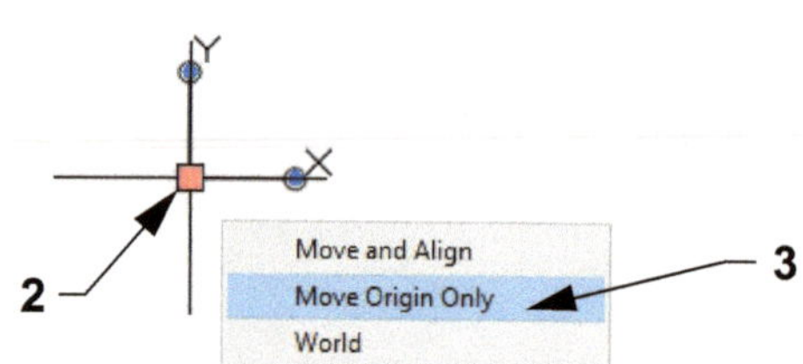

3. Select **Move Origin Only** from the Shortcut Menu.
4. Place the new Origin location by entering coordinates or pressing the left mouse button.

Continued on the next page...

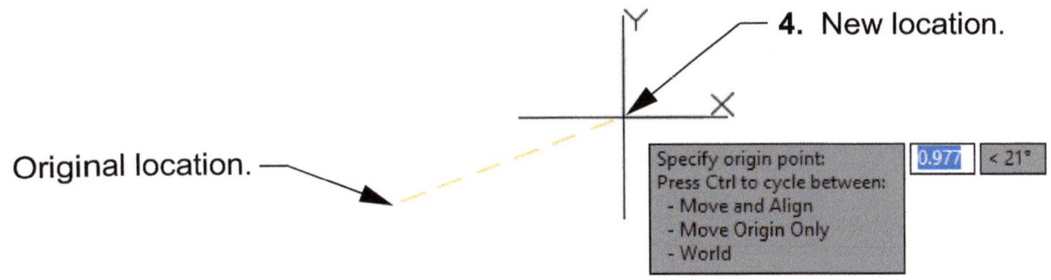

5. Press the ***<Esc>*** key to deselect the Origin icon.

CAD TIP! Press the ***<Ctrl>*** key to cycle between **Move and Align**, **Move Origin Only**, and World.

Moving the Origin to accurately place a Circle

If you move the Origin to the lower left corner of the rectangle, it will make it very easy to accurately place the center of a circle.

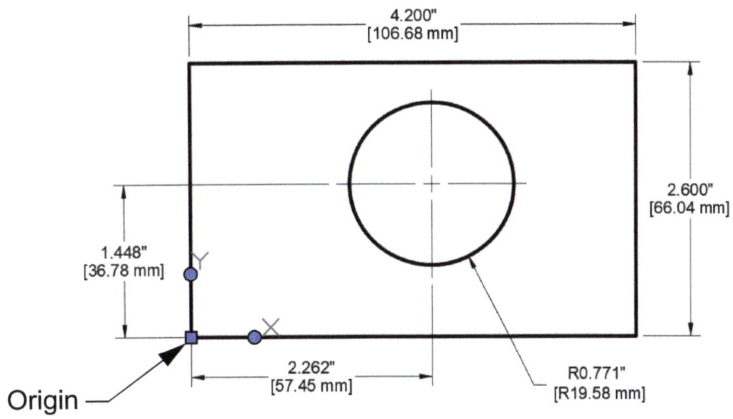

How to place the Circle accurately

1. Select **"Origin"** from the Shortcut Menu as shown on the previous page.
2. Snap to the lower left corner of the rectangle using Object Snap Endpoint. The UCS icon should now be displayed as shown above.
3. Select the **Circle** command.
4. Enter the coordinates to the center of the circle: ***2.262,1.448 <Enter>*** for inch users, or ***57.45,36.78 <Enter>*** for metric users.
5. Enter the radius: ***0.771 <Enter>*** for inch users, or ***19.58 <Enter>*** for metric users.
6. Select **"World"** from the Shortcut Menu to return the UCS icon to its default location. (Refer to page 10-2.)

More examples of why you would move the Origin

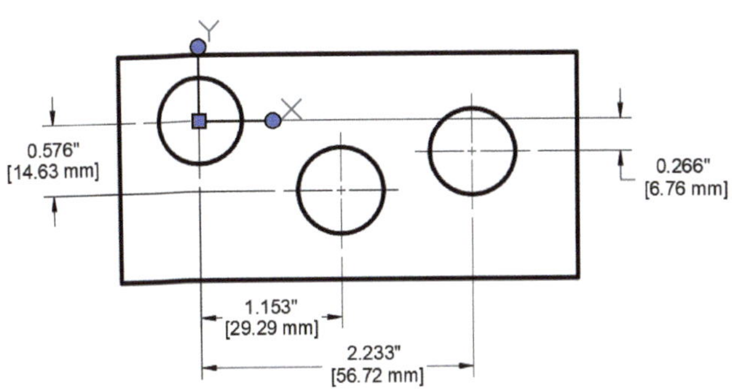

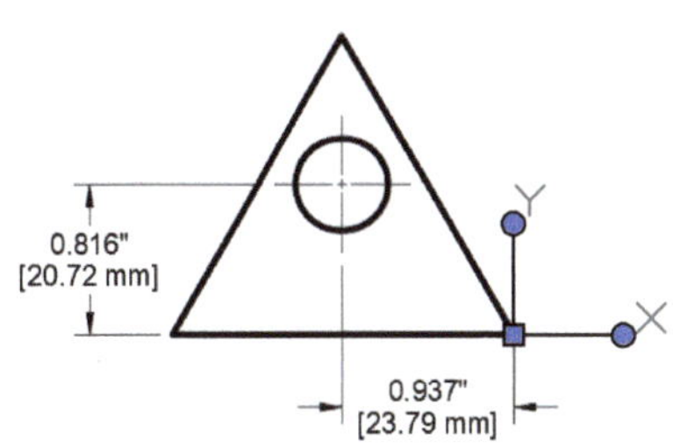

Displaying the UCS Icon

The **UCS icon** is merely a drawing aid that displays the location of the **Origin**. It can move with the Origin or stay in the default location. You can even change its appearance.

How to Show the UCS Icon at Origin

1. Right click on the Origin icon.
2. Select **UCS Icon Settings** from the Shortcut Menu.

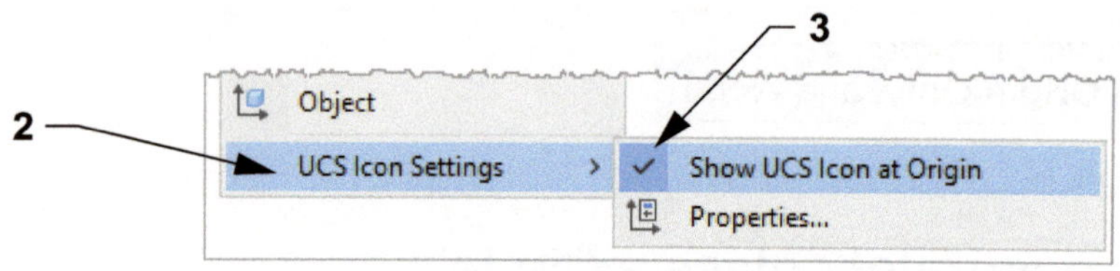

3. **Show UCS Icon at Origin**

☑ The UCS icon will follow the Origin as you move it. You will be able to see the Origin location at a glance.

☐ The UCS icon will not follow the Origin. It will stay in its default location.

How to change the UCS icon appearance

1. Right click on the Origin icon.
2. Select **UCS Icon Settings** from the Shortcut Menu.

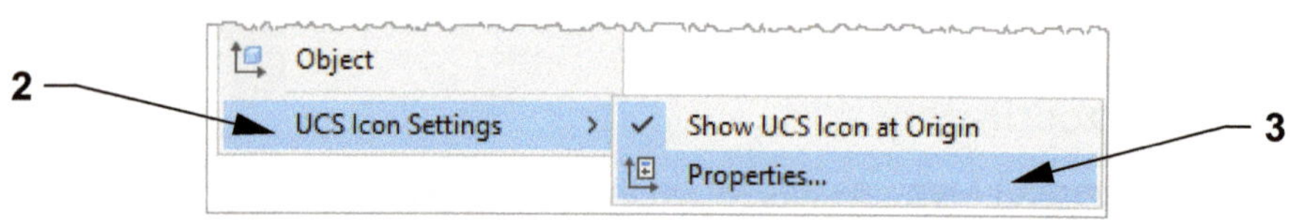

3. Select **Properties...**
When you select this option, the dialog box shown below will appear. You may change the style, size, and color of the icon at any time. Changing the appearance is a personal preference and will not affect the drawing or commands.

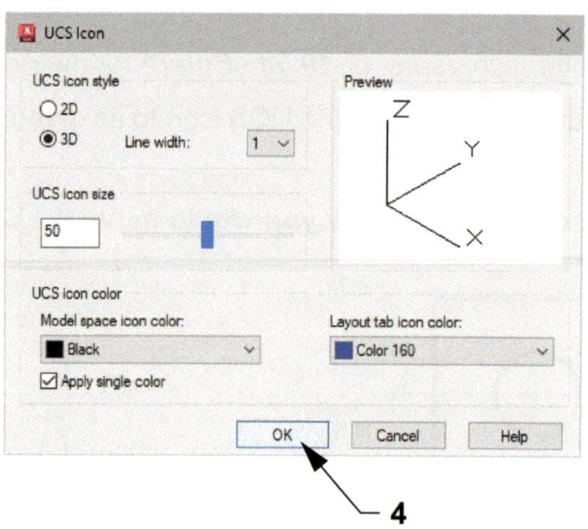

4. When complete, select the **OK** button.

Exercise 10A

Instructions:

1. Start a **New** file using the **Border A-Inch.dwt** or the **Border A-Metric.dwt**
2. Move the Origin to the locations noted before you draw the objects.
3. Use Layer Object Line.
4. Edit the **Title** and **Ex-XX** by double clicking on the text. Do not erase and replace.
5. Do not dimension.
6. **Save** as **Ex-10A**
7. Plot using Page Setup **Class Model A**. (Refer to pages 9-12 through 9-14.)

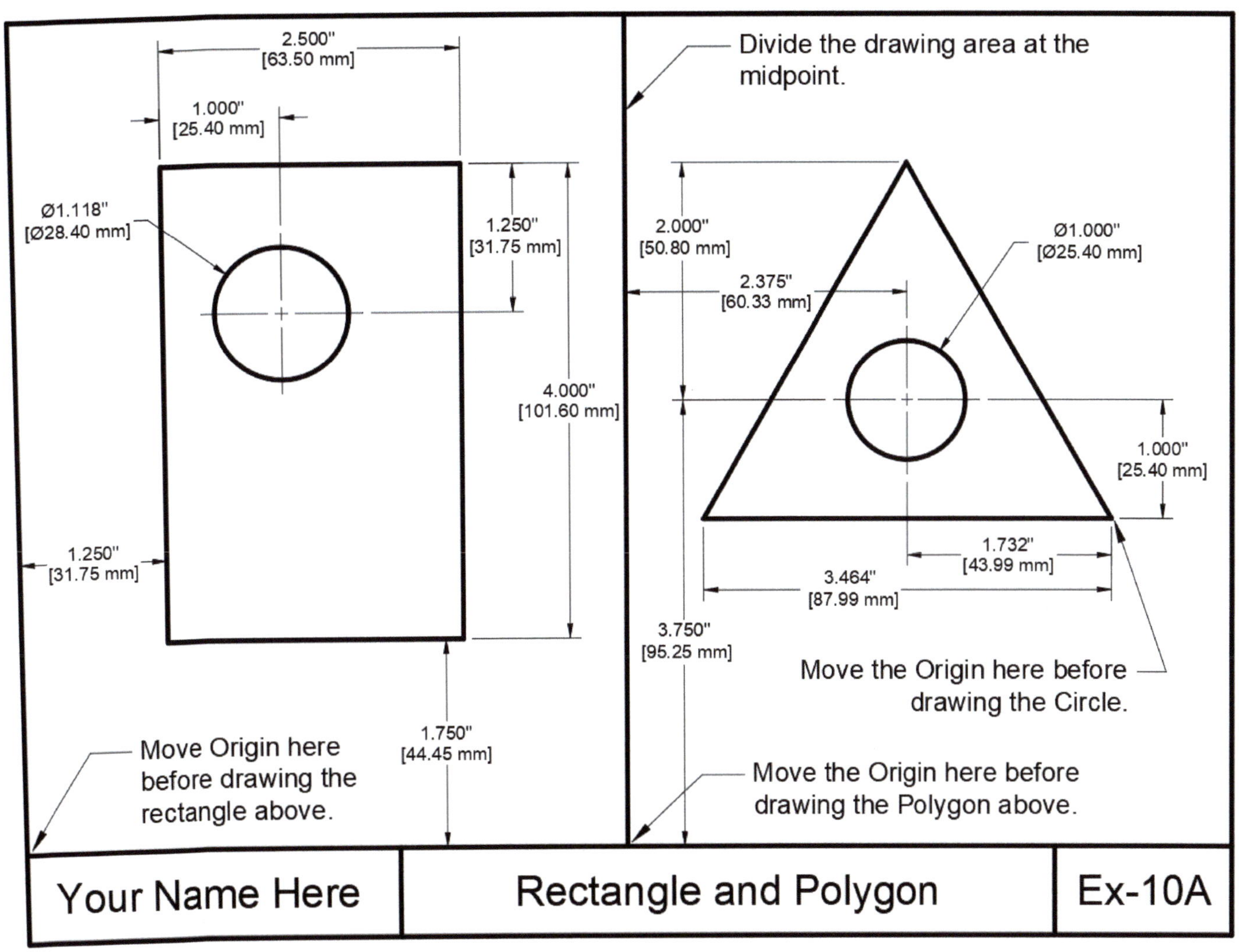

Exercise 10B

Instructions:

1. Start a **New** file using the **Border A-Inch.dwt** or the **Border A-Metric.dwt**
2. Move the Origin to the locations noted before you draw the objects.
3. Use Layer Object Line.
4. Edit the **Title** and **Ex-XX** by double clicking on the text. Do not erase and replace.
5. Do not dimension.
6. **Save** as **Ex-10B**
7. Plot using Page Setup **Class Model A**. (Refer to pages 9-12 through 9-14.)

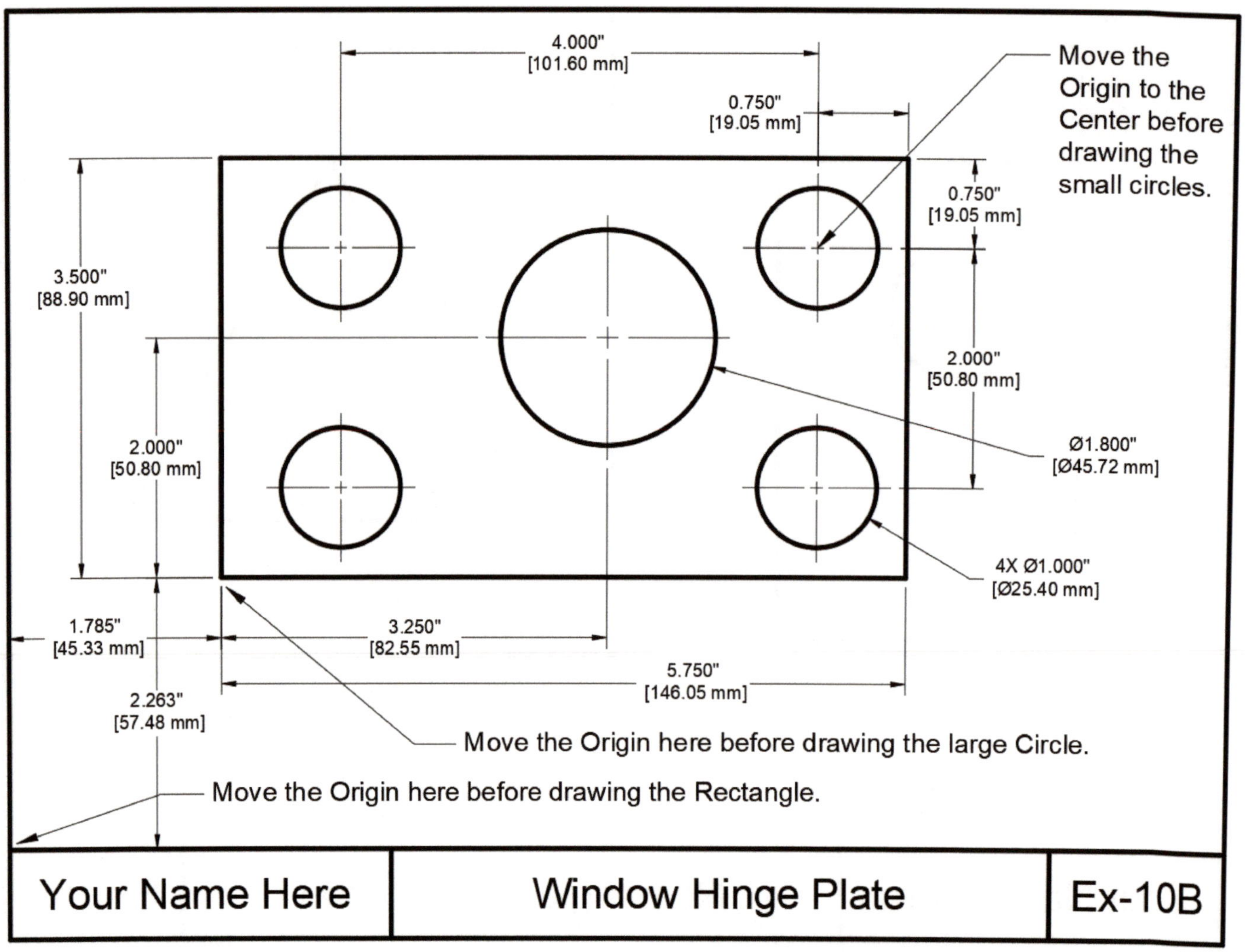

LESSON 11

LEARNING OBJECTIVES

After completing this lesson, you will be able to:

- Understand the Polar Degree Clock
- Draw Lines to a specific length and angle
- Draw objects using Polar coordinate input
- Use Dynamic Input
- Use Polar Tracking and Polar Snap

Polar Coordinate Input

In Lesson 9, you learned to control the length and direction of horizontal and vertical lines using Relative Input and Direct Distance Entry. Now you will learn how to control the length and **angle** of a line using **Polar coordinate** input.

Understanding the "Polar Degree Clock"

Previously when drawing horizontal and vertical lines, you controlled the direction using a **Positive** or **Negative** input. ***Polar coordinate input is different***. The angle of the line will determine the direction.

Example: If you wanted to draw a line at a 45 degree angle toward the upper right corner, you would use the angle 45. But if you wanted to draw a line at a 45 degree angle toward the lower left corner, you would use the angle 225.

You may also use Polar coordinate input for horizontal and vertical lines using the angles 0, 90, 180, and 270. No negative input is required.

Polar Degree Clock

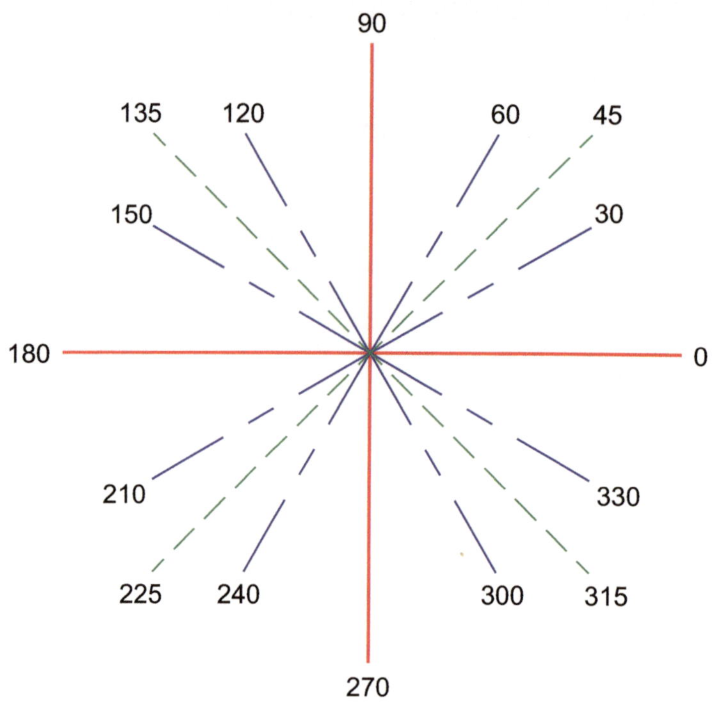

Drawing with Polar Coordinate Input

When entering Polar coordinates, the first number represents the **distance** and the second number represents the **angle**. The two numbers are separated by the **less than (<)** symbol. The input format is: **distance < angle**.

Note: If you are using Dynamic Input, refer to the next page.

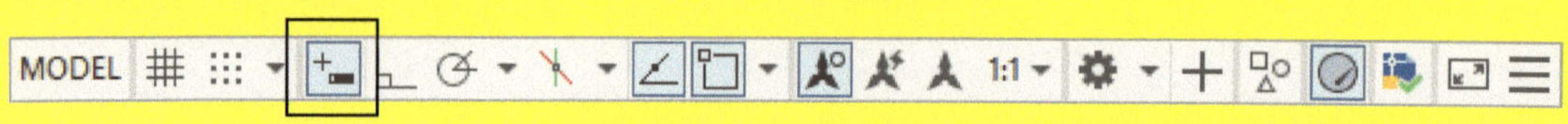

Continued on the next page...

A Polar coordinate of **@6<45** will be a distance of **6 units** and an angle of **45 degrees** from the last point entered.

Here is an example of Polar coordinate input for four line segments.

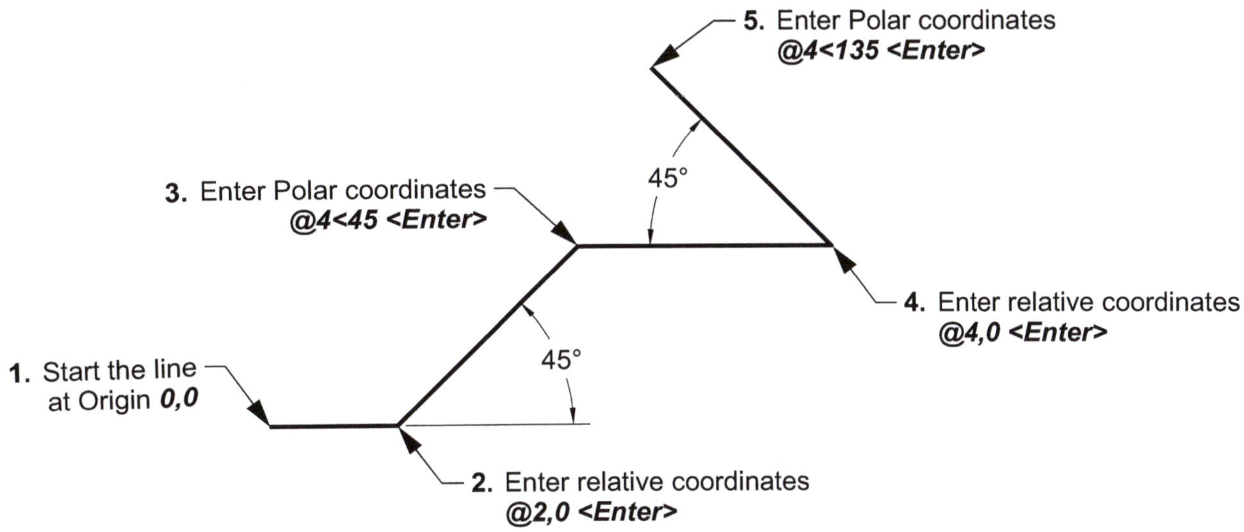

Dynamic Input

To help you keep your focus in the "drawing area", AutoCAD has provided a command interface called **Dynamic Input**. You may input information within the Dynamic Input tooltip box instead of on the Command Line.

When AutoCAD prompts you for the **first point**, the Dynamic Input tooltip displays the **absolute: X, Y** distance from the Origin.

Enter the **X** dimension, press the ***<Tab>*** key, enter the **Y** dimension, and then press ***<Enter>***.

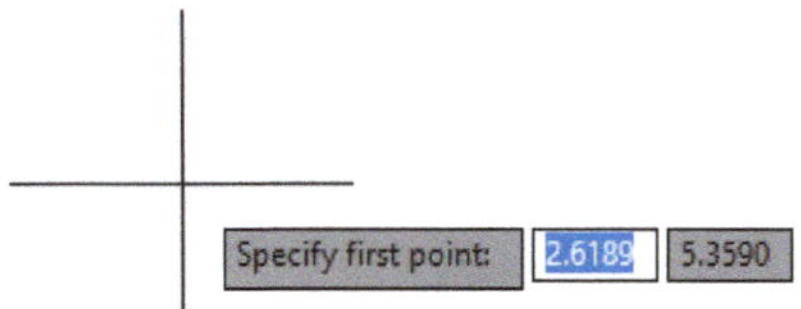

First point

When AutoCAD prompts you for the **second** and all **next points**, the Dynamic Input tooltip displays the **relative: distance and angle** from the last point entered.

Enter the **distance**, press the ***<Tab>*** key, move the cursor in the approximate desired angle, enter the **angle**, and then press ***<Enter>***.

Note: The @ is not necessary.

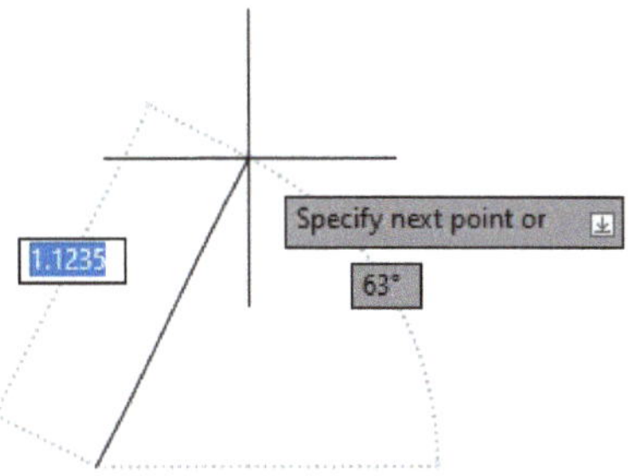

Second and all next points

How to turn Dynamic Input on or off

Select the **Dynamic Input** button on the Status Bar or use the ***<F12>*** key.

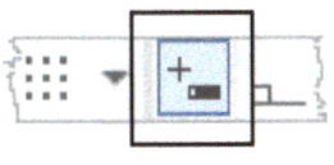

Continued on the next page...

Here is an example of Dynamic Input for four line segments.

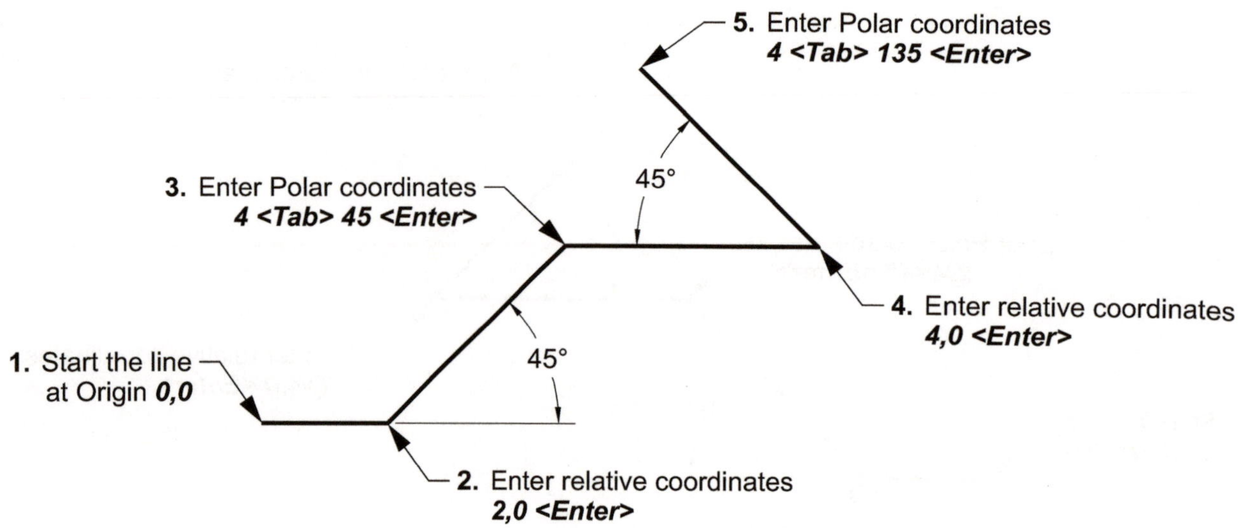

How to enter Cartesian coordinates (X and Y)

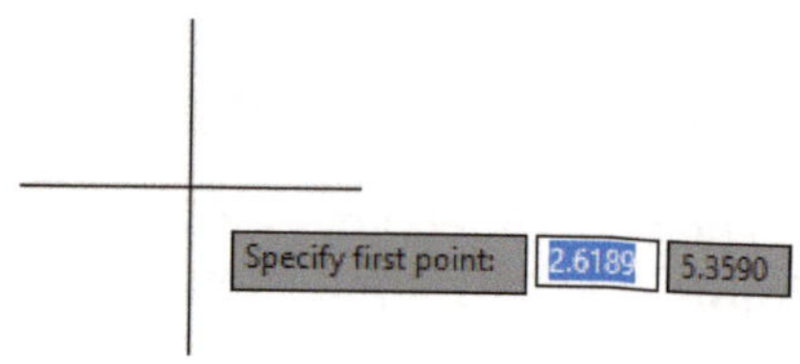

1. Enter an **"X"** coordinate value and a **comma**.
2. Enter a **"Y"** coordinate value ***<Enter>***.

How to enter Polar coordinates (from the last point entered)

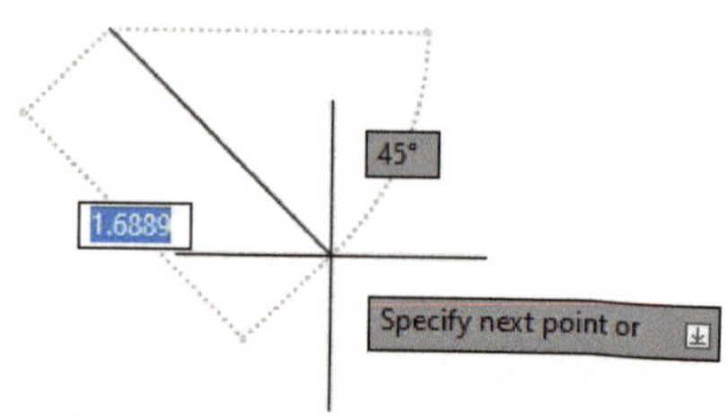

1. Enter the **distance** value from the last point entered.
2. Press the ***<Tab>*** key.
3. Move the cursor in the approximate direction and enter the **angle** value ***<Enter>***.

Move the cursor in the approximate direction and enter an angle value of **0-180 only**. Dynamic Input does not use 360 degrees. (Refer to the example on the next page.)

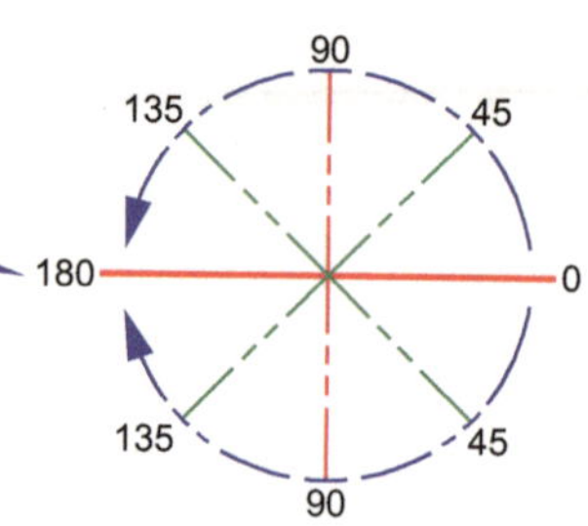

How to specify absolute or relative coordinates while using Dynamic Input

To enter **absolute** coordinates when relative coordinate format is displayed in the tooltip: Enter **#** to temporarily override the setting.

To enter **relative** coordinates when absolute coordinate format is displayed in the tooltip: Enter **@** to temporarily override the setting.

CAD TIP! You may toggle OrthoMode **on** and **off** by holding down the ***<Shift>*** key. This is an easy method for using Direct Distance Entry while using Dynamic Input.

Using Dynamic Input and Polar Coordinates

The following is a simple drawing to practice Dynamic Input and Polar coordinates. Think about how this differs from the basic Polar coordinate input on page 11-2 and 11-3.

See the example on the next page.

1. Set the Status Bar as follows:
   Dynamic Input = **on** All others = **off**

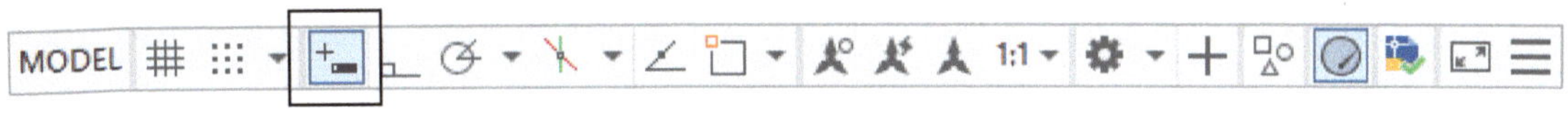

2. Select the **Line** command.

3. Start the line near the lower left corner of the drawing area.

Line A 4. Move the cursor to the right.
5. Type ***2 [50.8] <Tab> 0 <Enter>*** (**Note:** Sizes in **[....]** are for metric users.)

Line B 6. Move the cursor up and to the right.
7. Type ***3 [76.2] <Tab> 45 <Enter>***.

Line C 8. Move the cursor up.
9. Type ***2 [50.8] <Tab> 90 <Enter>***.

Line D 10. Move the cursor down and to the left.
11. Type ***4 [101.6] <Tab> 135 <Enter>***. (**Note**: 180 – 45 = **135**)

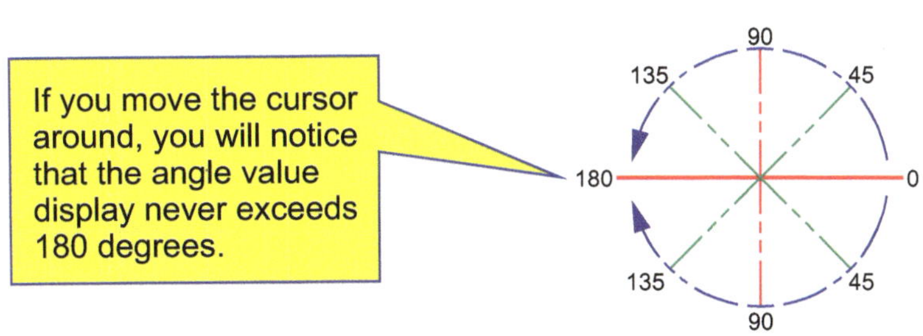

Line E 12. Move the cursor to the left.
13. Type ***1.293 [32.84] <Tab> 180 <Enter>***.

Line F 14. Move the cursor down.
15. Type ***1.293 [32.84] <Tab> 90 <Enter>***.
16. Press ***<Enter>*** to stop.

Continued on the next page...

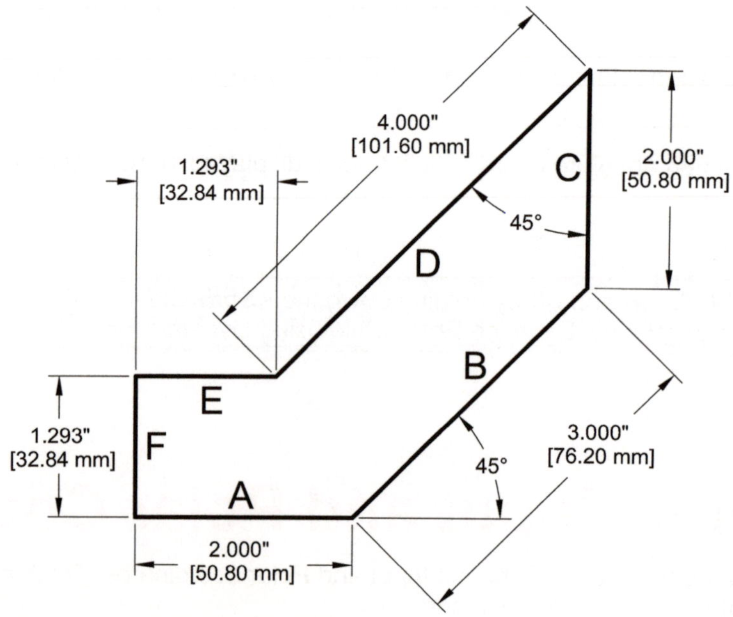

Note: For more on Dynamic Input, see Appendix B.

Polar Tracking

Polar Tracking can be used **instead** of **Dynamic Input**. When **Polar Tracking** is **"on"**, a dotted **"tracking"** line and a **"tooltip"** box appear. The tracking line **"snaps"** to a **preset angle increment** when the cursor approaches one of the preset angles. The word **"Polar"**, followed by the **"distance"** and **"angle"** from the last point appears in the box. (A step-by-step example is described on the next page.)

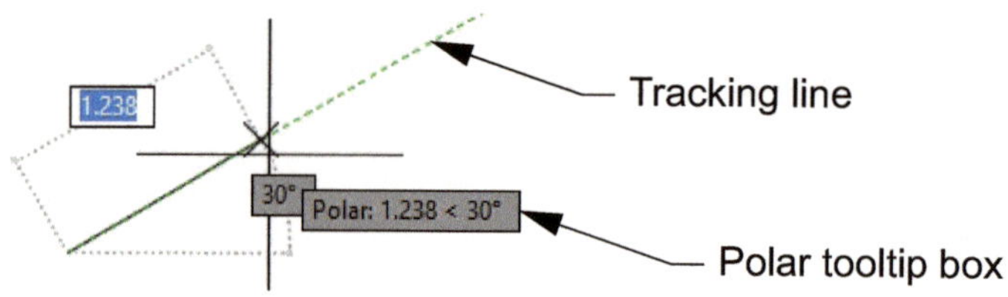

How to set the Increment Angle

1. Left click on the **Polar Tracking** button down arrow ▾ on the Status Bar and select **"Tracking Settings"**, or select an angle from the list.

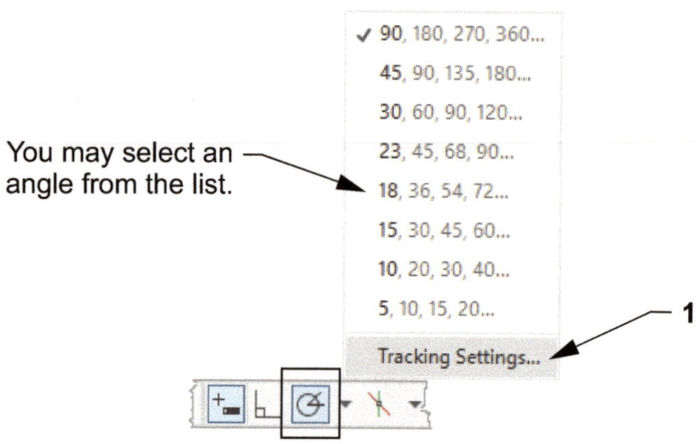

Continued on the next page...

2. The **Drafting Settings** dialog box will appear.

3. Select the **Polar Tracking** Tab.

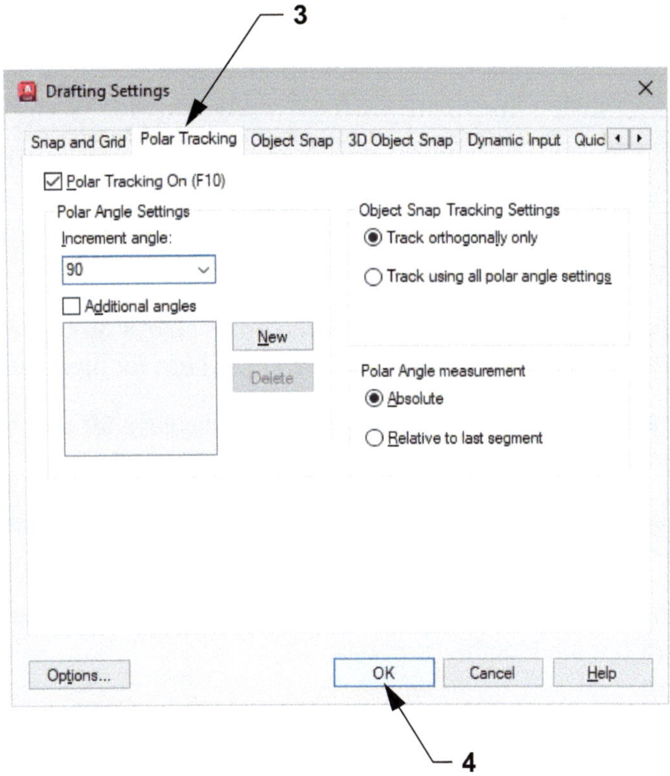

Polar Angle Settings

Increment angle: Choose from the Increment angle list including 90, 45, 30, 22.5, 18, 15,10, and 5. It will also snap to the selected angles multiples. For example, if you choose 30, it will snap to 30, 60, 90, 120, 150, 180, 210, 240, 270, 300, 330, and, 0.

Additional angles: Check this box if you would like to use an angle other than one in the Increment angle list. For example, 12.5.

New: You may add an angle by selecting the **"New"** button. You will be able to snap to this new angle in addition to the incremental angle selected. **But you will not be able to snap to its multiple**. For example, if you selected 7, you would not be able to snap to 14.

Delete: Deletes an Additional angle. Select the Additional angle to be deleted and then the **Delete** button.

Polar Angle Measurement Settings

Absolute: Polar Tracking **angles** are relative to the UCS.

Relative to last segment: Polar Tracking **angles** are relative to the last segment.

4. Make any necessary changes and then select the **OK** button to close the **Drafting Settings** dialog box.

Using Polar Tracking and DDE

1. Set the Polar Tracking Increment angle to **15**.

2. Turn all the Status Bar buttons to **off** except **Polar**. (**Note:** Dynamic Input should be **off**, but you may wish to leave it on.)

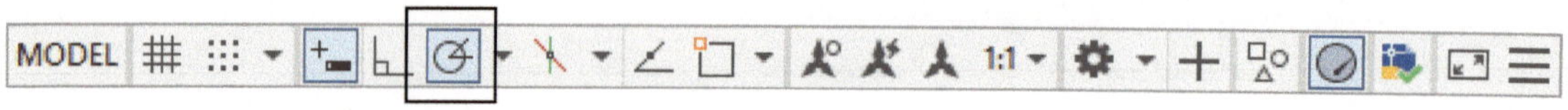

3. Select the **Line** command.

P1 4. Start the line in the lower left area of the drawing area.

P2 5. Move the cursor in the direction of **P2** until the tooltip box displays **30** degrees.
6. Type ***2 [50.8] <Enter>*** (for the length). (**Note:** Sizes in **[....]** are for metric users.)

P3 7. Move the cursor in the direction of **P3** until the tooltip box displays **90** degrees.
8. Type ***2 [50.8] <Enter>*** (for the length).

P4 9. Move the cursor in the direction of **P4** until the tooltip box displays **0** degrees.
10. Type ***2 [50.8] <Enter>*** (for the length).

P5 11. Move the cursor in the direction of **P5** until the tooltip box displays **150** degrees.
12. Type ***2 [50.8] <Enter>*** (for the length).

P6 13. Move the cursor in the direction of **P6** until the tooltip box displays **180** degrees.
14. Type ***2 [50.8] <Enter>*** (for the length).

15. Then type ***C*** for Close.

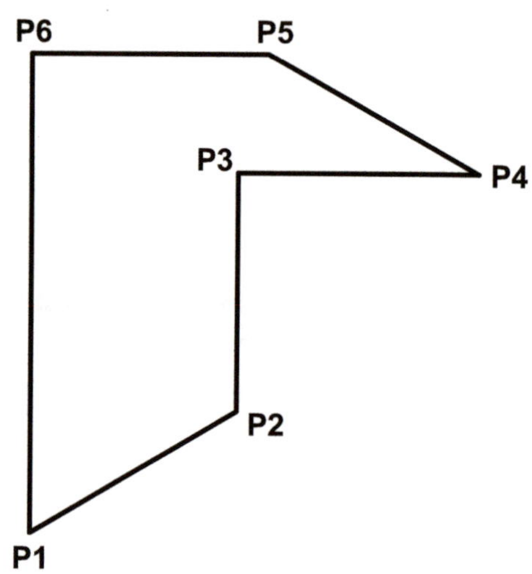

CAD TIP! You may toggle Polar Tracking **on** or **off** by using one of the following methods: either left click on the **Polar** button on the Status Bar or press ***<F10>***.

Polar Snap

Polar Snap is used with **Polar Tracking** to make the cursor snap to specific **distances** and **angles**. If you set **Polar Snap distance** to **1** and **Polar Tracking** to **angle 30**, you can draw lines 1, 2, 3, or 4 units long at an angle of 30, 60, 90, etc. without typing anything. You just move the cursor and watch the tooltips.

See the step-by-step example on the next page.

How to set the Polar Snap

1. Set the **Polar Tracking Increment angle** as shown on page 11-7.
2. Left click on the **Snap** button down arrow ▾ on the Status Bar and select **"Snap Settings"**.

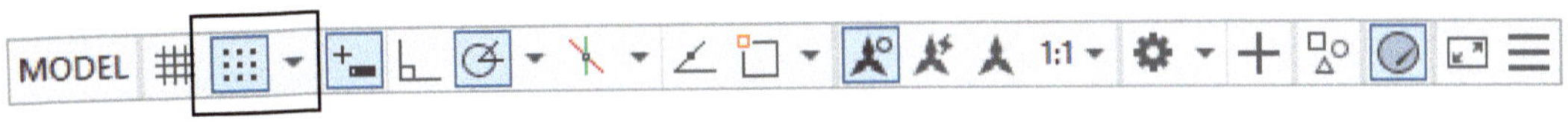

3. Select the **Snap and Grid** Tab.

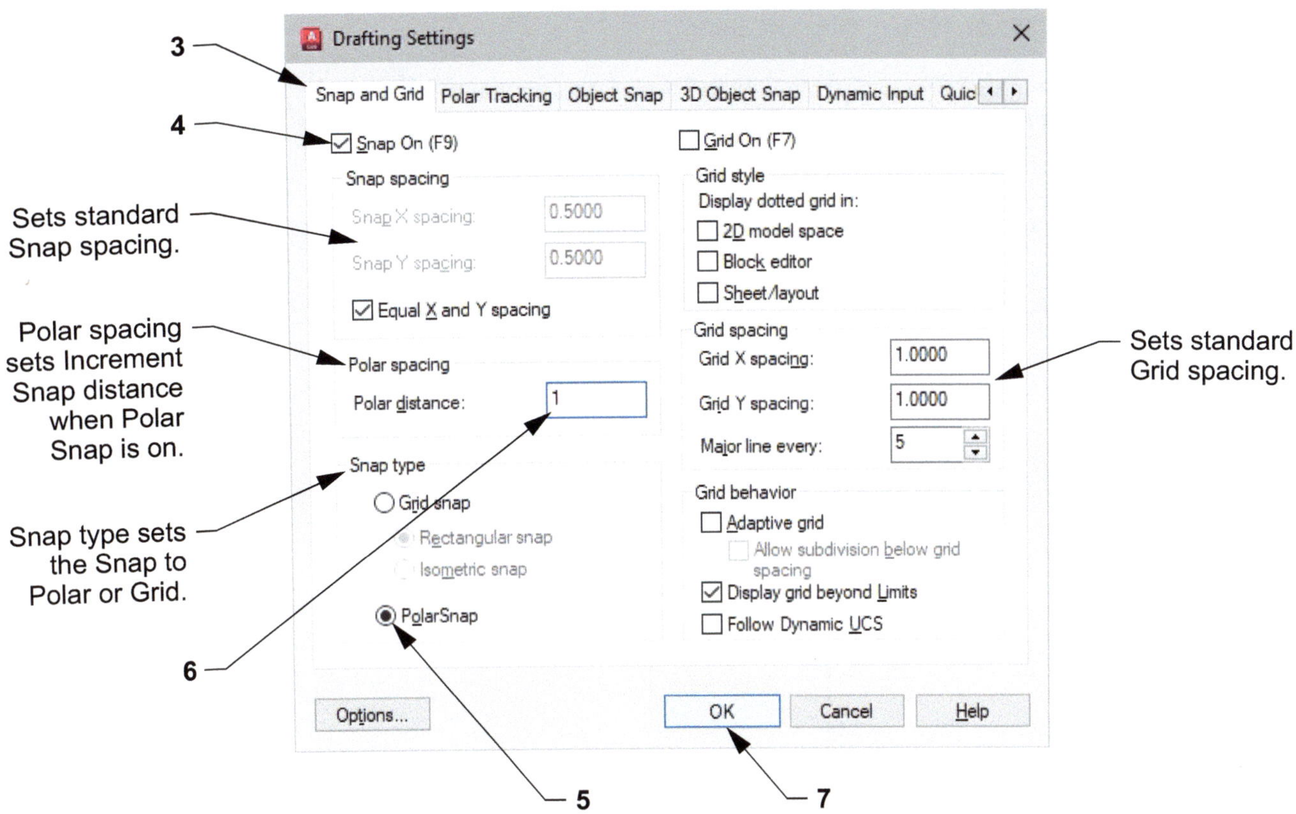

4. Select **Snap On**.
5. Select **PolarSnap**.
6. Set the **Polar distance**.
7. Select the **OK** button.

Using Polar Tracking and Polar Snap

Now let's draw the objects below again, but this time let's use "Polar Snap".

1. Set Polar Tracking Increment angle to **30** and Polar Snap to **1.000 [25.4]**

2. Turn all the Status Bar buttons to **off** except **Snap** and **Polar**.

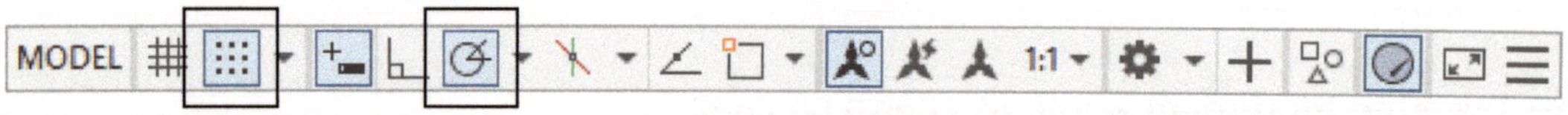

3. Select the **Line** command.

P1 4. Start the line in the lower left area of the drawing area.

P2 5. Move the cursor in the direction of **P2** until the tooltip box displays ***Polar 2.000 [50.8] <30°*** and then **left click**.

P3 6. Move the cursor in the direction of **P3** until the tooltip box displays ***Polar 2.000 [50.8] <90°*** and then **left click**.

P4 7. Move the cursor in the direction of **P4** until the tooltip box displays ***Polar 2.000 [50.8] <0°*** and then **left click**.

P5 8. Move the cursor in the direction of **P5** until the tooltip box displays ***Polar 2.000 [50.8] <150°*** and then **left click**.

P6 9. Move the cursor in the direction of **P6** until the tooltip box displays ***Polar 2.000 [50.8] <180°*** and then **left click**.

10. Then type ***C*** for Close.

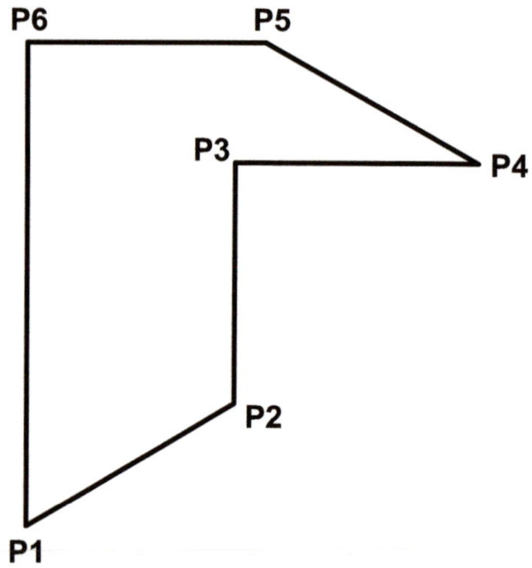

CAD TIP! You may **override** Polar settings at any time by typing: Polar coordinates (***@Length<Angle***) on the Command Line. For example, ***@2<30***

Exercise 11A-Inch

Instructions:

1. Start a **New** file using the **Border A-Inch.dwt**
2. After reviewing the lengths and angles below, set the Polar Tracking Increment angle and Polar Snap distance.

   Note: You may have to "override" a few of the lengths. (See bottom of 11-10.)
3. Use Layer Object Line.
4. Edit the **Title** and **Ex-XX** by double clicking on the text. Do not erase and replace.
5. Do not dimension.
6. **Save** as **Ex-11A-I**
7. Plot using Page Setup **Class Model A**

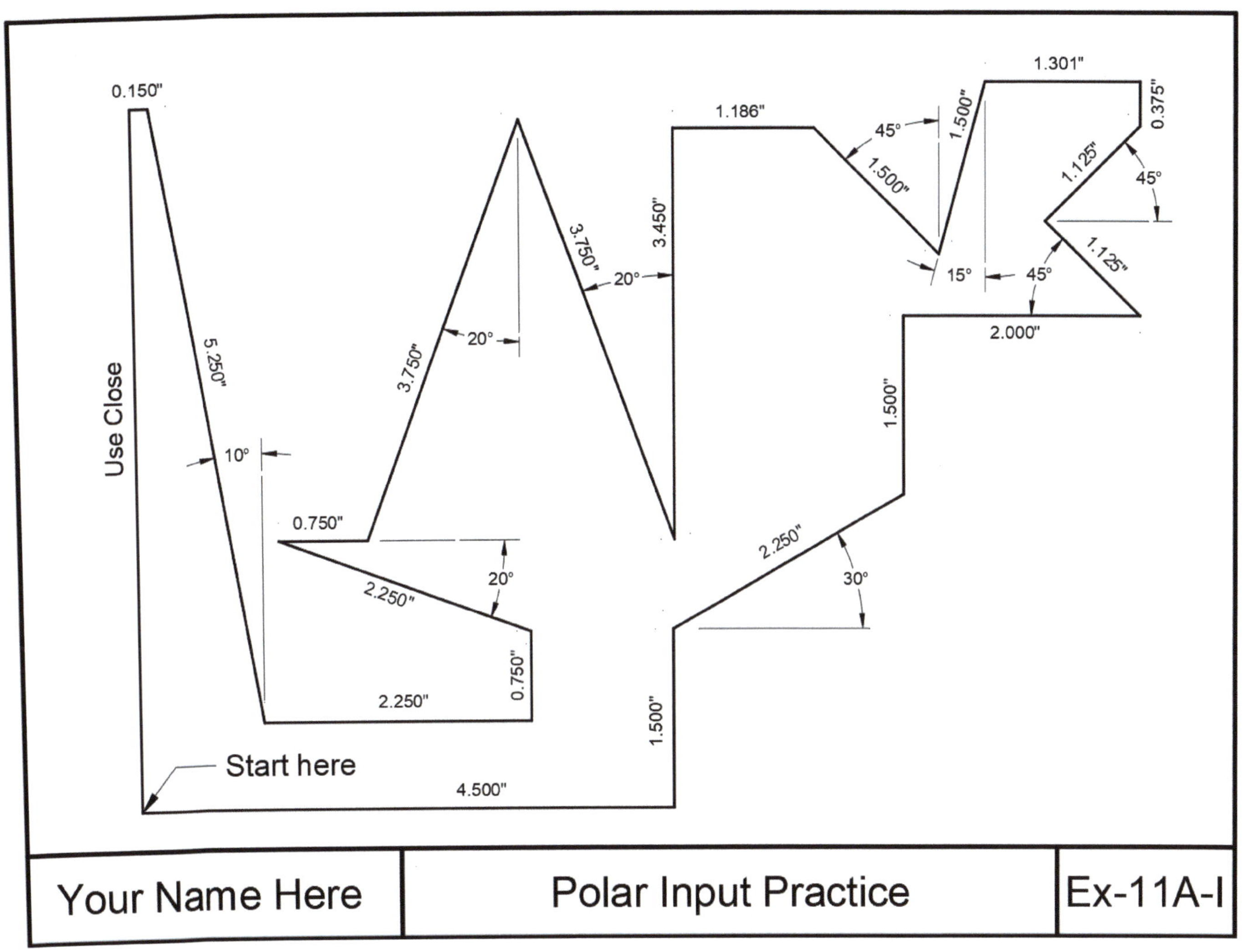

Exercise 11A-Metric

Instructions:

1. Start a **New** file using the **Border A-Metric.dwt**
2. After reviewing the lengths and angles below, set the Polar Tracking Increment angle and Polar Snap distance.

   Note: You may have to "override" a few of the lengths. (See bottom of 11-10.)
3. Use Layer Object Line.
4. Edit the **Title** and **Ex-XX** by double clicking on the text. Do not erase and replace.
5. Do not dimension.
6. **Save** as **Ex-11A-M**
7. Plot using Page Setup **Class Model A**

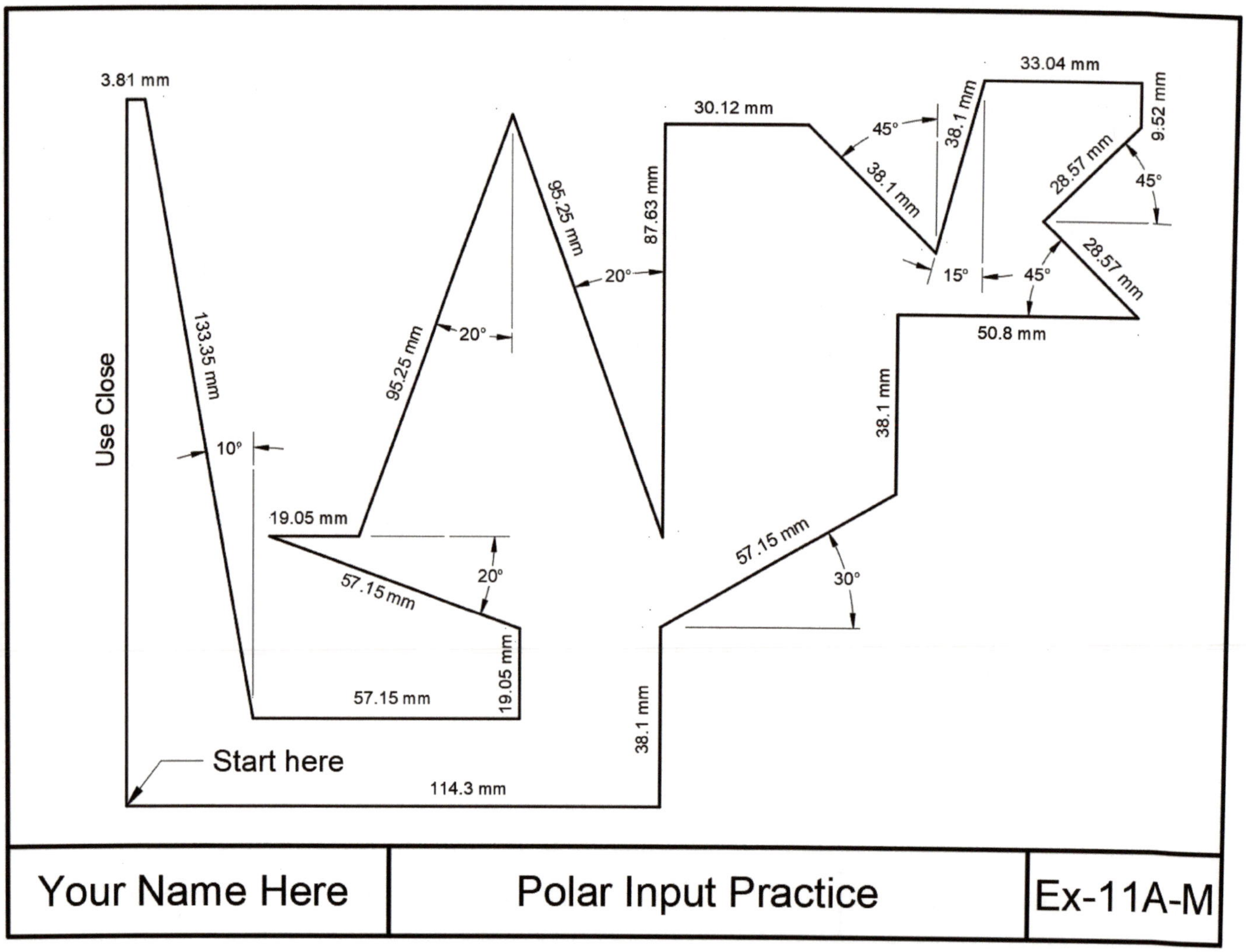

Exercise 11B-Inch

Instructions:

1. Start a **New** file using the **Border A-Inch.dwt**
2. After reviewing the lengths and angles below, set the Polar Tracking Increment angle to **30**, and the Polar Snap distance to **0.125**

   Note: The isometric lines are 30, 90, 150, 210, 270, and 330. (Refer to page 11-2.)
3. Use Layer Object Line.
4. Edit the **Title** and **Ex-XX** by double clicking on the text. Do not erase and replace.
5. Do not dimension.
6. **Save** as **Ex-11B-I**
7. Plot using Page Setup **Class Model A**

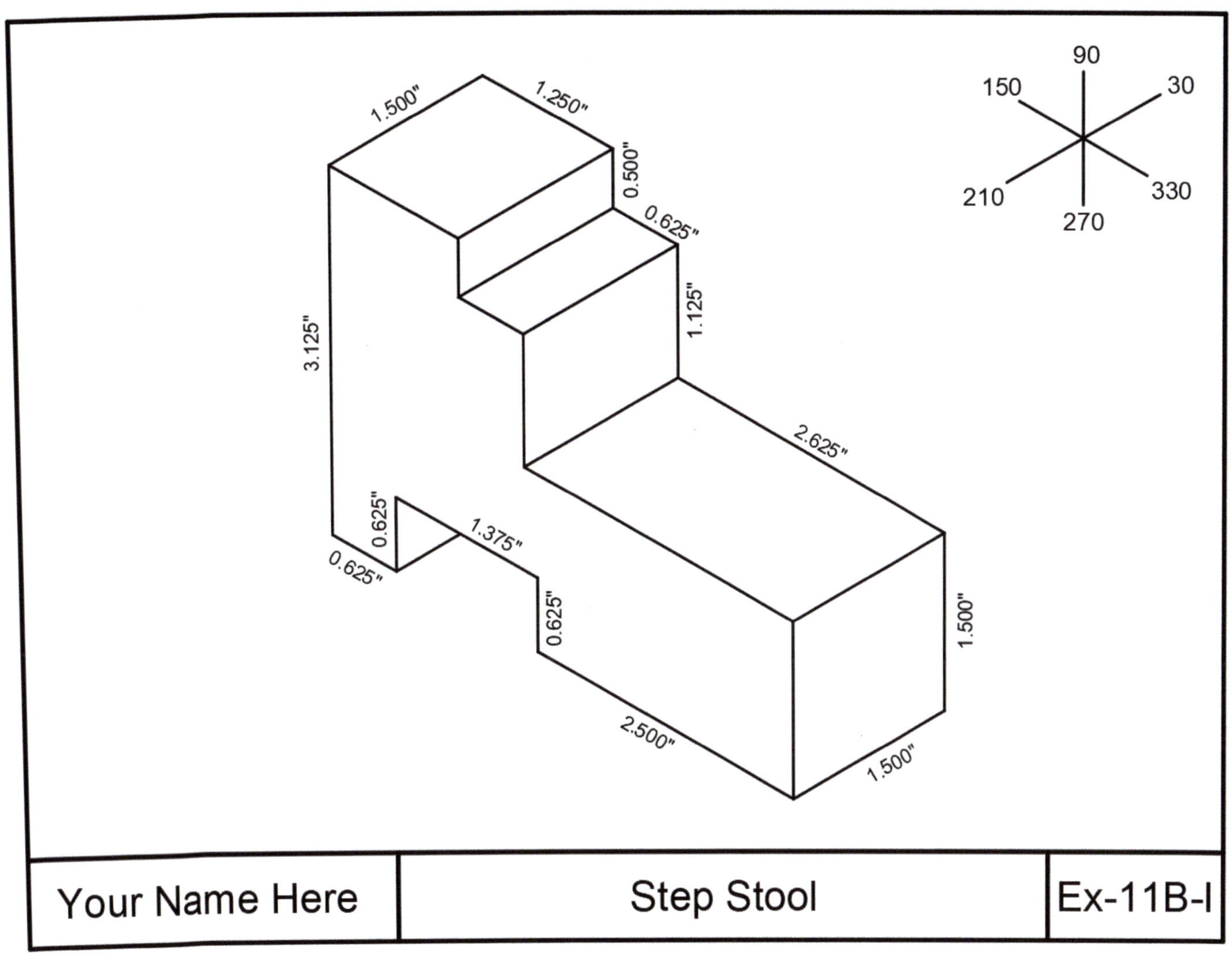

Exercise 11B-Metric

Instructions:

1. Start a **New** file using the **Border A-Metric.dwt**
2. After reviewing the lengths and angles below, set the Polar Tracking Increment angle to **30**, and the Polar Snap distance to **5**

   Note: The isometric lines are 30, 90, 150, 210, 270, and 330. (Refer to page 11-2.)
3. Use Layer Object Line.
4. Edit the **Title** and **Ex-XX** by double clicking on the text. Do not erase and replace.
5. Do not dimension.
6. **Save** as **Ex-11B-M**
7. Plot using Page Setup **Class Model A**

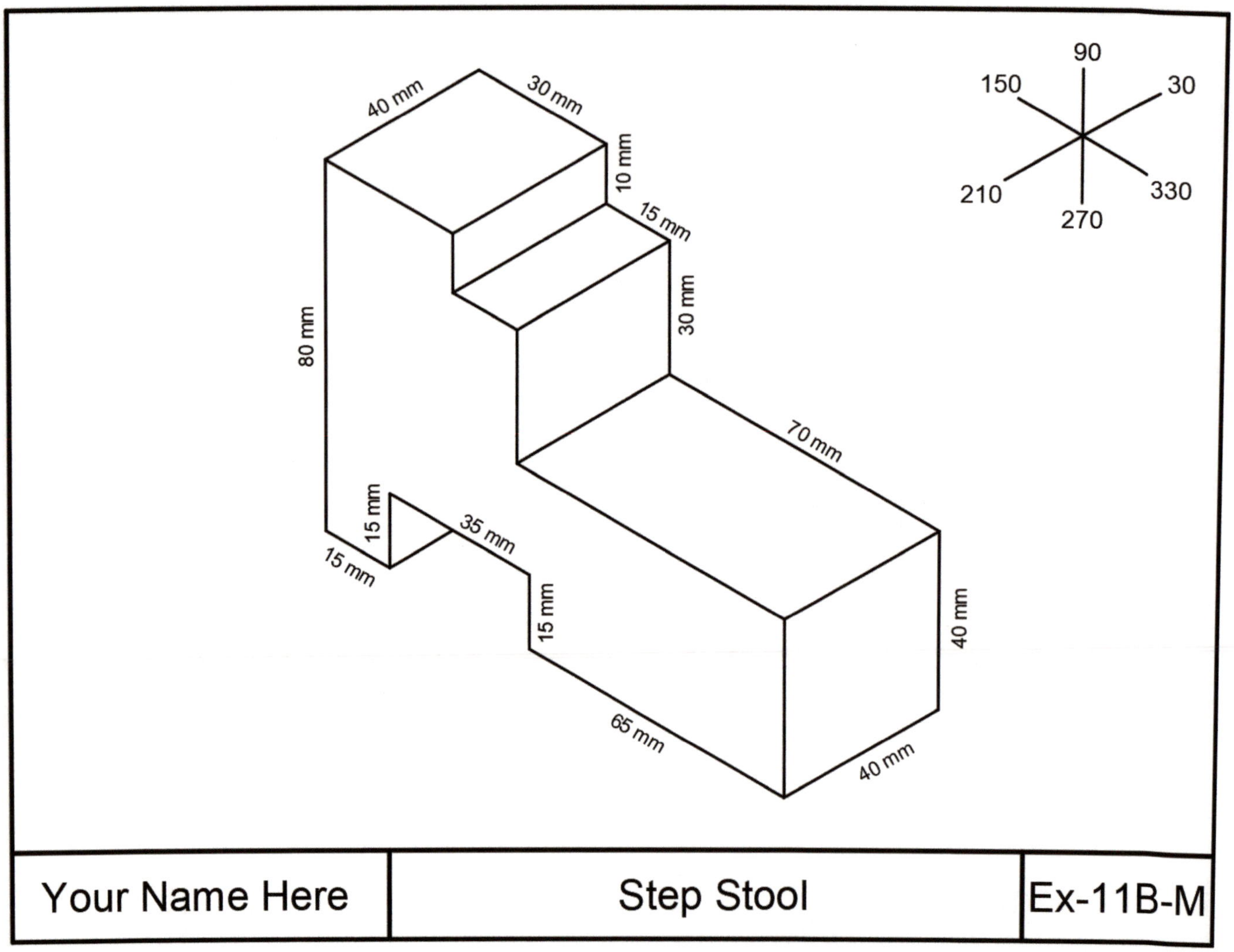

Exercise 11C

Instructions:

1. Start a **New** file using the **Border A-Inch.dwt** or the **Border A-Metric.dwt**

   Note: You will have to do some adding and subtracting on this one.

2. Draw the 5.000" [127 mm] diameter circle using Layer Object Line.
3. Draw the 4.000" [101.6 mm] diameter circle using Layer Centerline.
4. Draw each Polar line on Layer Centerline using the following example:
   For example, the line marked **"X"** is drawn as follows:

   A. Place first endpoint in the center of the circle.
   B. Enter Polar coordinates for second endpoint (distance 2.750" [69.85 mm] angle 138).
   (Refer to the Polar Degree Clock on page 11-2.)

5. Draw the 0.500" [12.7 mm] diameter circles. Locate their center using Object Snap Intersection.
6. Edit the **Title** and **Ex-XX** by double clicking on the text. Do not erase and replace.
7. Do not dimension.
8. **Save** as **Ex-11C**
9. Plot using Page Setup **Class Model A**

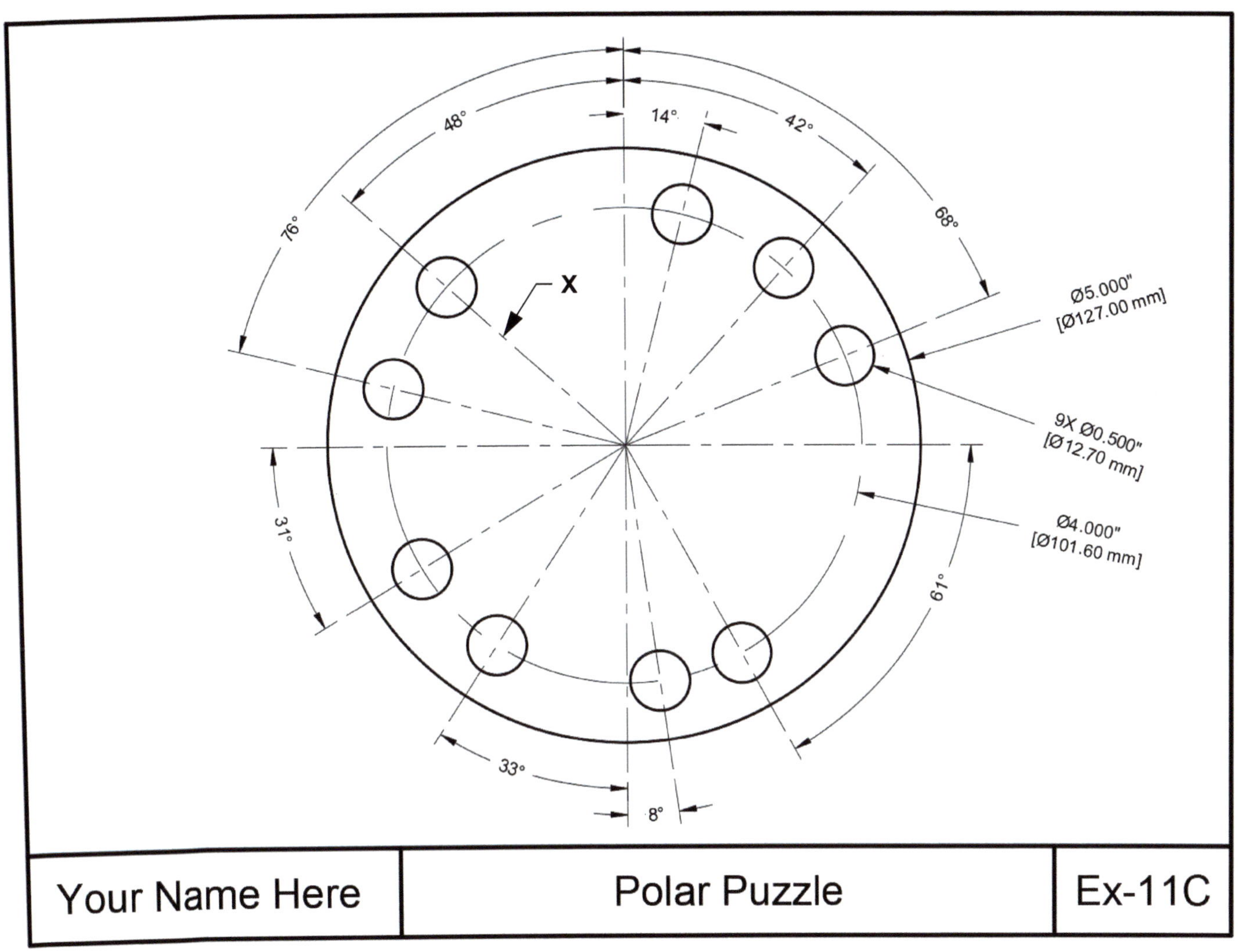

Notes:

LESSON 12

LEARNING OBJECTIVES

After completing this lesson, you will be able to:

- Duplicate an object at a specified distance away
- Make changes to an object's properties
- Use the Quick Properties Panel

Offset

The **Offset** command duplicates an object parallel to the original object at a specified distance. You can offset lines, arcs, circles, ellipses, 2D polylines, and splines. You may duplicate the original object or assign the offset copy to another layer.

Examples of Offset objects:

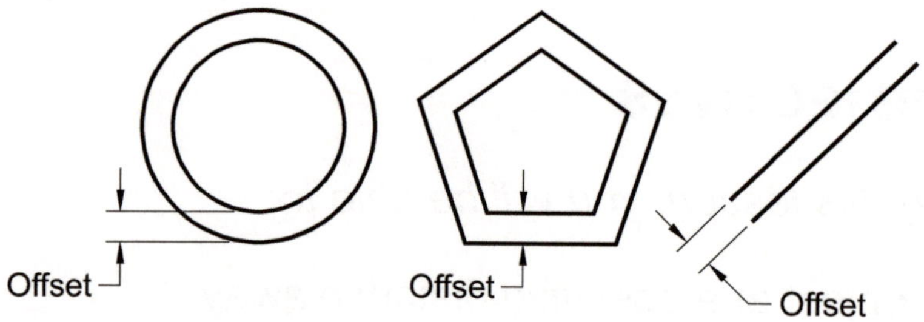

Method 1. Duplicate original object

1. Select the **Offset** command using one of the following:

   Ribbon = Home Tab / Modify Panel / 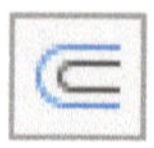
   or
   Keyboard = *offset <Enter>*

   The following will appear on the Command Line:

   ```
   OFFSET Specify offset distance or [Through Erase Layer] <Through>:
   ```

2. Type the offset distance you require and then press ***<Enter>***. (Also see **Offset Options** on the next page.)

   The following will appear on the Command Line:

   ```
   OFFSET Select object to offset or [Exit Undo] <Exit>:
   ```

3. Select the object to offset (left click).

   The following will appear on the Command Line:

   ```
   OFFSET Specify point on side to offset or [Exit Multiple Undo] <Exit>:
   ```

4. Select which side of the original you want the duplicate to appear on by placing your cursor and left clicking. (Also see **Offset Options** on the next page.)

5. Continue to offset objects or press ***<Enter>*** to end the command.

Method 2. Duplicate original object with Offset copy on a different layer

To automatically place the offset copy on a different layer than the original, you must first change the **"current"** layer to the layer you want the offset copy to be placed on.

1. Select the layer that you want the offset copy placed on from the list of layers.

2. Select the **Offset** command.

Continued on the next page...

The following will appear on the Command Line:

```
OFFSET Specify offset distance or [Through Erase Layer] <Through>:
```

3. Type ***L*** and then press ***<Enter>***.

   The following will appear on the Command Line:

```
OFFSET Enter layer option for offset objects [Current Source] <Source>:
```

4. Type ***C*** and then press ***<Enter>***.

5. Now follow **Step 2** through **Step 5** on the previous page to finish the Offset command.

Offset Options

Through: Creates an object passing through a specified point.

Erase: Erases the source object after it is offset.

Layer: Determines whether offset objects are created on the **current** layer or on the layer of the **source** object. Select **Layer** and then select **current** or **source**. (Source is the default.)

Multiple: Turns on the multiple offset mode, which allows you to continue creating duplicates of the original without reselecting the original.

Exit: Exits the Offset command.

Undo: Removes the previous offset copy.

Properties Palette

The **Properties Palette**, shown below, makes it possible to change an object's properties. You simply open the Properties Palette, select an object, and you can change any of the properties that are listed.

How to open the Properties Palette

1. Open the **Properties Palette** using one of the following:

   Ribbon = Home Tab / Properties Panel / ↘
   or
   Keyboard = *<Ctrl + 1>*

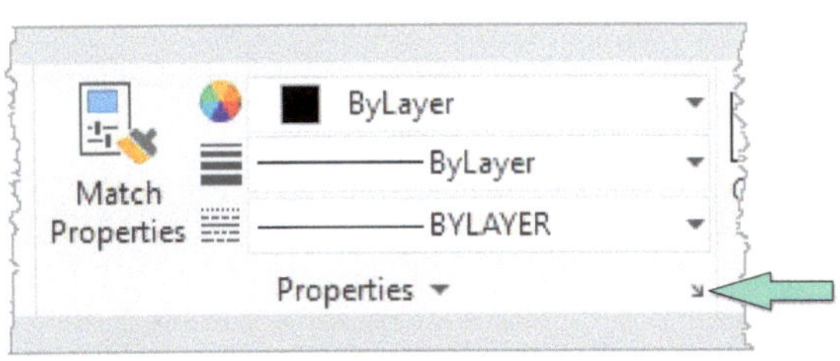

Continued on the next page...

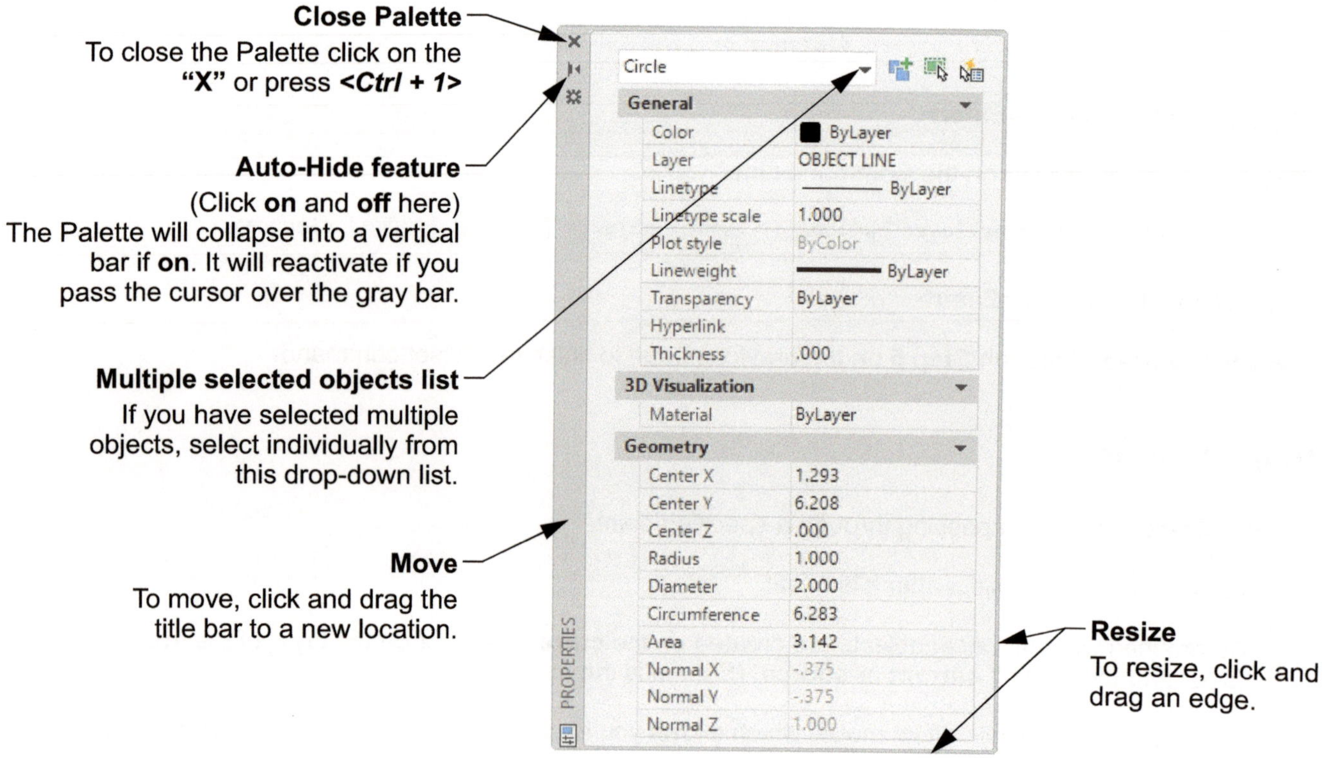

Example of editing an object using the Properties Palette

1. Draw a **2.000" [50.8 mm]** radius circle.

2. Open the **Properties Palette** and select the **Circle**. (The Properties for the circle should appear. You may change any of the properties listed in the Properties Palette for this object. When you press ***<Enter>***, the circle will change.)

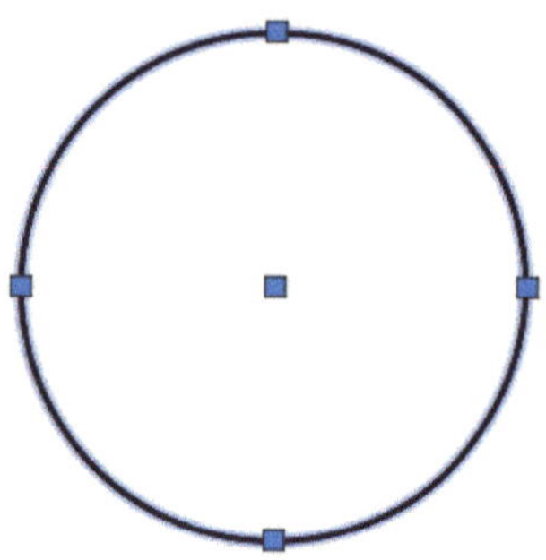

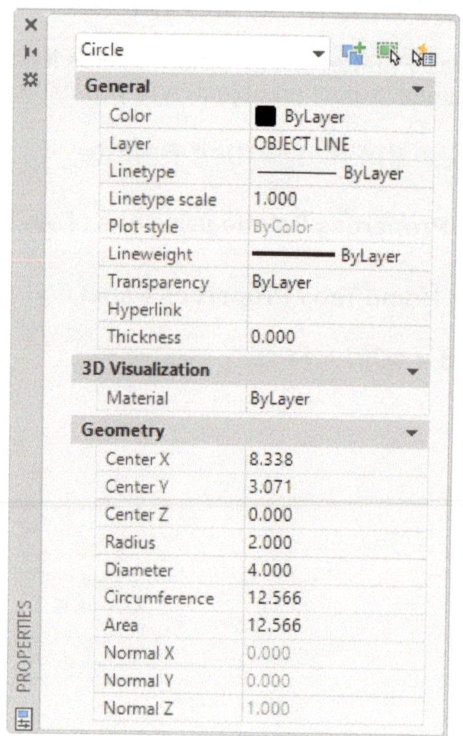

Continued on the next page...

3. Highlight and change the **"Radius"** to **1.000" [25.4 mm]** and the **"Layer"** to Hidden Line and then press ***<Enter>***. The circle got smaller and the layer changed as shown below.

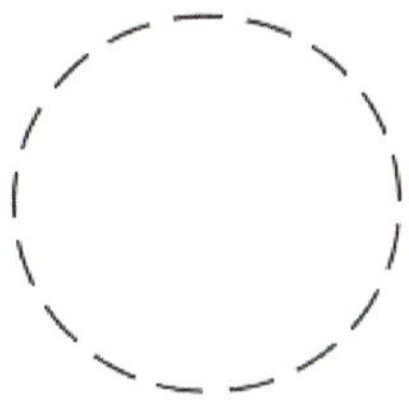

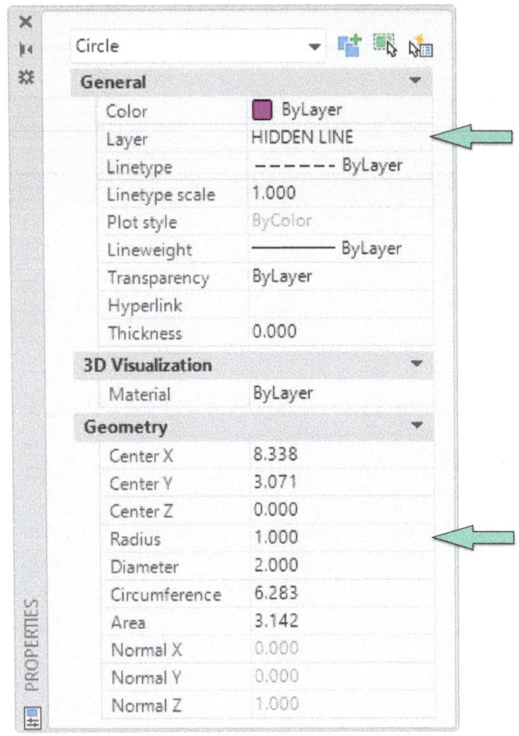

Note: The Properties Palette will show different sizes for metric users.

Quick Properties Panel

The **Quick Properties Panel**, shown below, will only appear if you have it set to **o**n. The Quick Properties Panel displays fewer properties and appears when you click once on an object. You may make changes to the objects properties using Quick Properties just as you would using the Properties Palette. (AutoCAD is just giving you another option.)

How to turn Quick Properties Panel on or off

Select the **Quick Properties** button on the Status Bar. **Note:** The Quick Properties button is not displayed by default, to display the button, refer to pages 1-12 and 1-16.

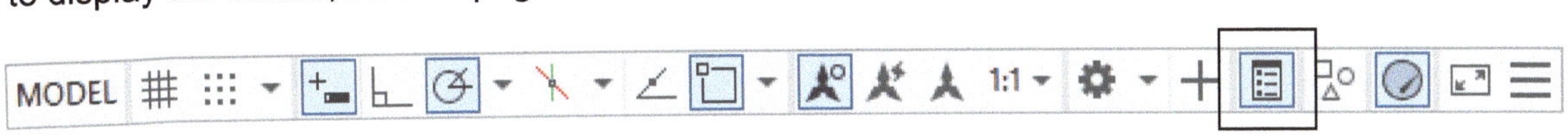

1. Select an object.

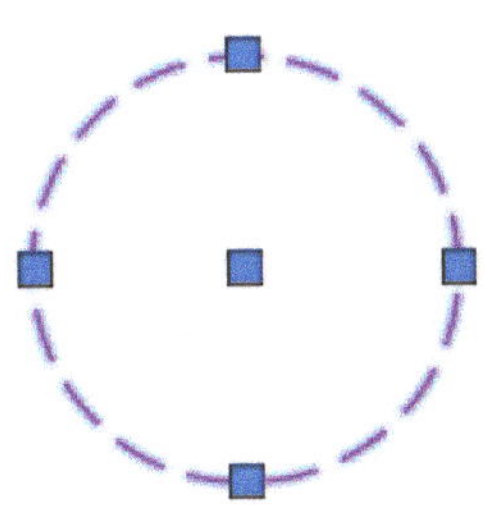

Continued on the next page...

2. The Quick Properties Panel appears.

Note: The list depends on the type of object you have selected, such as: circle, line, rectangle, etc.

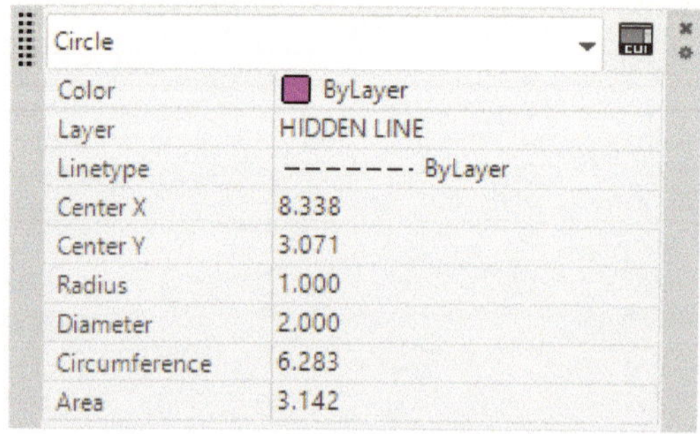

Customizing the Quick Properties Panel

You may add or remove properties from the Quick Properties Panel. And it is easy.

1. Select the **Customize User Interface** button.

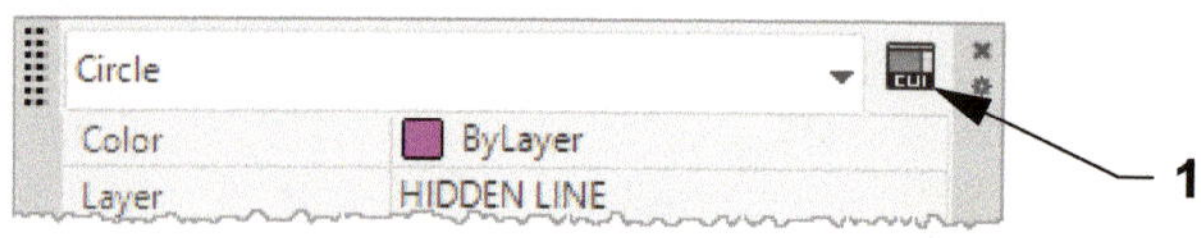

2. Select the object, from the list, that you would like to customize.

3. **Check** the boxes for the properties that you **want to appear**. **Uncheck** the boxes that you **do not want to appear**.

4. Select **Apply** and then **OK**

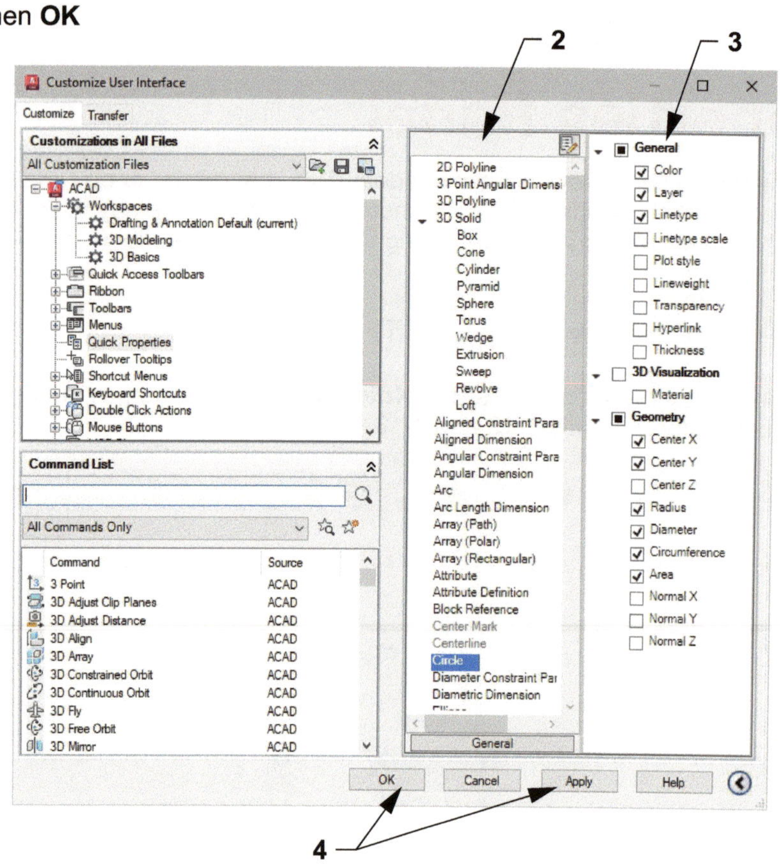

Note: The customizing is saved to the AutoCAD program settings, not the drawing file.

Offsetgaptype

When you offset a closed 2D object, such as a rectangle, to create a larger object, it results in potential gaps between the segments. The **offsetgaptype** system variable controls how these gaps are closed.

How to set the offsetgaptype

1. Type ***offsetgaptype*** and then press ***<Enter>***.

   The following will appear on the Command Line:

   ```
   OFFSETGAPTYPE Enter new value for OFFSETGAPTYPE <0>:
   ```

2. Enter one of the following variables:

   0 = Fills the gap by extending the polyline segments (default setting).

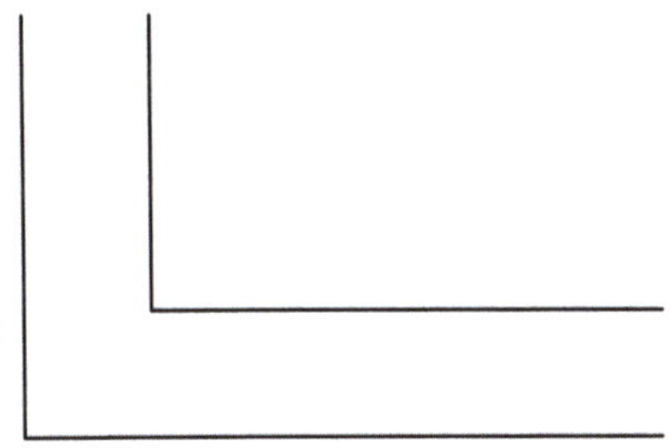

   1 = Fills the gap with **filleted arc segments**. The radius of each arc segment is equal to the offset distance.

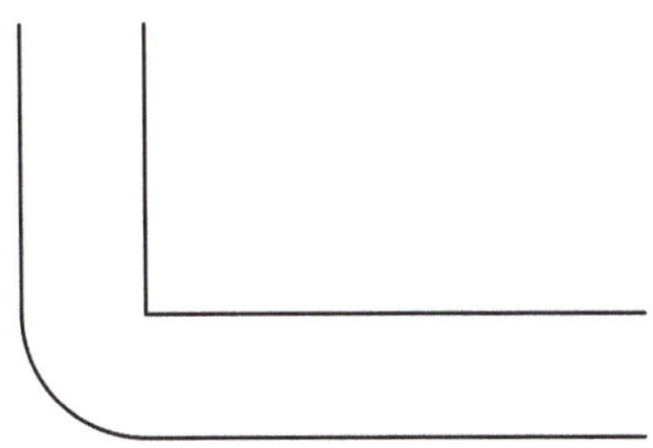

   2 = Fills the gap with **chamfered line segments**. The perpendicular distance to each chamfer is equal to the offset distance.

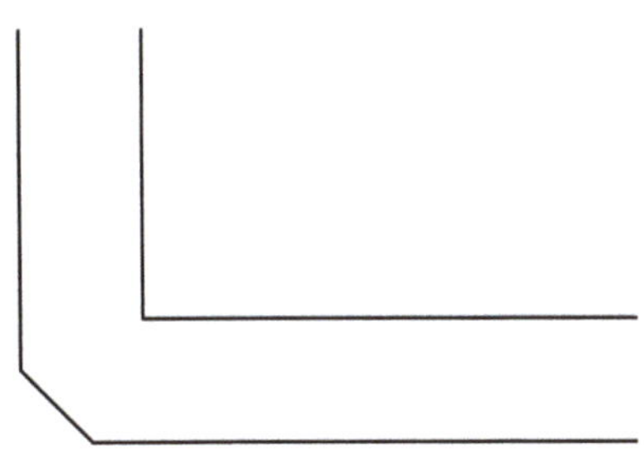

Exercise 12A

Instructions:

1. Start a **New** file using the **Border A-Inch.dwt** or the **Border A-Metric.dwt**
2. Draw the objects below using circles, rectangles, and lines.
3. Use the Offset command to offset the objects.
4. Use Layers Object Line and Hidden Line.
5. Edit the **Title** and **Ex-XX** by double clicking on the text. Do not erase and replace.
6. Do not dimension.
7. **Save** as **Ex-12A**
8. Plot using Page Setup **Class Model A**

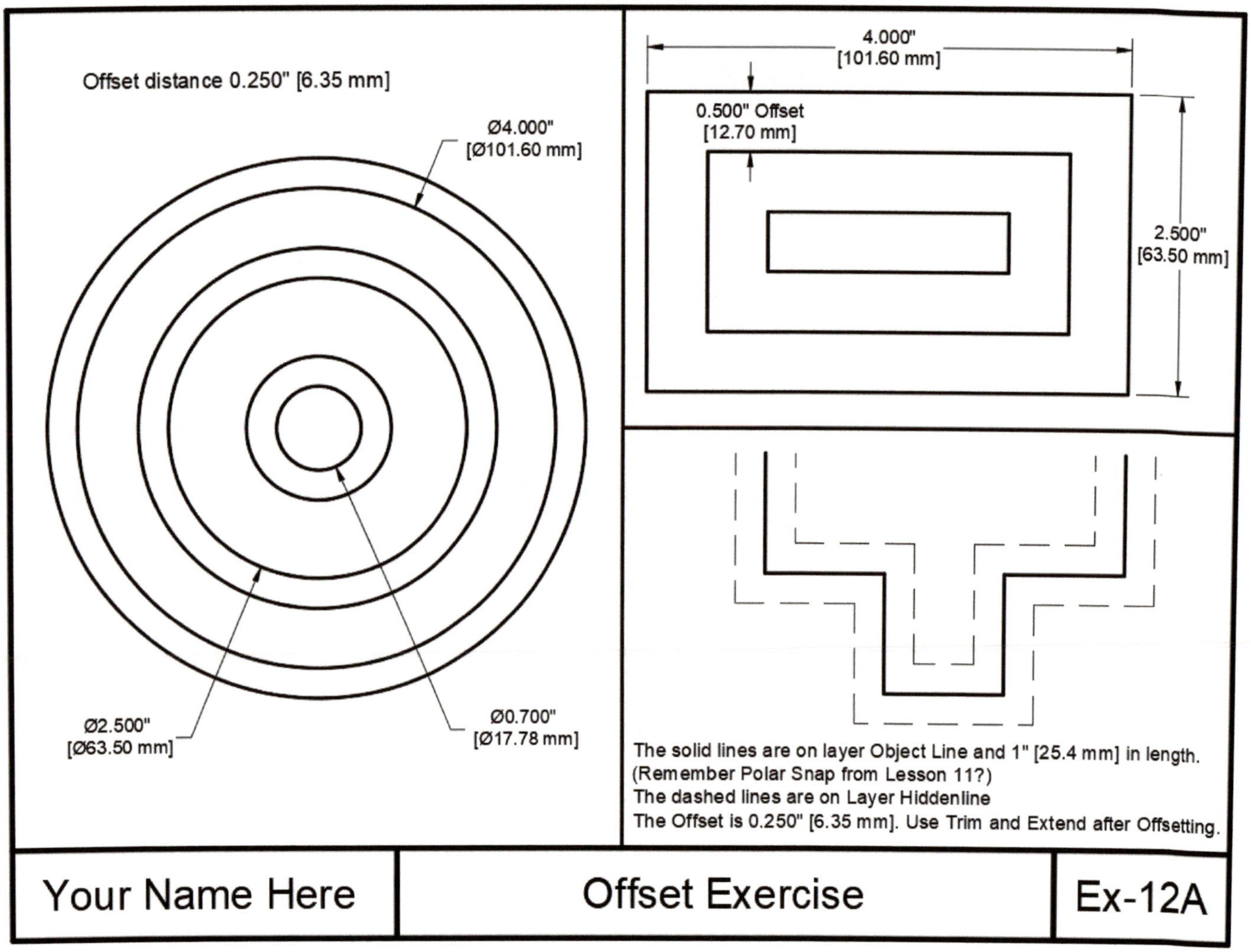

Exercise 12B-Inch

Instructions:

1. Start a **New** file using the **Border A-Inch.dwt**
2. Draw the Window Schedule shown below.
   Instructions for doing this are given on the next page.
3. Use Layers Object Line and Text (change the Object Line lineweight to **0.020"**).
4. Edit the **Title** and **Ex-XX** by double clicking on the text. Do not erase and replace.
5. Do not dimension.
6. **Save** as **Ex-12B-I**
7. Plot using Page Setup **Class Model A**

WINDOW SCHEDULE		
SYM	SIZE	TYPE
(A)	5'-0" X 4'-0"	WOOD FIXED
(B)	3'-0" X 4'-0"	WOOD FIXED
(C)	3'-0" X 3'-0"	WOOD FIXED
(D)	2'-0" X 3'-0"	ALUMINUM SLIDER

Your Name Here	Window Schedule	Ex-12B-I

Exercise 12B-Inch Helper

Step 1. Draw the Window Schedule using the **Offset** command.

Step 2. Draw four diagonal guidelines using the "Intersection" Snap. Then draw four circles with a diameter of **0.250"** using the "Midpoint" Snap. Then change the lineweight of the circles to **0.010"**.

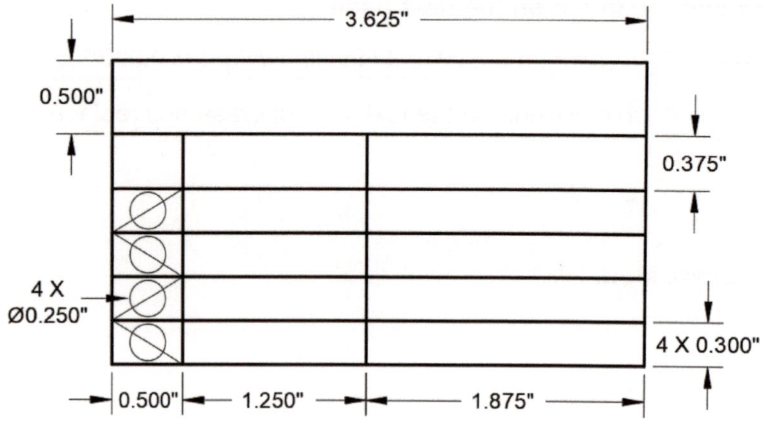

Step 3. Draw the guidelines as shown using Object Snap "Intersection" for the diagonal lines. Use the Offset command for the horizontal and vertical lines.

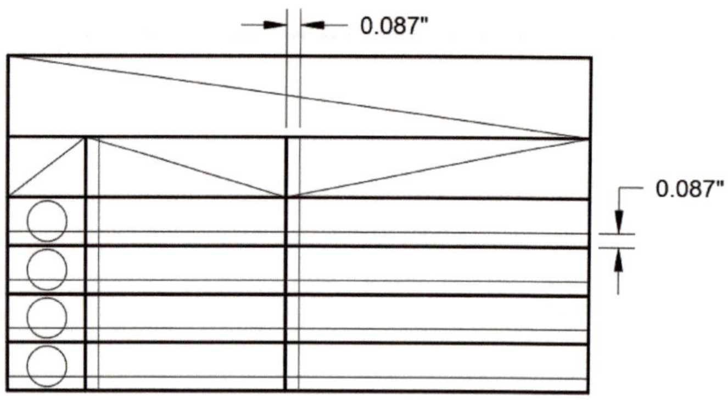

Step 4. Using **"Single Line Text"** and place text as shown. All text height is **0.125"** except the title. Title height is **0.188"**.

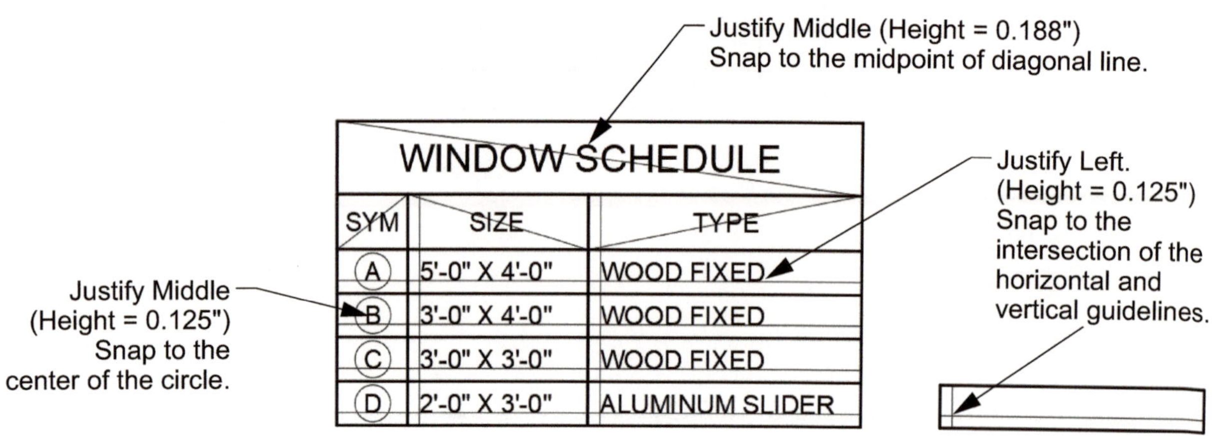

Step 5. Erase all the guidelines.

Exercise 12B-Metric

Instructions:

1. Start a **New** file using the **Border A-Metric.dwt**
2. Draw the Window Schedule shown below.
   Instructions for doing this are given on the next page.
3. Use Layers Object Line and Text (change the Object Line lineweight to **0.5 mm**).
4. Edit the **Title** and **Ex-XX** by double clicking on the text. Do not erase and replace.
5. Do not dimension.
6. **Save** as **Ex-12B-M**
7. Plot using Page Setup **Class Model A**

WINDOW SCHEDULE		
SYM	SIZE	TYPE
(A)	2M X 1.5M	WOOD FIXED
(B)	1M X 1.5M	WOOD FIXED
(C)	1M X 1M	WOOD FIXED
(D)	0.5M X 1M	ALUMINIUM SLIDER

Your Name Here	Window Schedule	Ex-12B-M

Exercise 12B-Metric Helper

Step 1. Draw the Window Schedule using the **Offset** command.

Step 2. Draw four diagonal guidelines using the “Intersection” Snap. Then draw four circles with a diameter of **6.35 mm** using the “Midpoint” Snap. Then change the lineweight of the circles to **0.25 mm**.

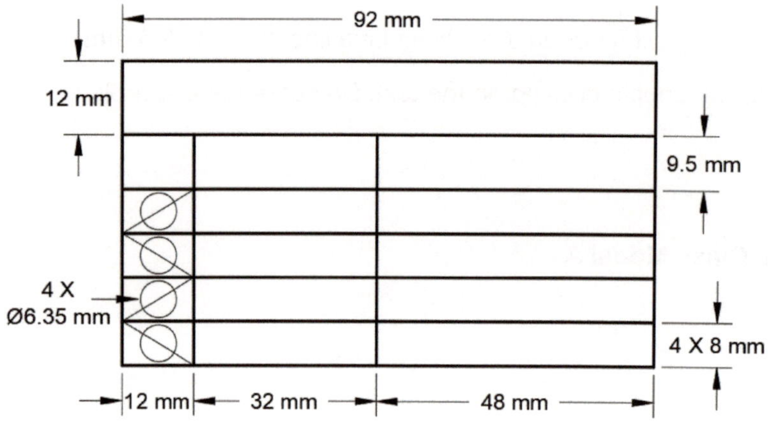

Step 3. Draw the guidelines as shown using Object Snap “Intersection” for the diagonal lines. Use the Offset command for the horizontal and vertical lines.

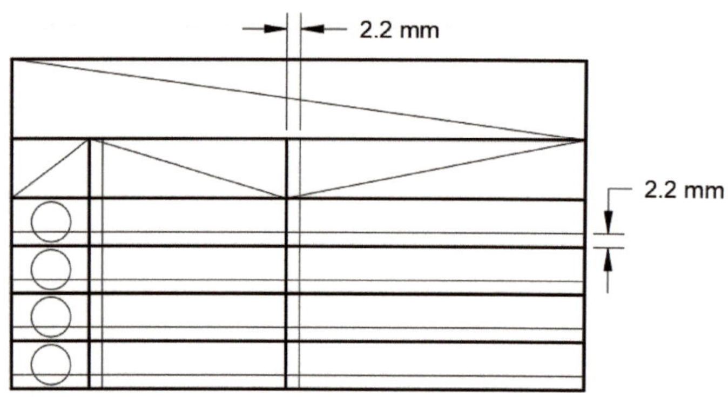

Step 4. Using **“Single Line Text”** and place text as shown. All text height is **3.2 mm** except the title. Title height is **5 mm**.

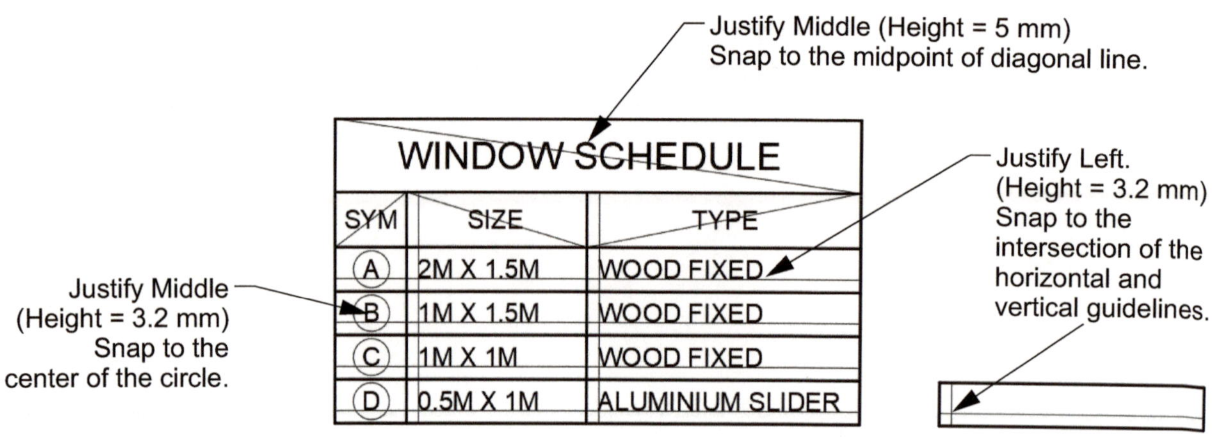

Step 5. Erase all the guidelines.

Exercise 12C

Instructions:

1. Start a **New** file using the **Border A-Inch.dwt** or the **Border A-Metric.dwt**
2. Draw a horizontal and vertical line on Layer Object Line.

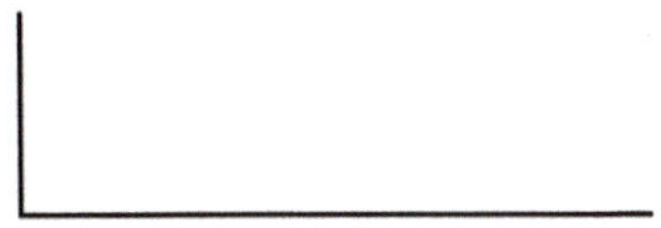

3. Draw the upper half using the Offset command.

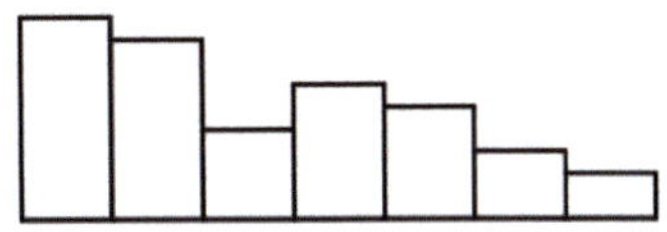

4. **Mirror** the upper half. (Do not include the horizontal Centerline.)
5. Change the horizontal line to Layer Centerline using Properties Palette or Quick Properties.
6. Edit the **Title** and **Ex-XX** by double clicking on the text. Do not erase and replace.
7. Do not dimension.
8. **Save** as **Ex-12C**
9. Plot using Page Setup **Class Model A**

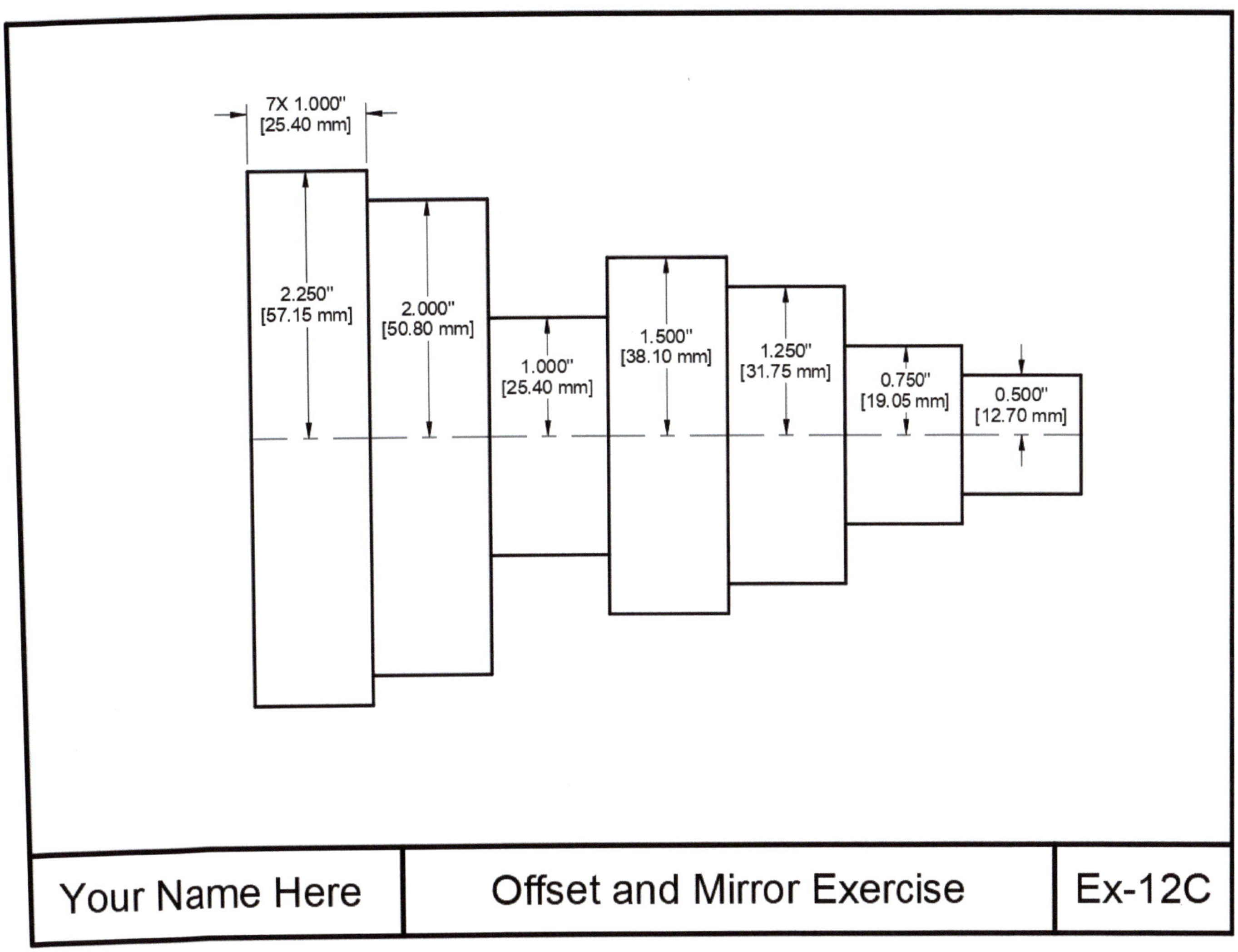

Exercise 12D

Instructions:

1. Start a **New** file using the **Border A-Inch.dwt** or the **Border A-Metric.dwt**
2. Draw the lamp base below.
3. Practice using the "Layer method" when offsetting the hidden lines. Or use the Properties Palette or Quick Properties.
4. Notice that the part is symmetrical. Consider the **Mirror** command.
5. Use Layers Object line, Hidden Line, and Centerline.
6. Edit the **Title** and **Ex-XX** by double clicking on the text. Do not erase and replace.
7. Do not dimension.
8. **Save** as **Ex-12D**
9. Plot using Page Setup **Class Model A**

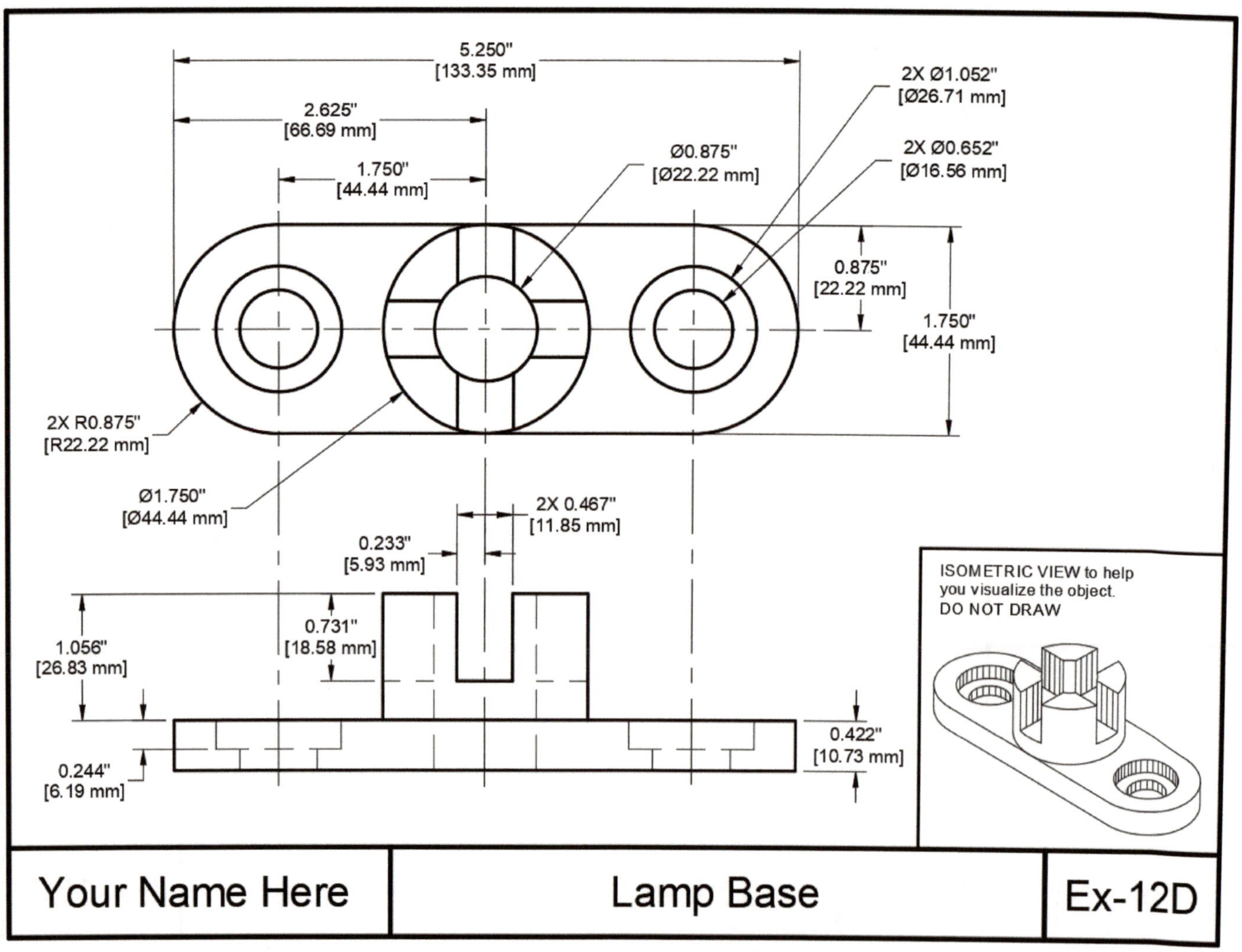

LESSON 13

LEARNING OBJECTIVES

After completing this lesson, you will be able to:

- Create multiple copies in a Rectangular or Circular pattern or Path
- Understand how to Array objects

Array

The **Array** command allows you to make multiple copies in a **Rectangular** or **Polar** (circular) pattern and even on a **Path**. The maximum limit of copies per array is 100,000. This limit can be changed but should accommodate most users. (Refer to the Autodesk Assistant or AutoCAD Help Menu if you choose to change the limit.)

Rectangular Array

This method allows you to make multiple copies of object(s) in a **rectangular pattern**. You specify the number of rows (horizontal), columns (vertical) and the spacing between the rows and columns. The spacing will be equally spaced between copies.

Spacing is sometimes tricky to understand. ***Read this carefully***. The spacing is the **distance from a specific location on the original to that same location** on the future copy. It is **not just the space in between** the two. Refer to the example below.

To use the Rectangular Array command, you will select the object(s) and specify how many rows and columns are desired, and the spacing for the rows and the columns.

Example of a Rectangular Array:

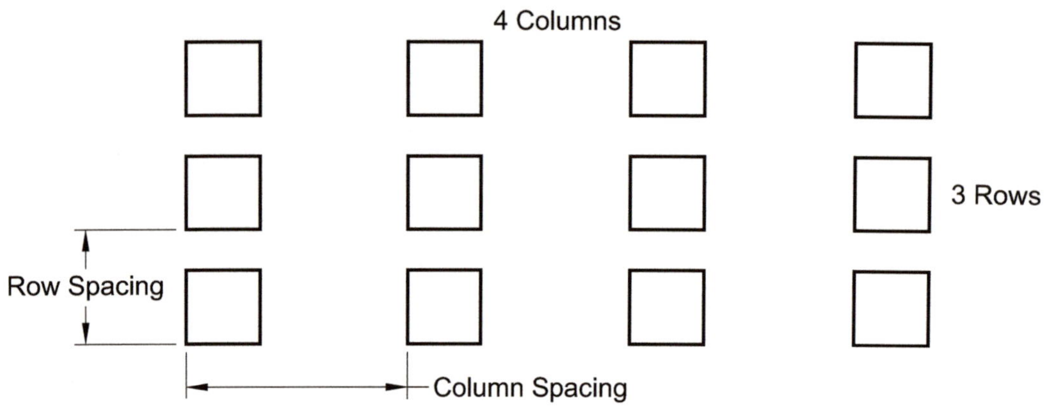

How to create a Rectangular Array

1. Draw a **1" [25.4 mm]** square rectangle. 
2. Select the **Array** command using one of the following:

   Ribbon = Home Tab / Modify Panel / ▾
   or
   Keyboard = *array <Enter>*

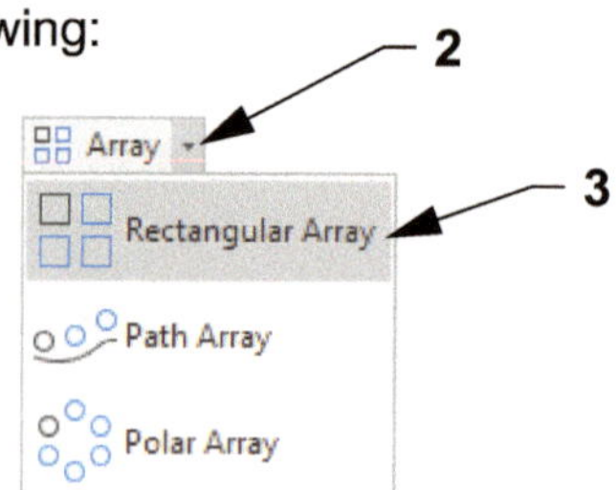

3. Select **Rectanguar Array**.

   The following will appear on the Command Line:

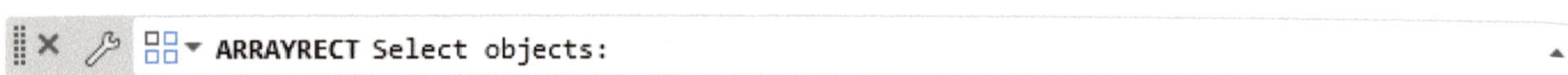

4. Select the object(s) to be arrayed and then press ***<Enter>***.

Continued on the next page...

The **Array Creation** Tab appears and a 3 x 4 default grid array of the object(s) selected.

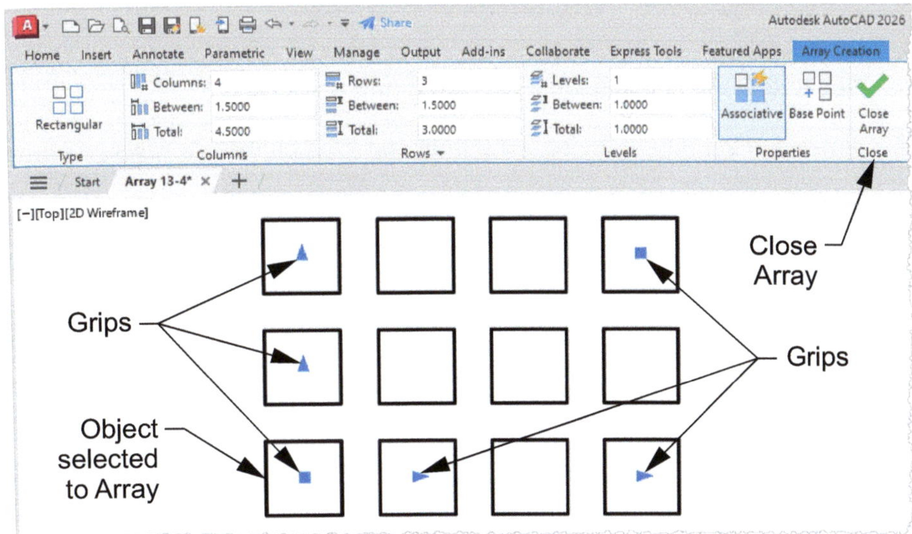

5. Make any changes necessary in the **Array Creation** Tab, then press ***<Enter>*** to display any changes.

6. If the display is correct, select **Close Array**.

How to edit a Rectangular Array

1. Select the Array to edit.

   The **Array** Tab is displayed. (The **Quick Properties** will also be displayed if you have the Quick Properties button **on** in the Status Bar.)

2. Make any changes necessary in the **Array** Tab, then press ***<Enter>*** to display any changes.

3. If the display is correct, select **Close Array**.

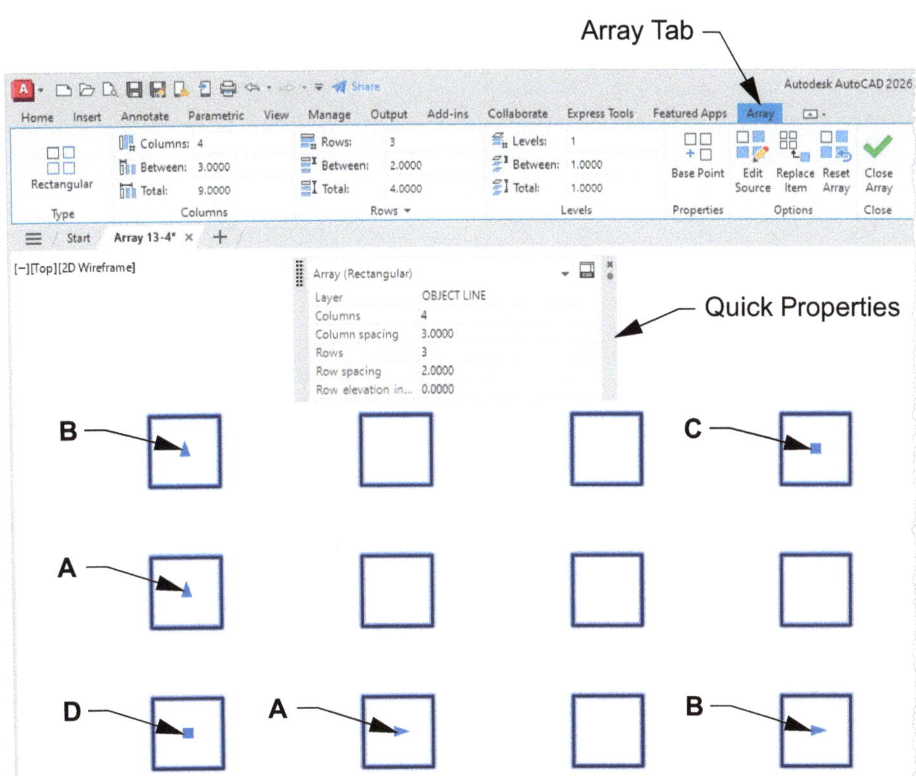

How to edit a Rectangular Array

Using Grips to edit.

You may also use the grips to edit the spacing. Just click on a grip and drag.

A. The first ► or ▲ allows you to change the spacing between the columns or rows.

B. The last ► or ▲ allows you to change the total spacing between the basepoint and the last ► or ▲ and also to add extra columns or rows, or change the axis angle.

C. The ■ allows you to change the total row and column spacing simultaneously, and also to add extra columns and rows simultaneously.

D. Use the basepoint grip ■ to **Move** the entire Array.

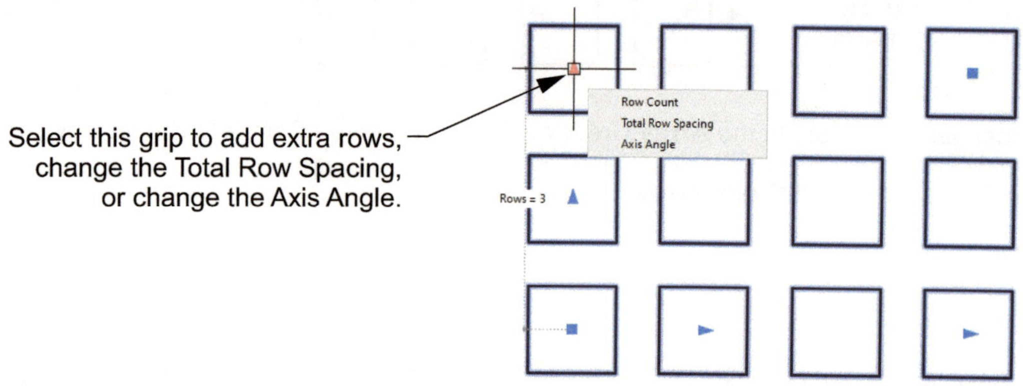

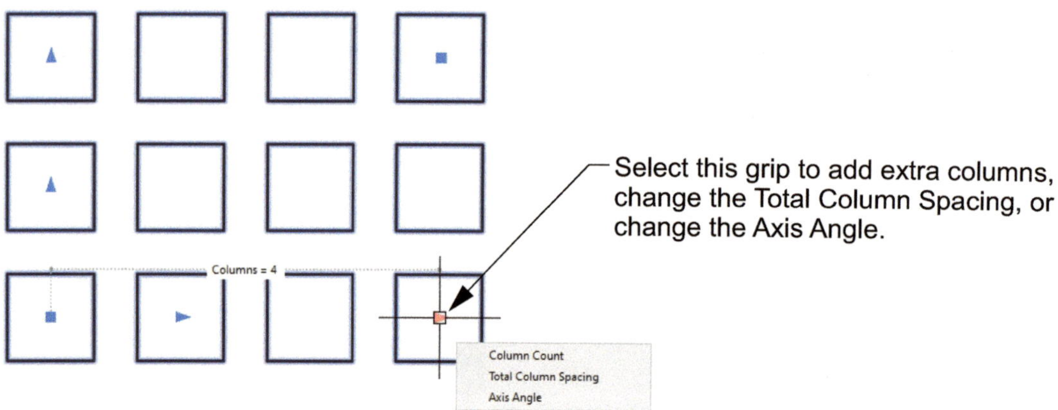

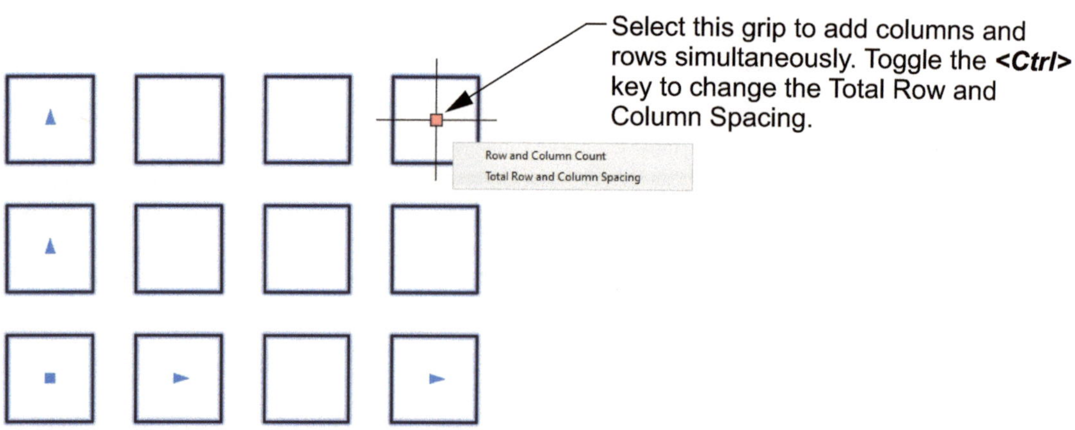

Continued on the next page...

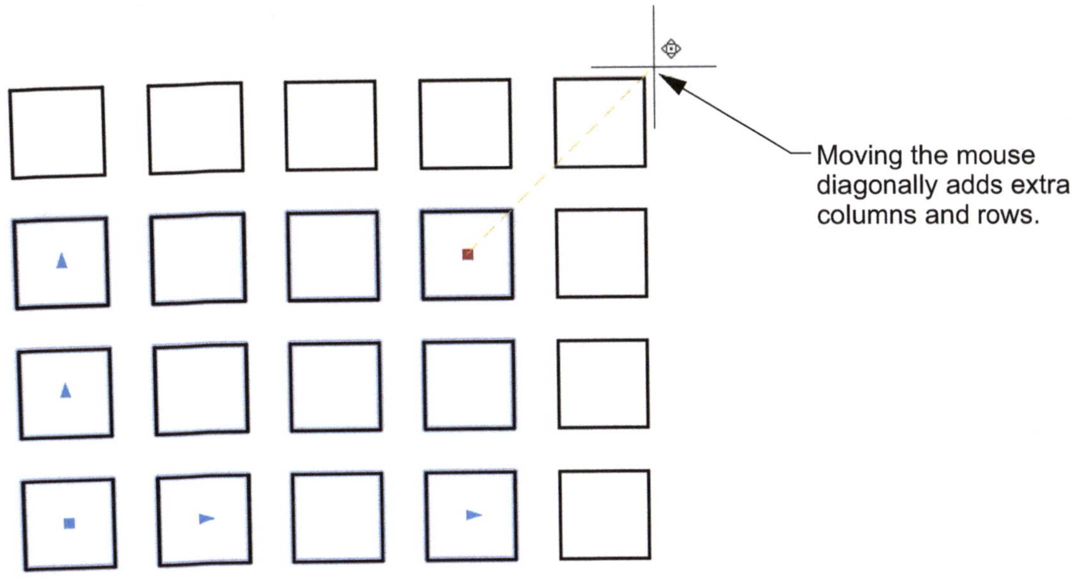

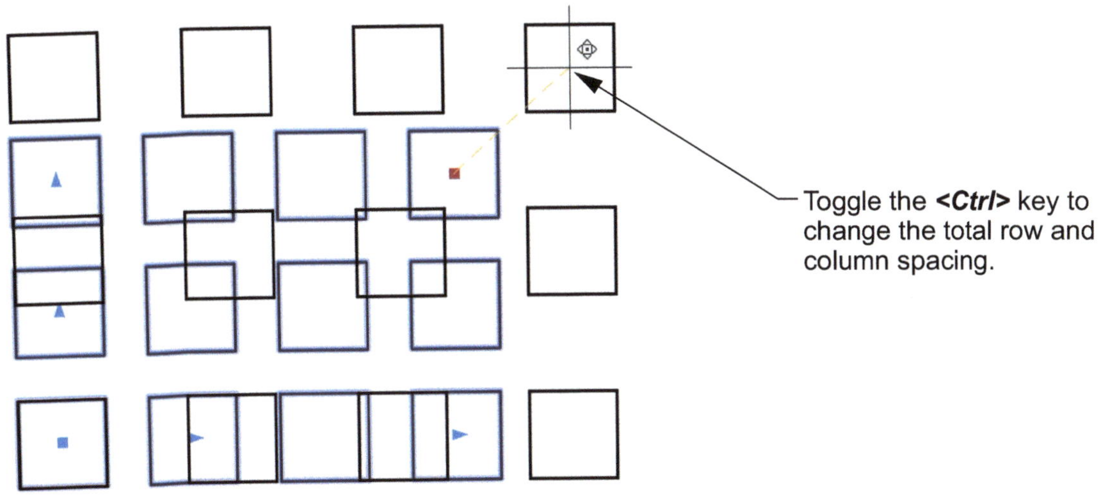

Polar Array

This method allows you to make multiple copies in a **circular pattern**. You specify the total number of copies to fill a specific angle or specify the angle between each copy and angle to fill.

To use the Polar Array command, you select the object(s) to array and specify the center of the array, the number of copies or the angle between the copies, the angle to fill, and if you would like the copies to rotate as they are copied, or not rotate as they are copied.

Example of a Polar Array:

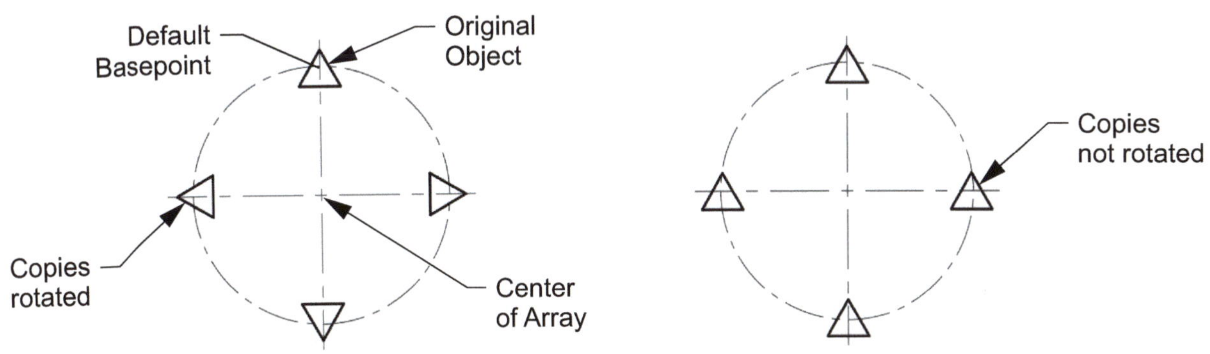

Note: The two examples shown on the previous page use the **objects default basepoint**. The examples below displays what happens if you **specify a basepoint**.

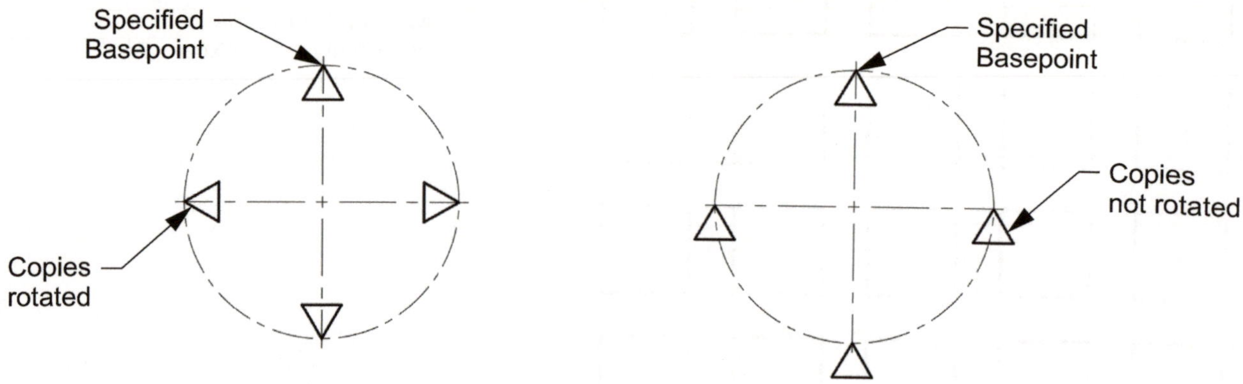

How to create a Polar Array

Using "Number of Items".

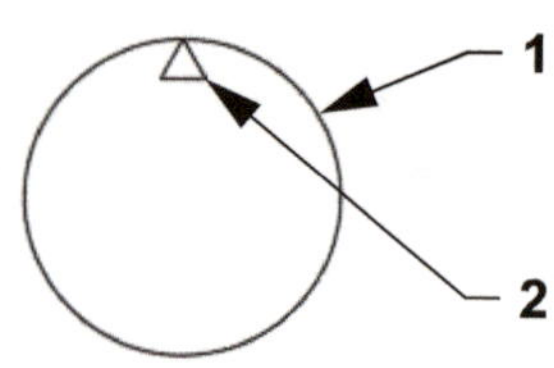

1. Draw a **3" [76.2mm]** radius circle.

2. Add a **0.50" [12.7 mm]** radius three-sided polygon and place as shown.

3. Select the **Array** command using one of the following:

   Ribbon = Home Tab / Modify Panel / ▾
   or
   Keyboard = *array <Enter>*

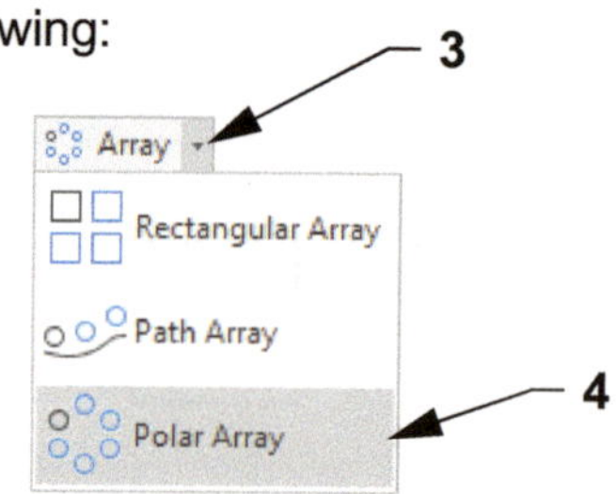

4. Select **Polar Array**.

   The following will appear on the Command Line:

```
ARRAYPOLAR Select objects:
```

5. Select the object to be arrayed (the polygon) and then press ***<Enter>***.

   The following will appear on the Command Line:

```
ARRAYPOLAR Specify center point of array or [Base point Axis of rotation]:
```

6. Select the center point of the circle.

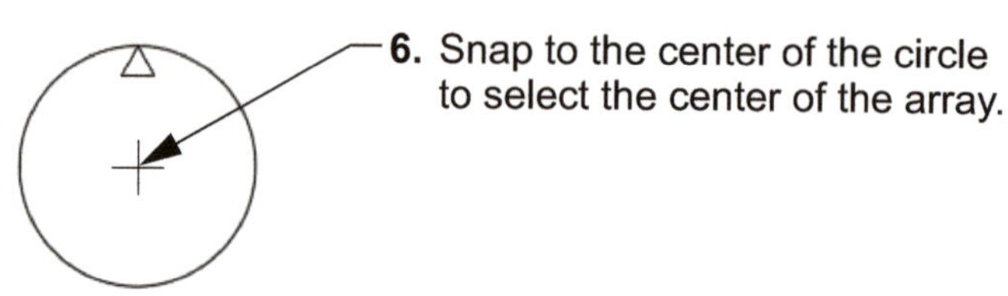

Continued on the next page...

The **Array Creation** Tab appears and the array defaults to 6 items.

7. Enter Items: **12**
8. Enter Fill: **360**
9. Press ***<Enter>*** to display the selections.
10. Select **Close Array** if display is correct.

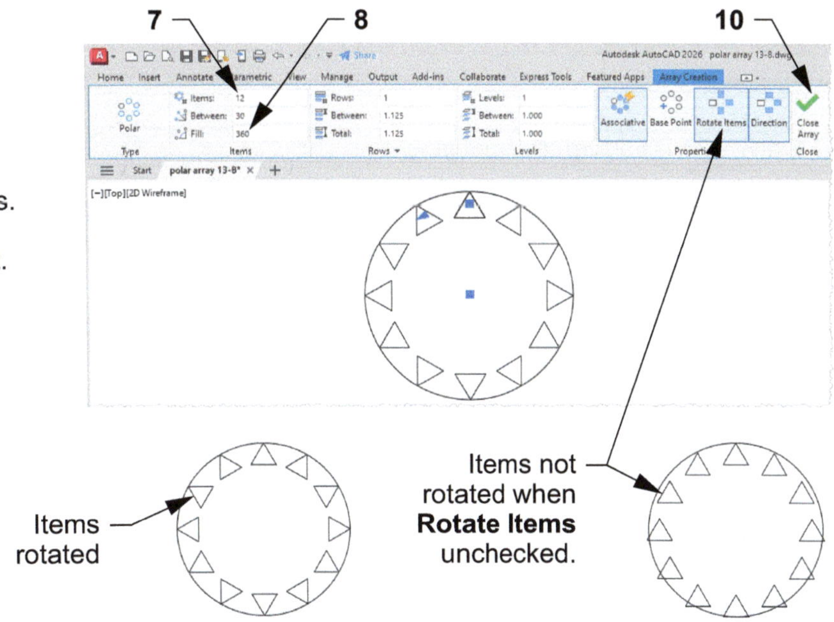

Note: 12 items were evenly distributed within 360 degrees.

How to create a Polar Array

Using "Angle Between".

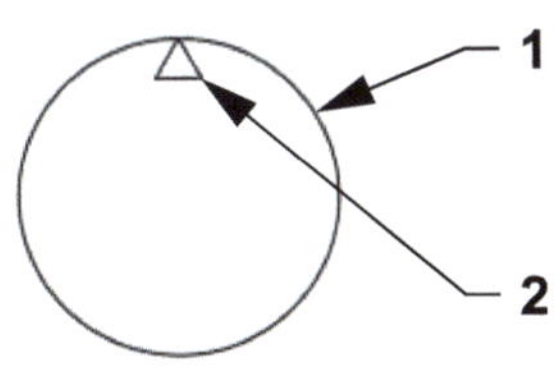

1. Draw a **3" [76.2mm]** radius circle.
2. Add a **0.50" [12.7 mm]** radius three-sided polygon and place as shown.
3. Select **Polar Array.**
4. Select the object to be arrayed (the polygon) and then press ***<Enter>***.
5. Select the center point of the circle.

The **Array Creation** Tab appears and the array defaults to 6 items.

6. If not already correct, enter Items: **6**
7. Enter Between: **45**
8. Press ***<Enter>*** to display the selections.
9. Select **Close Array** if display is correct.

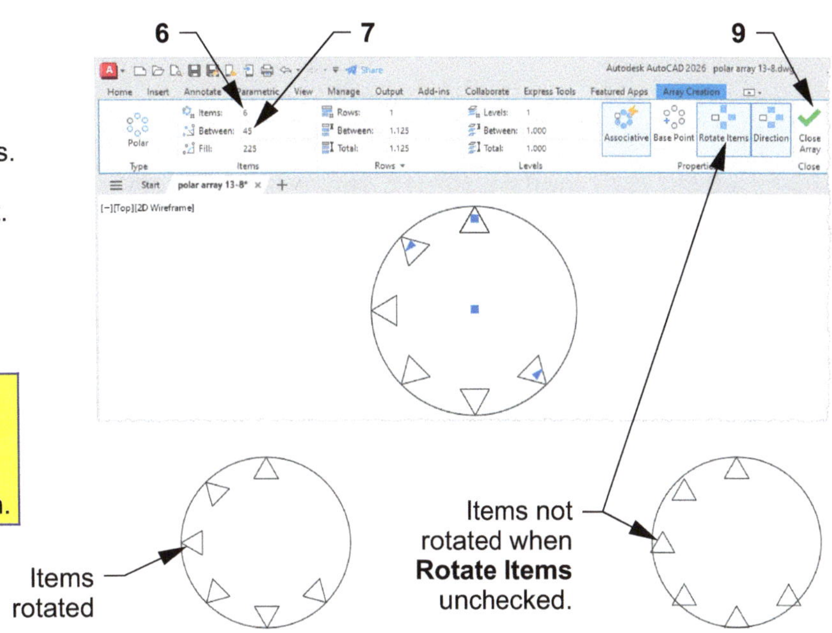

Note: 6 items were copied at each 45 degree counter-clockwise.

To have the items copied in a clockwise direction, uncheck the **"Direction"** button.

How to create a Polar Array

Using "Fill Angle".

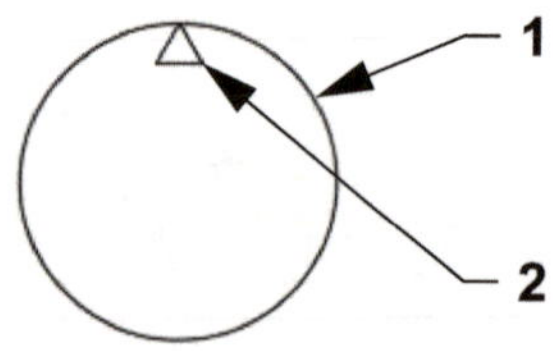

1. Draw a **3" [76.2mm]** radius circle.
2. Add a **0.50" [12.7 mm]** radius three-sided polygon and place as shown.
3. Select **Polar Array.**
4. Select the object to be arrayed (the polygon) and then press ***<Enter>***.
5. Select the center point of the circle.

The **Array Creation** Tab appears and the array defaults to 6 items.

6. Enter items: **8**
7. Enter Fill: **180**
8. Press ***<Enter>*** to display the selections.
9. Select **Close Array** if display is correct.

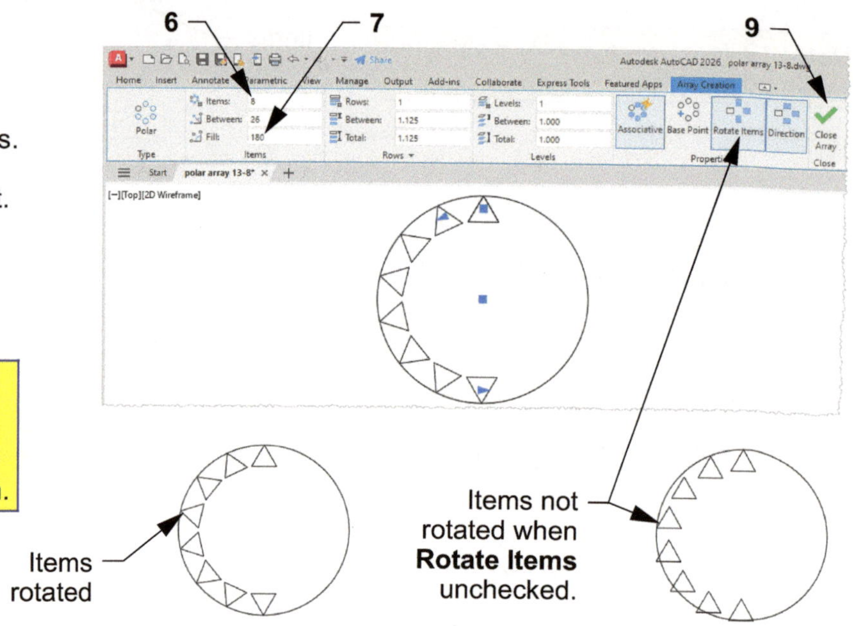

Note: 8 items were evenly distributed within 180 degree counter-clockwise.

To have the items copied in a clockwise direction, uncheck the **"Direction"** button.

How to create a Path Array

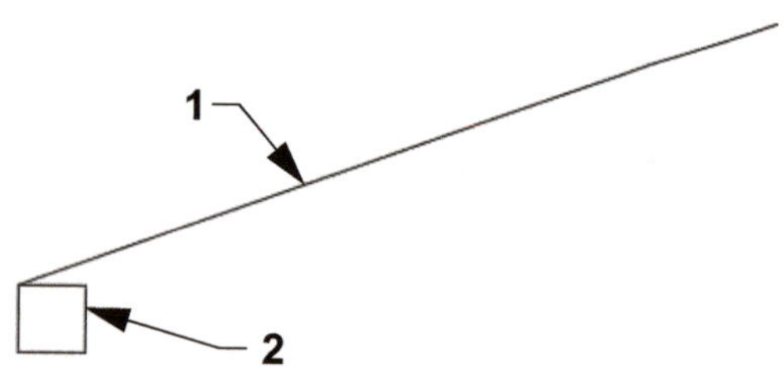

1. Draw a Line **6" [152.4 mm]** long at **20** degrees.
2. Add a **0.50" [12.7 mm] x 0.50" [12.7 m]** rectangle and place as shown.
3. Select the **Array** command using one of the following:

   Ribbon = Home Tab / Modify Panel / ▾
   or
   Keyboard = *array <Enter>*

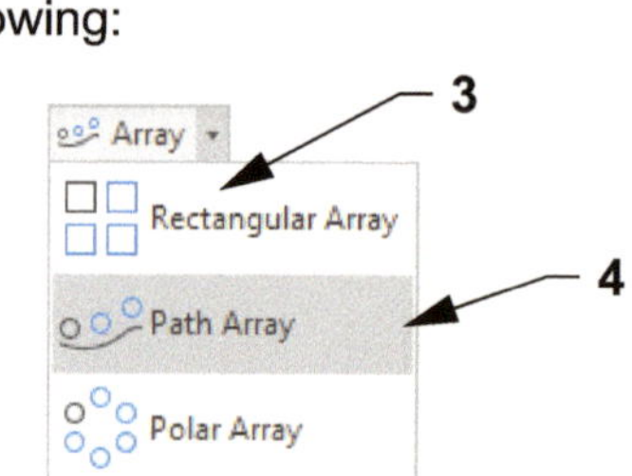

4. Select **Path Array**.

Continued on the next page...

The following will appear on the Command Line:

```
ARRAYPATH Select objects:
```

5. Select the object to be arrayed (the small rectangle) and then press ***<Enter>***.

   The following will appear on the Command Line:

```
ARRAYPATH Select path curve:
```

6. Select the path curve (the angled line).

CAD TIP! The **Path** can be a line, polyline, spline, helix, arc, circle, or ellipse.

The **Array Creation** Tab appears and the array defaults to 9 items.

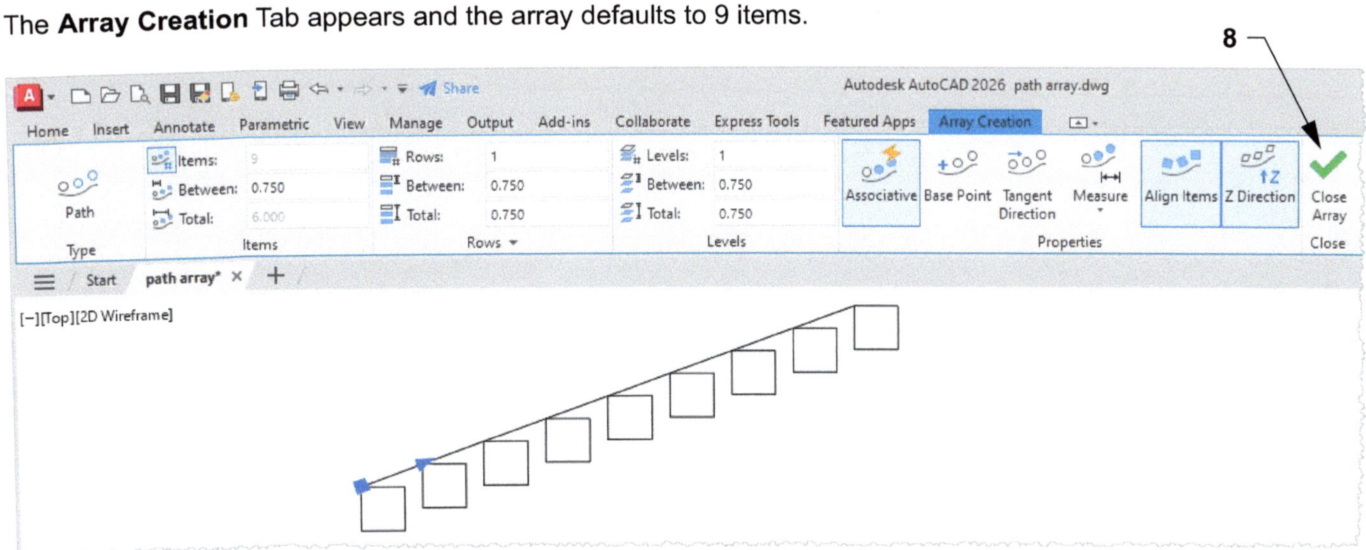

If you want to change the number of items on the path, you will need to select the **Measure** drop-down arrow on the **Properties Panel**, then select **Divide**. You can then change the number of items in the **Items Panel**.

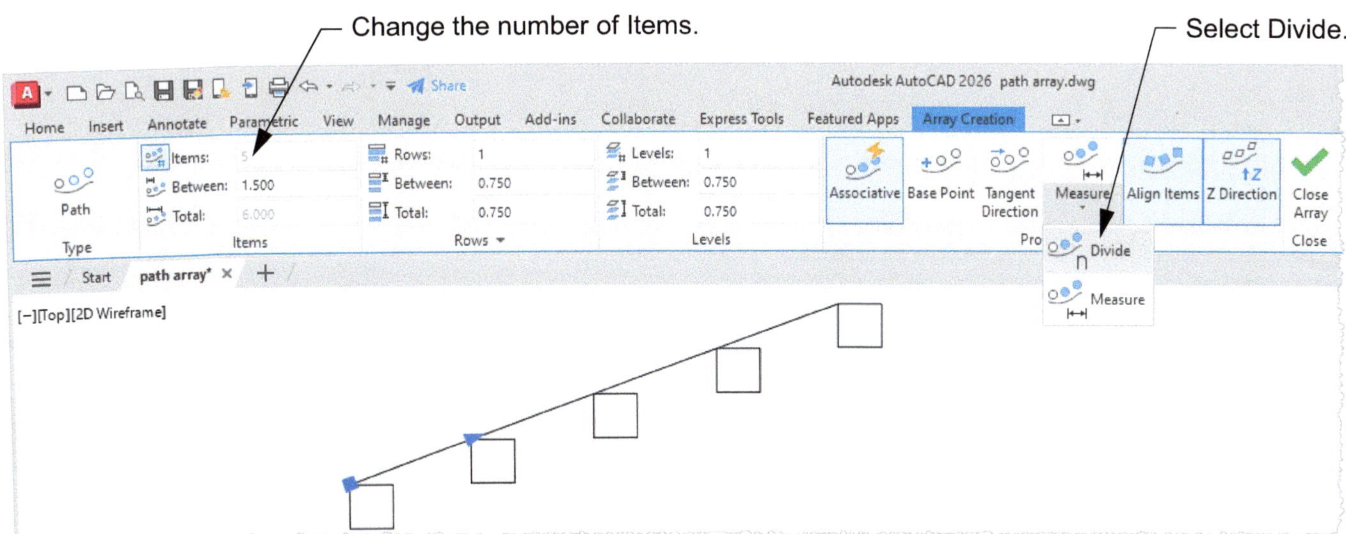

7. Make any alterations and press ***<Enter>*** to display.

8. If correct, select **Close Array**.

Exercise 13A

Instructions:

1. Start a **New** file using the **Border A-Inch.dwt** or the **Border A-Metric.dwt**
2. Draw the original **0.500" [12.7 mm]** square on Layer Object Line.
3. Array the original square as shown.
4. Edit the **Title** and **Ex-XX** by double clicking on the text. Do not erase and replace.
5. Do not dimension.
6. **Save** as **Ex-13A**
7. Plot using Page Setup **Class Model A**

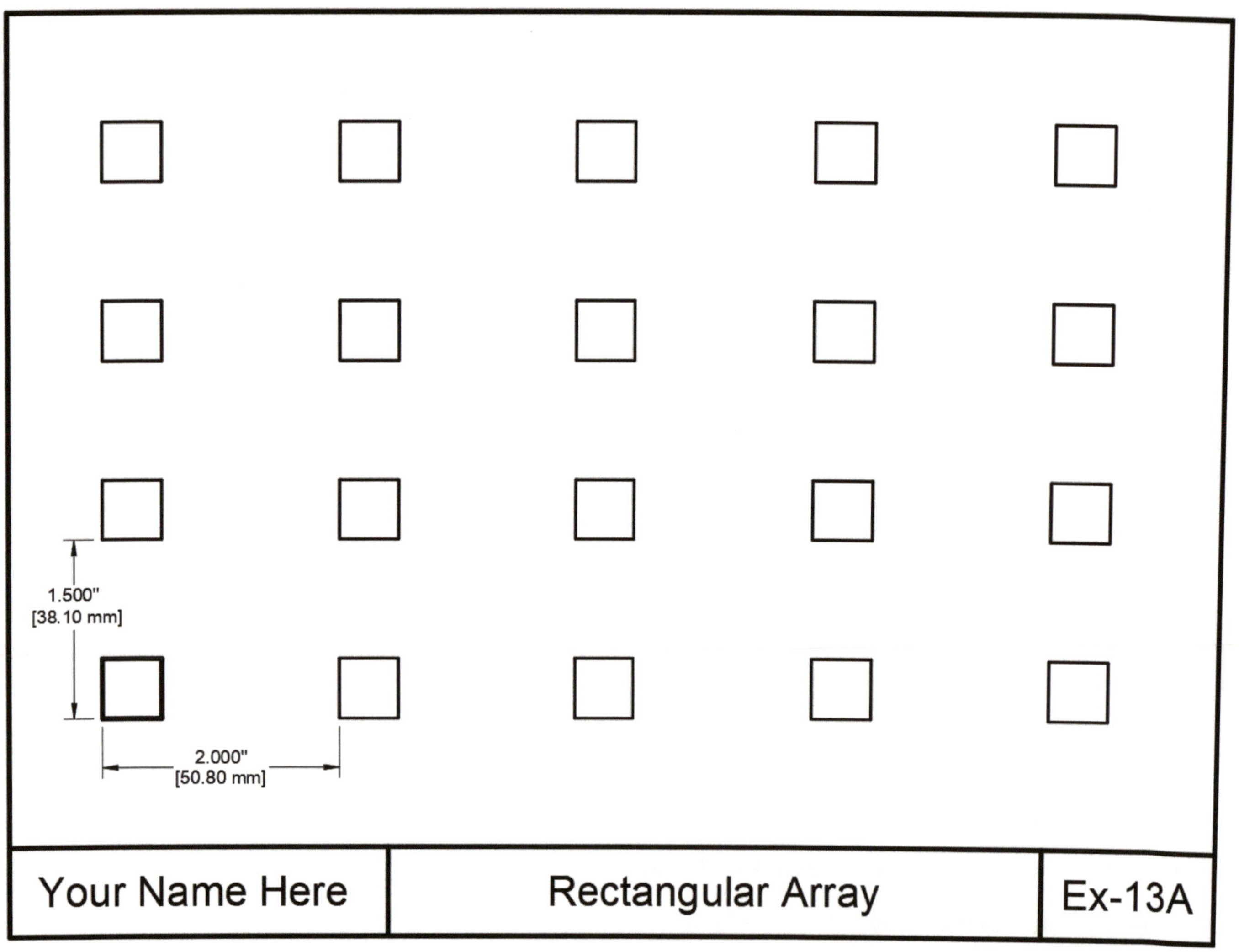

Exercise 13B

Instructions:

1. Start a **New** file using the **Border A-Inch.dwt** or the **Border A-Metric.dwt**
2. Draw the objects shown below using the most efficient methods.
3. Use Layer Object Line.
4. Edit the **Title** and **Ex-XX** by double clicking on the text. Do not erase and replace.
5. Do not dimension.
6. **Save** as **Ex-13B**
7. Plot using Page Setup **Class Model A**

Refer to the next page for helpful hints and more dimensions.

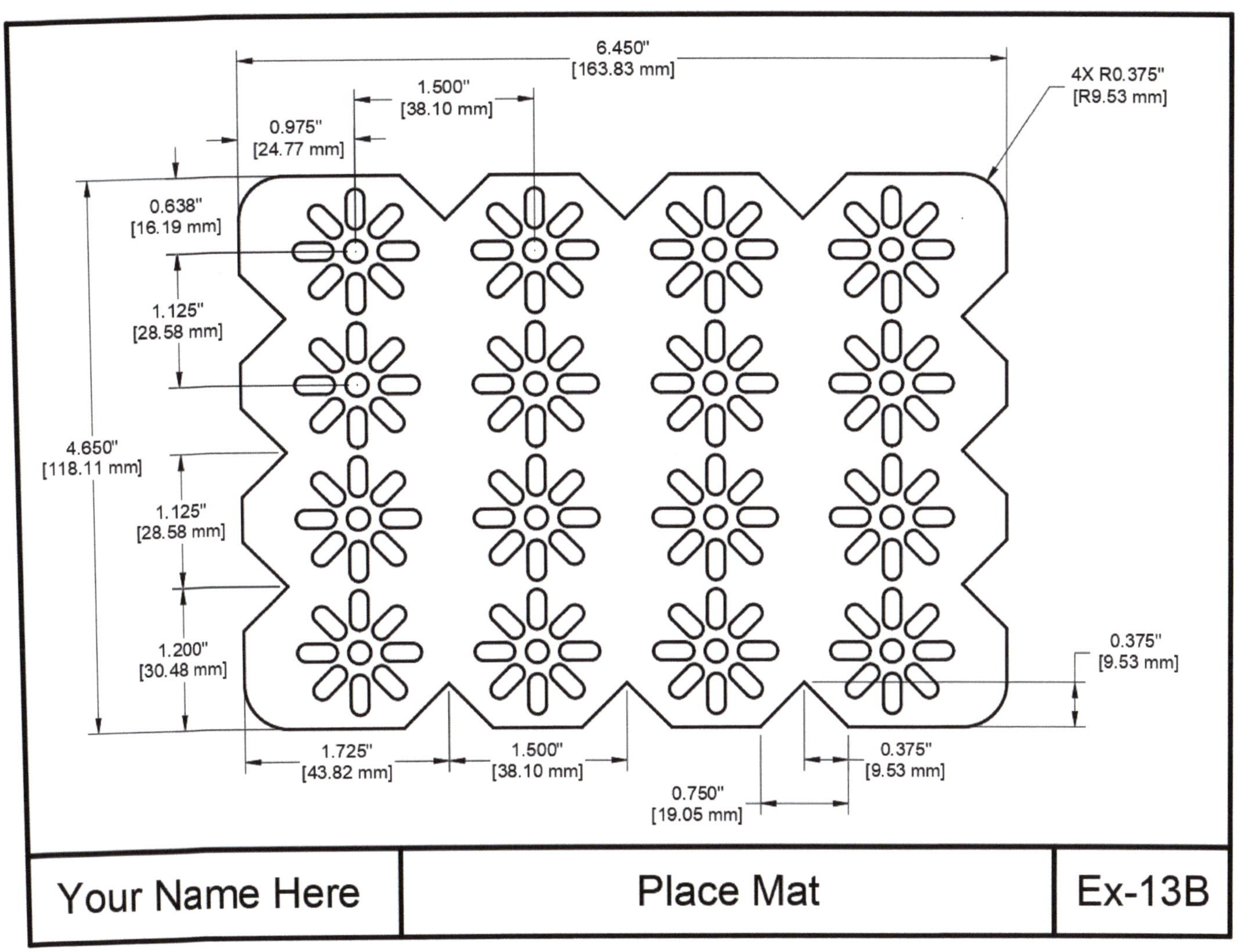

Exercise 13B-Helper

Notches

Step 1.
Draw the outside rectangle and then offset the location for the center of the notches.

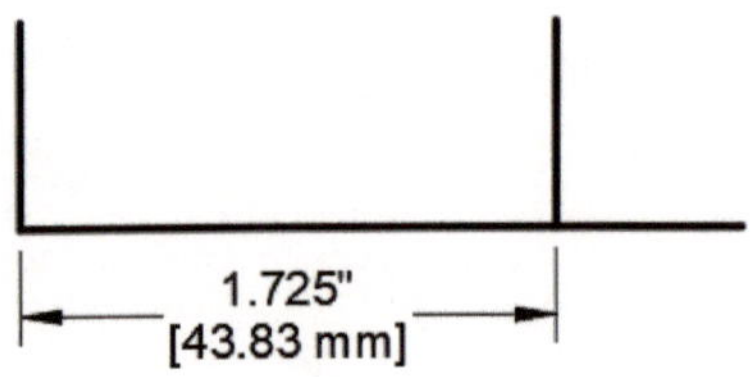

Step 2.
Use Offset to frame the **0.375" [9.53 mm]** width and height.

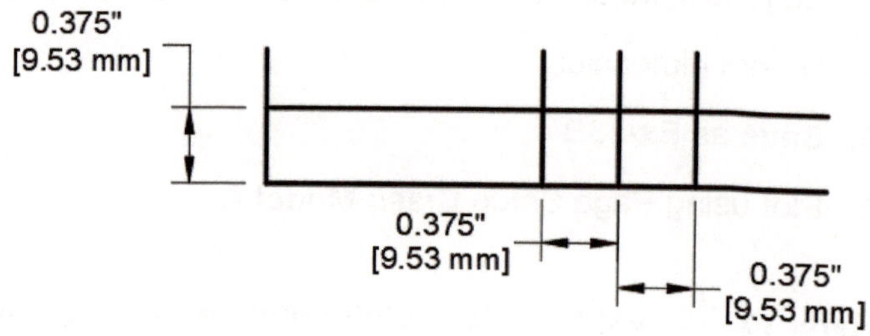

Step 3.
Draw the diagonal lines by snapping to intersections.

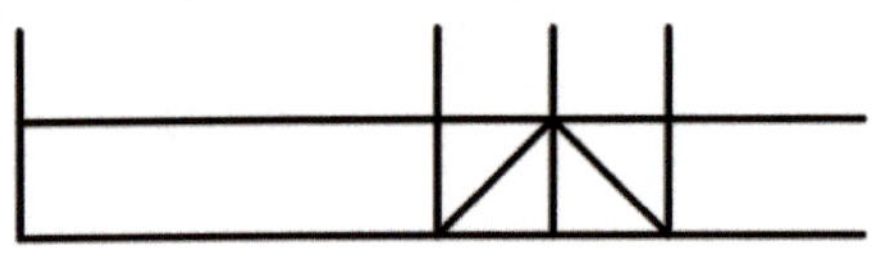

Step 4.
Trim and Erase the lines.

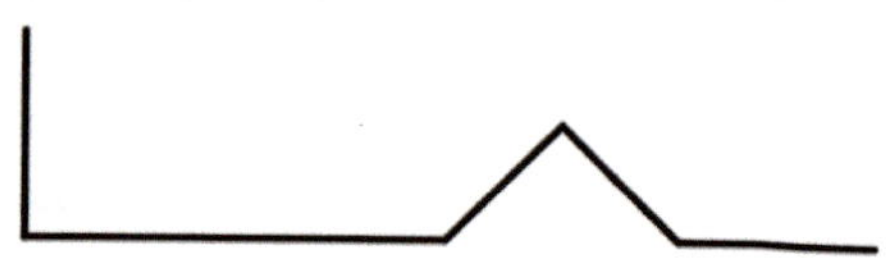

Flower

Step 1.
Offset.

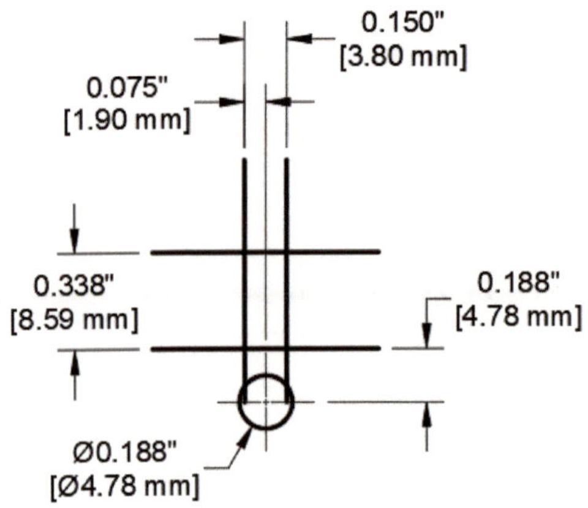

Step 2.
Fillet.

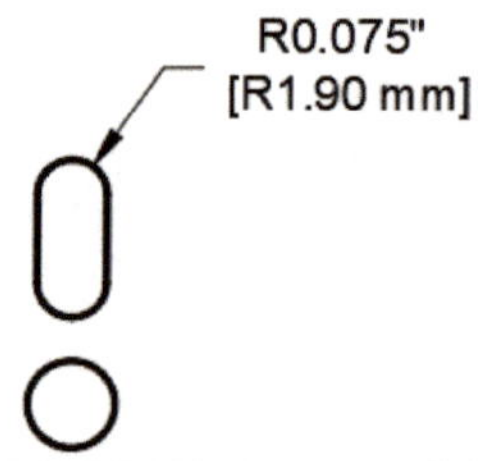

Step 3.
Array.

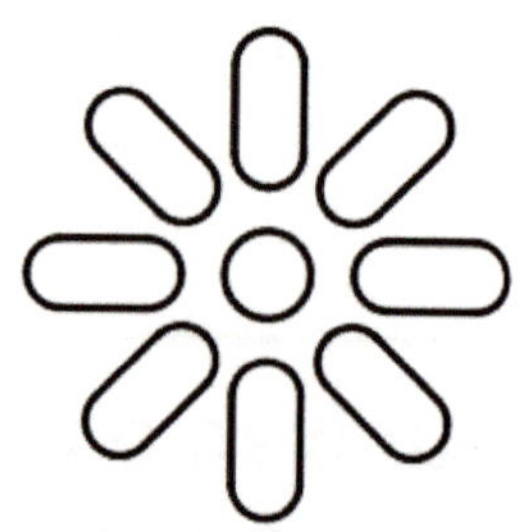

Exercise 13C

Instructions:

1. Start a **New** file using the **Border A-Inch.dwt** or the **Border A-Metric.dwt**
2. Draw the Centerlines and circles first.
3. Draw one polygon at the 12:00 location.
4. Array the original polygon as shown.
5. Edit the **Title** and **Ex-XX** by double clicking on the text. Do not erase and replace.
6. Do not dimension.
7. **Save** as **Ex-13C**
8. Plot using Page Setup **Class Model A**

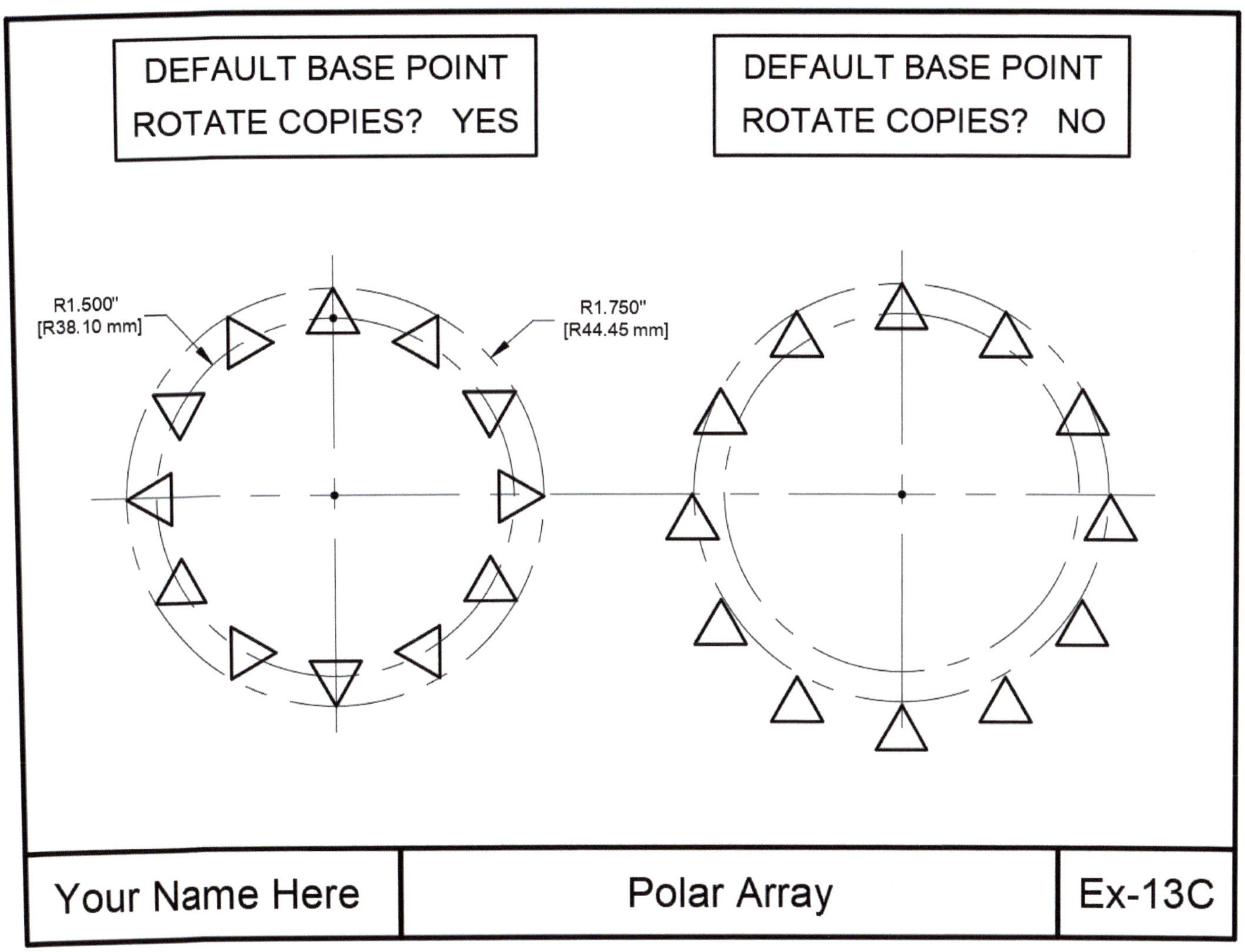

Exercise 13D

Instructions:

1. Start a **New** file using the **Border A-Inch.dwt** or the **Border A-Metric.dwt**
2. Draw the line first. **9.000" [228.6 mm]** long and at an angle of **30** degrees.
3. Draw one circle with a radius of **0.500" [12.7 mm]** at the lower left end.
4. Array **8** circles along the Path (Line), evenly divided.
5. Edit the **Title** and **Ex-XX** by double clicking on the text. Do not erase and replace.
6. Do not dimension.
7. **Save** as **Ex-13D**
8. Plot using Page Setup **Class Model A**

Note: When the Array Tab appears, you will need to click on the **Measure** drop-down arrow on the Properties Panel, and then click on **Divide**. This will enable you to change the number of items to **8**.

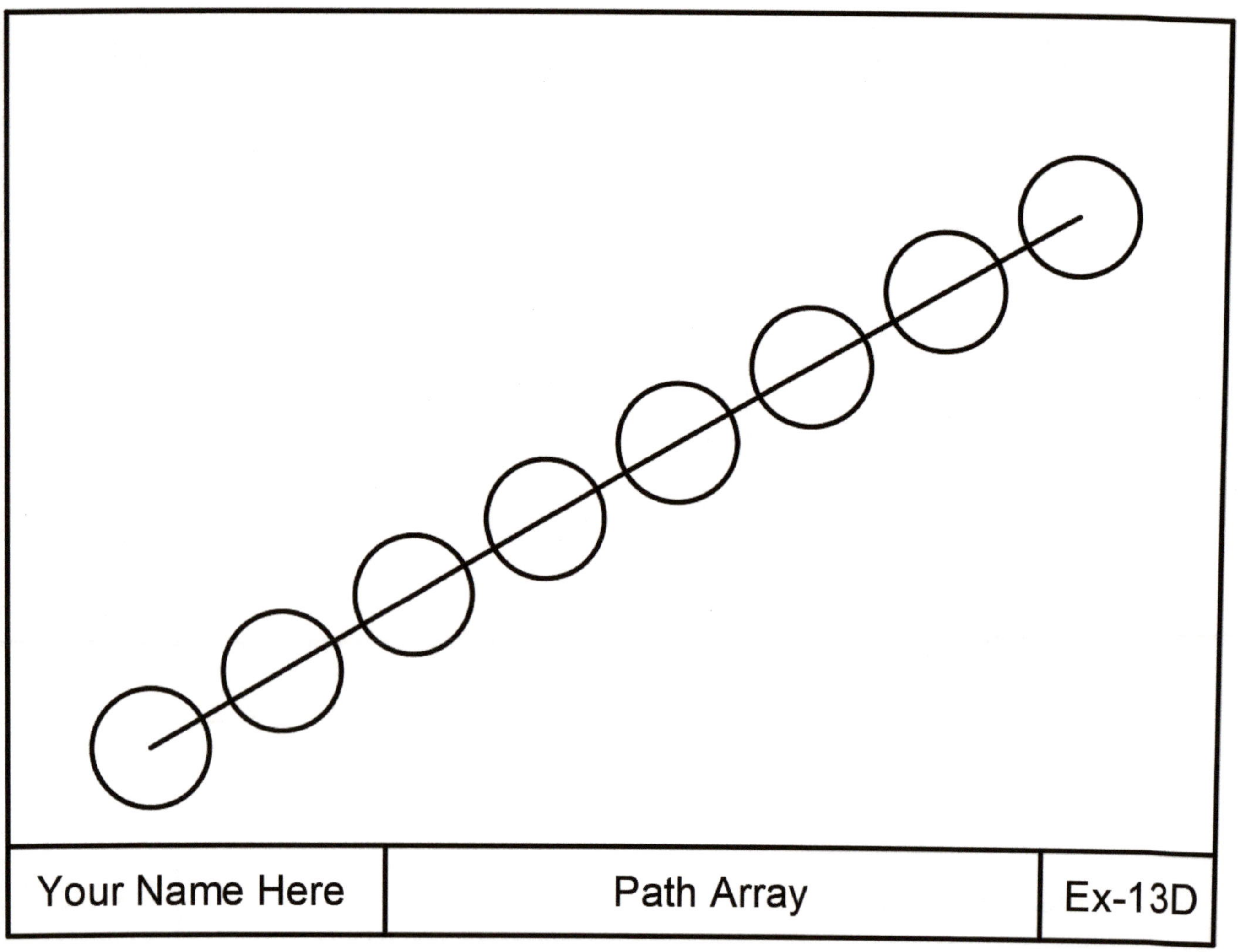

LESSON 14

LEARNING OBJECTIVES

After completing this lesson, you will be able to:

- Make an existing object larger or smaller proportionately
- Stretch or Compress an existing object
- Rotate an existing object to a specific Angle

Scale

The **Scale** command is used to make one or more objects larger or smaller **proportionately**. You may scale using a scale factor or a reference length. You must also specify a basepoint. The basepoint is a stationary point from which the objects scale.

1. Select the **Scale** command using one of the following:

   Ribbon = Home Tab / Modify Panel / 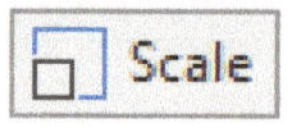

   or
   Keyboard = *scale <Enter>*

Scale Factor

The following will appear on the Command Line:

2. Select the objects and then press ***<Enter>***.

   The following will appear on the Command Line:

3. Select the stationary point on the objects (left click).

   The following will appear on the Command Line:

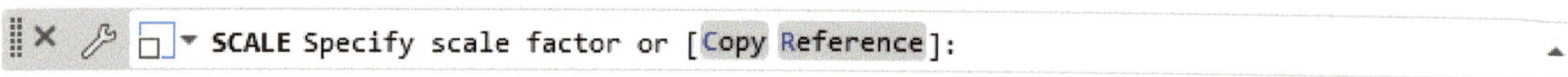

4. Type the **scale factor** and then press ***<Enter>***.

If the scale factor is greater than 1, the objects will increase in size.
If the scale factor is less than 1, the objects will decrease in size.

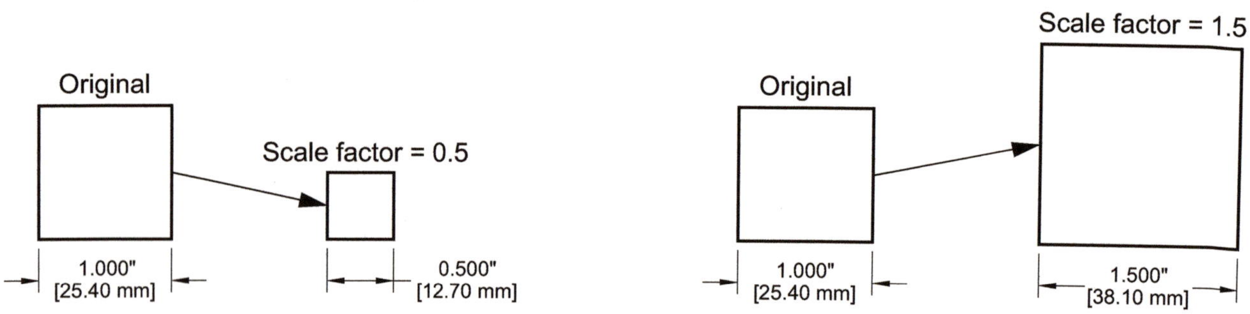

Reference Option

1. Select the **Scale** command.

   The following will appear on the Command Line:

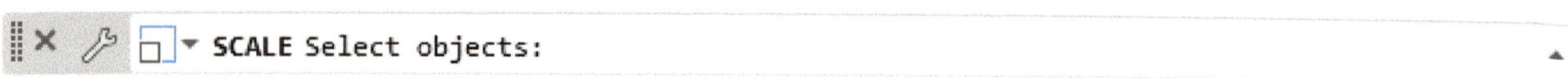

2. Select the objects and then press ***<Enter>***.

Continued on the next page...

The following will appear on the Command Line:

```
SCALE Specify base point:
```

3. Select the stationary point on the objects (left click).

   The following will appear on the Command Line:

```
SCALE Specify scale factor or [Copy Reference]:
```

4. Type ***R*** and then press ***<Enter>***.

   The following will appear on the Command Line:

```
SCALE Specify reference length <1.000>:
```

5. Specify a **reference length** and then press ***<Enter>***.

   The following will appear on the Command Line:

```
SCALE Specify new length or [Points] <1.500>:
```

6. Specify the new length and then press ***<Enter>***.

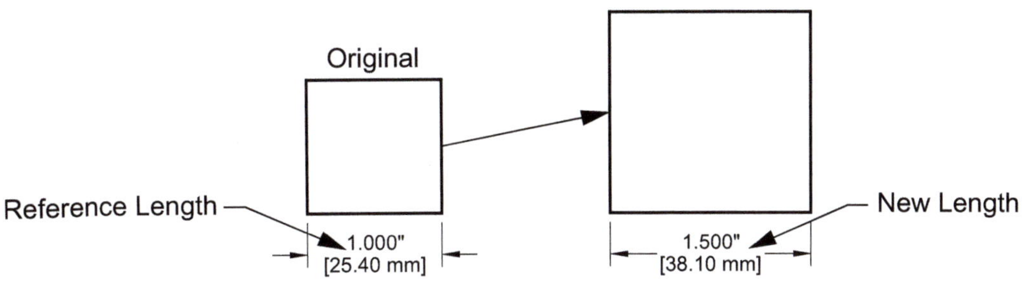

Copy Option

The **Copy Option** of the **Scale** command creates a duplicate of one or more selected objects. The duplicate is directly on top of the original. The duplicate will be scaled. The Original remains the same.

Stretch

The **Stretch** command allows you to stretch or compress one or more objects. Unlike the Scale command, you can alter an object's proportion with the Stretch command. In other words, you may increase the length without changing the width and vice versa.

Stretch is a very valuable tool. Take some time to really understand this command. It will save you hours when making corrections to drawings.

When selecting the objects, you must use a **Crossing Window**.

Objects that are crossed, will **stretch**.

Objects that are totally enclosed, will **move**.

Continued on the next page...

1. Select the **Stretch** command using one of the following:

   Ribbon = Home Tab / Modify Panel / 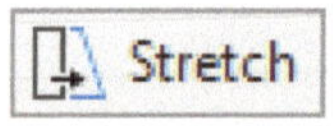

   or
   Keyboard = *S <Enter>*

   The following will appear on the Command Line:

```
STRETCH Select objects:
```

2. Select the first corner of the Crossing Window (left click).

   The following will appear on the Command Line:

```
STRETCH Select objects: Specify opposite corner:
```

3. Specify the opposite corner of the Crossing Window (left click) and then press ***<Enter>***.

   The following will appear on the Command Line:

```
STRETCH Specify base point or [Displacement] <Displacement>:
```

4. Select a basepoint (where it stretches from).

   The following will appear on the Command Line:

```
STRETCH Specify second point or <use first point as displacement>:
```

5. Type coordinates or place the location with the cursor.

Before Stretch

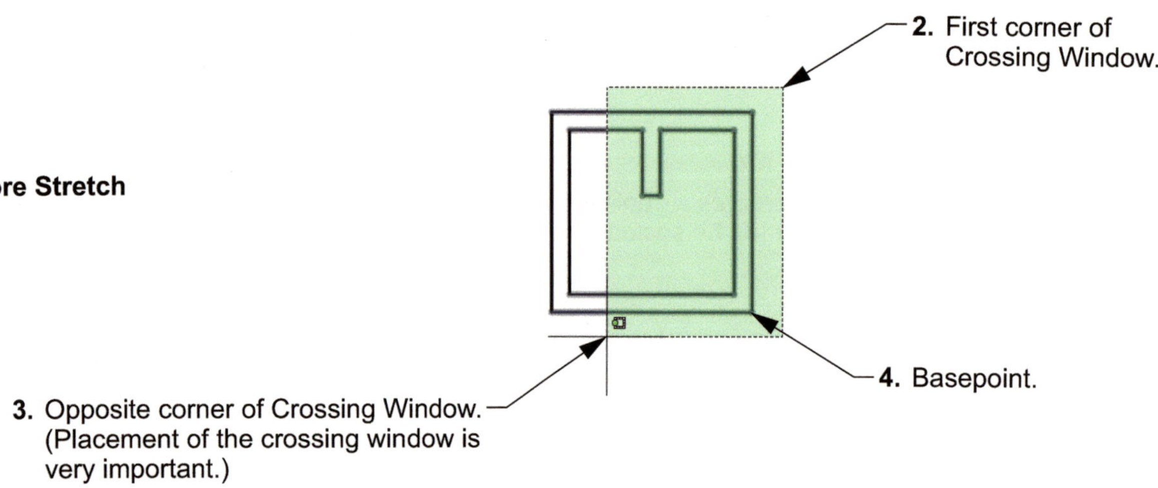

After Stretch

(Notice which objects stretched and which objects moved.)

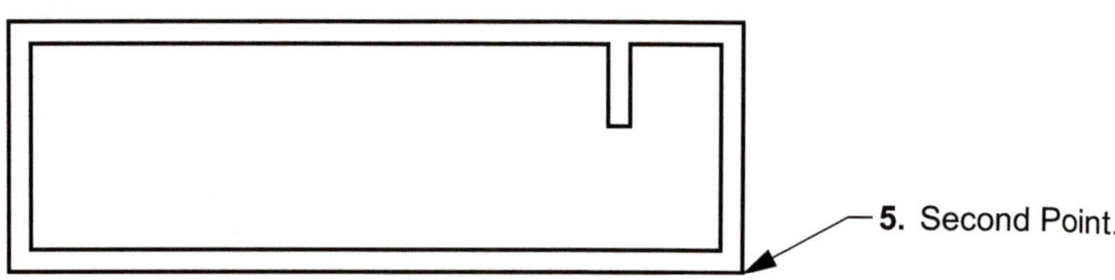

Rotate

The **Rotate** command is used to rotate objects around a basepoint (pivot point).

After selecting the objects and the basepoint, you will enter the rotation angle from its **current** rotation angle or select a reference angle followed by the new angle.

A **Positive** rotation angle revolves the objects counter-clockwise. A **Negative** rotation angle revolves the objects clockwise.

1. Select the **Rotate** command using one of the following:

   Ribbon = Home Tab / Modify Panel / 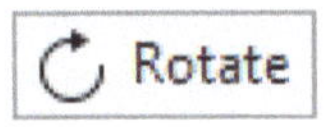

   or
   Keyboard = ***RO <Enter>***

Rotation Angle Option

The following will appear on the Command Line:

```
ROTATE Select objects:
```

2. Select the objects to rotate and then press ***<Enter>***.

   The following will appear on the Command Line:

```
ROTATE Specify base point:
```

3. Select the basepoint (pivot point).

   The following will appear on the Command Line:

```
ROTATE Specify rotation angle or [Copy Reference] <0>:
```

4. Type the angle of rotation and then press ***<Enter>***.

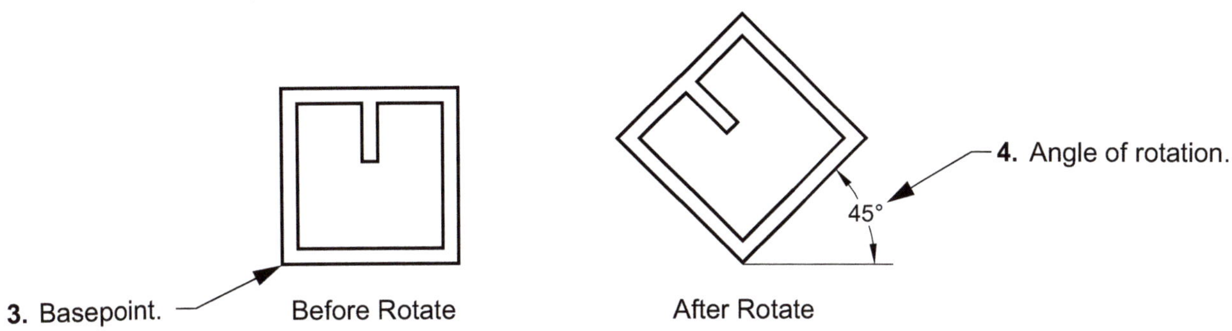

Reference Option

1. Select the **Rotate** command.

   The following will appear on the Command Line:

```
ROTATE Select objects:
```

2. Select the objects to rotate and then press ***<Enter>***.

Continued on the next page...

The following will appear on the Command Line:

```
ROTATE Specify base point:
```

3. Select the basepoint (pivot point).

The following will appear on the Command Line:

```
ROTATE Specify rotation angle or [Copy Reference] <0>:
```

4. Type ***R*** and then press ***<Enter>***.

The following will appear on the Command Line:

```
ROTATE Specify the reference angle <0>:
```

5. Snap to the first reference point on the objects.

The following will appear on the Command Line:

```
ROTATE Specify the reference angle <0>:  Specify second point:
```

6. Snap to the second reference point on the objects.

The following will appear on the Command Line:

```
ROTATE Specify the new angle or [Points] <0>:
```

7. Type ***P*** and then press ***<Enter>***.

The following will appear on the Command Line:

```
ROTATE Specify first point:
```

8. Select the first endpoint of the new angle.

The following will appear on the Command Line:

```
ROTATE Specify first point:  Specify second point:
```

9. Select the second endpoint of the new angle.

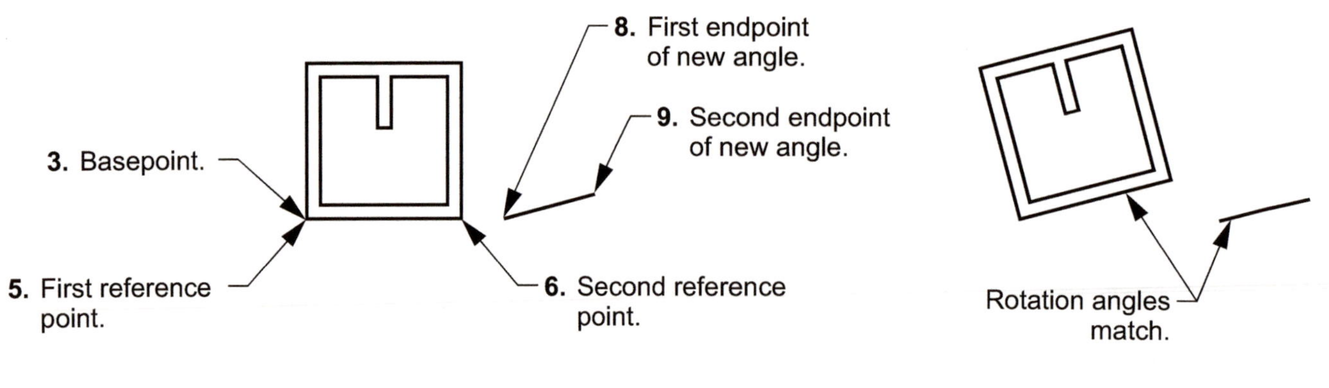

Before Rotate After Rotate

Copy Option

The **Copy Option** of the **Rotate** command creates a duplicate of selected objects. The duplicate is rotated at the angle you enter. The Original remains in the same position.

Exercise 14A

Instructions:

1. Start a **New** file using the **Border A-Inch.dwt** or the **Border A-Metric.dwt**
2. Draw both **2" [50.8 mm]** Squares shown below labelled **"Original"**. Use Layer Object Line.
3. Scale the original on the left using **"Scale Factor"** method.
4. Scale the original on the right using **"Scale Reference"** method.
5. Edit the **Title** and **Ex-XX** by double clicking on the text. Do not erase and replace.
6. Do not dimension.
7. **Save** as **Ex-14A**
8. Plot using Page Setup **Class Model A**

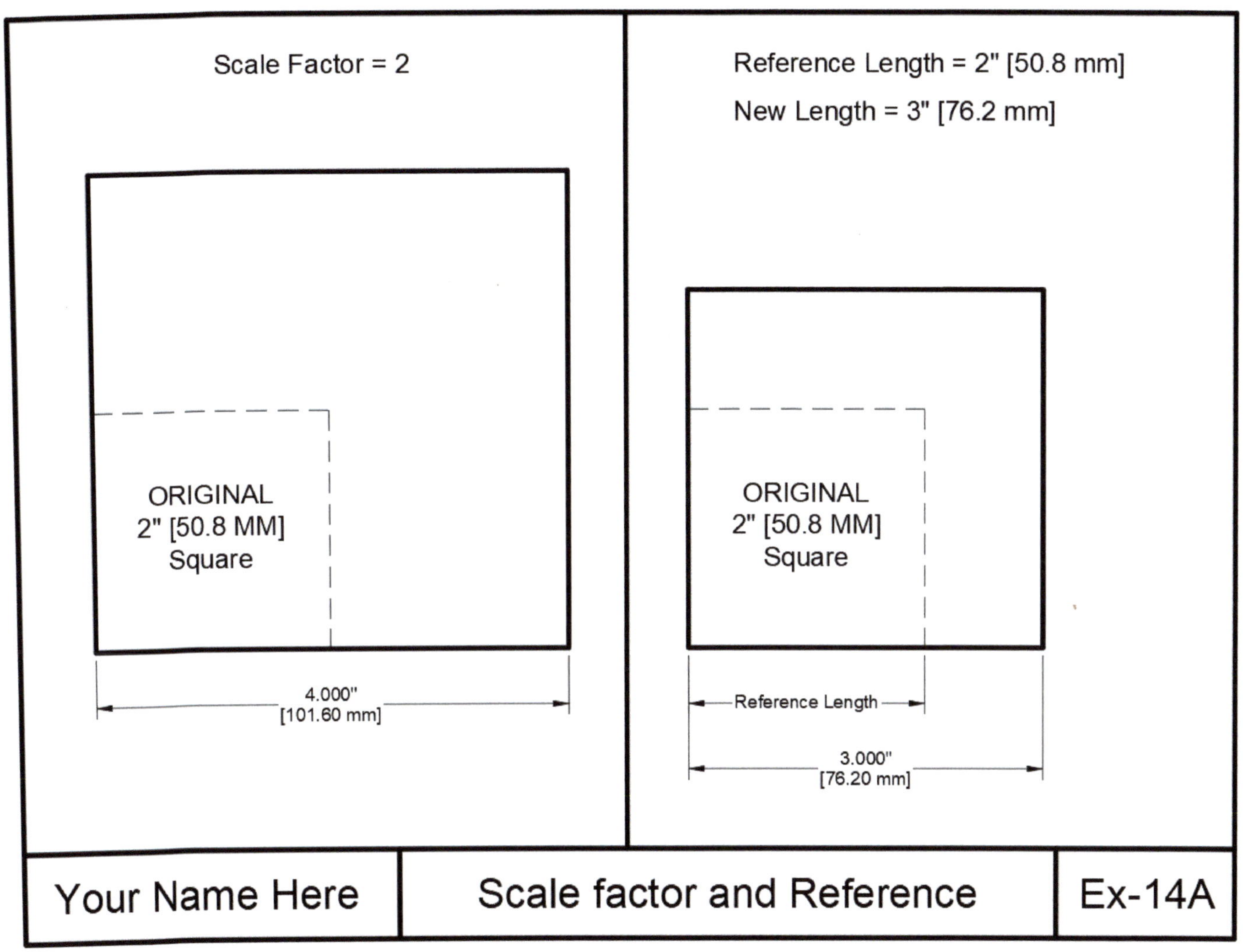

Exercise 14B

Instructions:

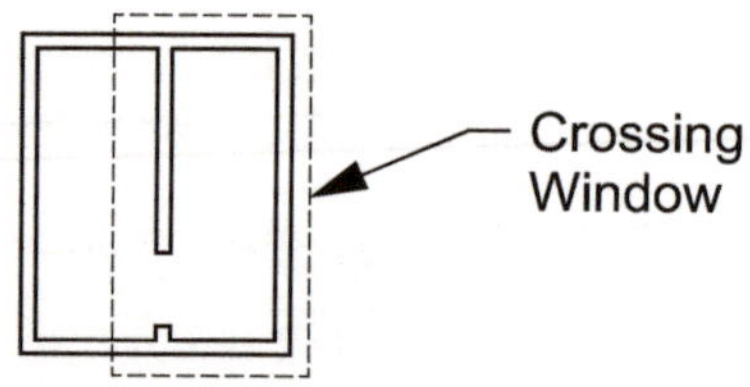

1. Start a **New** file using the **Border A-Inch.dwt** or the **Border A-Metric.dwt**
2. Draw the Original shape on the left side of the border. Use Layer Object Line.

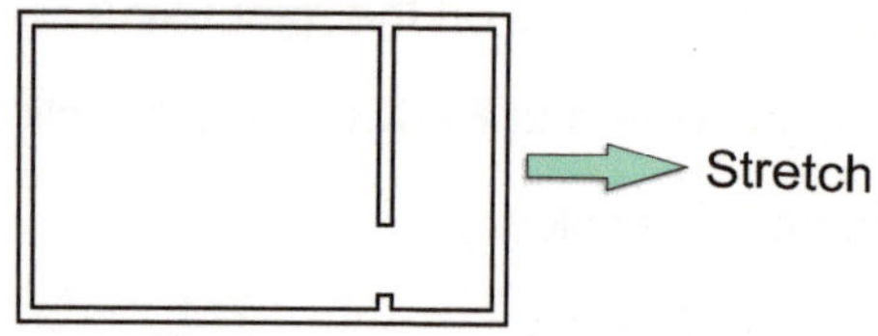

3. Stretch the overall length to **8.5" [215.9 mm]** as shown. Notice the placement of the crossing window.
4. Edit the **Title** and **Ex-XX** by double clicking on the text. Do not erase and replace.
5. Do not dimension.
6. **Save** as **Ex-14B**
7. Plot using Page Setup **Class Model A**

Note: If this was an actual floor plan, consider how the Stretch command could be very helpful.

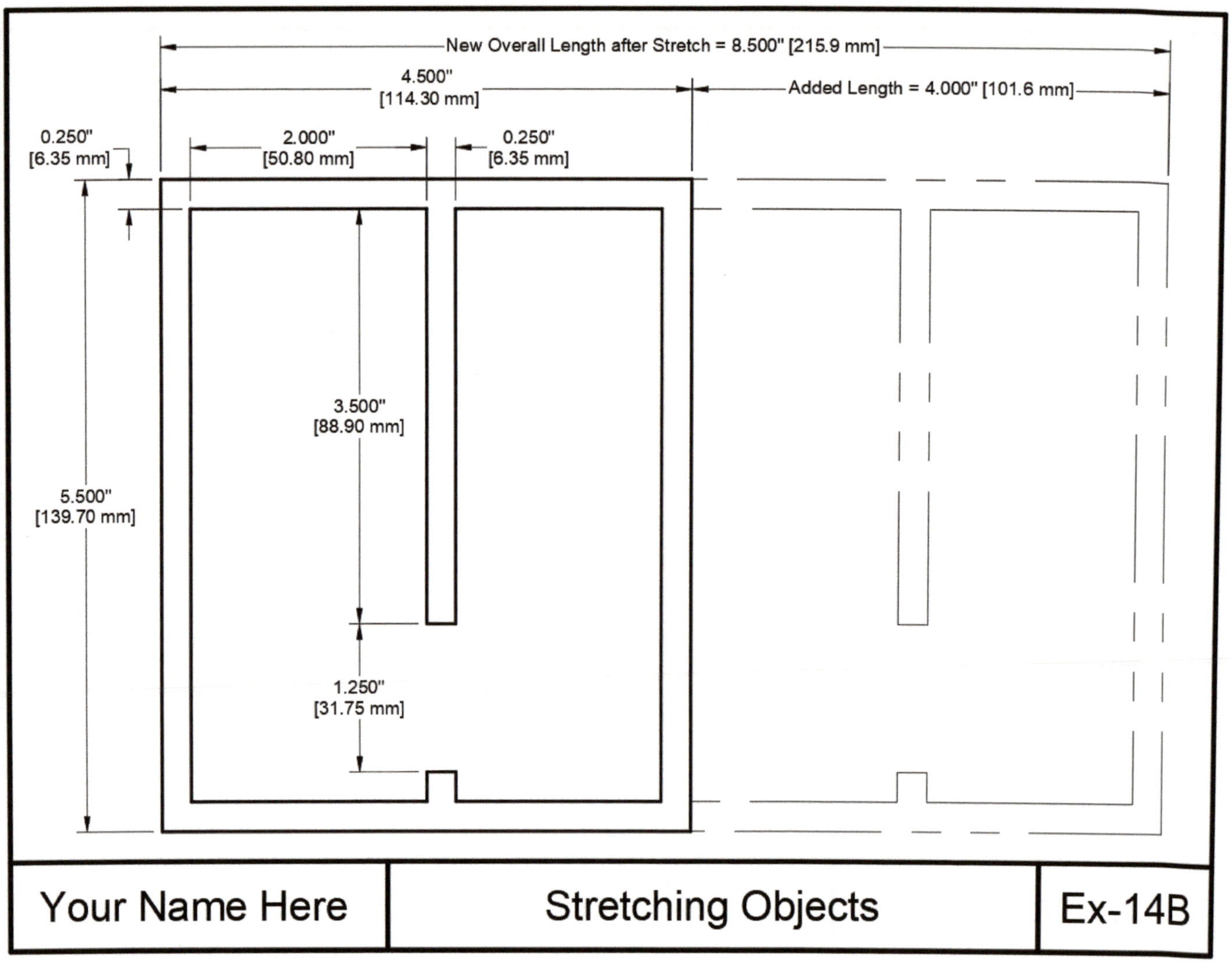

Exercise 14C

Instructions:

1. Open **Ex-14B**

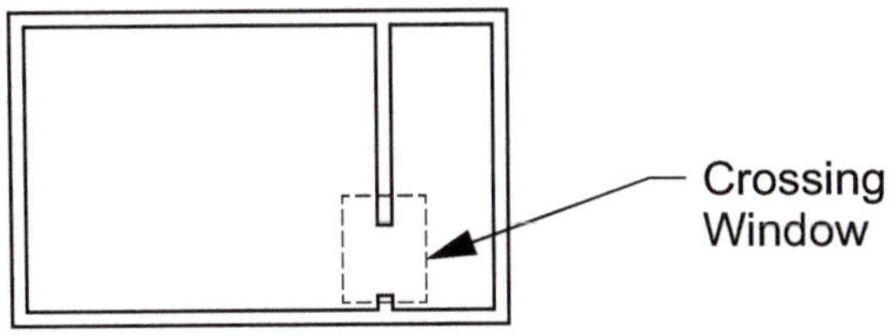

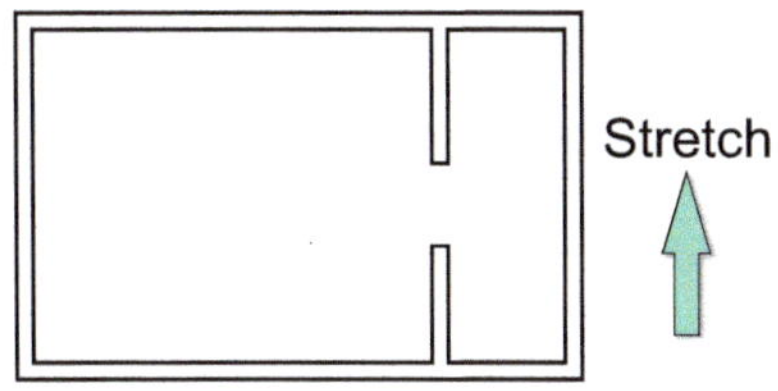

2. Move the opening to the new location using the Stretch command. Notice the placement of the Crossing Window.
3. Edit the **Title** and **Ex-XX** by double clicking on the text. Do not erase and replace.
4. Do not dimension.
5. **Save** as **Ex-14C**
6. Plot using Page Setup **Class Model A**

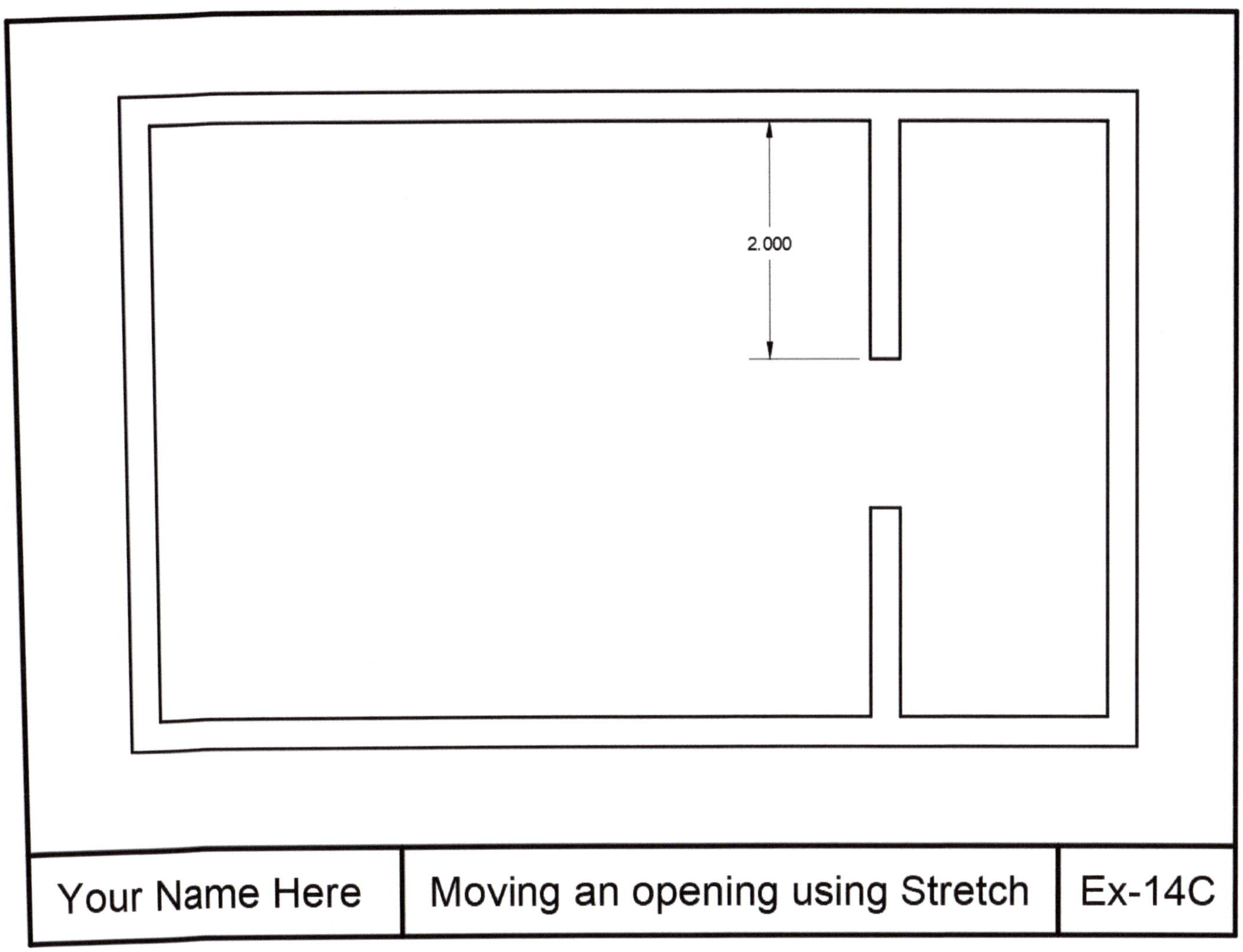

Exercise 14D

Instructions:

1. Start a **New** file using the **Border A-Inch.dwt** or the **Border A-Metric.dwt**
2. Draw the objects at position **A**. Use Layer Object Line.
3. Copy the objects to position **B** and **C**. (Shown as dashed lines.)
4. Rotate the copies as shown. **Note:** The basepoint is important.
5. Edit the **Title** and **Ex-XX** by double clicking on the text. Do not erase and replace.
6. Do not dimension.
7. **Save** as **Ex-14D**
8. Plot using Page Setup **Class Model A**

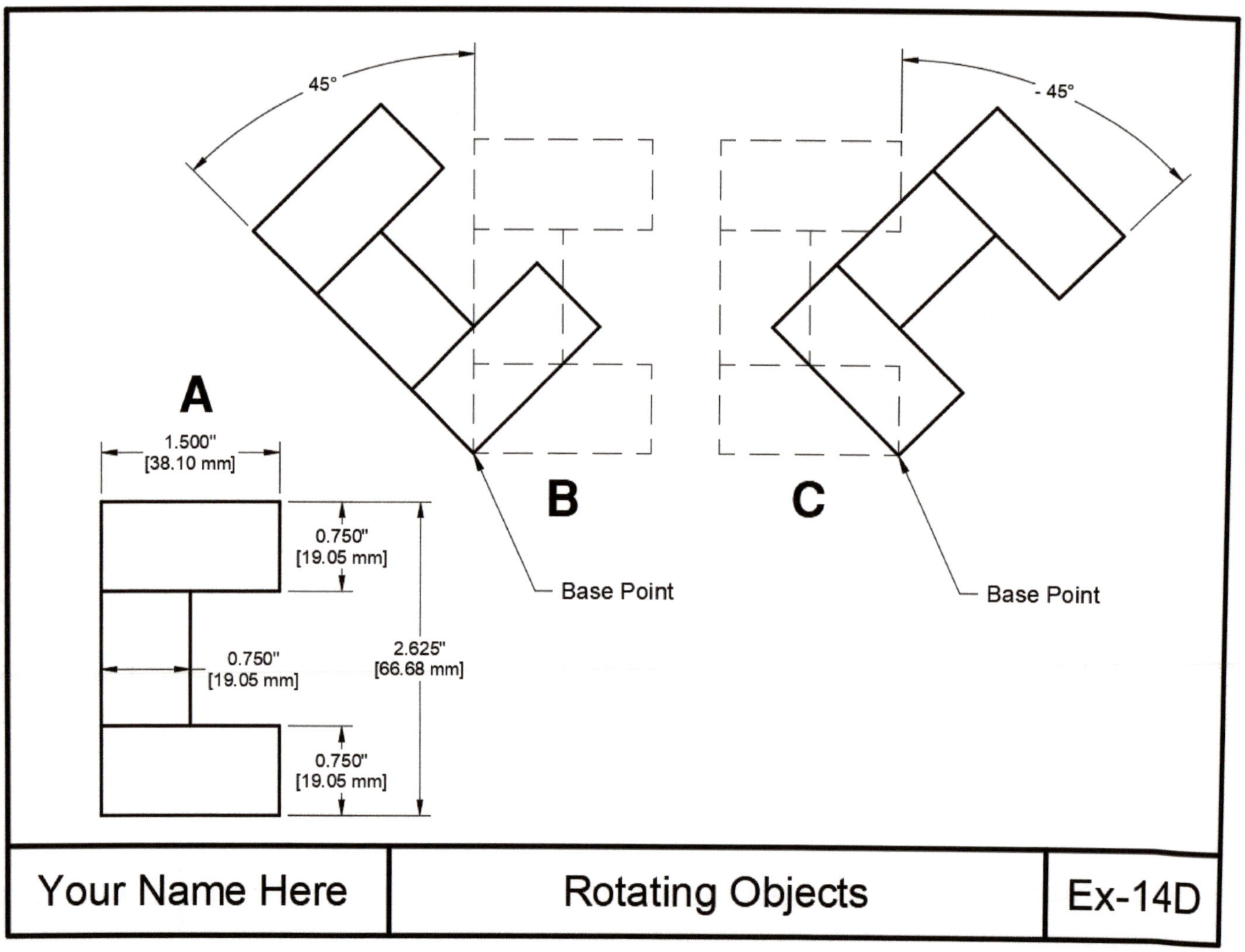

LESSON 15

LEARNING OBJECTIVES

After completing this lesson, you will be able to:

- Place a Hatch pattern in a Boundary
- Fill an Area with solid fill
- Add Gradient fills
- Make changes to existing Hatch sets
- Place a Hatch pattern without a Boundary

Hatch

The **Hatch** command is used to create hatch lines for section views or filling areas with specific patterns. To draw hatch, you usually start with a closed boundary. A closed boundary is an area completely enclosed by objects. A rectangle would be a closed boundary. You simply place the cursor inside the closed boundary or select objects.

Although most hatches are created using a closed boundary, the **Hatch** command now also can be used to create a hatch without the need for a closed boundary. (Refer to page 15-11.)

Note: A Hatch set is one object. It is good drawing management to always place the Hatch on its own layer. Use Layer Hatch. You may also make the Hatch appear or disappear with the **Fill** command. (Refer to page 5-6.)

How to place a Hatch

1. Draw a rectangle.

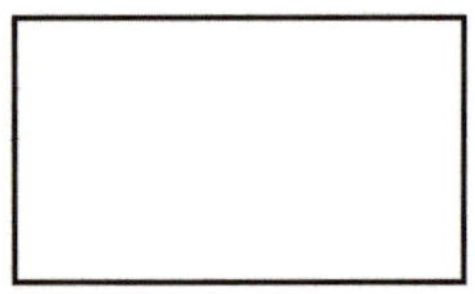

2. Select the **Hatch** command using one of the following:

Ribbon = Home Tab / Draw Panel / 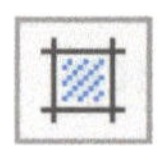
or
Keyboard = ***BH <Enter>***

The **Hatch Creation** Ribbon Tab appears automatically.

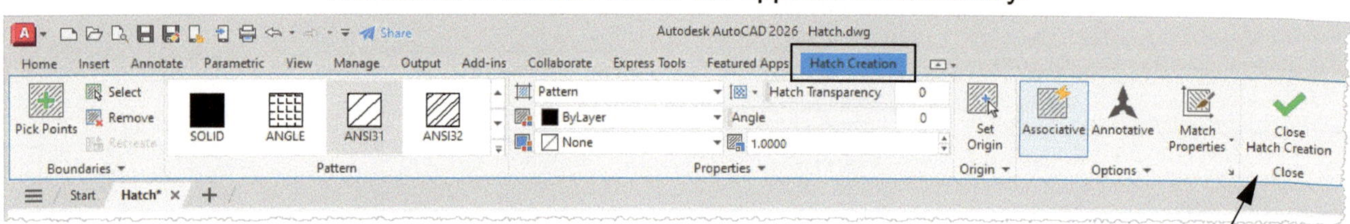

5

3. Place the cursor inside the rectangle (a closed boundary).

A Hatch pattern preview will appear.

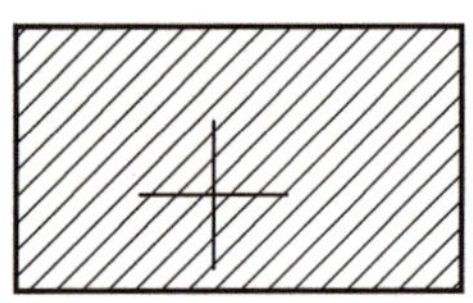

4. Press the left mouse button to accept the Hatch.

5. Select **Close Hatch Creation** or press ***<Enter>***.

Hatch Properties

When you select the Hatch command, the **Hatch Creation** Ribbon Tab appears automatically. The Panels on this Tab help set the properties of the Hatch. You should set the properties desired previous to placing the Hatch set, although you can easily edit an existing Hatch set.

Boundaries Panel

The Boundaries Panel allows you to choose what method you will use to select the Hatch boundary.

Continued on the next page...

Pick Point: Pick Point is the default selection. When you select the Hatch command, AutoCAD assumes that you want to use the Pick Points method. You merely place the cursor in the closed area to select the boundary. The Hatch set preview will appear. Press the left mouse button to accept.

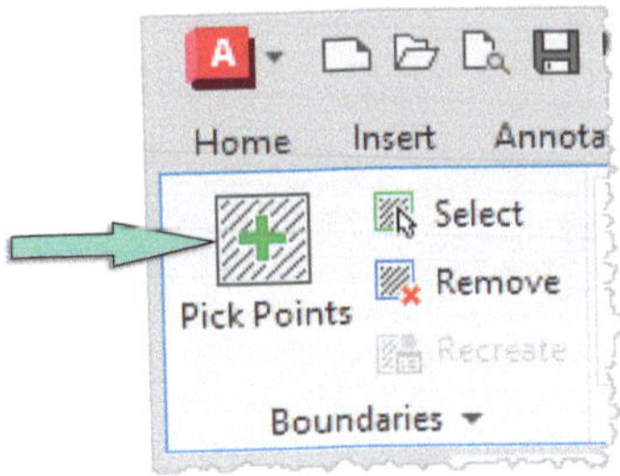

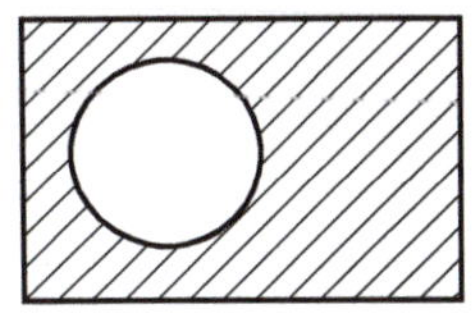

Select and Remove: You may select or remove objects to a boundary. **Note: Remove** will not be available unless you click on **Select**.

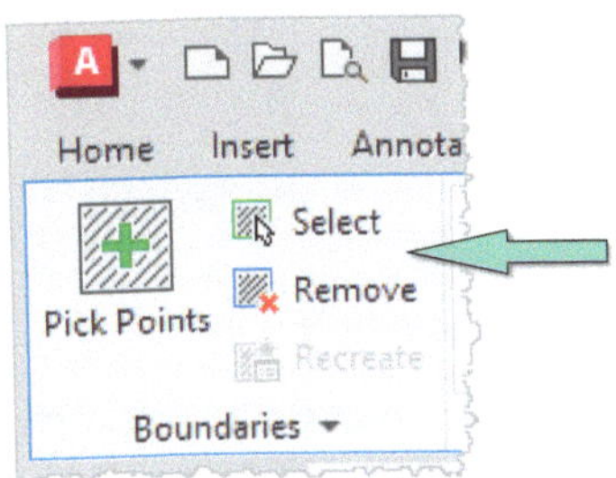

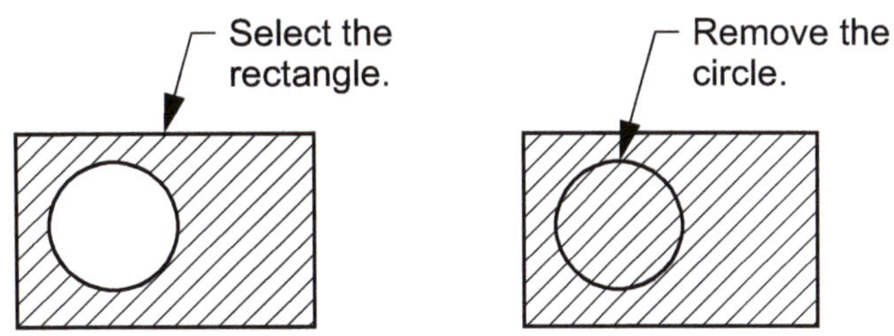

Pattern Panel

The Pattern Panel displays the Hatch swatches that relate to the Hatch Type that has been selected in the Properties Panel. Refer to Properties Panel below.

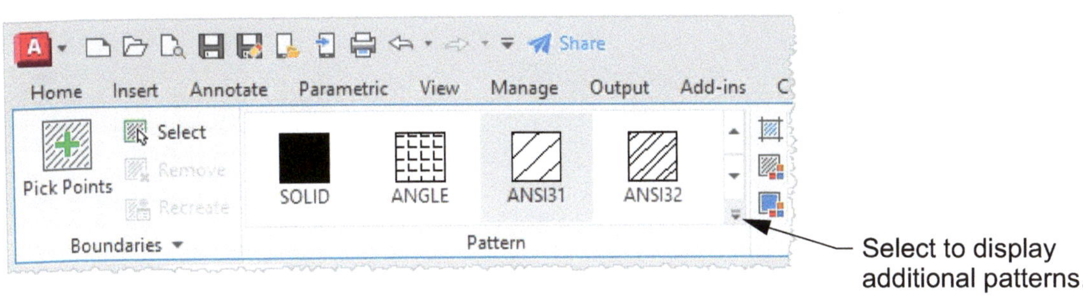

Properties Panel

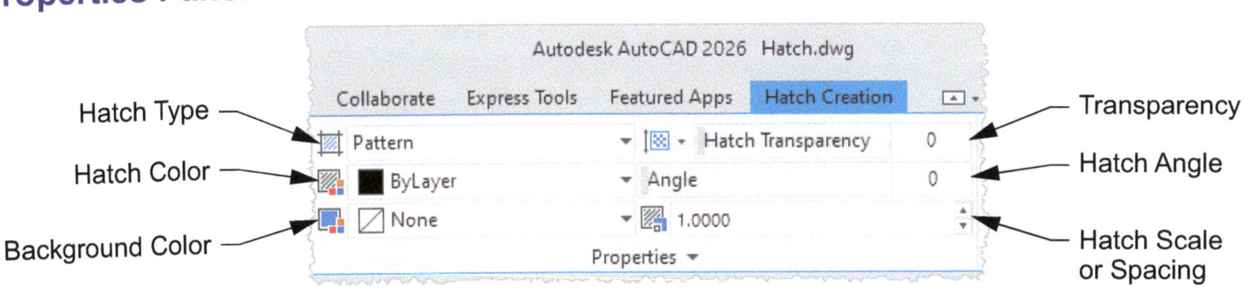

Hatch Type: When you select the drop-down arrow ▾, you may select one of the Hatch Types: **Solid**, **Gradient**, **Pattern**, or **User Defined**.

Note: Hatch Types will be explained in more detail on pages 15-7 through 15-9.

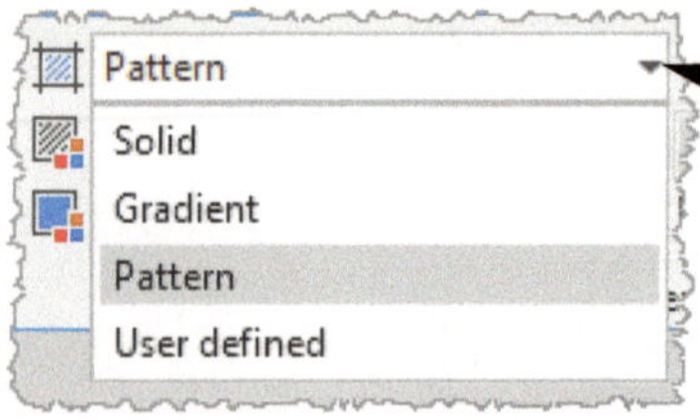

Select drop-down arrow.

When you select a Hatch Type from the drop down list, the **Pattern Panel** displays the related hatch swatches from which to select.

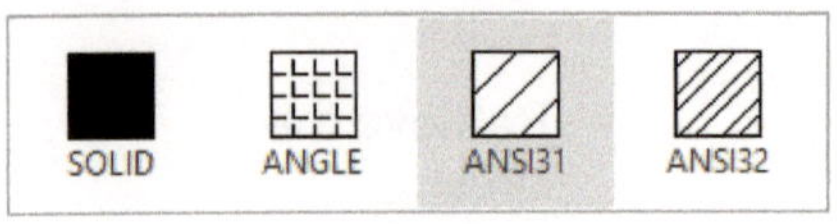

Pattern

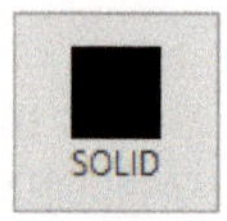

Solid

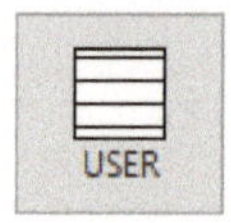

User Defined

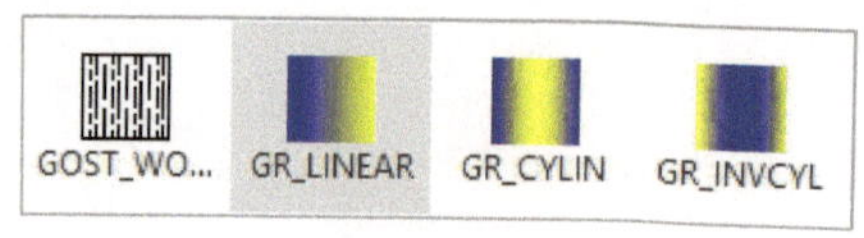

Gradient

Hatch Color: This color selection is specific to the Hatch and will not affect any other objects.

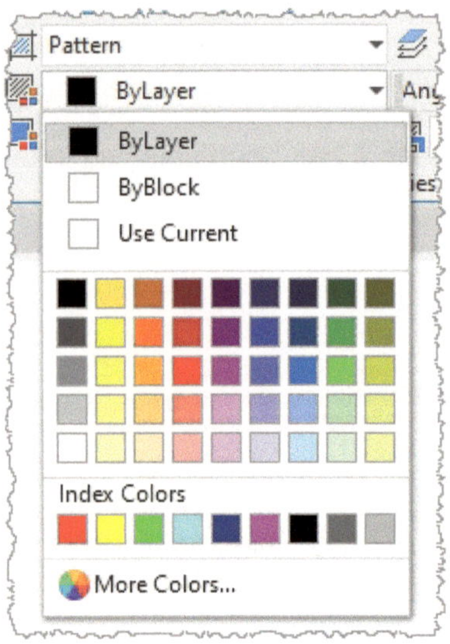

Background Color: You may select a background color for the Hatch area.

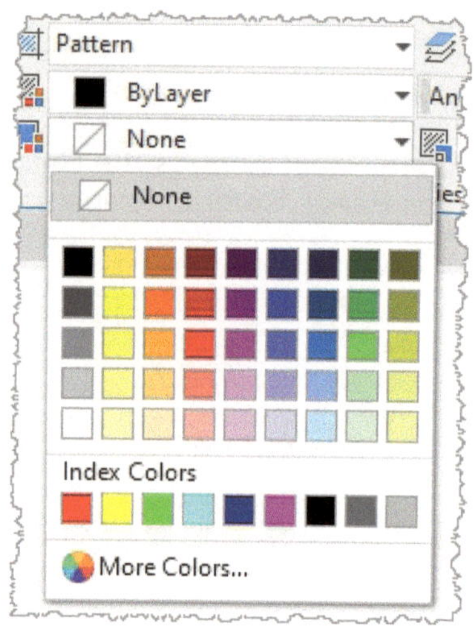

Hatch Transparency: Displays the selected transparency setting. You may select to use the current setting for the drawing or the current layer setting or to select a specific value.

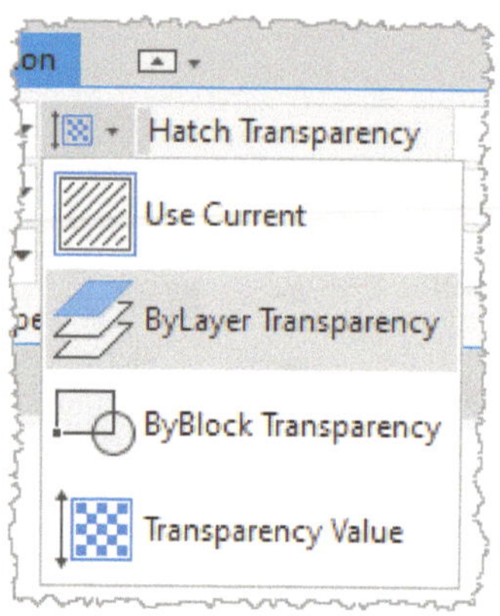

Hatch Angle

The default angle of a Pattern Hatch is **0**. If you change this angle, the pattern will rotate relative to its original design.

With a User defined Hatch, you will need to specify the actual angle of the hatch lines.

Hatch Scale or Spacing

If Hatch Type **Pattern** is selected, this value determines the scale of the Hatch pattern. A value greater than **1** will increase the scale. A value less than **1** will decrease the scale.

If Hatch Type **User Defined** is selected, this value determines the spacing between the hatch lines.

Origin Panel

You may specify where the Hatch will originate. Lower left, lower right, upper left, upper right, center, or even the at the UCS Origin.

The Origin locations shown on the right are displayed when Hatch Type, Pattern, Solid, or User Defined are selected.

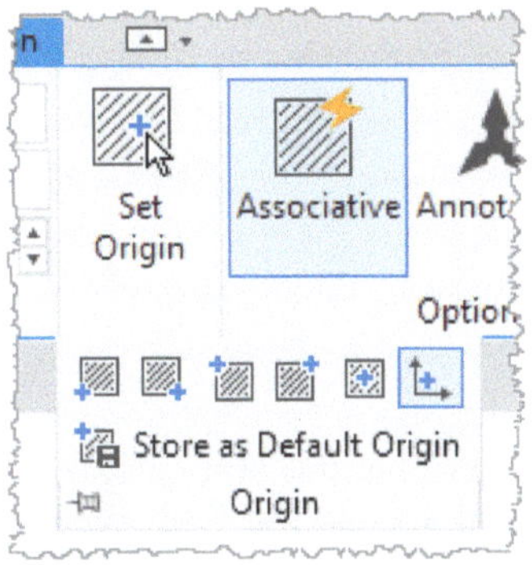

The Origin location **Centered** is displayed only when Hatch Type Gradient is selected.

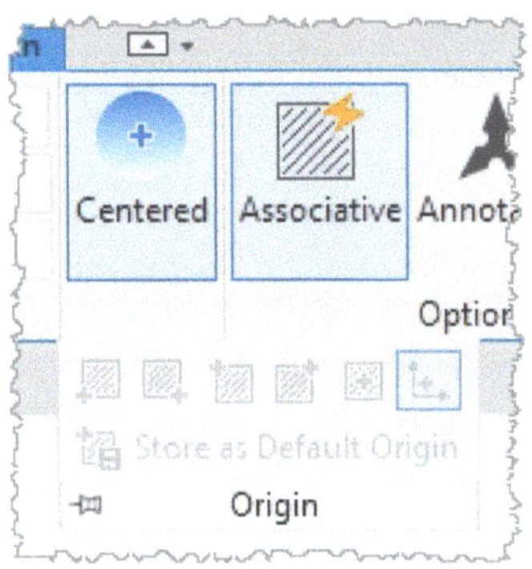

Options Panel

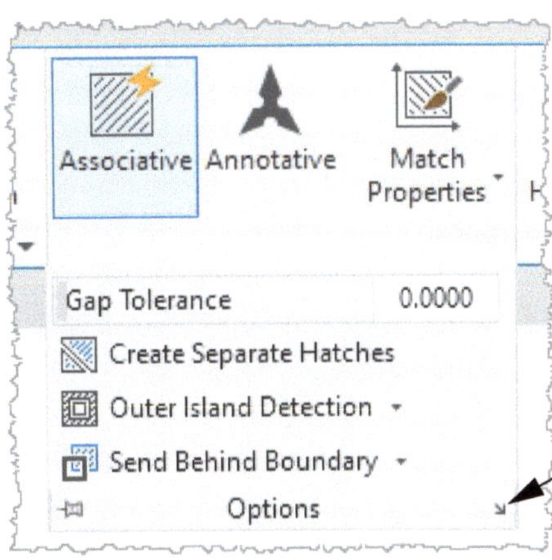

Opens the **Hatch and Gradient** Dialog Box. You may choose to use it instead of the Ribbon.

Continued on the next page...

Associative: If the Associative option is selected, the Hatch set is associated to the boundary. This means if the boundary size is changed, the hatch will automatically change to fit the new boundary shape. (Refer to page 15-10 for an example.)

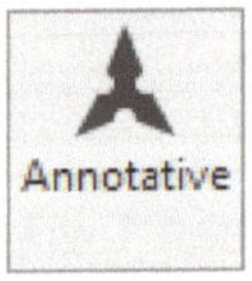

Annotative: AutoCAD will automatically adjust the scale to match the current Annotative scale. This option will be discussed in Lesson 27.

Match Properties

Match Properties allows you to set the properties of the new Hatch set by selecting an existing Hatch set. You may choose to **"use the current origin"** or **"use the source hatch origin"**.

Use current origin: Sets all properties except the hatch origin.

Use source hatch origin: Sets all properties including the hatch origin.

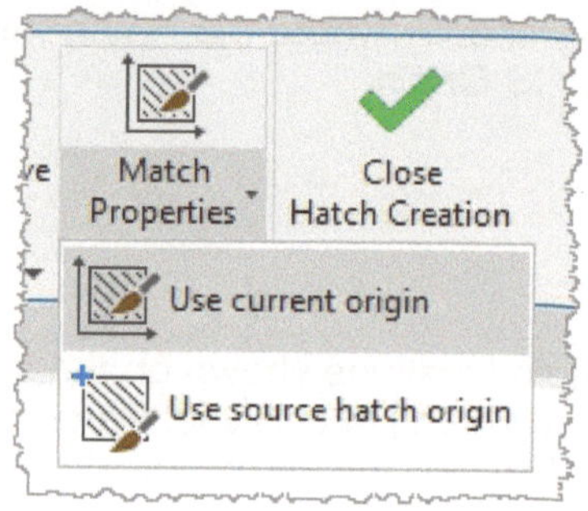

Gap Tolerance

If the area you selected to hatch is not completely closed (gaps), AutoCAD will bridge the gap depending on the Gap Tolerance. The Gap Tolerance can be set to a value from **0** to **5000**. Any gaps equal to or smaller than the value you specify are ignored and the boundary is treated as closed.

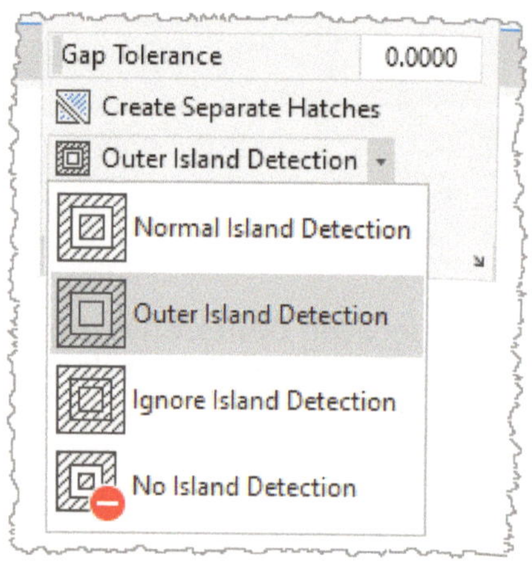

Create Separate Hatches

Controls whether Hatch creates a single hatch object or separate hatch objects when selecting several closed boundaries.

Outer Island Detection

These selections determine how Hatch recognizes internal objects.

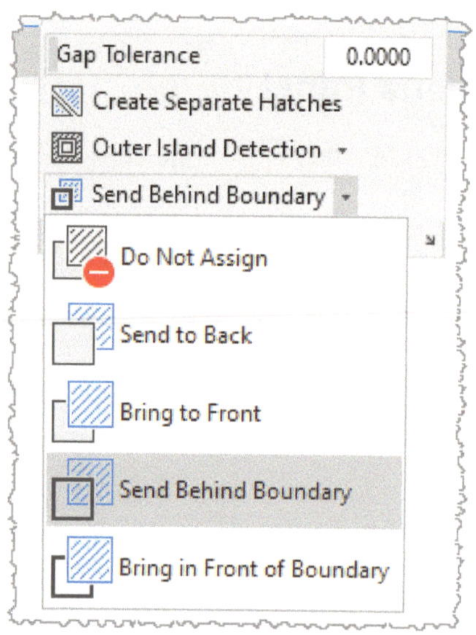

Send Behind Boundary

These selections determine the draw order of the Hatch set.

Hatch Types

Pattern

AutoCAD includes many previously designed Hatch patterns. **Note:** Using Hatch patterns will greatly increase the size of the drawing file. So use them conservatively. You may also purchase patterns from other software companies.

1. Select the Hatch **Pattern**.
2. Select a Pattern from the list of Hatch patterns displayed in the Pattern Panel.

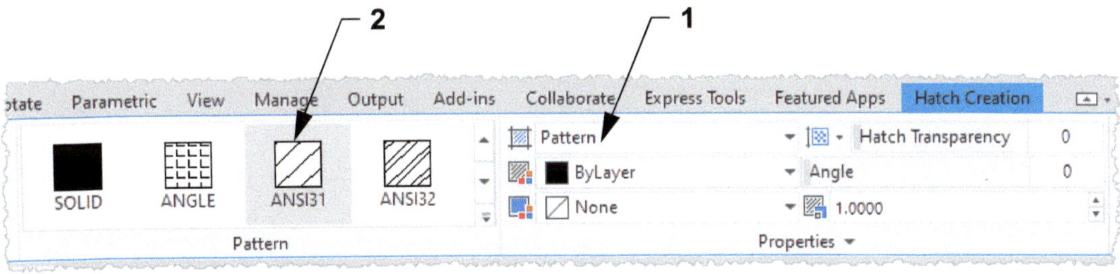

3. Select the Hatch Color, Boundary Background Color, and Hatch Transparency.
4. **Angle:** A previously designed pattern has a default angle of **0**. If you change this angle, the pattern will rotate relative to its original design. It is important that you understand how to control the angle.

   Example: If **ANSI31** Hatch pattern is used and the angle is set to **45** degrees, the pattern will rotate relative to its original design and the pattern will appear to be **90** degrees.

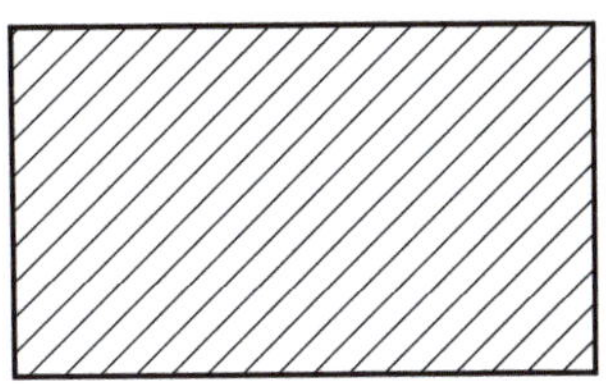

Original Pattern Design
Angle = **0**

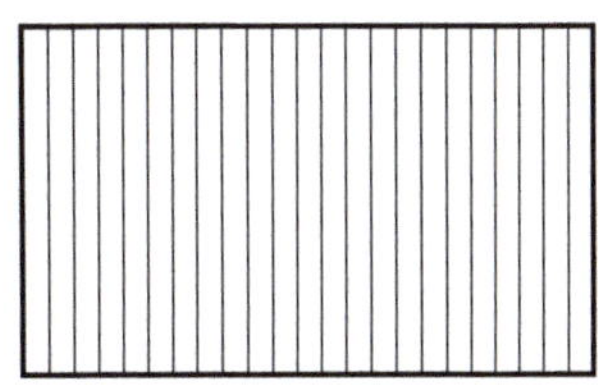

Pattern Design Rotated
Angle = **45**

5. **Scale:** A value greater than 1 will increase the scale. A value less than 1 will decrease the scale. If the Hatch set is Annotative, the scale will automatically adjust to the Annotative scale. But you might have to tweak the scale additionally to make it display exactly as you desire. (*This will be discussed more in Lesson 27. Don't be too concerned about it right now.*)

User Defined

This Hatch Type allows you to simply draw continuous lines. No special pattern. You specify the **angle** of the lines and the **spacing** between the lines.

Note: This Hatch Type does not significantly increase the size of the drawing file.

1. Select the Hatch Type **User defined**.
2. You also can select the **User** swatch on the Pattern Panel.

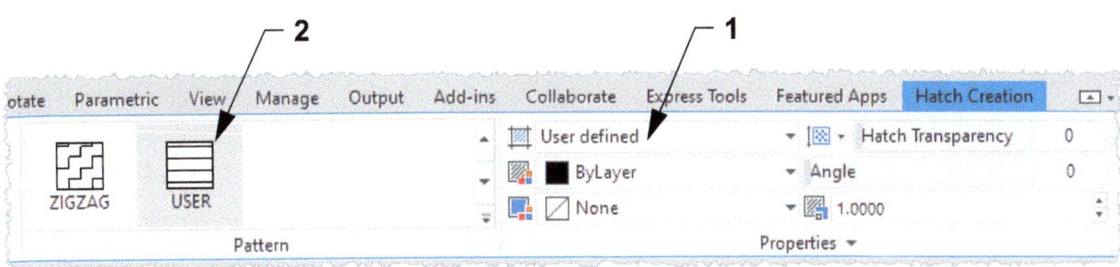

Continued on the next page...

3. Select the Hatch Color, Boundary Background Color, Hatch Transparency.
4. **Angle:** Specify the actual angle that you desire from 0 to 180.

   Example: If you want the lines to be on an angle of **45** degrees, you would enter **45**. If you want the lines to be on an angle of **90** degrees, you would enter **90**. (This is different from the angle for Patterns.)

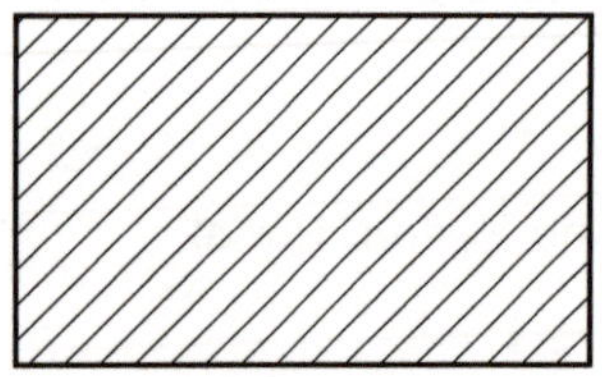

Angle = **45**

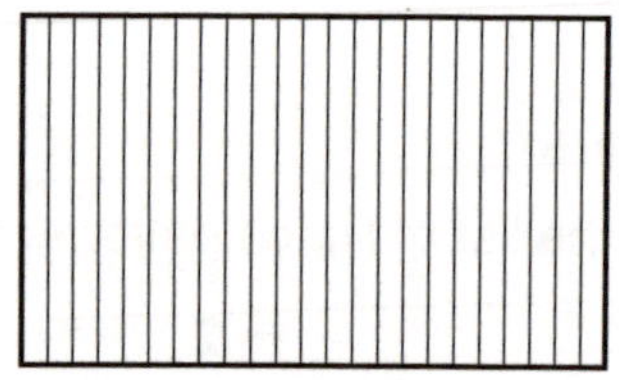

Angle = **90**

5. **Scale:** Specify the actual distance between each of the hatch lines.

Solid

If you would like to fill an area with a solid fill, you should use Hatch Type **Solid**.

1. Select **Solid** by selecting the Hatch Type **Solid** on the Properties Panel
2. You also can select the **Solid** swatch on the Pattern Panel.

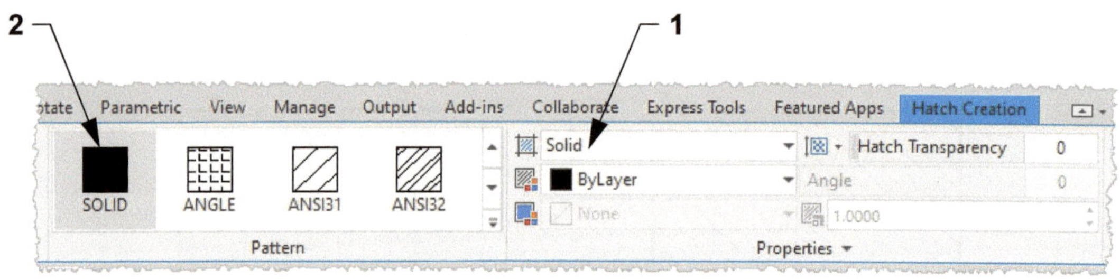

3. Select the Hatch **Color**. **Note:** The Boundary Background Color is not available when using Hatch Type Solid.
4. Select Hatch **Transparency**.

Transparency = **0**

Transparency = **75**

Note: Angle and Scale do not apply when using Hatch Type Solid.

Gradient

Gradients are fills that gradually change from dark to light or from one color to another. Gradient fills can be used to enhance presentation drawings, giving the appearance of light reflecting on an object, or creating interesting backgrounds for illustrations.

Gradients are definitely fun to experiment with, but you will have to practice to achieve complete control. They will also greatly increase the size of the drawing file.

Continued on the next page...

1. Select the Hatch Type **Gradient**.
2. Select a Gradient Pattern from the nine **GR_** patterns displayed in the Pattern Panel.

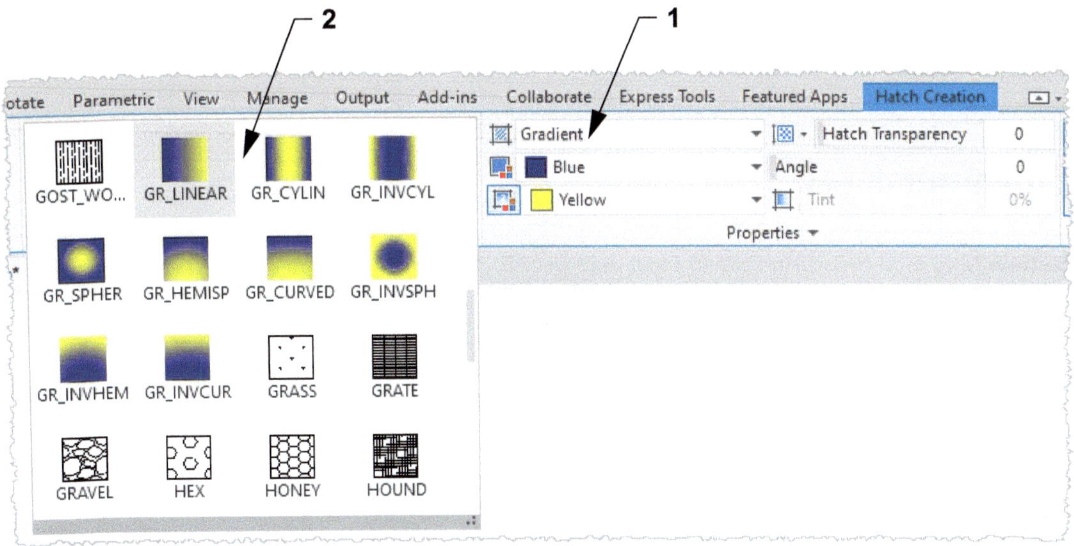

3. Select the Hatch **Color**. The Gradient can be in one color or two colors. If one color, you can select the Tint and Shade of that color. (See Step 6 below.)
4. Select the Hatch **Transparency**.
5. **Angle:** Specify the actual angle that you want from **0** to **180**. The pattern will rotate the pattern relative to its original design.

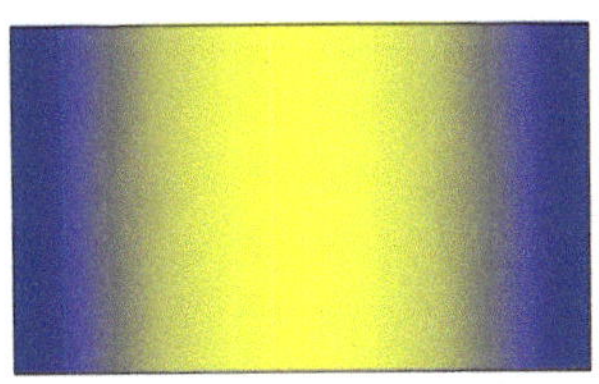
Angle = **0**

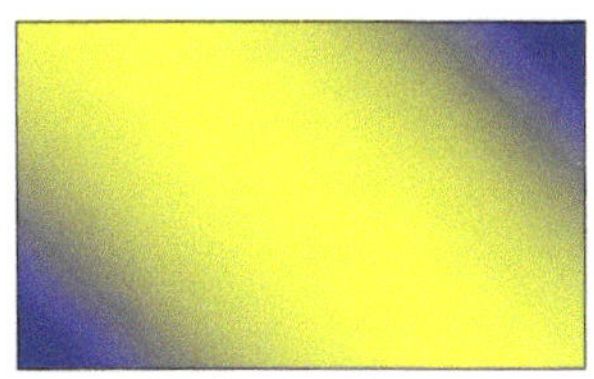
Angle = **45**

6. **Tint** and **Shade:** Tint and Shade is used when you are using only one-color gradient fill. Specify the tint or shade of the color selected in Step 3 above.

Further examples of the **Gradient Hatch** using two colors, black and white.

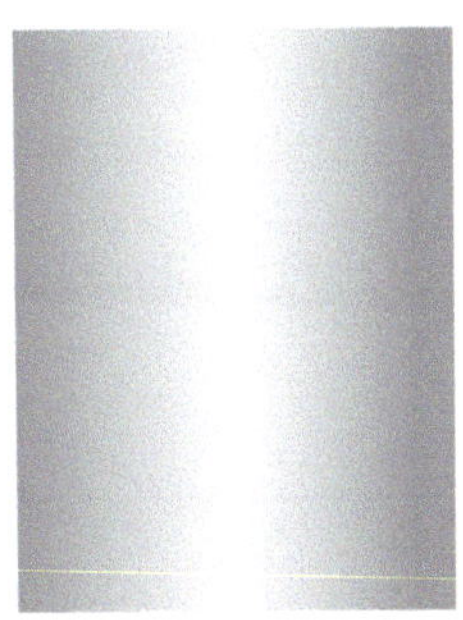
Cylinder using pattern
GR_CYLIN
Angle = **0**

Sphere using pattern
GR_SPHER
Angle = **0**

Editing Hatch Set Properties

Editing a Hatch set properties is easy.

1. Simply select the Hatch set to edit.
2. The Hatch Editor Ribbon Tab will appear.
3. Make new selections. Any changes are applied immediately.

Changing the Boundary

If the Hatch set is **Associative** (see 15-5), you may change the shape of the boundary and the Hatch set will conform to the new shape. If the Hatch set is **Non-Associative**, the Hatch set will not change.

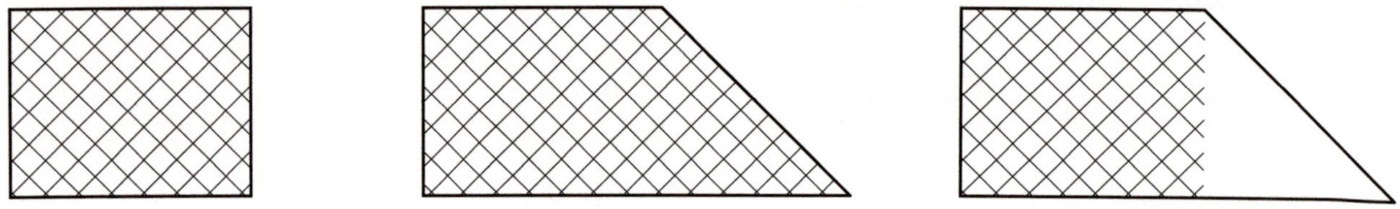

Original Boundary

New Boundary with Associative Hatch

New Boundary with Non-Associative Hatch

Trimming a Hatch

You may trim a Hatch set just like any other object.

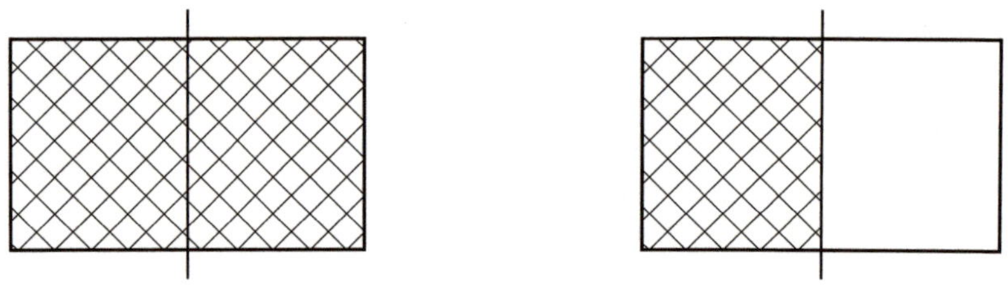

Before Trim

After Trim

Mirror a Hatch

An existing Hatch set can be **mirrored**. The boundary shape will automatically mirror. But you may control whether the Hatch pattern is mirrored or not.

To control the Hatch pattern mirror: (**Note:** Set prior to using the Mirror command.)

1. Type ***mirrhatch*** and then press ***<Enter>***.
2. Enter ***0*** or ***1*** and the press ***<Enter>***. **0** = Hatch not Mirrored. **1** = Hatch Mirrored.

Mirrhatch = **0**

Mirrhatch = **1**

Hatch without Boundaries

The **Hatch** command also has the ability to create a hatch without the need for a closed boundary. You can create a circle, rectangle, or polyline hatch. To create a circle or rectangle hatch to specific dimensions, follow the instructions for creating a circle or rectangle described in Lesson 3.

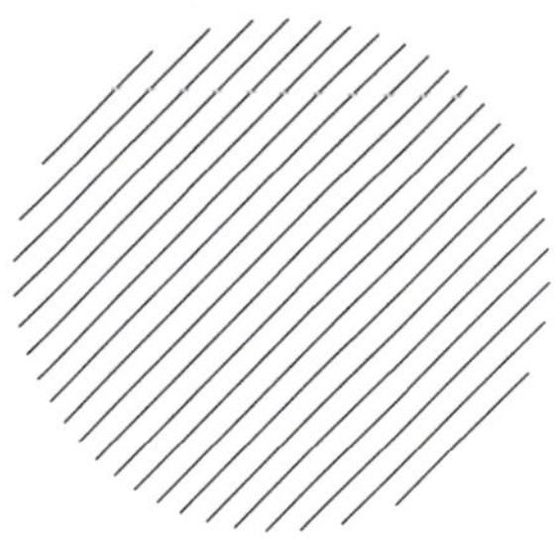

Circle Hatch

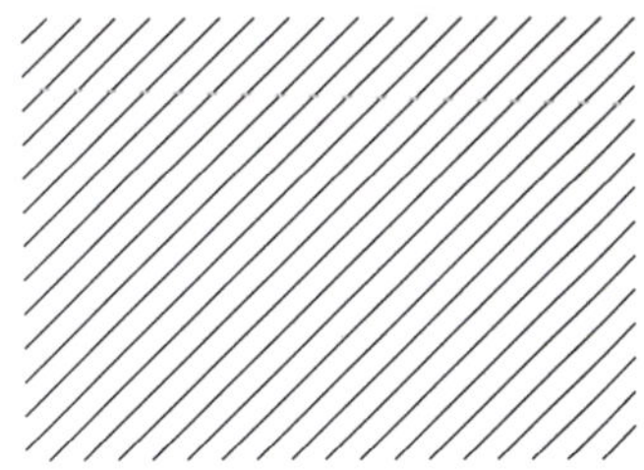

Rectangle Hatch

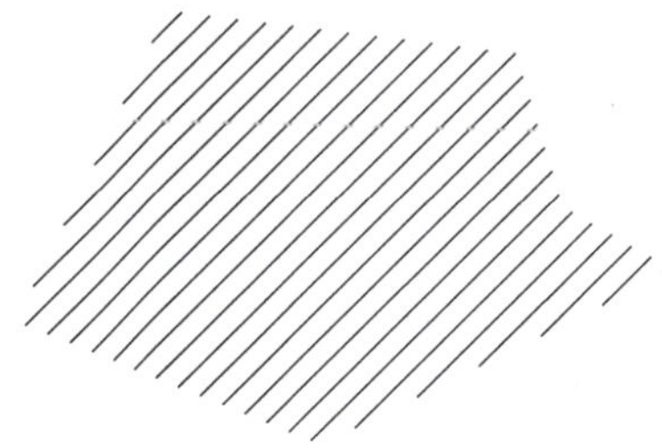

Polyline Hatch

How to create a circle Hatch without boundaries

1. Select the **Hatch** command.

   The following will appear on the Command Line:

   ```
   HATCH Pick internal point or [Select objects Draw Undo seTtings]:
   ```

2. Type ***D*** and then press ***<Enter>***.

   The following will appear on the Command Line:

   ```
   HATCH Specify start point or [picK internal point Select objects Rectangle Circle
   Mode seTtings]:
   ```

3. Type ***C*** and then press ***<Enter>***.

   The following will appear on the Command Line:

   ```
   HATCH Specify center point for circle or [3P 2P Ttr (tan tan radius)]:
   ```

4. Locate the center point for the circle Hatch by moving the cursor to the desired location in the drawing area (**P1**) and then press the left mouse button.

   The following will appear on the Command Line:

   ```
   HATCH Specify radius of circle or [Diameter]:
   ```

5. Now move the cursor away from the center point. You should see a circle forming.

6. When the circle is the size desired (**P2**), press the left mouse button, or type in the radius and then press ***<Enter> <Enter>***.

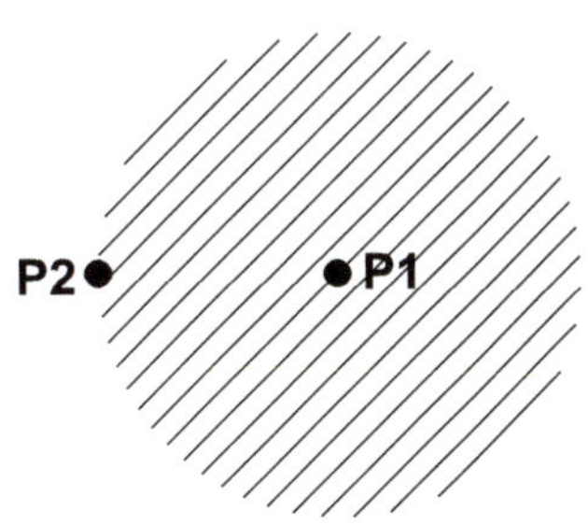

Note: Also see Lesson 3 for other methods of creating a circle for the Hatch.

How to create a rectangle Hatch without boundaries

1. Select the **Hatch** command.

   The following will appear on the Command Line:

   ```
   HATCH Pick internal point or [Select objects Draw Undo seTtings]:
   ```

2. Type ***D*** and then press ***<Enter>***.

   The following will appear on the Command Line:

   ```
   HATCH Specify start point or [picK internal point Select objects Rectangle Circle
   Mode seTtings]:
   ```

3. Type ***R*** and then press ***<Enter>***.

   The following will appear on the Command Line:

   ```
   HATCH Specify first corner point or [Chamfer Elevation Fillet]:
   ```

4. Specify the location of the first corner by moving the cursor to a location (**P1**) and then press the left mouse button.

   The following will appear on the Command Line:

   ```
   HATCH Specify other corner point or [Area Dimensions Rotation]:
   ```

5. Specify the location of the diagonal corner (**P2**) by moving the cursor diagonally away from the first corner (**P1**) and then pressing the left mouse button, followed by ***<Enter>***.

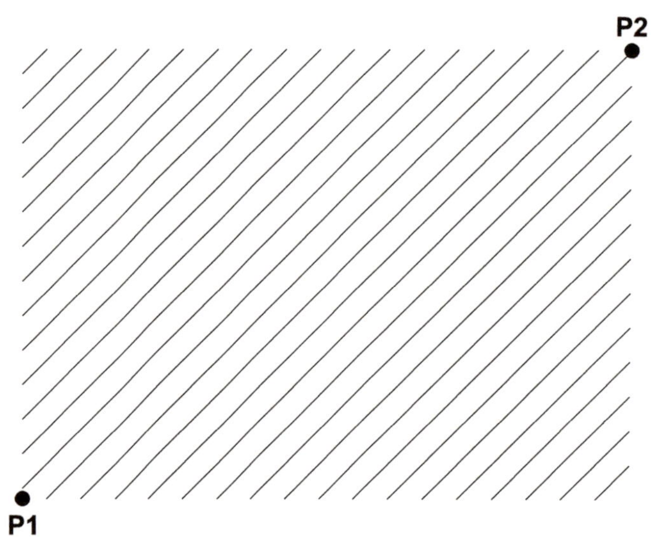

Note: Also see Lesson 3 for other methods of creating a rectangle for the Hatch.

How to create a polyline Hatch without boundaries

1. Select the **Hatch** command.

   The following will appear on the Command Line:

   ```
   HATCH Pick internal point or [Select objects Draw Undo seTtings]:
   ```

2. Type ***D*** and then press ***<Enter>***.

   The following will appear on the Command Line:

   ```
   HATCH Specify start point or [picK internal point Select objects Rectangle Circle
   Mode seTtings]:
   ```

3. Specify the location of the first point by moving the cursor to a location (**P1**) and then press the left mouse button.

   The following will appear on the Command Line:

   ```
   HATCH Specify the next point or [Arc Length Undo Exit] <Exit>:
   ```

4. Specify the location of the second point by moving the cursor to a location (**P2**) and then press the left mouse button.

   The following will appear on the Command Line:

   ```
   HATCH Specify the next point or [Arc Length Undo Exit] <Exit>:
   ```

5. Specify the location of the third point by moving the cursor to a location (**P3**) and then press the left mouse button.

   The following will appear on the Command Line:

   ```
   HATCH Specify the next point or [Arc Close Length Undo Exit] <Exit>:
   ```

6. Continue to enter more points or press ***<Enter> <Enter>*** to stop entering points.

Note: The example below uses a four-point polyline Hatch.

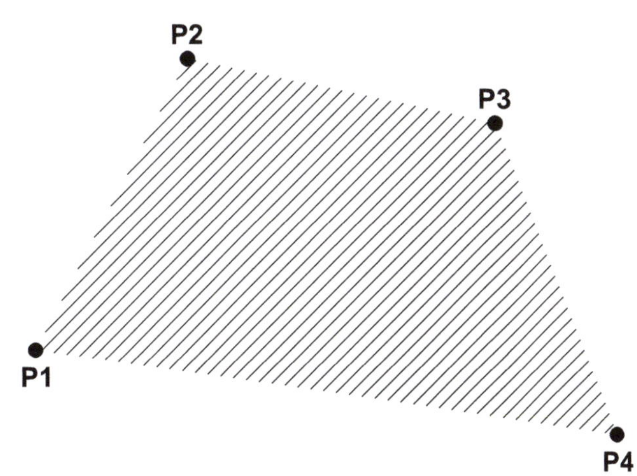

Exercise 15A

Instructions:

1. Start a **New** file using the **Border A-Inch.dwt** or the **Border A-Metric.dwt**
2. Draw the rectangle, circle, and polygon as shown.
3. Use Layers Object Line and Hatch.
4. Hatch the boundaries shown using the Patterns shown.
5. Edit the **Title** and **Ex-XX** by double clicking on the text. Do not erase and replace.
6. Do not dimension.
7. **Save** as **Ex-15A**
8. Plot using Page Setup **Class Model A**

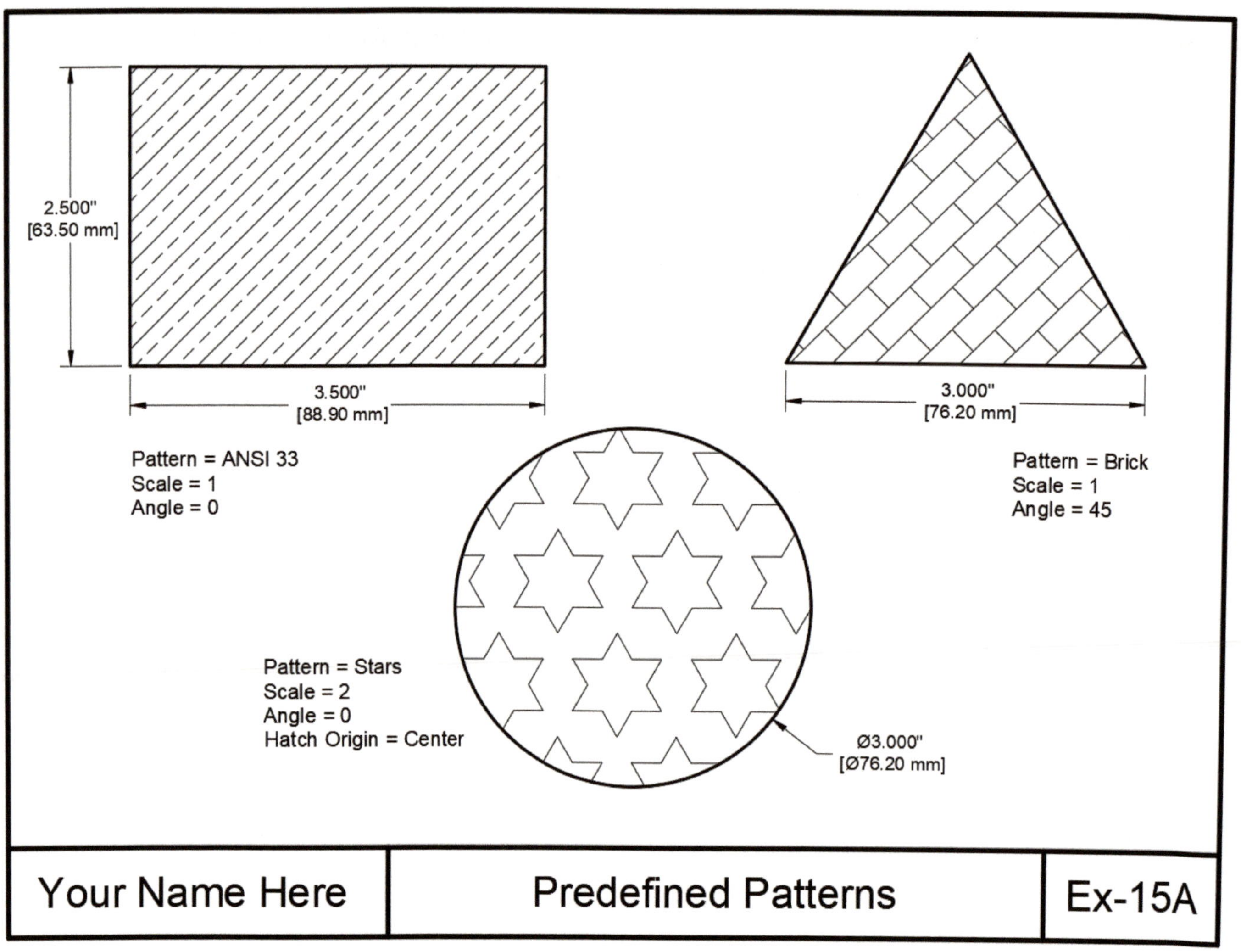

Exercise 15B

Instructions:

1. Start a **New** file using the **Border A-Inch.dwt** or the **Border A-Metric.dwt**
2. Draw the four rectangles as shown.
3. Use Layers Object Line and Hatch.
4. Hatch the boundaries shown using the Patterns shown.
   Notice the angles rotate the Hatch pattern from its original design. For example, angle 45 appears to be rotation 90 because the original design was drawn at a 45 degree...think about it.
5. Edit the **Title** and **Ex-XX** by double clicking on the text. Do not erase and replace.
6. Do not dimension.
7. **Save** as **Ex-15B**
8. Plot using Page Setup **Class Model A**

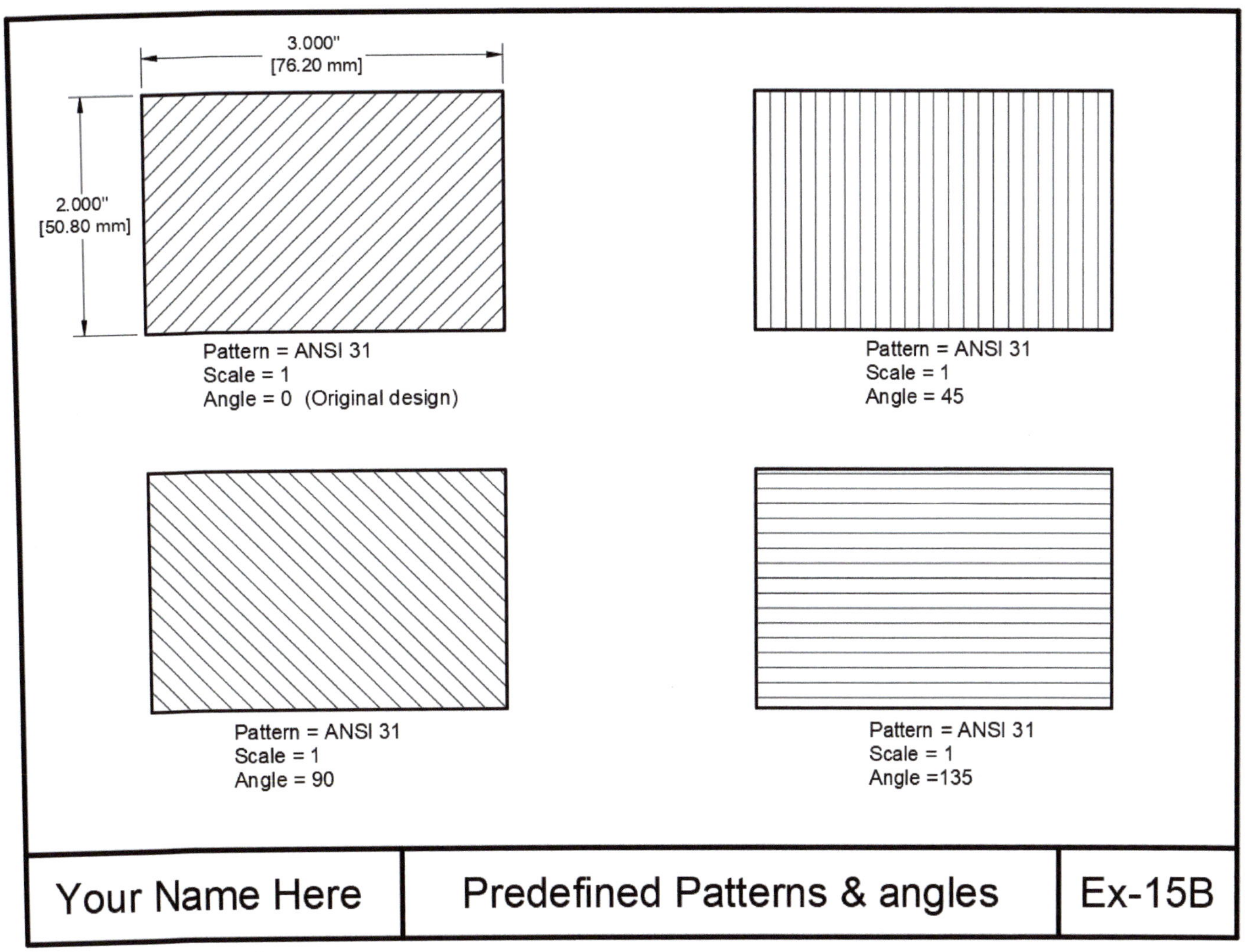

Exercise 15C

Instructions:

1. Start a **New** file using the **Border A-Inch.dwt** or the **Border A-Metric.dwt**
2. Draw the rectangle as shown using Layer Object Line.
3. Place the **1" [25.4 mm]** high text in the middle of the rectangle. Use Layer Text.
4. Use Hatch Pattern **ANSI 31**. Use Layer Hatch.
5. Select the boundary using **"Select Objects"** this time
6. Edit the **Title** and **Ex-XX** by double clicking on the text. Do not erase and replace.
7. Do not dimension.
8. **Save** as **Ex-15C**
9. Plot using Page Setup **Class Model A**

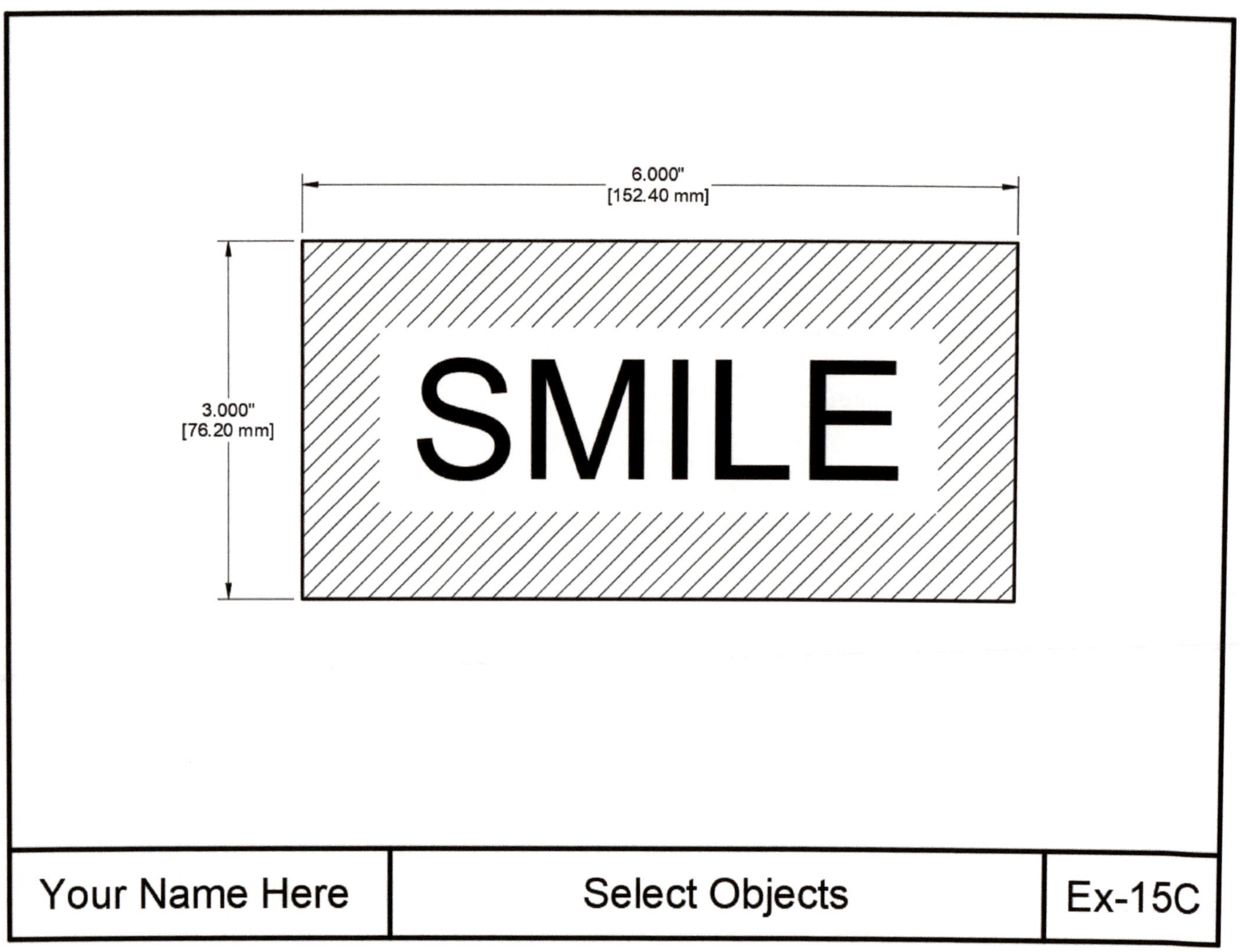

Exercise 15D

Instructions:

1. Start a **New** file using the **Border A-Inch.dwt** or the **Border A-Metric.dwt**
2. Draw the circle, rectangle, and polyline hatches approximately as shown using Layer Hatch.
3. Use Hatch pattern **ANSI 31** Scale = **1** and Angle = **0**.
4. Edit the **Title** and **Ex-XX** by double clicking on the text. Do not erase and replace.
5. **Save** as **Ex-15D**

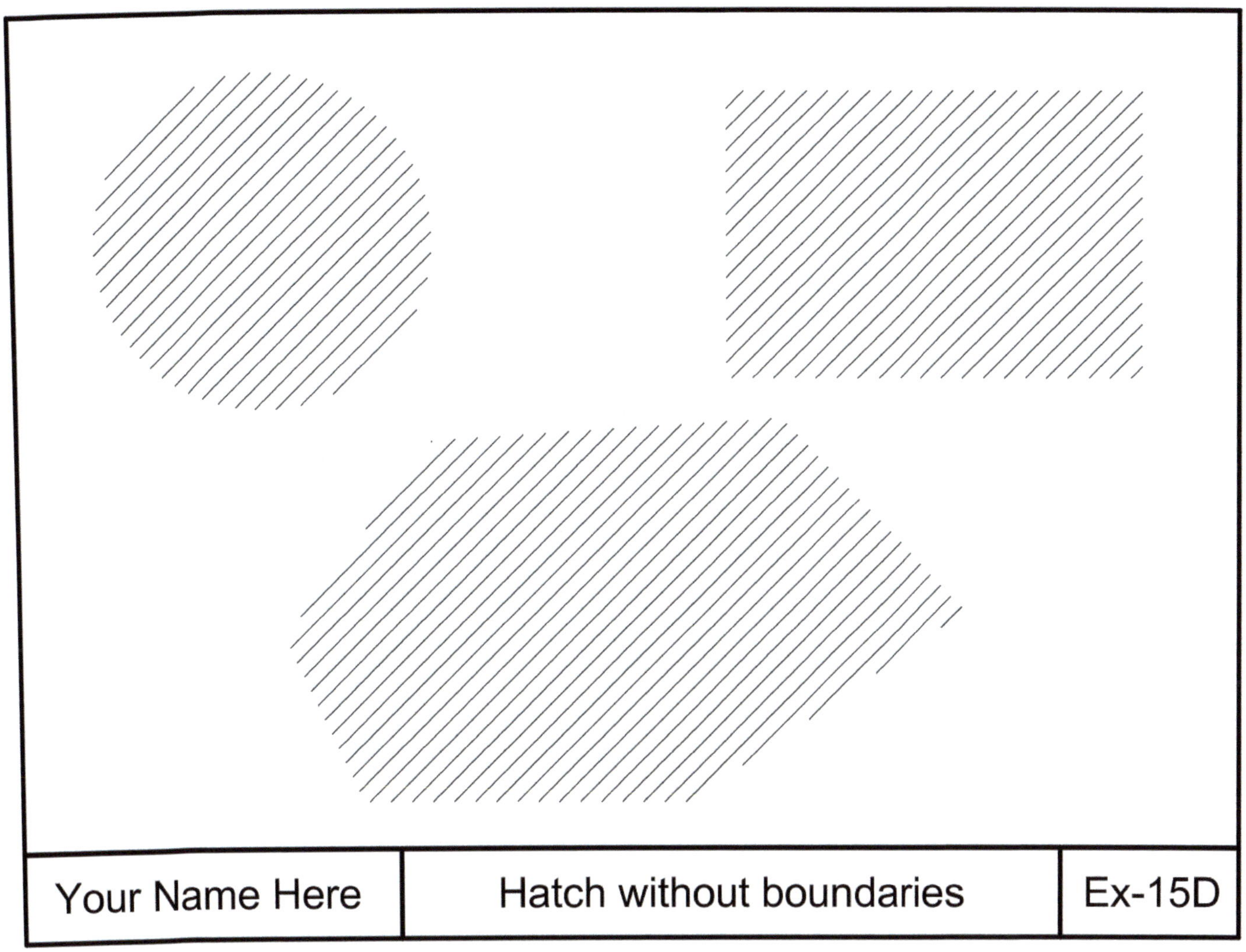

Exercise 15E

Instructions:

1. Start a **New** file using the **Border A-Inch.dwt** or the **Border A-Metric.dwt**
2. Draw the Hub as shown using Layer Object Line.
3. Use Hatch pattern **ANSI 31** Scale = **1** and Angle = **0**. Use Layer Hatch.
4. Draw the section arrowhead, approximately as shown. Fill with **"Solid"** hatch.
5. Edit the **Title** and **Ex-XX** by double clicking on the text. Do not erase and replace.
6. Do not dimension.
7. **Save** as **Ex-15E**
8. Plot using Page Setup **Class Model A**

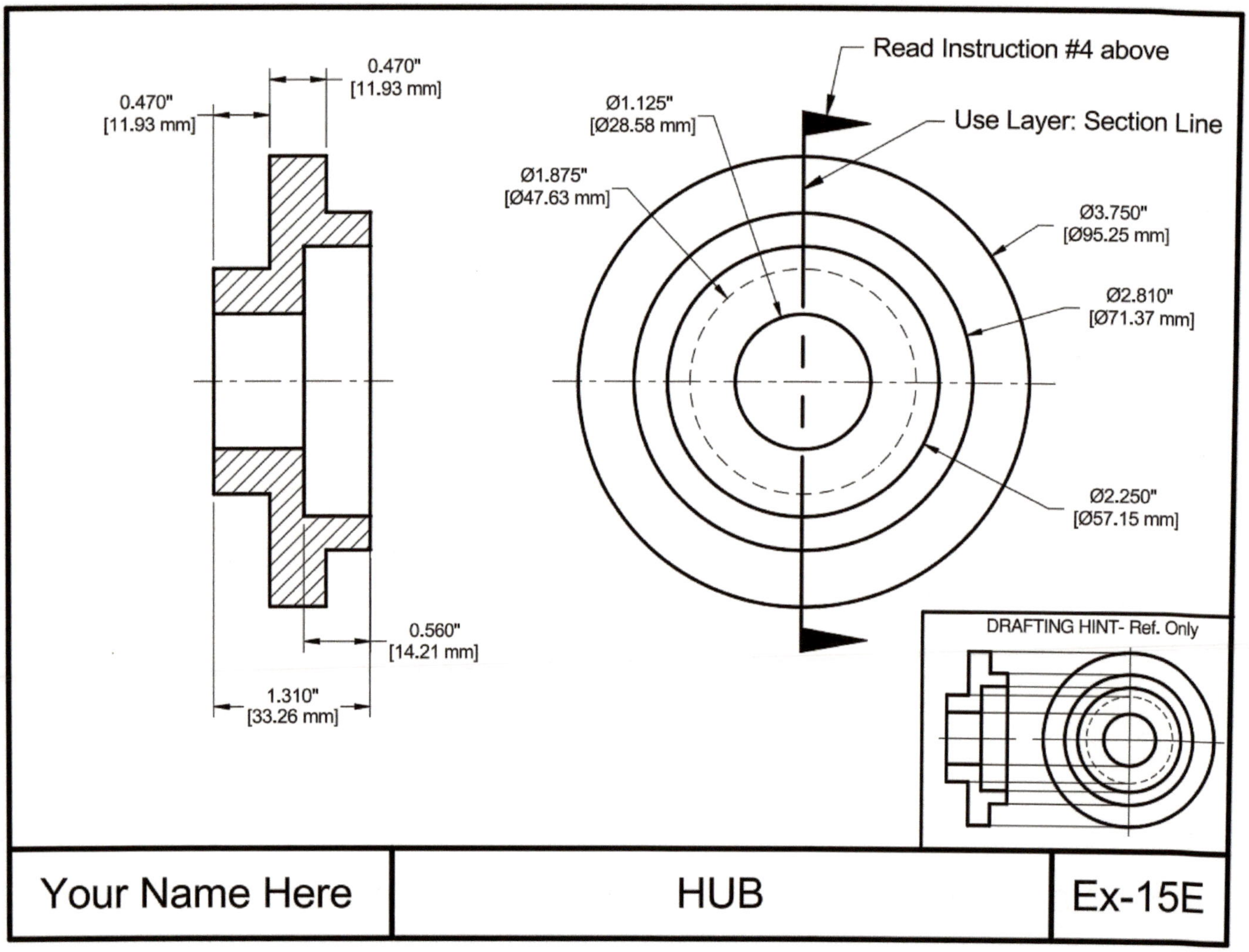

LESSON 16

LEARNING OBJECTIVES

After completing this lesson, you will be able to:

- Understand the importance of True Associative dimensioning
- Create Linear, Continue, and Baseline dimensions
- Draw, Select, and Erase objects
- Use Grips to manipulate dimensions
- Create a new Dimension Style
- Ignore Hatch Lines when dimensioning

Dimensioning

Dimensioning is basically easy, but as always, there are many options to learn. As a result, I have divided the dimensioning process into multiple lessons.

In addition, Annotative dimensioning will be discussed in Lesson 27. So relax and just take it one lesson at a time.

In this Lesson you will learn how to create a **dimension style** and how to create linear (horizontal and vertical) dimensions.

Dimensions can be **Associative**, **Non-Associative**, or **Exploded**. You need to understand what these are so you may decide which setting you want to use. Most of the time, you will use Associative, but you may have reasons to also use Non-Associative and Exploded.

Associative

Associative Dimensioning means that the dimension is actually associated to the objects that they dimension. If you move the object, the dimension will move with it. If you change the size of the object, the dimension text value will change also.

(**Note:** *This is not parametric. In other words, you cannot change the dimension text value and expect the object to change. That would be parametric dimensioning.*)

Non-Associative

Non-Associative means the dimension is not associated to the objects (disassociated) and will not change if the size of the object changes.

Exploded

Exploded means the dimension will be exploded into lines, text, and arrowheads and Non-Associative.

How to set dimensioning to Associative, Non-Associative or Exploded

1. On the Command Line type ***dimassoc*** and then press ***<Enter>***.

2. Enter the number ***2***, ***1*** or ***0*** and then press ***<Enter>***.

   2 = Associative

   1 = Non-Associative

   0 = Exploded

How to Re-Associate a dimension

If a dimension is **Non-Associative**, and you would like to make it **Associative**, you may use the **dimreassociate** command to change.

1. Select **Reassociate** using one of the following:

   Ribbon = Annotate Tab / Dimension Panel / 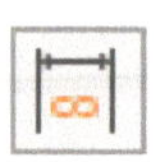
   or
   Keyboard = *dimreassociate <Enter>*

   The following will appear on the Command Line:

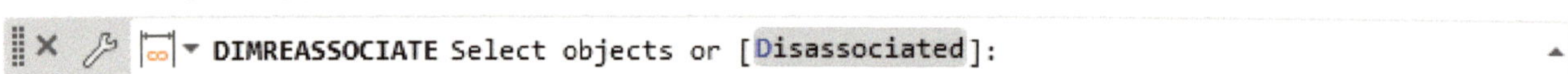

2. Select the dimensions to be re-associated and then press ***<Enter>***.

Continued on the next page...

The following will appear on the Command Line:

```
DIMREASSOCIATE Specify first extension line origin or [Select object] <next>:
```

3. An **"X"** will appear to identify which is the first extension; use Object Snap to select the exact location, on the object, for the extension line point.

   The following will appear on the Command Line:

```
DIMREASSOCIATE Specify second extension line origin <next>:
```

4. An **"X"** will appear on the second extension; use Object Snap to select the exact location, on the object, for the extension line point.

5. Continue until all extension line points are selected.

Note: You must use **Object Snap** to specify the exact location for the extension lines.

Regenerating Associative dimensions

Sometimes after panning and zooming, the Associative dimensions seem to be floating or not following the object. The **dimregen** command will move the Associative dimensions back into their correct location.

To move the Associative dimensions back into their correct location, type ***dimregen*** and then press ***<Enter>***.

Linear Dimensioning

Linear dimensioning allows you to create horizontal and vertical dimensions.

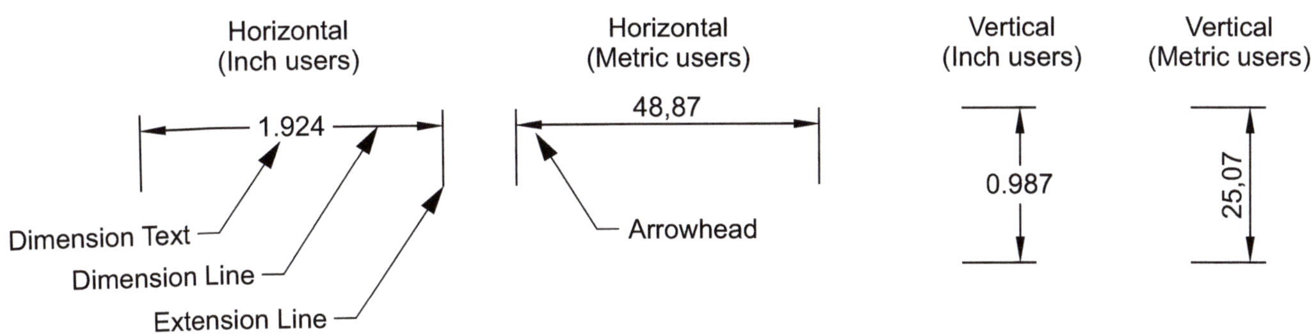

Note: A **decimal point** is used for **inch** dimensions and a **comma** is used for **metric** dimensions.

How to create a Linear Dimension

1. Select the **Linear** command using one of the following:

   Ribbon = Annotate Tab / Dimension Panel / 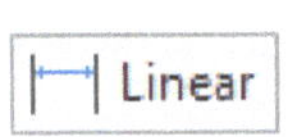

   or
   Keyboard = *dimlinear <Enter>*

   The following will appear on the Command Line:

```
DIMLINEAR Specify first extension line origin or <select object>:
```

Continued on the next page...

2. Snap to the first extension line origin (**P1**).

   The following will appear on the Command Line:

```
DIMLINEAR Specify second extension line origin:
```

3. Snap to the second extension line origin (**P2**).

   The following will appear on the Command Line:

```
DIMLINEAR [Mtext Text Angle Horizontal Vertical Rotated]:
```

4. Select where you want the dimension line placed (**P3**).

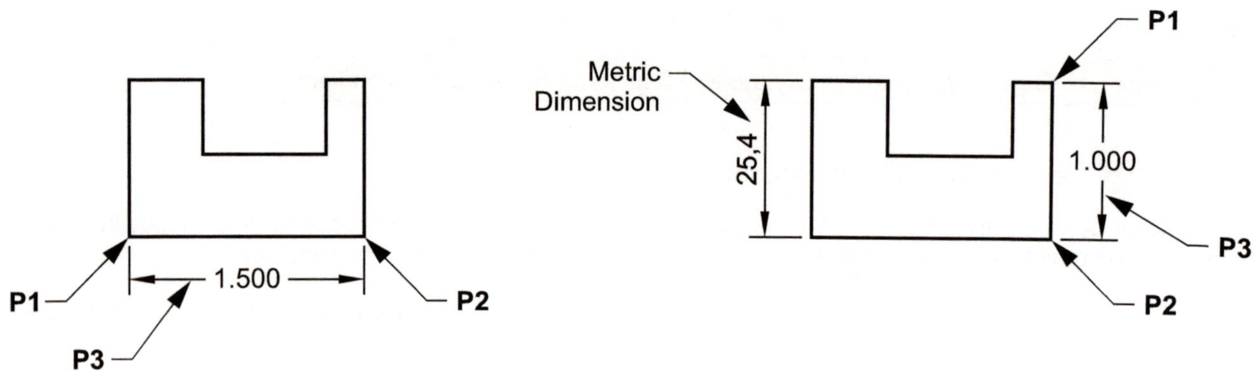

Continue Dimensioning

Continue Dimensioning creates a series of dimensions in-line with an existing dimension. If you use the continue dimensioning immediately after a Linear dimension, you do not have to specify the continue extension

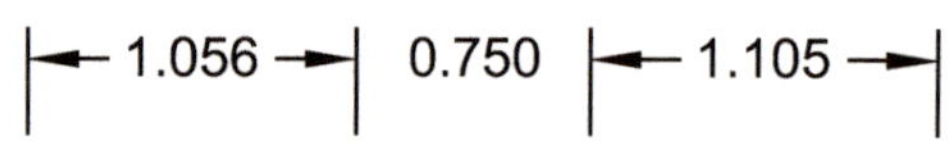

How to create Continue Dimensions

1. Create a Linear dimension first (**P1** and **P2** shown on the next page).

2. Select the **Continue** command using one of the following:

   Ribbon = Annotate Tab / Dimension Panel /
   or
   Keyboard = *dimcontinue <Enter>*

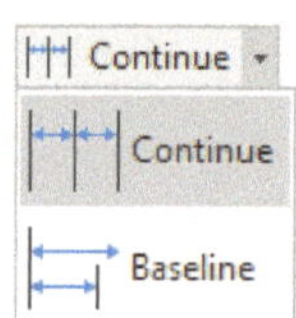

   The following will appear on the Command Line:

```
DIMCONTINUE Specify second extension line origin or [Select Undo] <Select>:
```

3. Snap to the second extension line origin (**P3**).

   The following will appear on the Command Line:

```
DIMCONTINUE Specify second extension line origin or [Select Undo] <Select>:
```

Continued on the next page...

4. Snap to the second extension line origin (**P4**).

   The following will appear on the Command Line:

```
DIMCONTINUE Specify second extension line origin or [Select Undo] <Select>:
```

5. Press ***<Enter> <Enter>*** (the second enter ends the command).

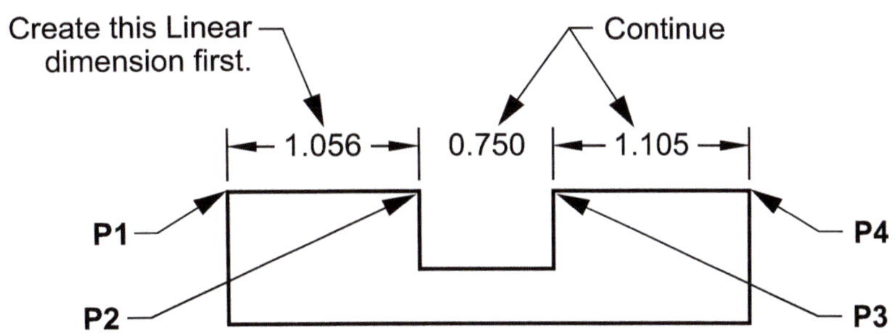

Baseline Dimensioning

Baseline Dimensioning allows you to establish a baseline for successive dimensions. The spacing between dimensions is automatic and should be set in the dimension style. (See the bottom figures on page 16-8, **Baseline spacing** setting.)

A Baseline dimension must be used with an existing dimension. If you use Baseline dimensioning immediately after a Linear dimension, you do not have to specify the baseline origin.

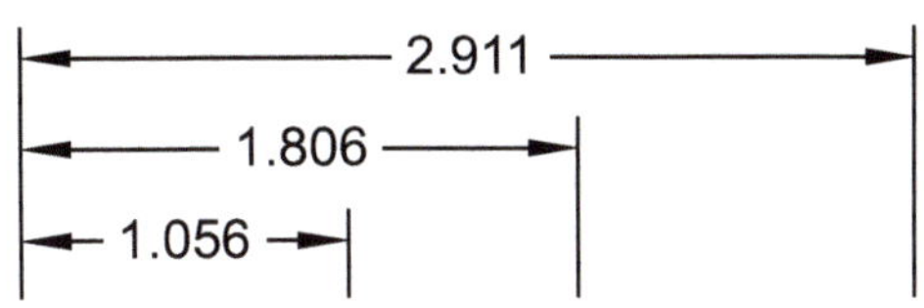

How to create Baseline Dimensions

1. Create a Linear dimension first (**P1** and **P2** shown on the next page).

2. Select the **Baseline** command using one of the following:

   Ribbon = Annotate Tab / Dimension Panel /
   or
   Keyboard = *dimbaseline <Enter>*

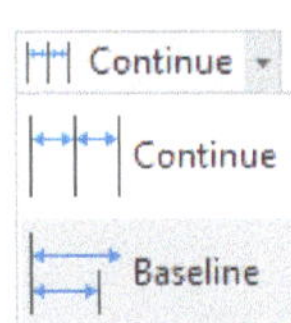

   The following will appear on the Command Line:

```
DIMBASELINE Specify second extension line origin or [Select Undo] <Select>:
```

3. Snap to the second extension line origin (**P3**).

   The following will appear on the Command Line:

```
DIMBASELINE Specify second extension line origin or [Select Undo] <Select>:
```

4. Snap to the second extension line origin (**P4**).

Continued on the next page...

The following will appear on the Command Line:

```
DIMBASELINE Specify second extension line origin or [Select Undo] <Select>:
```

5. Press ***<Enter> <Enter>*** (the second enter ends the command).

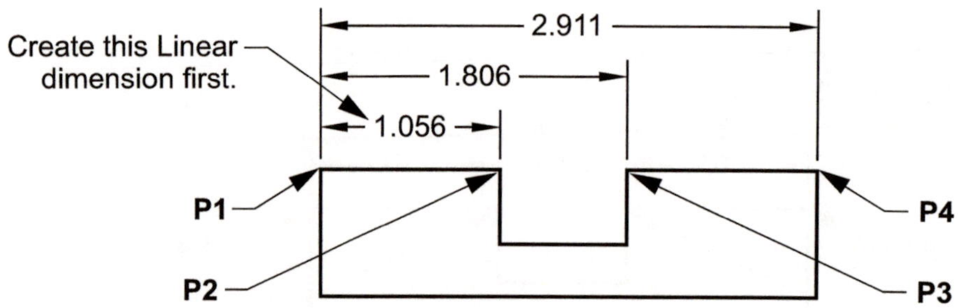

Dimension Styles

Using the **Dimension Style Manager** you can change the appearance of the dimension features, such as length of arrowheads, size of the dimension text, etc. There are over 70 different settings.

You can also Create New, Modify, Override, and Compare Dimension Styles. All of these are simple, by using the Dimension Style Manager described below.

1. Select the **Dimension Style Manager** using one of the following:

Ribbon = Annotate Tab / Dimension Panel / ↘
or
Keyboard = *dimstyle <Enter>*

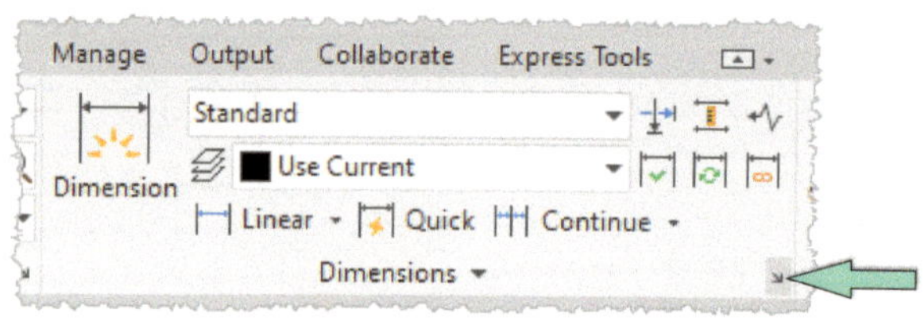

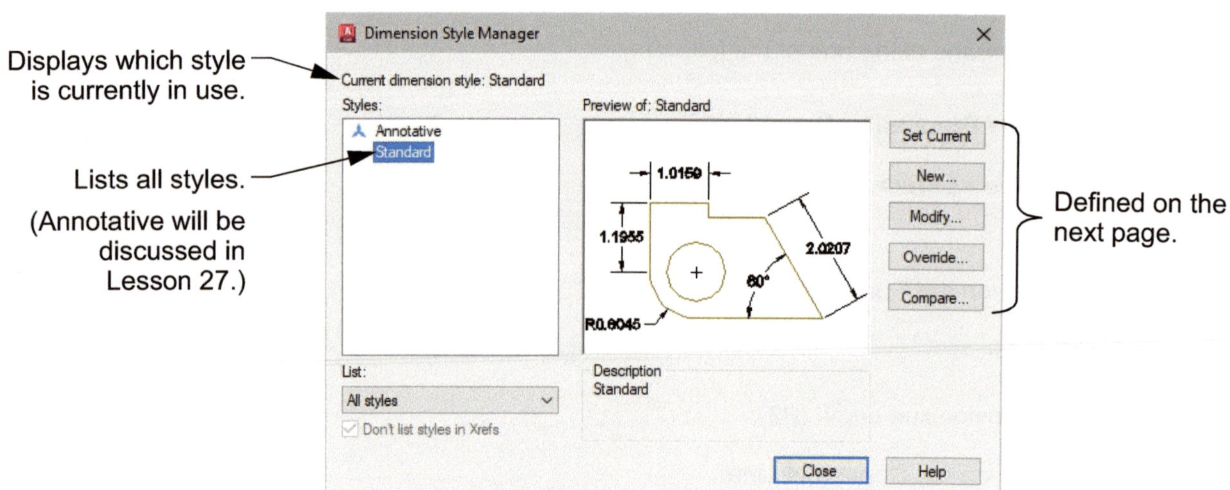

Continued on the next page...

Set Current: Select a style from the list of styles and select the **Set Current** button.

New: Select this button to create a new style. When you select this button, the **Create New Dimension Style** dialog box is displayed.

Modify: Selecting this button opens the **Modify Dimension Style** dialog box, which allows you to make changes to the style selected from the "list of Styles".

Override: An override is a temporary change to the current style. Selecting this button opens the **Override Current Style** dialog box.

Compare: Compares two dimension styles.

Creating a New Dimension Style

A dimension style is a group of settings that has been saved with a name you assign. When creating a new style, you must start with an existing style, such as **Standard**. Next, assign it a new name, make the desired changes, and when you select the **OK** button, the new style will have been successfully created.

How to create a new dimension style

1. Start a New file using the **Border A-Inch.dwt** or the **Border A-Metric.dwt**
2. Select the **Dimension Style Manager**. (Refer to the previous page.)
3. Select the **New...** button.

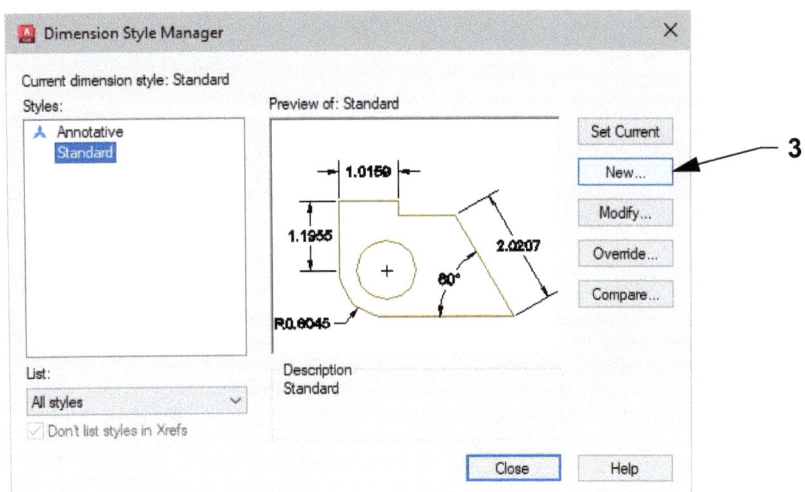

4. Enter **CLASS STYLE** in the **"New Style Name"** box.
5. Select **Standard** in the **"Start With:"** box.
6. The **"Use For:"** box will be discussed later. For now, leave it set to **"All dimensions"**.
7. Select the **Continue** button.

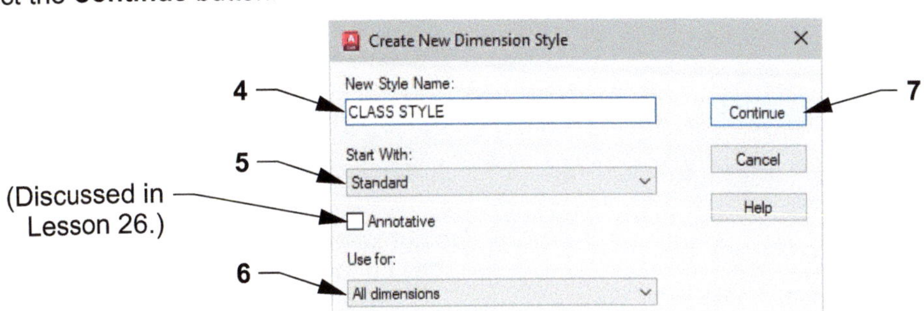

Continued on the next page...

8. Select the **Primary Units** Tab and change your settings to match the settings shown below.

Inch users

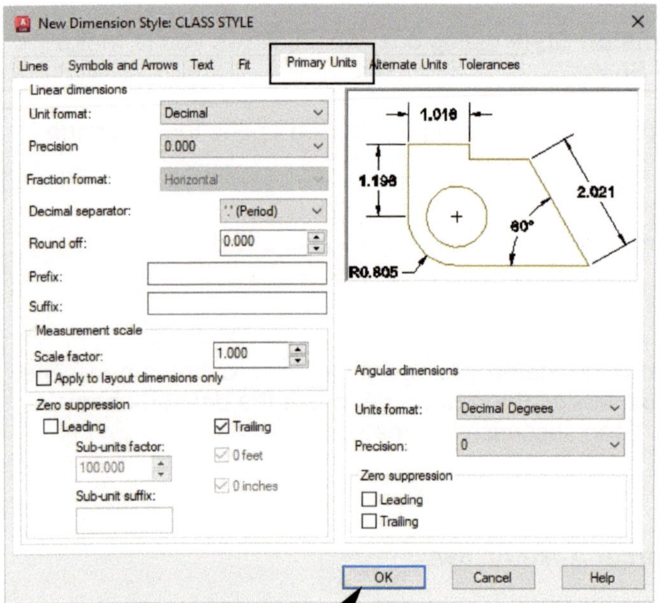

Do not select the **OK** button yet.

Metric users

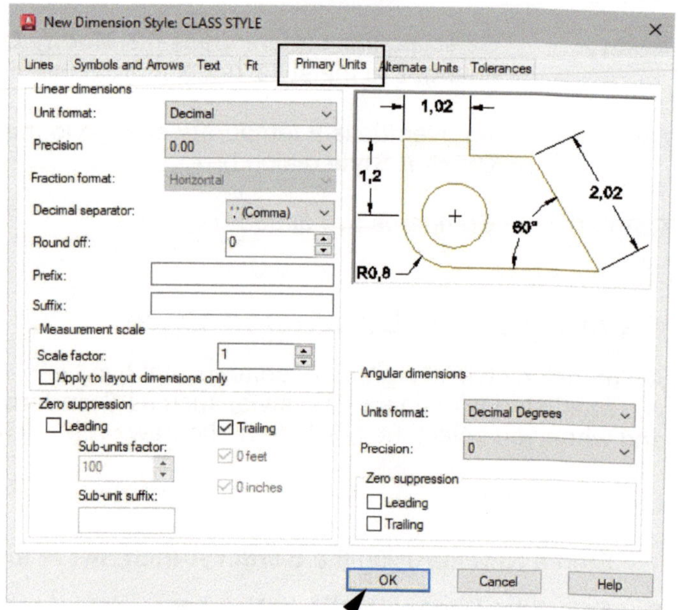

Do not select the **OK** button yet.

9. Select the **Lines** Tab and change your settings to match the settings shown below.

Inch users

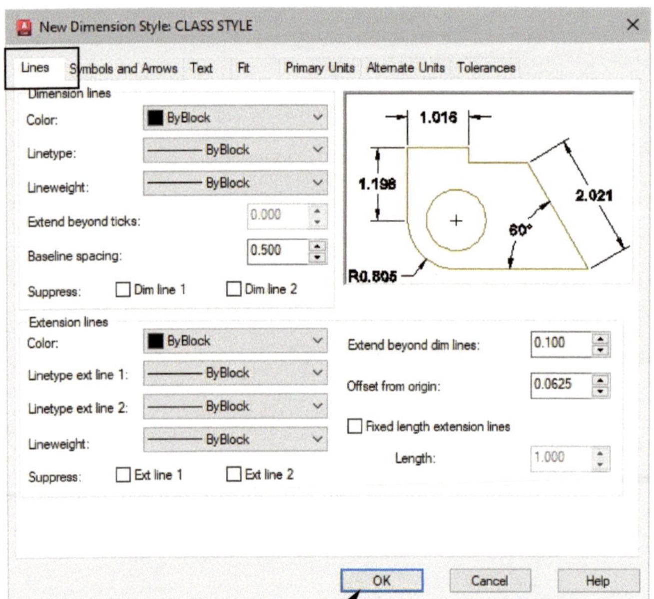

Do not select the **OK** button yet.

Metric users

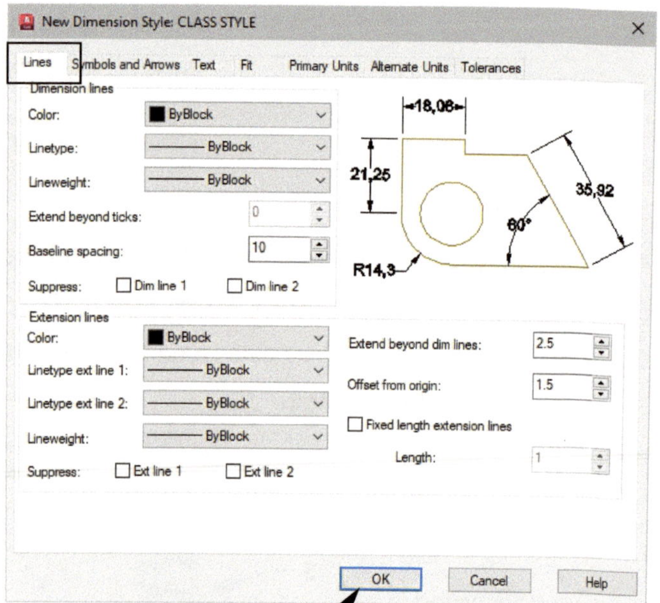

Do not select the **OK** button yet.

Continued on the next page...

10. Select the **Symbols and Arrows** Tab and change your settings to match the settings shown below.

Inch users

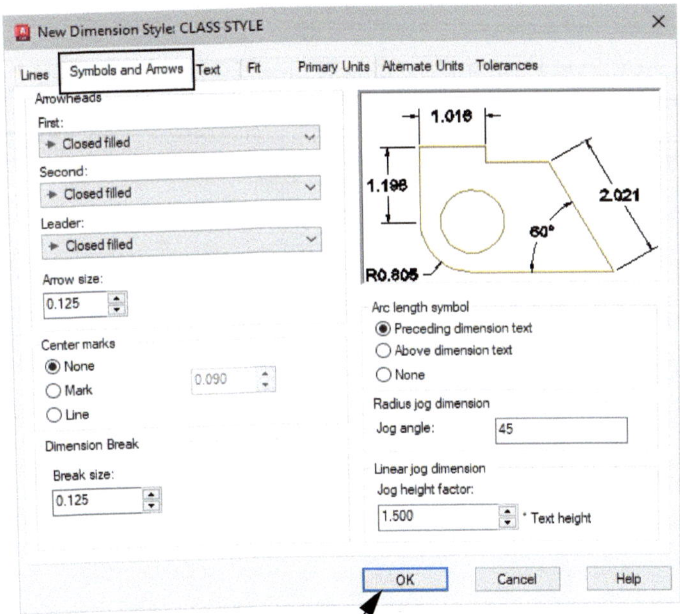

Do not select the **OK** button yet.

Metric users

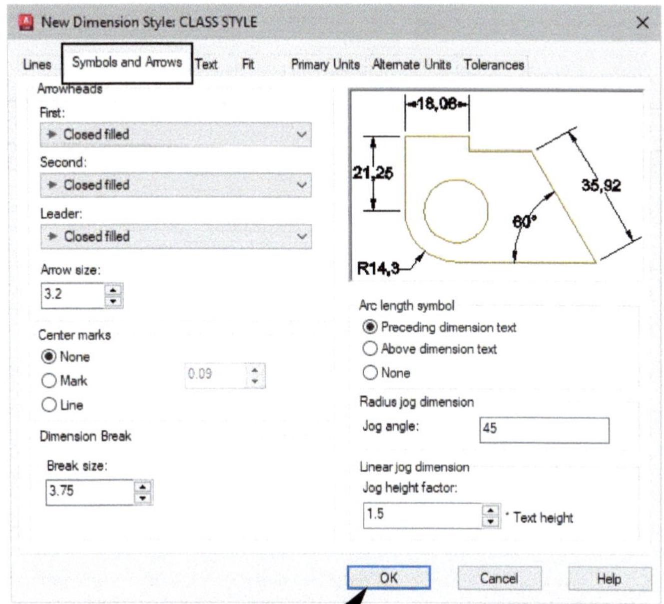

Do not select the **OK** button yet.

11. Select the **Text** Tab and change your settings to match the settings shown below.

Inch users

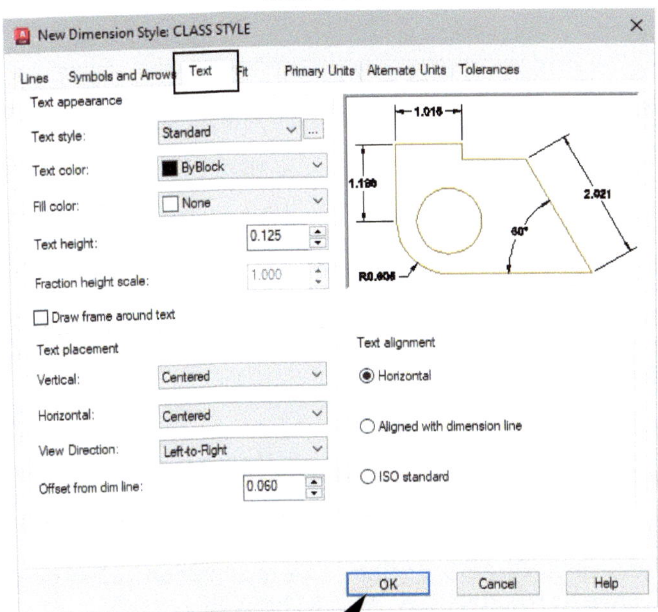

Do not select the **OK** button yet.

Metric users

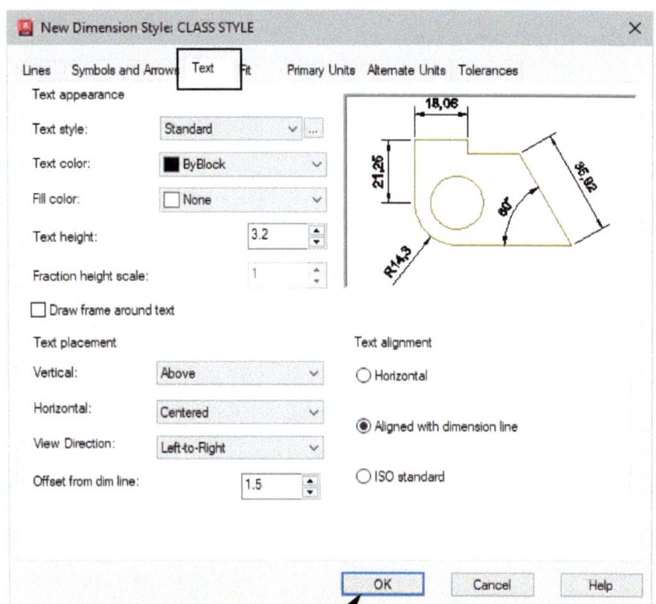

Do not select the **OK** button yet.

Continued on the next page...

12. Select the **Fit** Tab and change your settings to match the settings shown below.

Inch users

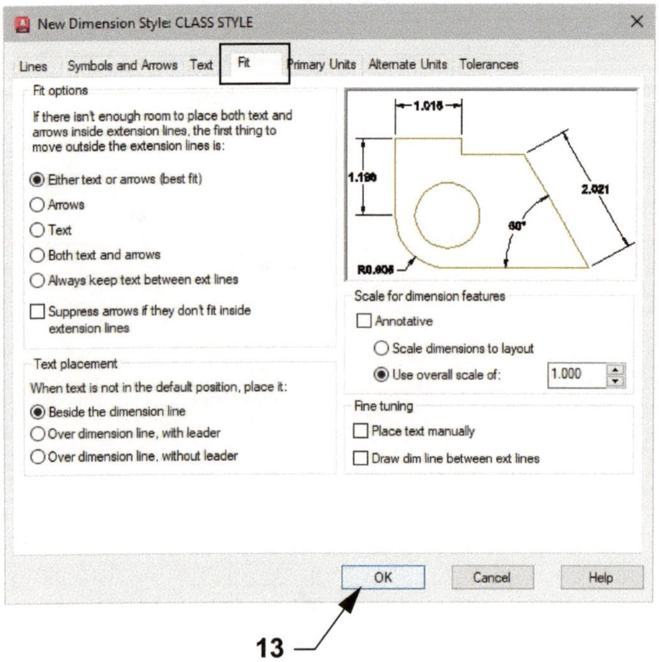

Metric users

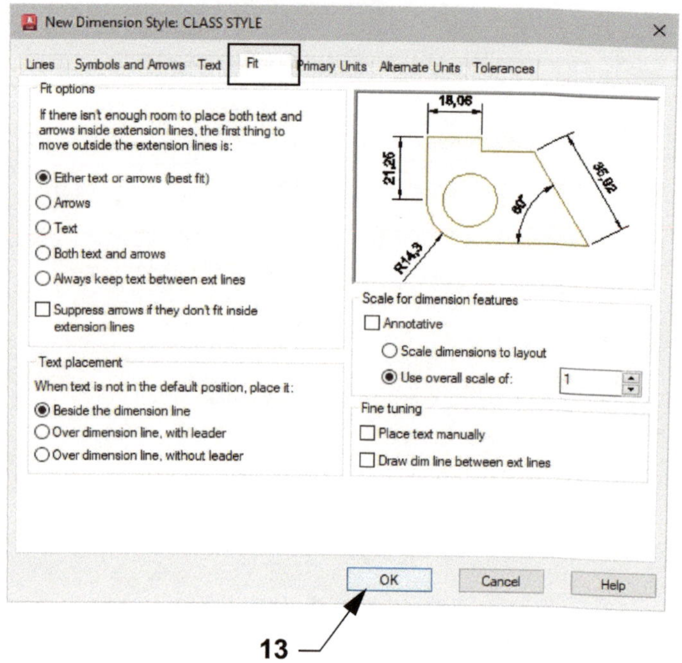

13. Now select the **OK** button.

Your new style **"CLASS STYLE"** should be listed.

14. Select the **"Set Current"** button to make your new style CLASS STYLE the style that will be used.

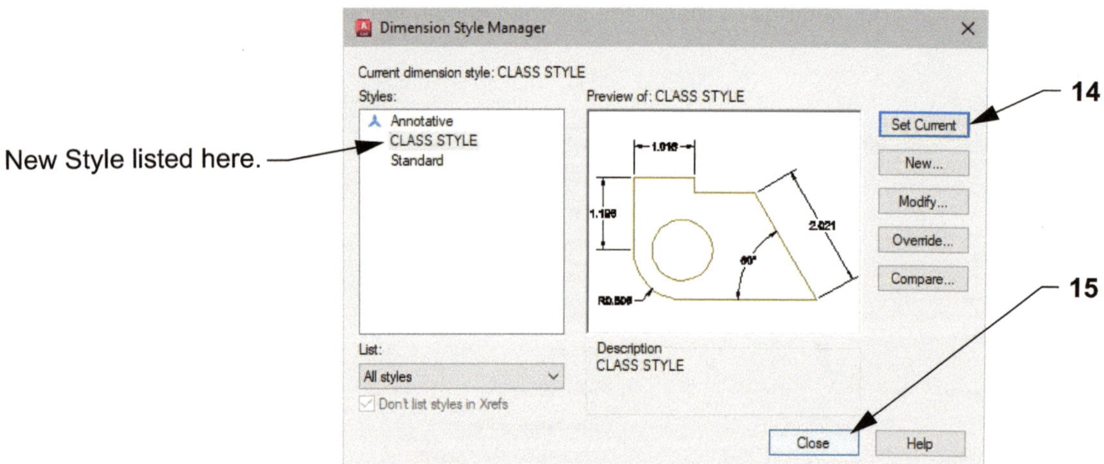

15. Select the **Close** button.

16. **Important:** Re-Save this "template" as **Border A-Inch.dwt** for inch users, or **Border A-Metric.dwt** for metric users. (Refer to page 2-3.)

Note: You have successfully created a new **Dimension Style** called **CLASS STYLE**. This style will be saved in your **Border A-Inch** or **Border A-Metric** template after you save the template. So every time you use this template, the dimension style will already be there, and you will not have to create it again.

It is important that you understand that this dimension style resides **only** in your **Border A-Inch** or **Border A-Metric** template. If you open another drawing, this dimension style will not be there.

Ignoring Hatch Objects

Occasionally, when you are dimensioning an object that has "Hatch Lines", your cursor will snap to the Hatch Line instead of the object that you want to dimension. To prevent this from occurring, select the option **"Ignore hatch objects".**

Example:

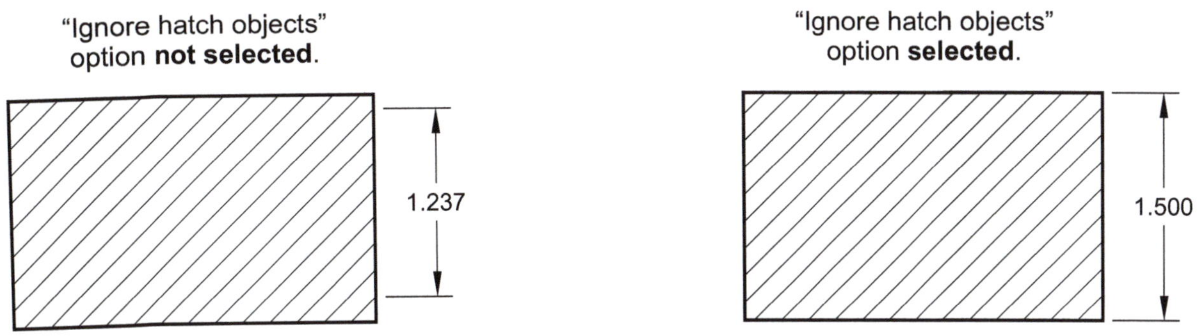

How to select the "Ignore hatch objects" option

1. Type ***Options*** and then press ***<Enter>***.
2. Select the **Drafting** Tab.
3. Check the **Ignore hatch objects** box.
4. Select the **OK** button to close the **Options** dialog box.

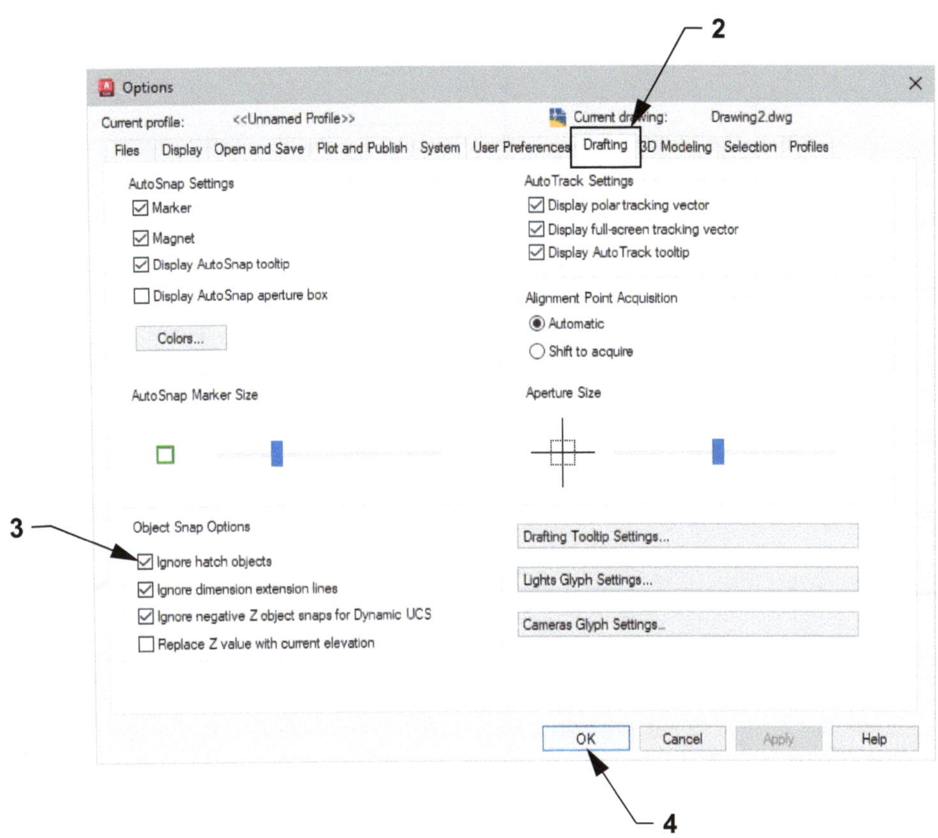

Exercise 16A-Inch

Instructions:

1. Start a **New** file using the **Border A-Inch.dwt**
2. Create a dimension style in your **Border A-Inch** template by following the instructions on pages 16-7 through 16-10 if you haven't already completed the task. Re-save the template as **Border A-Inch.dwt** before starting this drawing.
3. Draw the objects as shown using Layer Object Line.
4. Dimension as shown using Dimension Style CLASS STYLE on Layer Dimension.
5. Use Linear, Continue, and Baseline.
6. Edit the **Title** and **Ex-XX** by double clicking on the text. Do not erase and replace.
7. **Save** as **Ex-16A-I**
8. Plot using Page Setup **Class Model A**

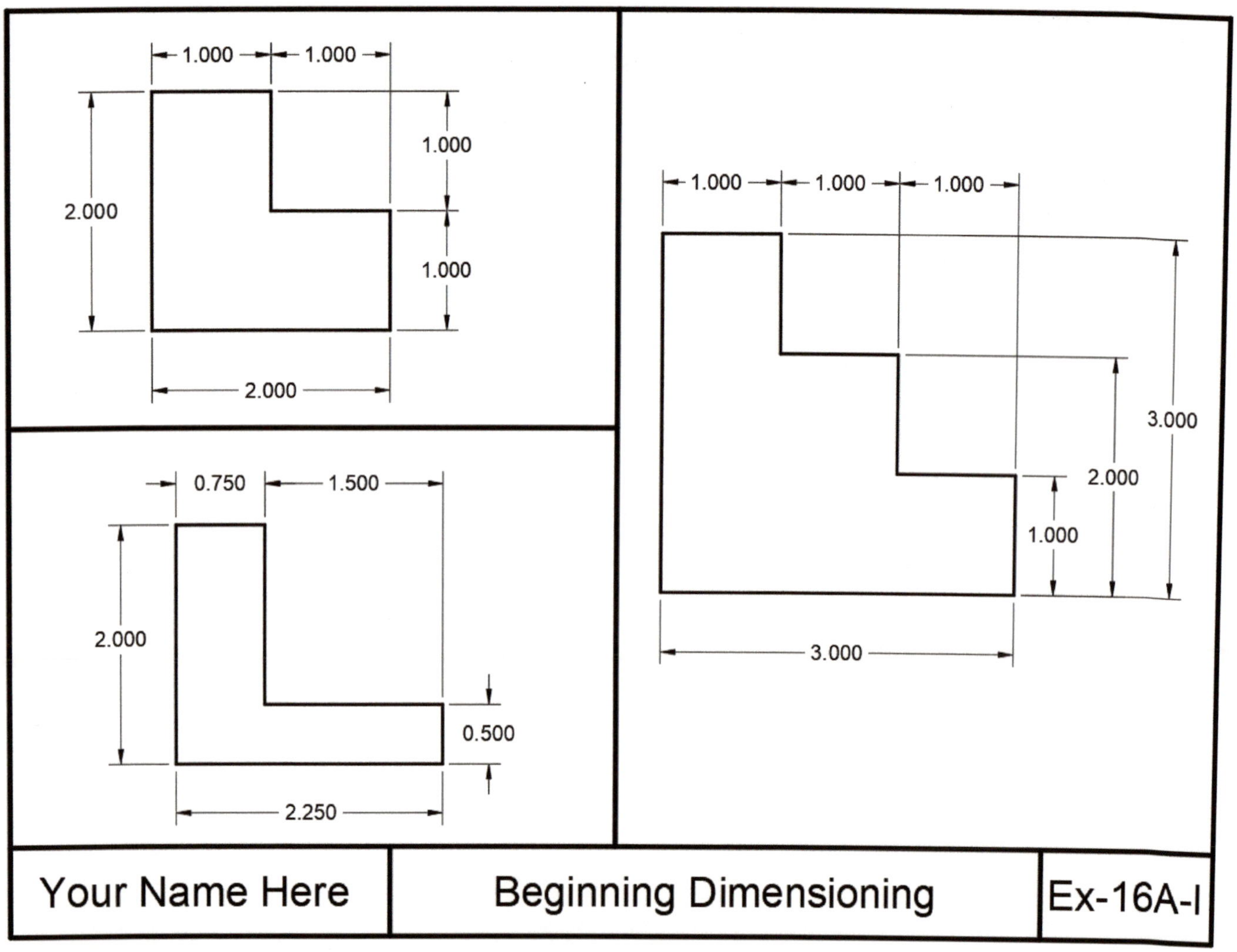

Exercise 16A-Metric

Instructions:

1. Start a **New** file using the **Border A-Metric.dwt**
2. Create a dimension style in your **Border A-Metric** template by following the instructions on pages 16-7 through 16-10 if you haven't already completed the task. Re-save the template as **Border A-Metric.dwt** before starting this drawing.
3. Draw the objects as shown using Layer Object Line.
4. Dimension as shown using Dimension Style CLASS STYLE on Layer Dimension.
5. Use Linear, Continue, and Baseline.
6. Edit the **Title** and **Ex-XX** by double clicking on the text. Do not erase and replace.
7. **Save** as **Ex-16A-M**
8. Plot using Page Setup **Class Model A**

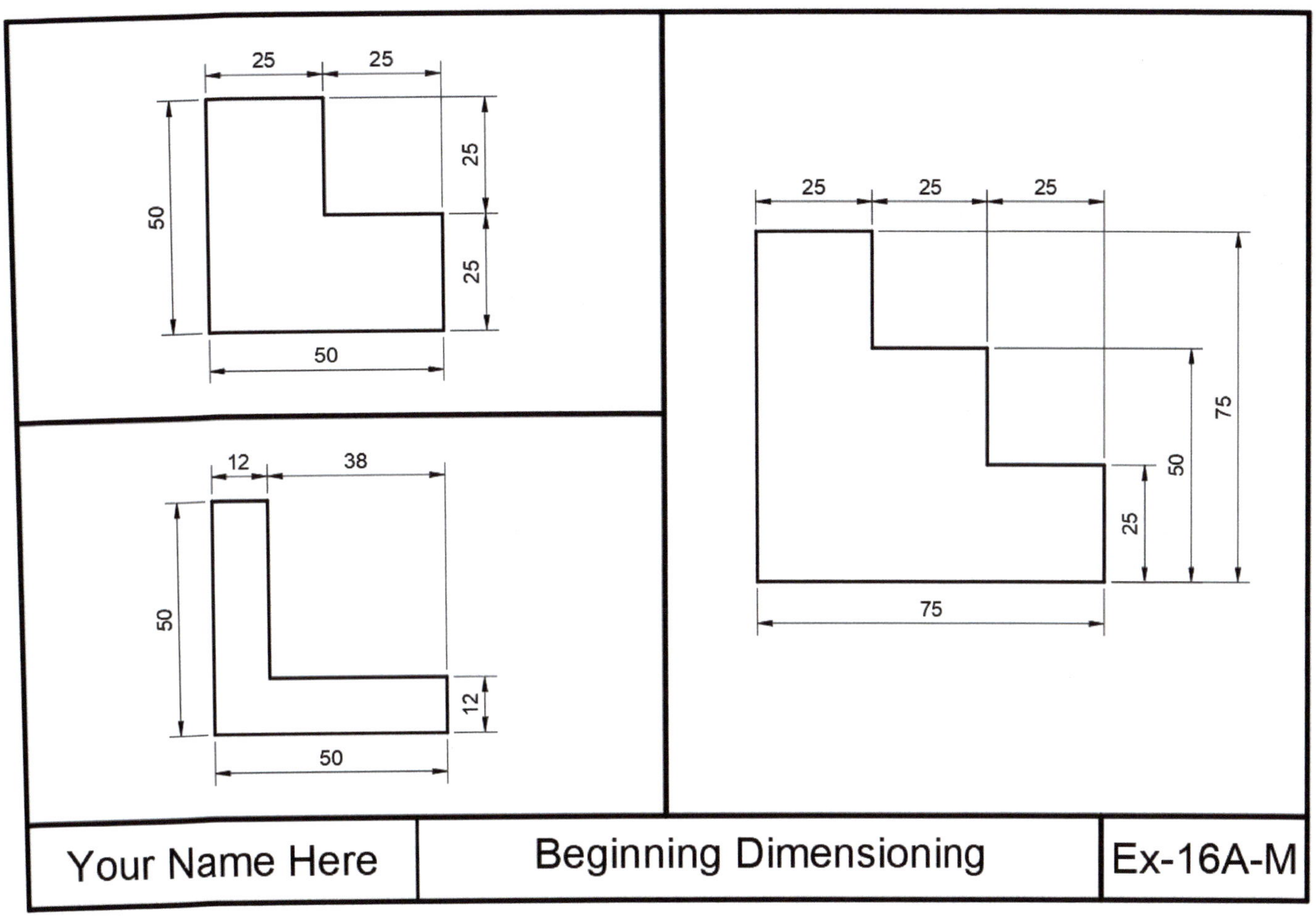

Exercise 16B-Inch

Instructions:

1. Start a **New** file using the **Border A-Inch.dwt**
2. Draw the objects as shown using Layer Object Line.
3. Dimension as shown using Dimension Style CLASS STYLE on Layer Dimension.
4. Use Linear, Continue, and Baseline.
5. Edit the **Title** and **Ex-XX** by double clicking on the text. Do not erase and replace.
6. **Save** as **Ex-16B-I**
7. Plot using Page Setup **Class Model A**

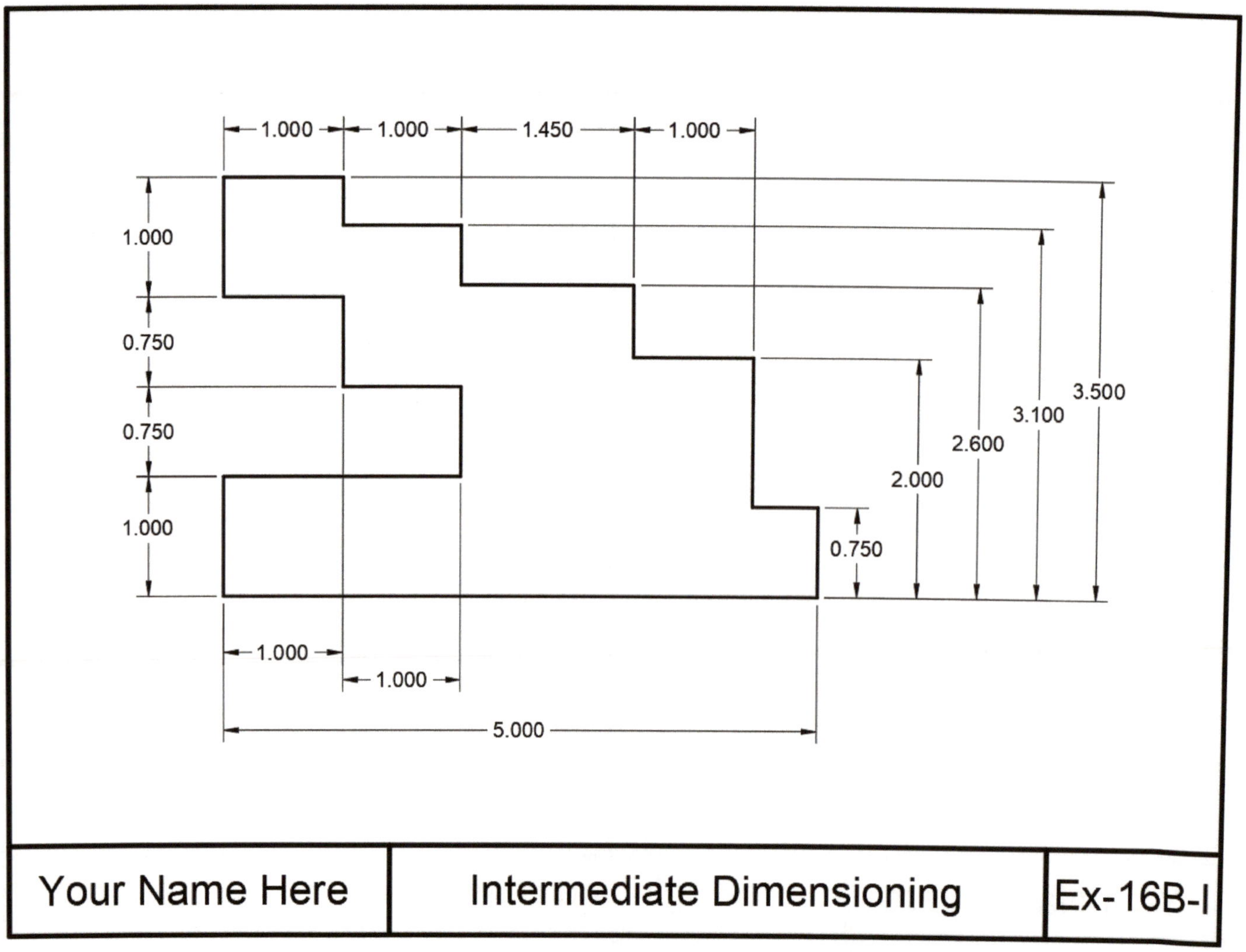

Exercise 16B-Metric

Instructions:

1. Start a **New** file using the **Border A-Metric.dwt**
2. Draw the objects as shown using Layer Object Line.
3. Dimension as shown using Dimension Style CLASS STYLE on Layer Dimension.
4. Use Linear, Continue, and Baseline.
5. Edit the **Title** and **Ex-XX** by double clicking on the text. Do not erase and replace.
6. **Save** as **Ex-16B-M**
7. Plot using Page Setup **Class Model A**

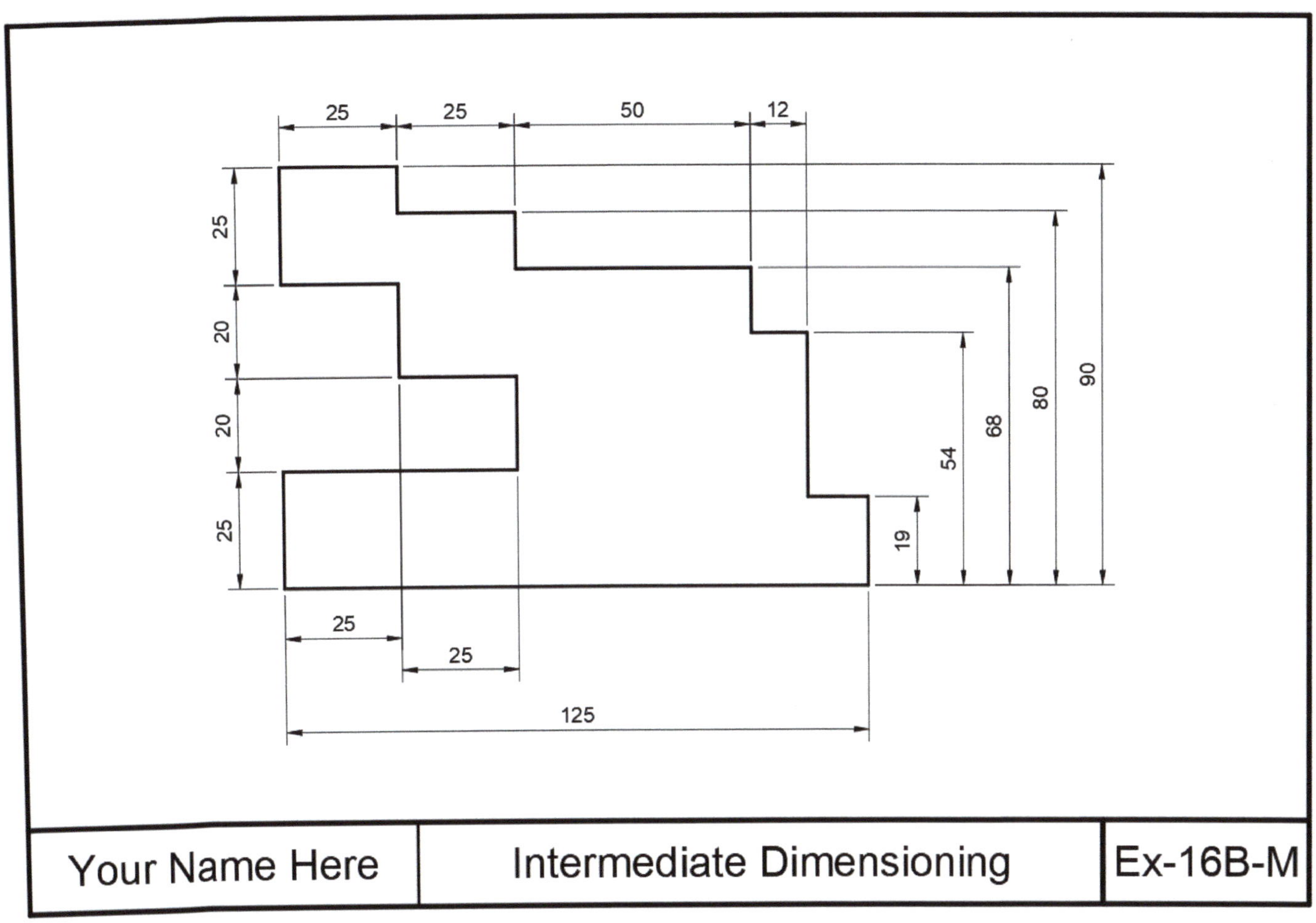

Exercise 16C-Inch

Instructions:

1. Start a **New** file using the **Border A-Inch.dwt**
2. Follow the instructions below.
3. Do not divide your drawing into four sections. Just draw one rectangle in the middle of your drawing area and follow the instructions.
4. Edit the **Title** and **Ex-XX** by double clicking on the text. Do not erase and replace.
5. **Save** as **Ex-16C-I**
6. Plot using Page Setup **Class Model A**

1. Draw a **2.250"** long by **1.500"** wide rectangle.

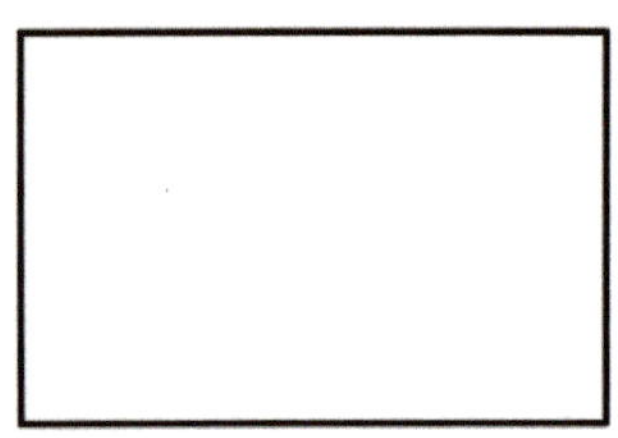

2. Set "Dimassoc" setting to **2**.
3. Dimension the rectangle as shown here.

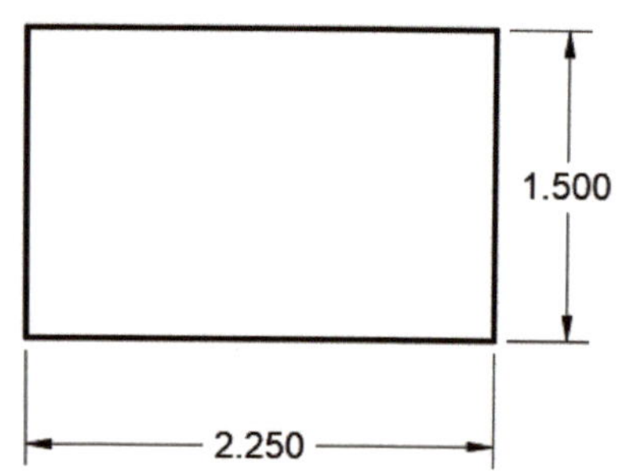

4. Stretch one corner of the rectangle using the grips approximately as shown (with Ortho and Osnap turned **off**).

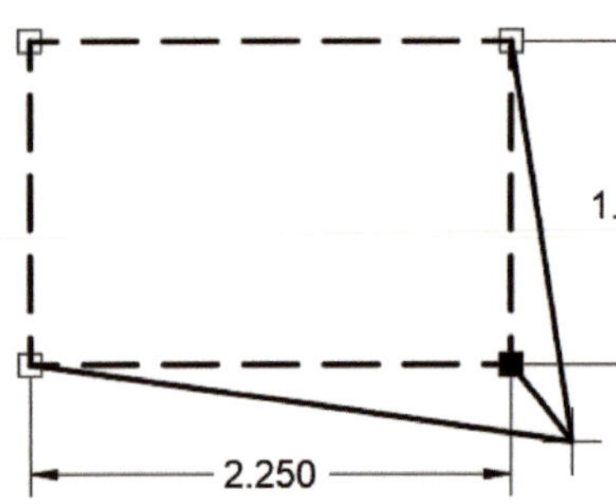

Did the dimension change?
Note: Your dimensions may not be exactly the same as shown. That's OK.

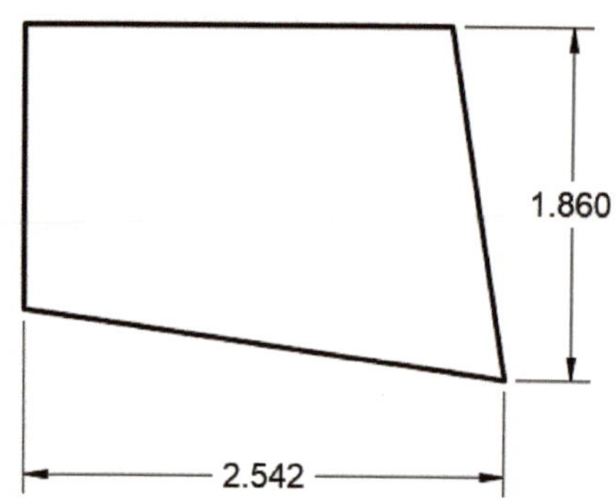

Your Name Here	Stretch and Associative Dims	Ex-16C-I

Exercise 16C-Metric

Instructions:

1. Start a **New** file using the **Border A-Metric.dwt**
2. Follow the instructions below.
3. Do not divide your drawing into four sections. Just draw one rectangle in the middle of your drawing area and follow the instructions.
4. Edit the **Title** and **Ex-XX** by double clicking on the text. Do not erase and replace.
5. **Save** as **Ex-16C-M**
6. Plot using Page Setup **Class Model A**

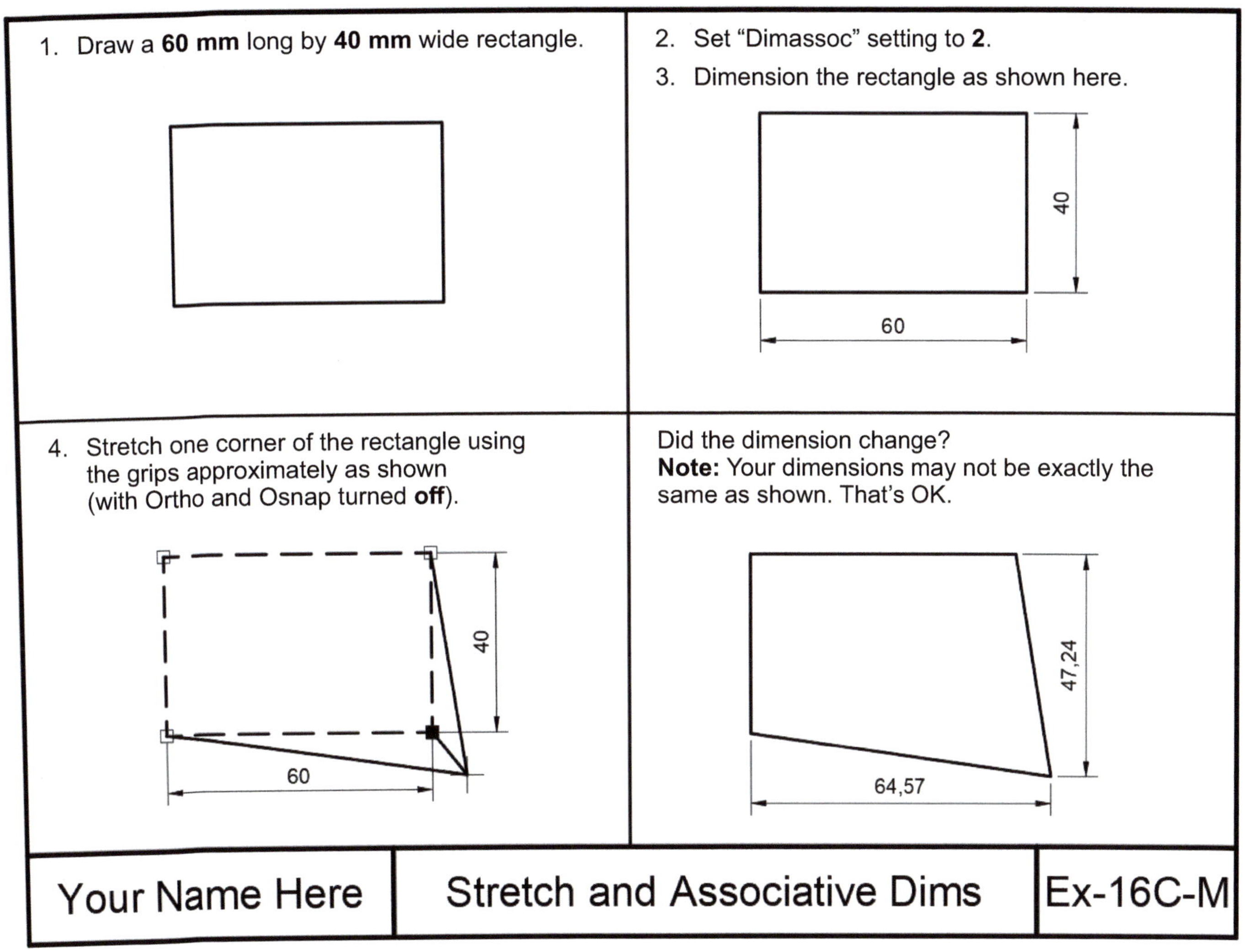

Exercise 16D-Inch

Instructions:

1. Start a **New** file using the **Border A-Inch.dwt**
2. Follow the instructions below.
3. Do not divide your drawing into four sections. Just draw one rectangle in the middle of your drawing area and follow the instructions.
4. Edit the **Title** and **Ex-XX** by double clicking on the text. Do not erase and replace.
5. **Save** as **Ex-16D-I**
6. Plot using Page Setup **Class Model A**

1. Draw a **2.250"** long by **1.500"** wide rectangle.

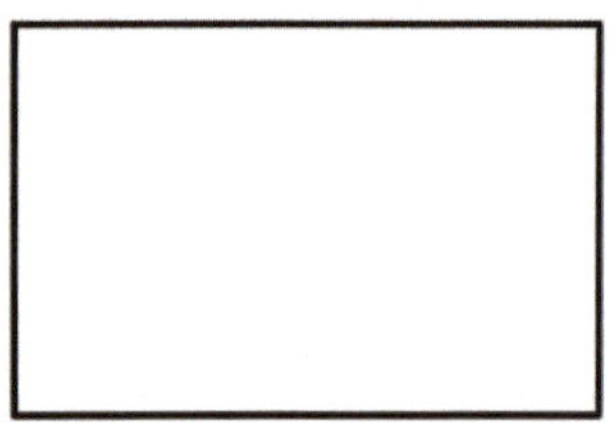

2. Set "Dimassoc" setting to **1**. ← **Note.**
3. Dimension the rectangle as shown here.

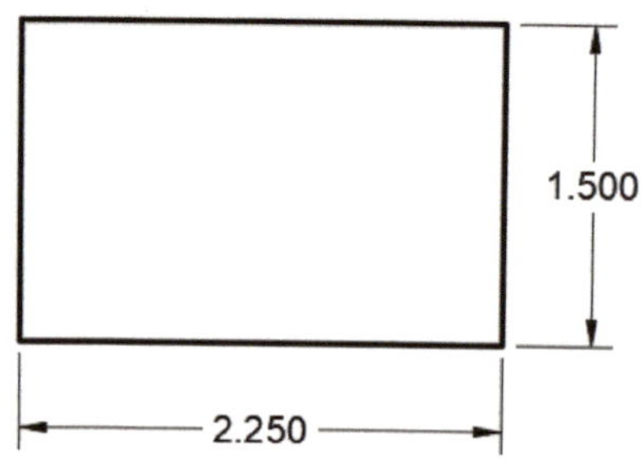

4. Stretch one corner of the rectangle using the grips approximately as shown (with Ortho and Osnap turned **off**).

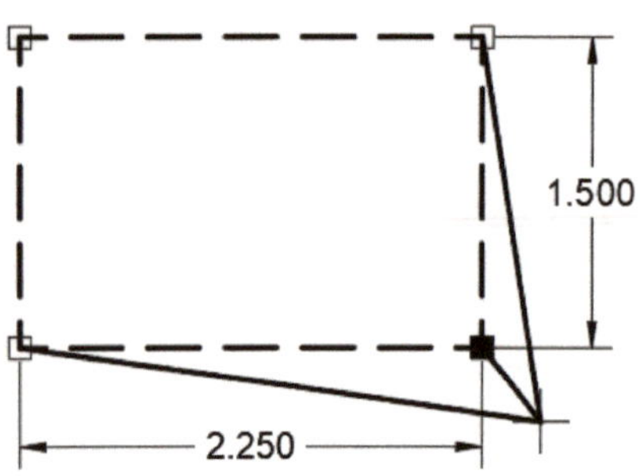

Notice the difference?
The dimensions are not associated to the object, so they did not respond to the stretch.

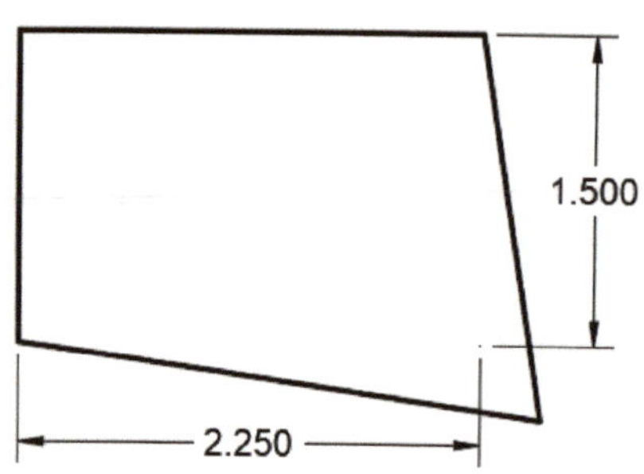

Your Name Here	Stretch and Non-Associative Dims	Ex-16D-I

Exercise 16D-Metric

Instructions:

1. Start a **New** file using the **Border A-Metric.dwt**
2. Follow the instructions below.
3. Do not divide your drawing into four sections. Just draw one rectangle in the middle of your drawing area and follow the instructions.
4. Edit the **Title** and **Ex-XX** by double clicking on the text. Do not erase and replace.
5. **Save** as **Ex-16D-M**
6. Plot using Page Setup **Class Model A**

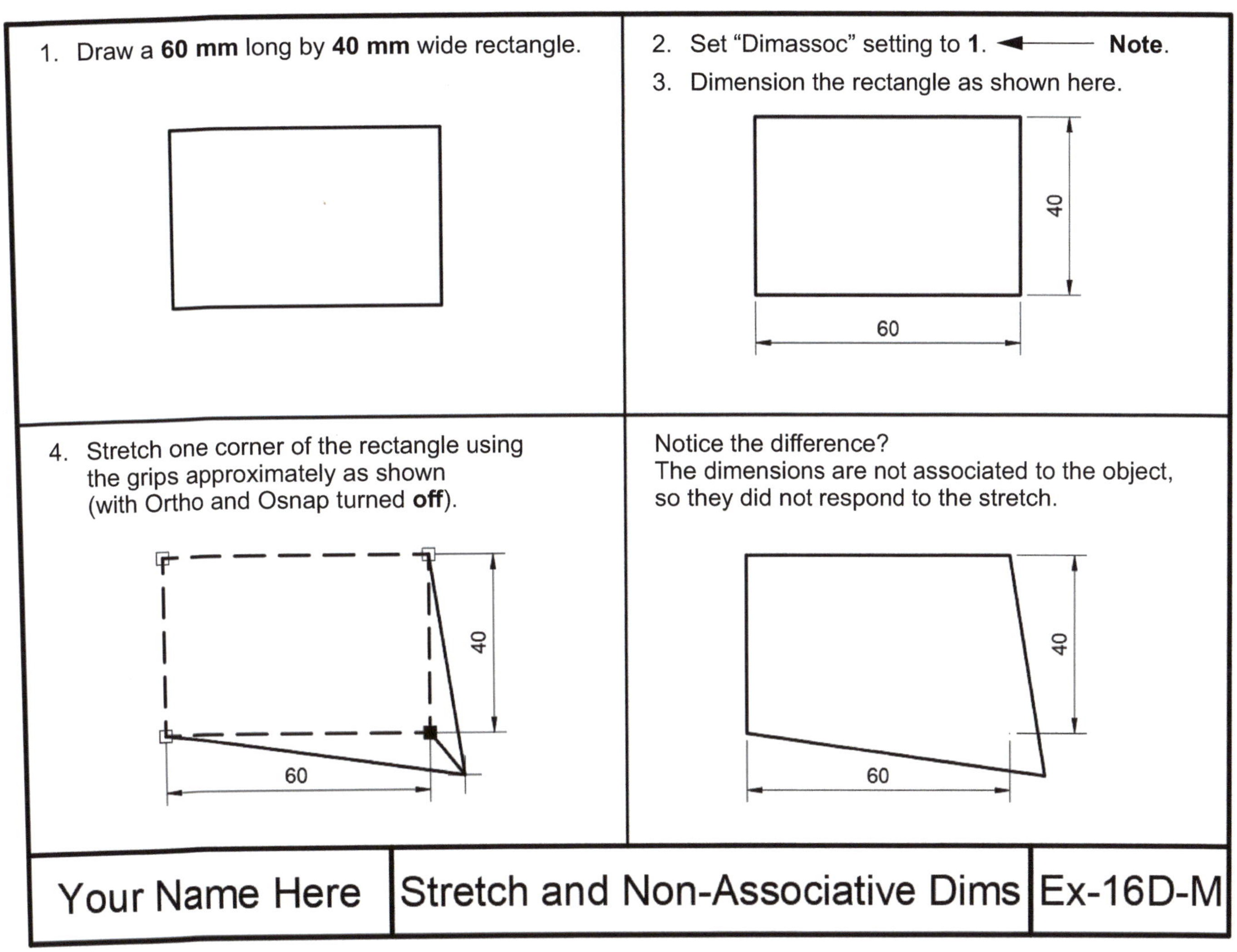

Notes:

LESSON 17

LEARNING OBJECTIVES

After completing this lesson, you will be able to:

- Edit the dimension text without changing the value
- Move the dimension text within the dimension lines
- Modify an entire dimension style
- Override a dimension style
- Use the Properties Palette to change a dimension
- Break intersecting extension and dimension lines
- Add a Jog to a dimension line
- Adjust the distance between dimensions

Editing Dimension Text Values

Sometimes you need to modify the dimension text value or text height. You may add a symbol, a note, or even change the text of an existing dimension.

Example: Add the word **"MAX"** to the existing dimension value text.

How to edit Dimension Text

1. Double left click on the dimension text to be edited.
2. The dimension text will be highlighted inside a text box area the same as Multiline Text.

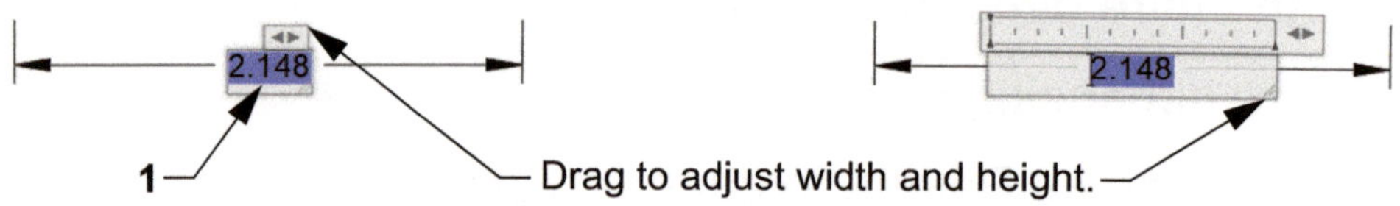

3. The **Text Editor Ribbon** will appear where you may change things like text height or color, add a symbol, or change the text style.

   Associative Dimension: If the dimension is **Associative**, the dimension text will appear highlighted.

   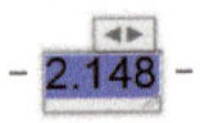

   Before Editing

   You may add text in front or behind the dimension text and it will remain Associative. Be careful not to disturb the dimension value text.

   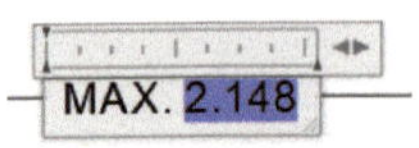

   After Editing

   Non-Associative or Exploded Dimension: If the dimension value has been changed or exploded, it will appear with a gray background and is not Associative.

   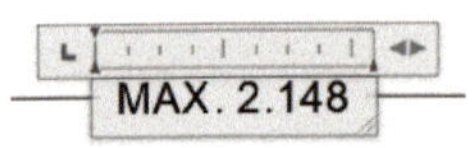

4. Make any changes.
5. Select the **Close Text Editor** button or left click outside the dimension text box area.

Editing the Dimension Position

Grips are great tools for repositioning dimensions. Grips are small, solid-filled squares that are displayed at strategic points on objects. You can drag these grips to stretch, move, rotate, scale, or mirror objects quickly.

How to use Grips

1. Select the object (no command can be in use while using grips).
2. Select one of the **blue** grips. It will turn to **red**. This indicates that grip is **"hot"**. The **"hot"** grip is the **basepoint**.

Continued on the next page...

3. Move the hot grip to the new location.

4. After editing you must press the ***<Esc>*** key to de-activate the grips.

The following is an example of how to use grips to quickly reposition dimensions.

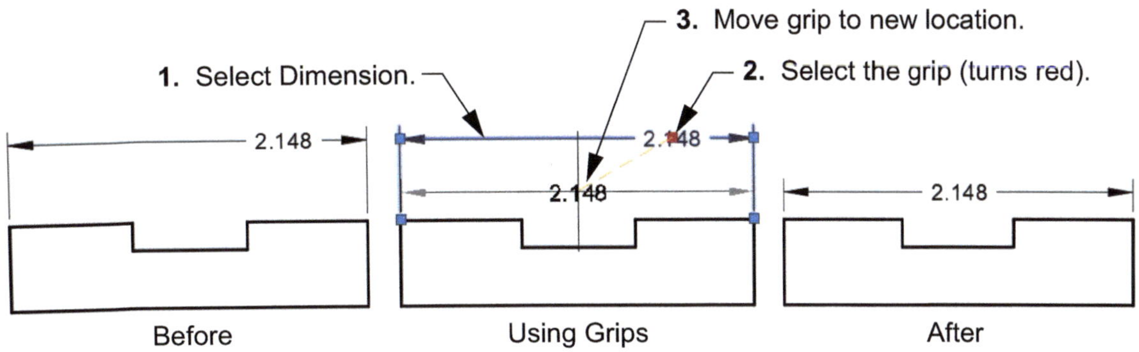

Additional editing options using the Shortcut Menu

1. Select the dimension that you want to change.

2. Place the cursor on one of the grips shown below. (Do not press the mouse button.)

3. Select an option from the Shortcut Menu that appears.

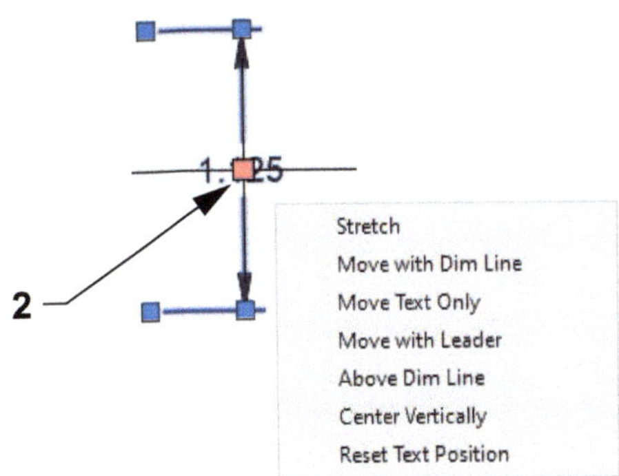

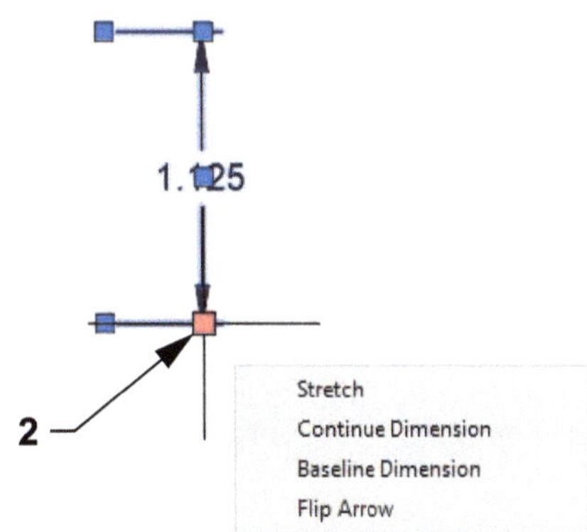

Modifying an Entire Dimension Style

After you have created a dimension style, you may find that you have changed your mind about some of the settings. You can easily change the entire style by using the **Modify...** button in the Dimension Style Manager dialog box. This will not only change the style for future use, but it will also update dimensions already in the drawing.

Note: If you do not want to update the dimensions already in the drawing but want to make a change to a new dimension, use **Override** (see the next page).

How to Modify a Dimension Style

1. Select the **Dimension Style Manager**. (Refer to page 16-7.)

2. Select the dimension style that you want to modify.

Continued on the next page...

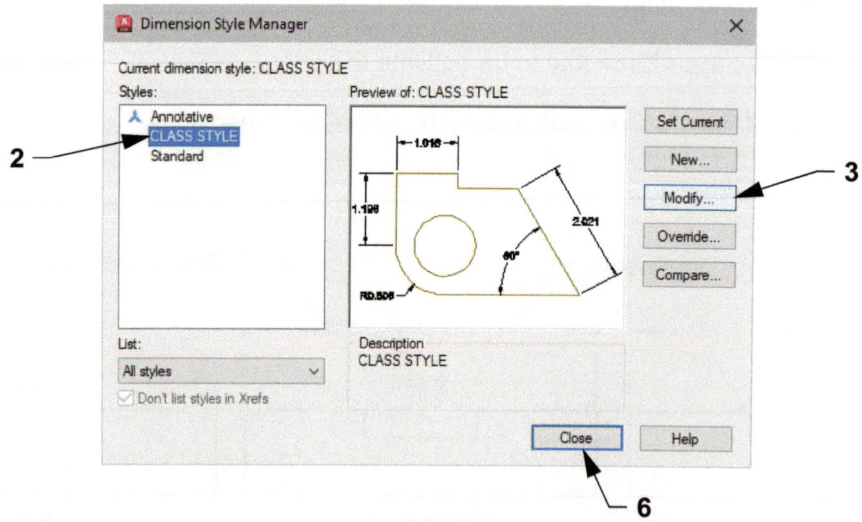

3. Select the **Modify...** button from the **Dimension Style Manager** dialog box.

4. Make the desired changes to the settings.

5. Select the **OK** button.

6. Select the **Close** button.

Now look at your drawing. Have your dimensions updated?

Note: The method above will not change other dimensions that were previously modified or exploded.

If some of the dimensions have not changed, do the following:

1. Type: ***-dimstyle <Enter>*** (notice the (-) **dash** in front of **"dimstyle"**).
2. Type: ***A <Enter>***.
3. Select dimensions to update and then press ***<Enter>***.

(Sometimes you have to give them a little nudge.)

Overriding a Dimension Style

A dimension Override is a temporary change to the dimension settings. An override **will not affect existing dimensions**. It will **only affect new dimensions**. Use this option when you want a **new** dimension just a little bit different but don't want to create a whole new dimension style and don't want the existing dimensions to change either.

Example: If you want the new dimension to have a text height of **0.500" [12.7 mm]** but you want the existing text to remain at **0.125" [3.17 mm]** height:

1. Select the **Dimension Style Manager**. (Refer to page 16-7.)

2. Select the **"style"** you want to override. (Such as: CLASS STYLE.)

3. Select the **Override...** button.

4. Make the desired changes to the settings. (Such as: Text height = **0.500" [12.7 mm]**.)

5. Select the **OK** button.

Continued on the next page...

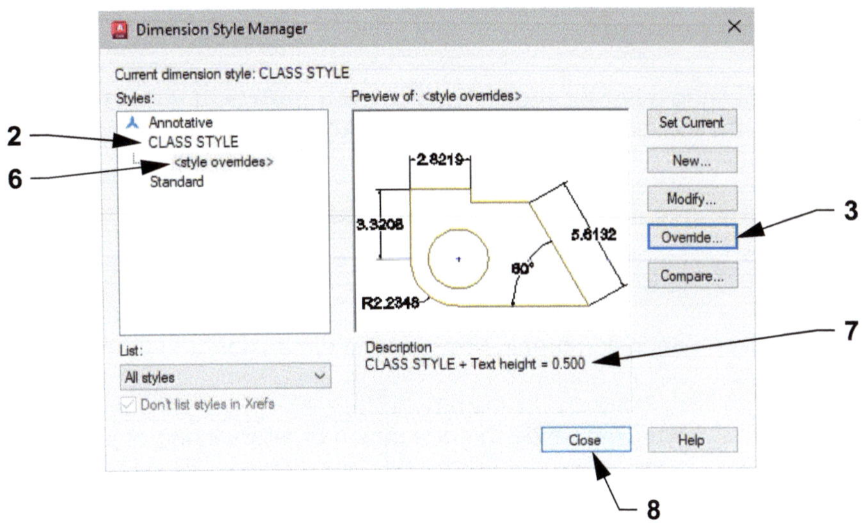

6. Confirm the Override.

   Look at the list of styles. Under the style name, a sub heading of **<style overrides>** should be displayed.

7. The description box should display the style name and the override settings.

8. Select the **Close** button.

When you want to return to style **CLASS STYLE**, select **CLASS STYLE** from the styles list, then select the **Set Current** button. Each time you select a different style, you must select the **Set Current** button to activate it.

Editing an Individual Existing Dimension

Sometimes you would like to modify the settings of an individual existing dimension. This can be achieved using the Properties Palette.

1. Open the **Properties Palette**. (Refer to page 12-3.)

2. Select the dimension to change.

3. Select and change the desired settings.

Example: Change the dimension text height to **0.500" [12.7 mm]**

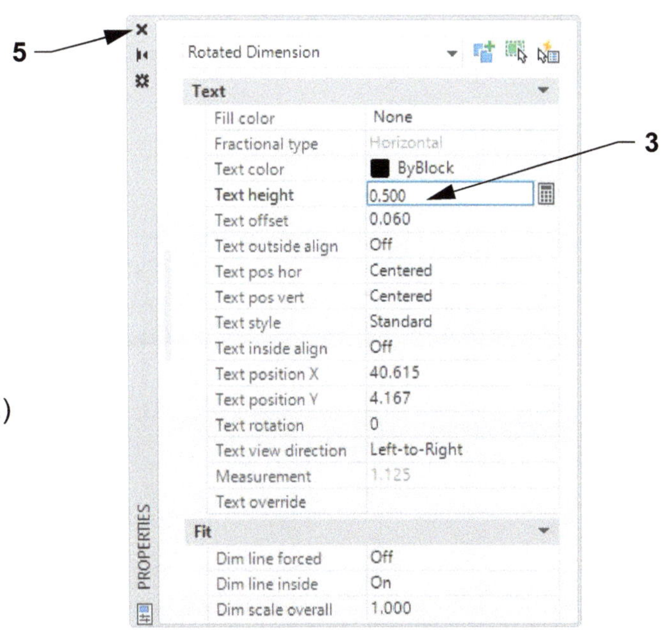

4. Press ***<Enter>***. (The dimension should have changed.)

5. Close the **Properties Palette**.

6. Press the ***<Esc>*** key to de-activate the grips.

Note: The dimension will remain Associative.

Dimension Breaks

Occasionally extension lines overlap other extension lines or even objects. If you do not like this, you may use the **Dimbreak** command to break the intersecting lines. Either the **Automatic** (described below) or **Manual** (described on the next page) method may be used.

Note: You also may use the **Remove** option to remove the break. Refer to page 17-8.

Automatic Dimension Break

To create an automatically placed dimension break, you select a dimension and then use the **Auto** option of the **Dimbreak** command.

Automatic dimension breaks are updated any time the dimension or intersecting objects are modified.

You control the size of automatically placed dimension breaks on the **Symbols and Arrows** Tab of the **Dimension Style Manager** dialog box.

The specified size is affected by the dimension break size, dimension scale, and current annotation scale for the current viewport. (Annotation is discussed in Lesson 27.)

1. Select the **Dimbreak** command using one of the following:

   Ribbon = Annotate Tab / Dimension Panel / 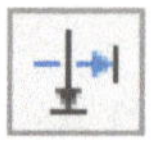
   or
   Keyboard = *dimbreak <Enter>*

   The following will appear on the Command Line:

   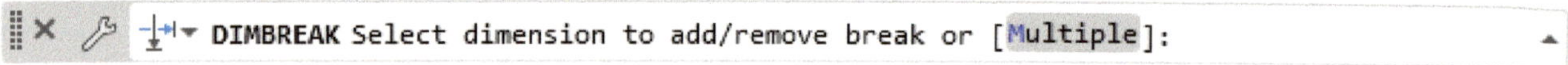

2. Type ***M*** and then press ***<Enter>***.

   The following will appear on the Command Line:

3. Select the dimension(s) you want to break and then press ***<Enter>***. (You may select more than one.)

   The following will appear on the Command Line:

   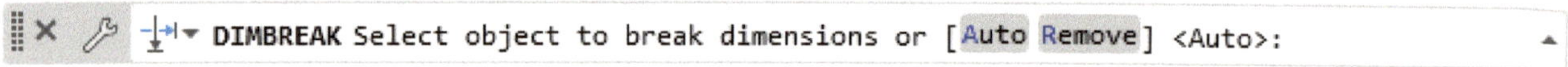

4. Press ***<Enter>***.

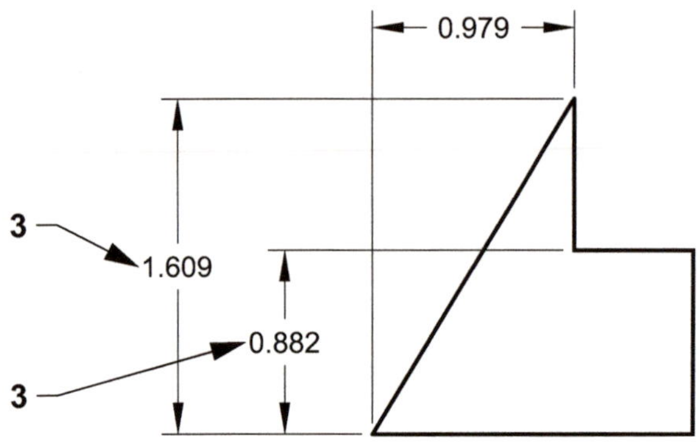

Before Dimbreak

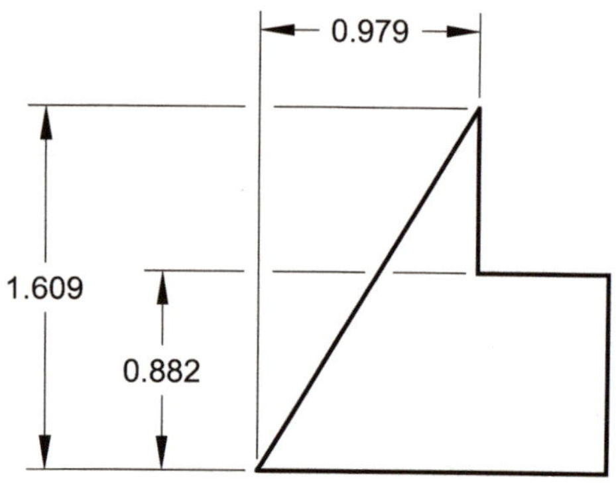

After Dimbreak

Manual Dimension Break

You can place a dimension break by **picking two points** on the dimension, extension, or leader line to determine the size and placement of the break.

Dimension breaks that are added manually by picking two points are not automatically updated if the dimension or intersecting object is modified. So if a dimension with a manually added dimension break is moved or the intersecting object is modified, you might have to restore the dimension and then add the dimension break again.

Note: The size of a dimension break that is created by picking two points is not affected by the current dimension scale or annotation scale value for the current viewport. You will learn Annotation in Lesson 27.

1. Select the **Dimbreak** command using one of the following:

   Ribbon = Annotate Tab / Dimension Panel / 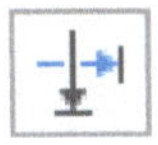
   or
   Keyboard = *dimbreak <Enter>*

   The following will appear on the Command Line:

   ```
   DIMBREAK Select dimension to add/remove break or [Multiple]:
   ```

2. Select the dimension you want to break.

   The following will appear on the Command Line:

   ```
   DIMBREAK Select object to break dimension or [Auto Manual Remove] <Auto>:
   ```

3. Type ***M*** and then press ***<Enter>***.

   The following will appear on the Command Line:

   ```
   DIMBREAK Specify first break point:
   ```

4. Select the first break point location (Object Snap **Nearest** is ideal for this operation).

   The following will appear on the Command Line:

   ```
   DIMBREAK Specify second break point:
   ```

5. Select the second break point location (Object Snap **Nearest** is ideal for this operation).

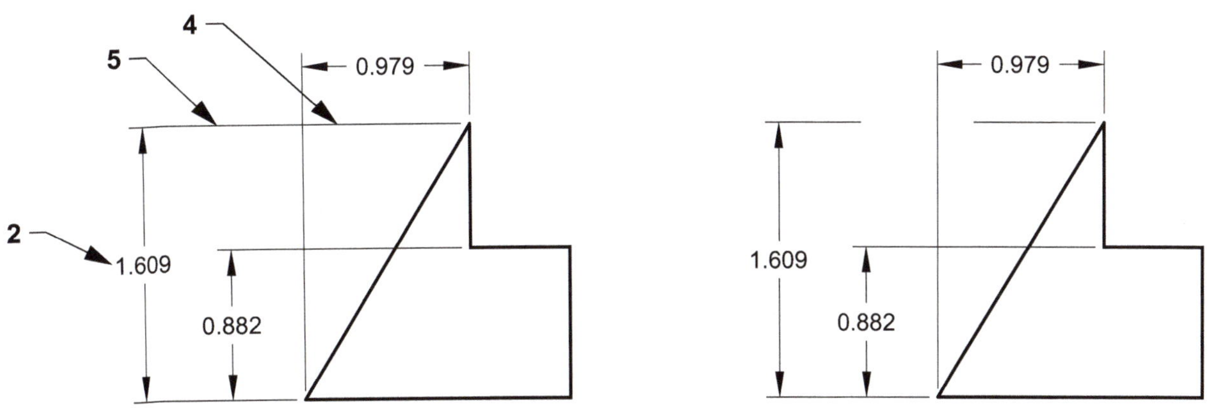

Before Dimbreak   After Dimbreak

CAD TIP! The following objects can be used as cutting edges when adding a dimension break: dimension, leader, line, circle, arc, spline, ellipse, polyline, text, and multiline text.

How to Remove a Dimension Break

Removing the break is easy using the **Remove** option.

1. Select the **Dimbreak** command.

   Ribbon = Annotate Tab / Dimension Panel /
   or
   Keyboard = dimbreak <Enter>

   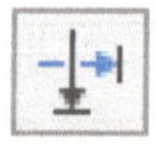

   The following will appear on the Command Line:

```
DIMBREAK Select dimension to add/remove break or [Multiple]:
```

2. Type ***M*** and then press ***<Enter>***.

   The following will appear on the Command Line:

```
DIMBREAK Select dimensions:
```

3. Select the dimension breaks(s) you want to remove and then press ***<Enter>***. (You may select more than one.)

   The following will appear on the Command Line:

```
DIMBREAK Select object to break dimensions or [Auto Remove] <Auto>:
```

4. Type ***R*** and then press ***<Enter>***.

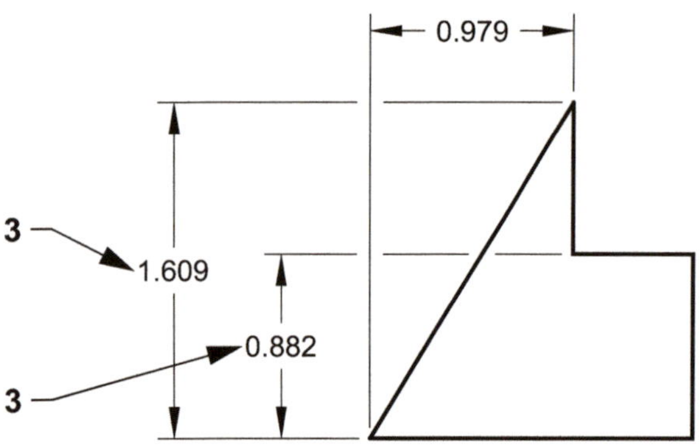

Before Remove

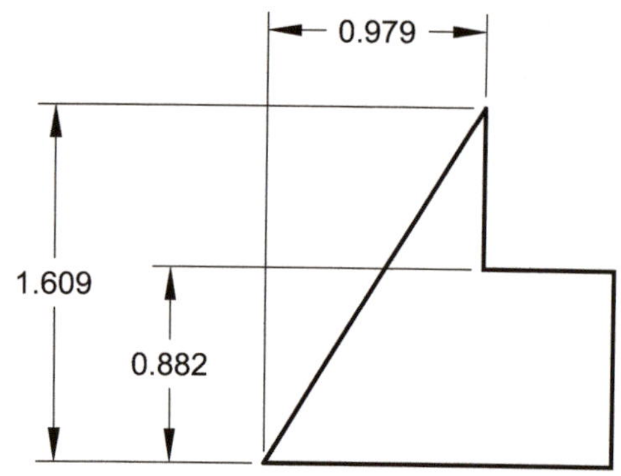

After Remove

Jogging a Dimension Line

Jog lines can be added to Linear dimensions. Jog lines are used to represent a dimension value that does not display the actual measurement. Before you add a jog, the jog angle and the height factor of the jog should be set in the **Symbols and Arrows** Tab within the dimension style.

Note: The jog height is calculated as a factor of the text height.

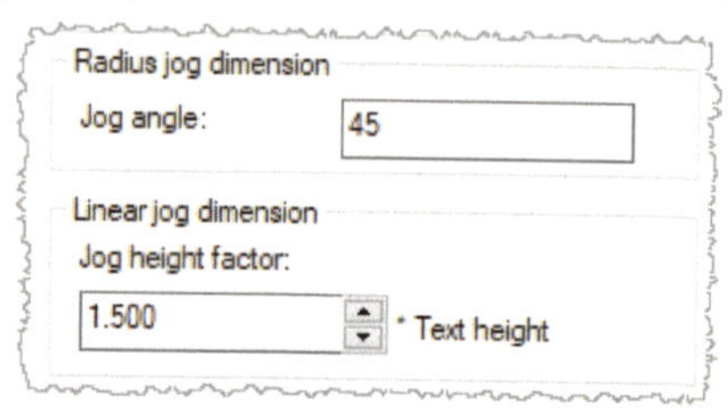

Example:

If the text height was 0.250" [6.35 mm] and the jog height factor was 1.500, the jog would be 0.375" [9.52 mm]

Formula: (0.250" height. X 1.500 jog height. factor = 0.375")
(6.35 mm height. X 1.500 jog height. factor = 9.52 mm)

Continued on the next page...

How to Jog a Dimension Line

1. Select the **Jogged Linear** command using one of the following:

   Ribbon = Annotate Tab / Dimension Panel / 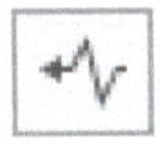
   or
   Keyboard = *dimjogline <Enter>*

   The following will appear on the Command Line:

```
DIMJOGLINE Select dimension to add jog or [Remove]:
```

2. Select a dimension to jog.

   The following will appear on the Command Line:

```
DIMJOGLINE Specify jog location (or press ENTER):
```

3. Press ***<Enter>***.

4. After you have added the jog, you can re-position it by using **grips** and adjust the height of the jog symbol using the **Properties Palette**.

Before Jog Line

After Jog Line

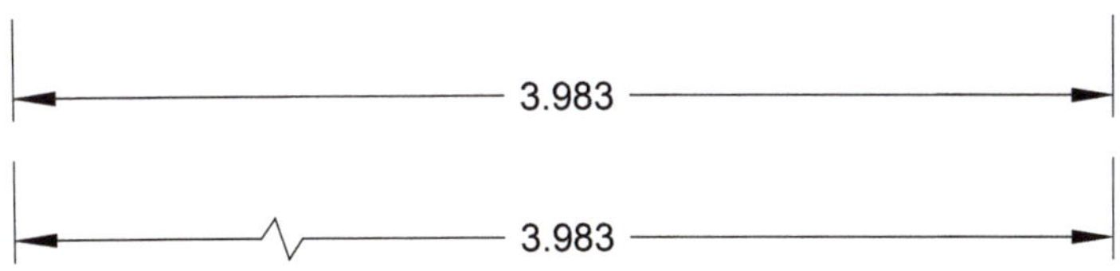

How to Remove a Jog Line

1. Select the **Jogged Linear** command.

   The following will appear on the Command Line:

```
DIMJOGLINE Select dimension to add jog or [Remove]:
```

2. Type ***R*** and then press ***<Enter>***.

   The following will appear on the Command Line:

```
DIMJOGLINE Select jog to remove:
```

3. Select the Jogged dimension.

Adjusting the Distance Between Dimensions

The **Adjust Space** command allows you to adjust the distance between existing parallel linear and angular dimensions, so they are equally spaced. You may also align the dimensions to create a string.

1. Select the **Dimension Space** command using one of the following:

   Ribbon = Annotate Tab / Dimension Panel / 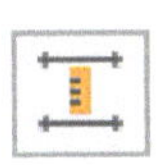
   or
   Keyboard = *dimspace <Enter>*

Continued on the next page...

The following will appear on the Command Line:

```
DIMSPACE Select base dimension:
```

2. Select the dimension that you want to use as the base dimension when equally spacing dimensions (**P1**).

   The following will appear on the Command Line:

```
DIMSPACE Select dimensions to space:
```

3. Select the next dimension to be spaced (**P2**).

   The following will appear on the Command Line:

```
DIMSPACE Select dimensions to space:
```

4. Select the next dimension to be spaced (**P3**).

   The following will appear on the Command Line:

```
DIMSPACE Select dimensions to space:
```

5. Continue to select more dimensions or press ***<Enter>*** to stop selecting dimensions.

   The following will appear on the Command Line:

```
DIMSPACE Enter value or [Auto] <Auto>:
```

6. Enter a value or press ***<Enter>*** for **Auto**.

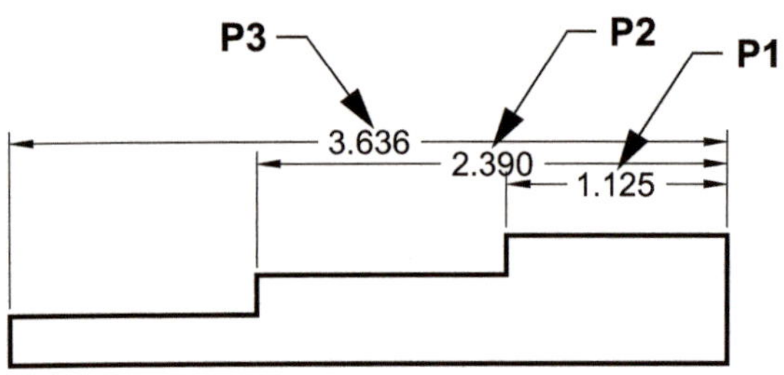

Before Spacing

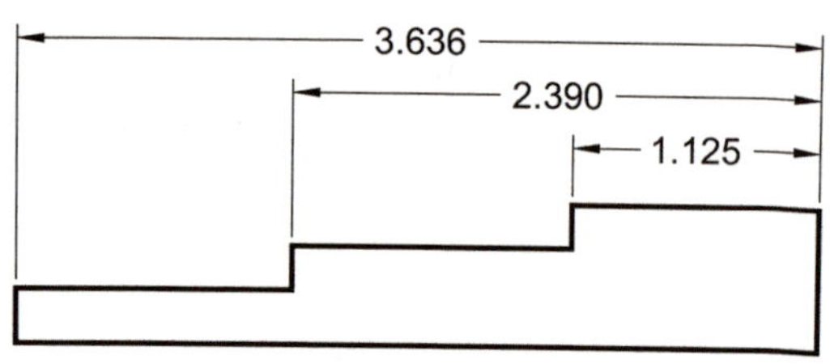

After Spacing

Auto creates a spacing value of twice the height of the dimension text.

Example:

If the dimension text is 1/8" [3.17 mm],
the spacing will be 1/4" [6.35 mm].

How to Align dimensions

Follow the steps shown above, but when asked for the value, enter ***0*** and then press ***<Enter>***.

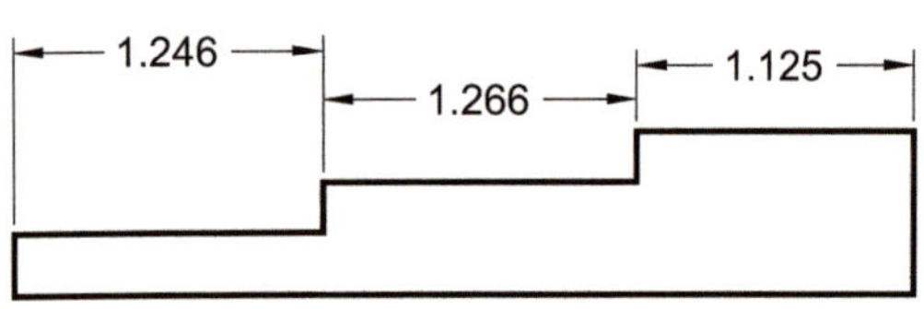

Before Aligning

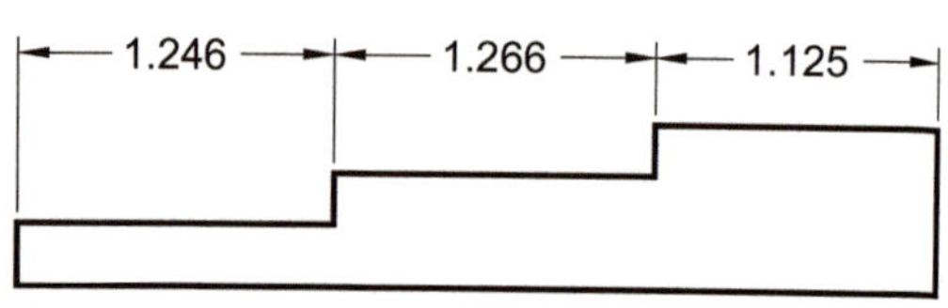

After Aligning

Exercise 17A-Inch

Instructions:

1. Start a **New** file using the **Border A-Inch.dwt**
2. Draw the **6" x 4"** rectangle shown below. Use Layer Object Line.
3. **Upper** and **right dimension**: Use Dimension Style CLASS STYLE. Use Layer Dimension.
4. **Lower** and **left dimension**: Use **Override**. Change the text height setting to **0.500"**. If the upper and right dimension also changed, you did not use "Override". Try again.
5. Edit the **Title** and **Ex-XX** by double clicking on the text. Do not erase and replace.
6. **Save** as **Ex-17A-I**
7. Plot using Page Setup **Class Model A**

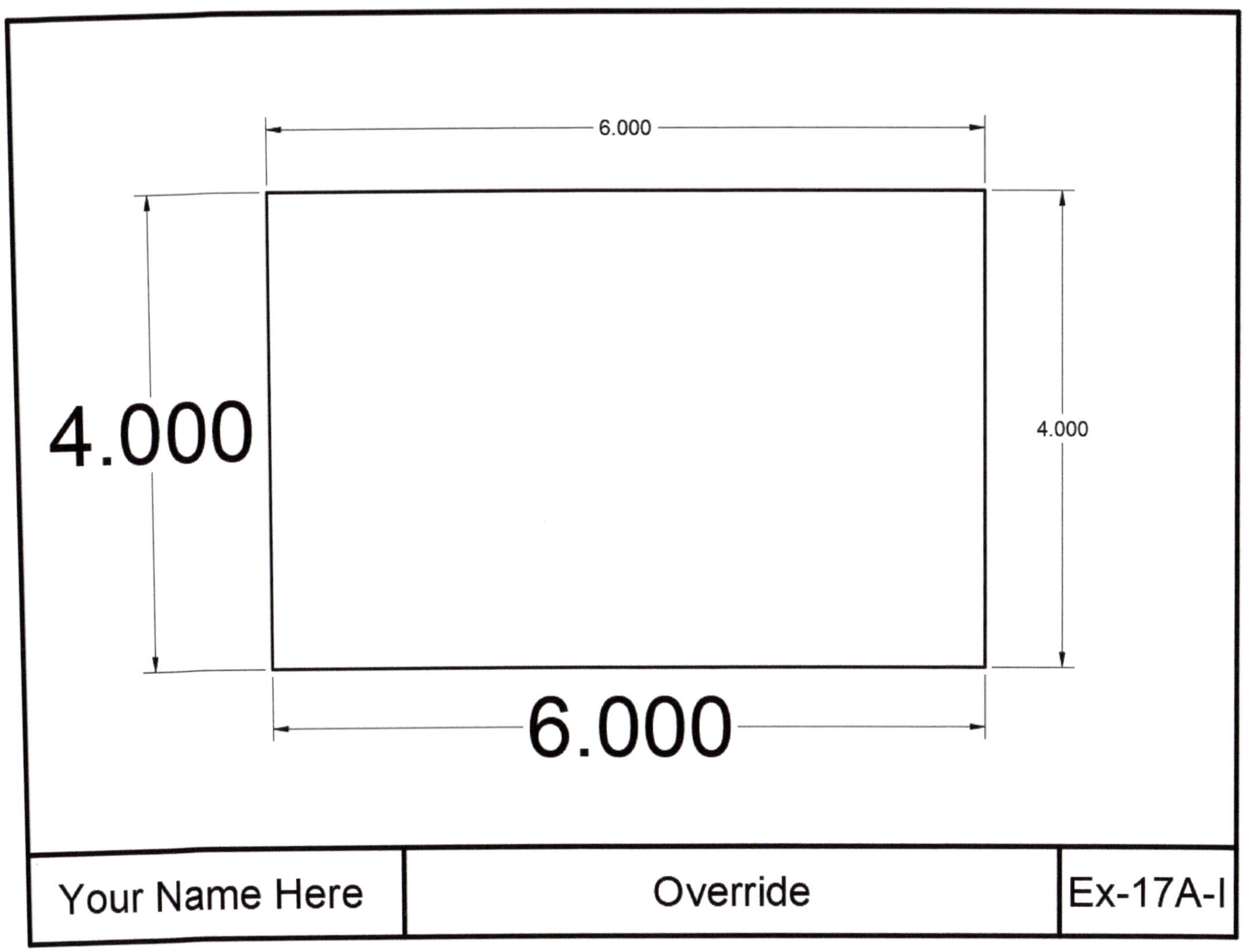

Exercise 17A-Metric

Instructions:

1. Start a **New** file using the **Border A-Metric.dwt**
2. Draw the **150 mm x 100 mm** rectangle shown below. Use Layer Object Line.
3. **Upper** and **right dimension**: Use Dimension Style CLASS STYLE. Use Layer Dimension.
4. **Lower** and **left dimension**: Use **Override**. Change the text height setting to **12 mm**. If the upper and right dimension also changed, you did not use "Override". Try again.
5. Edit the **Title** and **Ex-XX** by double clicking on the text. Do not erase and replace.
6. **Save** as **Ex-17A-M**
7. Plot using Page Setup **Class Model A**

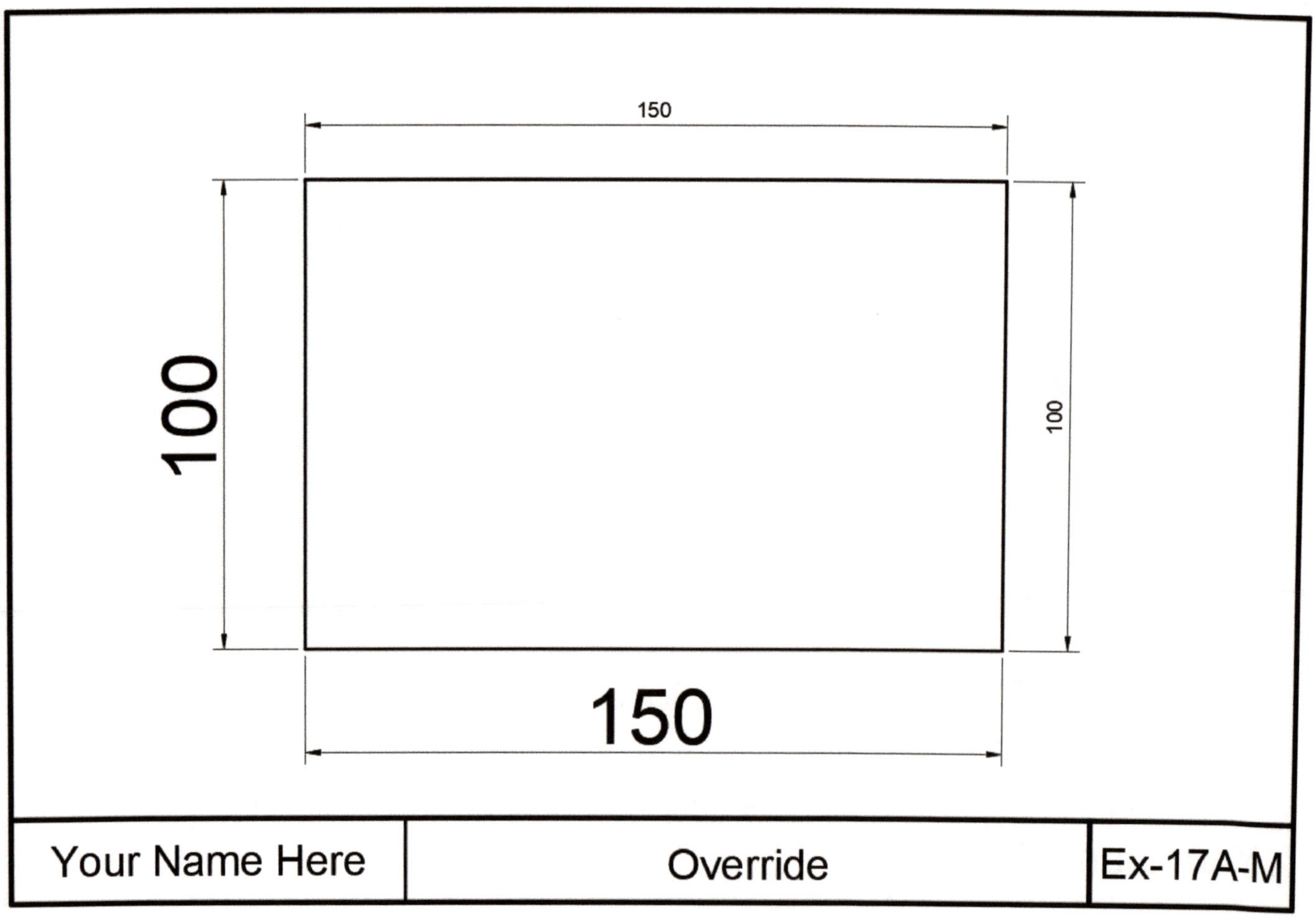

Exercise 17B-Inch

Instructions:

1. Open **Ex-16B-I**
2. Re-space the Baseline dimensions on the right side. Use **Auto** for spacing.
3. Align the two **1.000"** dimensions.
4. Edit the **Title** and **Ex-XX** by double clicking on the text. Do not erase and replace.
5. **Save** as **Ex-17B-I**
6. Plot using Page Setup **Class Model A**

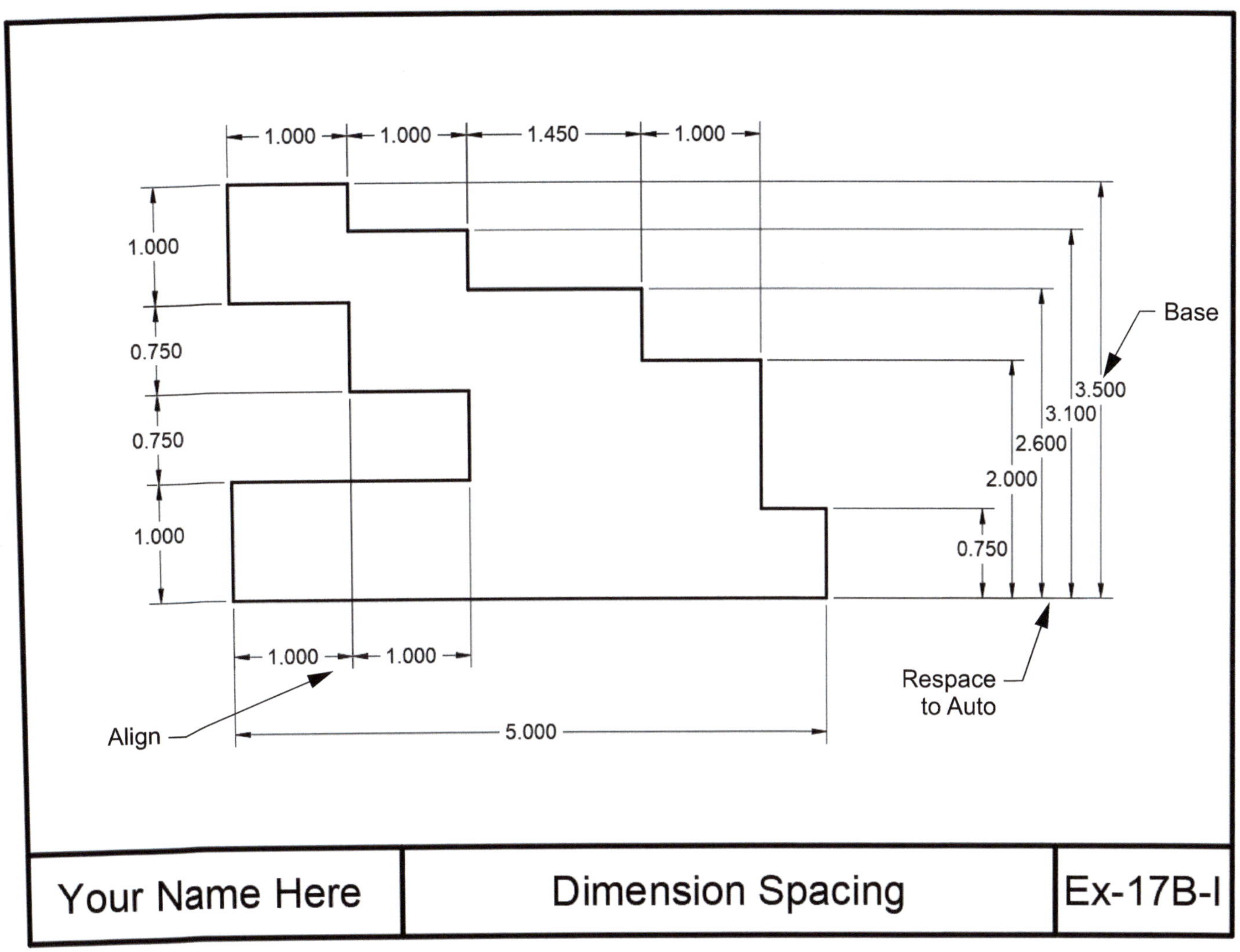

Exercise 17B-Metric

Instructions:

1. Open **Ex-16B-M**
2. Re-space the Baseline dimensions on the right side. Use **Auto** for spacing.
3. Align the two **25 mm** dimensions.
4. Edit the **Title** and **Ex-XX** by double clicking on the text. Do not erase and replace.
5. **Save** as **Ex-17B-M**
6. Plot using Page Setup **Class Model A**

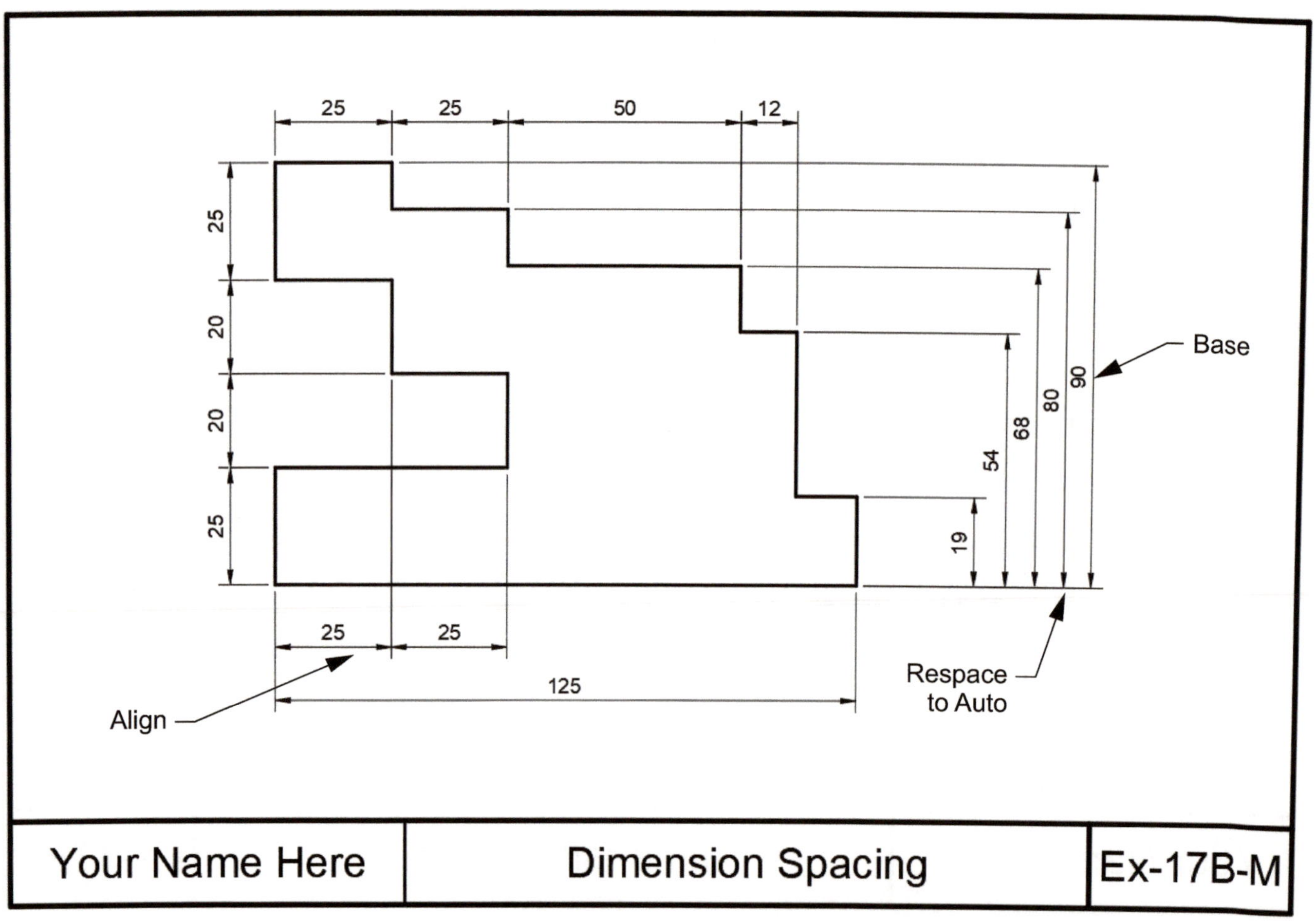

Exercise 17C-Inch

Instructions:

1. Open **Ex-17B-I**
2. Make the changes shown below. Use the grips and the Properties Palette.
3. Notice: Not all dimensions change, so you can't make changes to the dimension style.
4. Edit the **Title** and **Ex-XX** by double clicking on the text. Do not erase and replace.
5. **Save** as **Ex-17C-I**
6. Plot using Page Setup **Class Model A**

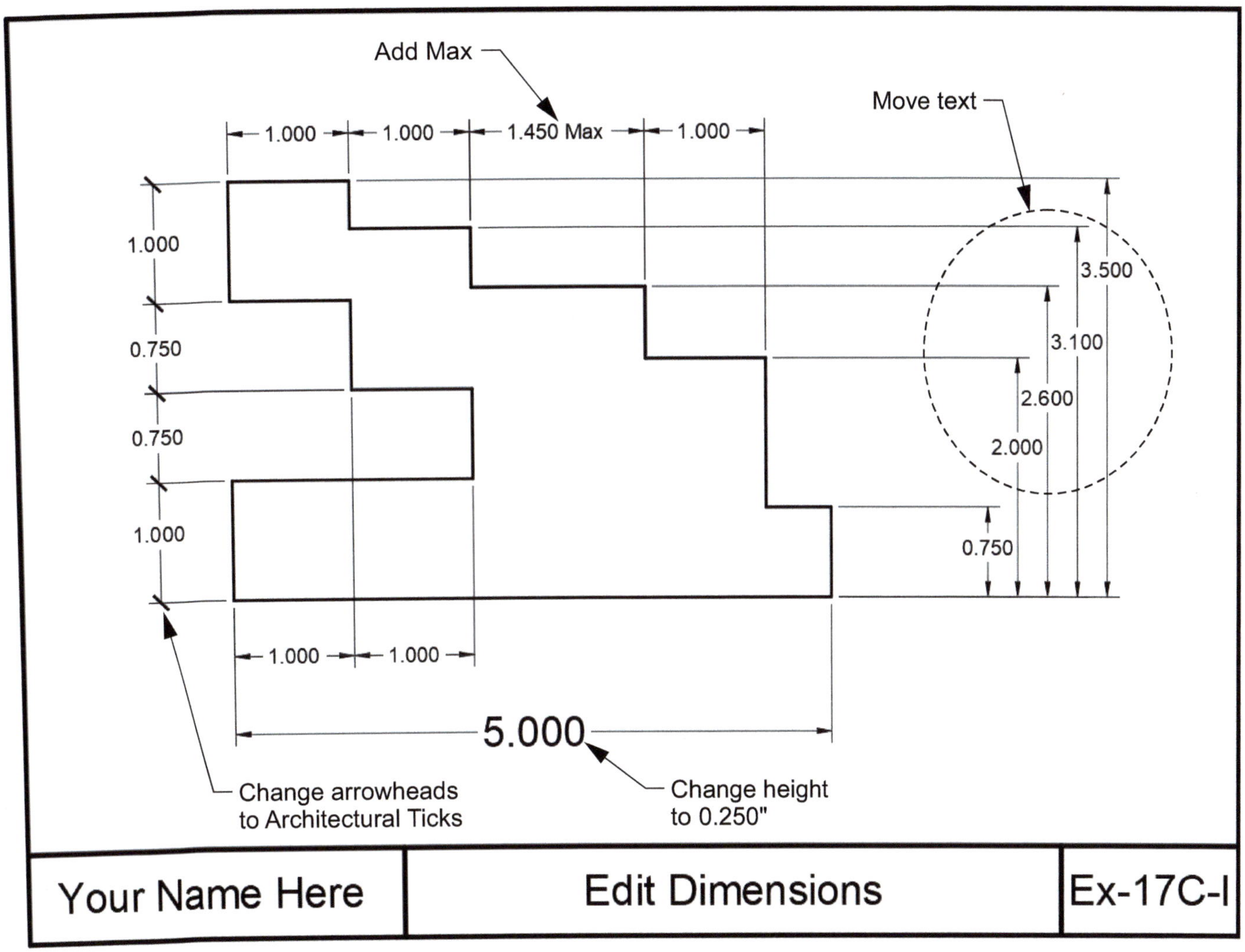

Exercise 17C-Metric

Instructions:

1. Open **Ex-17B-M**
2. Make the changes shown below. Use the grips and the Properties Palette.
3. Notice: Not all dimensions change, so you can't make changes to the dimension style.
4. Edit the **Title** and **Ex-XX** by double clicking on the text. Do not erase and replace.
5. **Save** as **Ex-17C-M**
6. Plot using Page Setup **Class Model A**

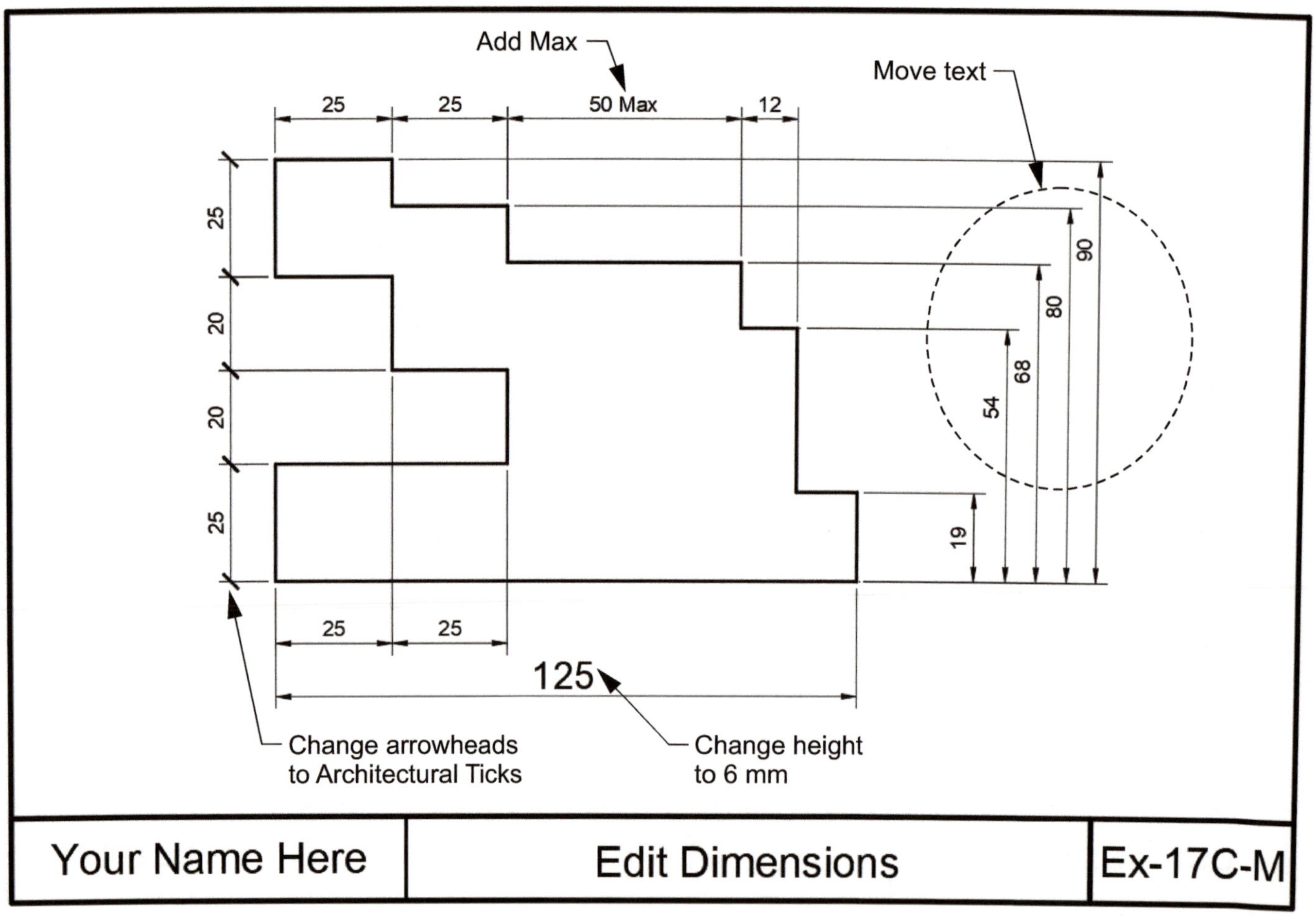

Exercise 17D-Inch

Instructions:

1. Start a **New** file using **Border A-Inch.dwt**
2. Draw the objects shown below.
3. Dimension as shown. Be very careful because there is just enough drawing area.
4. Use Dimension Break to break the extension lines as shown.
5. Use Dimension Jog to create a jog in the lower dimension.
6. Edit the **Title** and **Ex-XX** by double clicking on the text. Do not erase and replace.
7. **Save** as **Ex-17D-I**
8. Plot using Page Setup **Class Model A**

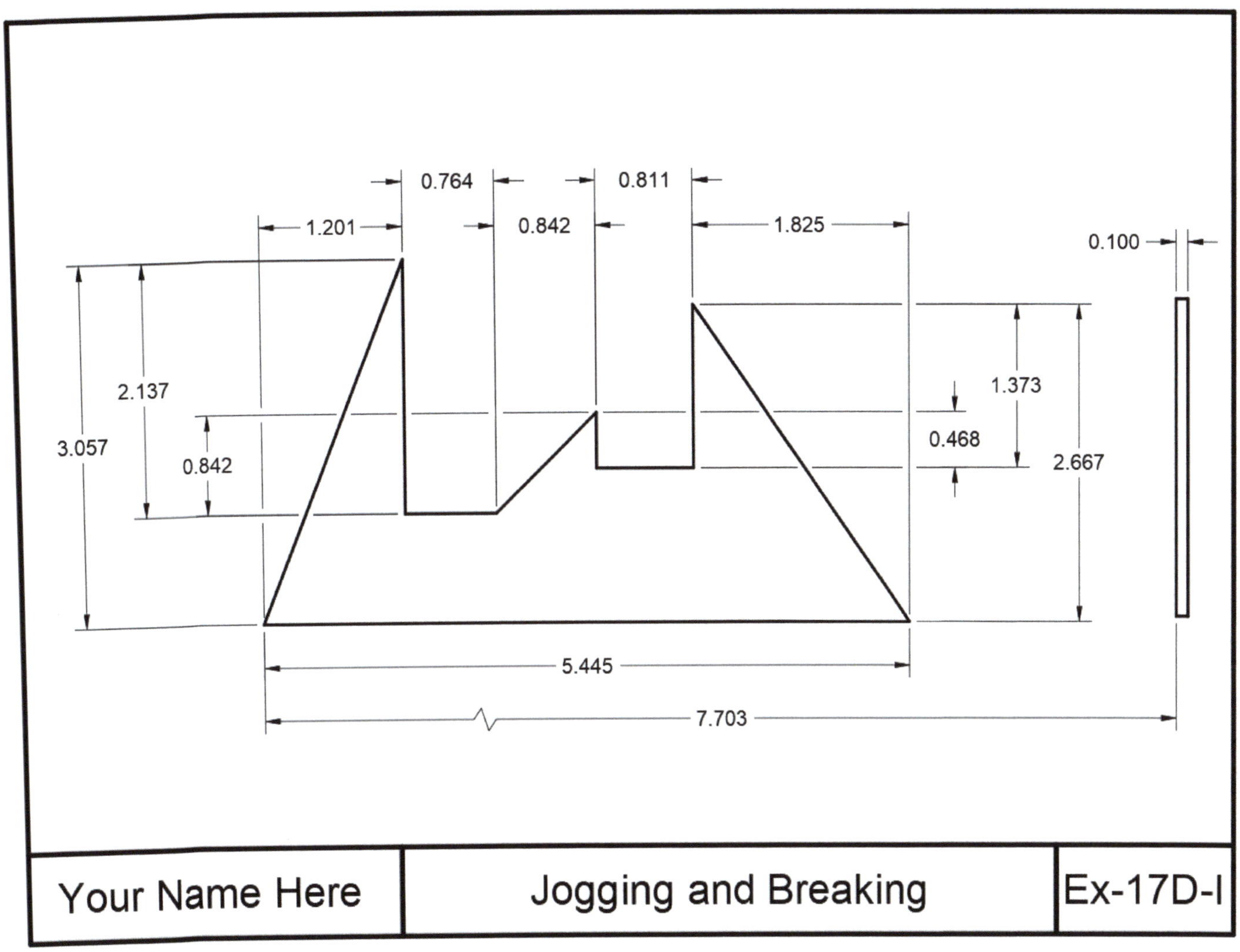

Exercise 17D-Metric

Instructions:

1. Start a **New** file using **Border A-Metric.dwt**
2. Draw the objects shown below.
3. Dimension as shown. Be very careful because there is just enough drawing area.
4. Use Dimension Break to break the extension lines as shown.
5. Use Dimension Jog to create a jog in the lower dimension.
6. Edit the **Title** and **Ex-XX** by double clicking on the text. Do not erase and replace.
7. **Save** as **Ex-17D-M**
8. Plot using Page Setup **Class Model A**

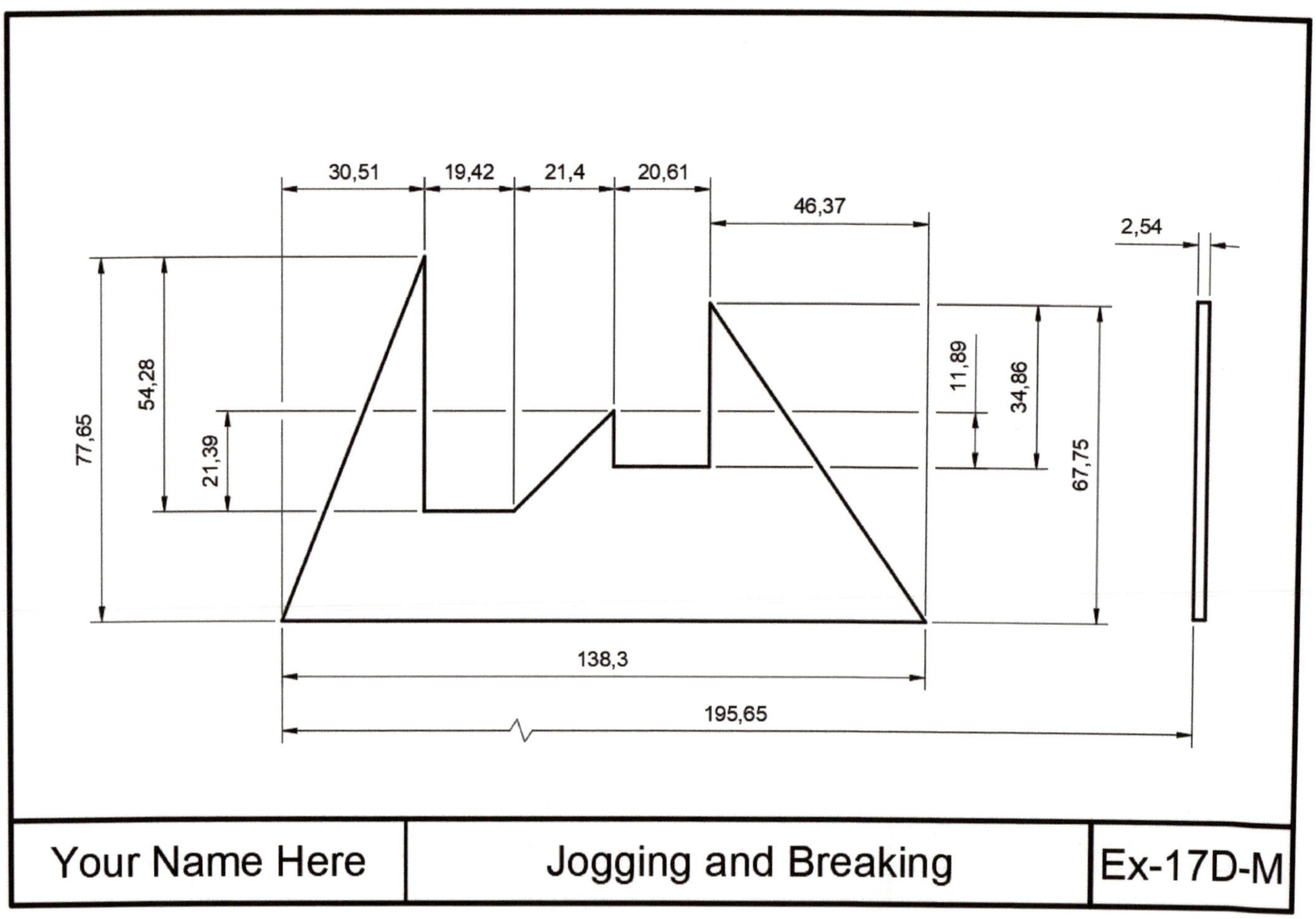

LESSON 18

LEARNING OBJECTIVES

After completing this lesson, you will be able to:

- Add Diameter, Radius and Angular dimensions to a drawing
- Draw and control Center Marks
- Manually place Center Marks and Centerlines
- Flip the direction of an arrowhead
- Understand the need for sub-styles
- Create a sub-style

Dimensioning Diameters

The **Diameter** dimensioning command should be used when dimensioning circles and arcs of **more than 180 degrees**. AutoCAD measures the selected circle or arc and displays the dimension text with the diameter symbol (**Ø**) in front of it.

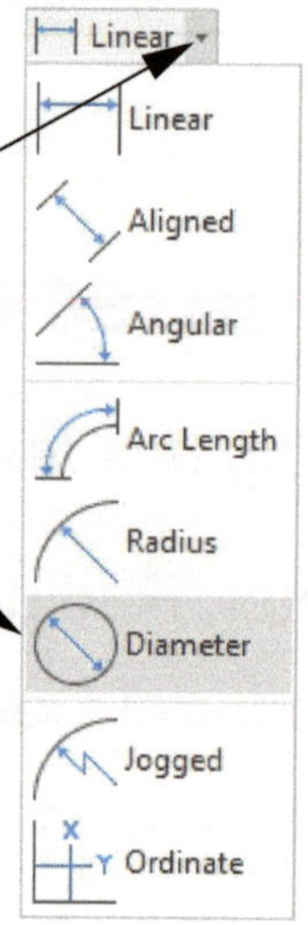

1. Select the **Diameter** command using one of the following:

   Ribbon = Annotate Tab / Dimension Panel / ▾
   or
   Keyboard = *dimdiameter <Enter>*

   The following will appear on the Command Line:

2. Select the arc or circle (**P1**).

   The following will appear on the Command Line:

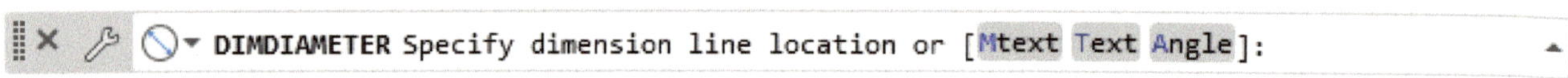

3. Place the dimension text location (**P2**).

Inch users

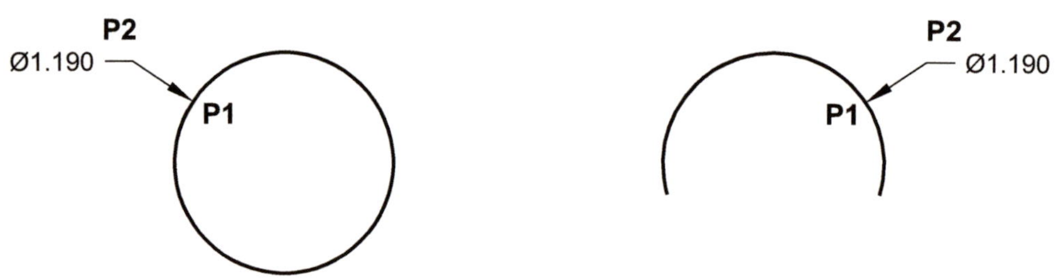

Metric users

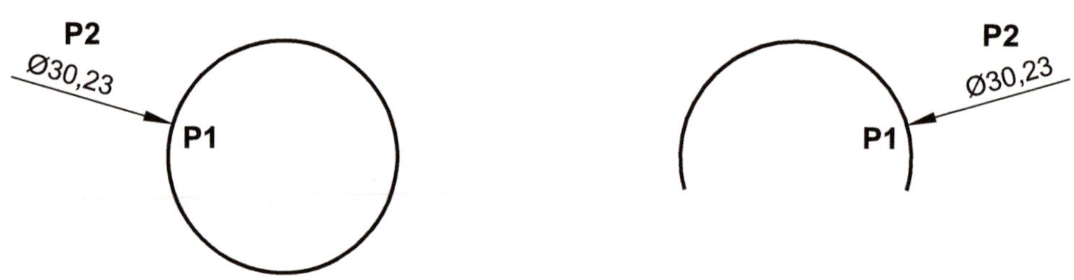

Note: Refer to page 18-6 if you require Center Marks to be included automatically with a diameter dimension, or if you want to manually place a Center Mark on a circle.

Controlling the Diameter dimension appearance

If you would like your diameter dimensions to appear as shown in the two examples below, you must change some settings in the Dimension Style Manager **Fit** Tab.

Inch users

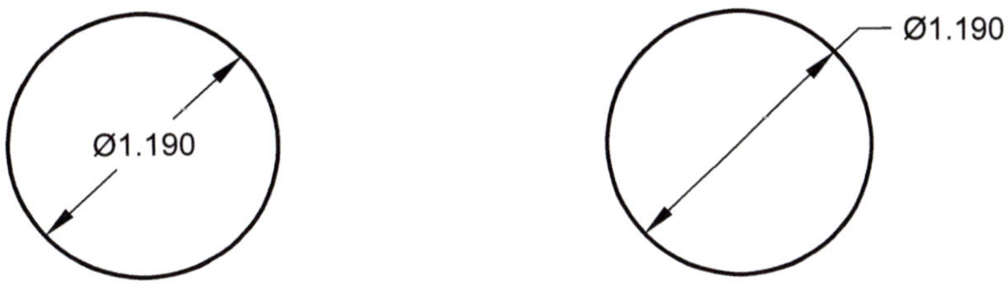

Metric users

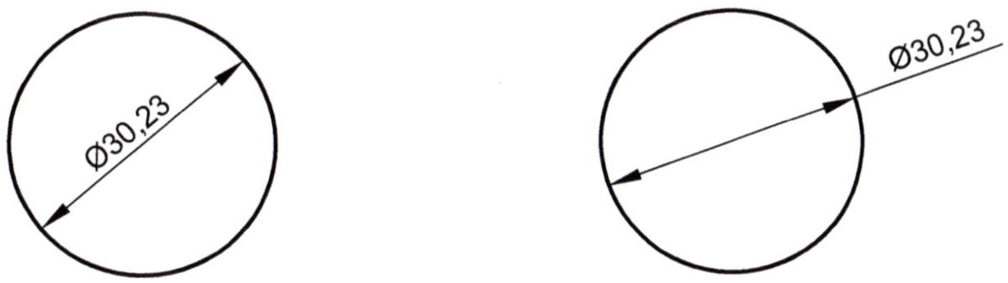

How to change the Diameter dimension appearance

1. Open the **Dimension Style Manager** dialog box.
2. Select the **Fit** Tab.
3. Fit options: Select **Text**.
4. Fine tuning: Check **Place text manually**.

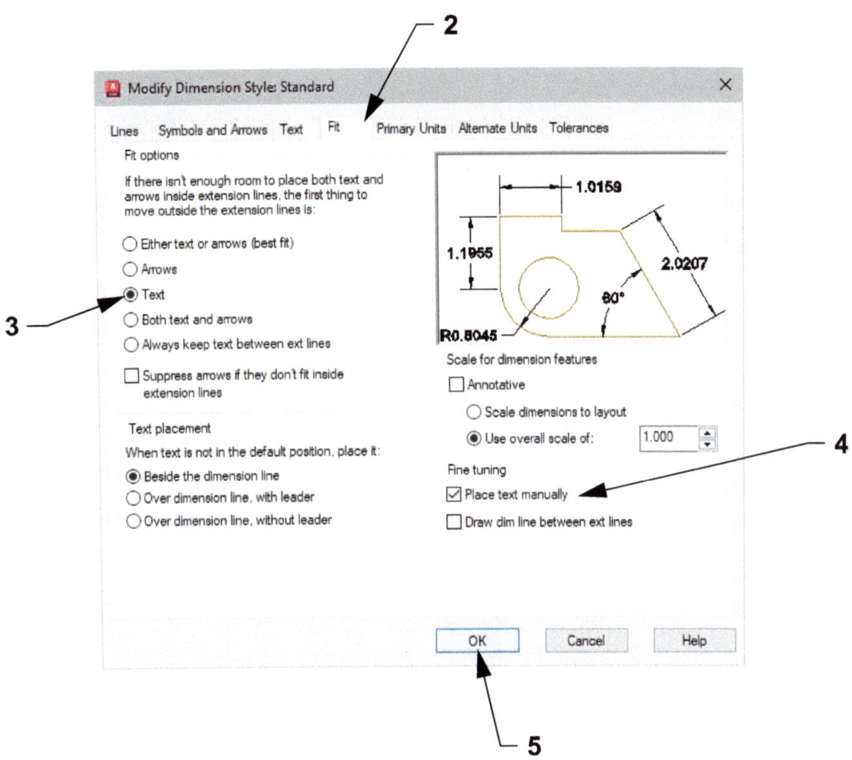

5. Select the **OK** button.

Dimensioning Radii

The **Radius** dimensioning command should be used when dimensioning arcs of **less than 180 degrees**. AutoCAD measures the selected arc and displays the dimension text with the radius symbol (**R**) in front of it.

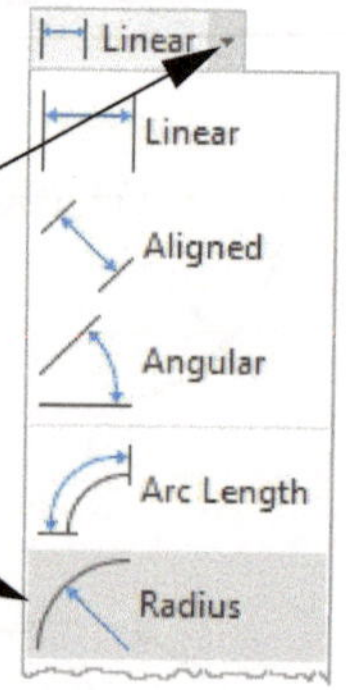

1. Select the **Radius** command using one of the following:

   Ribbon = Annotate Tab / Dimension Panel / ▾
   or
   Keyboard = *dimradius <Enter>*

   The following will appear on the Command Line:

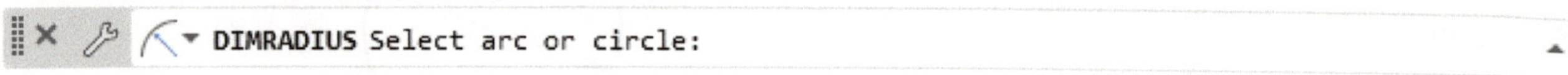

2. Select the arc or circle (**P1**).

   The following will appear on the Command Line:

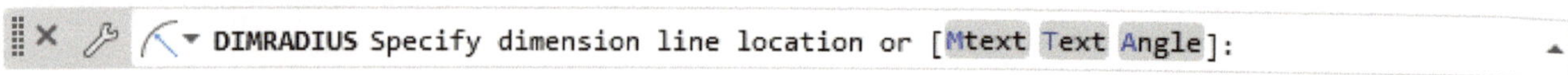

3. Place the dimension text location (**P2**).

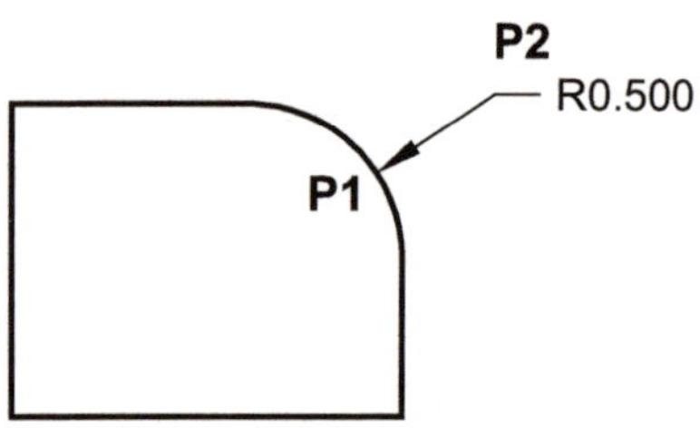

Inch users

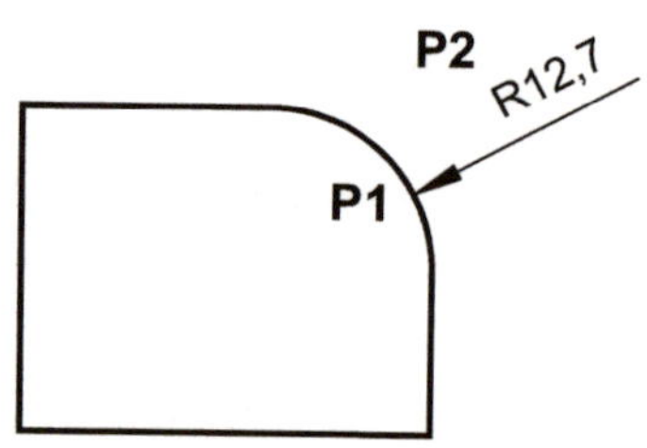

Metric users

Note: Refer to page 18-6 if you require Center Marks to be included automatically with a radius dimension, or if you want to manually place a Center Mark on a circle.

Controlling the Radius dimension appearance

If you would like your radius dimensions to appear as shown in the two examples below, you must change some settings in the Dimension Style Manager **Fit** Tab. Change the settings to be the same as those shown for diameter dimensions on page 18-3

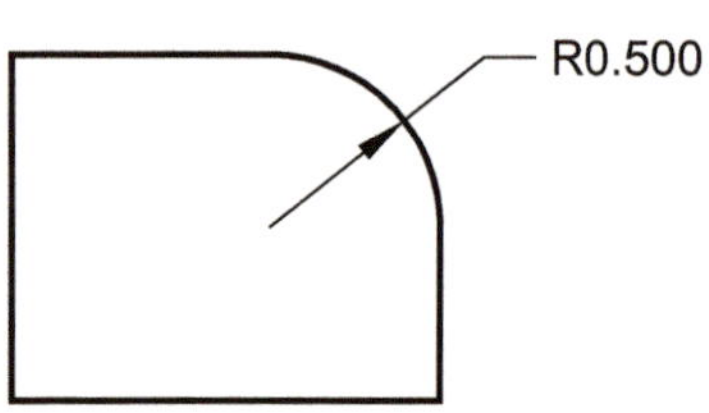

Inch users

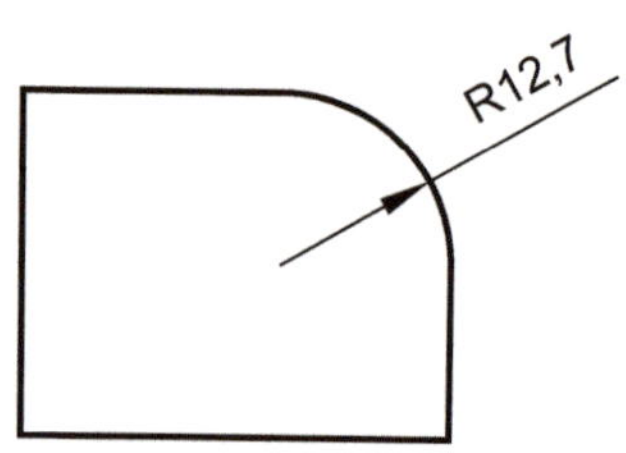

Metric users

Angular Dimensioning

The **Angular** dimension command is used to create an angular dimension between two lines that form an angle. AutoCAD determines the angle between the selected lines and displays the dimension text followed by a degree (°) symbol.

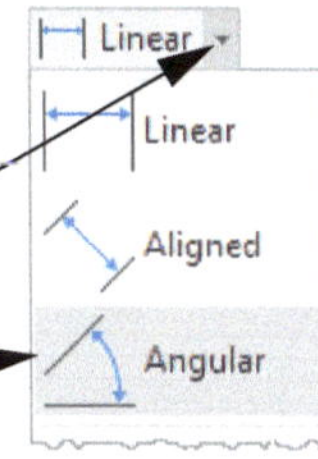

1. Select the **Angular** command using one of the following:

   Ribbon = Annotate Tab / Dimension Panel / ▾
   or
   Keyboard = *dimangular <Enter>*

   The following will appear on the Command Line:

```
DIMANGULAR Select arc, circle, line, or <specify vertex>:
```

2. Select the first line that forms the angle (**P1**). Location is not important, do not use Object Snap.

   The following will appear on the Command Line:

```
DIMANGULAR Select second line:
```

3. Select the first line that forms the angle (**P2**).

   The following will appear on the Command Line:

```
DIMANGULAR Specify dimension arc line location or [Mtext Text Angle Quadrant]:
```

4. Place the dimension text location (**P3**).

   Any of the four angular dimensions shown below can be displayed by moving the cursor in the direction of the dimension after selecting the two lines (**P1** and **P2**) that form the angle.

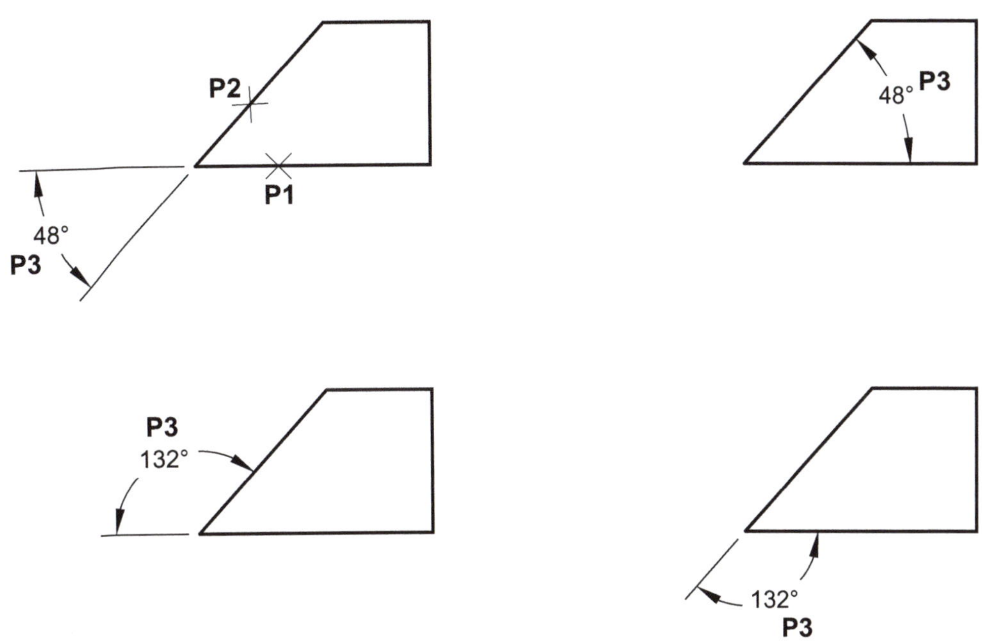

Center Mark – Automatic

Center Marks can automatically be drawn when you dimension circular objects like circles and arcs. You set the size and type.

The Center Mark has three types, **None**, **Mark**, and **Line** as shown below.

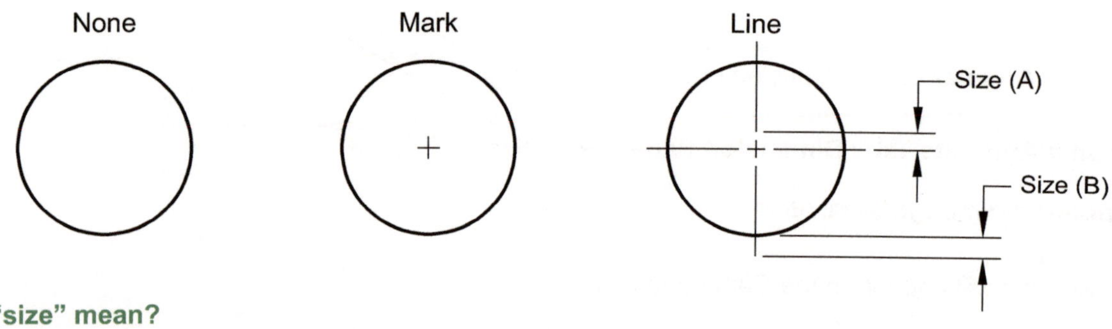

What does "size" mean?

The **size** setting determines both, (A) the length of half of the intersection line and (B) the length extending beyond the circle. (See above right.)

Where do you set the Center Mark "type" and "size"?

1. Open the **Dimension Style Manager** dialog box.
2. Select **Modify** or **Override**.
3. Select the **Symbols and Arrows** Tab.
4. Select the **Center marks** type.
5. Set the **size**.

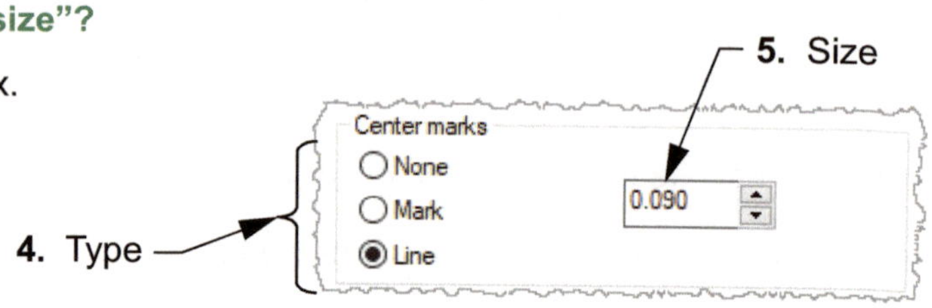

Center Mark – Manual

Center Marks can be manually placed on circles and arcs.

1. Select the **Centermark** command using one of the following:

   Ribbon = Annotate Tab / Centerlines Panel / 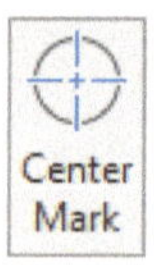

   or
   Keyboard = *centermark <Enter>*

   The following will appear on the Command Line:

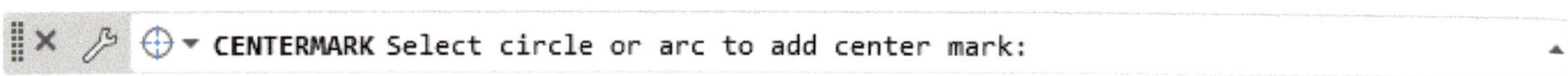

2. Select a circle or arc.
3. Continue to select more circles or arcs, or press ***<Enter>*** to end the command.

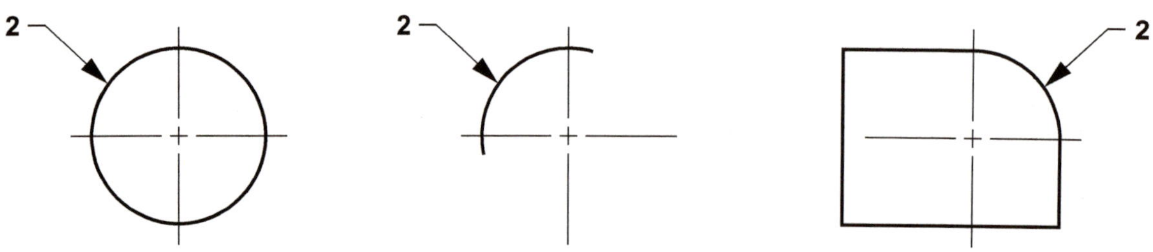

Modifying Center Marks

You can modify Center Marks by using the grips after first selecting a Center Mark. You can choose to lengthen each extension line individually by selecting an **arrow grip**, or you can lengthen all four extension lines equally at the same time by selecting the **square center grip** then selecting **"Change Extension Length"** from the menu.

Selecting an arrow grip to extend an individual line

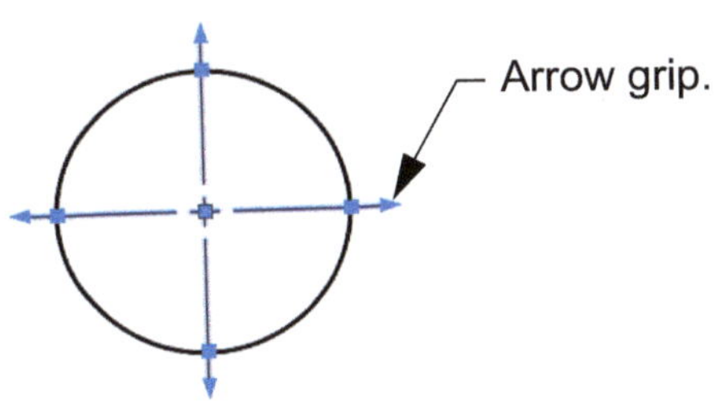

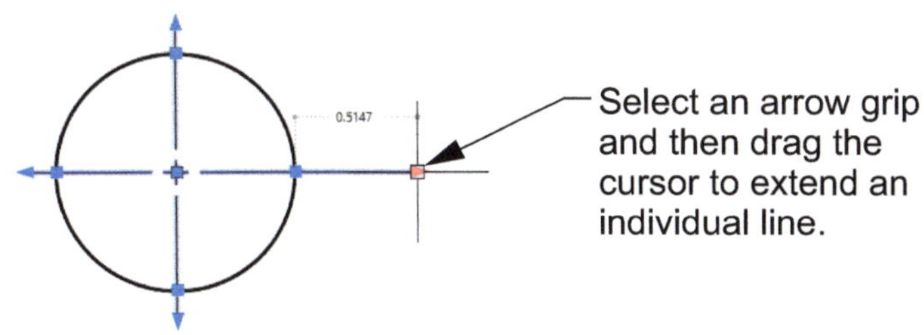

Selecting the square grip to change the extension length of all four lines equally

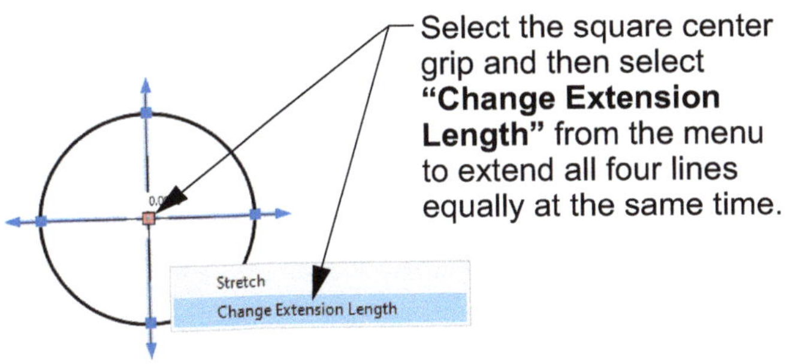

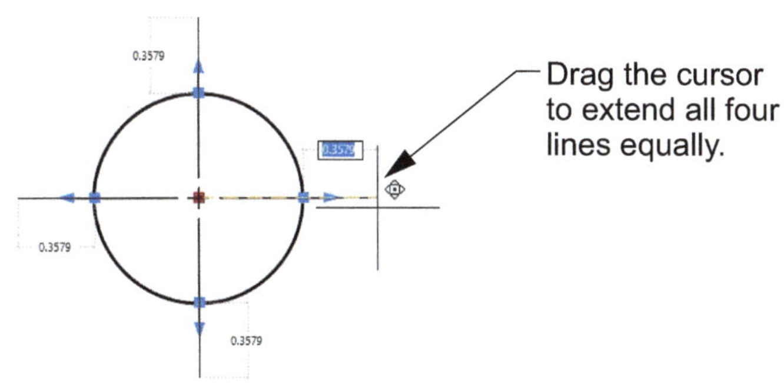

Note: Center Marks are **Associative**, so if you choose to move or change the size of a circle or arc, the Center Mark will automatically move or resize with it.

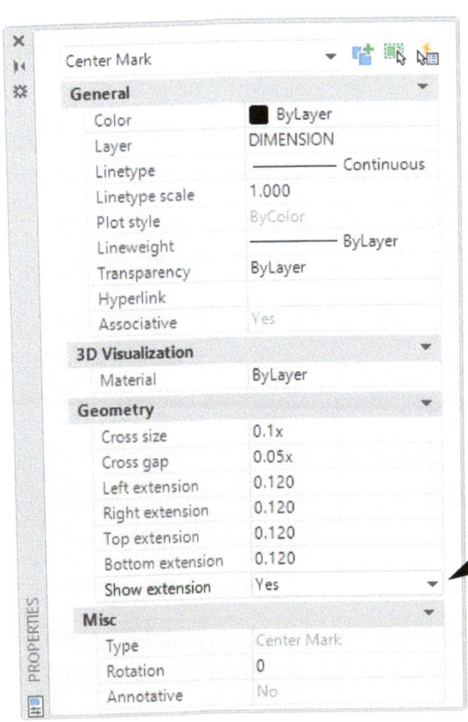

You may also change the properties of Center Marks by selecting a Center Mark then right clicking and choosing **Properties** from the menu.

You can change individual extension lines, the color, layer, and linetype, as well as the center cross size and the rotation of the Center Mark.

If you select the **Show extension** drop-down list and select **No**, you can choose to show just the center cross without the extension lines.

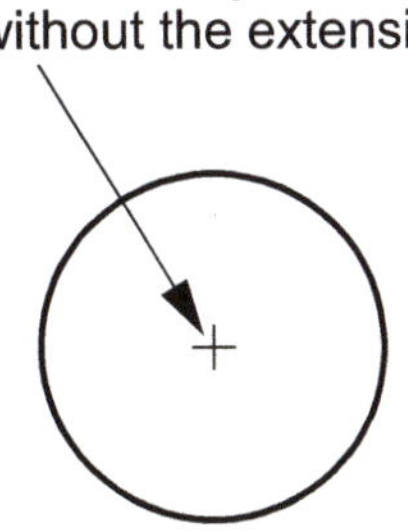

Centerline

Centerlines are created by selecting two lines or polylines.

1. Select the **Centerline** command using one of the following:

   Ribbon = Annotate Tab / Centerlines Panel /
   or
   Keyboard = *centerline <Enter>*

   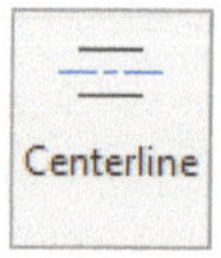

   The following will appear on the Command Line:

   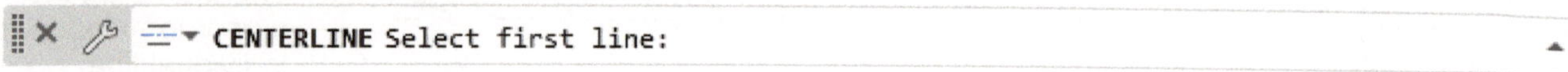

2. Select the first line.

   The following will appear on the Command Line:

3. Select the second line.

4. The Centerline will be created and the command will end automatically.

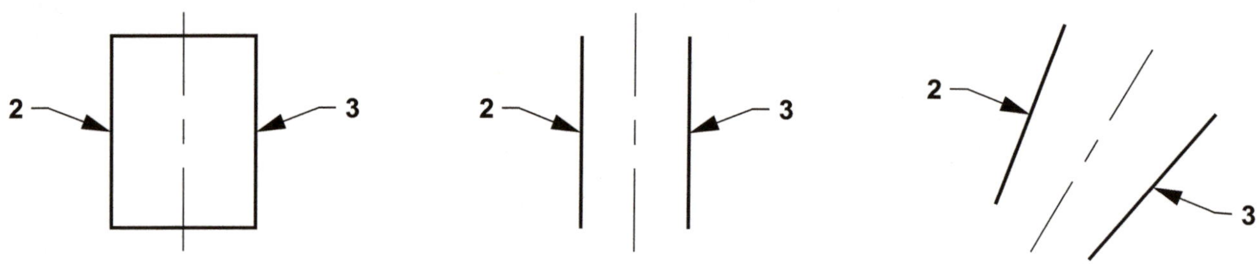

Modifying Centerlines

You can modify Centerlines by using the grips after first selecting a Centerline. You can choose to lengthen either end of the Centerline by dragging an arrow grip, or you may choose to extend the Centerline to another object as shown in the example below.

Before extending the Centerline

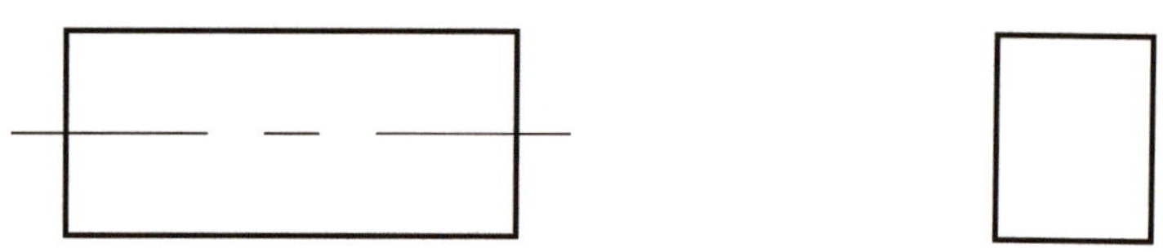

After extending the Centerline

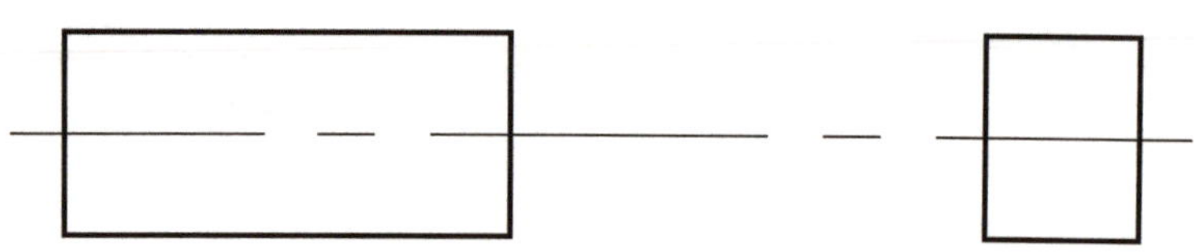

Note: Centerlines are **Associative**, so if you choose to move an object or one of the lines, the Centerline will automatically adjust to the new center position. You may also change the properties of Centerlines by selecting a Centerline and then right clicking and choosing **"Properties"** from the menu.

Flip Arrow

You can easily **flip** the direction of the arrowhead using the **Flip Arrow** option.

How to Flip an arrowhead

1. Select the dimension that you want to flip.
2. Rest your cursor on the arrow grip. (Do not press the button, just rest the cursor.)
3. Select **Flip Arrow** from the Shortcut Menu.
4. The arrowhead flips.
5. Press ***<Esc>*** and the grips will disappear. (Do not press ***<Enter>***.)

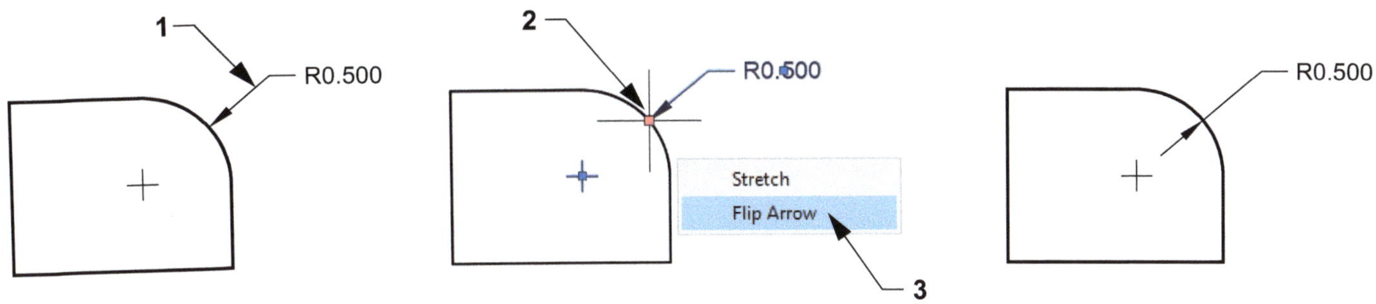

Creating a Dimension Sub-Style

In Lesson 16, you learned how to create a dimension style named CLASS STYLE. All of the dimensions created with that style appear identical because they have the same settings. Now you are going to learn how to create a **"sub-style"** of the CLASS STYLE.

Example:

If you want all of the **diameter** dimensions to have a **Center Mark** automatically displayed but do not want a **Center Mark** displayed when using the **Radius** dimension command, you must create a **"sub-style"** for all **diameter** dimensions.

Sub-styles have also been called "children" of the "parent" dimension styles. As a result, they form a family.

A sub-style is permanent, unlike the Override command, which is temporary.

Note: This sounds much more complicated than it is. Just follow the steps below. It is very easy.

How to create a sub-style for Diameter dimensions

You will set the Center Mark to **Line** for the Diameter command only. The Radius command Center Mark will not change.

1. Start a **New** file using either the **Border A-Inch.dwt** or the **Border A-Metric.dwt**
2. Select the **Dimension Style Manager**. (Refer to page 16-7.)

Continued on the next page...

3. Select **CLASS STYLE** from the Styles list.

4. Select the **New...** button.

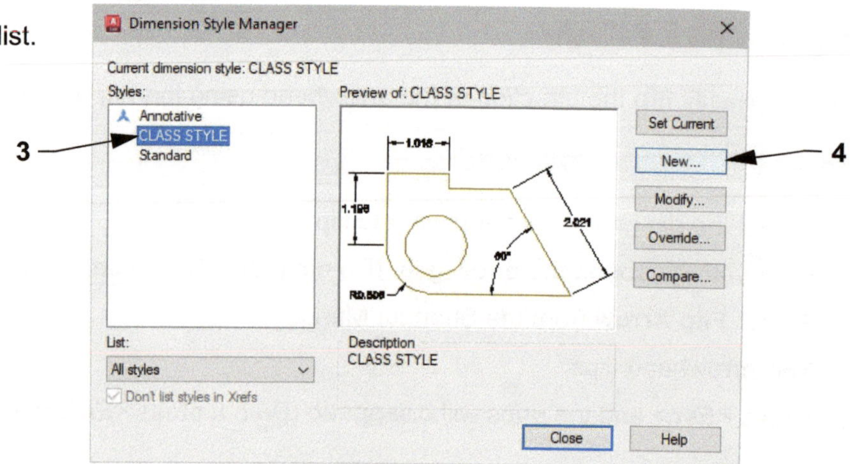

5. Change the "Use for" to: **Diameter dimensions**.

6. Select **Continue**.

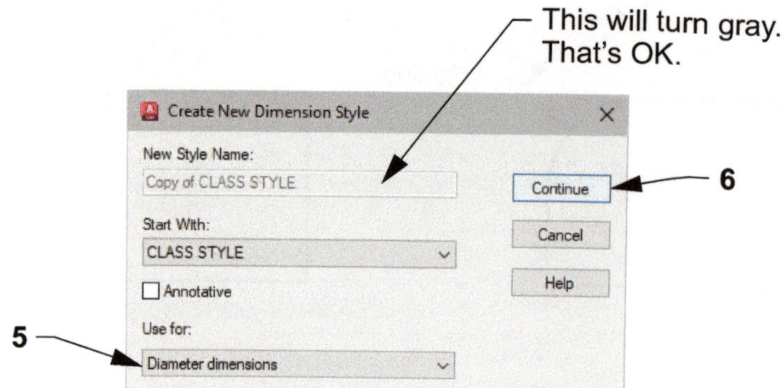

7. Select the **Symbols and Arrows** Tab.

8. Change the "Center marks" to **Line**.

9. Select the **OK** button.

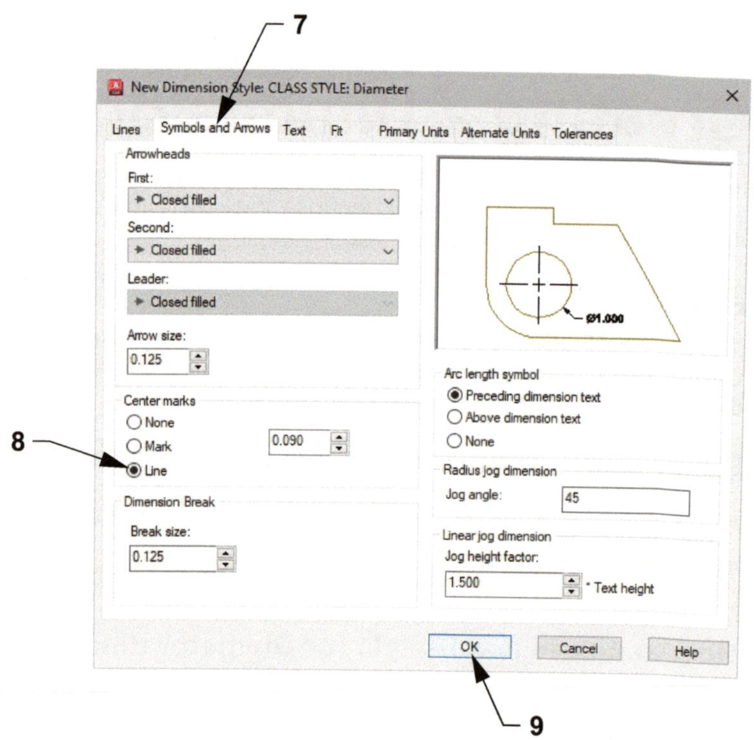

10. You now have a sub-style that will **automatically override** the basic CLASS STYLE whenever you use a Diameter command. Diameter dimensions will have Center Marks and radius dimensions will not.

    If you would like to keep this sub-style, re-save the template.

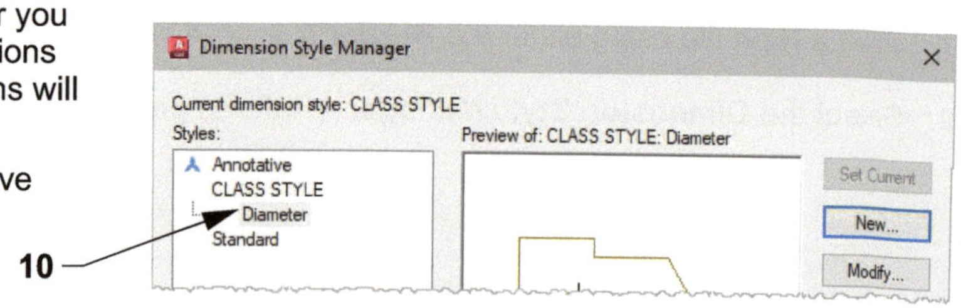

Exercise 18A-Inch

Instructions:

1. Start a **New** file using the **Border A-Inch.dwt**
2. Draw the objects below using Layer Object Line.
3. Dimension as shown using Linear, Diameter, and Radius. Use Layer Dimension.
4. Edit the **Title** and **Ex-XX** by double clicking on the text. Do not erase and replace.
5. **Save** as **Ex-18A-I**
6. Plot using Page Setup **Class Model A**

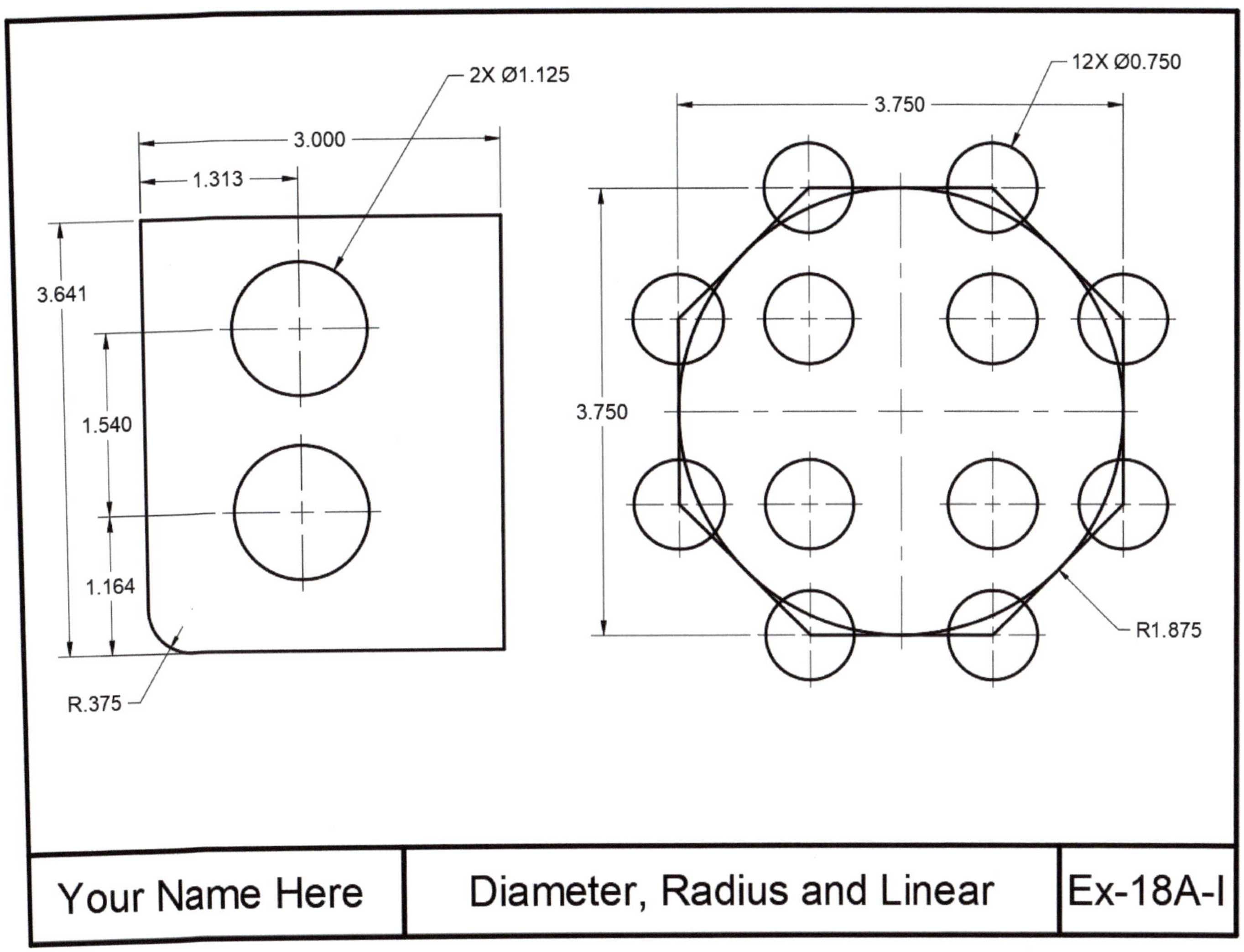

Exercise 18A-Metric

Instructions:

1. Start a **New** file using the **Border A-Metric.dwt**
2. Draw the objects below using Layer Object Line.
3. Dimension as shown using Linear, Diameter, and Radius. Use Layer Dimension.
4. Edit the **Title** and **Ex-XX** by double clicking on the text. Do not erase and replace.
5. **Save** as **Ex-18A-M**
6. Plot using Page Setup **Class Model A**

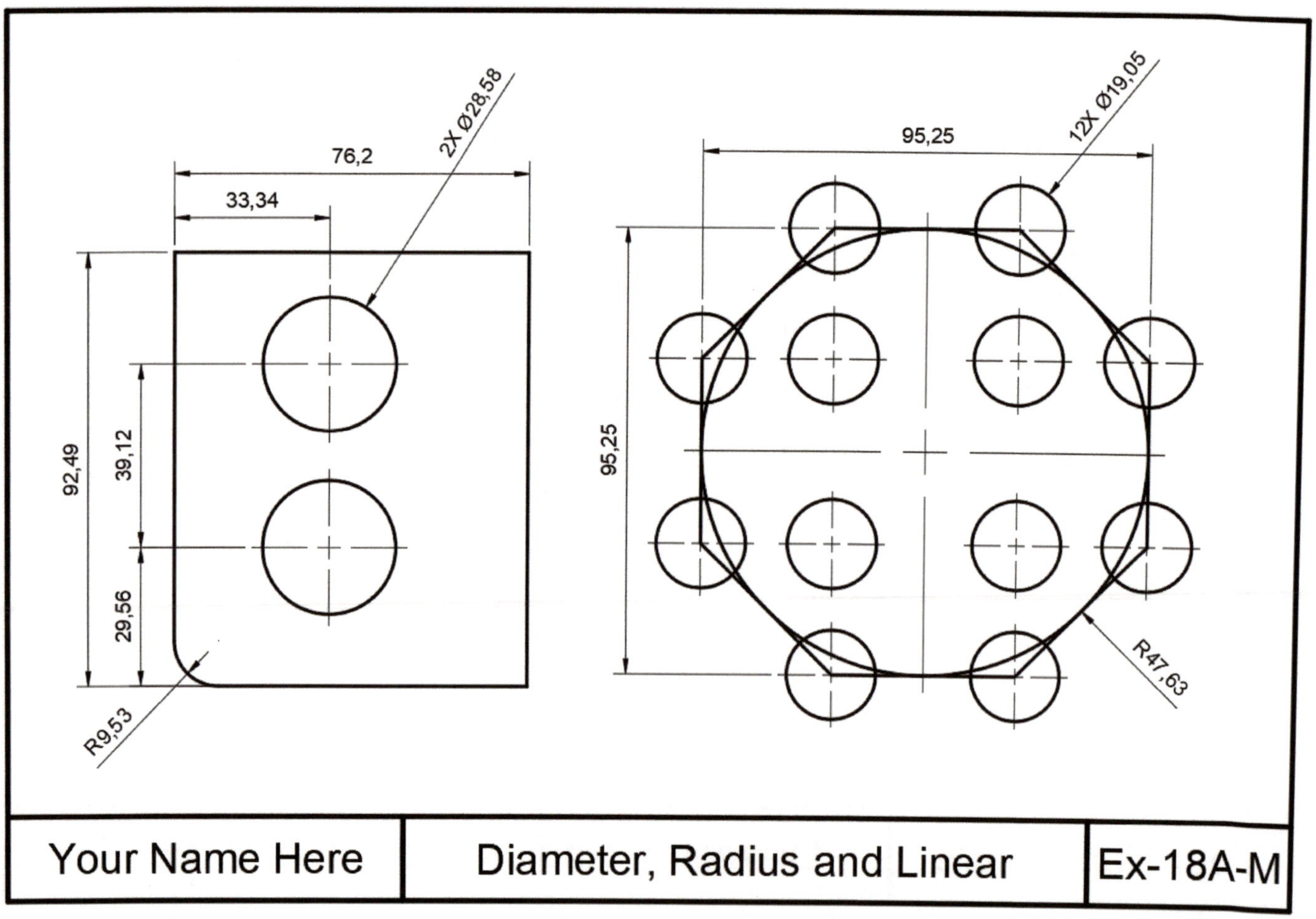

Exercise 18B-Inch

Instructions:

1. Start a **New** file using the **Border A-Inch.dwt**
2. Draw the objects below using Layer Object Line.
   Refer to the next page for drawing hints.
3. Dimension as shown using Linear, Diameter, and Angular. Use Layer Dimension.
4. Edit the **Title** and **Ex-XX** by double clicking on the text. Do not erase and replace.
5. **Save** as **Ex-18B-I**
6. Plot using Page Setup **Class Model A**

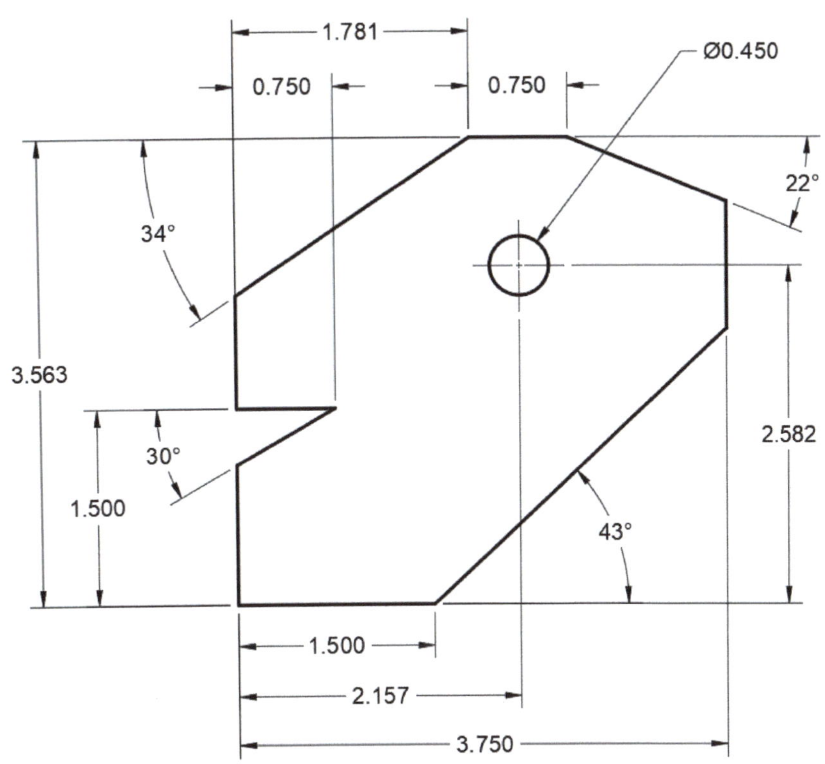

Refer to the next page for drawing hints

Your Name Here	Linear, Diameter and Angular	Ex-18B-I

Exercise 18B-Inch Helper

Follow the step-by-step instructions to complete the exercise.

Step 1. Draw horizontal and vertical lines.

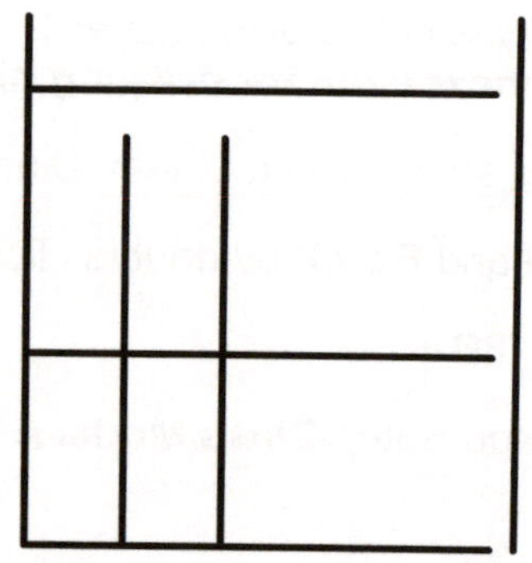

Step 2. Offset some lines.

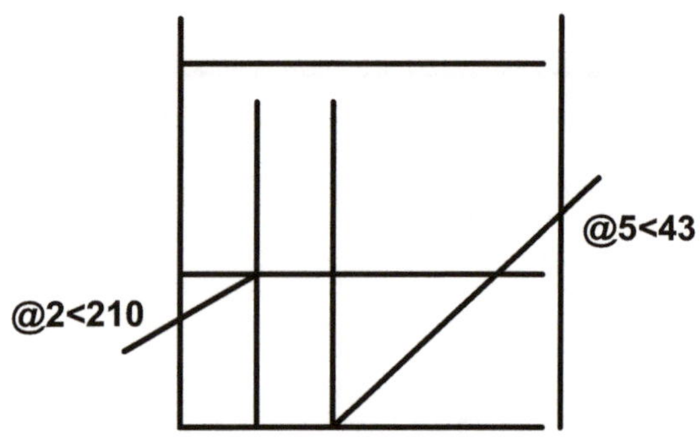

Step 3. Use Polar coordinate input to draw the two angled lines.

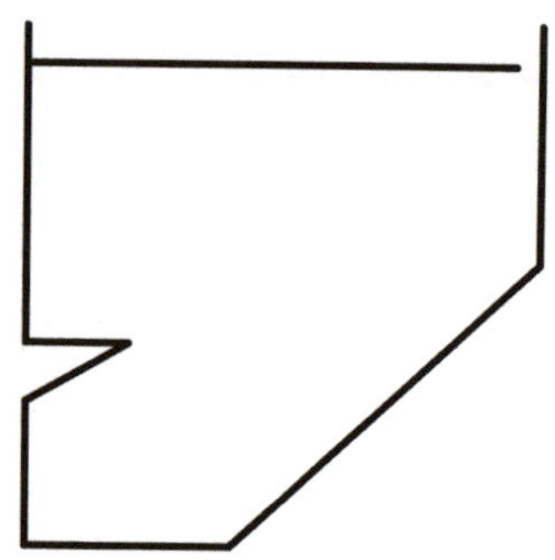

Step 4. Trim the lines so it doesn't get confusing.

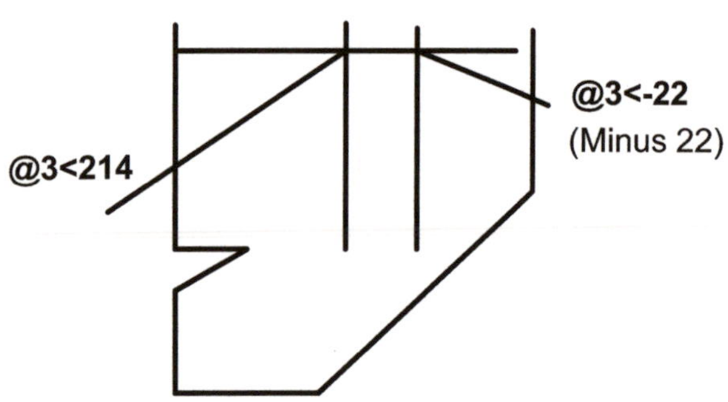

Step 5. Offset some more lines. Use Polar coordinate input again.

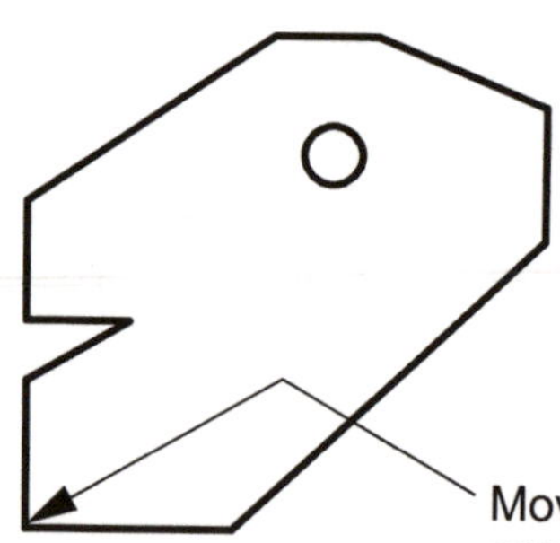

Step 6. Trim.

Exercise 18B-Metric

Instructions:

1. Start a **New** file using the **Border A-Metric.dwt**
2. Draw the objects below using Layer Object Line.
   Refer to the next page for drawing hints.
3. Dimension as shown using Linear, Diameter, and Angular. Use Layer Dimension.
4. Edit the **Title** and **Ex-XX** by double clicking on the text. Do not erase and replace.
5. **Save** as **Ex-18B-M**
6. Plot using Page Setup **Class Model A**

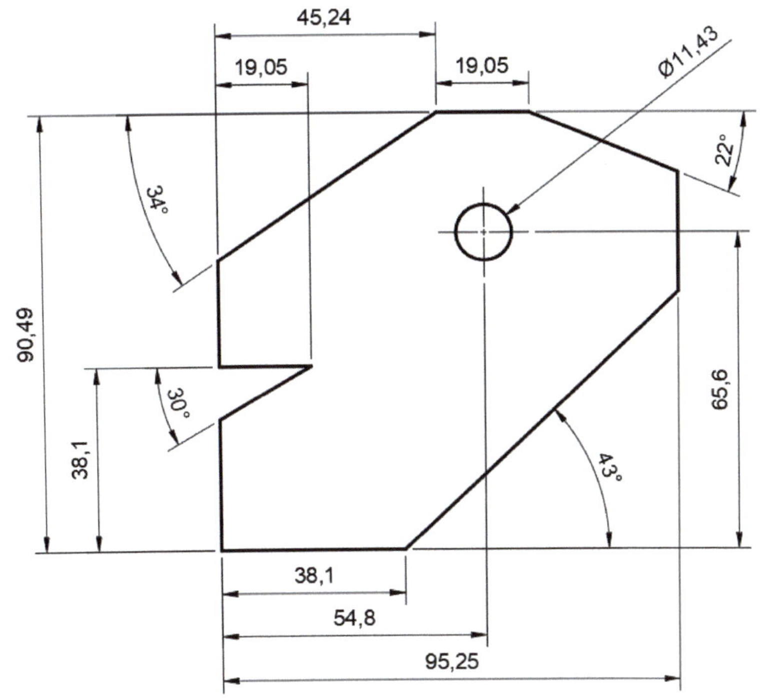

Refer to the next page for drawing hints

Your Name Here	Linear, Diameter and Angular	Ex-18B-M

Exercise 18B-Metric Helper

Follow the step-by-step instructions to complete the exercise.

Step 1. Draw horizontal and vertical lines.

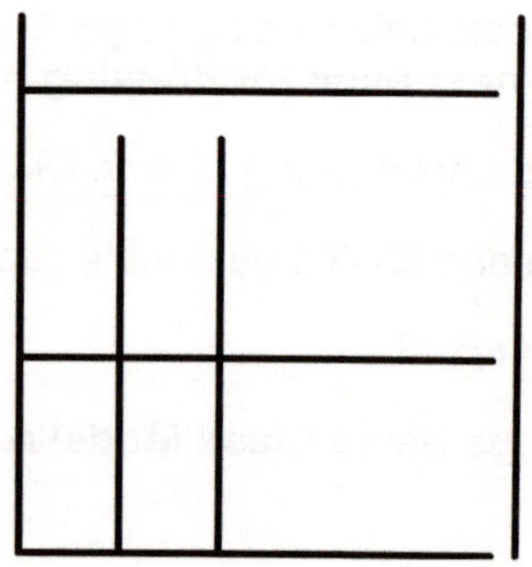

Step 2. Offset some lines.

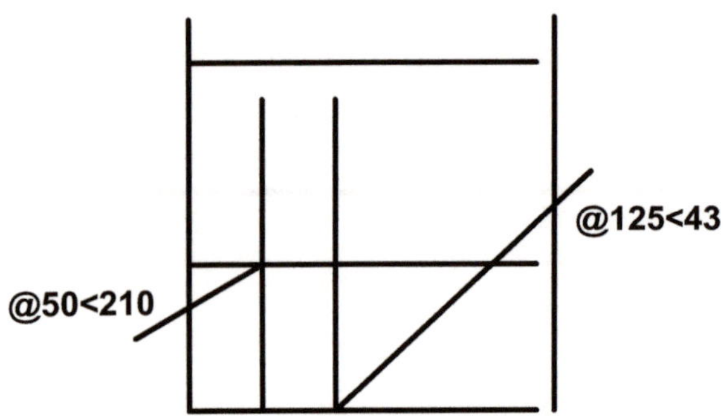

Step 3. Use Polar coordinate input to draw the two angled lines.

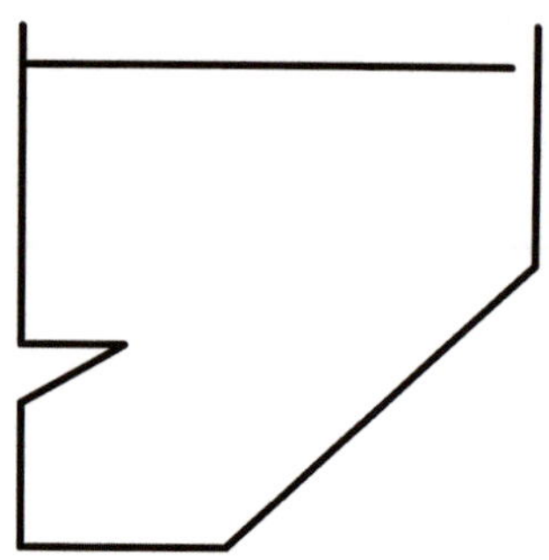

Step 4. Trim the lines so it doesn't get confusing.

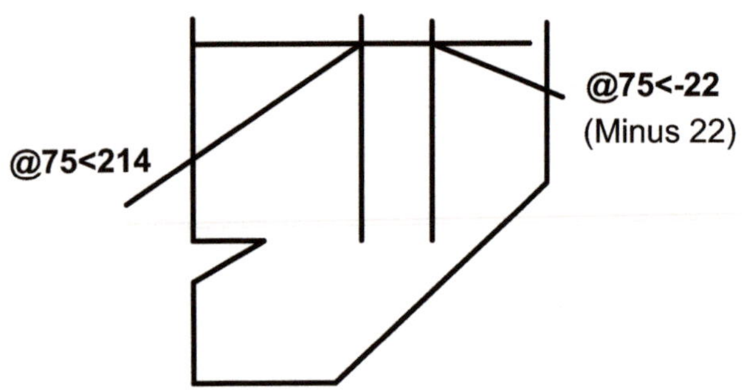

Step 5. Offset some more lines. Use Polar coordinate input again.

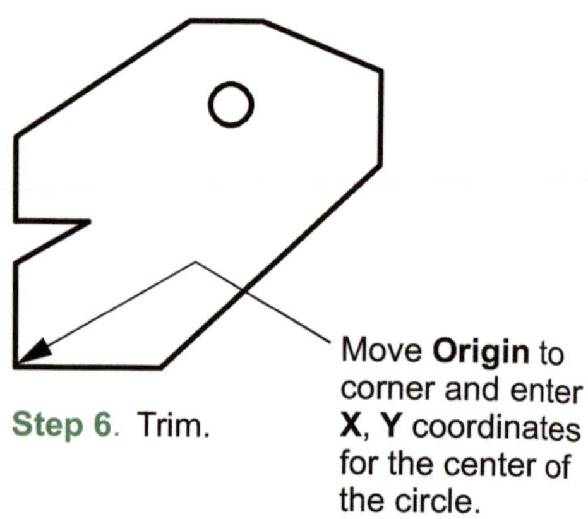

Step 6. Trim.

Exercise 18C-Inch

Instructions:

1. Start a **New** file using the **Border A-Inch.dwt**
2. Draw the objects below using Layer Object Line.
3. Manually place the Center Marks and Centerlines. Use Layer Centerline and change the line color to black.
4. Dimension as shown using Linear and Diameter. Use Layer Dimension.
5. Edit the **Title** and **Ex-XX** by double clicking on the text. Do not erase and replace.
6. **Save** as **Ex-18C-I**
7. Plot using Page Setup **Class Model A**

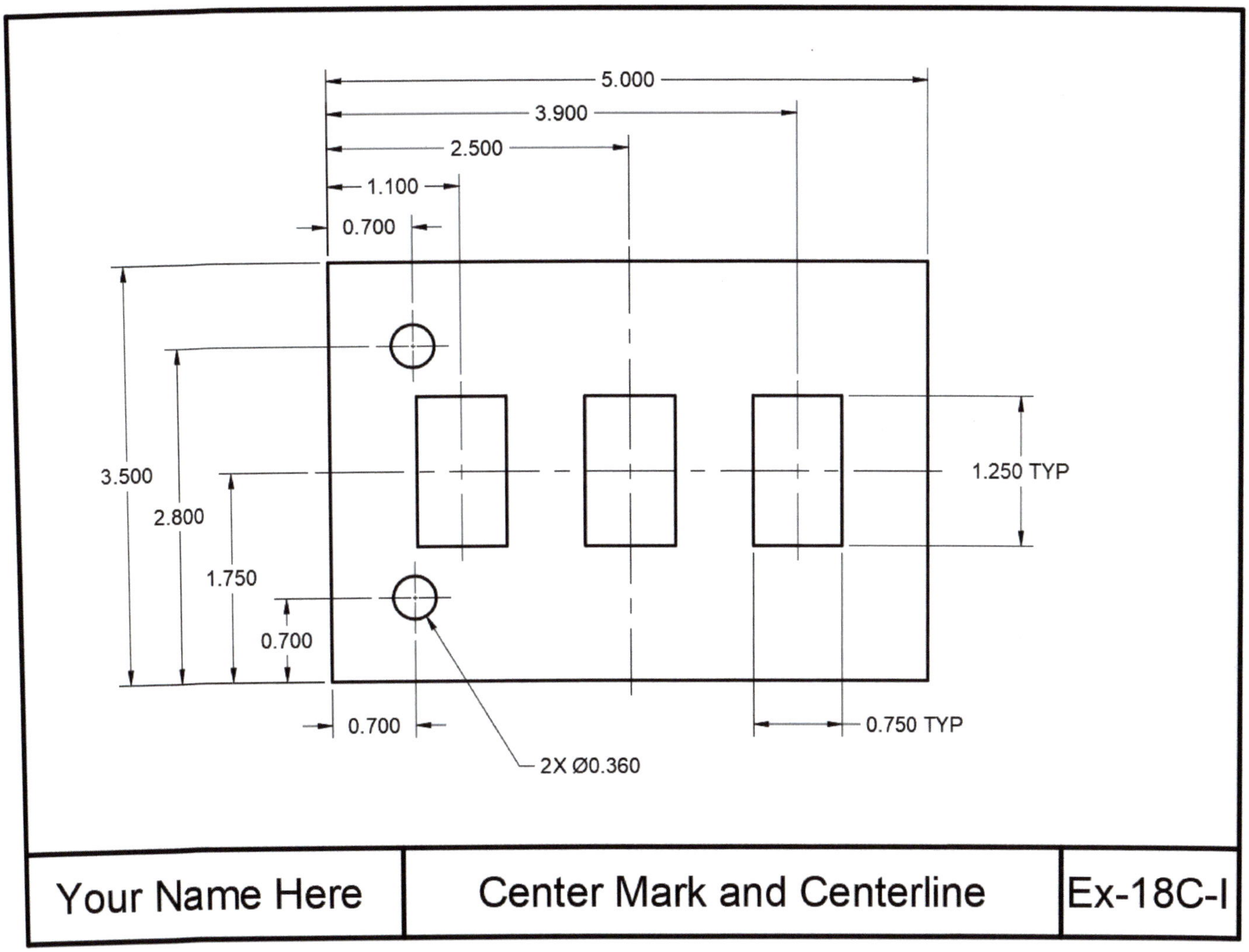

Exercise 18C-Metric

Instructions:

1. Start a **New** file using the **Border A-Metric.dwt**
2. Draw the objects below using Layer Object Line.
3. Manually place the Center Marks and Centerlines. Use Layer Centerline and change the line color to black.
4. Dimension as shown using Linear and Diameter. Use Layer Dimension.
5. Edit the **Title** and **Ex-XX** by double clicking on the text. Do not erase and replace.
6. **Save** as **Ex-18C-M**
7. Plot using Page Setup **Class Model A**

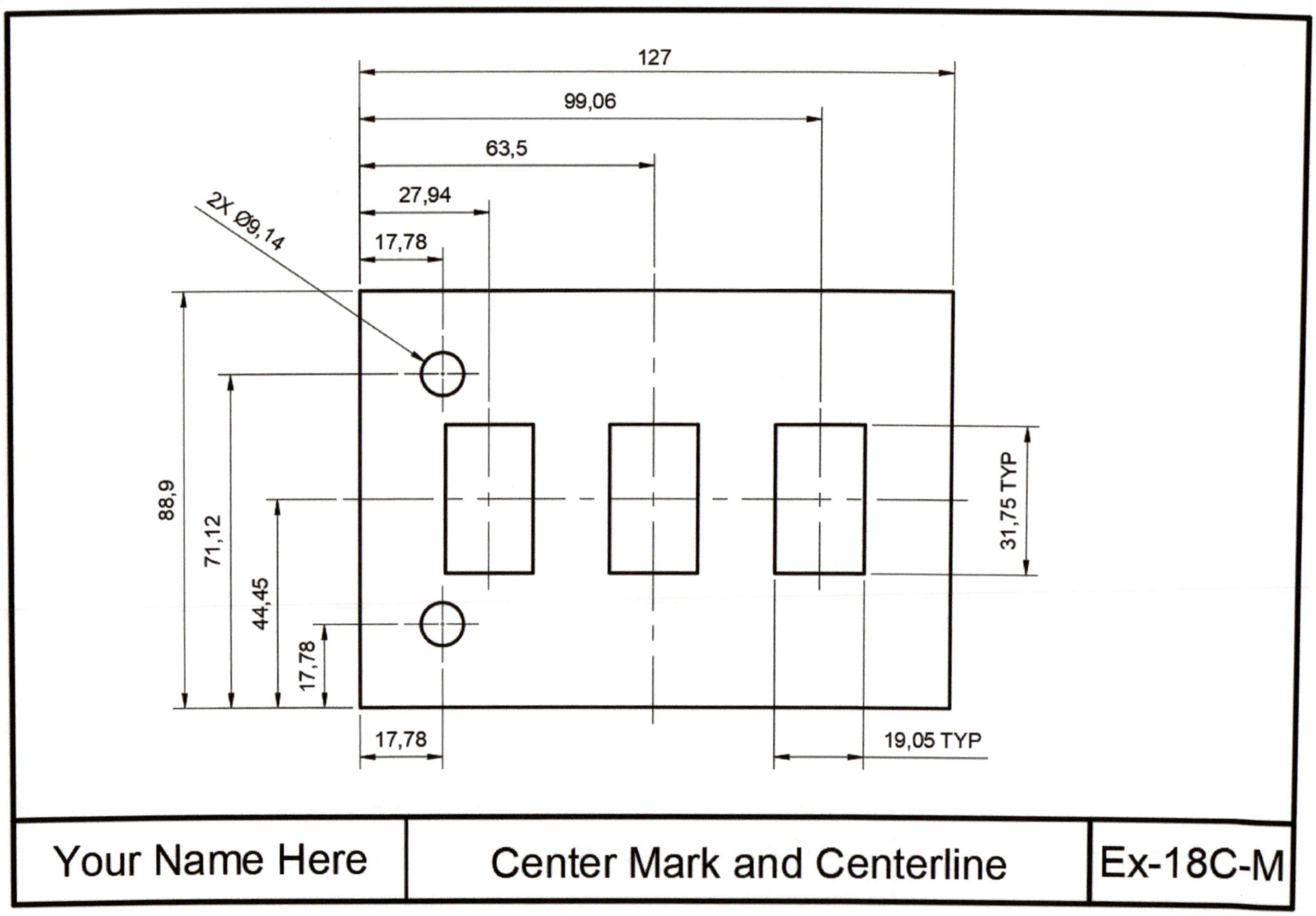

LESSON 19

LEARNING OBJECTIVES

After completing this lesson, you will be able to:

- Create Multileaders
- Modify Multileader lines
- Create a new Multileader style
- Dimension angular surfaces
- Add Special Characters
- Pre-assign a Prefix or Suffix to a dimension

Multileader

A **multileader** is a single object consisting of a pointer, leading line, landing line, and content. You may draw a multileader, pointer first, line first, or content first.

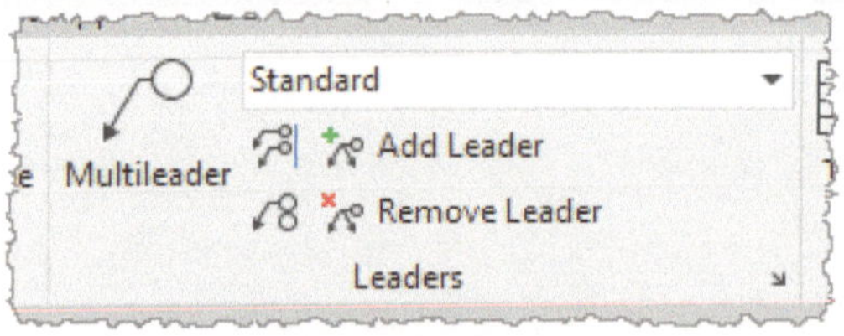

Leaders Panel

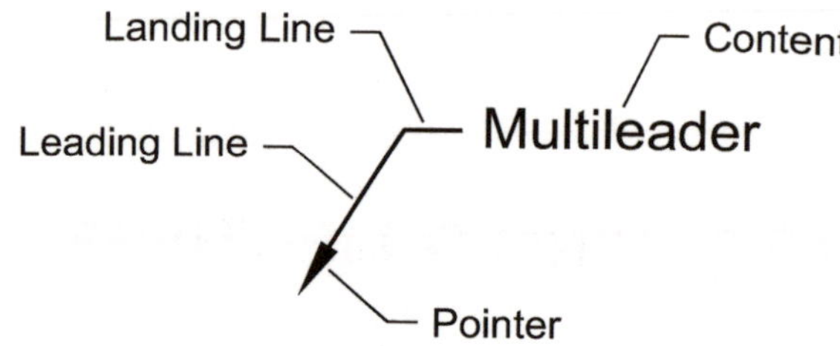

How to create a single Multileader

1. To select the **Multileader** command use one of the following:

   Ribbon = Annotate Tab / Leaders Panel /
   or
   Keyboard = ***MLD <Enter>***

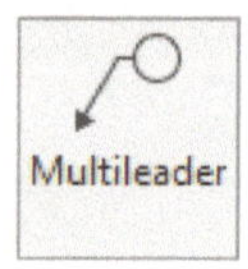

   The following will appear on the Command Line:

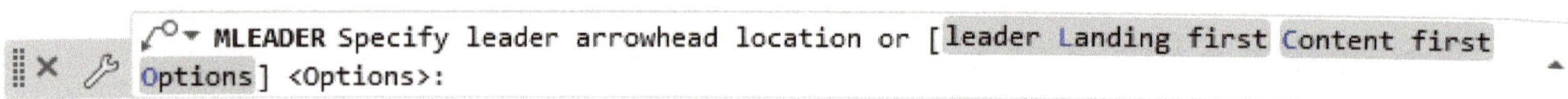

2. Specify the arrowhead location (**P1**).

   The following will appear on the Command Line:

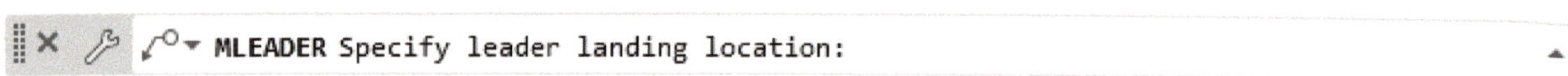

3. Specify the leader landing location (**P2**).

4. The Multiline Text Editor Ribbon will appear: Enter the text for the leader and then select **Close Text Editor**.

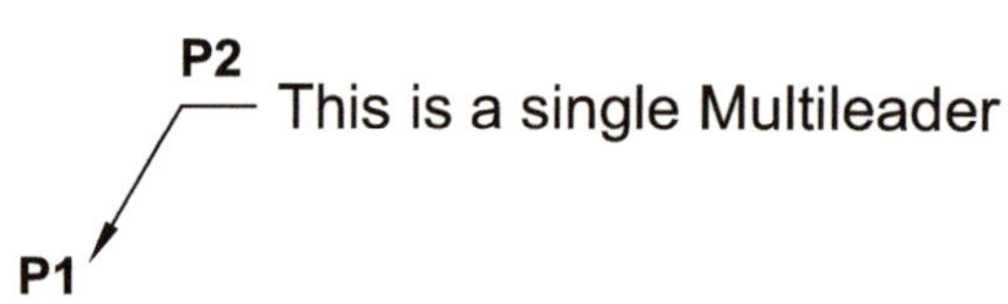

Add leader lines to an existing Multileader

1. Select the **Add Leader** tool.

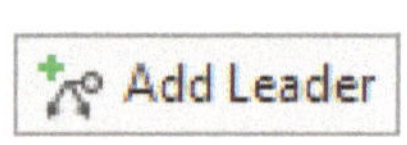

   The following will appear on the Command Line:

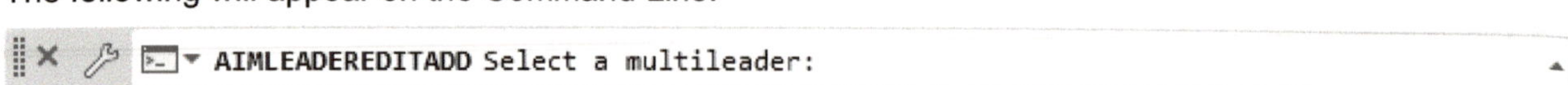

2. Select an existing multileader (**P1**).

   The following will appear on the Command Line:

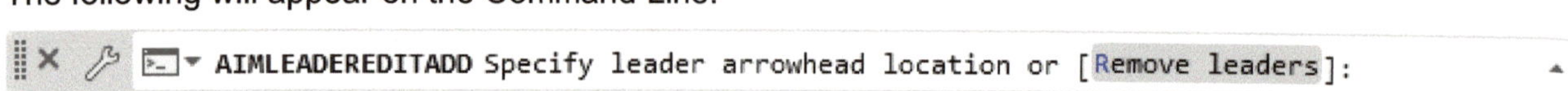

3. Specify the arrowhead location (**P2**).

Continued on the next page...

The following will appear on the Command Line:

```
AIMLEADEREDITADD Specify leader arrowhead location or [Remove leaders]:
```

4. Specify the arrowhead location (**P3**).

   The following will appear on the Command Line:

```
AIMLEADEREDITADD Specify leader arrowhead location or [Remove leaders]:
```

5. Specify the arrowhead location (**P4**).

6. Press ***<Enter>*** to stop.

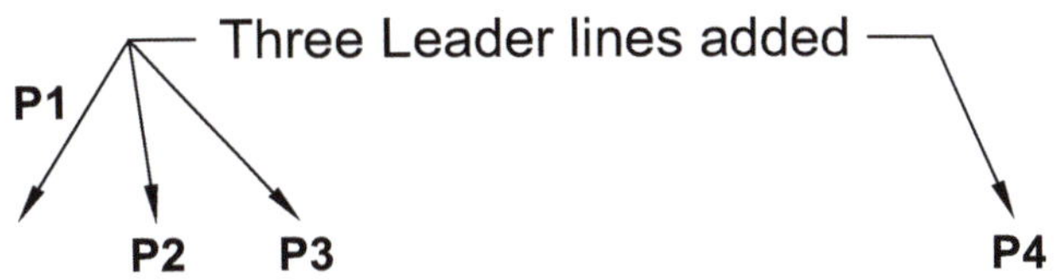

Remove a leader line from an existing Multileader

1. Select the **Remove Leader** tool. 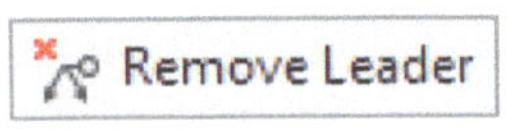

   The following will appear on the Command Line:

```
AIMLEADEREDITREMOVE Select a multileader:
```

2. Select an existing multileader (**P1**). (You can click anywhere on the multileader.)

   The following will appear on the Command Line:

```
AIMLEADEREDITREMOVE Specify leaders to remove or [Add leaders]:
```

3. Select the leader line to remove (**P2**).

4. You can select more leader lines to remove or press ***<Enter>*** to end the command.

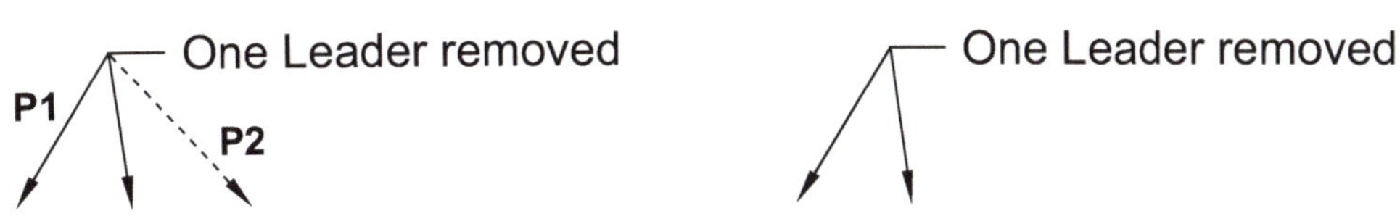

Align Multileaders

1. Select the **Align Multileader** tool. 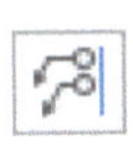

   The following will appear on the Command Line:

```
MLEADERALIGN Select multileaders:
```

2. Select the leaders to align (Note 1 and Note 3 as shown on the next page) and then press ***<Enter>*** to stop selecting.

   The following will appear on the Command Line:

```
MLEADERALIGN Select multileader to align to or [Options]:
```

Continued on the next page...

3. Select the leader to align to (Note 2).

   The following will appear on the Command Line:

```
MLEADERALIGN Specify direction:
```

4. Move the cursor in the direction you want the leaders aligned, and then press the left mouse button.

CAD TIP! If you want the leaders perfectly aligned vertically or horizontally, make sure **Ortho Mode <*F8*>** is **on**.

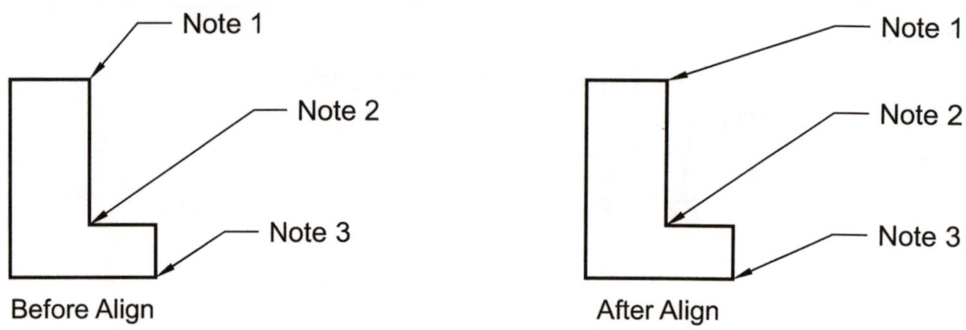

Note: Collect Multileader works best with Blocks and will be discussed in Lesson 29.

Creating a Multileader Style

You may create **multileader styles** to control its appearance. This will be similar to creating a dimension style.

1. Start a **New** file using either the **Border A-Inch.dwt** or the **Border A-Metric.dwt**

2. Select the **Multileader Style Manager** by selecting the diagonal arrow ↘ on the **Leaders Panel**.

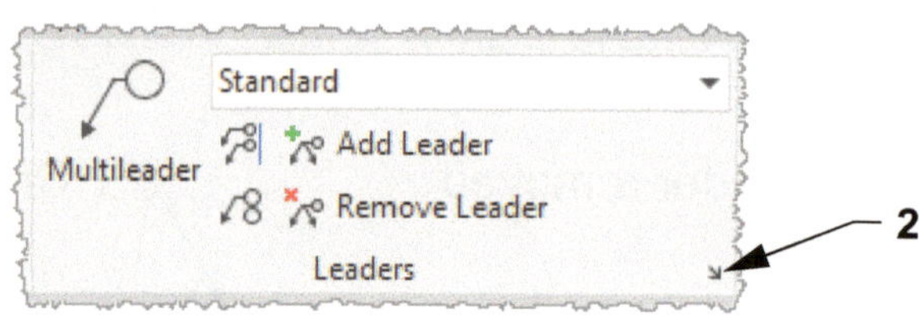

3. Select the **New...** button.

4. Enter the New Style name: **Class ML Style**

5. Select the **Continue** button.

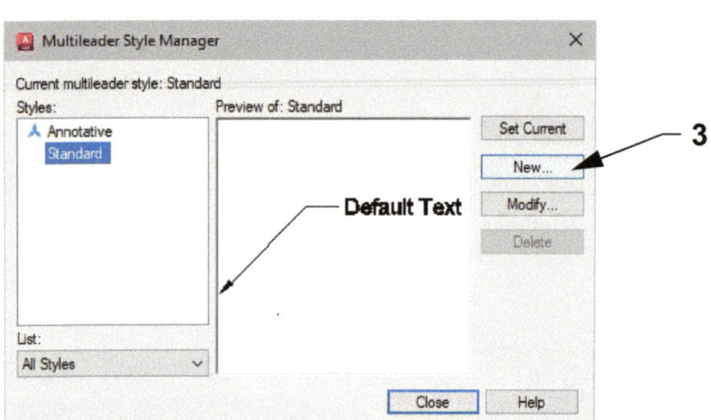

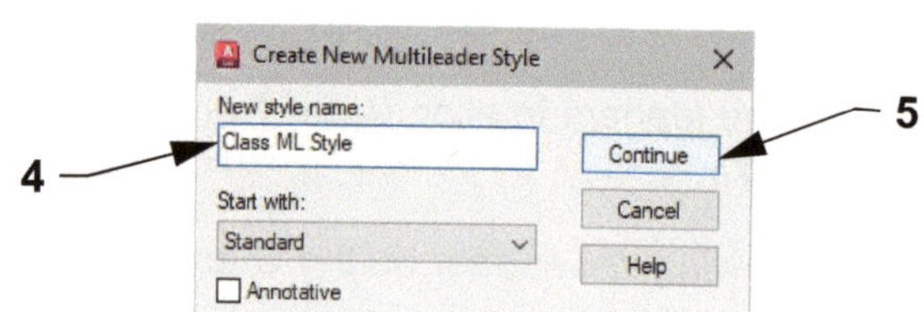

Continued on the next page...

6. Select the **Leader Format** Tab and change the settings as shown.

   Metric users: Arrowhead Size = **3.2**
   Break Size = **3.2**

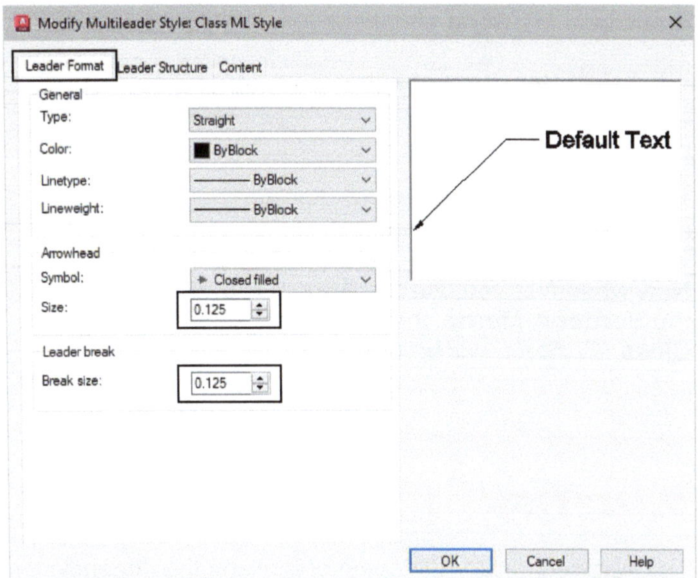

General: You may set the leader line to straight or spline (curved). You may change the color, linetype, and lineweight.

Arrowhead: Select the symbol for the pointer and size of pointer.

Leader break: Specify the size of the gap in the leader line if Dimension break is used.
(Refer to page 17-6 for Dimension break reminder.)

7. Select the **Leader Structure** Tab and change the settings as shown.

   Metric users: Set landing distance = **3.2**

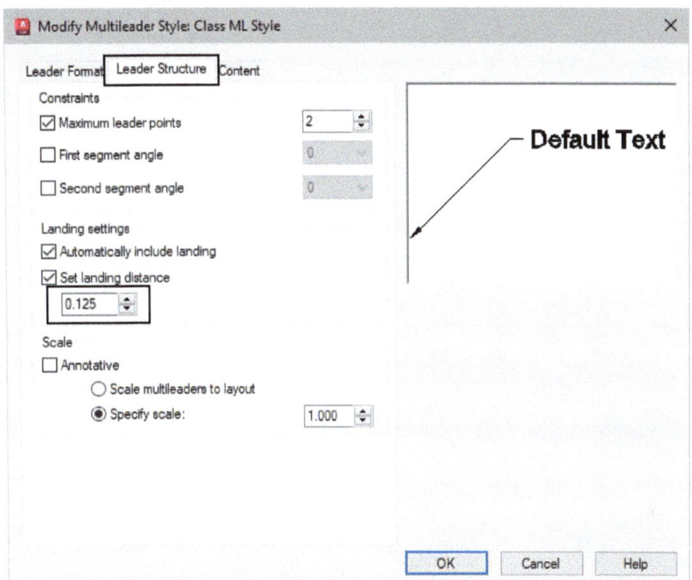

Constraints: Specify how many line segments to allow and on what angle.

Landing settings: Specify the length of the landing (the horizontal line next to the note) or turn it off by unchecking the "Automatically include landing" box.

Scale: "Annotative" will be discussed in Lesson 27

8. Select the **Content** Tab and change the settings as shown.

   Metric users: Text height = **3.2**
   Landing gap = **1.5**

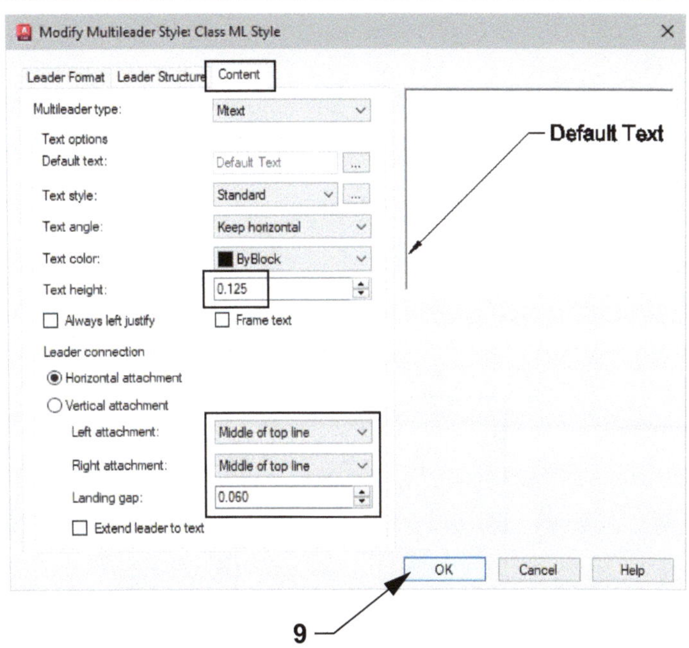

Multileader type: Select what to attach to the landing.

Text options: Specify the leader note appearance.

Leader connection: Specify how the note attaches to the landing.

9. Select the **OK** button.

Continued on the next page...

The **Class ML Style** should be listed in the "Styles" area.

10. Select **Set Current** button.

11. Select **Close** button.

12. Re-save your **Border A-Inch.dwt** or **Border A-Metric.dwt** again.

Now whenever you use the **Border A-Inch** or the **Border A-Metric**, the multileader style **Class ML Style**, will be there.

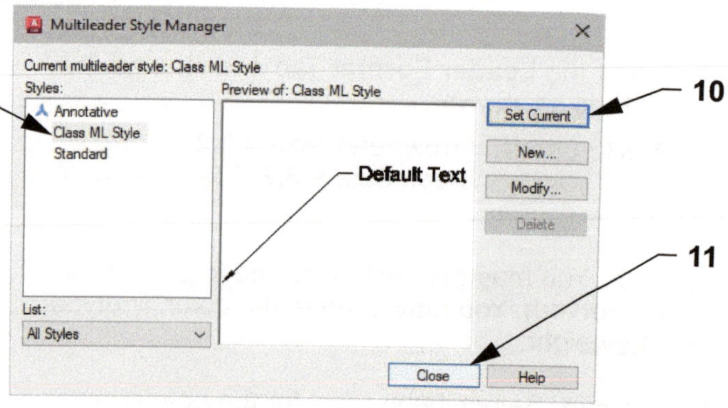

Aligned Dimensioning

The **Aligned** dimension command aligns the dimension with the angle of the object that you are dimensioning. The process is the same as Linear dimensioning. It requires two extension line origins and the placement of the dimension text location. (See the example below.)

1. Select the **Aligned** command using one of the following:

   Ribbon = Annotate Tab / Dimensional Panel / ▼
   or
   Keyboard = *DAL <Enter>*

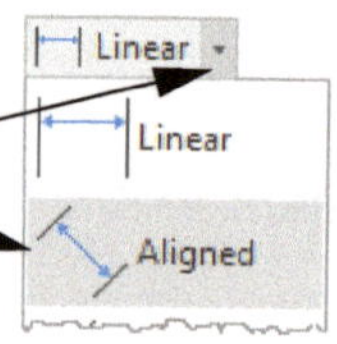

   The following will appear on the Command Line:

```
DIMALIGNED Specify first extension line origin or <select object>:
```

2. Select the first extension line origin (**P1**).

   The following will appear on the Command Line:

```
DIMALIGNED Specify second extension line origin:
```

3. Select the second extension line origin (**P2**).

   The following will appear on the Command Line:

```
DIMALIGNED [Mtext Text Angle]:
```

4. Place the dimension text (**P3**).

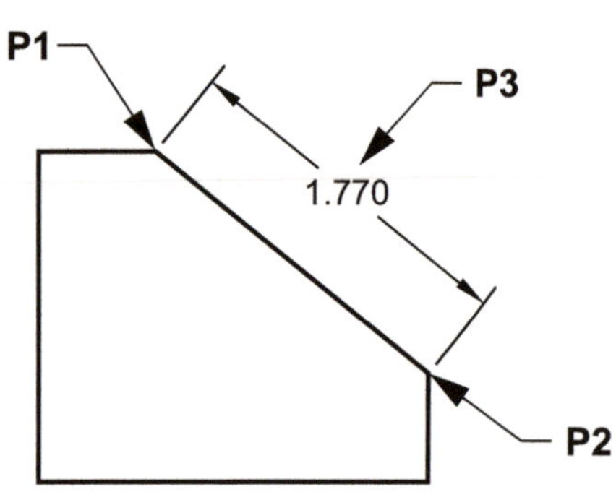

Inch users

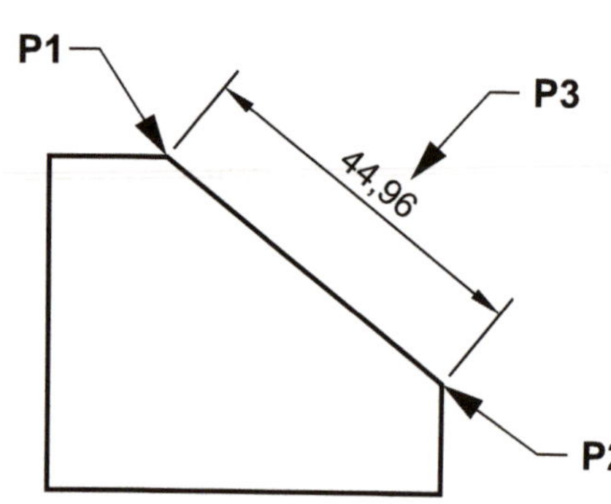

Metric users

Special Text Characters

Characters such as the **Degrees symbol (°)**, **Plus/Minus symbols (±)**, and the **Diameter symbol (Ø)**, are created by typing **%%** and then the appropriate **"code"** letter.

Symbol		Code
°	Degrees	%%d
±	Plus/Minus	%%p
Ø	Diameter	%%c

Single Line Text

When using **Single Line Text**, type the code in the sentence. After you enter the code, the symbol will appear.

Example:

Entering: ***350%%d*** will create: **350°**. The **"*d*"** is the **"code"** letter for degree.

Multiline Text

When using **Multiline Text**, you may enter the code in the sentence or you may select a symbol using the **Symbol** tool located on the **Insert Panel** of the **Text Editor** Ribbon.

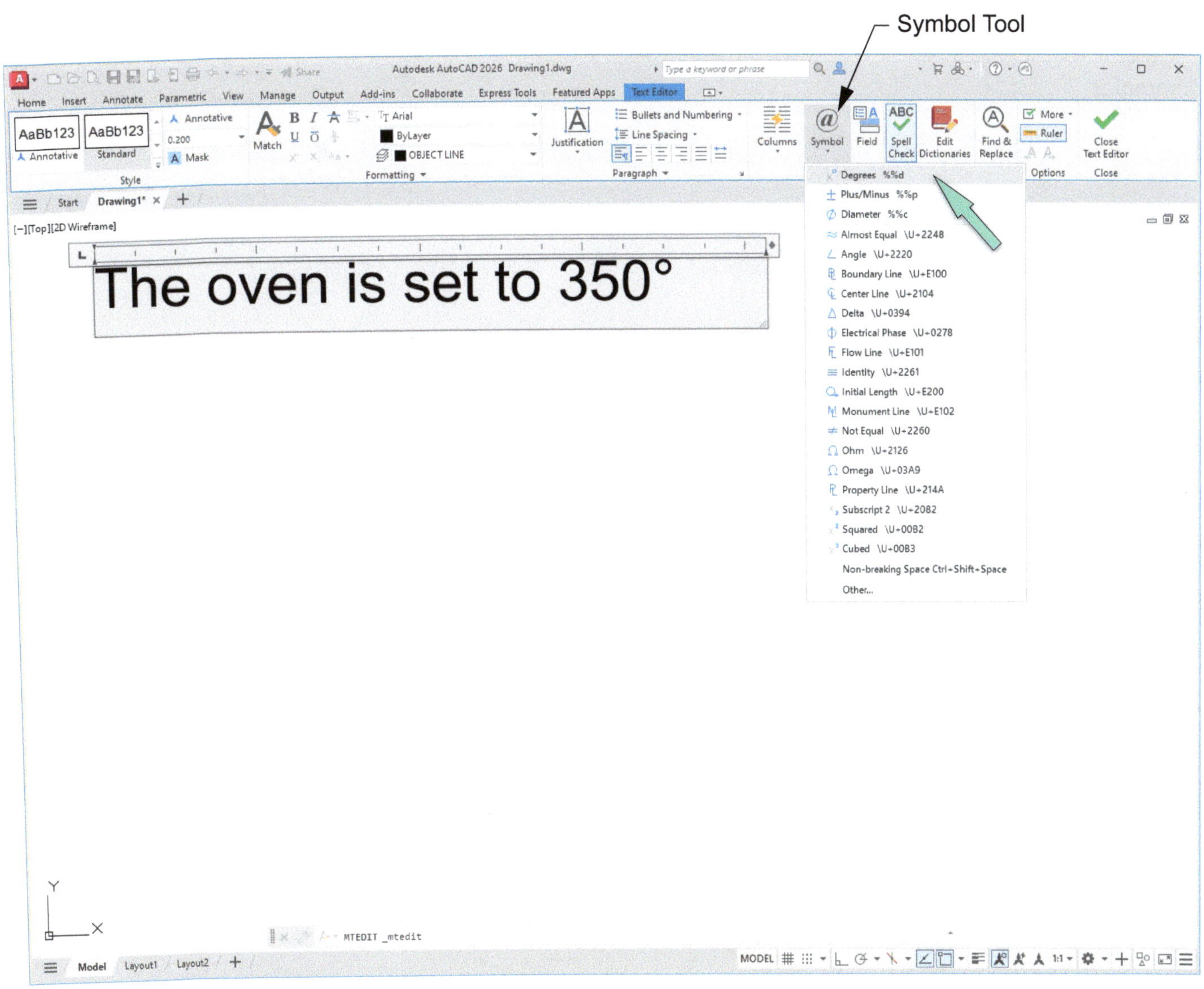

Prefix and Suffix

The **prefix** (before) and the **suffix** (after) allows you to preset text to be inserted automatically as you dimension.

Note: You would use this if you had multiple dimensions that included a prefix or a suffix. If you had only one or two, you would merely edit the dimension text as shown on page 17-2.

1. Select the **Dimension Style Manager**.
2. Select Dimension Style CLASS STYLE.
3. Select the **Primary Units** Tab.
4. Enter the **Prefix** or **Suffix** content.

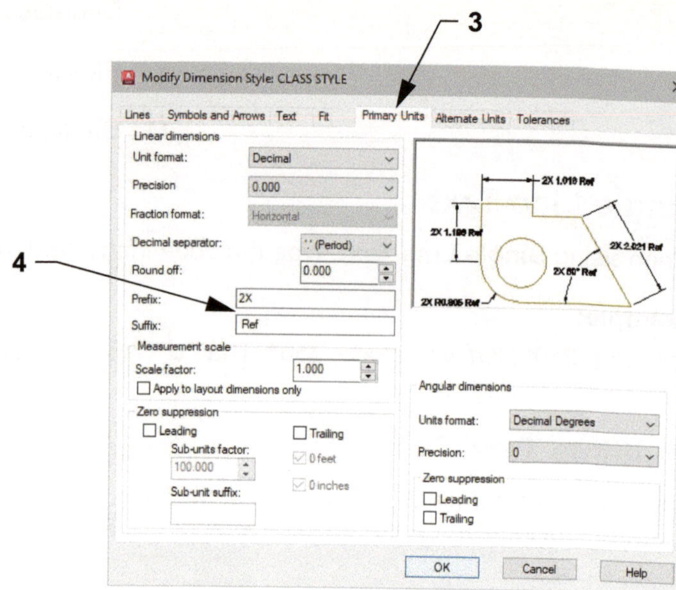

Example:

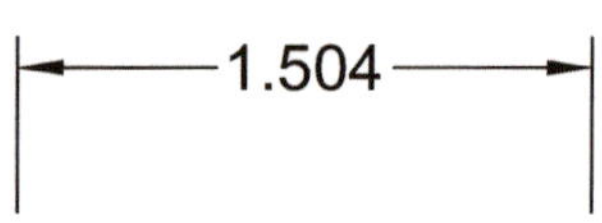

No Prefix or Suffix

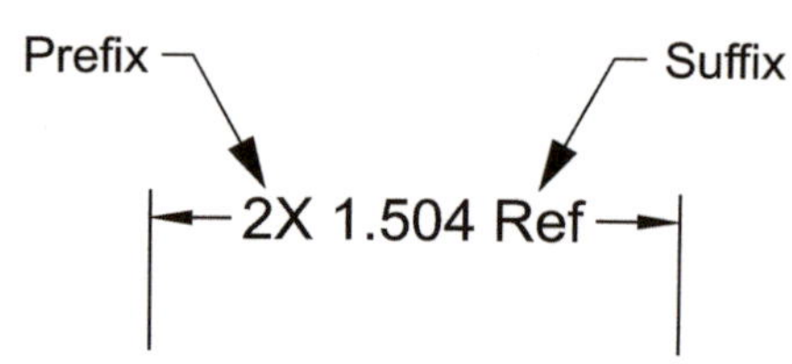

Prefix = 2X and Suffix = Ref

Make sure you add a space **after "2X"** in the **Prefix** box, and a space **before "Ref"** in the **Suffix** box. If you don't add those spaces, the dimension text will look like the example below.

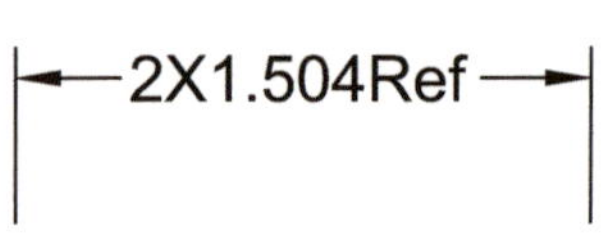

No spaces in Prefix and Suffix box

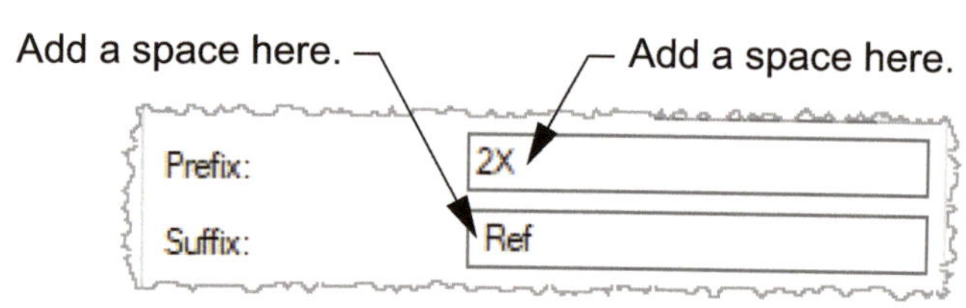

If you enter text in the **Prefix** box and you create **radius** or **diameter** dimensions, the **"R"** for radius or the **"Ø"** for the diameter symbol, will not be automatically drawn. You must create a **dimension sub-style** for both radius and diameter dimensions, and add the examples shown below to the **Prefix** box for each sub-style.

2X R

Prefix for a Radius Dimension Sub-Style

2X %%C

Prefix for a Diameter Dimension Sub-Style

Note: Refer to page 18-9 for information on creating dimension sub-styles.

Exercise 19A-Inch

Instructions:

1. Start a **New** file using the **Border A-Inch.dwt**
2. Draw the lines as shown using Polar coordinates or Dynamic Input. (Refer to Lessons 9 and 11.)
3. Use Layer Object Line.
4. Dimension using Linear, Aligned, and Angular.
5. Use Dimension Style CLASS STYLE and Layer Dimension.
6. Edit the **Title** and **Ex-XX** by double clicking on the text. Do not erase and replace.
7. **Save** as **Ex-19A-I**
8. Plot using Page Setup **Class Model A**

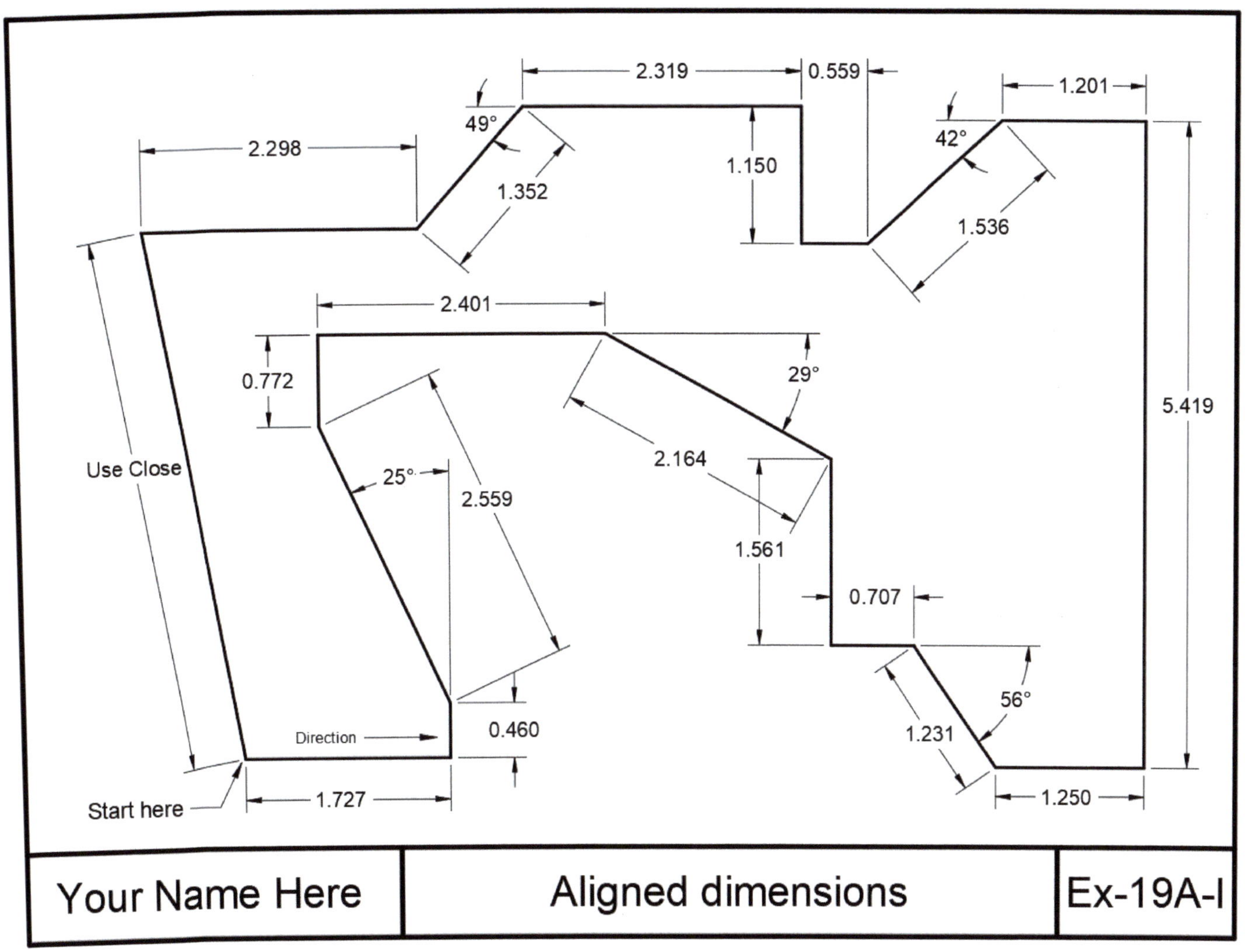

Exercise 19A-Metric

Instructions:

1. Start a **New** file using the **Border A-Metric.dwt**
2. Draw the lines as shown using Polar coordinates or Dynamic Input. (Refer to Lessons 9 and 11.)
3. Use Layer Object Line.
4. Dimension using Linear, Aligned, and Angular.
5. Use Dimension Style CLASS STYLE and Layer Dimension.
6. Edit the **Title** and **Ex-XX** by double clicking on the text. Do not erase and replace.
7. **Save** as **Ex-19A-M**
8. Plot using Page Setup **Class Model A**

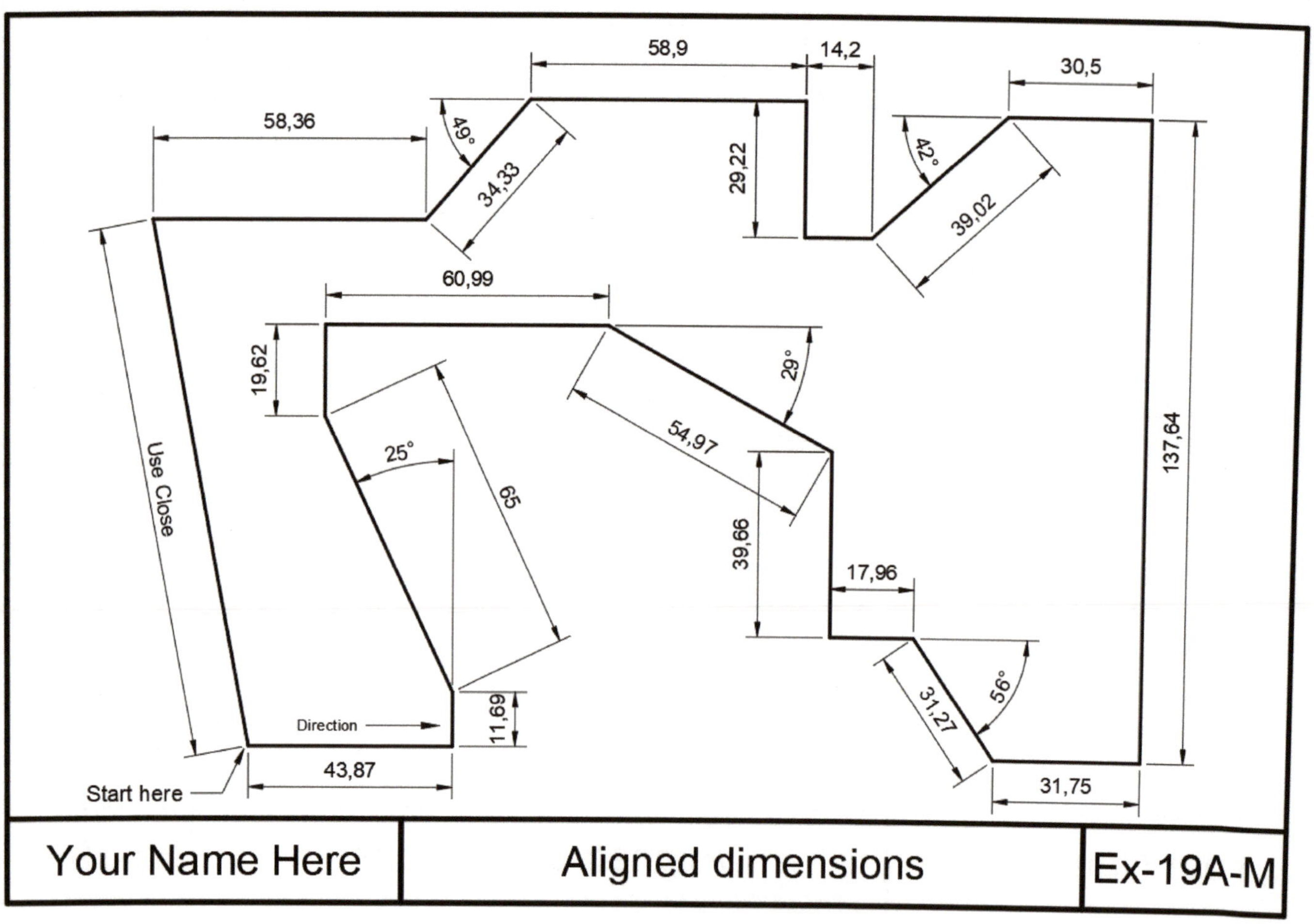

Exercise 19B-Inch

Instructions:

1. Start a **New** file using the **Border A-Inch.dwt**
2. Draw the object shown below.
3. Use Layer Object Line.
4. Dimension as shown.
5. Use Dimension Style CLASS STYLE and Layer Dimension.
6. For the note **"Special Instructions to Baker"** use Text Height = **0.125** Layer = Text.
7. Edit the **Title** and **Ex-XX** by double clicking on the text. Do not erase and replace.
8. **Save** as **Ex-19B-I**
9. Plot using Page Setup **Class Model A**

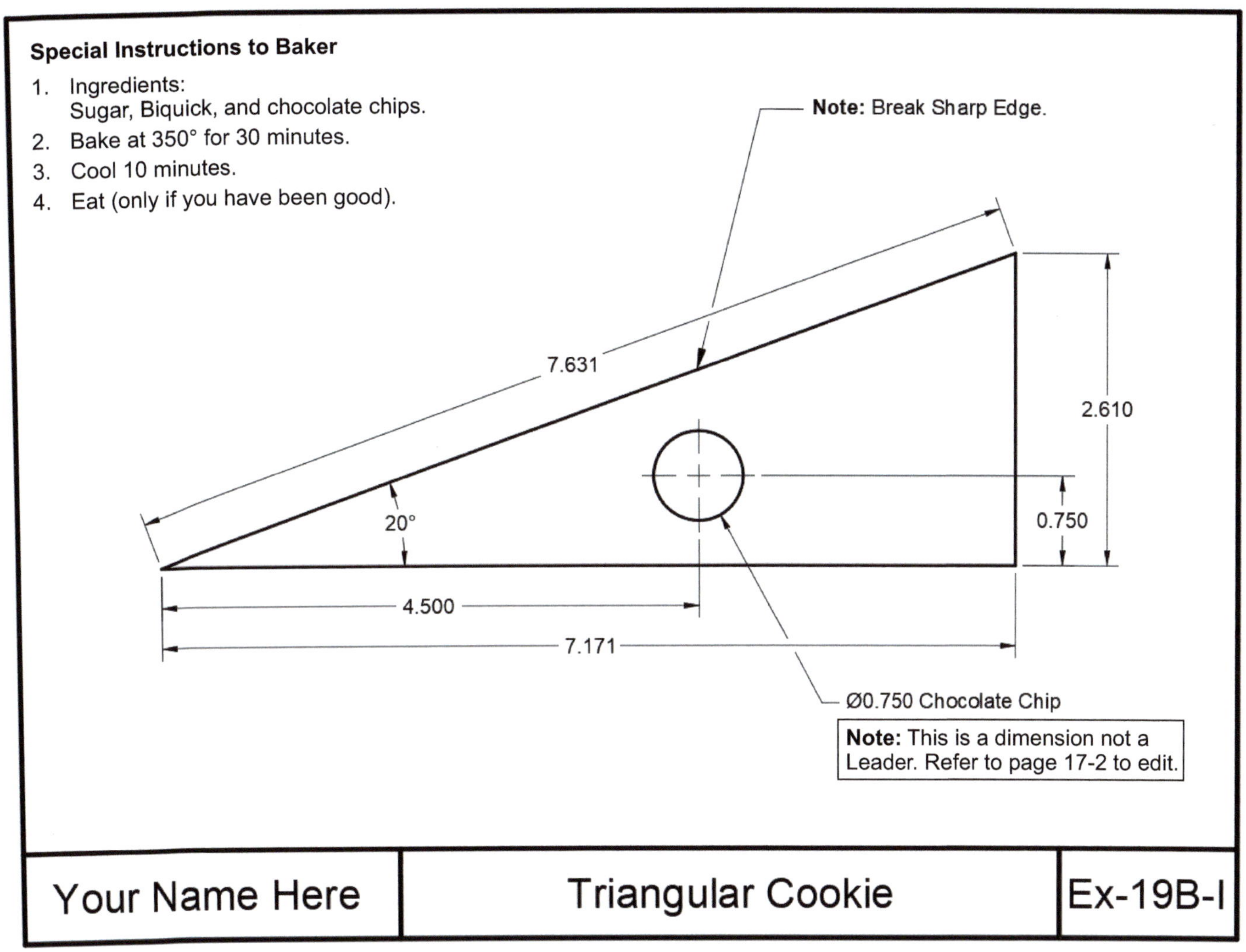

Exercise 19B-Metric

Instructions:

1. Start a **New** file using the **Border A-Metric.dwt**
2. Draw the object shown below.
3. Use Layer Object Line.
4. Dimension as shown.
5. Use Dimension Style CLASS STYLE and Layer Dimension.
6. For the note **"Special Instructions to Baker"** use Text Height = **3.2** Layer = Text.
7. Edit the **Title** and **Ex-XX** by double clicking on the text. Do not erase and replace.
8. **Save** as **Ex-19B-M**
9. Plot using Page Setup **Class Model A**

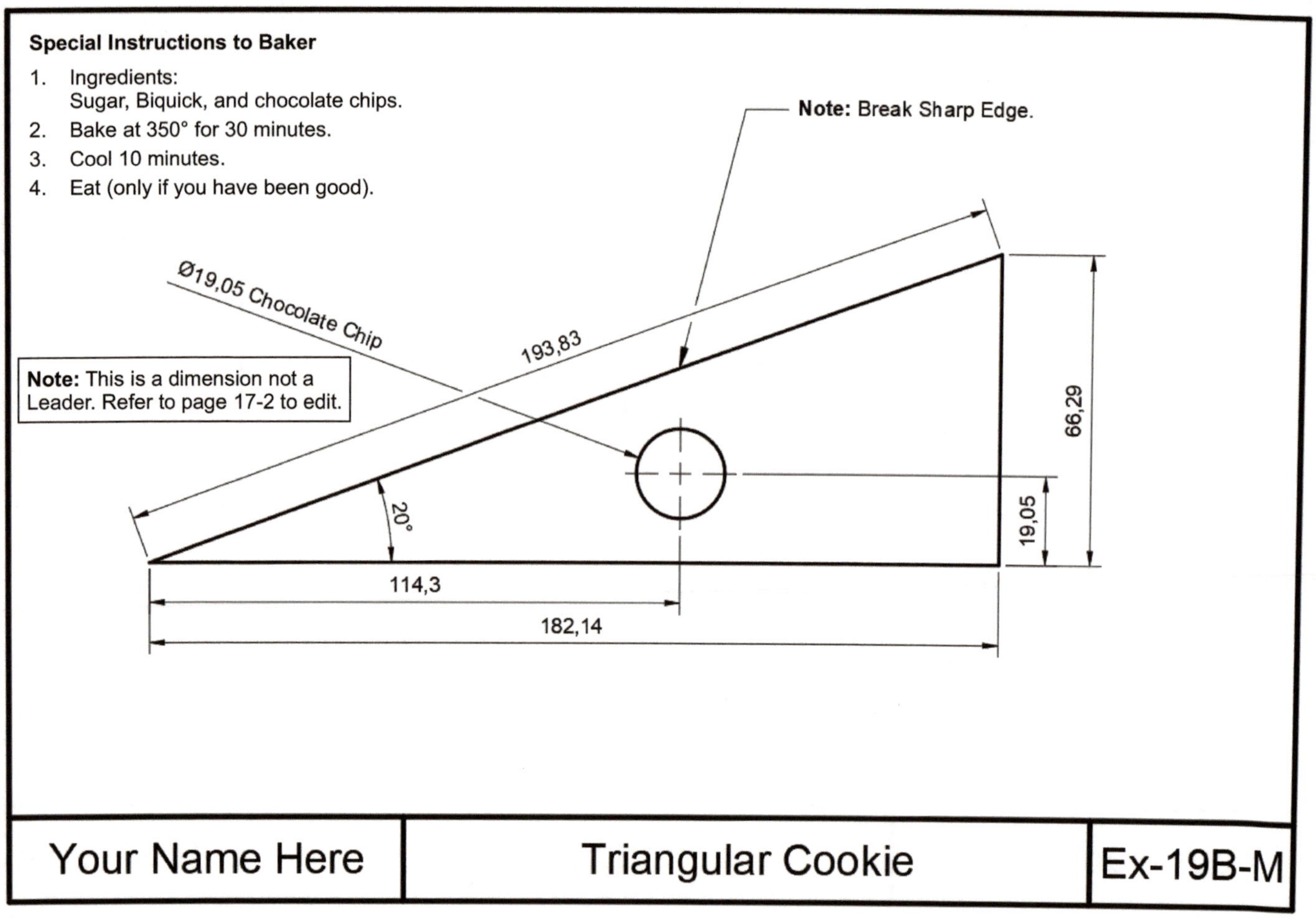

Exercise 19C-Inch

Instructions:

1. Start a **New** file using the **Border A-Inch.dwt**
2. Create the multileader style shown on page 19-4 if you haven't already.
3. Draw the object shown below.
4. Use Layer Object Line.
5. Dimension as shown.
6. Use Dimension Style CLASS STYLE and Class ML Style and Layer Dimension.
7. Edit the **Title** and **Ex-XX** by double clicking on the text. Do not erase and replace.
8. **Save** as **Ex-19C-I**
9. Plot using Page Setup **Class Model A**

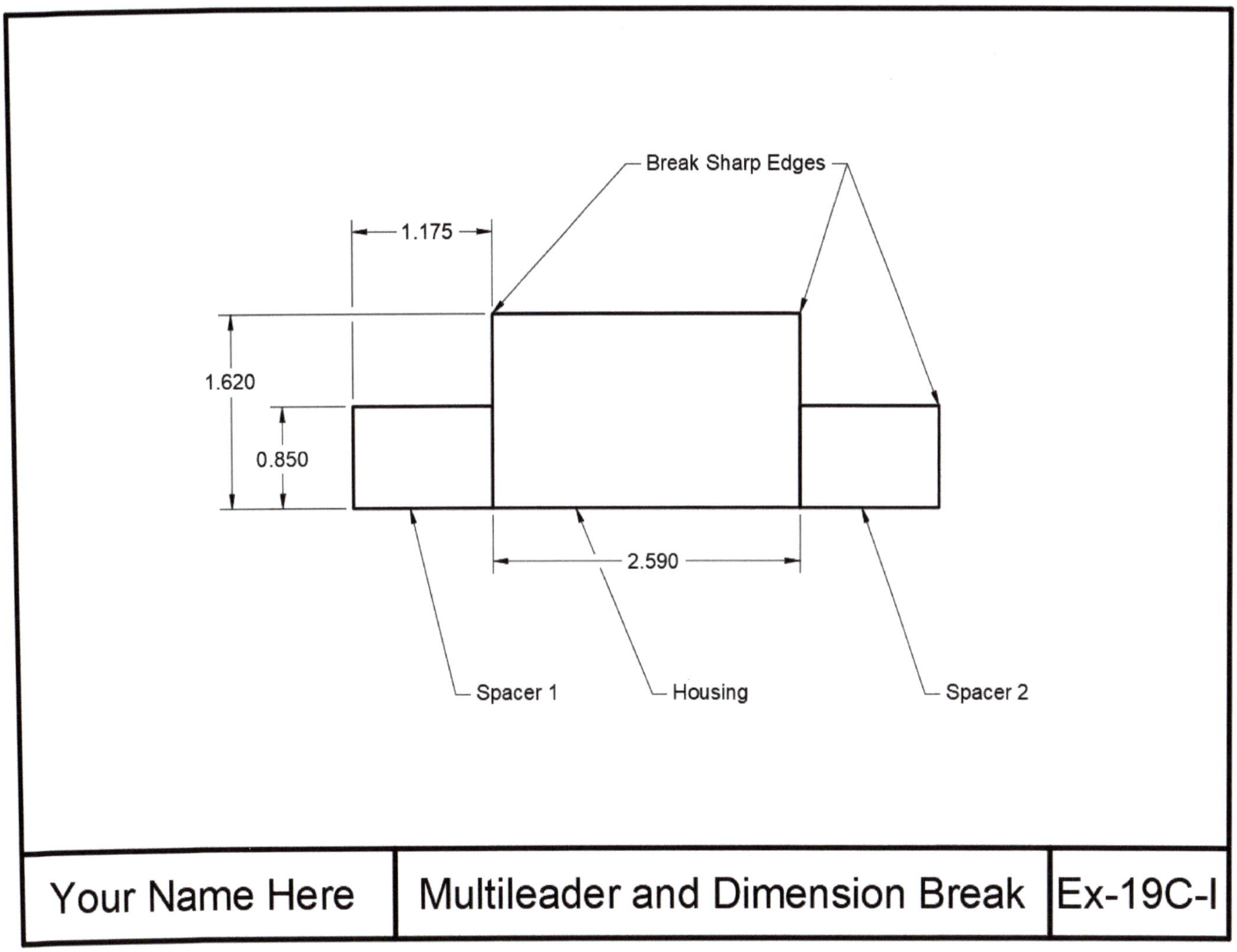

Exercise 19C-Metric

Instructions:

1. Start a **New** file using the **Border A-Metric.dwt**
2. Create the multileader style shown on page 19-4 if you haven't already.
3. Draw the object shown below.
4. Use Layer Object Line.
5. Dimension as shown.
6. Use Dimension Style CLASS STYLE and Class ML Style and Layer Dimension.
7. Edit the **Title** and **Ex-XX** by double clicking on the text. Do not erase and replace.
8. **Save** as **Ex-19C-M**
9. Plot using Page Setup **Class Model A**

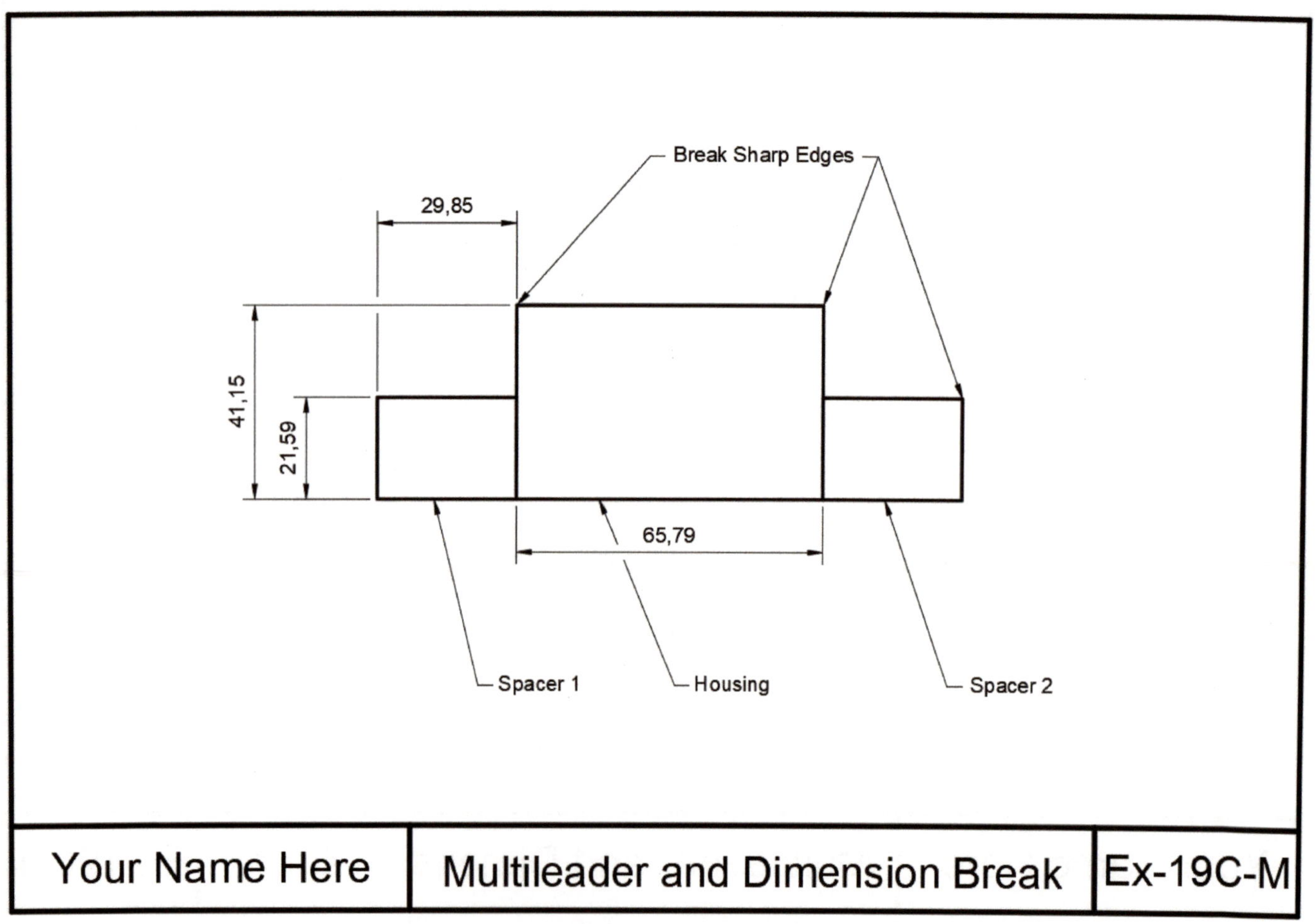

LESSON 20

LEARNING OBJECTIVES

After completing this lesson, you will be able to:

- Use the Dim command to create multiple dimensions all within a single command

Dim Command

Dim command allows you to automatically create dimensions based on the object you select. After selecting the Dim command, you simply hover the cursor over an object to see a preview of the dimension before you create it.

Dim command allows you to create multiple dimensions all within the one command by selecting an object or by using the appropriate prompts on the Command Line. In previous lessons, you had to select a command for each dimension you wanted to create.

1. Select the **Dim** command using one of the following:

   Ribbon = Annotate Tab / Dimensions Panel /
   or
   Keyboard = *DIM* *<Enter>*

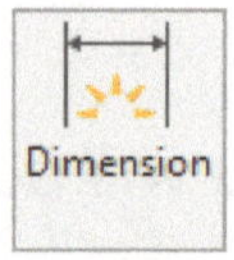

The following examples show you the previews when creating dimensions for Linear objects. **Note:** Metric users will see the dimension text aligned with the dimension line.

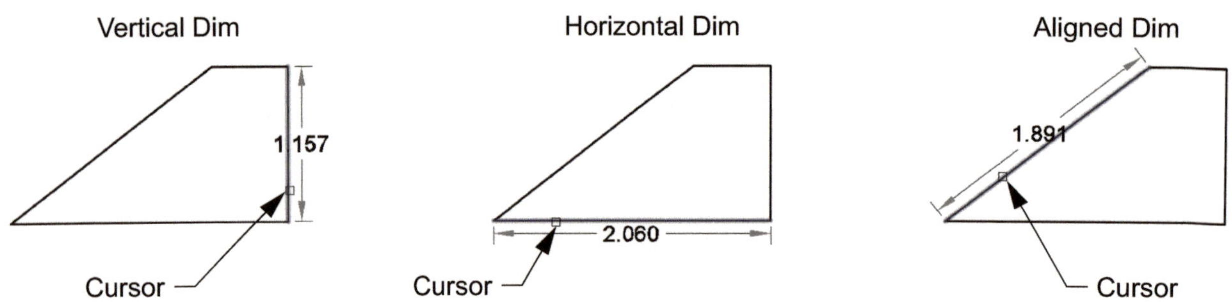

2. If you are satisfied with the preview of the dimension required, you simply left click to select the object and then drag the cursor to place the dimension in the position you require. Then left click again to accept the position.

Note: It is best to turn off **Ortho Mode *<F8>***, especially for **Aligned** dimensions.

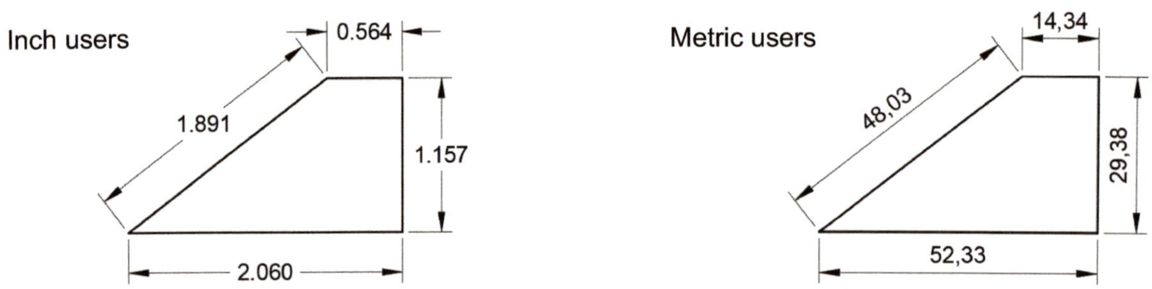

Here are some examples showing the previews when selecting arcs and circles:

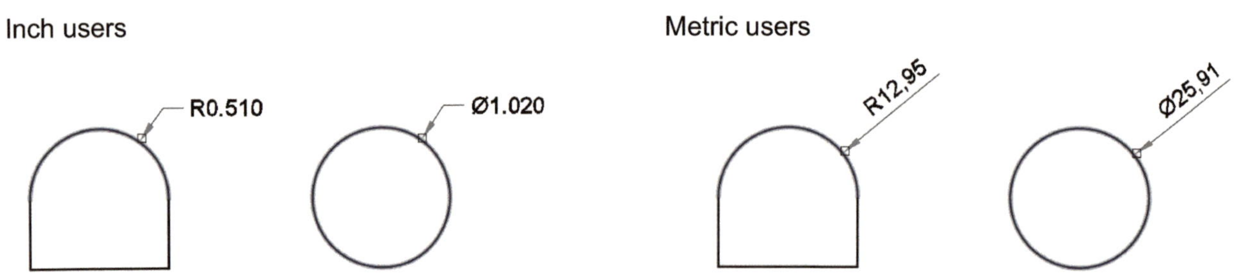

Baseline Dimensioning

The following is an example of how to use the **Baseline** option of the **Dim** command.

1. Select the **Dim** command.
2. Left click on the line (**P1**), drag the cursor down to the position you require for the first dimension, and then left click again to place the dimension (**P2**).

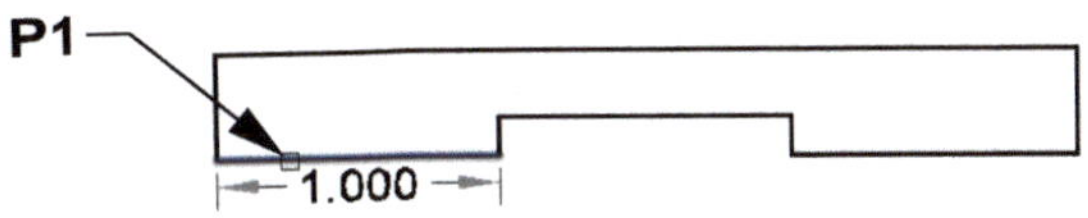

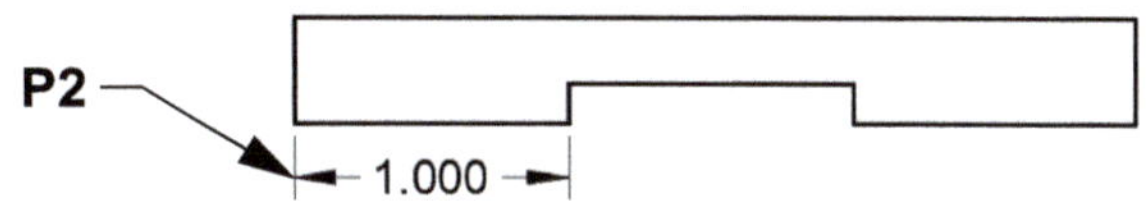

The following will appear on the Command Line:

```
DIM Select objects or specify first extension line origin or [Angular Baseline Continue Ordinate aliGn Distribute Layer Undo]:
```

3. Select **Baseline** on the Command Line or type ***B <Enter>***.

The following will appear on the Command Line:

```
DIM Specify first extension line origin as baseline or [Offset]:
```

4. Left click on the first extension line (**P3**).

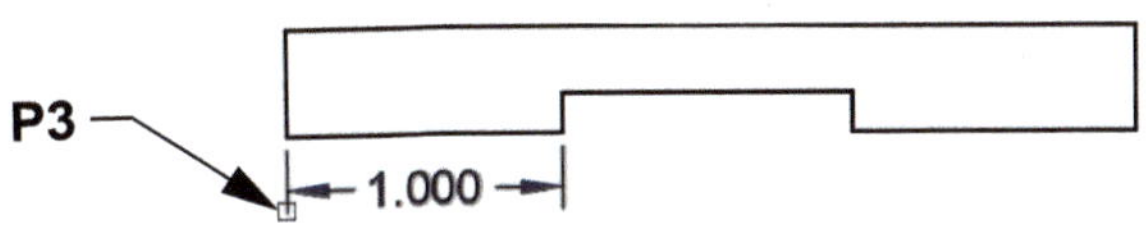

The following will appear on the Command Line:

```
DIM Specify second extension line origin or [Select Offset Undo] <Select>:
```

5. Left click on the endpoints at **P4** and **P5** (make sure **Object Snap** is turned on).

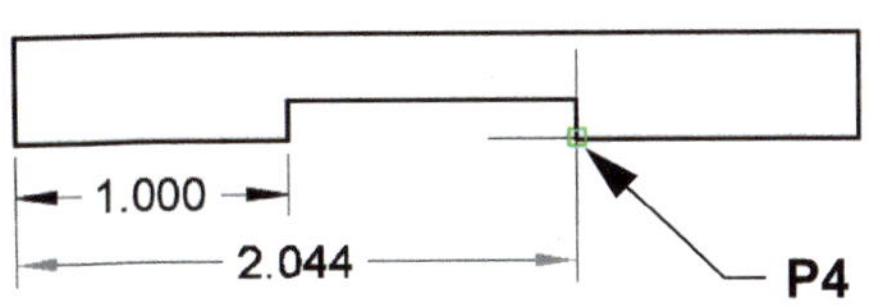

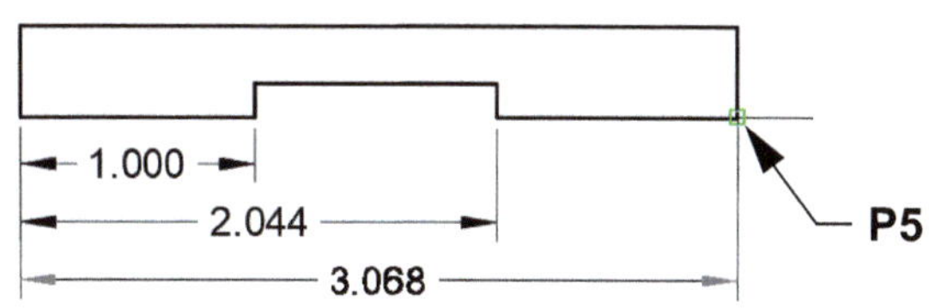

6. Press ***<Enter> <Enter> <Enter>*** to end the **Dim** command.

Continuous Dimensioning

The following is an example of how to use the **Continue** option of the **Dim** command.

1. Select the **Dim** command.
2. Place the first dimension **P1** and **P2** the same as shown for the **Baseline** option of the **Dim** command.

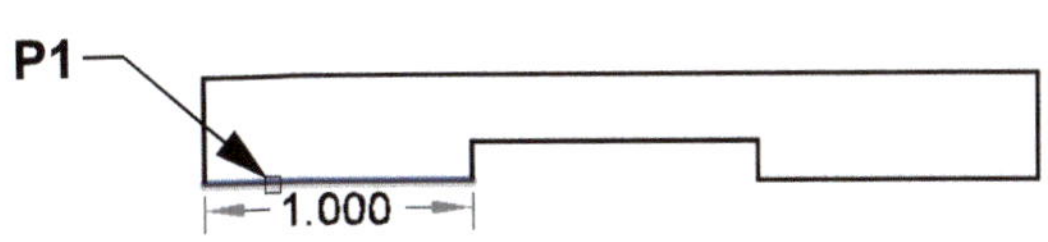

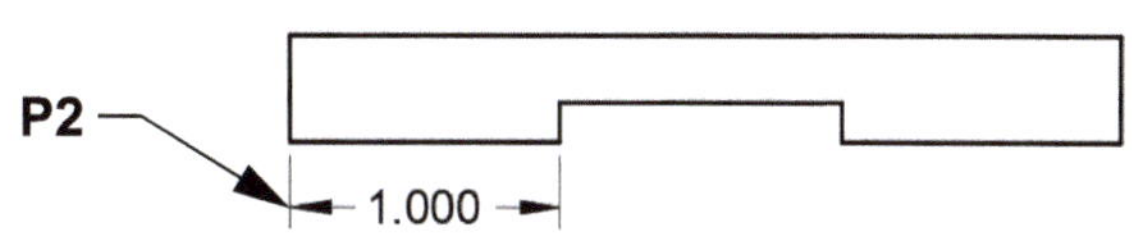

Continued on the next page...

The following will appear on the Command Line:

```
DIM Select objects or specify first extension line origin or [Angular Baseline
Continue Ordinate aliGn Distribute Layer Undo]:
```

3. Select **Continue** on the Command Line or type ***C <Enter>***.

The following will appear on the Command Line:

```
DIM Specify first extension line origin to continue:
```

4. Left click on the first extension line (**P3**).

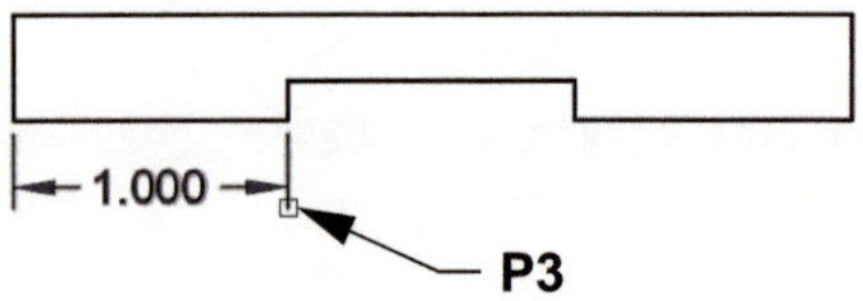

The following will appear on the Command Line:

```
DIM Specify second extension line origin or [Select Undo] <Select>:
```

5. Left click on the endpoints at **P4** and **P5** (make sure **Object Snap** is turned on).

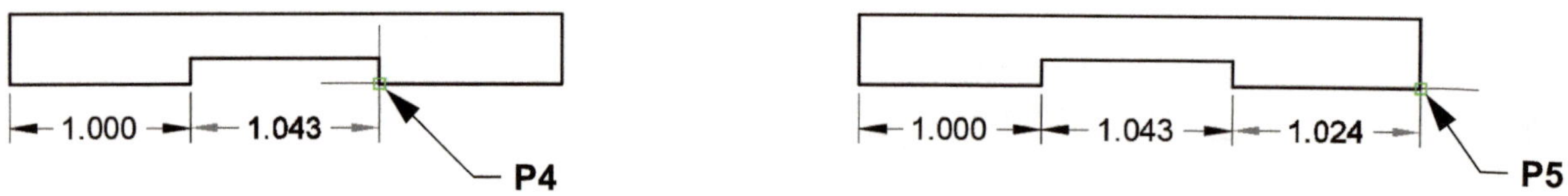

6. Press ***<Enter> <Enter> <Enter>*** to end the **Dim** command.

Angular Dimensioning

The following is an example of how to dimension **angles** with the **Dim** command. (Metric users will see the dimension text aligned with the dimension line.)

1. Select the **Dim** command.
2. Left click on the first line (**P1**) and then on the second line (**P2**).
3. Drag the cursor to the position you require for the dimension and then left click (**P3**).

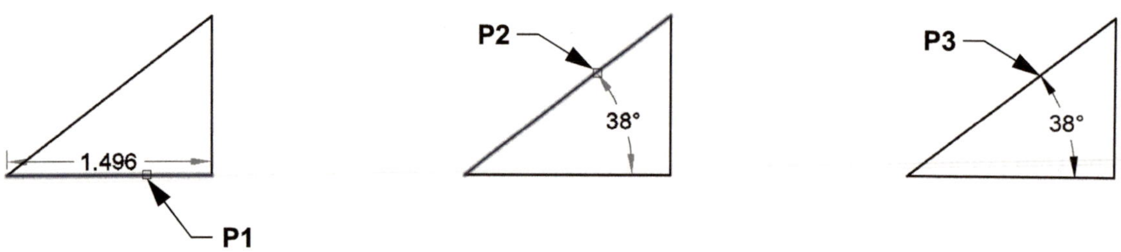

4. Press ***<Enter>*** to end the **Dim** command.

Continued on the next page...

You can place the Angular dimension in various positions as shown in the examples below.

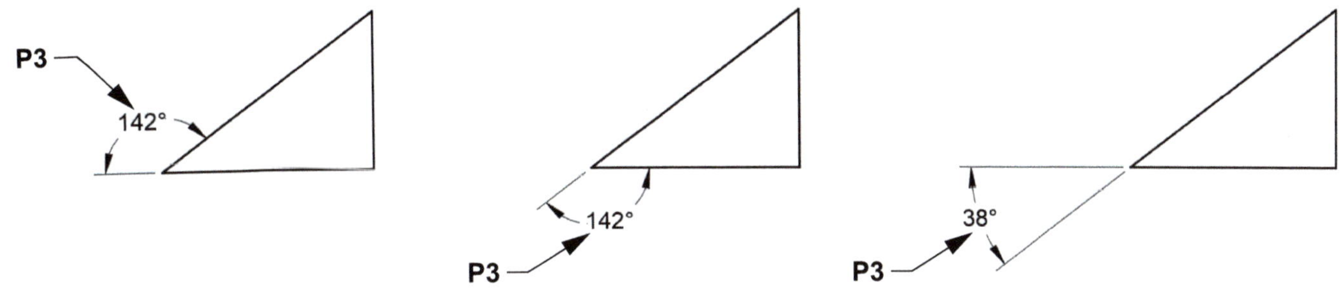

Linear Dimensioning

You also may wish to dimension the horizontal and vertical distances of an angled line as shown in the inch and metric examples below.

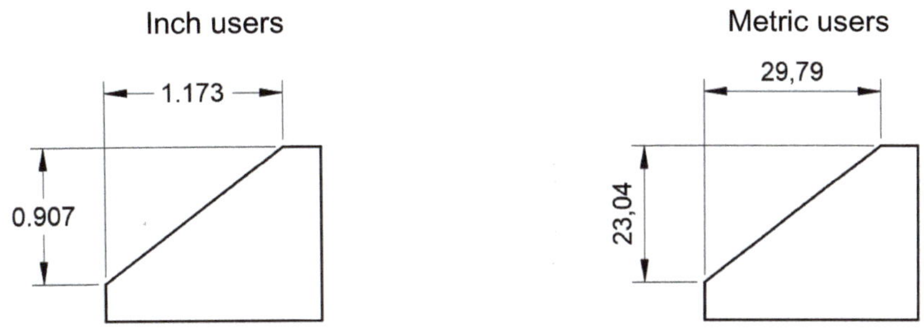

1. Select the **Dim** command.
2. Left click on the angled line (**P1**) and then drag the cursor down and to the left (**P2**). Left click again to place the dimension.

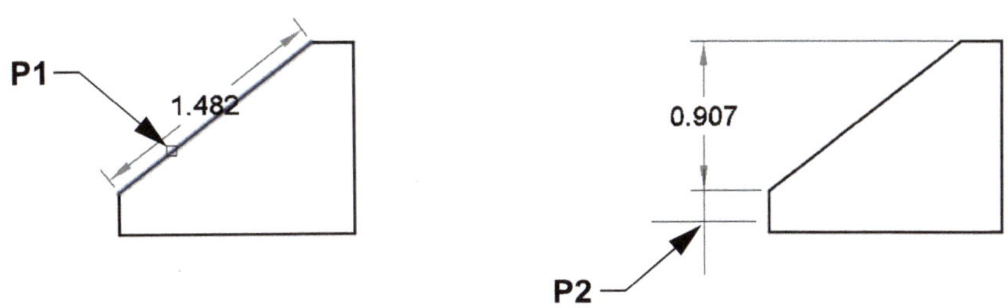

3. Left click on the angled line (**P3**) and then drag the cursor up and to the right (**P4**). Left click again to place the dimension.

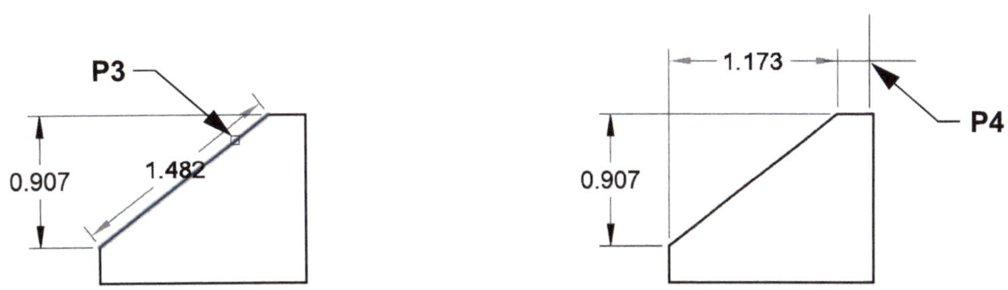

4. Press ***<Enter>*** to end the **Dim** command.

Note: Quick Dimension is still an option, but the **Dim** command is a better choice as it is more effective and has eliminated some of the challenges users had with Quick Dimension.

Exercise 20A-Inch

Instructions:

1. Start a **New** file using the **Border A-Inch.dwt**
2. Draw the object shown below.
3. Dimension as shown using a mixture of **Baseline** and **Linear** dimensions with the **Dim** command.
4. Use Dimension Style CLASS STYLE and Layer Dimension.
5. Edit the **Title** and **Ex-XX** by double clicking on the text. Do not erase and replace.
6. **Save** as **Ex-20A-I**
7. Plot using Page Setup **Class Model A**

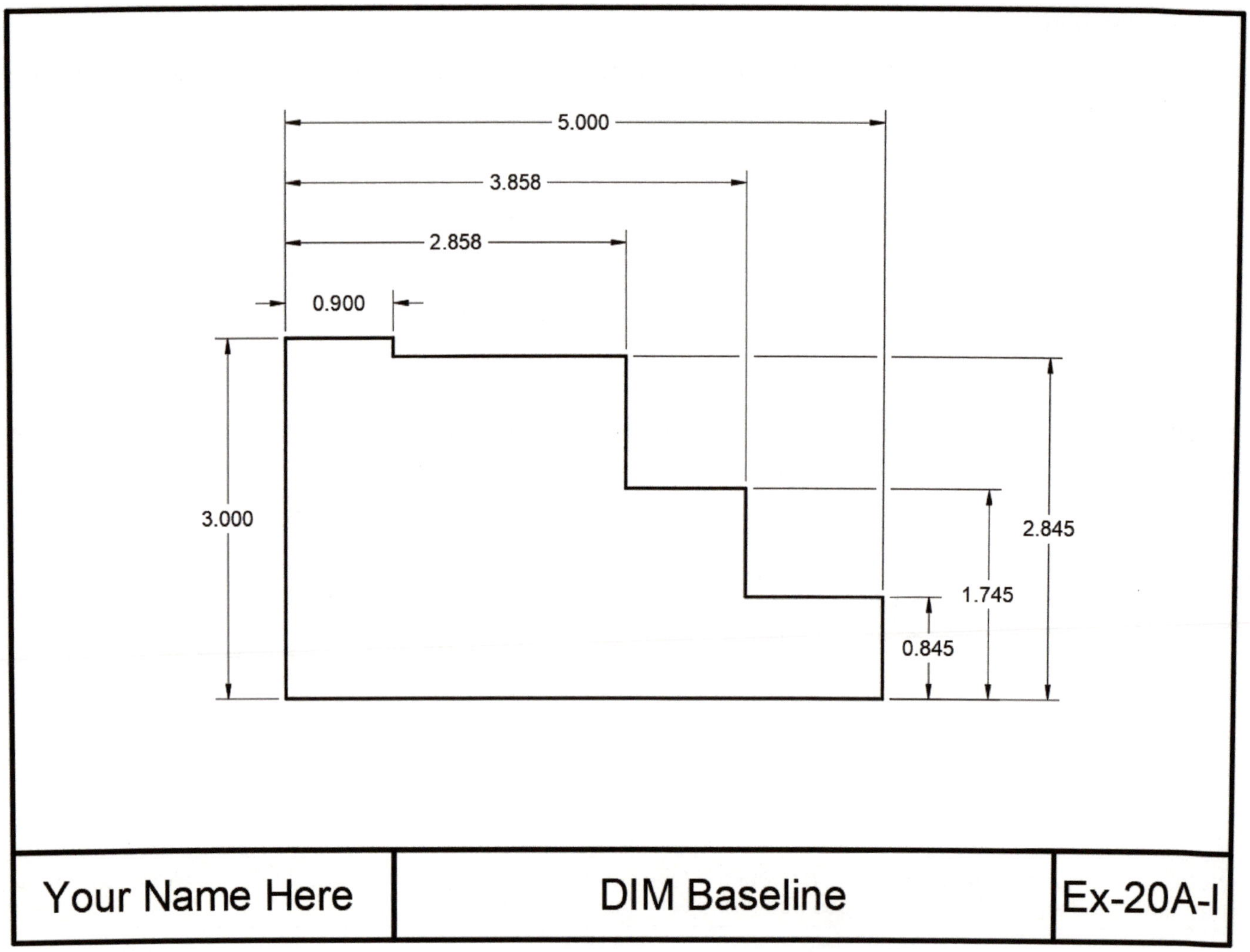

Exercise 20A-Metric

Instructions:

1. Start a **New** file using the **Border A-Metric.dwt**
2. Draw the object shown below.
3. Dimension as shown using a mixture of **Baseline** and **Linear** dimensions with the **Dim** command.
4. Use Dimension Style CLASS STYLE and Layer Dimension.
5. Edit the **Title** and **Ex-XX** by double clicking on the text. Do not erase and replace.
6. **Save** as **Ex-20A-M**
7. Plot using Page Setup **Class Model A**

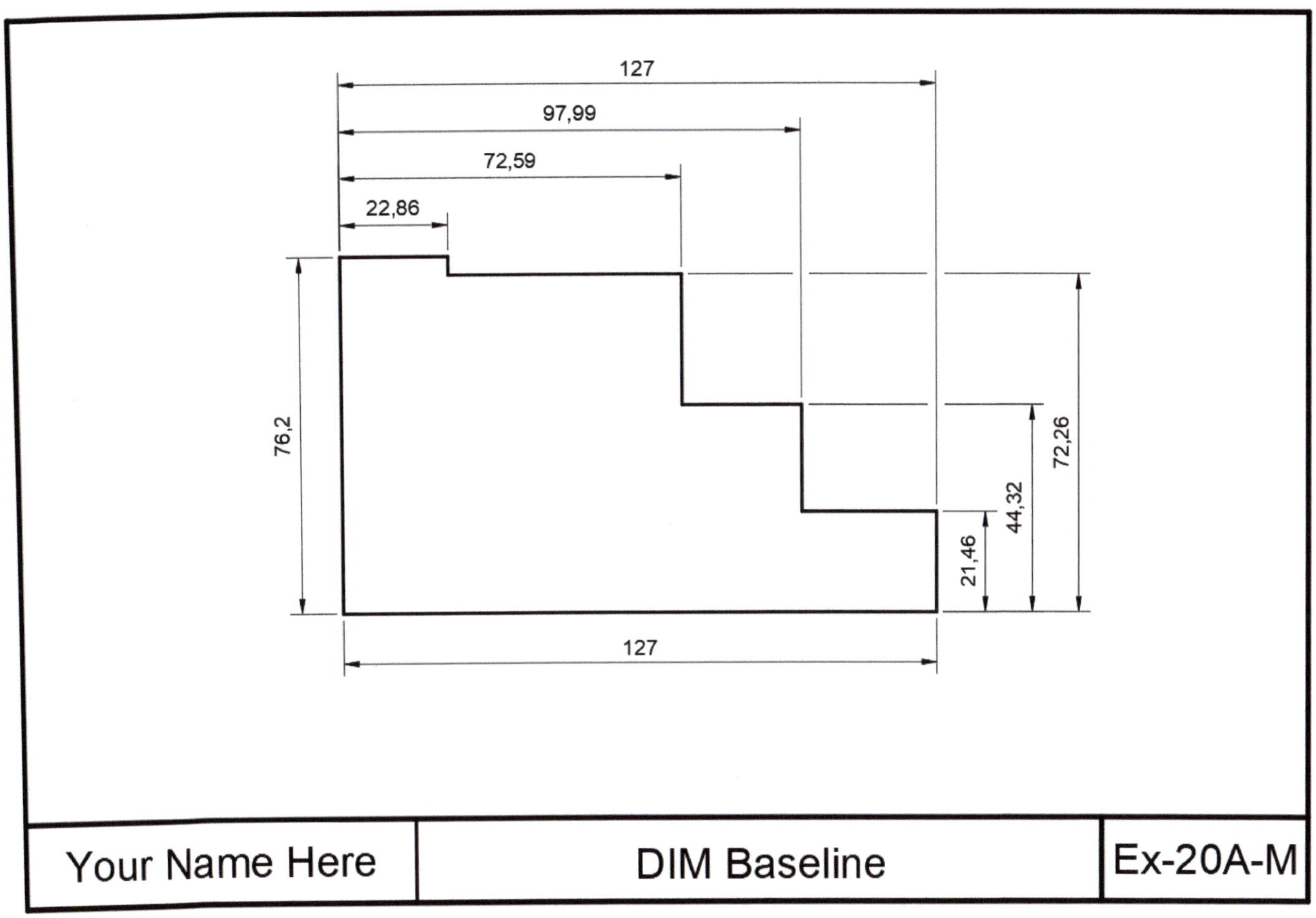

Exercise 20B-Inch

Instructions:

1. Open **Ex-20A-I**
2. Erase the original dimensions.
3. Dimension as shown using a mixture of **Continuous** and **Linear** dimensions with the **Dim** command.
4. Use Dimension Style CLASS STYLE and Layer Dimension.
5. Edit the **Title** and **Ex-XX** by double clicking on the text. Do not erase and replace.
6. **Save** as **Ex-20B-I**
7. Plot using Page Setup **Class Model A**

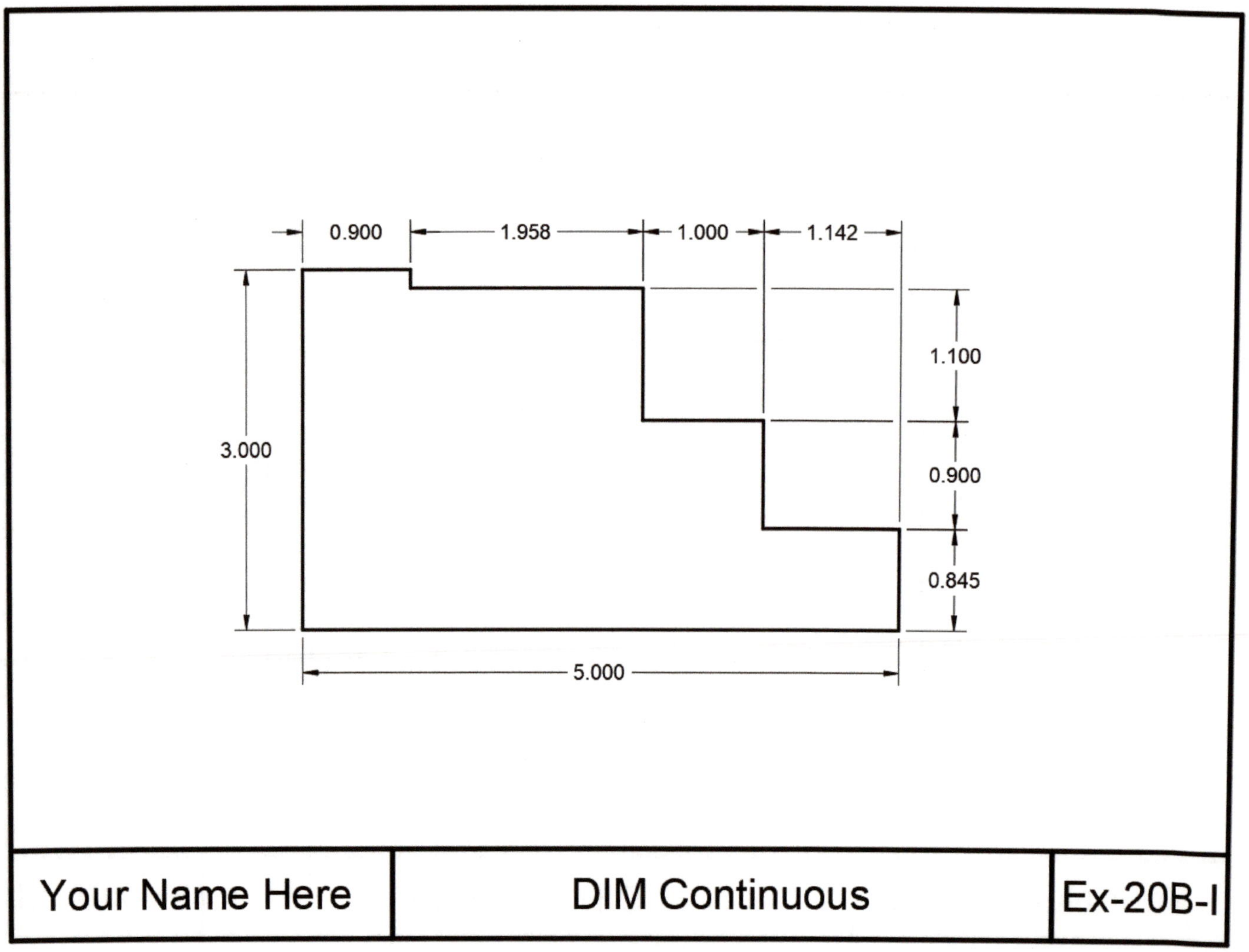

Exercise 20B-Metric

Instructions:

1. Open **Ex-20A-M**
2. Erase the original dimensions.
3. Dimension as shown using a mixture of **Continuous** and **Linear** dimensions with the **Dim** command.
4. Use Dimension Style CLASS STYLE and Layer Dimension.
5. Edit the **Title** and **Ex-XX** by double clicking on the text. Do not erase and replace.
6. **Save** as **Ex-20B-M**
7. Plot using Page Setup **Class Model A**

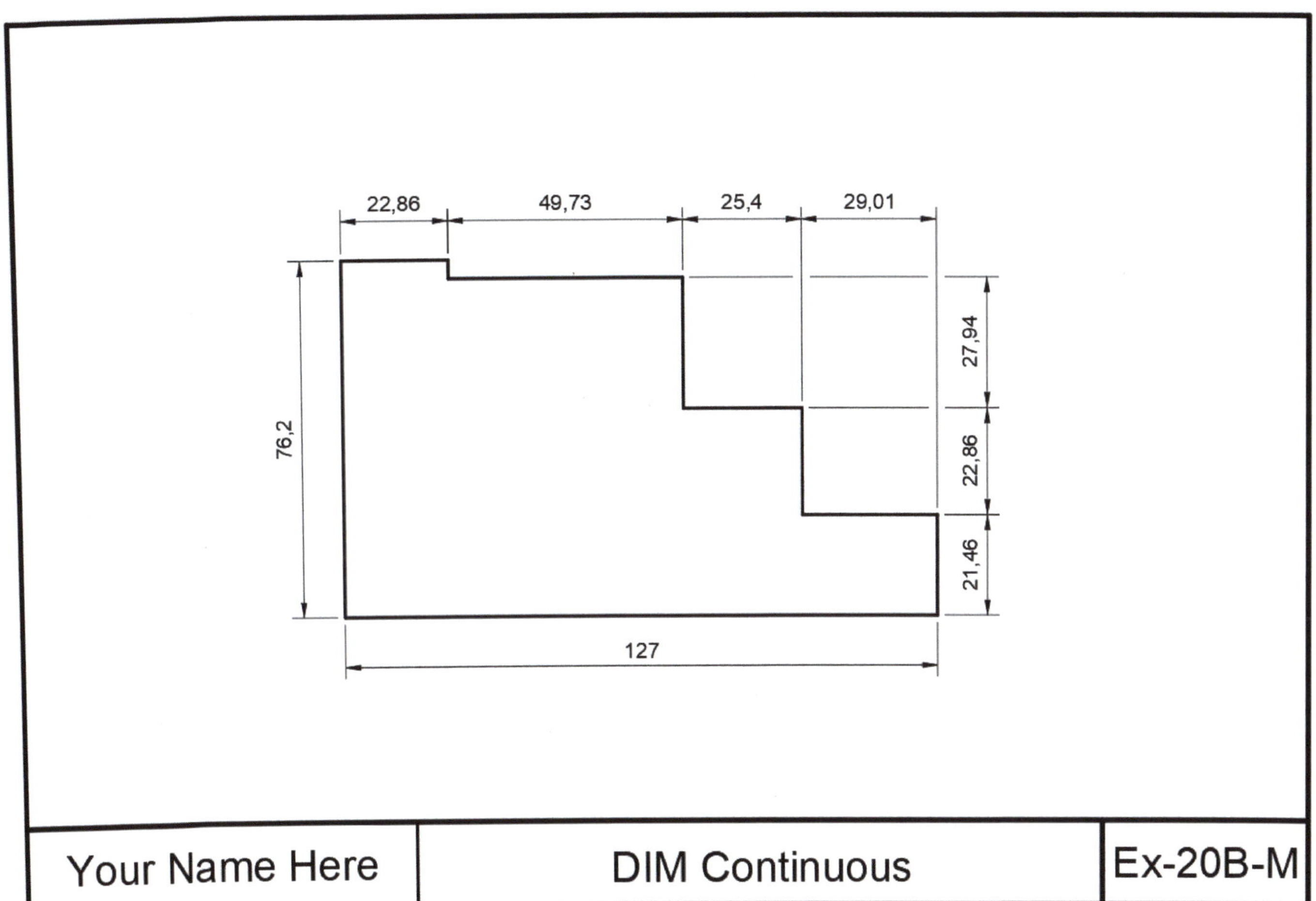

Exercise 20C-Inch

Instructions:

1. Start a **New** file using the **Border A-Inch.dwt**
2. Draw the object shown below.
3. Dimension as shown using a mixture of **Linear**, **Diameter**, **Radius**, and **Angular** dimensions with the **Dim** command.
4. Use Dimension Style CLASS STYLE and Layer Dimension.
5. Edit the **Title** and **Ex-XX** by double clicking on the text. Do not erase and replace.
6. **Save** as **Ex-20C-I**
7. Plot using Page Setup **Class Model A**

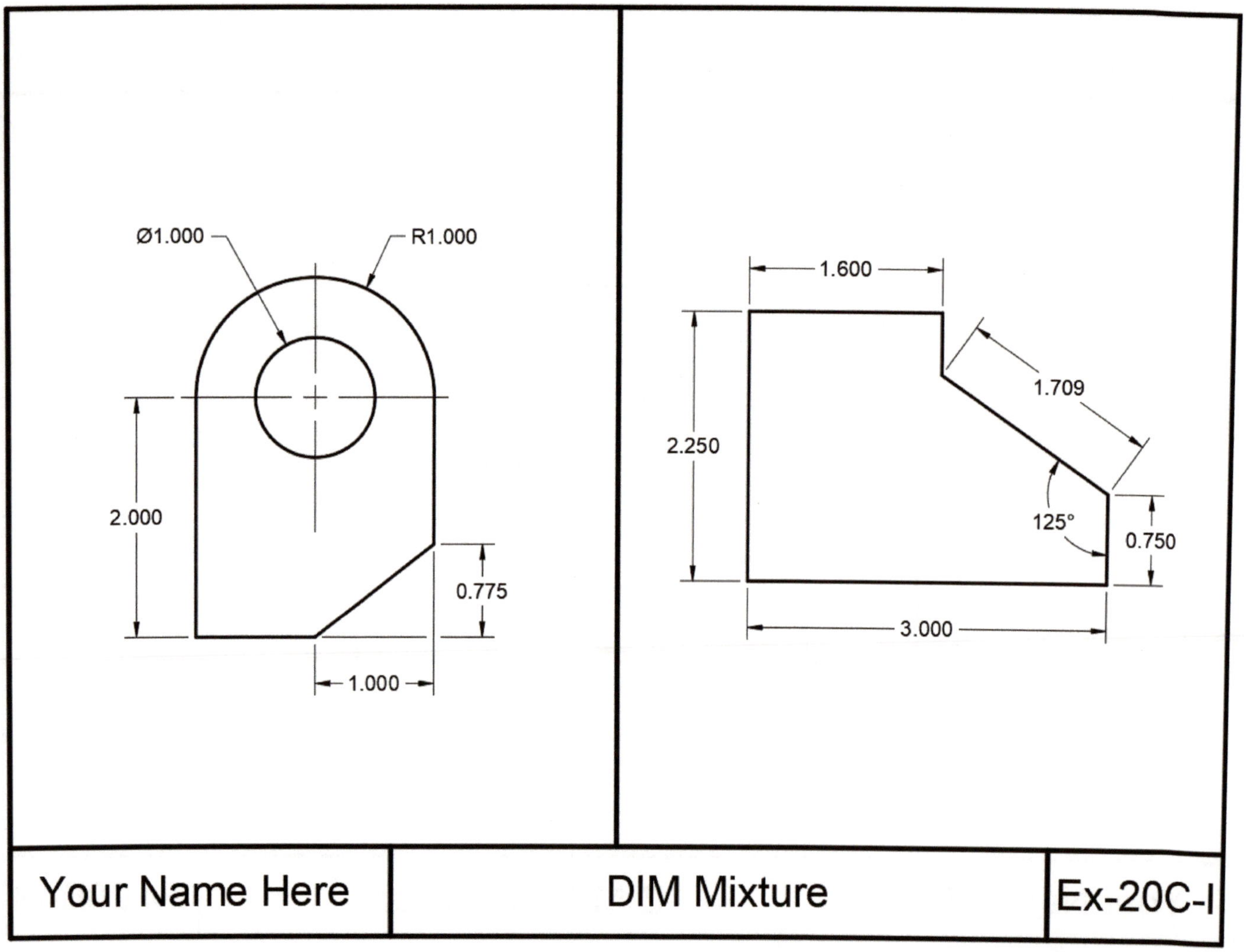

Exercise 20C-Metric

Instructions:

1. Start a **New** file using the **Border A-Metric.dwt**
2. Draw the object shown below.
3. Dimension as shown using a mixture of **Linear**, **Diameter**, **Radius**, and **Angular** dimensions with the **Dim** command.
4. Use Dimension Style CLASS STYLE and Layer Dimension.
5. Edit the **Title** and **Ex-XX** by double clicking on the text. Do not erase and replace.
6. **Save** as **Ex-20C-M**
7. Plot using Page Setup **Class Model A**

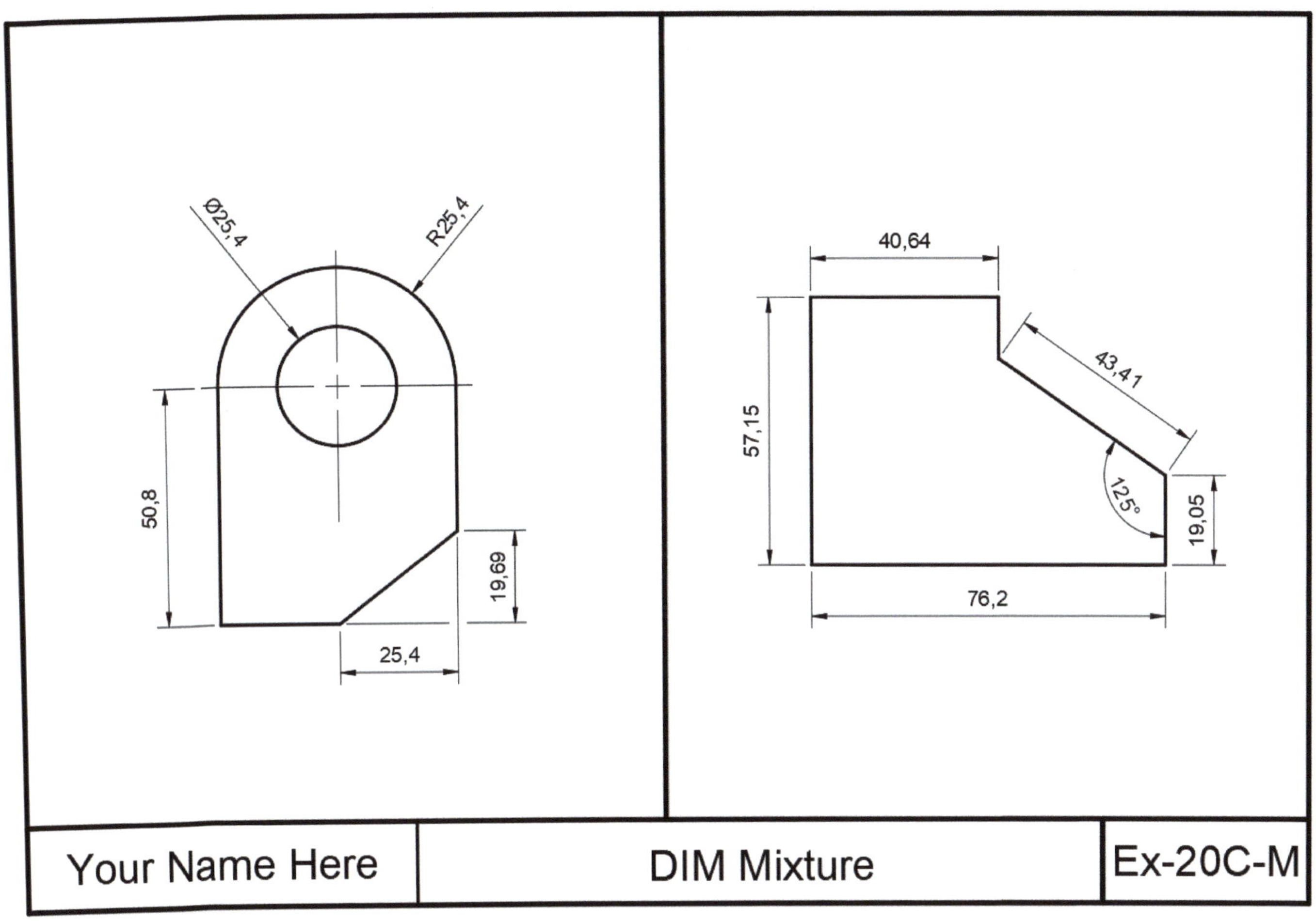

Notes:

LESSON 21

LEARNING OBJECTIVES

After completing this lesson, you will be able to:

- Match an object's properties with another object
- Create a Revision Cloud
- Select a Revision Cloud style
- "Wipe Out" an area of an object

Match Properties

Match Properties is used to "paint" the properties of one object to another. This is a simple and useful command. You first select the object that has the desired properties (the source object) and then select the object you want to "paint" the properties to (destination object).

Only one "source object" can be selected. But its properties can be painted to any number of "destination objects".

1. Select the **Match Properties** command using one of the following:

   Ribbon = Home Tab / Properties Panel / 

   or
   Keyboard = *MA <Enter>*

   The following will appear on the Command Line:

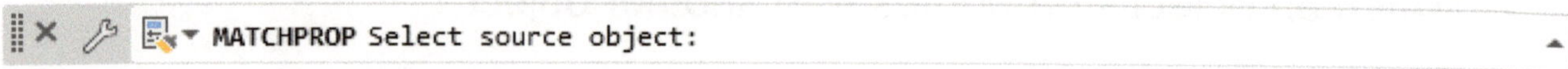

2. Select the object with the desired properties to match..

   The following will appear on the Command Line:

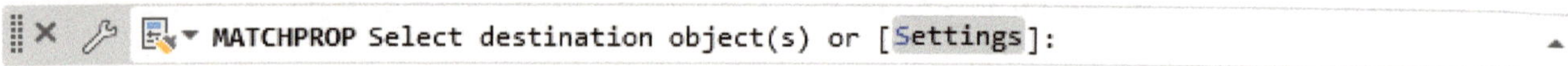

3. Select the object(s) you want to receive the matching properties and then press ***<Enter>***.

Note: If you do not want to match all of the properties, after you have selected the source object, right click and select **"Settings"** from the Shortcut Menu, before selecting the destination object.

In the **Property Settings** dialog box, uncheck all the properties you do not want to match and then select the **OK** button. Then select the destination object(s).

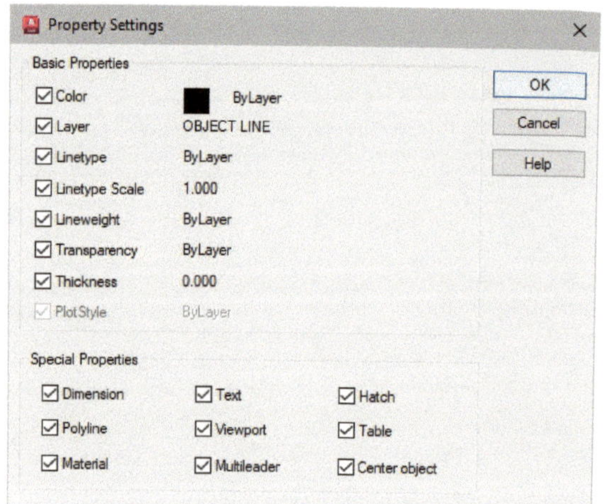

Match Layer

If you draw an object on the wrong layer, you can easily change it to the desired layer using the Layer Match command. You first select the object that needs to be changed and then select an object that is on the correct layer (object on destination layer).

1. Select the **Layer Match** command using one of the following:

   Ribbon = Home Tab / Layers Panel / 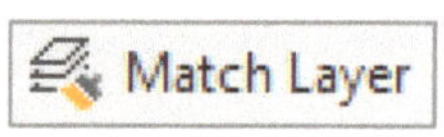

   or
   Keyboard = *Laymch <Enter>*

Continued on the next page...

The following will appear on the Command Line:

```
LAYMCH Select objects:
```

2. Select the object(s) that need to be changed and then press ***<Enter>***.

   The following will appear on the Command Line:

```
LAYMCH Select object on destination layer or [Name]:
```

3. Select the object that is on the layer that you want to change.

You may also easily change the layer of an object to another layer as follows:

1. Select the object that you want to change.
2. Select the Layer drop-down menu and select the layer that you want the object to be placed on.

Example:

If you had some text on the OBJECT LINE layer by mistake:

1. Select the text that is mistakenly on the OBJECT LINE layer.
2. Select the TEXT layer from the drop-down menu.

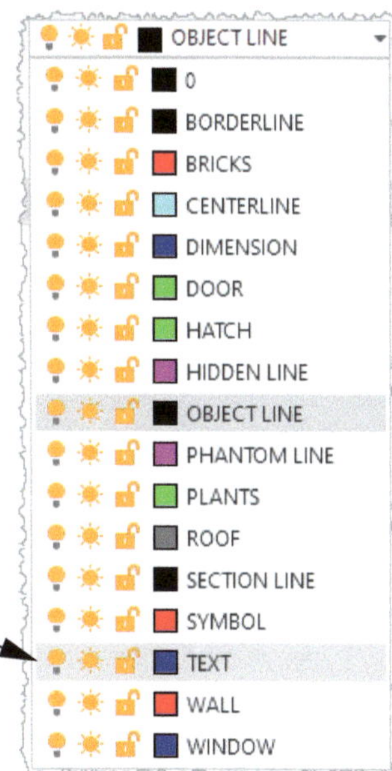

Creating a Revision Cloud

When you make a revision to a drawing, it is sometimes helpful to highlight the revision for someone viewing the drawing. A common method to highlight the area is to draw a "Revision Cloud" around the revised area. This can be accomplished easily with the **Revision Cloud** command.

The Revision Cloud command creates a series of sequential arcs to form a cloud shaped object. The size of the arcs are measured by an approximate chord length which is the distance between the endpoints of each arc segment. You can set the arc chord lengths when you first start the Revision Cloud command, or you can change the arc chord lengths in the Properties Palette.

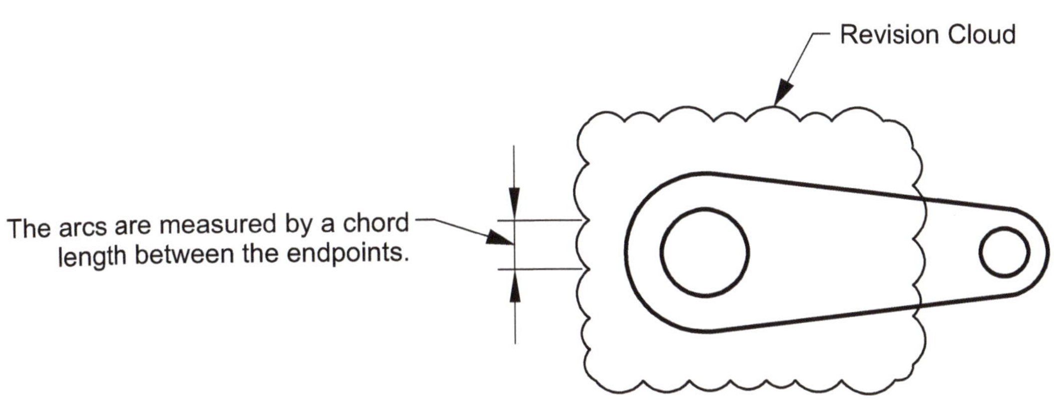

Continued on the next page...

Freehand Revision Cloud

To draw a **Freehand Revision Cloud**, specify the start point with a left click and then drag the cursor to form the outline. AutoCAD automatically draws the arcs. When the cursor gets very close to the start point, AutoCAD snaps the last arc to the first arc and closes the shape.

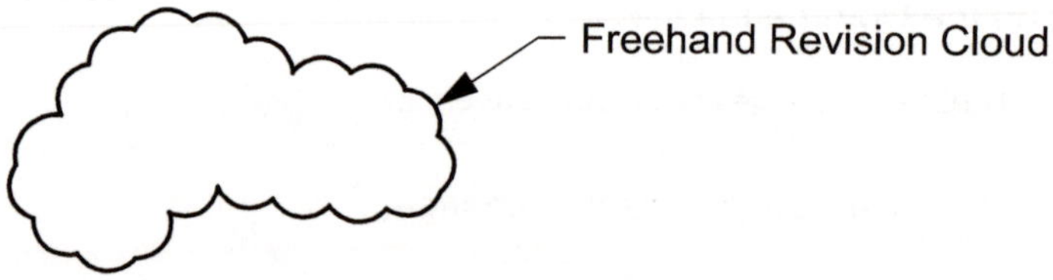

1. Select the **Freehand Revision Cloud** command using one of the following:

Ribbon = Annotate Tab / Markup Panel / ▾
or
Keyboard = *revcloud <Enter>*

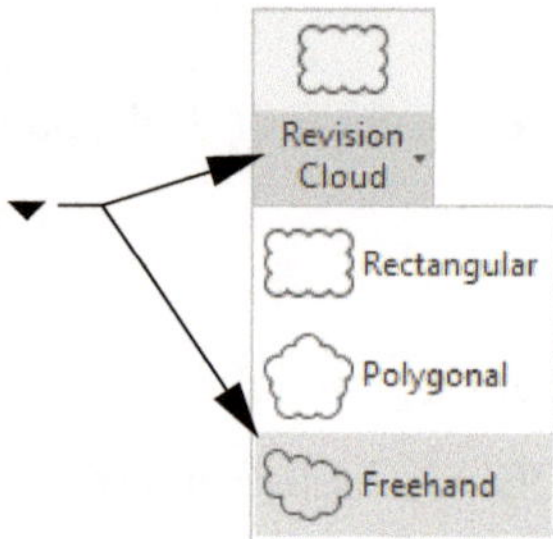

The following will appear on the Command Line:

```
REVCLOUD Specify first point or [Arc length Object Rectangular Polygonal Freehand Style Modify] <Object>:
```

2. Type ***A*** and then press ***<Enter>***.

The following will appear on the Command Line:

```
REVCLOUD Specify approximate length of arc <0.3000>:
```

3. Enter the arc chord length you require and then press ***<Enter>***.

4. Place the cursor at the start location and then left click.

The following will appear on the Command Line:

```
REVCLOUD Guide crosshairs along cloud path...
```

5. Move the cursor to create the cloud outline.

6. When the cursor approaches the start point, the cloud closes automatically and the command ends.

Rectangular Revision Cloud

To draw a **Rectangular Revision Cloud**, specify the start point with a left click and then drag the cursor to form the rectangular outline. AutoCAD automatically draws the arcs. When you are satisfied with the size of the rectangular shape, left click again to finish the command.

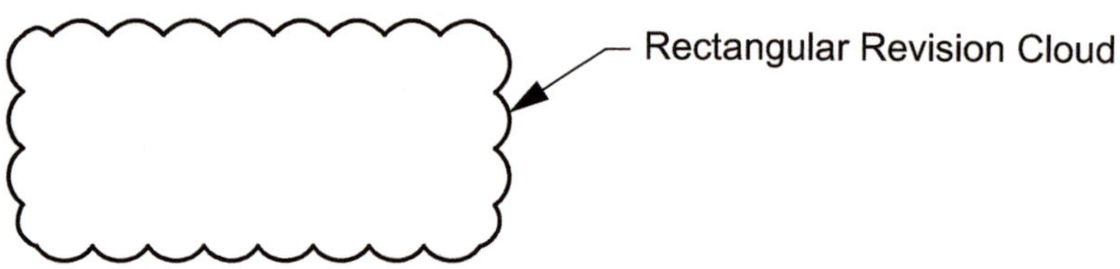

Continued on the next page...

1. Select the **Rectangular Revision Cloud** command using one of the following:

Ribbon = Annotate Tab / Markup Panel / ▼
or
Keyboard = *revcloud <Enter>*

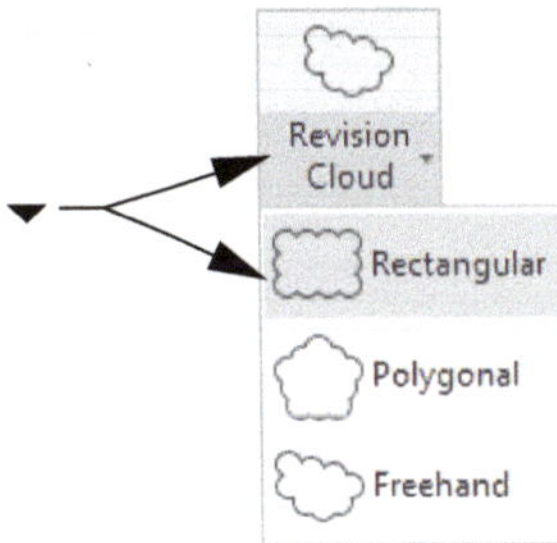

The following will appear on the Command Line:

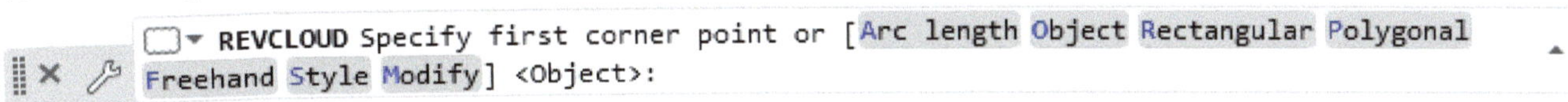

2. Type ***A*** and then press ***<Enter>***.

The following will appear on the Command Line:

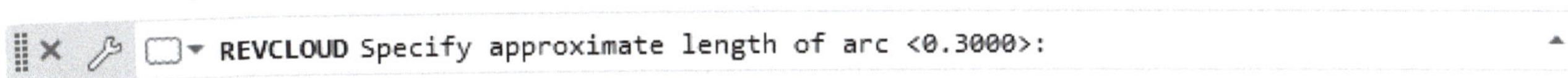

3. Enter the arc chord length you require and then press ***<Enter>***.
4. Place the cursor at the start location and then left click.

The following will appear on the Command Line:

5. Move the cursor to the opposite corner to create the cloud outline.
6. When you are satisfied with the size of the shape, left click to finish the command.

Polygonal Revision Cloud

To draw a **Polygonal Revision Cloud**, specify the start point with a left click and then drag the cursor to form the polygonal outline, using a left click to specify each point of the shape. AutoCAD automatically draws the arcs. When you are satisfied with the size of the polygonal shape, press ***<Enter>*** to finish the command.

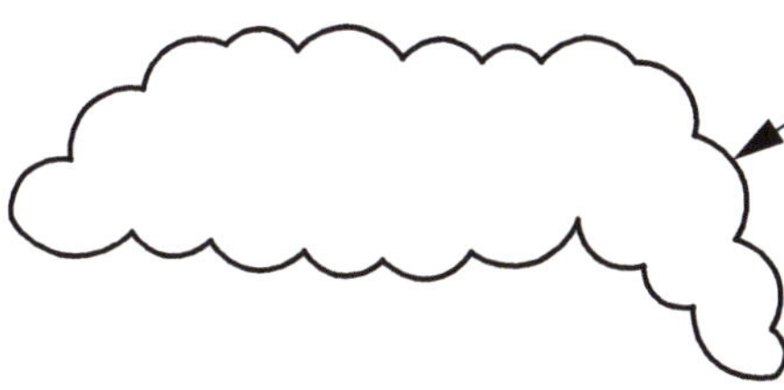

Polygonal Revision Cloud

1. Select the **Polygonal Revision Cloud** command using one of the following:

Ribbon = Annotate Tab / Markup Panel / ▼
or
Keyboard = *revcloud <Enter>*

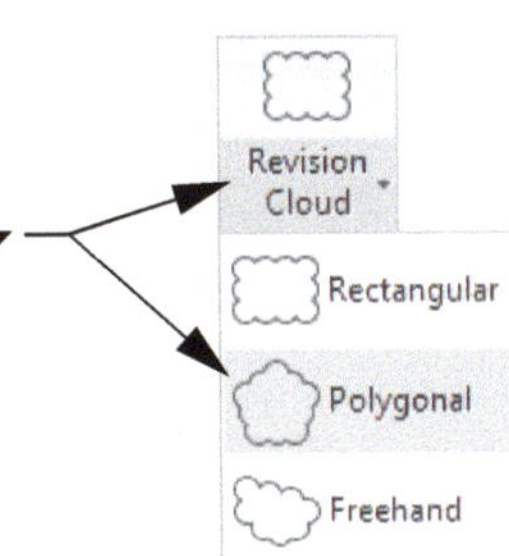

Continued on the next page...

The following will appear on the Command Line:

```
REVCLOUD Specify start point or [Arc length Object Rectangular Polygonal Freehand Style Modify] <Object>:
```

2. Type ***A*** and then press ***<Enter>***.

The following will appear on the Command Line:

```
REVCLOUD Specify approximate length of arc <0.3000>:
```

3. Enter the arc chord length you require and then press ***<Enter>***.
4. Place the cursor at the start location and then left click.

The following will appear on the Command Line:

```
REVCLOUD Specify next point:
```

5. Move the cursor to the next point and left click.

The following will appear on the Command Line:

```
REVCLOUD Specify next point or [Undo]:
```

6. Continue to left click for the next points until you are satisfied with the shape.
7. When you are satisfied with the size of the shape, press ***<Enter>*** to finish the command.

Converting an Object to a Revision Cloud

You can convert a closed object, such as a circle, ellipse, rectangle, or closed polyline to a Revision Cloud. The original object is deleted when it is converted.

If you want the original object to remain, in addition to the new Revision Cloud, set the variable **"delobj"** to **"0"**. The default setting is **"3"**.

1. Draw a closed object such as a circle.
2. Select the **Revision Cloud** command using one of the following:

Ribbon = Annotate Tab / Markup Panel /
or
Keyboard = *revcloud <Enter>*

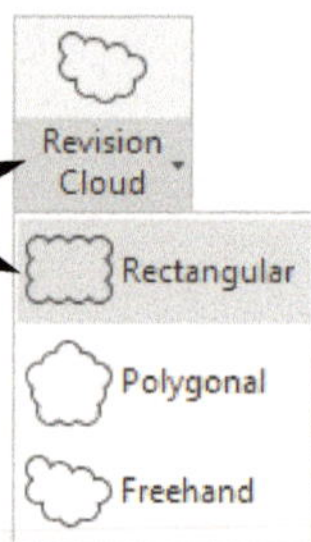

The following will appear on the Command Line:

```
REVCLOUD Specify first corner point or [Arc length Object Rectangular Polygonal Freehand Style Modify] <Object>:
```

3. Type ***A*** and then press ***<Enter>***.

Continued on the next page...

The following will appear on the Command Line:

```
REVCLOUD Specify approximate length of arc <0.3000>:
```

4. Enter the arc chord length you require and then press ***<Enter>***.

   The following will appear on the Command Line:

```
REVCLOUD Specify first corner point or [Arc length Object Rectangular Polygonal
Freehand Style Modify] <Object>:
```

5. Type the letter ***O*** and then press ***<Enter>***, or select **Object** on the Command Line.

   The following will appear on the Command Line:

```
REVCLOUD Select object:
```

6. Select the object to convert.

   The following will appear on the Command Line:

```
REVCLOUD Reverse direction [Yes No] <No>:
```

7. Select **Yes** or **No**.

Reverse direction?

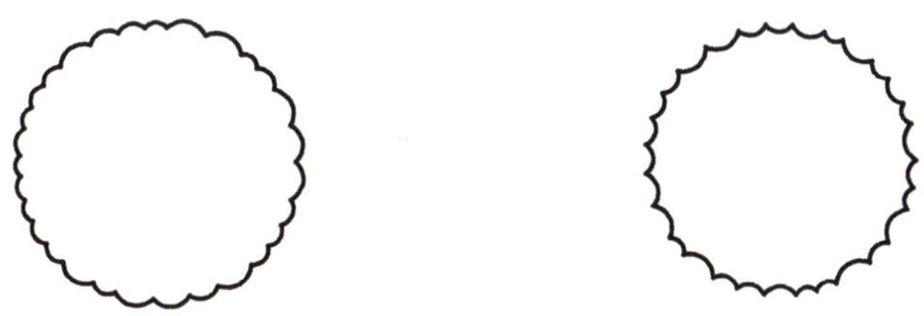

No Yes

Note: The **Match Properties** command will not match the arc length from the source cloud to the destination cloud.

Selecting the Revision Cloud Style

You may select one of two styles for the Revision Cloud: **Normal** or **Calligraphy**.

Normal will draw the cloud with one line width.

Calligraphy will draw the cloud with variable line widths to appear as though you used a chiseled calligraphy pen.

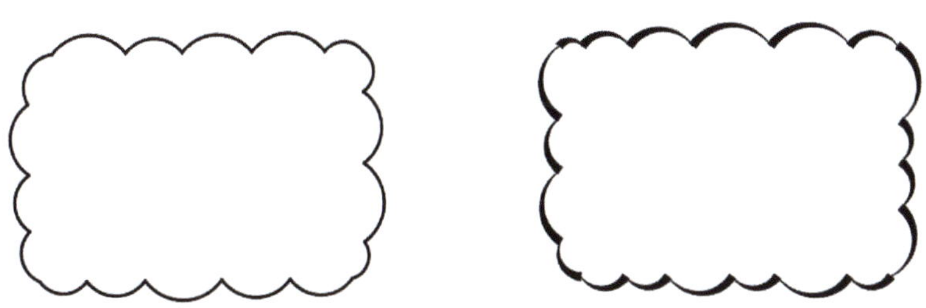

Normal Calligraphy

Continued on the next page...

1. Select the **Revision Cloud** command using one of the following:

Ribbon = Annotate Tab / Markup Panel /
or
Keyboard = *revcloud <Enter>*

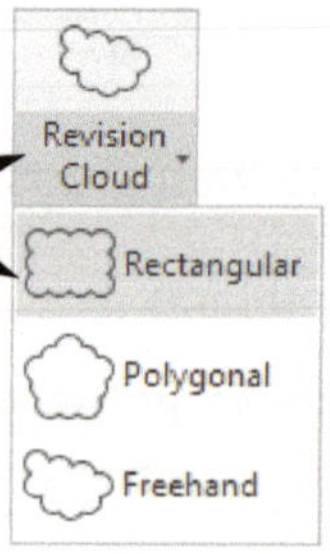

The following will appear on the Command Line:

```
REVCLOUD Specify first corner point or [Arc length Object Rectangular Polygonal Freehand Style Modify] <Object>:
```

2. Type ***S*** and then press ***<Enter>***, or select **Style** on the Command Line.

The following will appear on the Command Line:

```
REVCLOUD Select arc style [Normal Calligraphy] <Normal>:
```

3. Type ***N*** (for Normal) or ***C*** (for Calligraphy) and then press ***<Enter>***, or select from the Command Line.

4. Continue to create the Revision Cloud as shown in previous examples.

CAD TIP! You can use the **Calligraphy** style on any of the Revision Cloud commands.

Wipeout

The **Wipeout** command creates a blank area that covers existing objects. The area has a background that matches the background of the drawing area. This area is bounded by the wipeout frame, which you can turn on or off.

1. Select the **Wipeout** command using one of the following:

Ribbon = Annotate Tab / Markup Panel /
or
Keyboard = *wipeout <Enter>*

The following will appear on the Command Line:

```
WIPEOUT _wipeout Specify first point or [Frames Polyline] <Polyline>:
```

2. Specify the first point of the shape (**P1**).

The following will appear on the Command Line:

```
WIPEOUT Specify next point:
```

3. Specify the next point (**P2**).

Continued on the next page...

The following will appear on the Command Line:

```
WIPEOUT Specify next point or [Undo]:
```

4. Specify the next point (**P3**).

   The following will appear on the Command Line:

```
WIPEOUT Specify next point or [Close Undo]:
```

5. Specify the next point or press ***<Enter>*** to end the command.

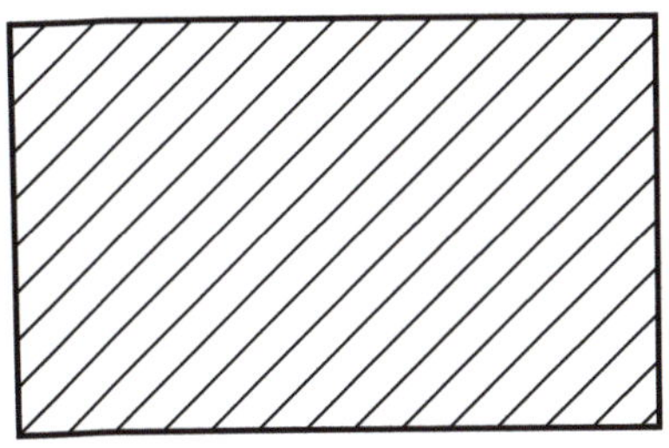

Before Wipeout

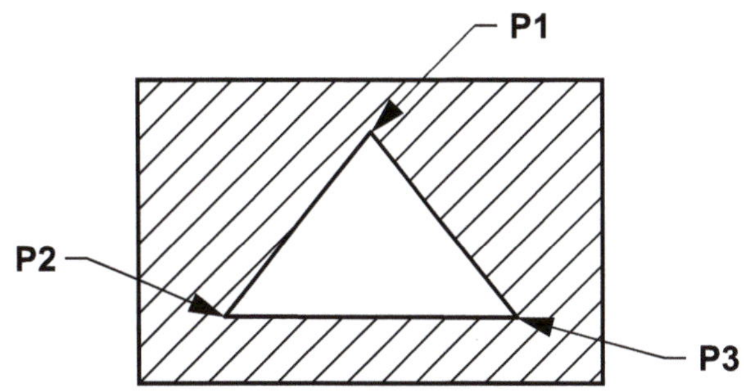

After Wipeout

How to Turn Wipeout Frames On or Off

1. Select the **Wipeout** command.

   The following will appear on the Command Line:

```
WIPEOUT _wipeout Specify first point or [Frames Polyline] <Polyline>:
```

2. Type ***F*** and then press ***<Enter>***, or select **Frames** on the Command Line.

   The following will appear on the Command Line:

```
WIPEOUT Enter mode [ON OFF Display but not plot] <ON>:
```

3. Type ***on*** or ***off*** and then press ***<Enter>***, or select from the Command Line.

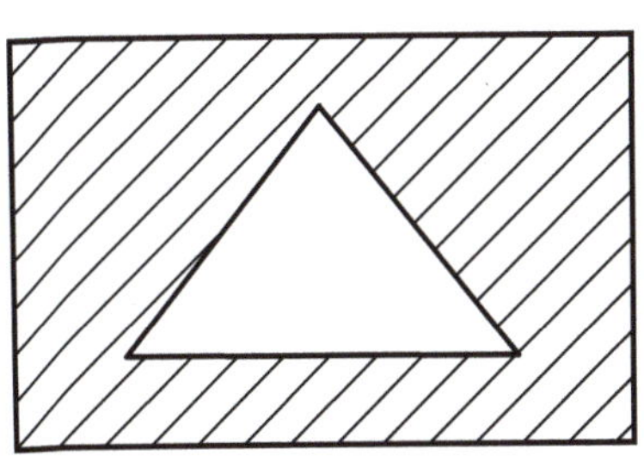

Frames on

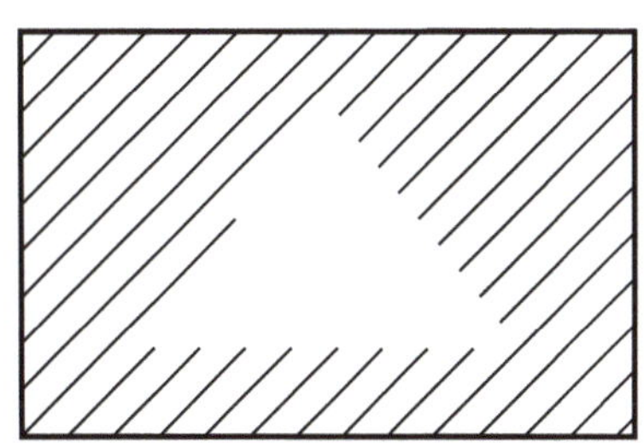

Frames off

Note: If you want to move the objects and the wipeout area, you must select both and move them at the same time. Do not move them separately.

Exercise 21A

Instructions:

1. Open **Ex-5F.dwg**
2. Using **Match Properties**, change the properties of the polygon to the same properties as the donut.
3. Use **Wipeout**, block out the bottom donut approximately as shown.
4. **Save** as **Ex-21A**

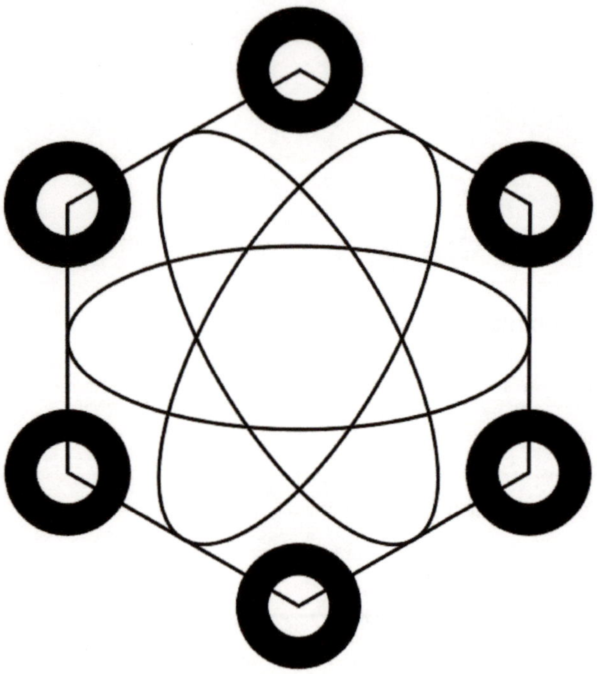

Before

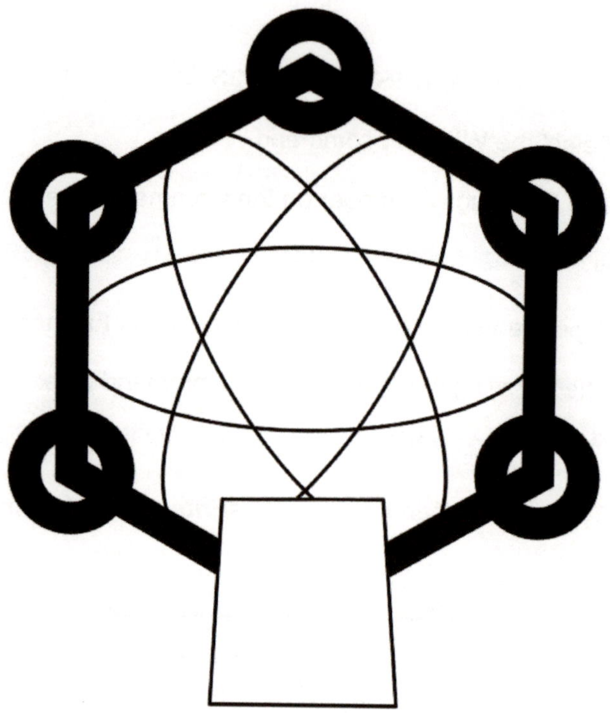

After

Exercise 21B

Instructions:

1. Start a **New** file using the **Border A-Inch.dwt** or the **Border A-Metric.dwt**
2. Draw the Revision Clouds shown below:
   #1: Set the arc chord length to **0.500" [12.7 mm]**.
   #2: Set the arc chord length to **0.500" [12.7 mm]**.
   #3: Set the arc chord length to **0.750" [19.05 mm]** and change the Style to **Calligraphy**.
3. Draw (2) **2.500" [63.5 mm]** diameter circles and **Convert** to Revision Clouds.
   Set the arc chord length to **0.300" [7.62 mm]**
   #4: Reverse = No. #5: Reverse = Yes.
4. Use Layer Object Line.
5. Edit the **Title** and **Ex-XX** by double clicking on the text. Do not erase and replace.
6. **Save** as **Ex-21B**
7. Plot using Page Setup **Class Model A**

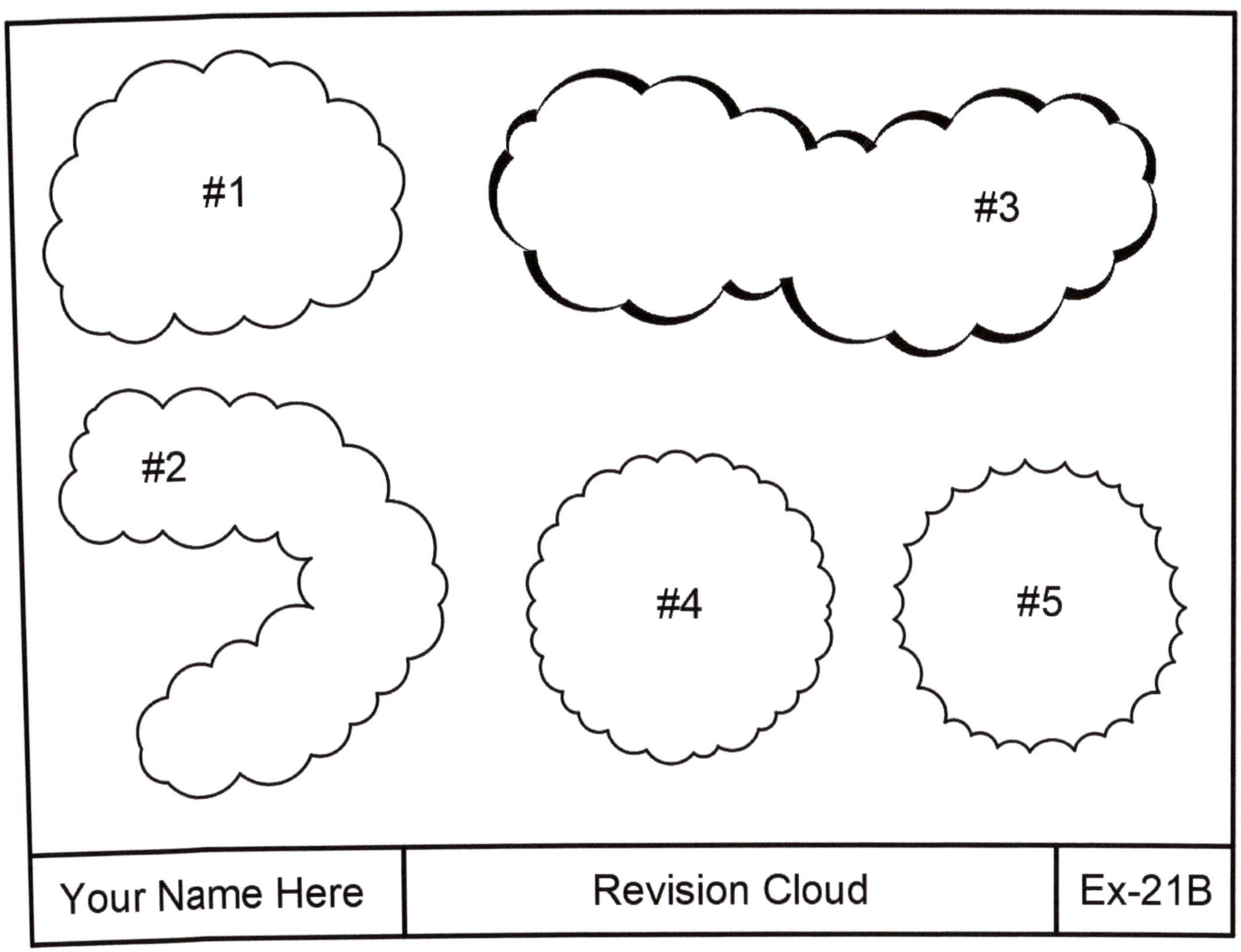

Exercise 21C-Inch

Instructions:

1. Start a **New** file using the **Border A-Inch.dwt**
2. The drawing below is designed to give you more practice drawing objects and dimensioning. Use any method that you have learned in the previous lessons.
3. Edit the **Title** and **Ex-XX** by double clicking on the text. Do not erase and replace.
4. **Save** as **Ex-21C-I**
5. Plot using Page Setup **Class Model A**

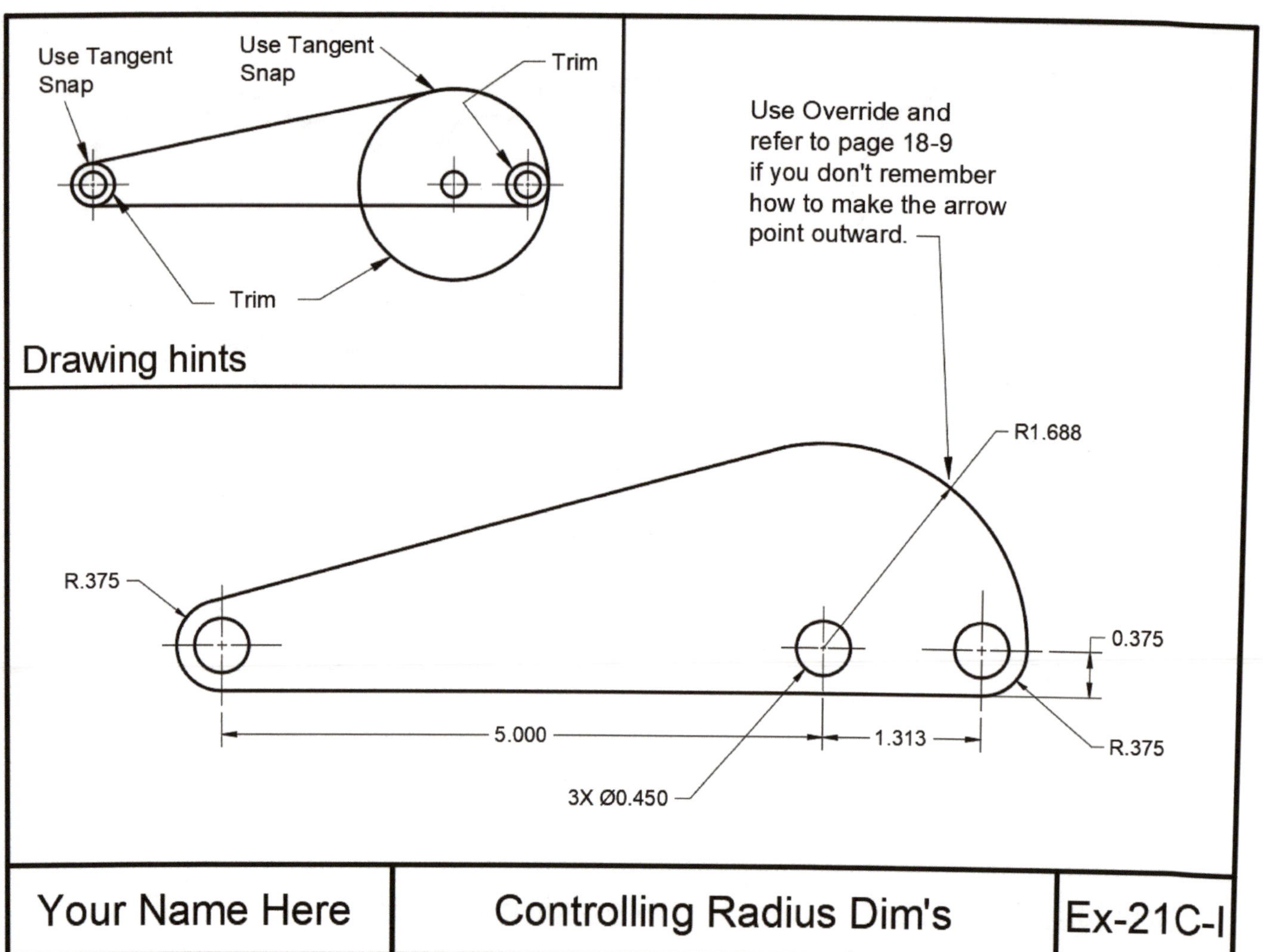

Exercise 21C-Metric

Instructions:

1. Start a **New** file using the **Border A-Metric.dwt**
2. The drawing below is designed to give you more practice drawing objects and dimensioning. Use any method that you have learned in the previous lessons.
3. Edit the **Title** and **Ex-XX** by double clicking on the text. Do not erase and replace.
4. **Save** as **Ex-21C-M**
5. Plot using Page Setup **Class Model A**

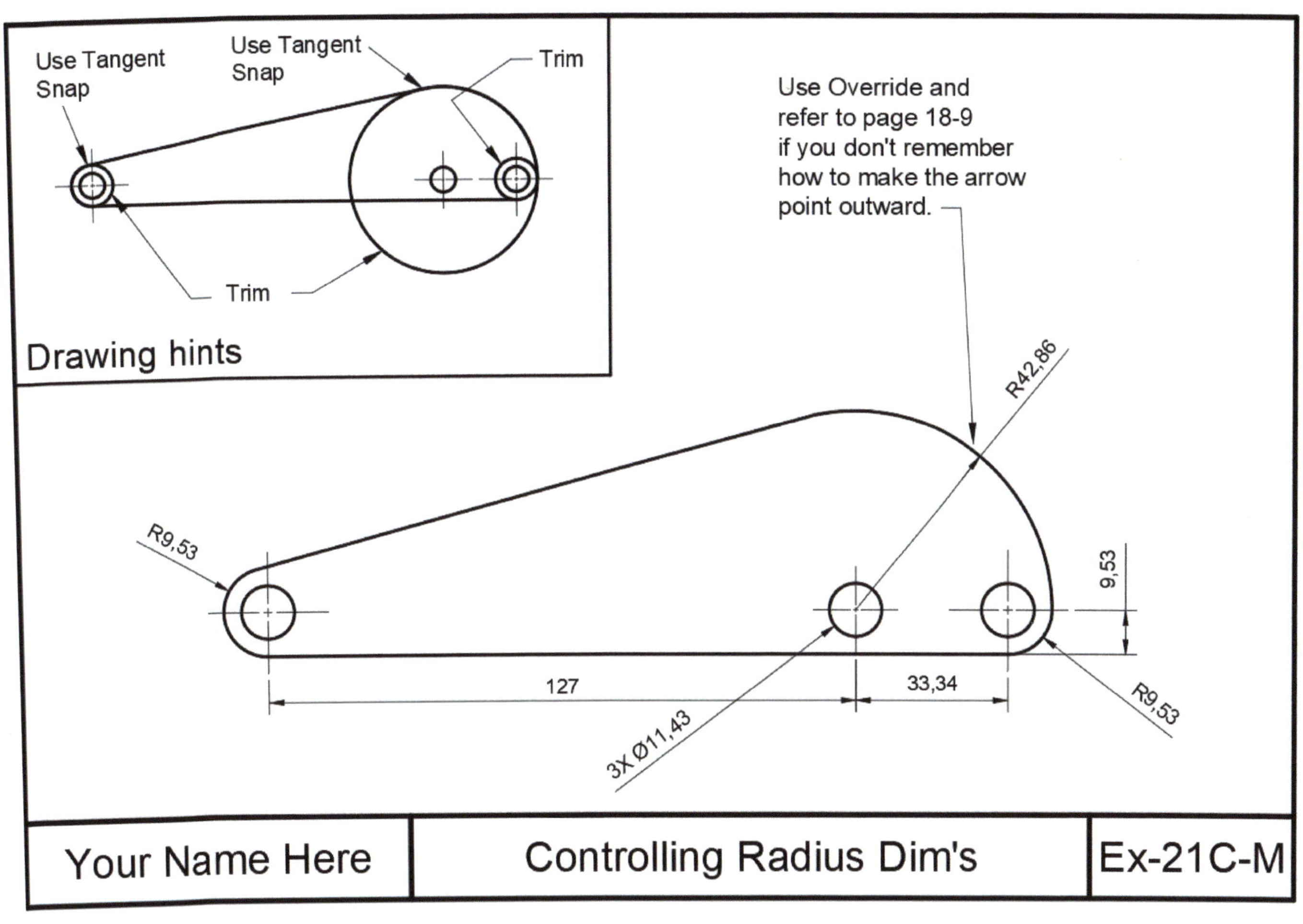

Exercise 21D-Inch

Instructions:

1. Start a **New** file using the **Border A-Inch.dwt**
2. The drawing below is designed to give you more practice drawing objects and dimensioning. Use any method that you have learned in the previous lessons.
3. Edit the **Title** and **Ex-XX** by double clicking on the text. Do not erase and replace.
4. **Save** as **Ex-21D-I**
5. Plot using Page Setup **Class Model A**

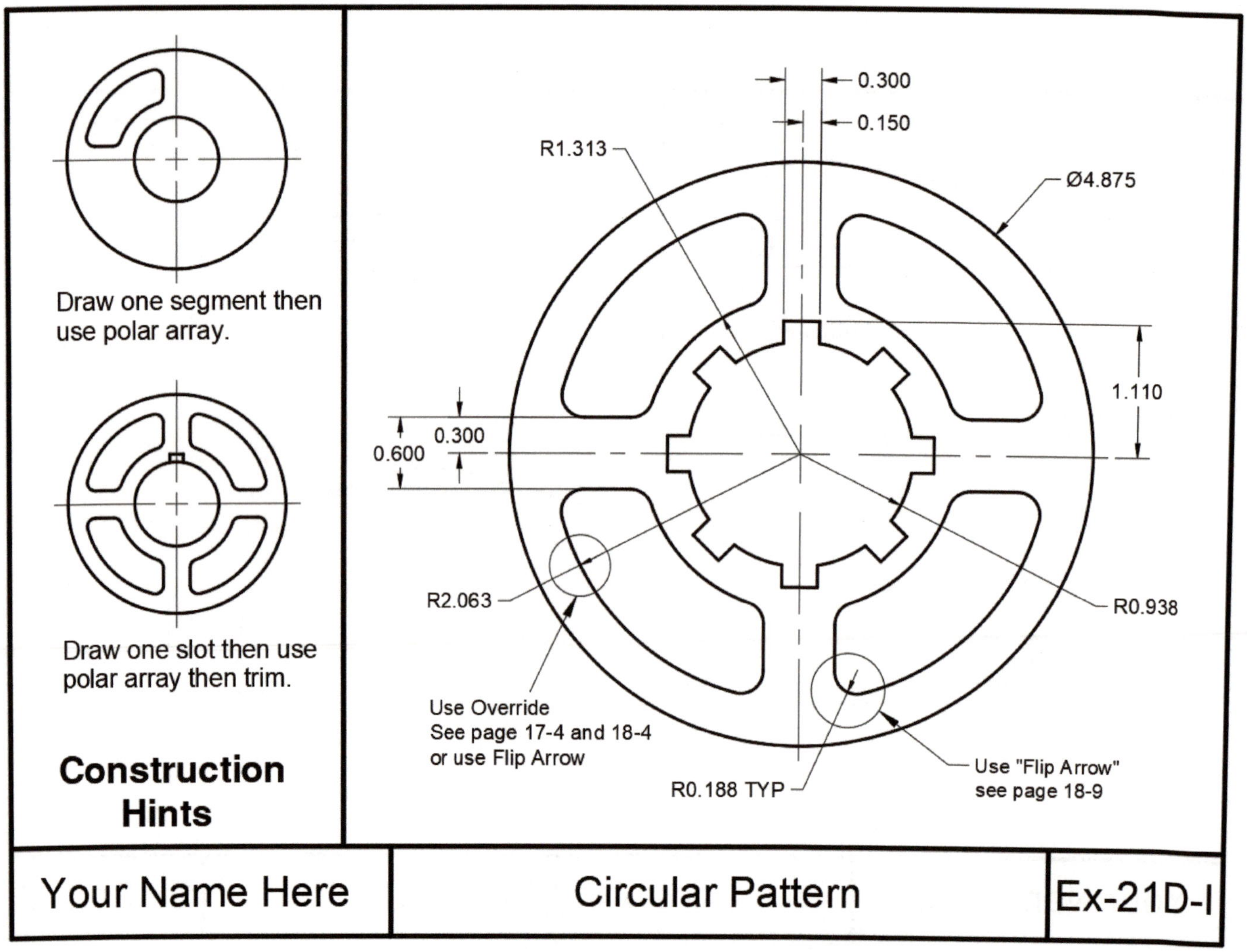

Exercise 21D-Metric

Instructions:

1. Start a **New** file using the **Border A-Metric.dwt**
2. The drawing below is designed to give you more practice drawing objects and dimensioning. Use any method that you have learned in the previous lessons.
3. Edit the **Title** and **Ex-XX** by double clicking on the text. Do not erase and replace.
4. **Save** as **Ex-21D-M**
5. Plot using Page Setup **Class Model A**

Note: You will notice that the **diameter** and **radius** dimension text is horizontal in the drawing below. By default the diameter and radius text is aligned with the dimension arrow for metric dimensions.

If you prefer to have your diameter and radius dimensions horizontal, you must set up a **Dimension Sub-Style** for both the diameter and radius dimensions (see page 18-9). Select the **Text** Tab, then in the **Text Alignment Panel**, change it to **"Horizontal"**.

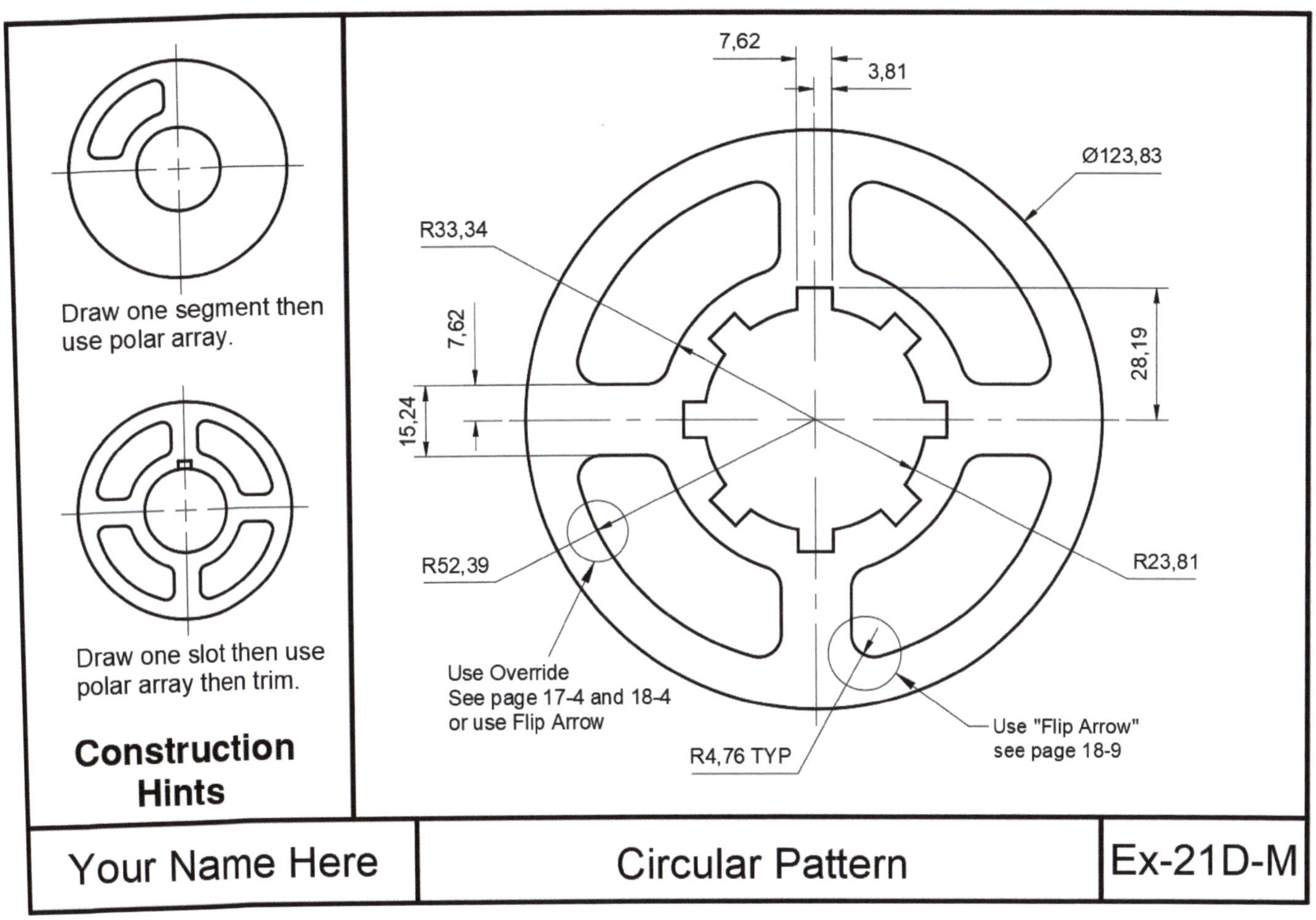

Notes:

LESSON 22

LEARNING OBJECTIVES

After completing this lesson, you will be able to:

- Draw an Arc using 10 different methods
- Dimension Arc segments

Drawing an Arc

There are 10 ways to draw an **arc** in AutoCAD. Not all of the arcs options are easy to create, so you may find it is often easier to **trim a circle** or use the **Fillet** command.

On the job, you will probably only use two of these methods. Which of these you choose to use will depend on the application.

An **arc** is a **segment** of a circle and **must be less than 360 degrees**.

By default, arcs are drawn counter-clockwise. You can change the direction by holding down the ***<Ctrl>*** key to draw in a clockwise direction. Or in some cases, you can enter a negative input to draw in a clockwise direction.

1. Select the **Arc** command using one of the following:

   Ribbon = Home Tab / Draw Panel / ▾
   or
   Keyboard = *A <Enter>*

2. Refer to Exercises 22A and 22B for examples of two of the most commonly used methods of creating an arc.

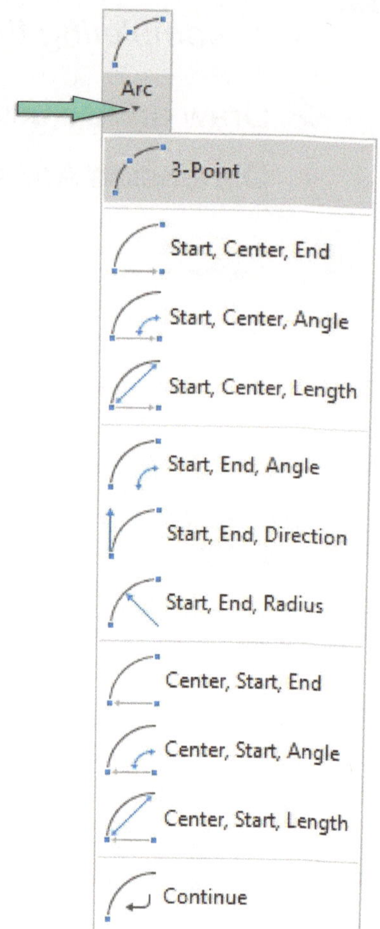

Dimensioning Arc Lengths

You may dimension the distance along an arc. This is known as the **arc length**. Arc length is an **Associative** dimension.

Example:

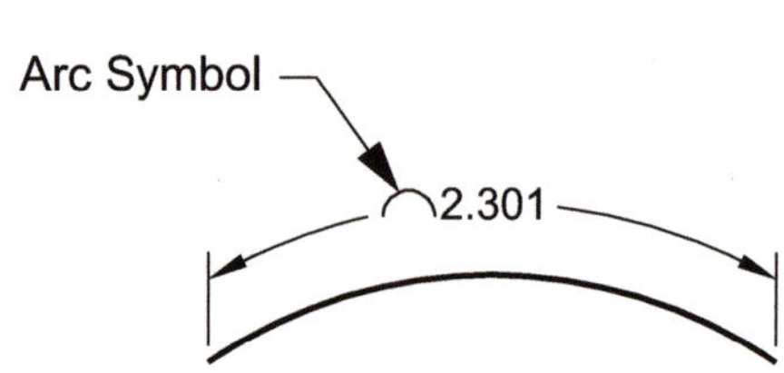

Continued on the next page...

1. Select the **Arc Length** dimension command using one of the following:

Ribbon = Annotate Tab / Dimension Panel / ▾
or
Keyboard = *dimarc* <*Enter*>

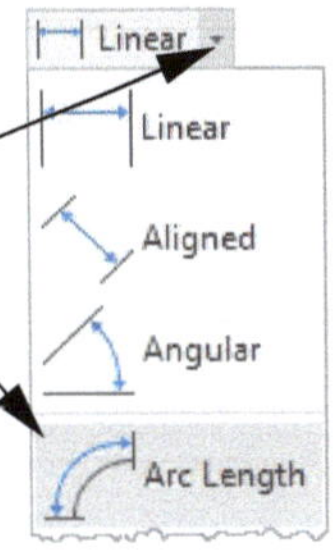

The following will appear on the Command Line:

```
DIMARC Select arc or polyline arc segment:
```

2. Select the arc.

The following will appear on the Command Line:

```
DIMARC Specify arc length dimension location, or [Mtext Text Angle Partial]:
```

3. Place the dimension line and text location.

To differentiate the arc length dimensions from linear or angular dimensions, arc length dimensions display an arc (⌒) symbol by default. (Also called a "hat" or "cap".)

The arc symbol may be displayed either **above** or **preceding** the dimension text. You may also choose **not** to display the arc symbol.

Example:

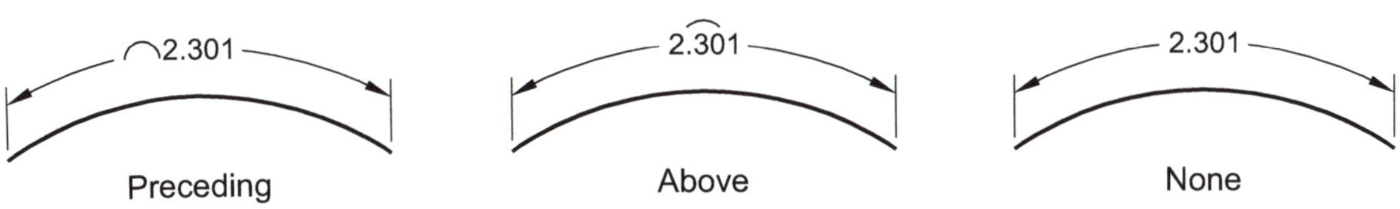

Specify the placement of the arc length symbol in the **Dimension Style Manager / Symbols and Arrows** Tab, or you may edit its position using the **Properties Palette**.

Dimension Style Manager Symbols and Arrows Tab

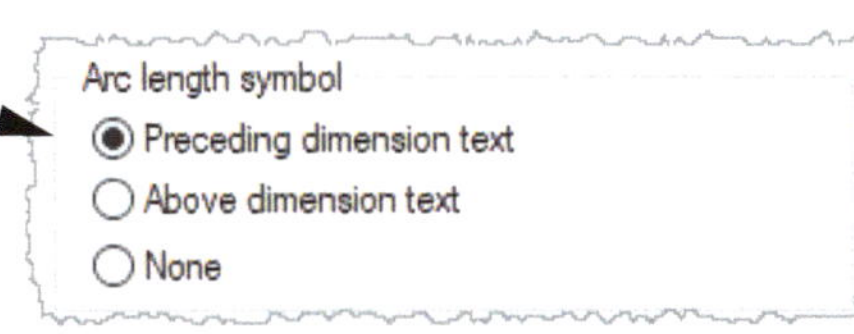

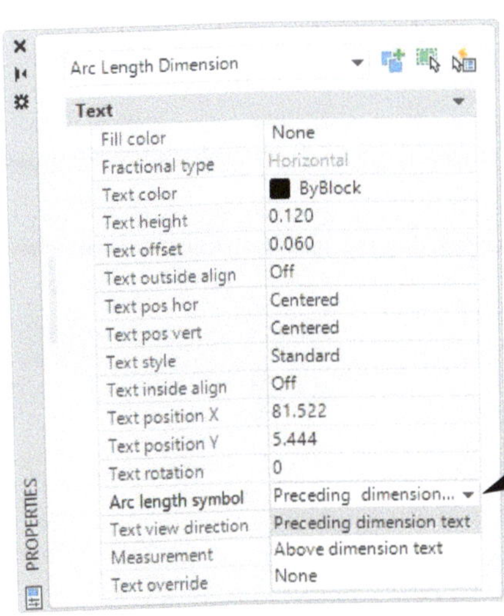

Properties Palette

Continued on the next page...

The extension lines of an arc length dimension are displayed as **radial** if the included angle is **greater than 90 degrees**.

The extension lines of an arc length dimension are displayed as **orthogonal** if the included angle is **less than 90 degrees**.

Example:

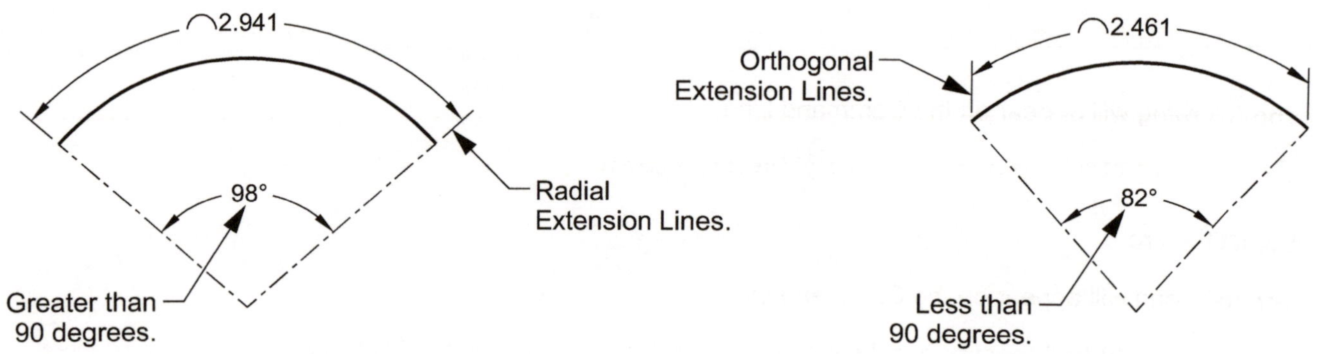

Dimensioning a Large Curve

When dimensioning an arc, the dimension line should pass through the center of the arc. However, for large curves, the true center of the arc could be very far away, even off the sheet.

When the true center location cannot be displayed, you can create a **"Jogged"** radius dimension.

Example:

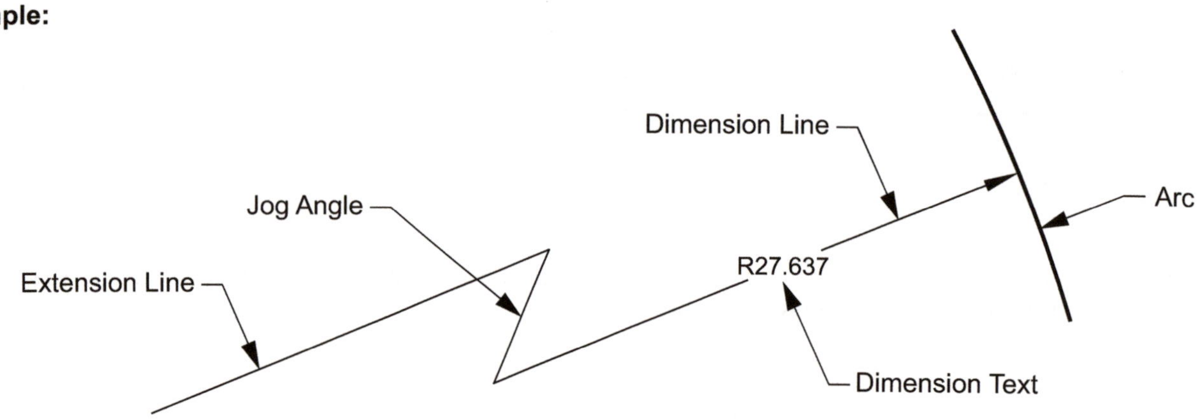

Controlling the Jog

You can specify the **Jog angle** and **Jog height factor** in the **Dimension Style Manager / Symbols and Arrows** Tab.

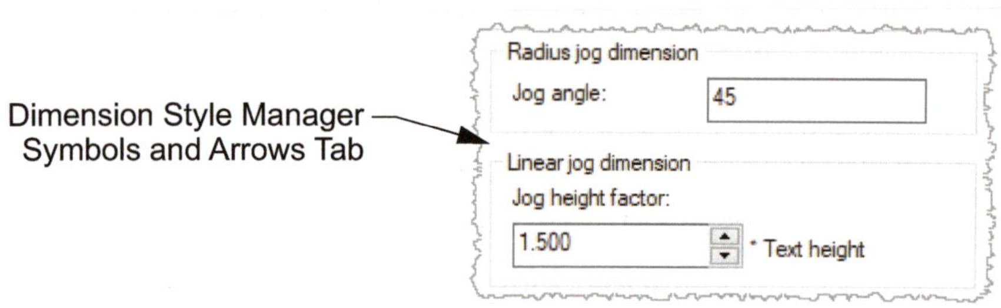

Continued on the next page...

1. Select the **Jogged** radius dimension command using one of the following:

Ribbon = Annotate Tab / Dimensions Panel / ▼
or
Keyboard = *dimjogged <Enter>*

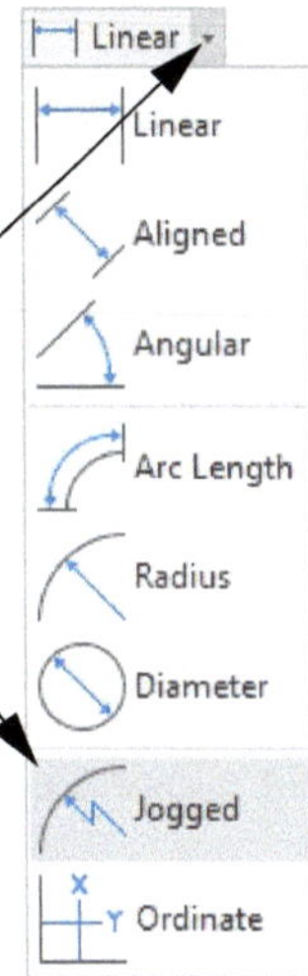

The following will appear on the Command Line:

```
DIMJOGGED Select arc or circle:
```

2. Select the large arc or circle (**P1**), anywhere on the arc or circle.

The following will appear on the Command Line:

```
DIMJOGGED Specify center location override:
```

3. Move the cursor and left click to specify the "fake" center location (**P2**).

The following will appear on the Command Line:

```
DIMJOGGED Specify dimension line location or [Mtext Text Angle]:
```

4. Move the cursor and left click to specify the location for the dimension text (**P3**).

The following will appear on the Command Line:

```
DIMJOGGED Specify jog location:
```

5. Move the cursor and left click to specify the location for the jog (**P4**).

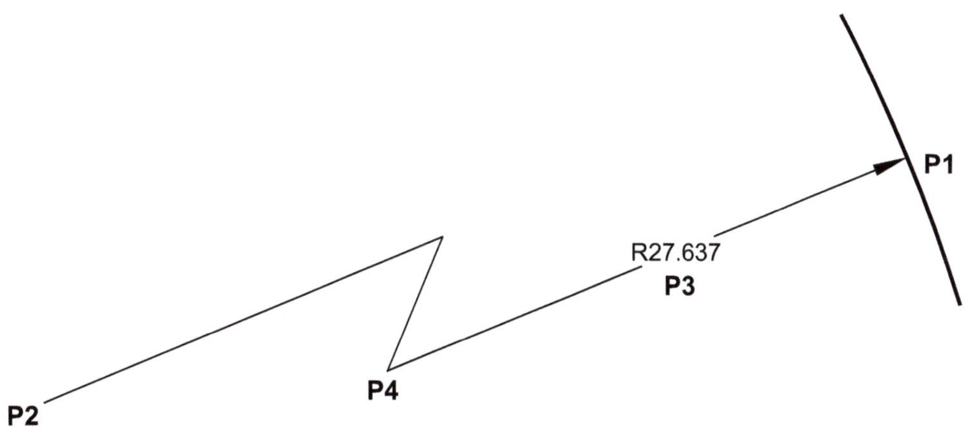

Exercise 22A

Instructions:

1. Start a **New** file using either the **Border A-Inch.dwt** or the **Border A-Metric.dwt**
2. Draw the Centerlines first on Layer Centerline.
3. Draw the **3-Point** arcs using Layer Object Line. Place the **first point** at location **1**, **second point** at location **2**, and **third point** at location **3**.
4. Dimension as shown using Dimension Style CLASS STYLE and Layer Dimension. (*See note below.*)
5. Edit the **Title** and **Ex-XX** by double clicking on the text. Do not erase and replace.
6. **Save** as **Ex-22A**
7. Plot using Page Setup **Class Model A**

Note: You will notice that I have used both inch and metric dimensions for Exercises **22A** through **22D**. Each of the exercises will be suitable for both inch and metric users. Your dimensions will either show inch or metric, depending on which template you use.

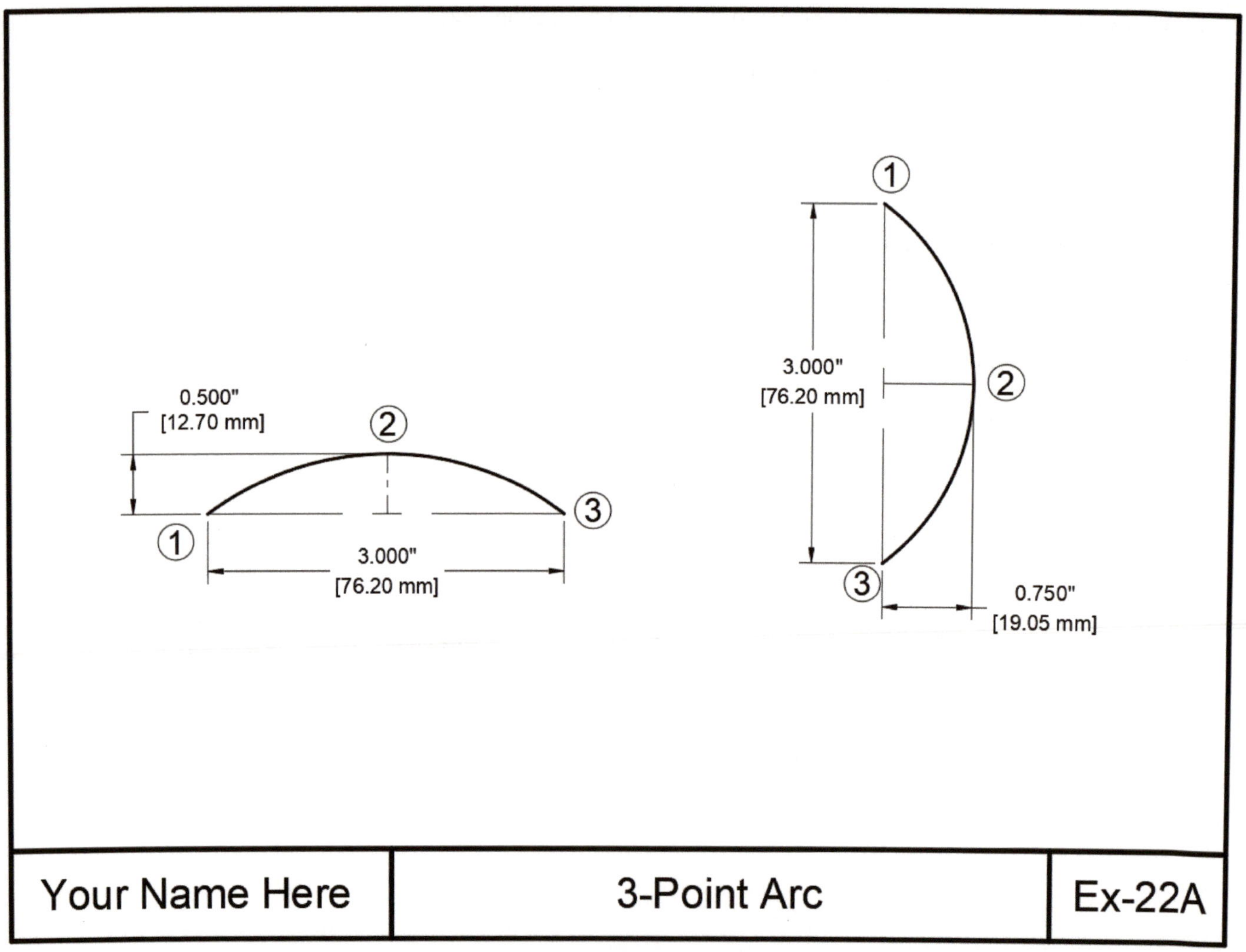

Exercise 22B

Instructions:

1. Start a **New** file using either the **Border A-Inch.dwt** or the **Border A-Metric.dwt**
2. Draw the Centerlines first on Layer Centerline.
3. Draw the two arcs using method **Start**, **End**, **Radius** on Layer Object Line. Place the **Start** at location **1**, **End** at location **2** and enter the **Radius** of **3**.
4. Dimension as shown using Dimension Style CLASS STYLE and Layer Dimension.
5. Edit the **Title** and **Ex-XX** by double clicking on the text. Do not erase and replace.
6. **Save** as **Ex-22B**
7. Plot using Page Setup **Class Model A**

Note: Positive radius draws the small segment and negative radius draws the large segment. To reverse the directions hold down the ***<Ctrl>*** key.

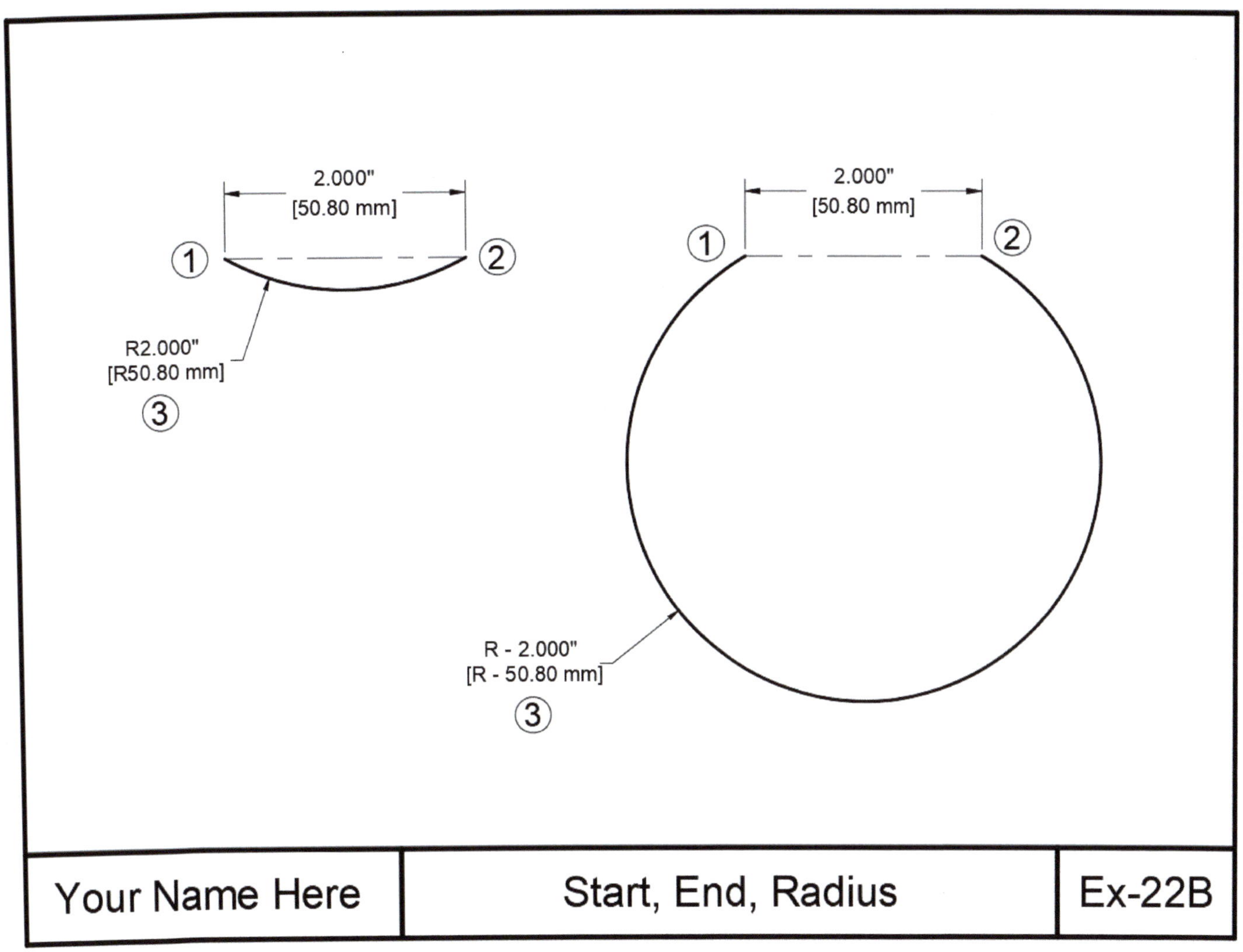

Exercise 22C

Instructions:

1. Start a **New** file using either the **Border A-Inch.dwt** or the **Border A-Metric.dwt**
2. Draw the Centerlines first on Layer Centerline.
3. Draw the arc using Layer Object Line. Refer to the previous pages to select method.
4. Dimension as shown using Dimension Style CLASS STYLE and Layer Dimension.
5. Edit the **Title** and **Ex-XX** by double clicking on the text. Do not erase and replace.
6. **Save** as **Ex-22C**
7. Plot using Page Setup **Class Model A**

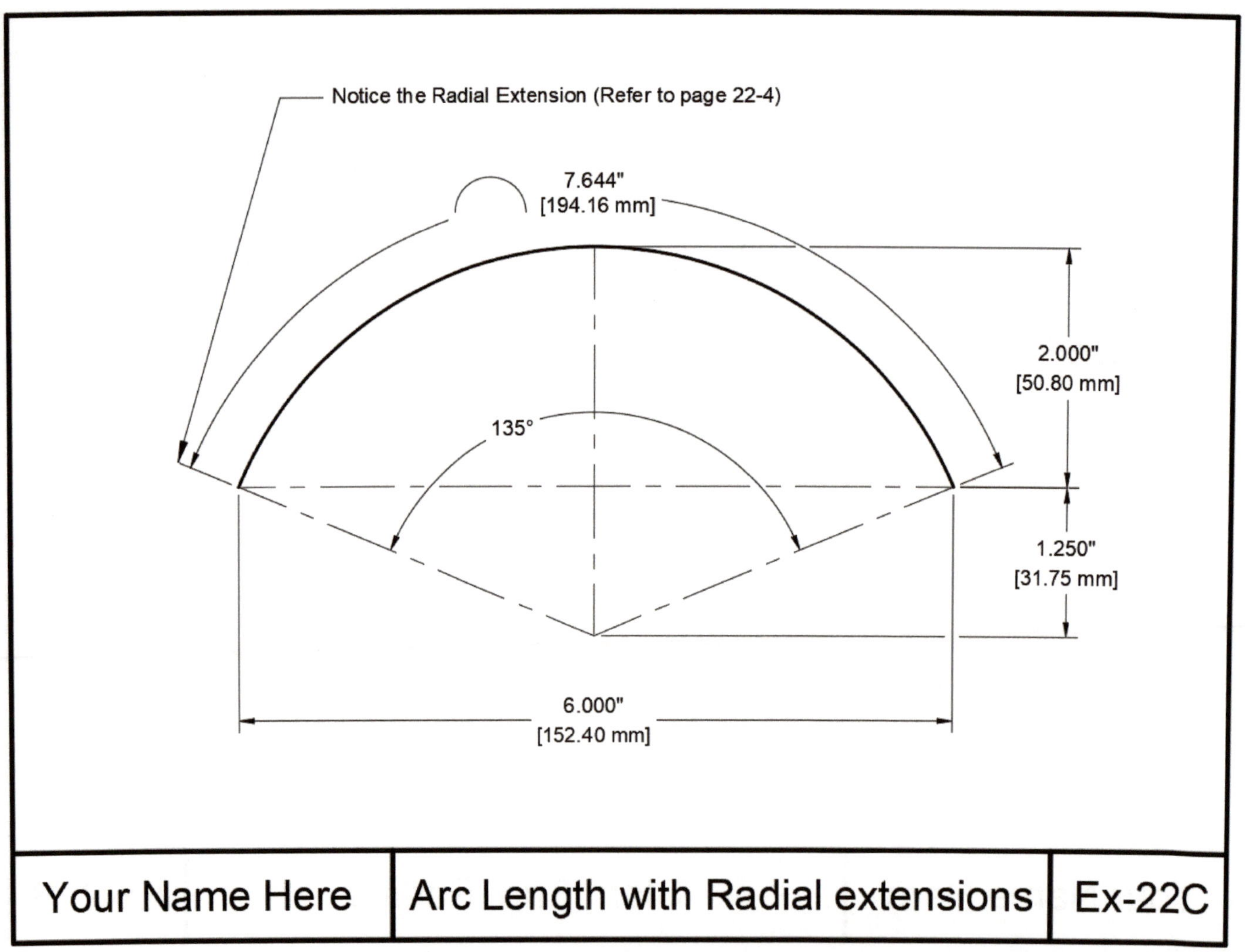

Exercise 22D

Instructions:

1. Start a **New** file using either the **Border A-Inch.dwt** or the **Border A-Metric.dwt**
2. Draw the Centerlines first on Layer Centerline.
3. Draw the arc using Layer Object Line. Refer to the previous pages to select method.
4. Dimension as shown using Dimension Style CLASS STYLE and Layer Dimension.
5. Edit the **Title** and **Ex-XX** by double clicking on the text. Do not erase and replace.
6. **Save** as **Ex-22D**
7. Plot using Page Setup **Class Model A**

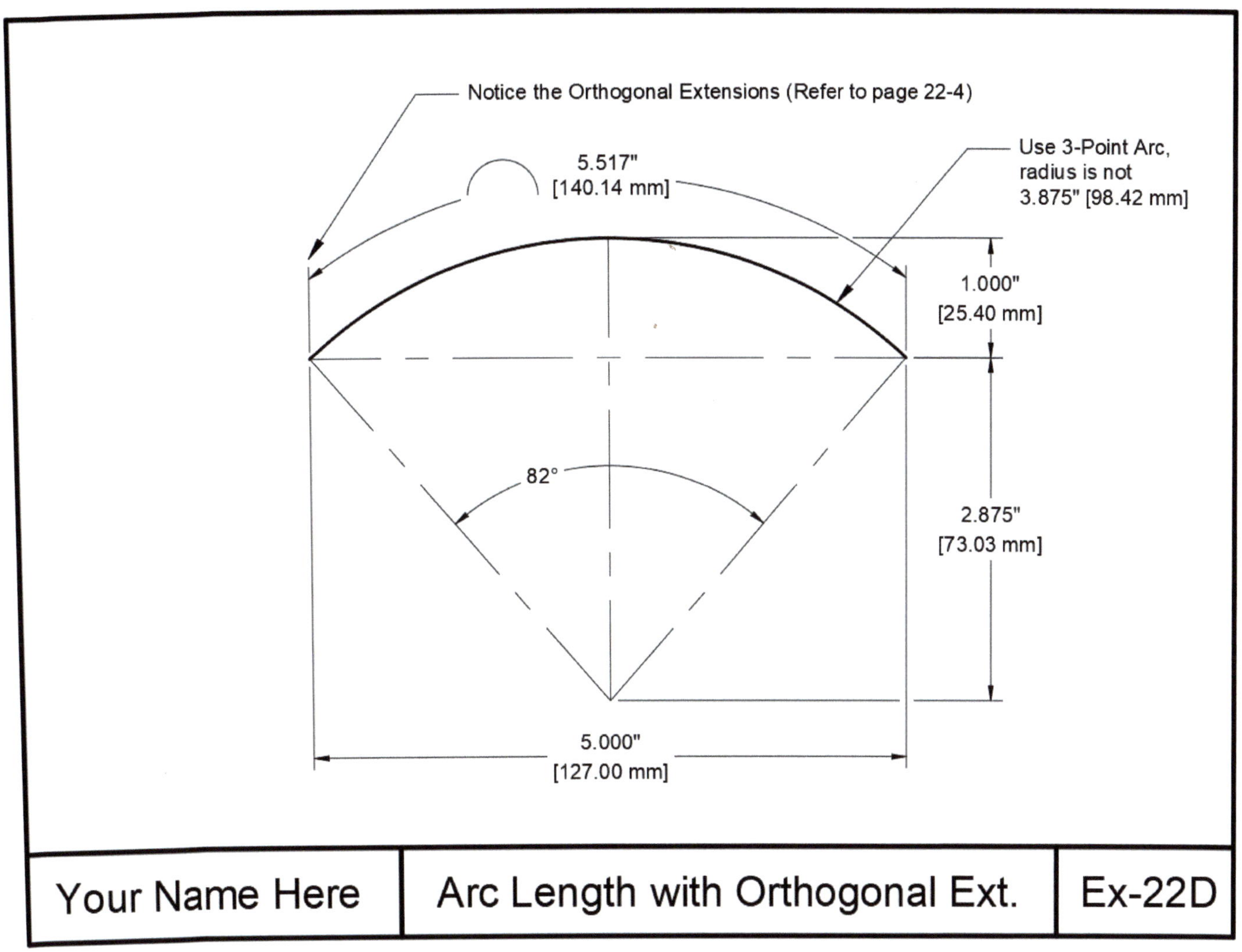

Exercise 22E-Inch

Instructions:

1. Start a **New** file using the **Border A-Inch.dwt**
2. Draw the objects below using circles, arcs, lines, and fillets on Layer Object Line.
   See the next page for drawing assistance.
3. Dimension as shown using Dimension Style CLASS STYLE and Layer Dimension. (Refer to pages 18-3 and 18-4 for diameter and radius dimension settings.)
4. Edit the **Title** and **Ex-XX** by double clicking on the text. Do not erase and replace.
5. **Save** as **Ex-22E-I**
6. Plot using Page Setup **Class Model A**

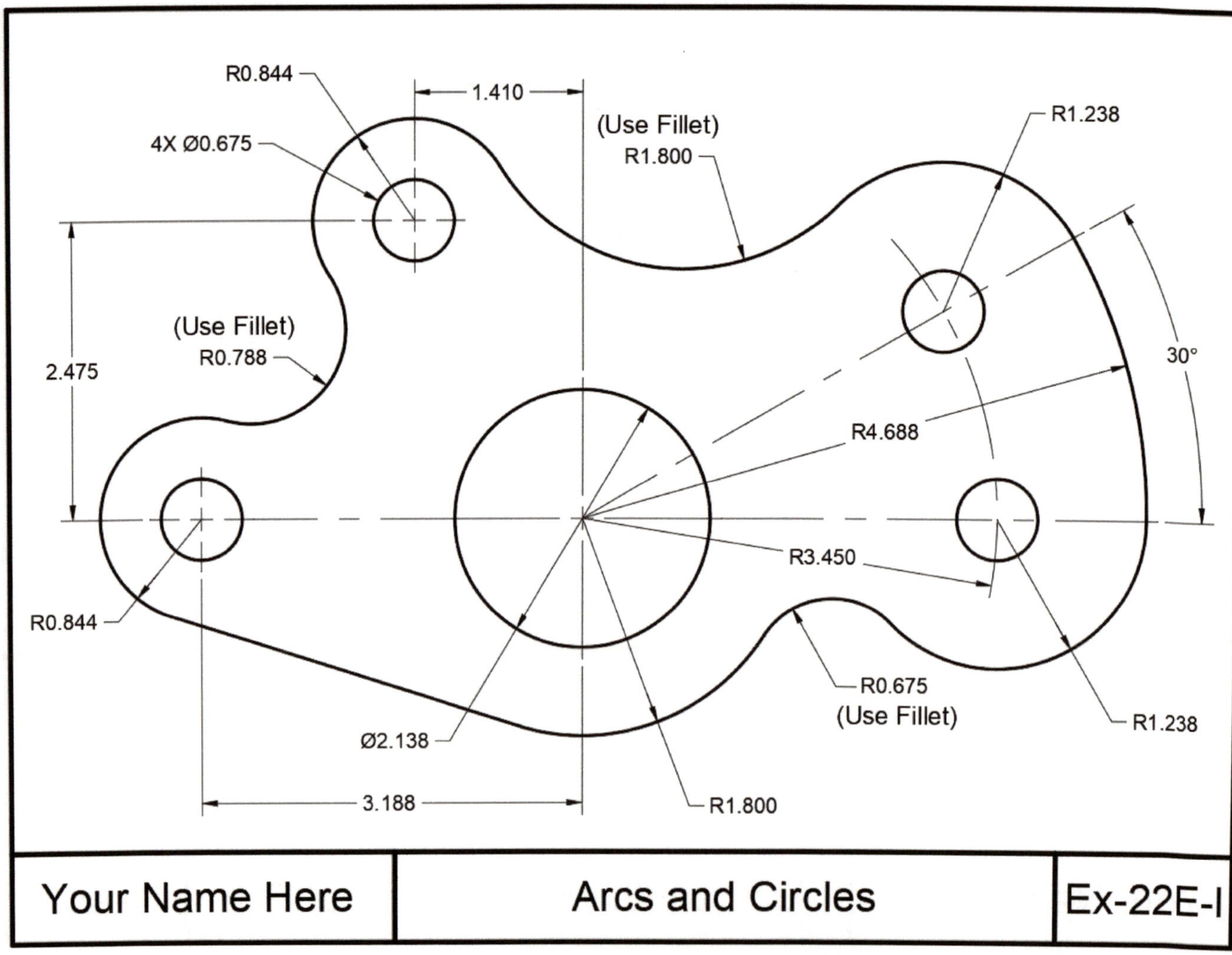

Exercise 22E-Inch Helper

Instructions:

Step 1.

Draw the horizontal and vertical lines, then offset to create intersections for circles. Draw the 30 degree line.

Step 2.

Draw the circles.

Step 3.

Draw the arcs and draw the circles.

Step 4.

Fillet, draw the tangent line, and dimension.

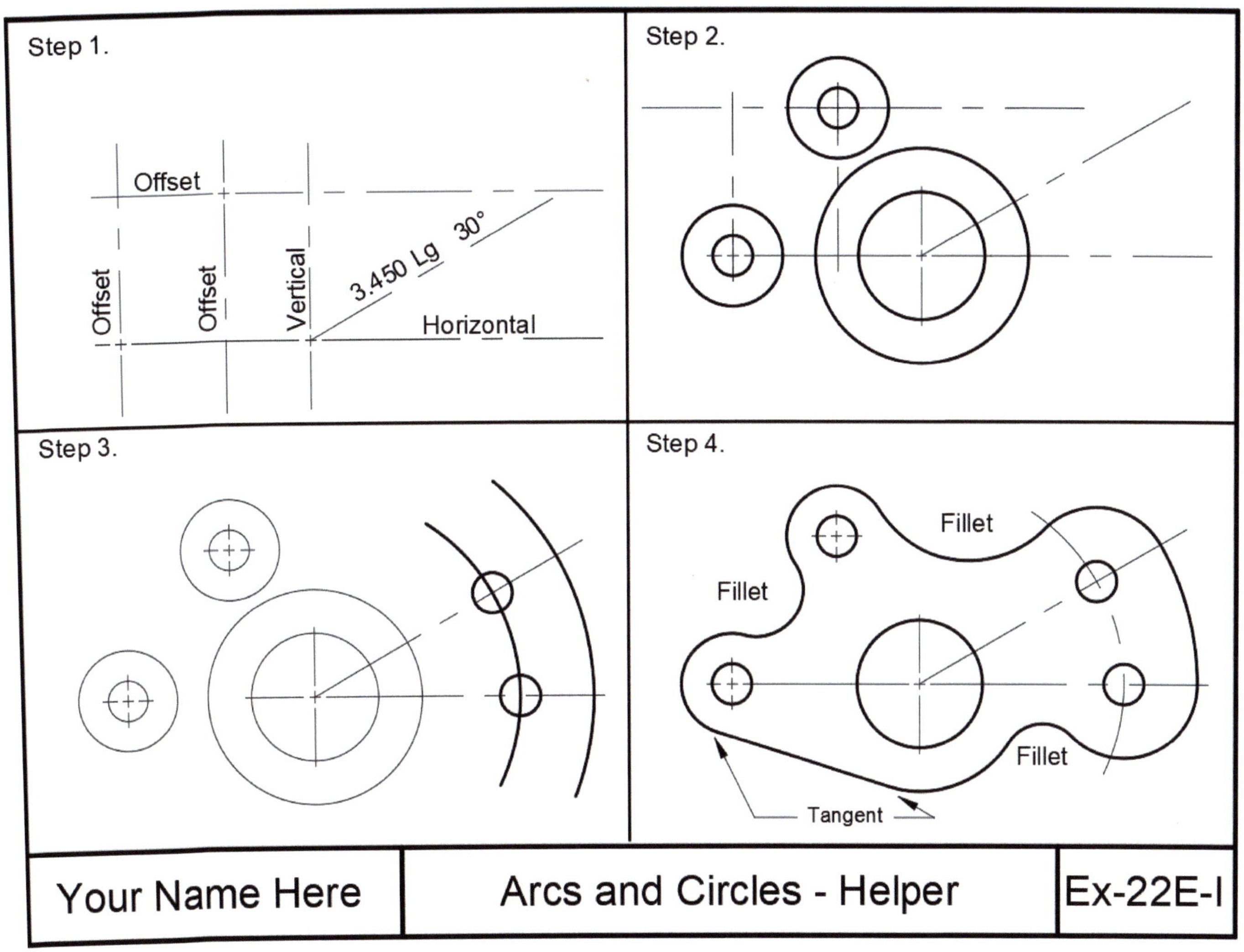

Exercise 22E-Metric

Instructions:

1. Start a **New** file using the **Border A-Metric.dwt**
2. Draw the objects below using circles, arcs, lines, and fillets on Layer Object Line. **See the next page for drawing assistance.**
3. Dimension as shown using Dimension Style CLASS STYLE and Layer Dimension. (Refer to pages 18-3 and 18-4 for diameter and radius dimension settings.)
4. Edit the **Title** and **Ex-XX** by double clicking on the text. Do not erase and replace.
5. **Save** as **Ex-22E-M**
6. Plot using Page Setup **Class Model A**

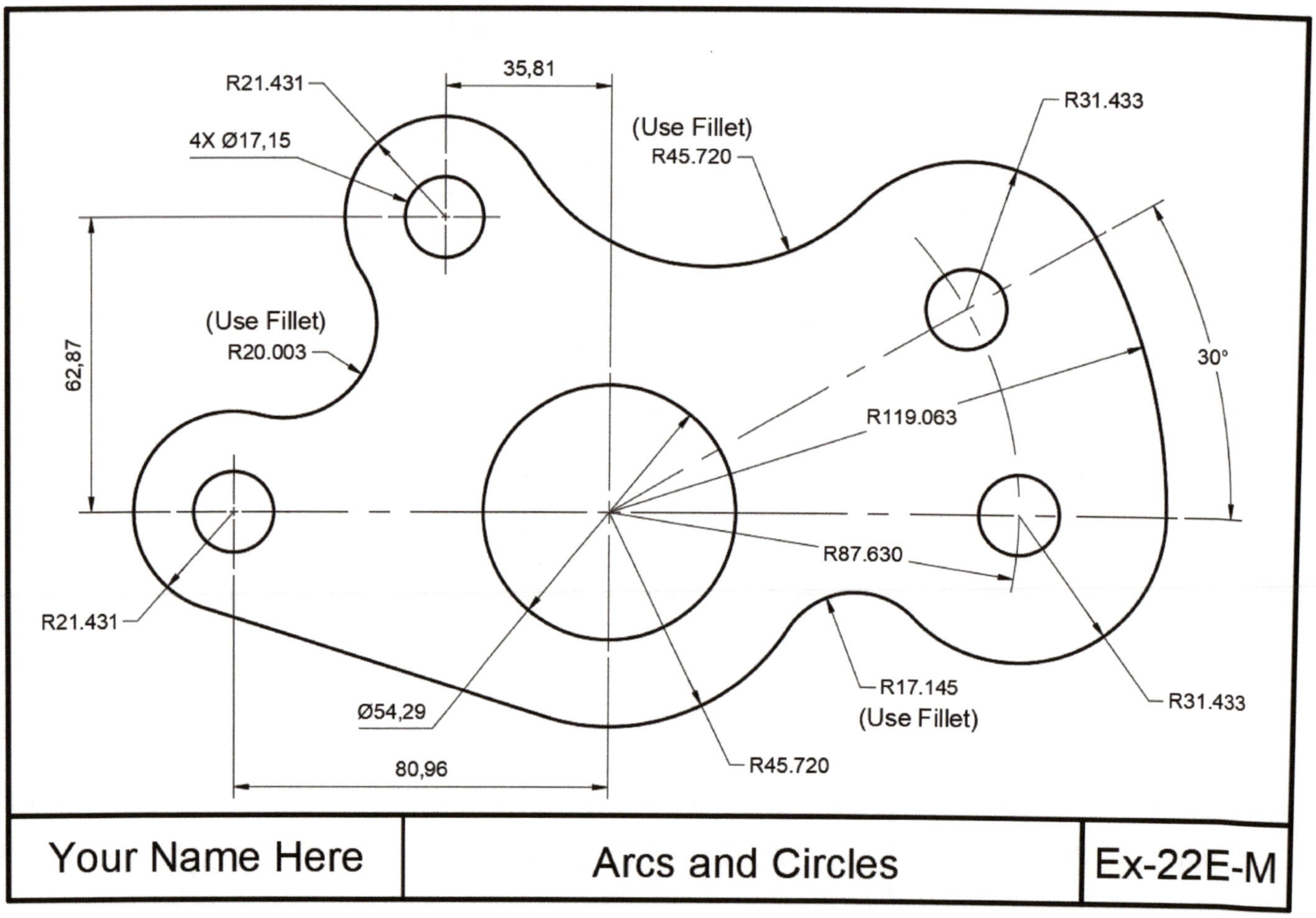

Exercise 22E-Metric Helper

Instructions:

Step 1.

Draw the horizontal and vertical lines, then offset to create intersections for circles. Draw the 30 degree line.

Step 2.

Draw the circles.

Step 3.

Draw the arcs and draw the circles.

Step 4.

Fillet, draw the tangent line, and dimension.

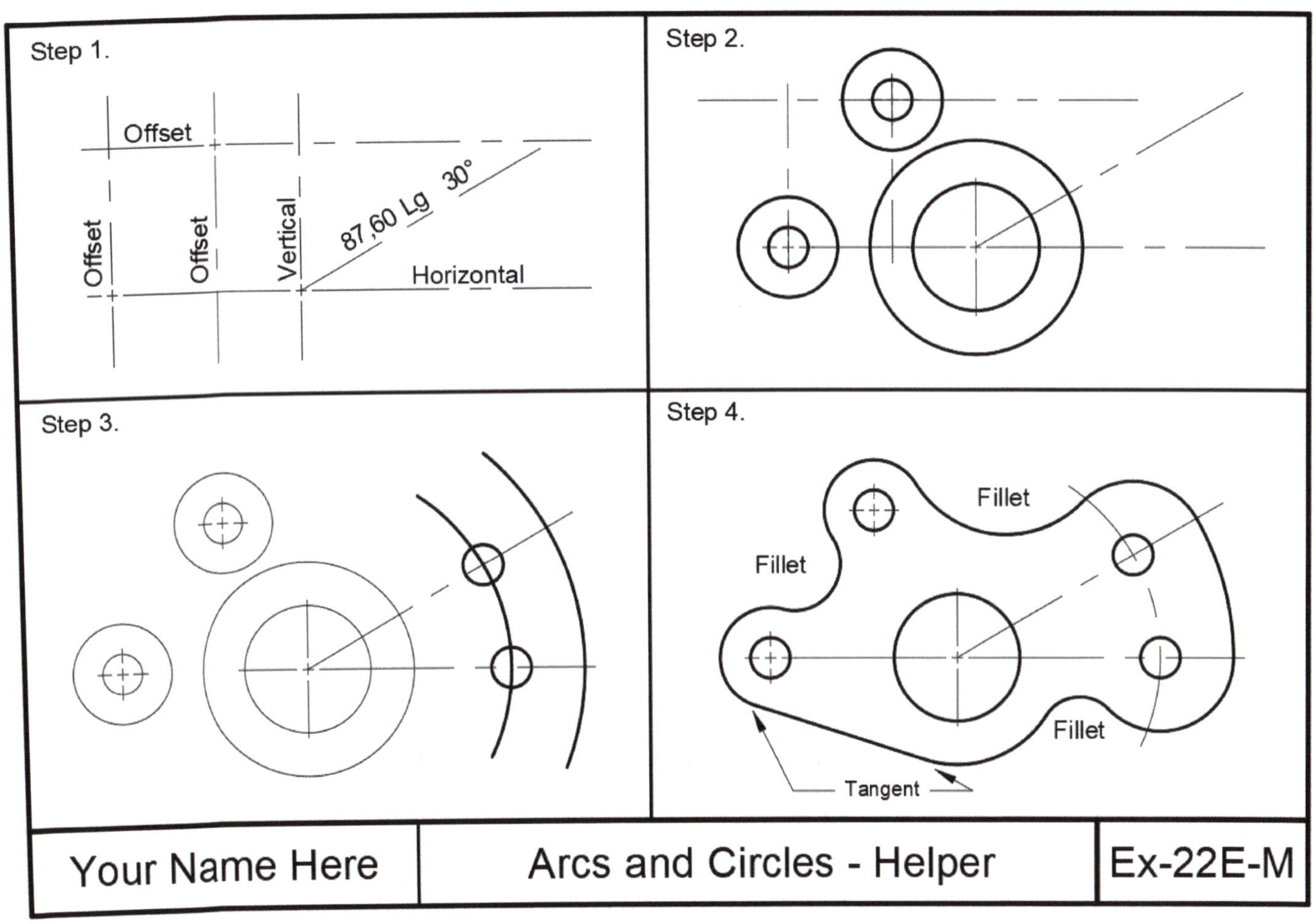

Exercise 22F

Instructions:

1. Start a **New** file using either the **Border A-Inch.dwt** or the **Border A-Metric.dwt**
2. Draw **9.000" [228.60 mm]** radius circle on Layer Object Line, as shown below.
3. Dimension as shown using Dimension Style CLASS STYLE and Layer Dimension. (Refer to page 22-4.)
4. Edit the **Title** and **Ex-XX** by double clicking on the text. Do not erase and replace.
5. **Save** as **Ex-22F**
6. Plot using Page Setup **Class Model A**

Note: You will notice that I again have used dual dimensions for this exercise. Your dimension will either show inch or metric, depending on which template you use.

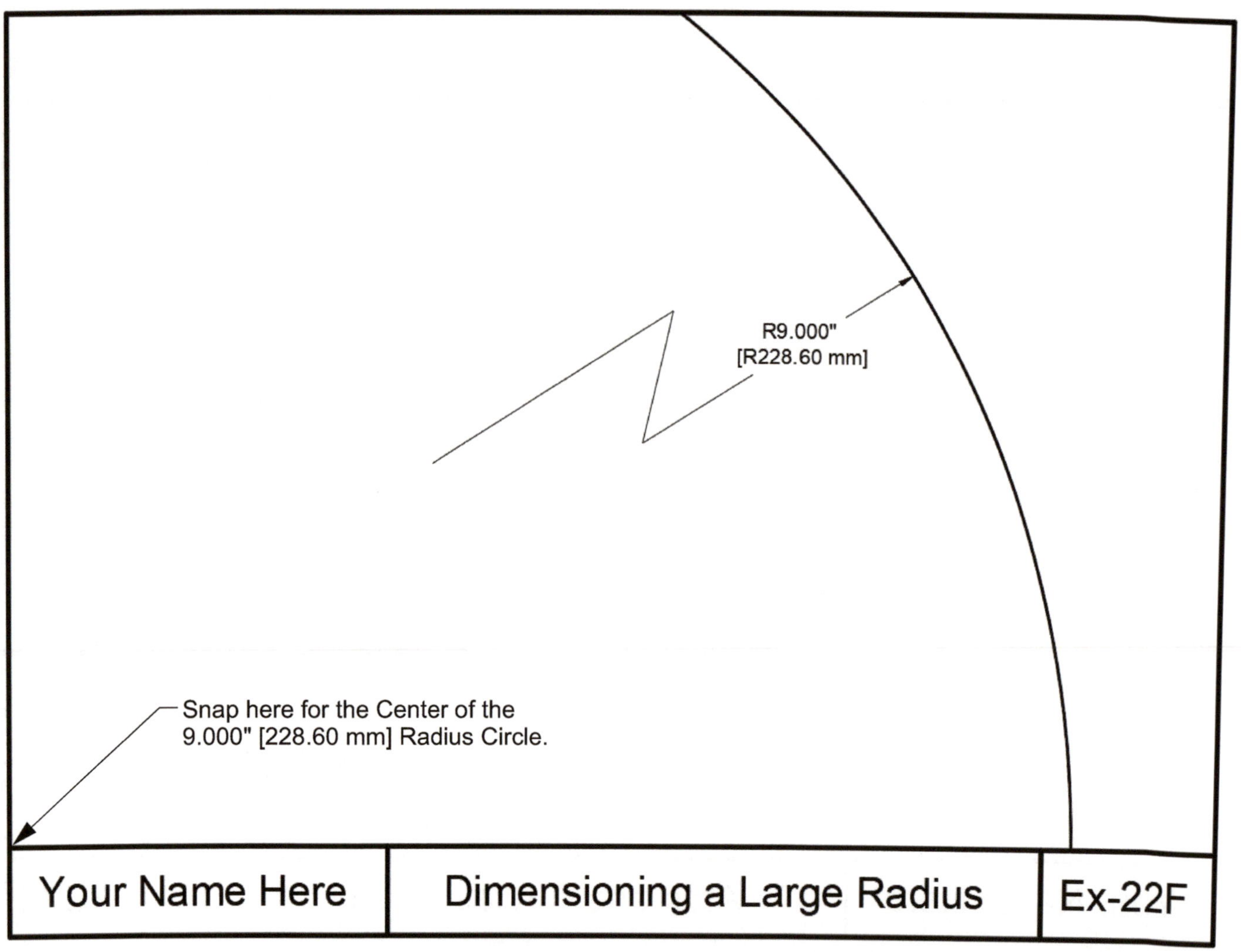

LESSON 23

LEARNING OBJECTIVES

After completing this lesson, you will be able to:

- Understand the Polyline command
- Draw a Polyline and a Polyarc
- Assign widths to Polylines
- Set the Fill mode to On or Off
- Explode a Polyline

Understanding and Creating Polylines

A **polyline** is very similar to a line. It is created in the same way a line is drawn. It requires first and second endpoints. But a polyline has additional features, as follows:

A. A polyline is one object, even though it may have many segments.

B. You may specify a specific width to each segment.

C. You may specify a different width to the start and end of a polyline segment.

1. Select the **Polyline** command using one of the following:

Ribbon = Home Tab / Draw Panel /
or
Keyboard = *PL <Enter>*

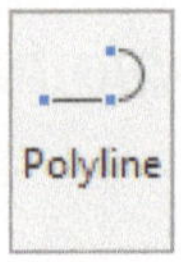

The following are examples of polylines with widths assigned.

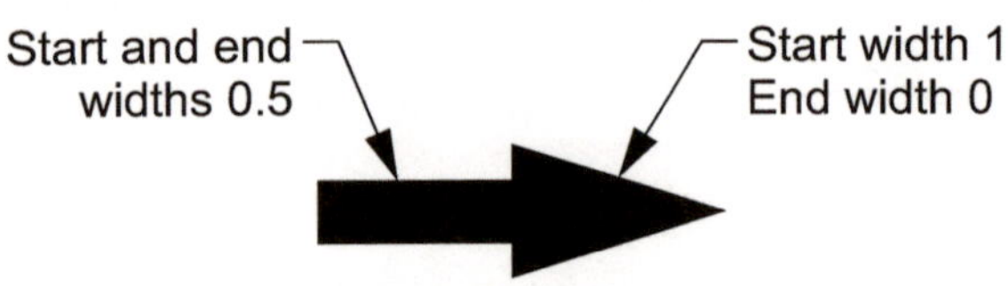

2 polyline segments

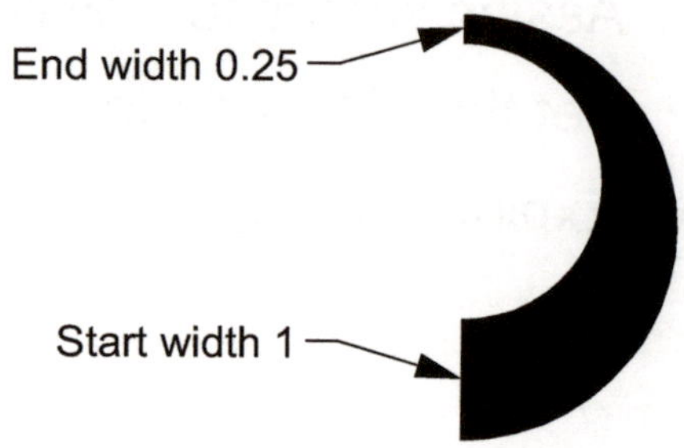

Single polyline arc

Options

Width: Specify the start and end width.

You can create a tapered polyline by specifying different starting and ending widths.

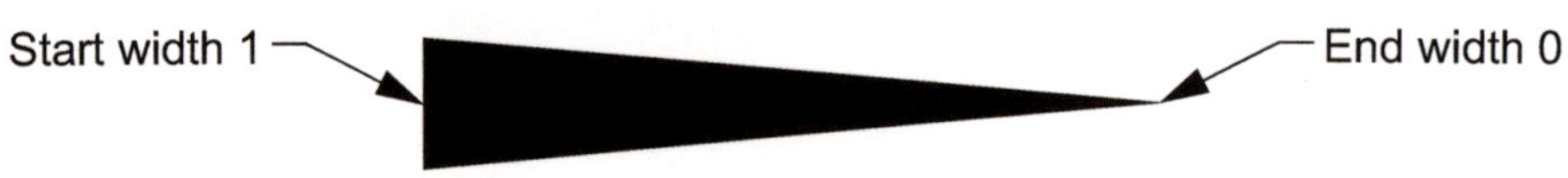

Halfwidth: The same as width except the starting and ending halfwidth specifies half the width rather than the entire width.

Continued on the next page...

Arc: This option allows you to create a circular polyline less than 360 degrees. You may use two 180 degree arcs to form a full circular shape.

Close: The close option is the same as in the Line command. Close attaches the last segment to the first segment.

Length: This option allows you to draw a polyline at the same angle as the last polyline drawn. This option is very similar to the **Offset** command. You specify the first endpoint and the length. The new polyline will automatically be drawn at the same angle as the previous polyline.

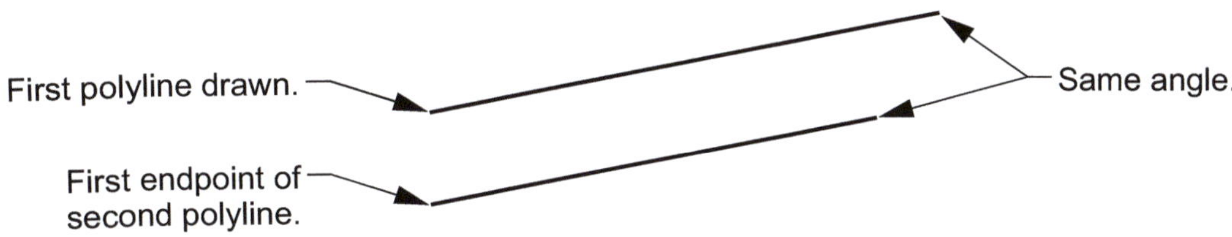

Controlling the Fill Mode

If you turn **Fill Mode off**, the polylines will appear as shown below.

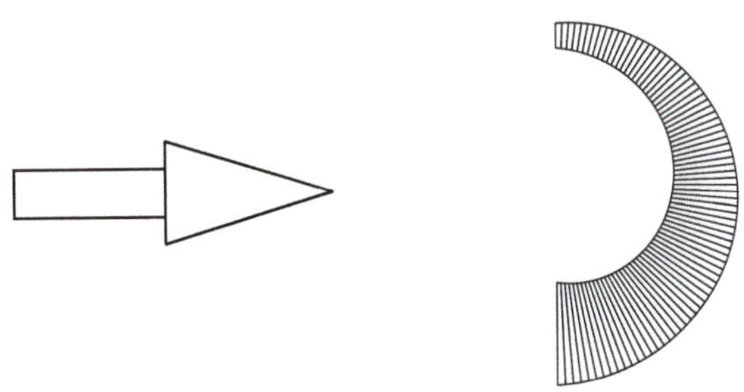

How to turn Fill Mode on or off

1. Type ***fill*** and then press ***<Enter>***.

   The following will appear on the Command Line:

2. Type ***on*** or ***off*** and then press ***<Enter>***.
3. Type ***regen*** and then press ***<Enter>***.

Exploding a Polyline

If you explode a polyline, it loses its width and turns into a regular line as shown below.

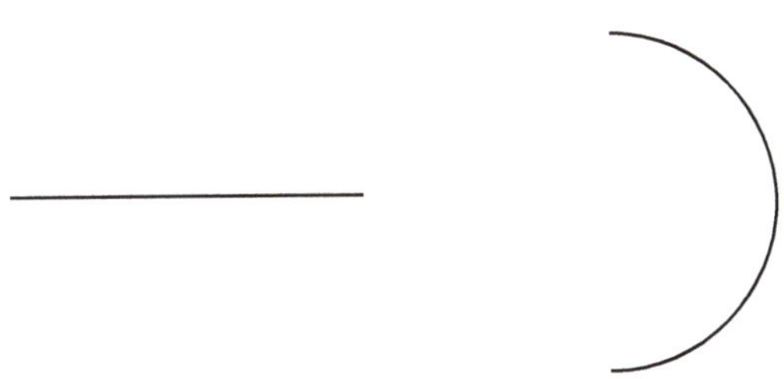

The following is an example of how to draw a polyline using **Width**.

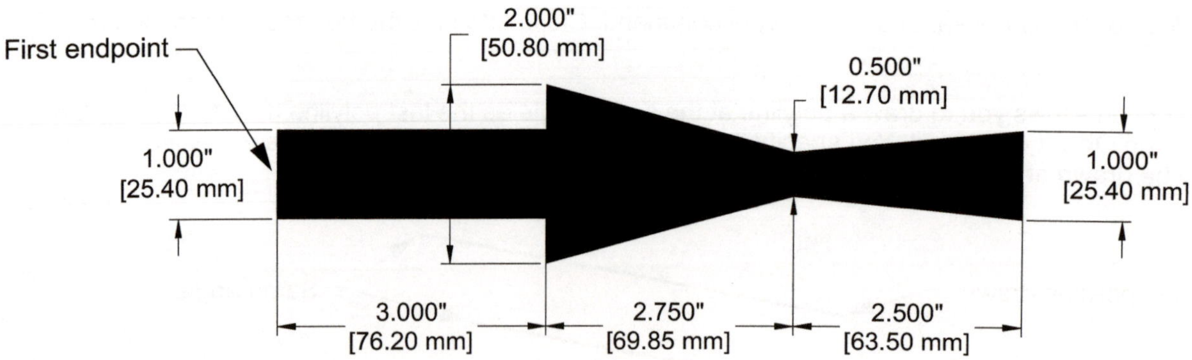

Note: Metric users, type in the sizes inside the brackets **[....]**

1. Select the **Polyline** command.

   The following will appear on the Command Line:

   ```
   PLINE Specify start point:
   ```

2. Place the first endpoint of the polyline.

   The following will appear on the Command Line:

   ```
   PLINE Specify next point or [Arc Halfwidth Length Undo Width]:
   ```

3. Type in ***W*** and then press ***<Enter>***, or select **Width** on the Command Line.

   The following will appear on the Command Line:

   ```
   PLINE Specify starting width <0.000>:
   ```

4. Type in ***1 [25.4]*** and then press ***<Enter>***.

   The following will appear on the Command Line:

   ```
   PLINE Specify ending width <1.000>:
   ```

5. Type in ***1 [25.4]*** and then press ***<Enter>***.

   The following will appear on the Command Line:

   ```
   PLINE Specify next point or [Arc Halfwidth Length Undo Width]:
   ```

6. Type in ***L*** and then press ***<Enter>***, or select **Length** on the Command Line.

   The following will appear on the Command Line:

   ```
   PLINE Specify length of line:
   ```

7. Type in ***3 [76.2]*** and then press ***<Enter>***.

   The following will appear on the Command Line:

   ```
   PLINE Specify next point or [Arc Halfwidth Length Undo Width]:
   ```

Continued on the next page...

8. Type in ***W*** and then press ***<Enter>***, or select **Width** on the Command Line.

   The following will appear on the Command Line:

```
PLINE Specify starting width <1.000>:
```

9. Type in ***2 [50.8]*** and then press ***<Enter>***.

   The following will appear on the Command Line:

```
PLINE Specify ending width <2.000>:
```

10. Type in ***0.5 [12.7]*** and then press ***<Enter>***.

    The following will appear on the Command Line:

```
PLINE Specify next point or [Arc Halfwidth Length Undo Width]:
```

11. Type in ***L*** and then press ***<Enter>***, or select **Length** on the Command Line.

    The following will appear on the Command Line:

```
PLINE Specify length of line:
```

12. Type in ***2.75 [69.85]*** and then press ***<Enter>***.

    The following will appear on the Command Line:

```
PLINE Specify next point or [Arc Halfwidth Length Undo Width]:
```

13. Type in ***W*** and then press ***<Enter>***, or select **Width** on the Command Line.

    The following will appear on the Command Line:

```
PLINE Specify starting width <0.500>:
```

14. Type in ***0.5 [12.7]*** and then press ***<Enter>***.

    The following will appear on the Command Line:

```
PLINE Specify ending width <0.500>:
```

15. Type in ***1 [25.4]*** and then press ***<Enter>***.

    The following will appear on the Command Line:

```
PLINE Specify next point or [Arc Halfwidth Length Undo Width]:
```

16. Type in ***L*** and then press ***<Enter>***, or select **Length** on the Command Line.

    The following will appear on the Command Line:

```
PLINE Specify length of line:
```

17. Type in ***2.5 [63.5]*** and then press ***<Enter>***.

    The following will appear on the Command Line:

```
PLINE Specify next point or [Arc Halfwidth Length Undo Width]:
```

18. Press ***<Enter>*** to stop the Polyline command.

Exercise 23A

Instructions:

1. Start a **New** file using either the **Border A-Inch.dwt** or the **Border A-Metric.dwt**
2. Draw the polylines as shown using Layer Object Line.
3. Do not dimension.
4. Edit the **Title** and **Ex-XX** by double clicking on the text. Do not erase and replace.
5. **Save** as **Ex-23A**
6. Plot using Page Setup **Class Model A**

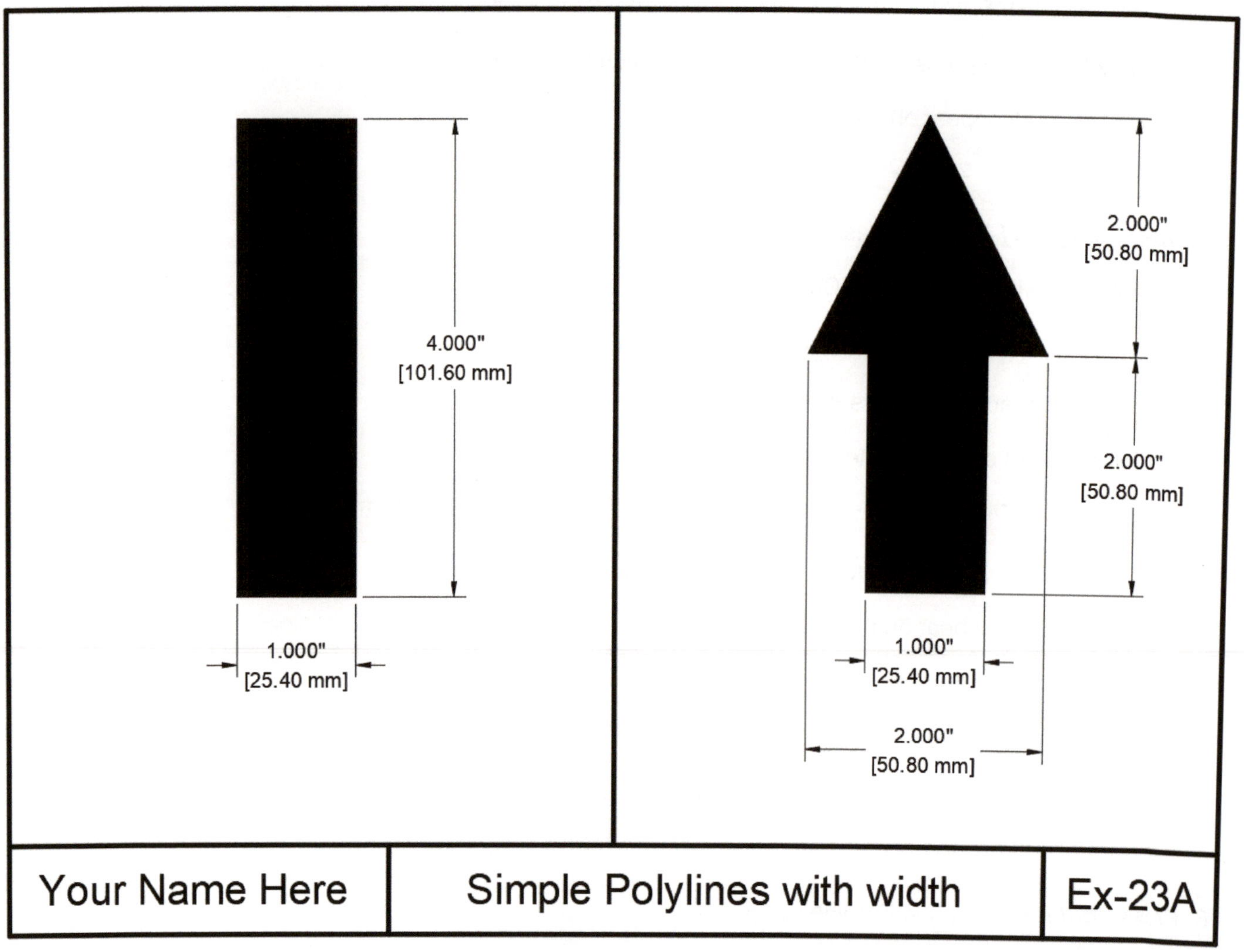

Exercise 23B

Instructions:

1. Open **Ex-23A**
2. Turn off the **Fill Mode**. (Refer to page 23-3.)
3. Do not dimension.
4. Edit the **Title** and **Ex-XX** by double clicking on the text. Do not erase and replace.
5. **Save** as **Ex-23B**
6. Plot using Page Setup **Class Model A**

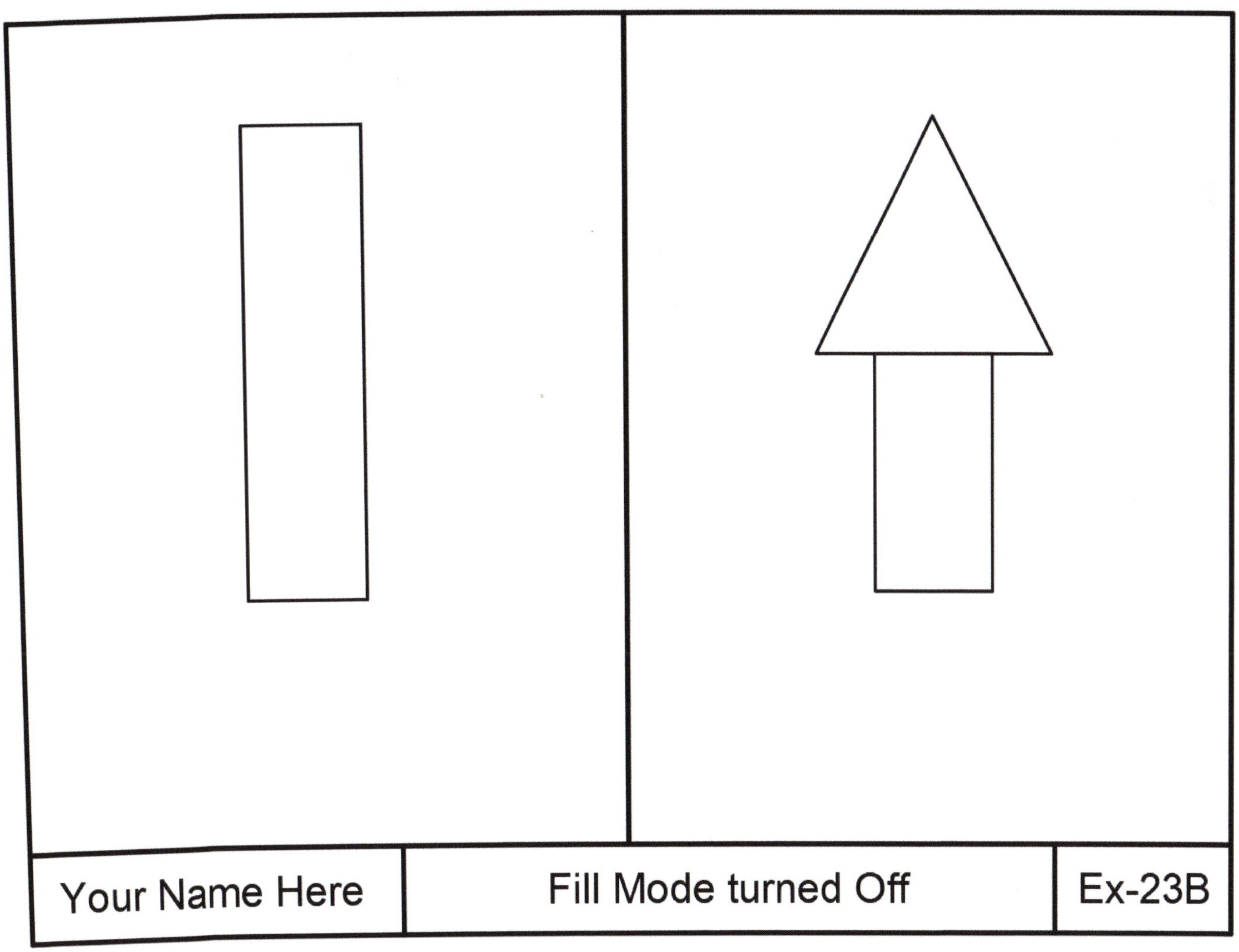

Exercise 23C

Instructions:

1. Open **Ex-23B**
2. Explode each of the polylines.
3. Edit the **Title** and **Ex-XX** by double clicking on the text. Do not erase and replace.
4. **Save** as **Ex-23C**
5. Plot using Page Setup **Class Model A**

Note: Architectural students may be thinking that the Polyline command would be good to use when creating floor plan walls. But notice what happens if you explode the polylines. Consider instead using the Multiline or Double line (LT) commands for floor plan walls. You can find these commands by referring to the Autodesk Assistant or AutoCAD Help Menu.

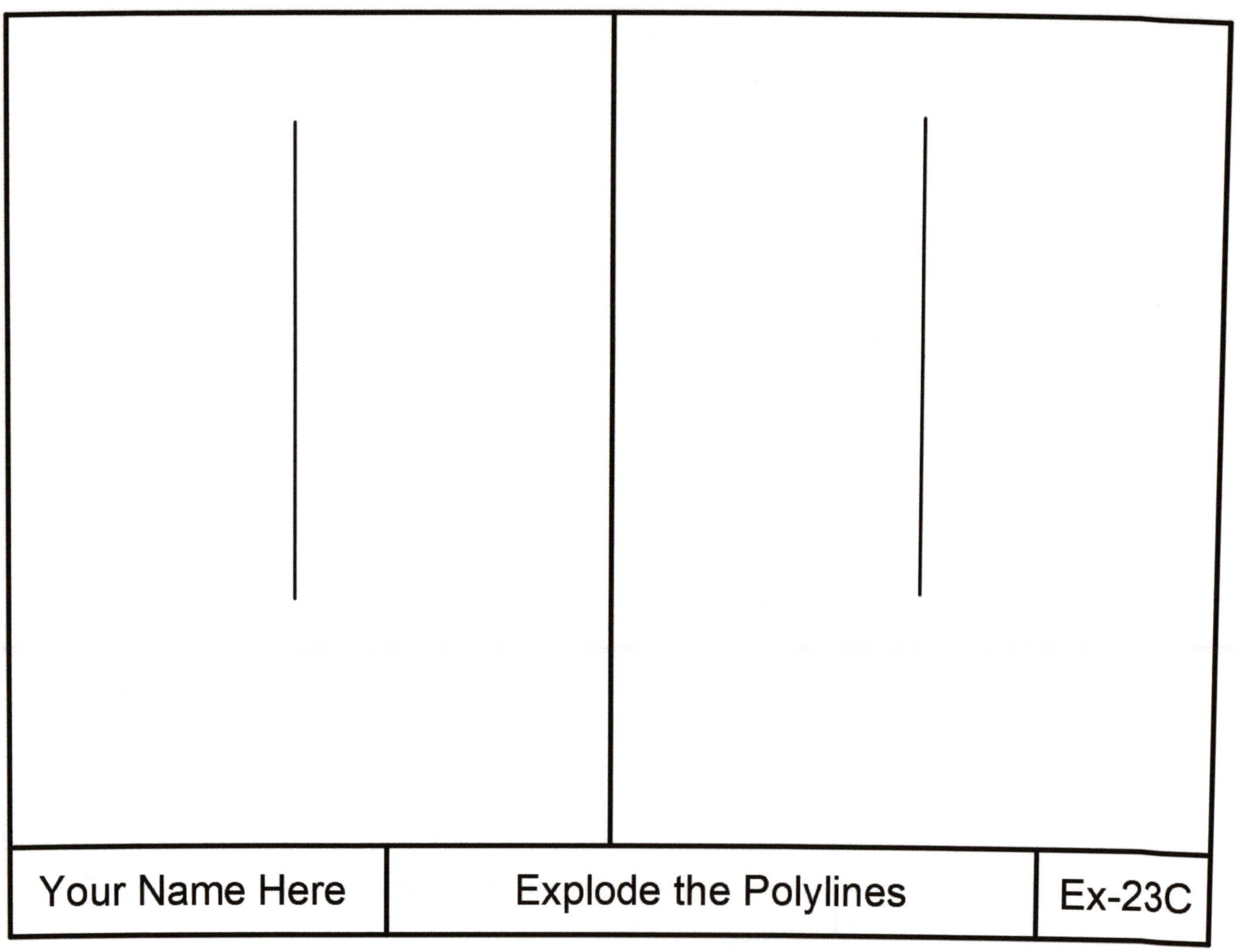

Exercise 23D

Instructions:

1. Start a **New** file using either the **Border A-Inch.dwt** or the **Border A-Metric.dwt**
2. Draw the polyline as shown using Layer Object Line.
3. Do not dimension.
4. Edit the **Title** and **Ex-XX** by double clicking on the text. Do not erase and replace.
5. **Save** as **Ex-23D**
6. Plot using Page Setup **Class Model A**

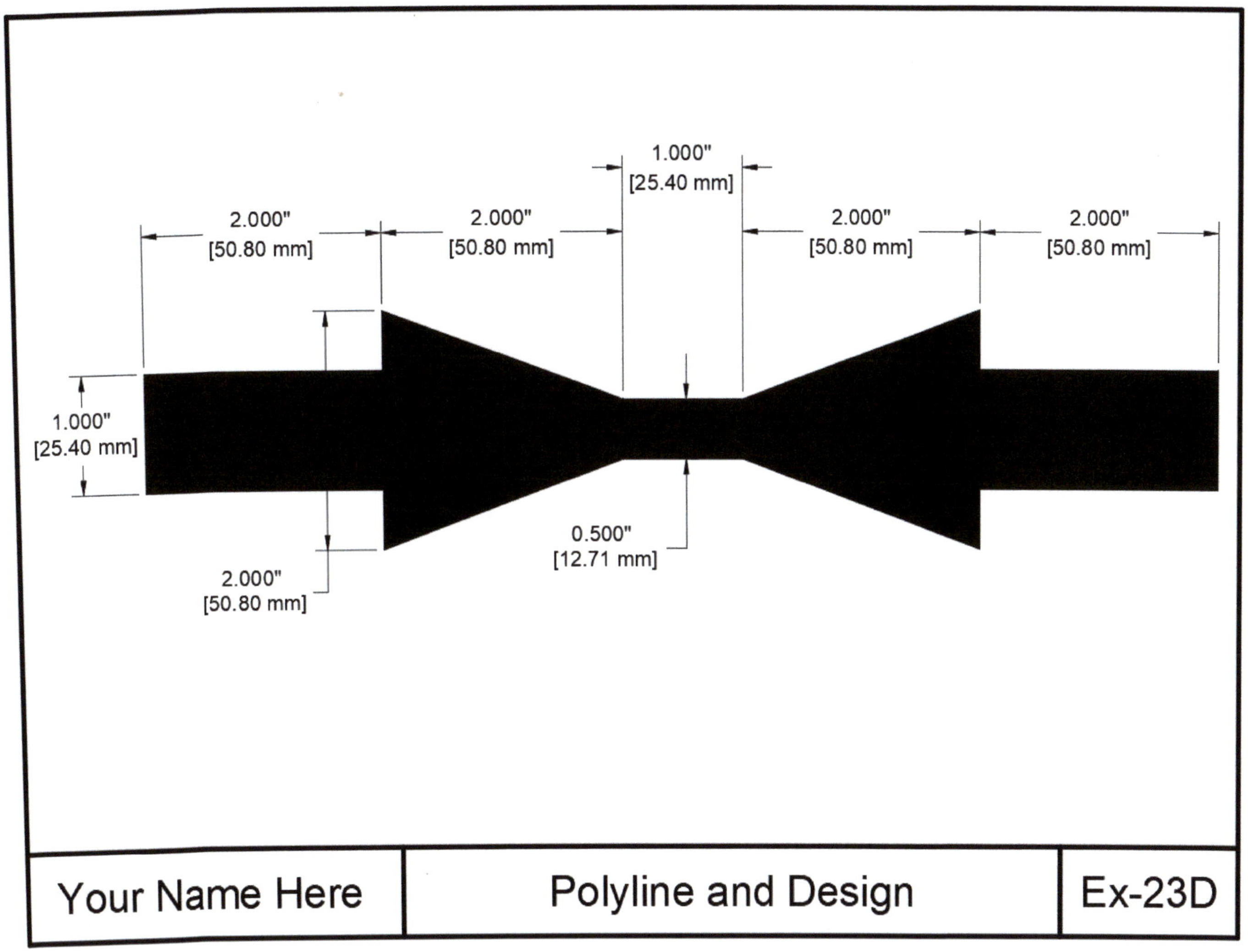

Exercise 23E

Instructions:

1. Start a **New** file using either the **Border A-Inch.dwt** or the **Border A-Metric.dwt**
2. Select the **Polyline** command.
3. Place the start point approximately as shown.
4. Set the **Widths**: Starting width = **0.100 [2.54]** Ending width = **0.500 [12.7]**.
5. Select option **Arc**.
6. Select option **Radius** and enter **2.50 [63.5]**.
7. Select option **Angle** and enter **180**.
8. Enter **90** for the **Direction of the Chord**.
9. Press ***<Enter>*** to end the command.
10. Do not dimension.
11. Edit the **Title** and **Ex-XX** by double clicking on the text. Do not erase and replace.
12. **Save** as **Ex-23E**
13. Plot using Page Setup **Class Model A**

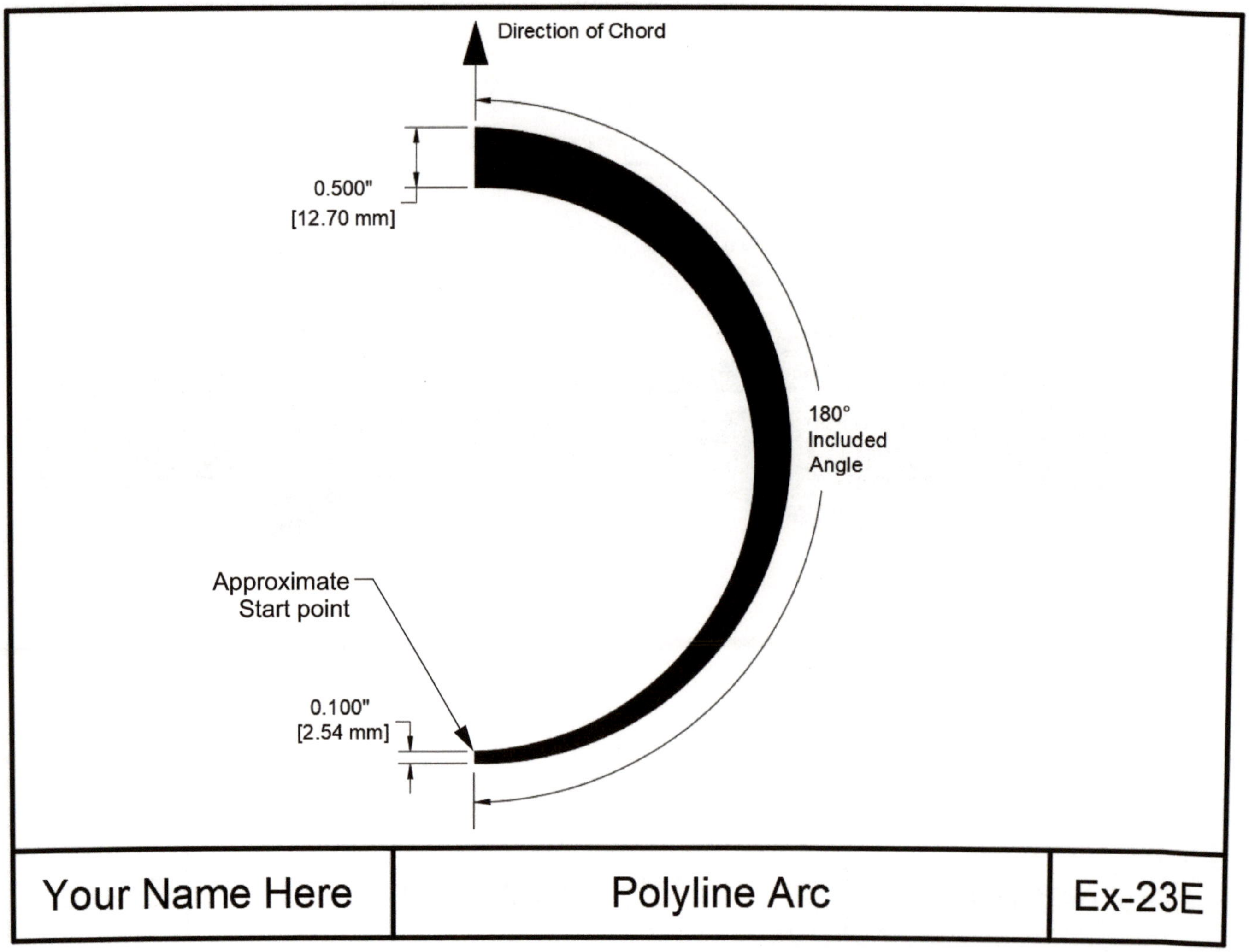

LESSON 24

LEARNING OBJECTIVES

After completing this lesson, you will be able to:

- Edit the width of Polylines
- Join Polylines
- Convert Polylines to Curves
- Convert a basic Line to a Polyline
- Join Lines, Polylines, Arcs, and Splines

Editing Polylines

The **Edit Polyline** command allows you to make changes to a polyline's option, such as the width. You can also change a regular line into a polyline and **Join** individual polyline segments.

Note: If you select a line that is not a polyline, the prompt will ask if you would like to turn it into a polyline.

1. Select the **Edit Polyline** command using one of the following:

   Ribbon = Home Tab / Modify Panel / ▾
   or
   Keyboard = *PE <Enter>*

   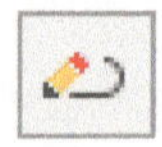

   The following will appear on the Command Line:

```
PEDIT Select polyline or [Multiple]:
```

Note: You may modify **"Multiple"** polylines simultaneously.

2. Select the polyline to be edited or type ***M*** (for multiple polylines).

   The following will appear on the Command Line:

```
PEDIT Enter an option [Close Join Width Edit vertex Fit Spline Decurve Ltype gen
Reverse Undo]:
```

3. Select an option (descriptions of each option are listed below).

Options

Close: This option connects the last segment with the first segment of an **Open** polyline. AutoCAD considers a polyline open unless you use the **Close** option to connect the segments originally.

Open Close

Open: This option removes the closing segment, but only if the **Close** option was used to close the polyline originally.

Join: The **Join** option allows you to join individual polyline segments into one polyline. The segments must have matching endpoints.

Four individual Polylines Joined one Polyline

Although the result will not look any different, it will become a single polyline which can be edited.

Continued on the next page...

Width: The **Width** option allows you to change the width of the polyline. But the entire polyline will have the same width.

Edit Vertex: This option allows you to change the starting and ending width of each segment individually.

Spline: This option allows you to change straight polylines to curves.

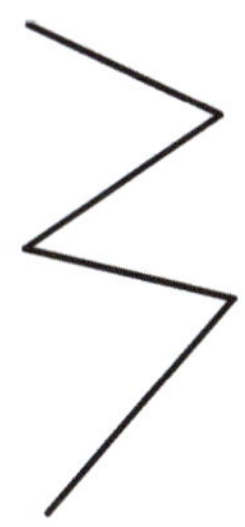

Polyline

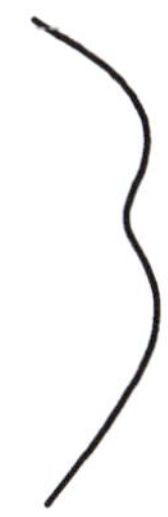

Splined

Decurve: This option removes the spline curves and returns the polyline to its original straight line segments.

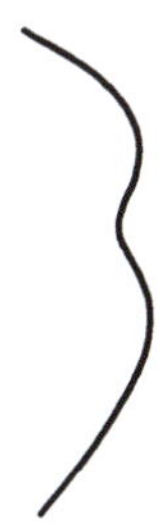

Splined

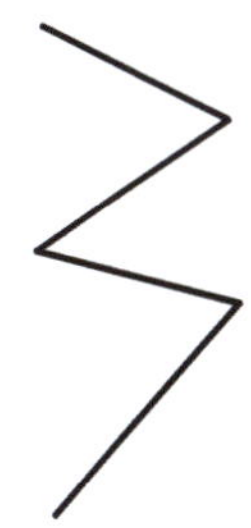

Polyline

Reverse: This option reverses the direction. The start point becomes the endpoint and vice versa.

Join Command

This might seem confusing, but this **Join** command is not the same as the Join "option" within the Polyedit tool. The Polyedit option can only be used to join individual polyline segments. This Join command joins similar objects to form a single unbroken object.

1. Select the **Join** command using one of the following:

   Ribbon = Home Tab / Modify Panel / ▾ 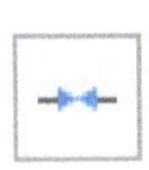
   or
   Keyboard = *J <Enter>*

   The following will appear on the Command Line:

   `JOIN Select source object or multiple objects to join at once:`

2. Select the **source** object.

Continued on the next page...

The following will appear on the Command Line:

3. Select the object(s) to **join** to the **source** object and then press ***<Enter>***.

Rules for Using the Join Command

Lines: The lines must be collinear (on the same plane) but can have gaps between them.

Before Join

After Join

Polylines: Same rules as lines except no gaps allowed between them.

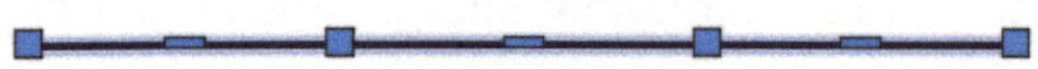

Before Join

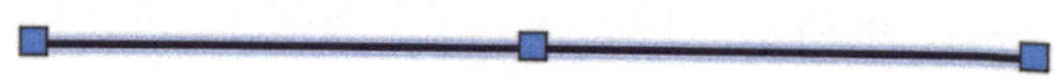

After Join

Arc: The arcs must lie on the same imaginary circle but can have gaps.

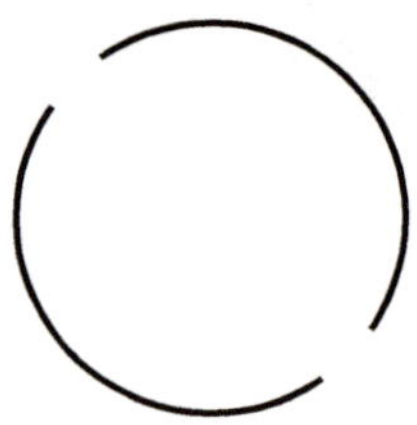

Before Join

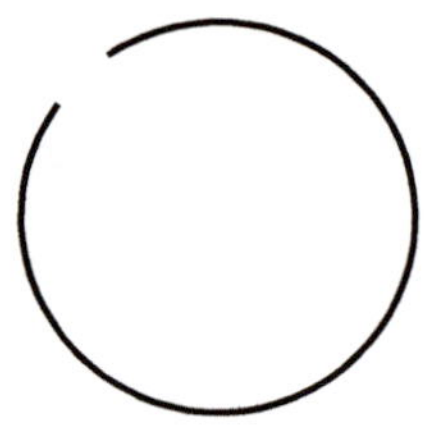

After Join

The **Close** option in Polyedit can be used to turn it into a circle.

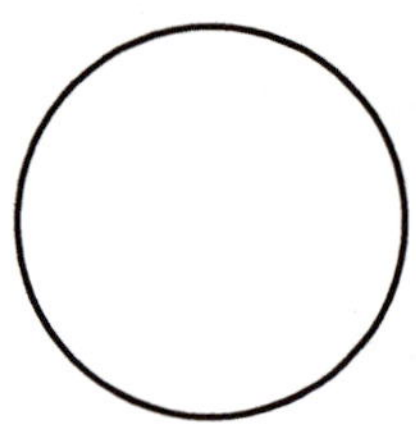

Spline: The spline must lie in the same plane and be contiguous (end to end). (You haven't learned this command yet.)

Exercise 24A

Instructions:

1. Start a **New** file using either the **Border A-Inch.dwt** or the **Border A-Metric.dwt**

Step 1.

2. Draw the polyline **4" [101.6 mm]** Long and **0** width using Layer Object Line.

Step 2.

3. Change the width of the polyline to **0.500 [12.7]**.
4. Do not dimension.
5. Edit the **Title** and **Ex-XX** by double clicking on the text. Do not erase and replace.
6. **Save** as **Ex-24A**
7. Plot using Page Setup **Class Model A**

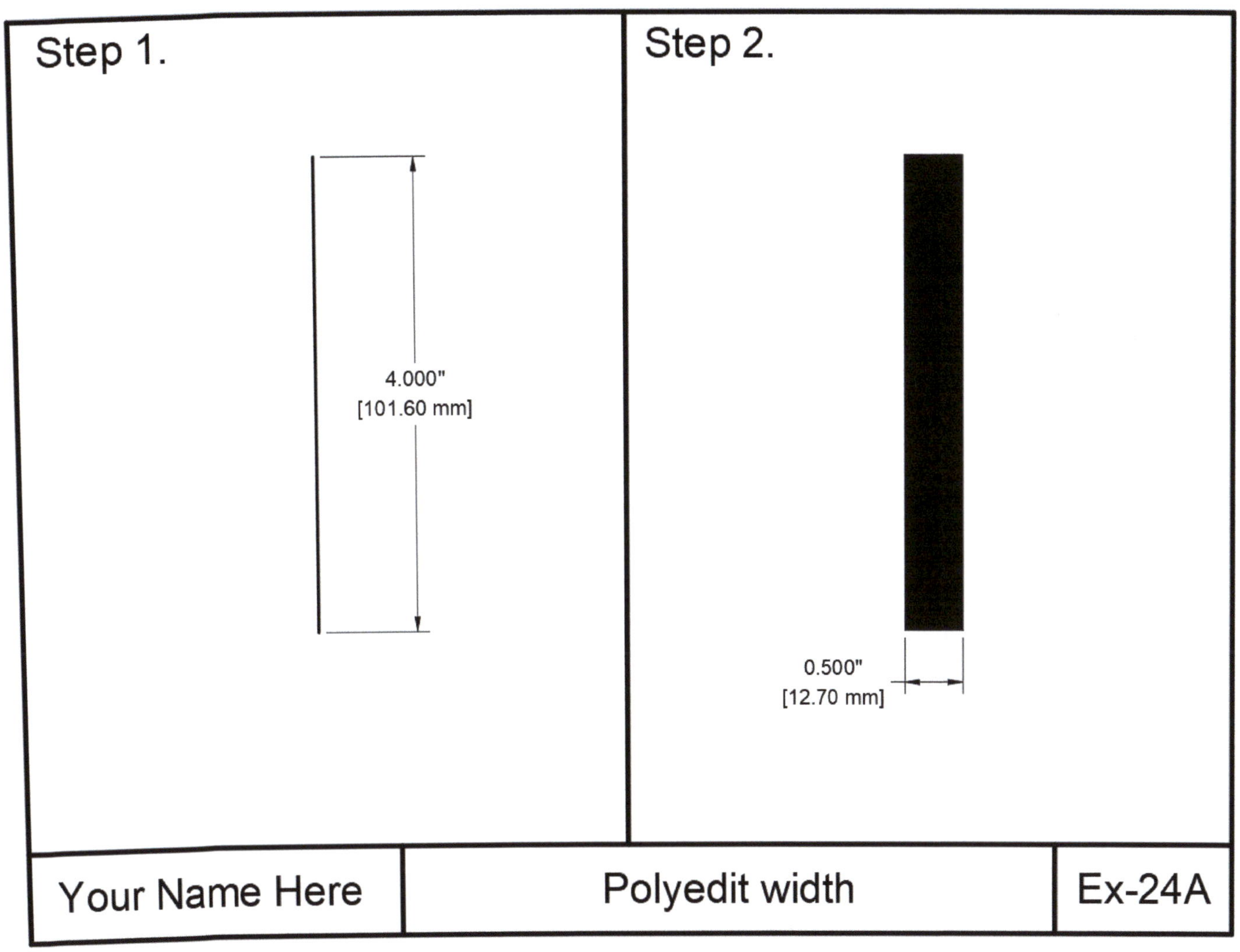

Exercise 24B

Instructions:

1. Start a **New** file using either the **Border A-Inch.dwt** or the **Border A-Metric.dwt**

Step 1.

2. Draw one continuous polyline with (3) **2" [50.8 mm]** Long segments.

Step 2.

3. Select the Polyedit command and select the first segment.
4. Select the Edit Vertex option. (Notice the "X" marking the starting point.
5. Select the Width option and set Start to **1 [25.4]** and Ending to **0.5 [12.7]**.
6. Select Next option. (The "X" should have moved to the next segment.)
7. Select the Width option and set Start to **0.5 [12.7]** and Ending to **0.5 [12.7]**.
8. Select the Next option (The "X" should have moved to the next segment).
9. Select the Width option and set Start to **2 [50.8]** and Ending to **0**.
10. Select the Exit option and press ***<Enter>*** to stop.
11. Edit the **Title** and **Ex-XX**. **Save** as **Ex-24B**. Plot using Page Setup **Class Model A**

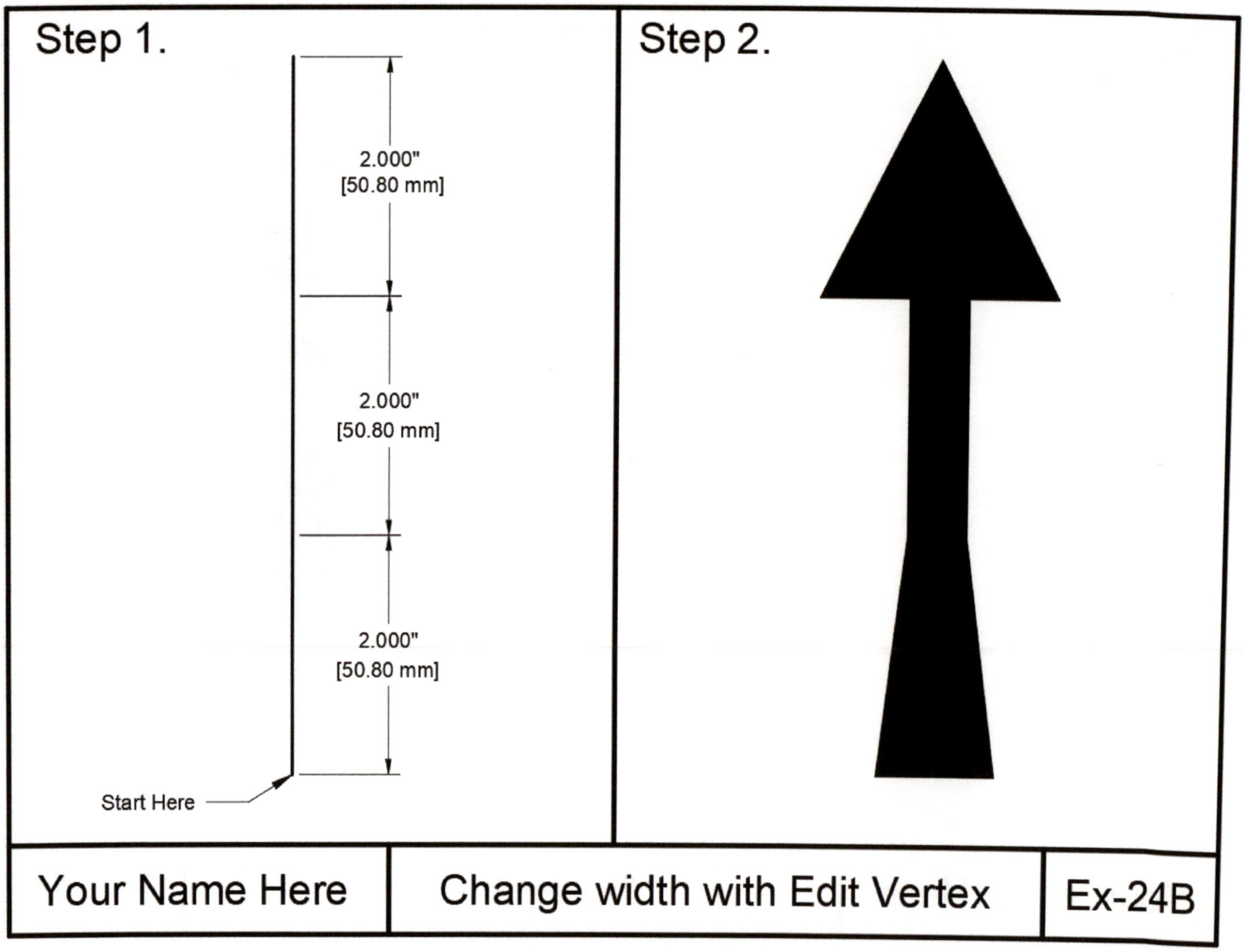

Exercise 24C

Instructions:

1. Start a **New** file using either the **Border A-Inch.dwt** or the **Border A-Metric.dwt**

Step 1.

2. Draw the objects below using the **Line** command not **Polyline**.

Step 2.

3. Select the Polyedit command and select the first segment drawn. (Answer **Yes** to *"Not a polyline, do you want to turn it into one?"*)
4. Select the Width option and set the width to **0.1 [2.54]**. (Notice only one segment changed.)
5. Select the Join option.
6. Select the remaining lines using a crossing window. (Now the width should be constant for all lines.)
7. Edit the **Title** and **Ex-XX** by double clicking on the text. Do not erase and replace.
8. **Save** as **Ex-24C**
9. Plot using Page Setup **Class Model A**

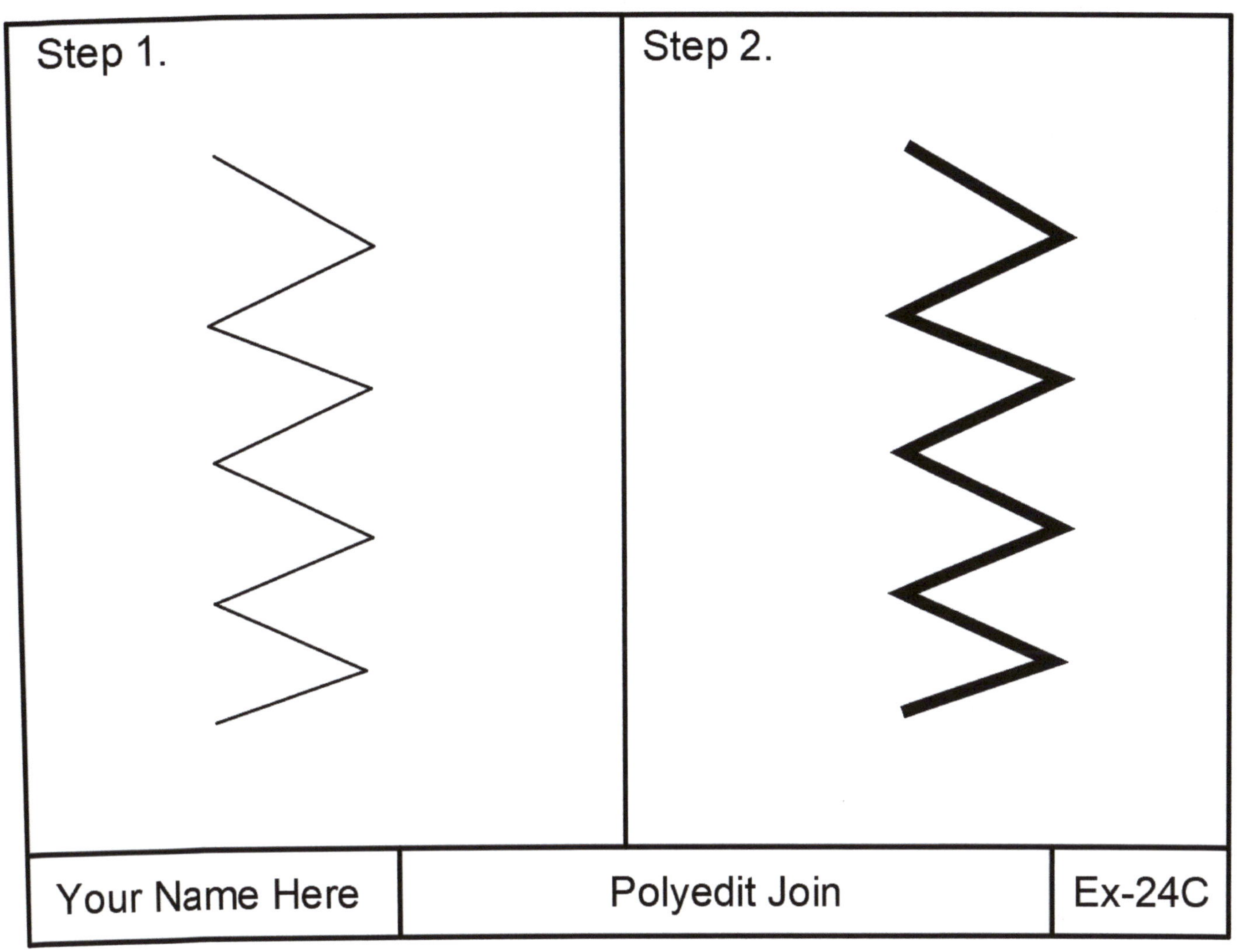

Exercise 24D

Instructions:

1. Open **Ex-24C**
2. Select the Polyedit command.
3. Select the polyline.
4. Select the Spline option
5. Edit the **Title** and **Ex-XX** by double clicking on the text. Do not erase and replace.
6. **Save** as **Ex-24D**
7. Plot using Page Setup **Class Model A**

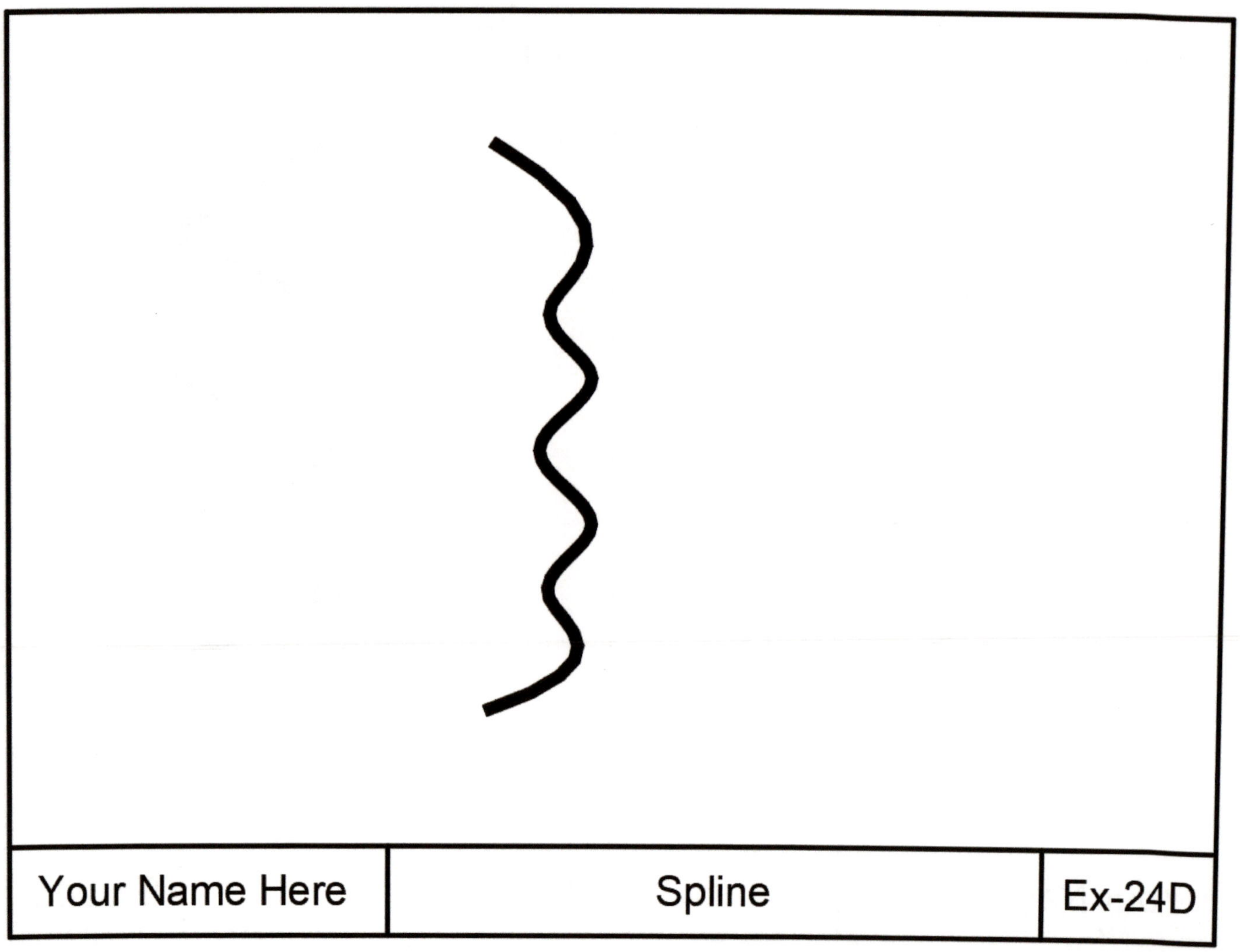

LESSON 25

LEARNING OBJECTIVES

After completing this lesson, you will be able to:

- Create a new Text style
- Change an existing Text style
- Delete a Text style
- Select a Point style
- Divide an object into a specified number of segments
- Divide an object into segments based on a specified measurement

Creating New Text Styles

AutoCAD provides you with two preset text styles named "Standard" and "Annotative". (Annotative will be discussed in Lesson 27.) You may want to create a new text style with a different font and effects. The following illustrates how to create a new text style.

1. Select the **Text Style** command using one of the following:

   Ribbon = Annotate Tab / Text Panel / ↘
   or
   Keyboard = *ST <Enter>*

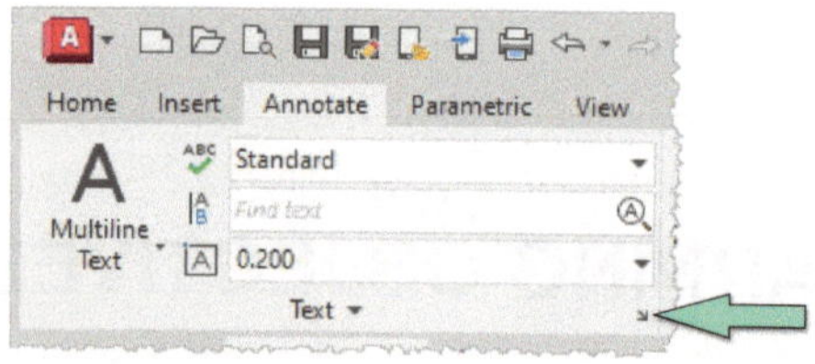

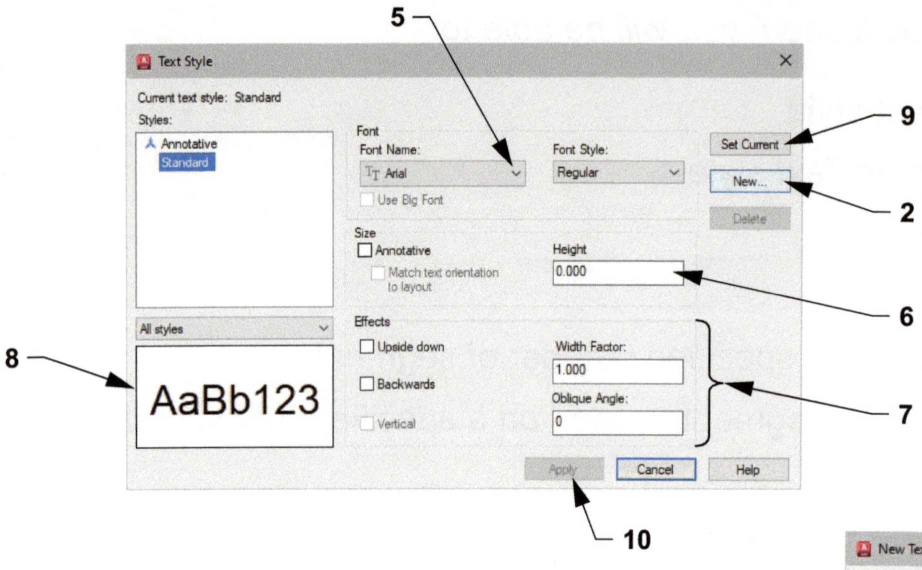

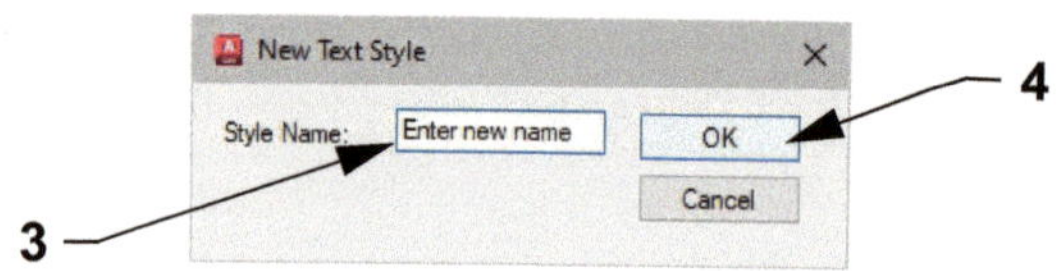

2. Select the **New...** button.

3. Type the new style name in **Style Name** box. Styles names can have a maximum of 31 characters.

4. Select the **OK** button.

5. Select the **Font**.

6. Enter the value of the **Height**.

Note: If the value is **0**, AutoCAD will always prompt you for a height. If you enter a number, the new text style will have a fixed height and AutoCAD will not prompt you for the height.

7. Assign **Effects**.

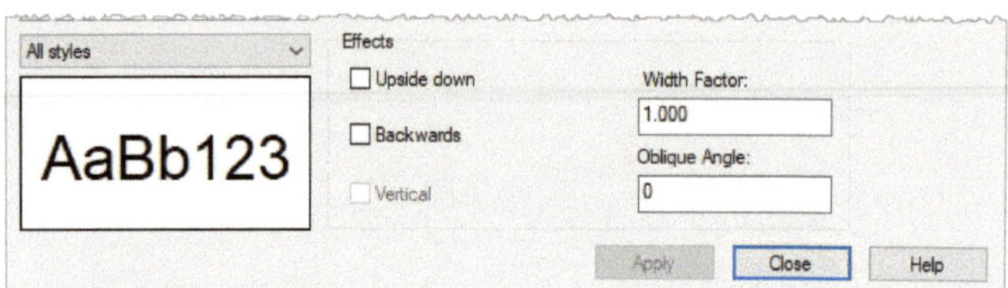

Upside down: Each letter will be created upside-down in the order in which it was typed. (**Note:** This is different from rotating text 180 degrees.)

Continued on the next page...

Backwards: The letters will be created backwards as typed.

Vertical: Each letter will be inserted directly under the other. Only **".shx"** fonts can be used. **Vertical** text will not display in the **Preview** box.

Width Factor: This effect compresses or extends the width of each character. A value less than **1** compresses each character. A value greater than **1** extends each character.

Oblique Angle: Creates letter with a slant, like italic. An angle of **0** creates a vertical letter. A positive angle will slant the letter forward. A negative angle will slant the letter backward.

8. The **Preview** box displays the text with the selected settings. In the example below, it shows what the text will look like if you select **Upside down** text.

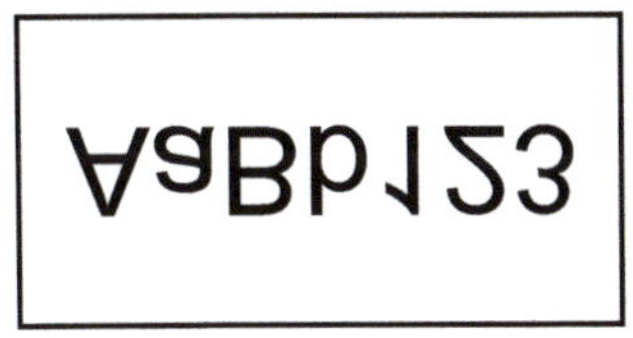

9. Select the **Set Current** button.

10. Select the **Apply** or **Close** button.

Selecting a Text Style

After you have created text styles, you will want to use them. You must select the text style before you use it.

Below are the methods when using Single Line or Multiline text.

Single Line Text

Select the style before selecting **Single Line Text** command.

1. Select the **Annotate** Tab.

2. Using the **Text Panel**, select the style down arrow ▼

3. Select the **text style**.

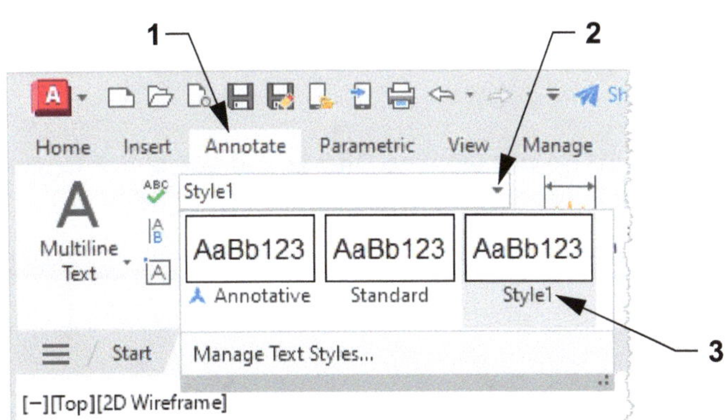

Continued on the next page...

Multiline Text

1. Select **Multiline Text** and place the **first corner** and **opposite corner**.
2. Find the **Style** Panel and scroll through the text styles available using the up and down arrows.

2 — Select the Text Style within the Text Editor.

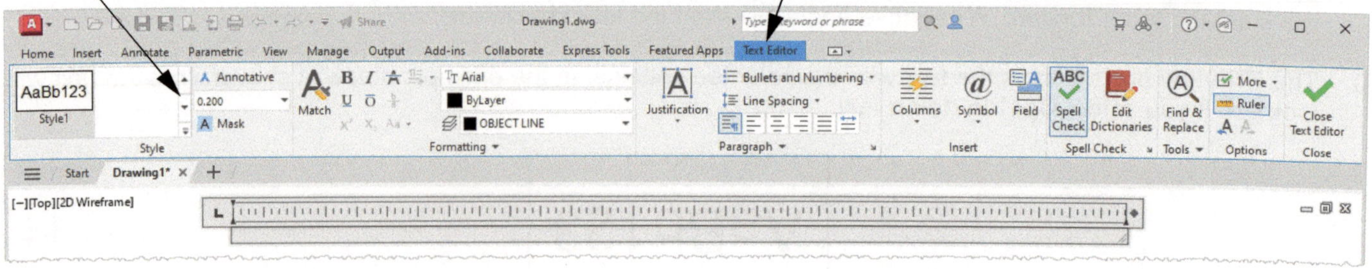

Deleting a Text Style

1. Select the **Text Style** command using one of the following:

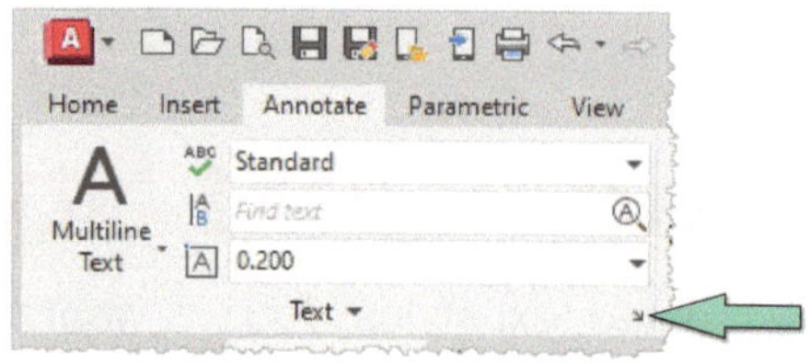

Ribbon = Annotate Tab / Text Panel / ↘
or
Keyboard = *ST <Enter>*

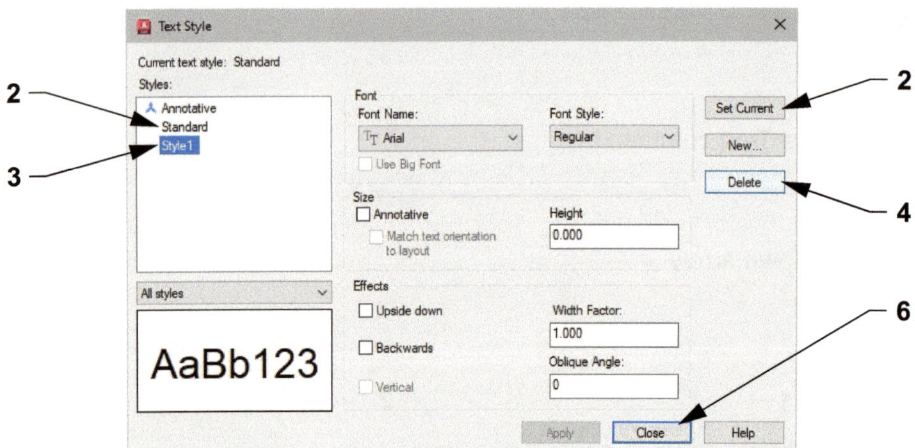

2. First, select a **text style** that you **do not want to Delete** and select **Set Current** button. (You can't Delete a Text Style that is in use.)
3. Select the **text style** that you want to Delete.
4. Select the **Delete** button.
5. A warning appears, select **OK** or **Cancel**.
6. Select the **Close** button.

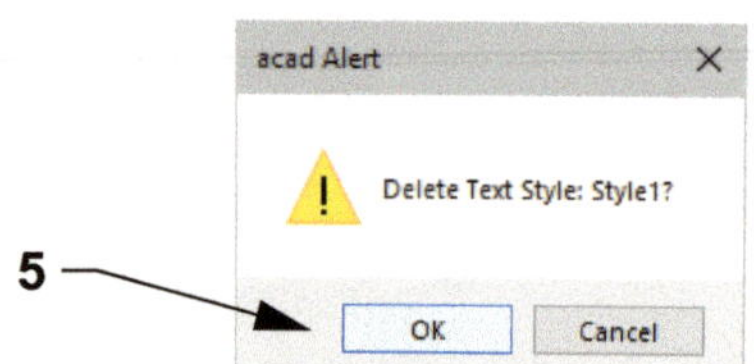

Note: Also refer to page 29-12 for the **Purge** command. The Purge command will remove any unused text styles, dimension styles, layers, and linetypes.

Changing the Effects of a Text Style

1. Select the **Text Style** command using one of the following:

Ribbon = Annotate Tab / Text Panel / ↘
or
Keyboard = *ST <Enter>*

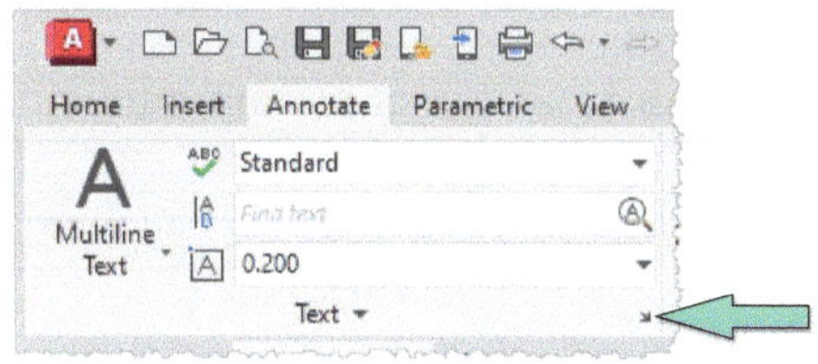

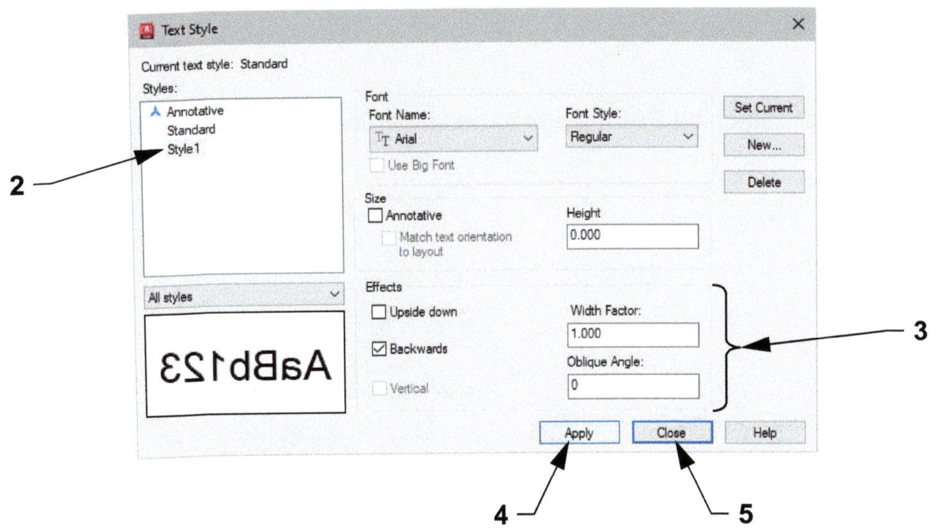

2. Select the **text style** you want to change.
3. Make the changes in the Effects boxes.

   Note about Vertical text: Only **.shx** fonts can be vertical. Vertical text will not display in the **Preview** box.

4. Select the **Apply** button. (Apply will stay gray if you did not change a setting.)
5. Select the **Close** button.

Divide Command

The **Divide** command divides an object mathematically by the **number** of segments you specify. A **Point** (object) is placed at each interval on the object.

Note: The object selected is not broken into segments. The Points are simply drawn on the object.

Example:

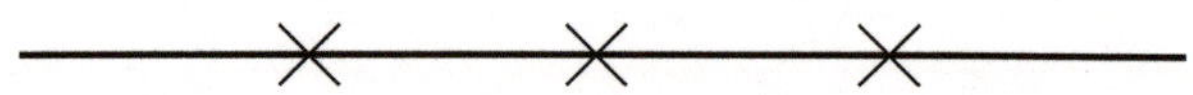

This **Line** has been divided into four **equal** lengths.
But remember, the line is not broken into segments.
The Points are simply drawn on the object.

1. First open the **Point Style** dialog box and select the **Point Style** to be placed on the object, using one of the following:

   Ribbon = Home Tab / Utilities Panel ▾ / Point Style
   or
   Keyboard = *ddptype <Enter>*

Select either of these two Point Styles only when you want to make the points disappear yet still be there.

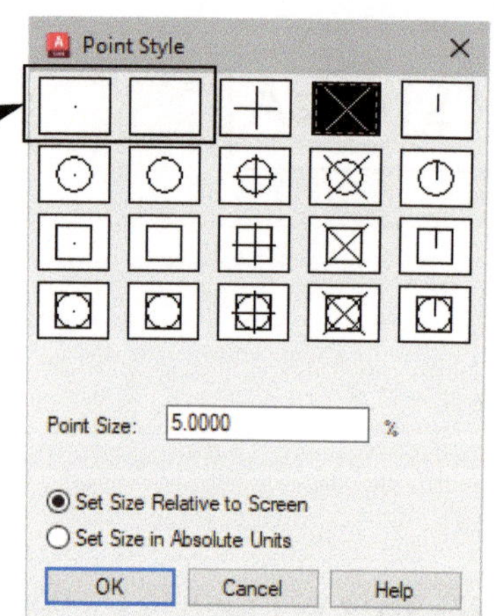

Note: Refer to page 5-7 if you need a refresher on Points.

2. Next select the **Divide** command using one of the following:

   Ribbon = Home Tab / Draw Panel ▾ / 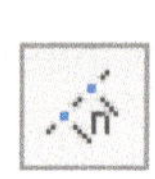
   or
   Keyboard = *div <Enter>*

   The following will appear on the Command Line:

   `DIVIDE Select object to divide:`

3. Select the object to divide.

   The following will appear on the Command Line:

   `DIVIDE Enter the number of segments or [Block]:`

4. Type the number of segments and then press ***<Enter>***.

Measure Command

The **Measure** command is very similar to the **Divide** command because point objects are drawn at intervals on an object. However, the **Measure** command allows you to specify the **length** of the segments rather than the number of segments.

Note: The object selected is not broken into segments. The Points are simply drawn on the object.

Example:

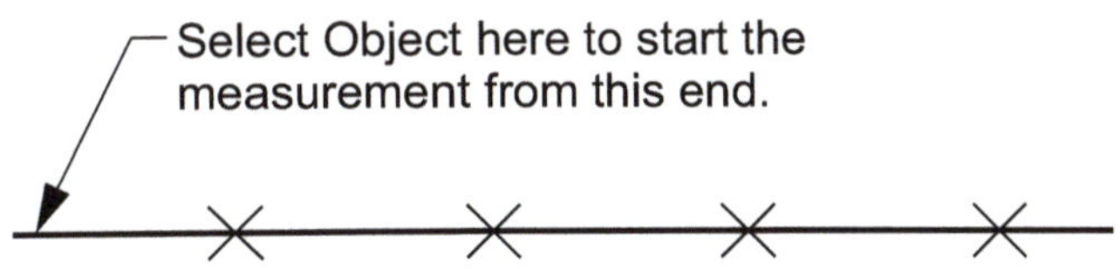

The **Measurement** was started at the left endpoint, and ended just short of the right end of the line. The remainder is less than the measurement length specified. You designate which end you want the measurement to start at by selecting the end when prompted to select the object.

1. First open the **Point Style** dialog box and select the **Point Style** to be placed on the object, using one of the following:

   Ribbon = Home Tab / Utilities Panel ▾ / Point Style
   or
   Keyboard = *ddptype <Enter>*

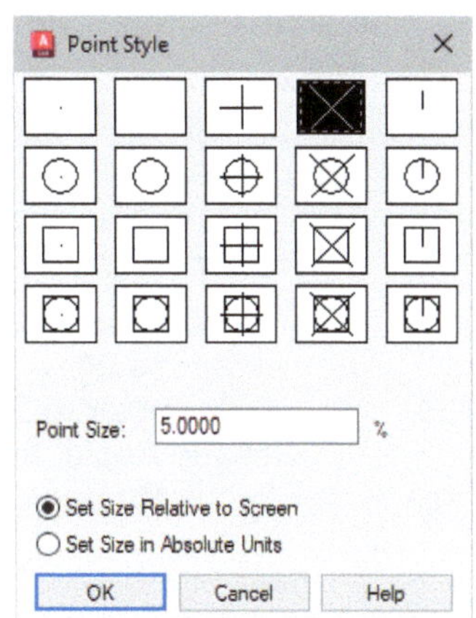

Note: Refer to page 5-7 if you need a refresher on Points.

2. Next select the **Measure** command using one of the following:

   Ribbon = Home Tab / Draw Panel ▾ / 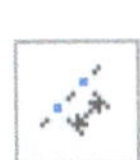
   or
   Keyboard = *me <Enter>*

   The following will appear on the Command Line:

3. Select the object to measure.

   The following will appear on the Command Line:

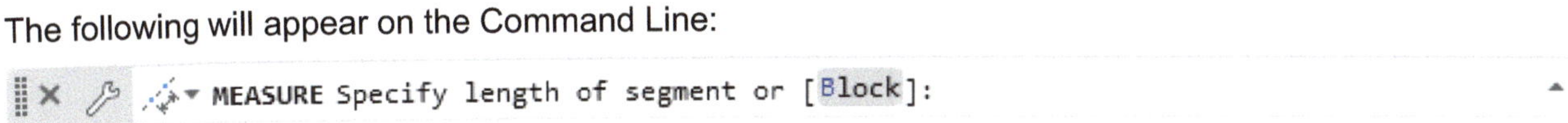

4. Type the length of one segment and then press ***<Enter>***.

Exercise 25A

Instructions:

1. Start a **New** file using the **Border A-Inch.dwt** or the **Border A-Metric.dwt**
2. Create the three text styles shown below. (Style 1, Style 2 and Style 3.)
3. Select all of the options shown.
4. **Save** as **Ex-25A**

Style 1

Inch users

Font: **romand.shx**
Height: **0.500**
Width Factor: **1.000**
Oblique Angle: **30**
Effects: **Vertical**

Metric users

Font: **romand.shx**
Height: **12.70**
Width Factor: **1.000**
Oblique Angle: **30**
Effects: **Vertical**

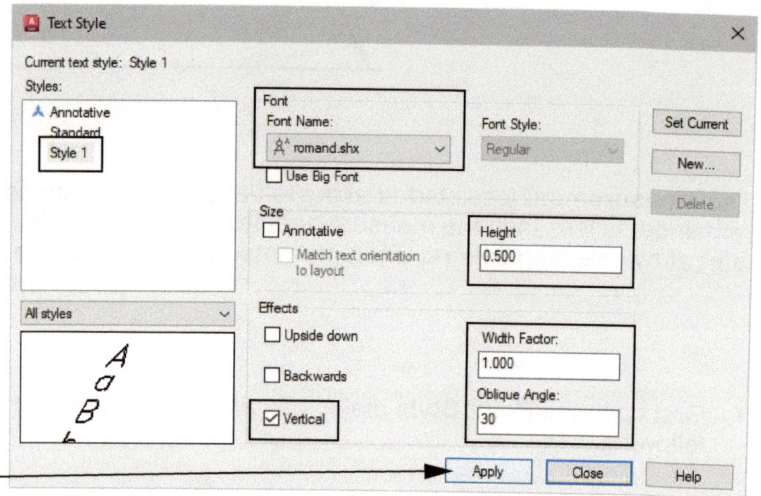

After selecting all the text features shown, select **Apply**, then the **New...** button again to start the new style.

Style 2

Inch users

Font: **gothice.shx**
Height: **0.750**
Width Factor: **0.750**
Oblique Angle: **0**
Effects: **Backwards**

Metric users

Font: **gothice.shx**
Height: **19.00**
Width Factor: **0.75**
Oblique Angle: **0**
Effects: **Backwards**

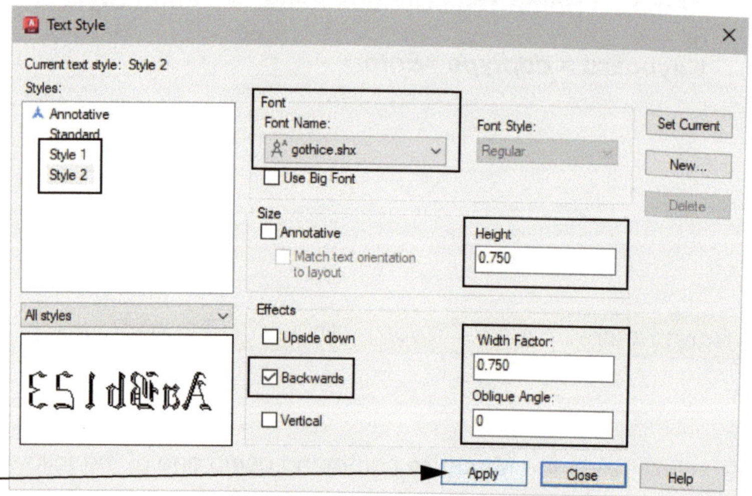

After selecting all the text features shown, select **Apply**, then the **New...** button again to start the new style.

Style 3

Inch users

Font: **italic.shx**
Height: **1.500**
Width Factor: **0.750**
Oblique Angle: **0**
Effects: **Upside down**

Metric users

Font: **italic.shx**
Height: **38.10**
Width Factor: **0.75**
Oblique Angle: **0**
Effects: **Upside down**

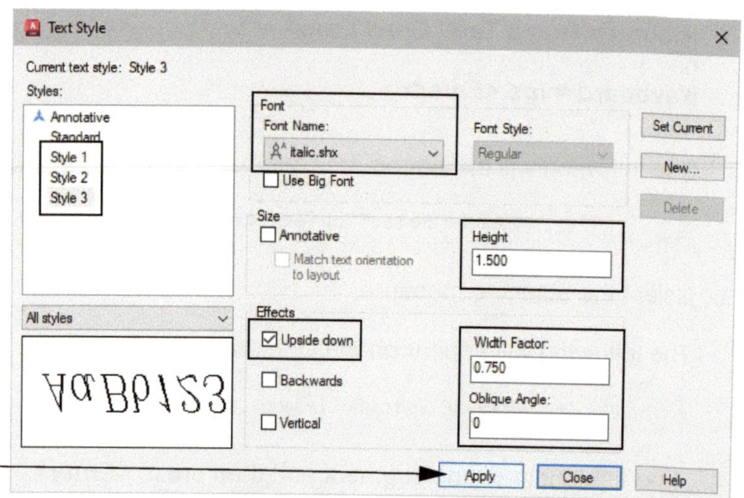

Select **Apply** and then **Close**.

Exercise 25B

Instructions:

1. Open **Ex-25A** (if not already open).
2. Before entering text, move the **Origin** to the lower left corner as shown.
3. Draw the three text styles using “Single Line Text”.
   Remember to select the text style before selecting the Single Line Text command.
4. Place the start point exactly as shown. (**Enter coordinates and the Rotation angle**.)
5. Use Layer Text.
6. Edit the **Title** and **Ex-XX** by double clicking on the text. Do not erase and replace.
7. **Save** as **Ex-25B**
8. Plot using Page Setup **Class Model A**

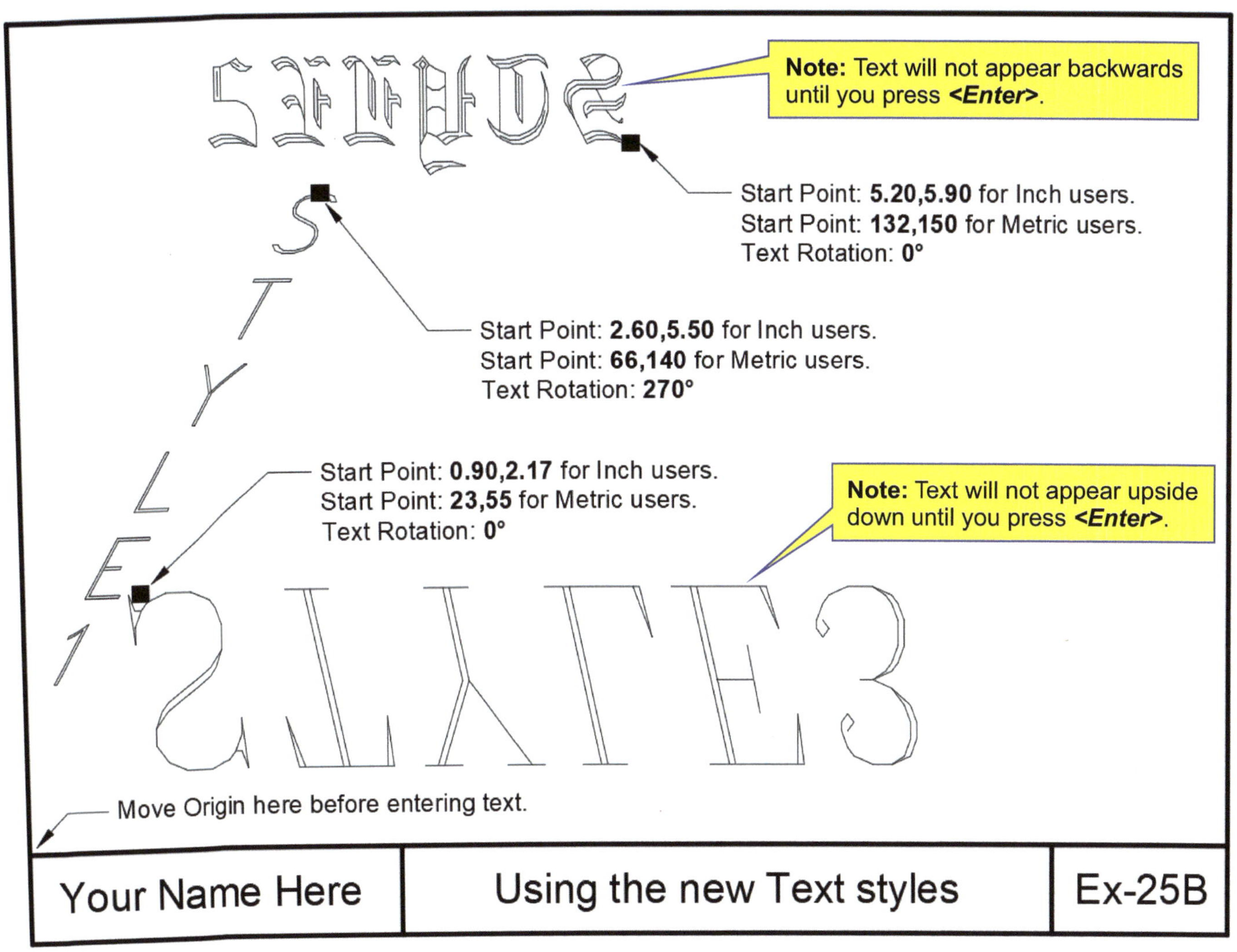

Exercise 25C

Instructions:

1. Open **Ex-25B** (if not already open).
2. Change the text to Text Style **Standard**.
   A. Select all the text.
   B. Right click and choose **"Quick Properties"** from the menu.
   C. Change text **"VARIES"** to **Standard**.
3. Edit the **Title** and **Ex-XX** by double clicking on the text. Do not erase and replace.
4. **Save** as **Ex-25C**
5. Plot using Page Setup **Class Model A**

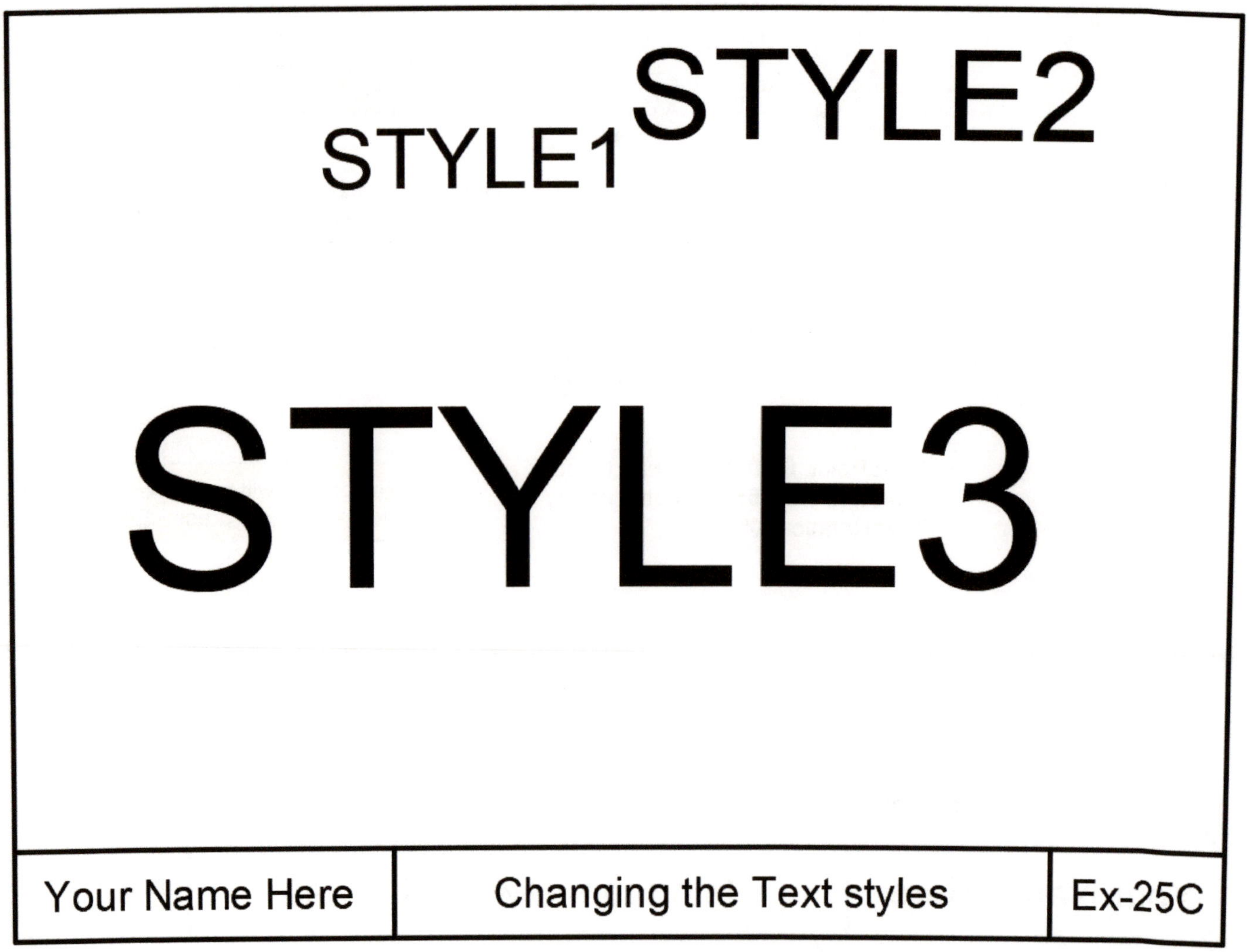

Exercise 25D

Instructions:

1. Open **Ex-25C** (if not already open).
2. Delete Text Styles 1, 2, and 3 as follows.
   A. Select the **Text Style** dialog box.
   B. Select the **Standard** text style.
   C. Select the **Set Current** button. (This makes Standard the current text style.)

Note: AutoCAD will not allow you to delete an active or current text style. That is why you must make the Standard text style current before you can delete a text style.

   D. Select the text style you want to **delete**.
   E. Click on the **Delete** button.

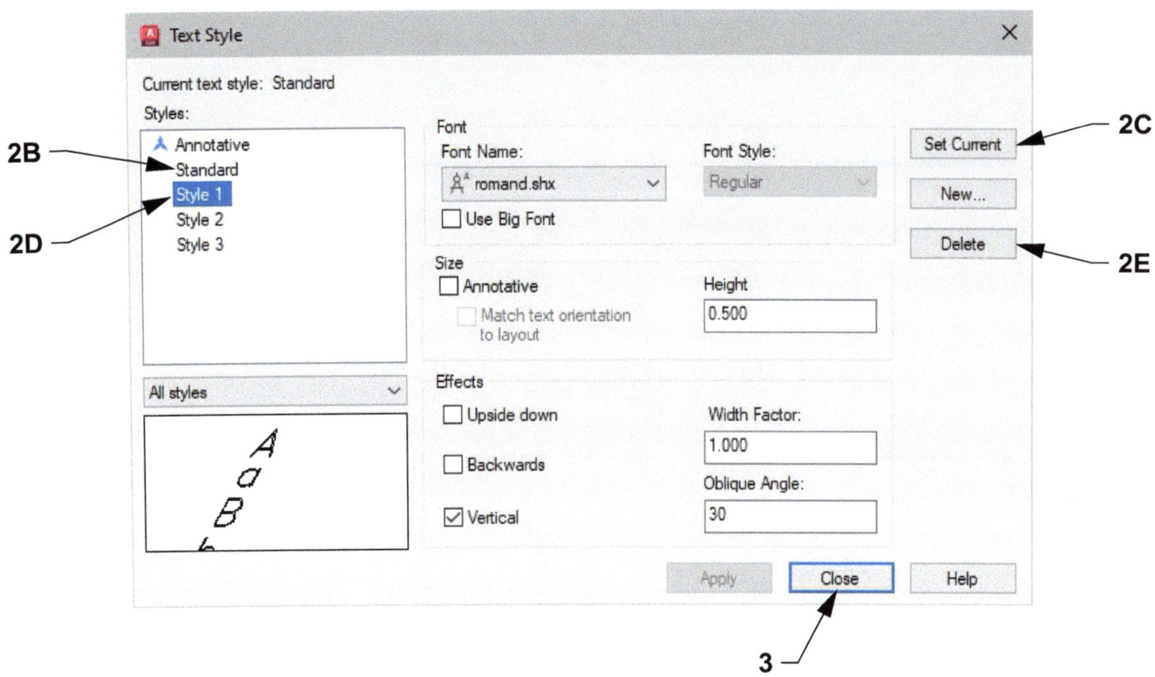

3. Select the **Close** button.
4. **Save** as **Ex-25D**

Exercise 25E

Instructions:

1. Start a **New** file using the **Border A-Inch.dwt** or the **Border A-Metric.dwt**
2. Draw the two **8" [203.2 mm]** long lines shown below on Layer Object Line.
3. **Divide** the upper line into 10 equal segments.
4. Using **Measure** create five **1.500" [38.1 mm]** long segments on the lower line as shown.
5. Dimension as shown. (Remember **"Node"** is the Object Snap to use for Points.)
6. Edit the **Title** and **Ex-XX** by double clicking on the text. Do not erase and replace.
7. **Save** as **Ex-25E**
8. Plot using Page Setup **Class Model A**

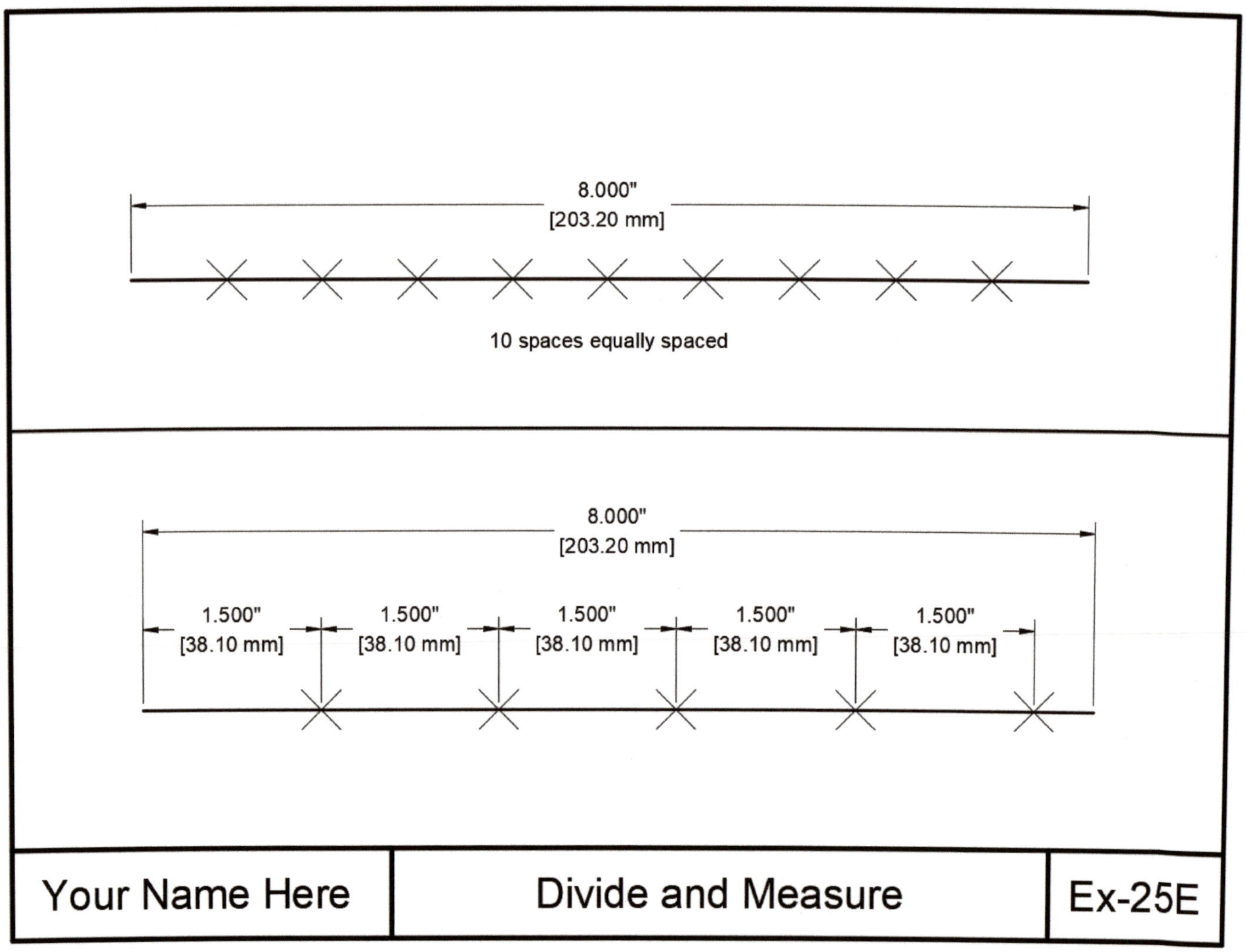

Exercise 25F

Instructions:

1. Start a **New** file using the **Border A-Inch.dwt** or the **Border A-Metric.dwt**
2. Draw the "pretend" house roof plan with the fence line shown below.
3. **Refer to the next page for instructions and helpful hints**.
4. Dimension using Dimension Style CLASS STYLE and Layer Dimension.
5. Edit the **Title** and **Ex-XX** by double clicking on the text. Do not erase and replace.
6. **Save** as **Ex-25F**
7. Plot using Page Setup **Class Model A**

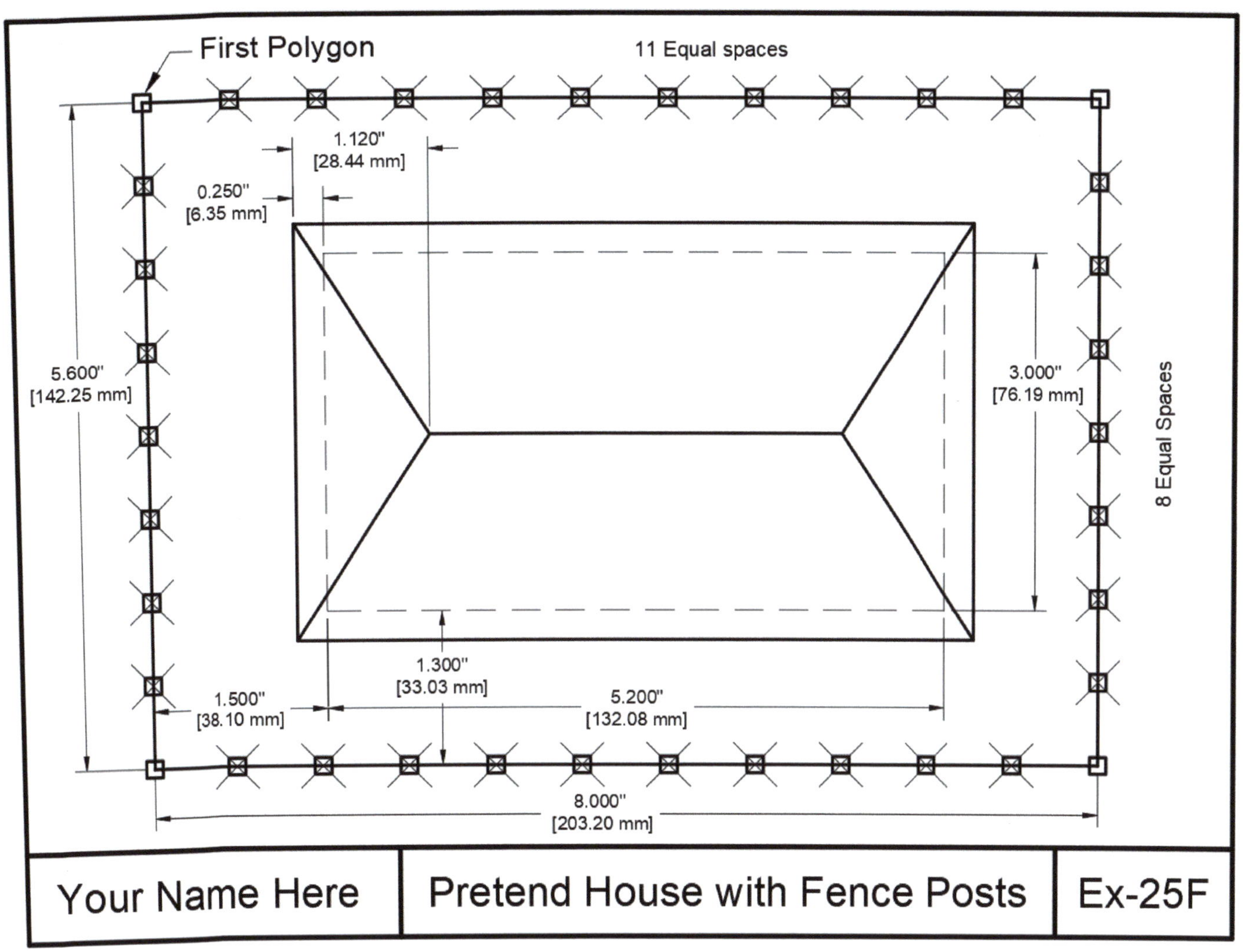

Exercise 25F Helper

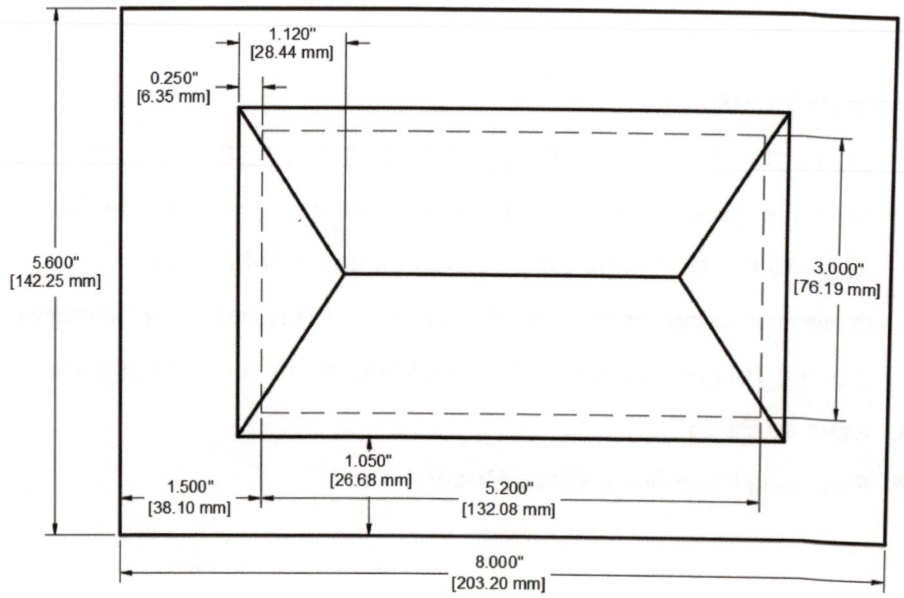

Step 1.

Draw the "pretend" house roof plan with the fence line.

Use Layers Roof, Walls, and Hidden Lines.

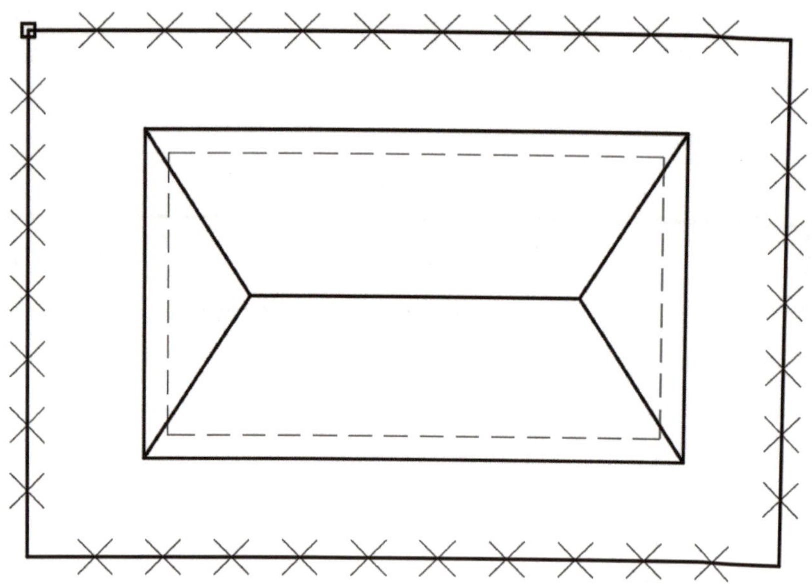

Step 2.

Divide the fence lines: 11 equal spaces horizontally, 8 equal spaces vertically.

Step 3.

Draw a fence post at upper-left corner using an inscribed, four-sided polygon, **0.100" [2.54 mm]** radius.

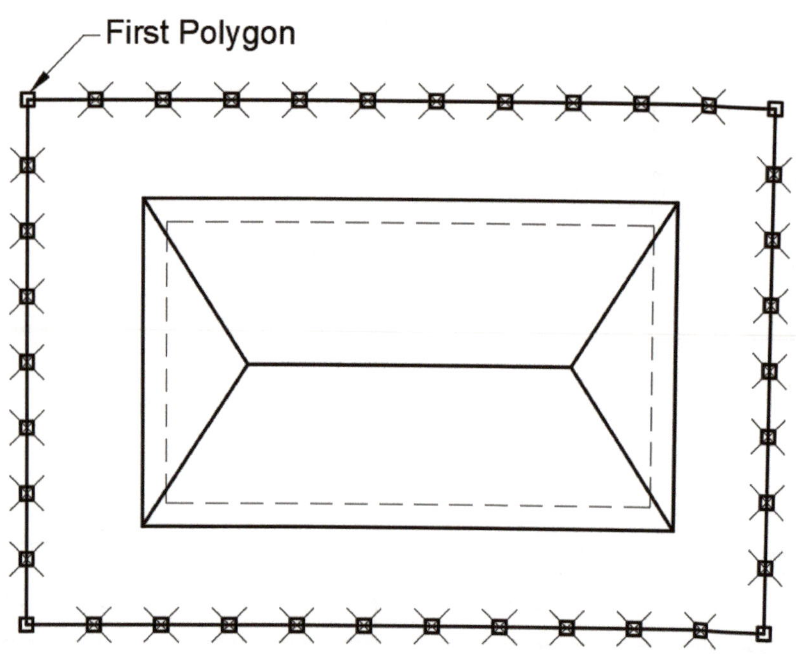

Step 4.

Copy the fence post to the "Points" and corner. Use Object Snap "Node" to snap to the "Points".

Step 5.

Change the Point style so the Points will not be displayed.

LESSON 26

LEARNING OBJECTIVES

After completing this lesson, you will be able to:

- Understand the difference between Model and Layout Tabs
- Create Viewports
- Create a Page Setup
- Create a Plot Page Setup
- Create a Master Decimal Setup template
- Create a new Border
- Experiment with multiple Viewports

Serious Business

In the previous lessons, you have been having fun learning most of the basic commands that AutoCAD offers and you have been using a template that included preset layers and drawing settings, etc. But now it is time to get down to the "serious business" of setting up your own drawing from scratch.

Starting from scratch means you will need to set or create the following:

Items 1 through 5 you have learned in previous lessons.

1. Drawing Units (Lesson 4).
2. Snap and Grid (Lesson 2).
3. Create text styles (Lesson 25).
4. Create dimension styles (Lesson 16).
5. Create new layers and load linetypes.

Items 6 through 9 will be learned in this lesson.

6. Create a "Layout" for plotting.
7. Create a "Floating Viewport" in the Layout.
8. Create a "Page Setup" to save plot settings.
9. Plot the drawing from Paper space.

After reading pages 26-3 through 26-17, start Exercise 26A and work your way through 26D. When you have completed Exercise 26E, you will have created a master template named, **"My Decimal Setup"**. This master template will have everything set, created, and prepared, ready to use each time you want to create a drawing using decimal units and to be plotted on an **11" X 8-1/2"** or **297 mm X 210 mm** sheet.

This means, for future drawings, you merely select **File / New** and **"My Decimal Setup.dwt"** and start drawing. No time consuming setups. It is all ready to go.

In Lesson 27, you will create a master template for **"feet and inches"**. Although metric users no longer use feet and inches, it is good practice and you can print your results out on an **A4** size sheet of paper.

So take it one page at a time and really concentrate on understanding the process.

Note: I am using sheet sizes of **11" X 8-1/2"** and **297 mm X 210 mm** so you can print the exercises on your printer. After you understand the Page Setup concept, you will be able to assign a larger sheet size to conform to any large format printer that you might use in the future.

Model and Layout Options

Very important: Before I discuss Model and Layout, I need you to confirm the Model and Layout Tabs are displayed. This will just take a minute.

How to turn on display of the Model and Layout Tabs

1. Type ***options*** and then press ***<Enter>***.

   or

   Right click and select **Options**.

2. Select the **Display** Tab.

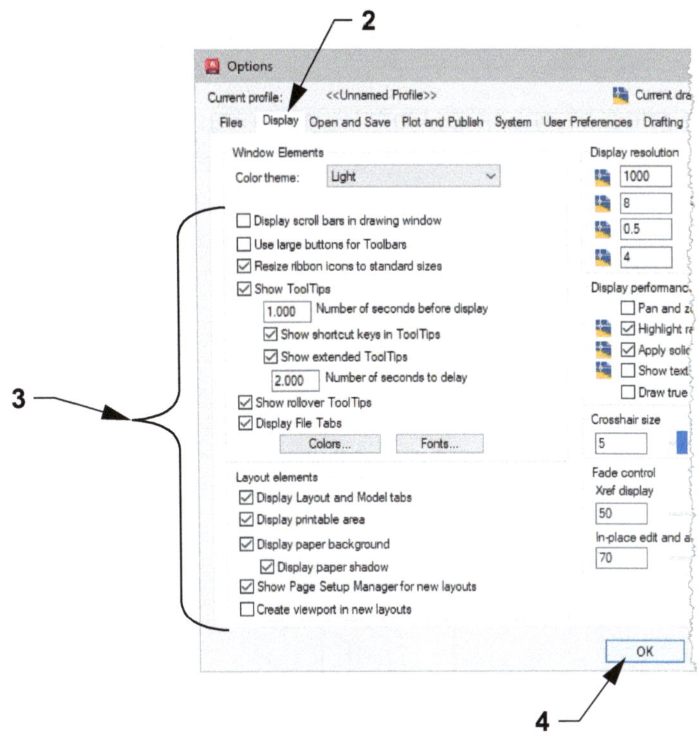

3. Check and un-check boxes as shown.

4. Select the **OK** button.

5. The lower left corner of the drawing area should display the three Tabs: Model, Layout1, and Layout2. And a few tools should be displayed in the lower right corner on the Status Bar.

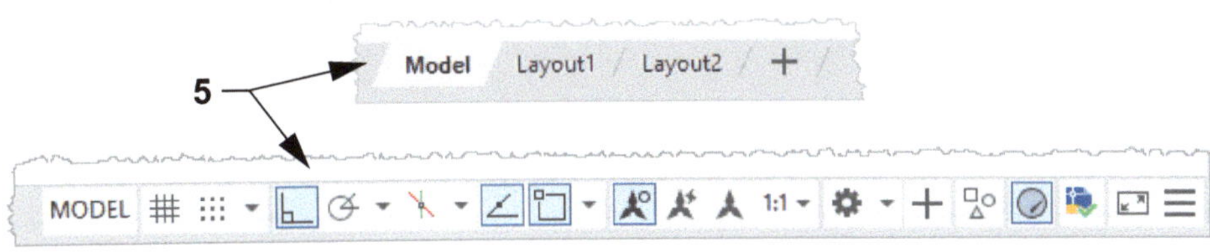

Model and Layout Tabs

Read this information carefully: It is very important that you understand this concept and the information on the following pages.

AutoCAD provides two drawing spaces, Model and Layout. You move into one or the other by selecting either the Model or Layout Tabs, located at the bottom left of the drawing area. (If you do not have these displayed, follow the instructions above.)

Model Tab (also called **Model Space**).

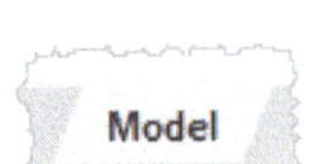

When you select the **Model** Tab, you enter **Model Space**. (This is where you have been drawing and plotting from for the last 25 lessons.) Model Space is where you **create** and **modify** your drawings.

Continued on the next page...

Layout Tabs (also called **Paper Space**). 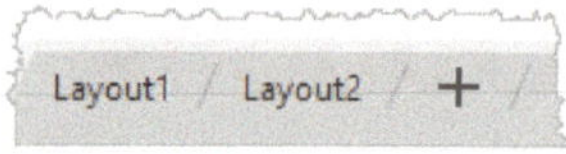

When you select a Layout Tab, you enter **Paper Space**. The primary function of Paper Space is to **prepare the drawing for plotting or printing**.

When you select the Layout Tab for the first time, the **"Page Setup Manager"** dialog box will appear. The Page Setup Manager allows you to specify the printing device and paper size to use. (More information on this is in "Creating a Page Setup" on page 26-9.)

When you select a Layout Tab, Model Space will seem to have disappeared, and a **blank sheet of paper** is displayed on the screen. This sheet of paper is basically **in front of the Model Space**. (Refer to the illustration below.)

To see the drawing in Model Space, while still in Paper Space, you must cut a hole in this sheet. This hole is called a **"viewport"**. (Refer to "Viewports" on page 26-5.)

Note: Try to think of this as a picture frame (**Paper Space**) in front of a photograph (**Model Space**).

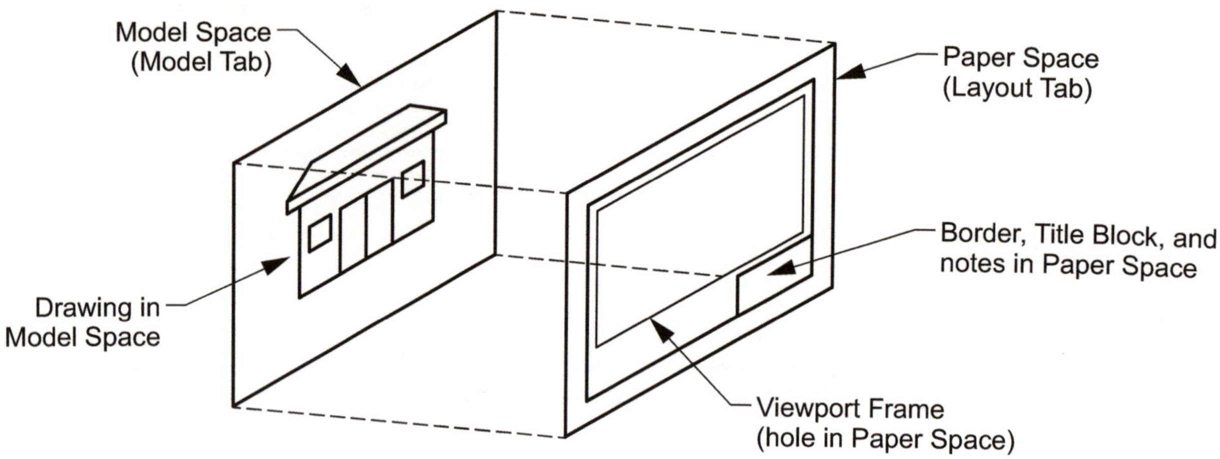

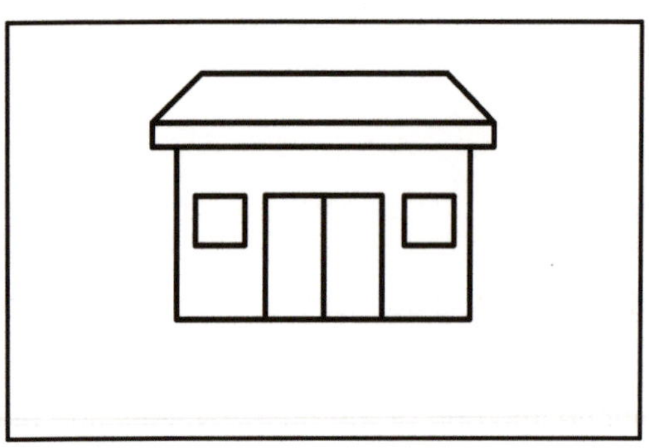

This is what you see when you select the 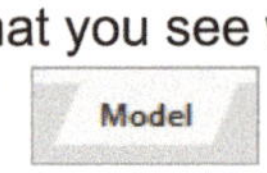

Tab.

You see Model Space.

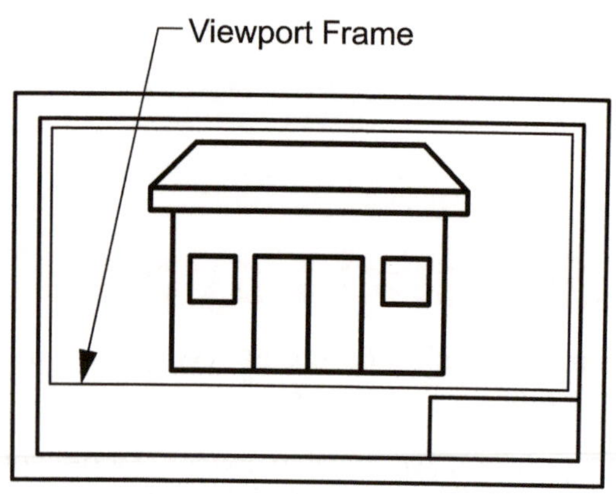

This is what you see when you select the 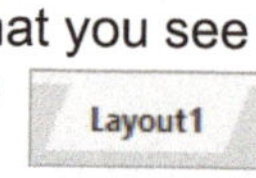

Tab.

You see through the viewport to Model Space.

Now hold this thought…more explanation is coming on the next few pages.

Viewports

Viewports are only used in Paper Space (Layout Tab). Viewports are holes cut into the sheet of paper displayed on the screen in Paper Space. **Viewport frames** are objects. They can be moved, stretched, scaled, copied, and erased. You can have multiple viewports, and their size and shape can vary.

Note that the following is just to present the concept. The actual step-by-step instructions follow in the exercises.

Note: It is considered good drawing management to create a layer for the viewport "frames" to reside on. This will allow you to control them separately: such as setting the Viewport layer to **"No plot"** so it will not be plotted out.

How to create a Viewport

1. First, create a drawing in **Model Space** (**Model** Tab 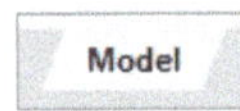) and save it.

2. Select the **Layout1** Tab. 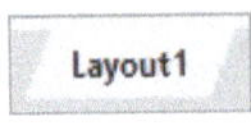

   When the **"Page Setup Manager"** dialog box appears, select the **New...** button. Then you will select the printing device and paper size to plot on. (Refer to *"Creating a Page Setup"* on page 26-9.)

3. You are now in **Paper Space**. Model Space appears to have disappeared, because a blank paper is now in front of Model Space, preventing you from seeing your drawing. You designated the size of this sheet in the "Page Setup" mentioned in Step 2 above. (The border, Title Block, and notes will be drawn on this paper.)

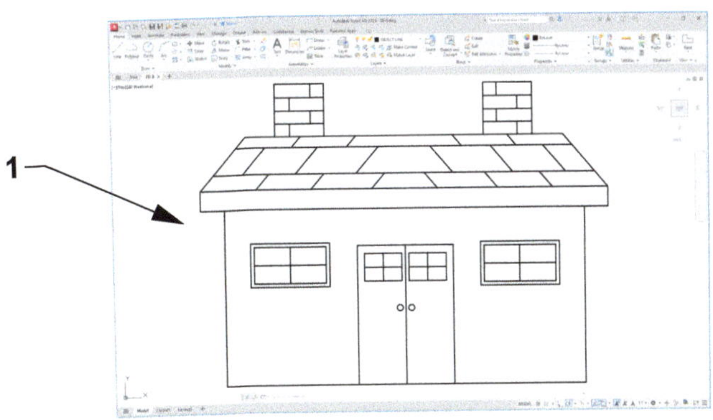

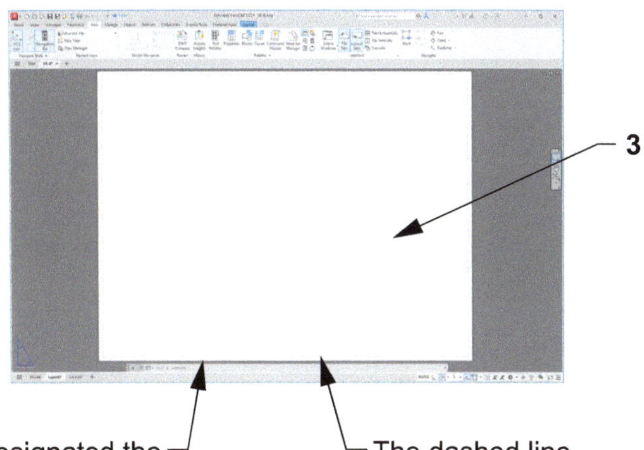

4. Draw a border, Title Block, and notes in Paper Space (Layout).

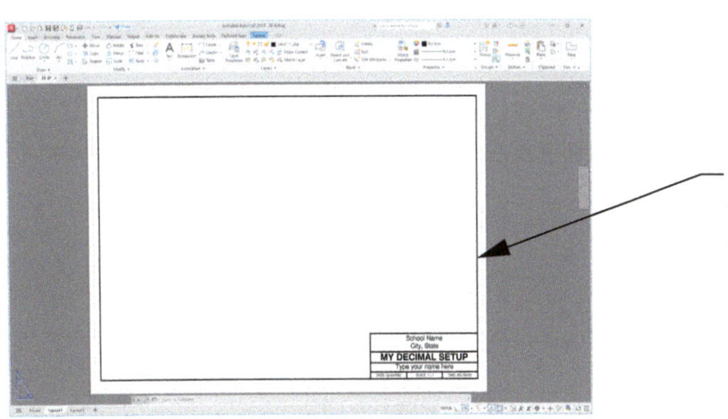

Continued on the next page...

*Now you will want to see the drawing that is in **Model Space**.*

5. Select the **"Viewport"** layer (you want the viewport frame to be on this layer).

6. Select the **Viewport** command using one of the following:

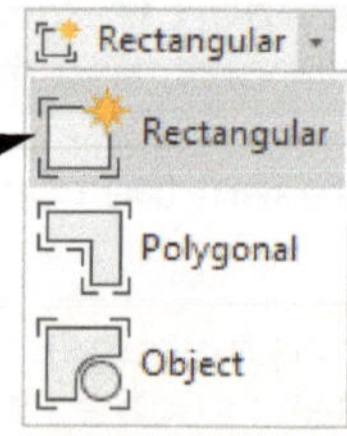

Ribbon = Layout Tab / Layout Viewports Panel / ▾ /
or
Keyboard = *MV <Enter>*

7. Draw a rectangular shaped viewport "frame" by placing the location for the "first corner" and then the "opposite corner" using the cursor. (While this is similar to drawing a rectangle, **do not** use the Rectangle command. You must remain in the **MV** command.)

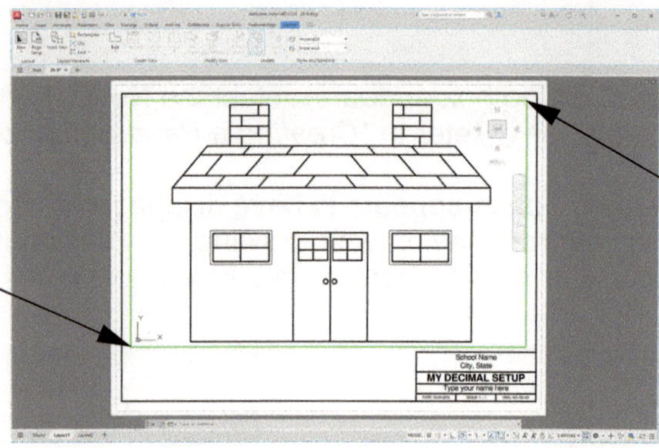

First corner of viewport.

Opposite corner of viewport.

You should now be able to look through the Paper Space sheet to Model Space and see your drawing because you just cut a rectangular-shaped hole in the sheet.

Note: You may go back to Model Space or return to Paper Space, simply by selecting the Model Tab or the Layout Tab.

Why Layouts are Useful

You are probably wondering why you should bother with Layouts. A Layout (Paper Space) is a great method to manipulate your drawing for plotting or printing.

Notice the drawing below with **multiple viewports**. Each viewport is a hole in the paper. You can see through each viewport (hole) to Model Space. Using Zoom and Pan, you can manipulate the display of model space in each viewport. To manipulate the display, you must be inside the viewport (see instructions on the next page).

The dashed line indicates the maximum printing area for the printer and paper selected. Any object outside of this area will not print.

Active viewport indicated by heavier viewport frame.

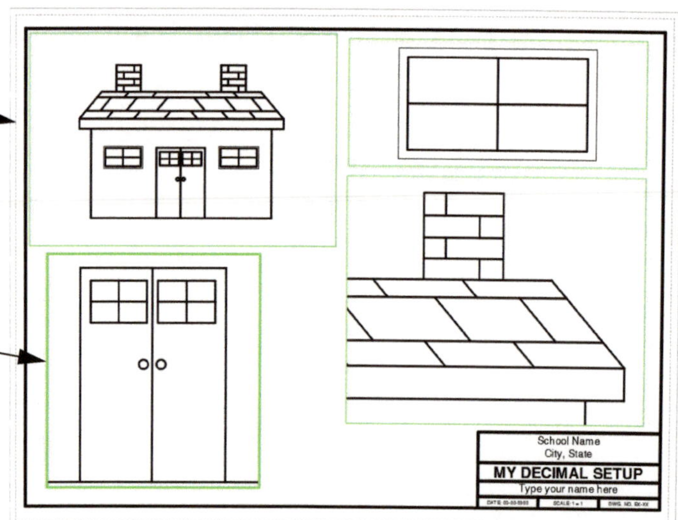

Reach Into a Viewport

To reach into a viewport, the rules are:

1. You have to be in Paper Space (Layout Tab) and at least one viewport must have been created.
2. You have to be **inside a viewport** to manipulate the scale or position of the drawing that you see in that Viewport.

How to reach into a viewport to manipulate the display

First, select a Layout Tab and cut a viewport.

At the bottom right of the screen on the Status Bar, there is a button that either says Model or Paper. This button displays which space you are in currently. 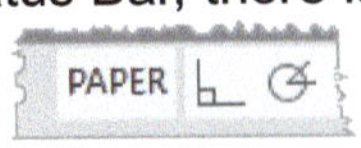

When the button is **Paper**, you are working on the Paper sheet that is in front of Model Space. (Refer to the illustration on page 26-4.) You may cut a viewport, draw a border, add a Title Block, and place notes.

If you want to reach into a viewport to manipulate the display, double click inside of the viewport frame. Only one viewport can be activated at one time. The active viewport is indicated by a heavier viewport frame. (Refer to the illustration on the previous page. The viewport displaying the doors is active.) Also, the Paper button changed to Model. 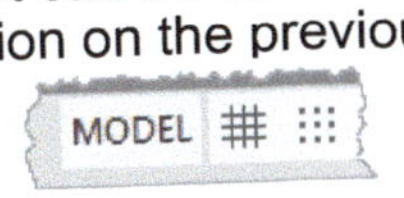

While you are inside a viewport, you may manipulate the scale and position of the drawing displayed. To return to the Paper surface, click on the word Model and it will change to Paper 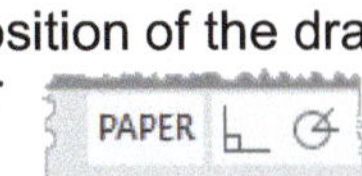

You may now work on the paper surface.

Model / Layout Tabs versus MODEL / PAPER button

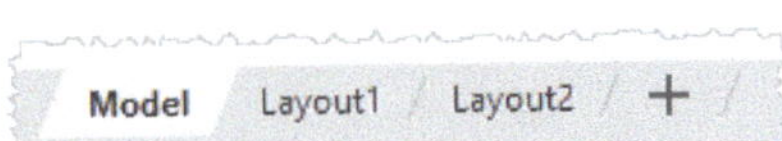

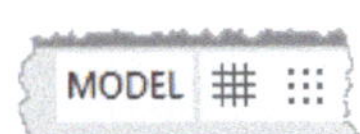

Do not confuse the Model / Layout Tabs with the MODEL / PAPER button. Here is the difference:

Model / Layout Tabs shuffle you from the actual drawing area (Model Space) to the Layout area (Paper Space). It is sort of like having two stacked pieces of paper and when you select the Model Tab, the drawing would come to the front and you could not see the layout. When you select the Layout Tab, a blank sheet would come to the front and you would not see Model Space, unless you have a viewport cut.

The **MODEL / PAPER** button allows you to work in Model Space or paper space without leaving the Layout Tab. No flipping of sheets. You are either on the paper surface or in the viewport reaching through to Model Space.

Don't worry it will get easier. This is the concept…but it will become clearer when you have completed the exercises in this lesson.

Pan

After you zoom in and out or adjust the scale of a viewport, the drawing within the viewport frame may not be positioned as you would like it. This is where **Pan** comes in handy. **Pan** will allow you to move the drawing around, within the viewport, without affecting the size or scale.

Note: Do not use the **Move** command. You do not want to actually move the original drawing. You only want to slide the viewport image, of the original drawing, around within the viewport.

Continued on the next page...

How to use the Pan command

1. Select a Layout Tab (Paper Space).
2. Unlock the viewport if it is locked. (Refer to page 26-9.)
3. Click inside a viewport.
4. Select the **Pan** command using one of the following:

   Ribbon = View Tab / Navigate Panel / 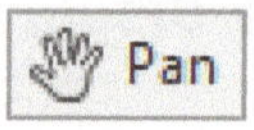 (Refer to pages 1-9 and 4-7.)

   or
   Keyboard = *P <Enter>*
   or
   Navigation Bar =  (Turned off on page 1-19.)

   (Consider adding the **Pan** tool to the **Quick Access Toolbar**. See page 1-8.)

5. Place the cursor inside the viewport and hold the left mouse button down while moving the cursor. (Click and drag.) When the drawing is in the desired location, release the mouse button.
6. Press the ***<Esc>*** key or press ***<Enter>*** to end the **Pan** command.
7. Lock the viewport. (Refer to page 26-9.)

Refer to the example below.

CAD TIP! If you have a wheel mouse, you may also use the **Pan** command by holding down the center wheel while moving the cursor.

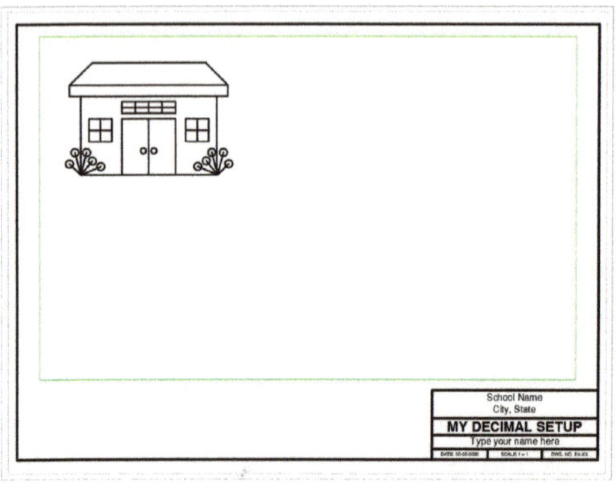

Before Pan.

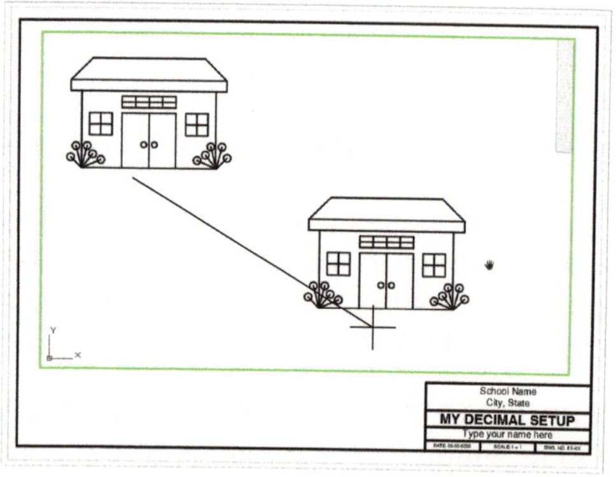

Double click in the viewport to active it.
Use Pan (Click, Drag, Release).

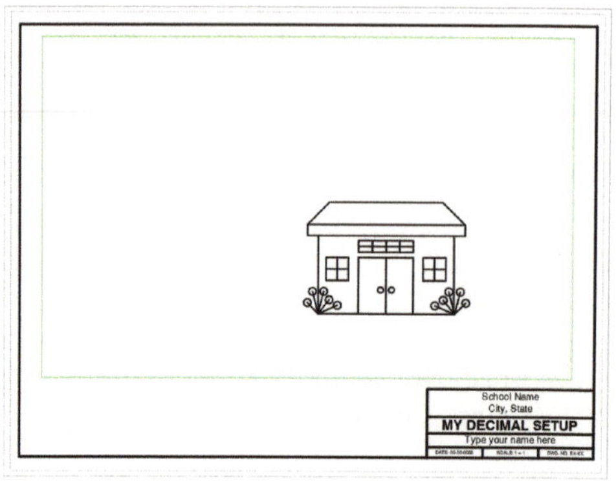

After Pan.

Locking a Viewport

After you have manipulated the drawing within each viewport to suit your display needs, you will want to **Lock** the viewport so the display can't be changed accidentally. Then you may zoom in and out, knowing that you will not disturb the display.

Note: Accurately adjusting the scale of a viewport will be discussed in detail in Lesson 27.

1. Make sure you are in **Paper Space**. 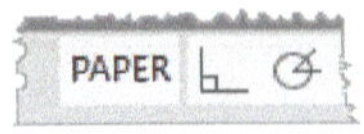

2. Click once on a **viewport frame**.

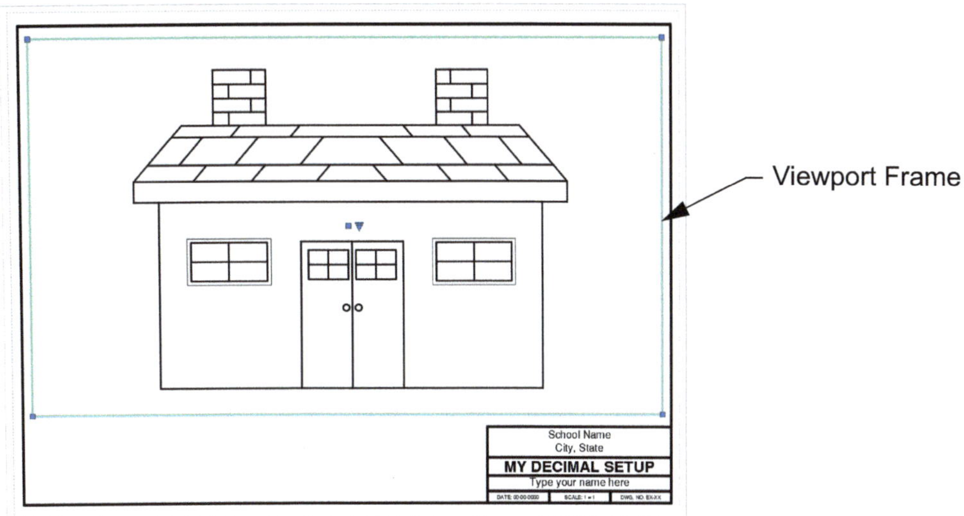

3. Click on the **Open Lock** tool located in the lower right corner of drawing area.

   The icon will change to a **Closed Lock** tool.

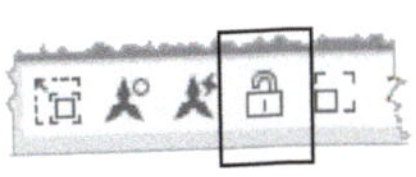

Viewport Unlocked

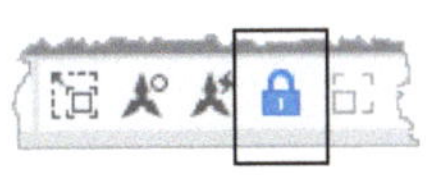

Viewport Locked

4. Press the **<*Esc*>** key to deactivate the grips on the viewport frame.

Note: Now, any time you want to know if a viewport is locked or unlocked, just glance down to the **Lock** tool shown above.

Creating a Page Setup

When you select a Layout Tab for the first time, the Page Setup Manager will appear. The Page Setup Manager allows you to select the **printer / plotter** and **paper size**. These specifications are called the **"Page Setup"**. This Page Setup will be saved to that Layout Tab so it will be available whenever you use that Layout Tab.

Note that the following is just to present the concept. The actual step-by-step instructions start in Exercise 26A.

1. Open the drawing you want to plot. (The drawing must be displayed on the screen.)

2. Select a Layout Tab. (Refer to page 26-3 if you do not have a Layout Tab.)

Continued on the next page...

Note: If the **"Page Setup Manager"** dialog box shown below does not appear automatically, right click on the Layout Tab and select **"Page Setup Manager"**.

3. Select the **New…** button.

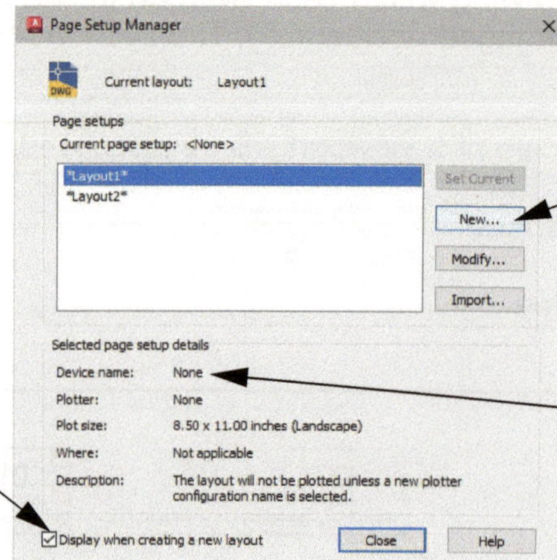

3

Yours will be different. It may even display "None". That's OK for now.

Check this box if the Page Setup Manager did not appear automatically when you selected a Layout Tab.

4. Select **<Default output device>** in the **Start with:** list.

5. Enter the New Page Setup name: **Setup A**

6. Select the **OK** button.

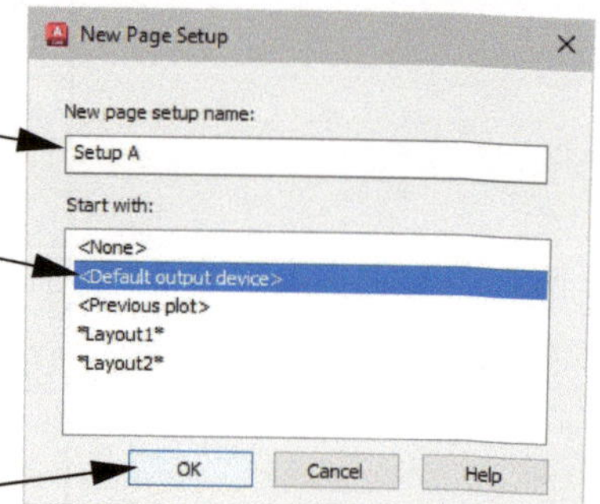

5

4

(I am assuming that your computer is attached to a printer. If not, select **Layout1**.)

6

This is where you will select the **printer / plotter**, **paper size**, and the **plot offset**.

7. Select the **Printer / Plotter**
   Your current system printer should already be displayed here. If you prefer another, select the down arrow and select from the list. If the preferred printer is not in the list, you must configure the printer. (Refer to **Appendix-A** for instructions.)

8. Select the **Paper size**.

9. Select **Plot offset**.

Notice the name you entered is now displayed as the Page setup name.

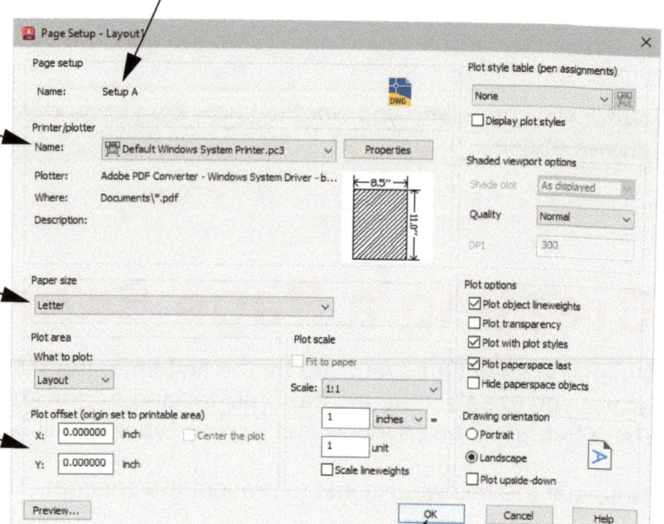

7
(Yours will be different.)

8
(Yours may state 8-1/2 x 11 or A4.)

9
(Should remain at 0.000000 0.000000)

10. Select the **OK** button.

10

Continued on the next page...

11. Select the Page Setup.

12. Select the **Set Current** button.

13. Select the **Close** button.

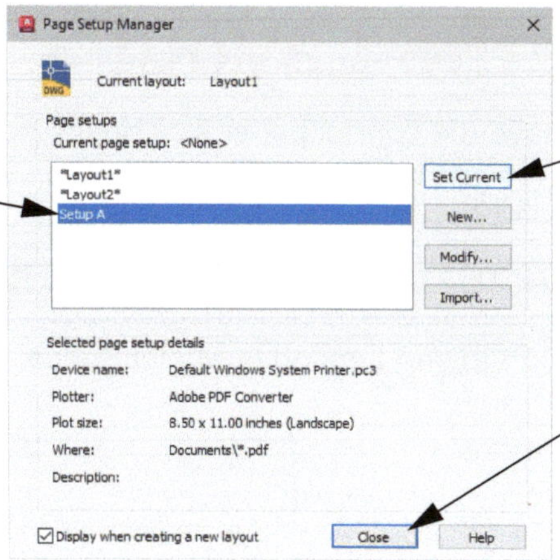

You should now have a sheet of paper displayed on the screen.
This sheet is the size you specified in the "Page Setup".
This sheet is in front of Model Space.

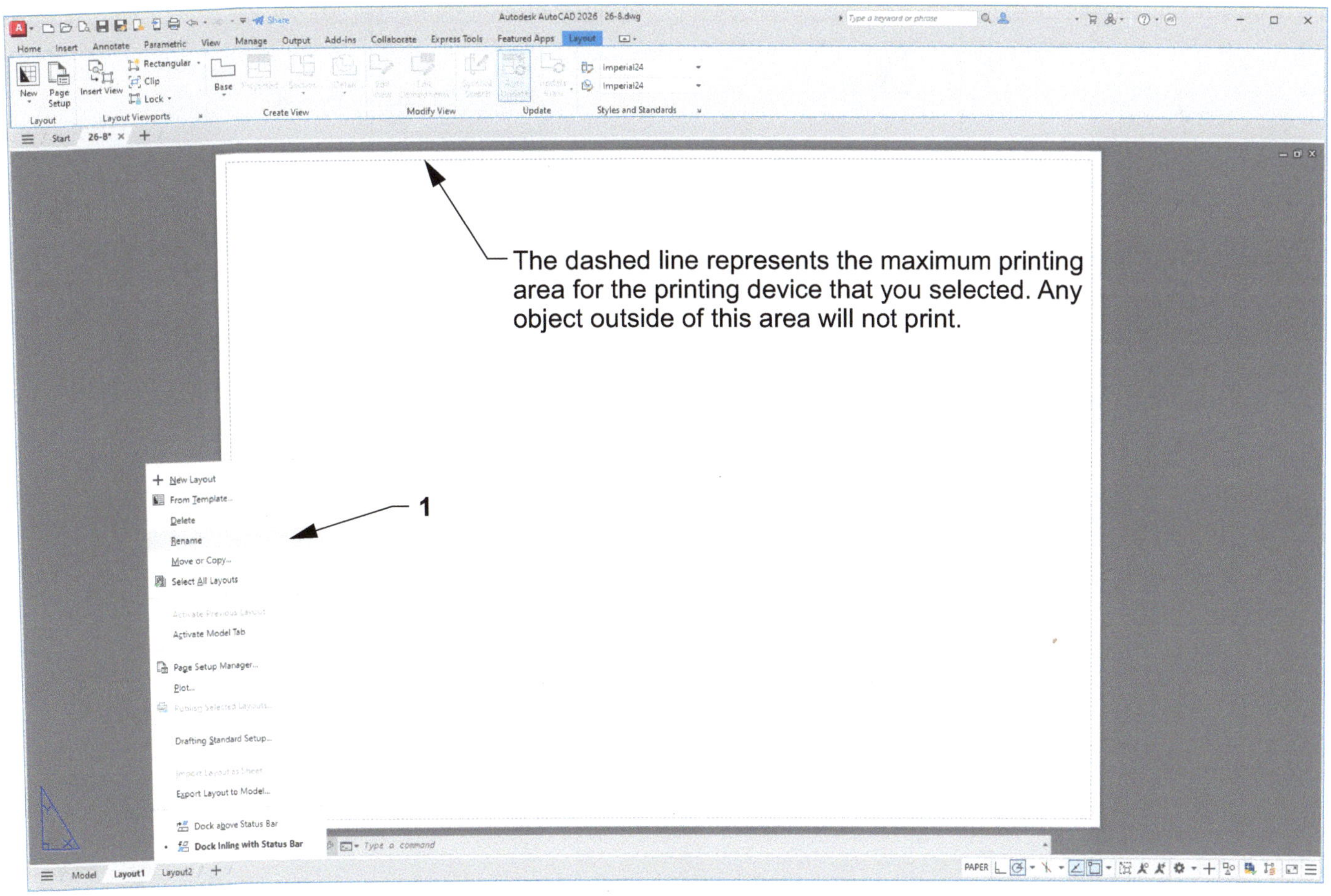

How to rename the Layout Tab

1. Right click on the active Layout Tab and select **Rename** from the list.

2. Enter the new Layout name **A Size** and then press ***<Enter>***.

This was just the concept of Page Setup. You will actually create one in Exercise 26B.

Using the Layout

Now that you have the correct paper size on the screen, you need to do a little bit more to make it useful.

The next step is to:

1. Add a border, Title Block, and notes in Paper Space.
2. Cut a viewport to see through to Model Space. (Refer to page 26-6.)

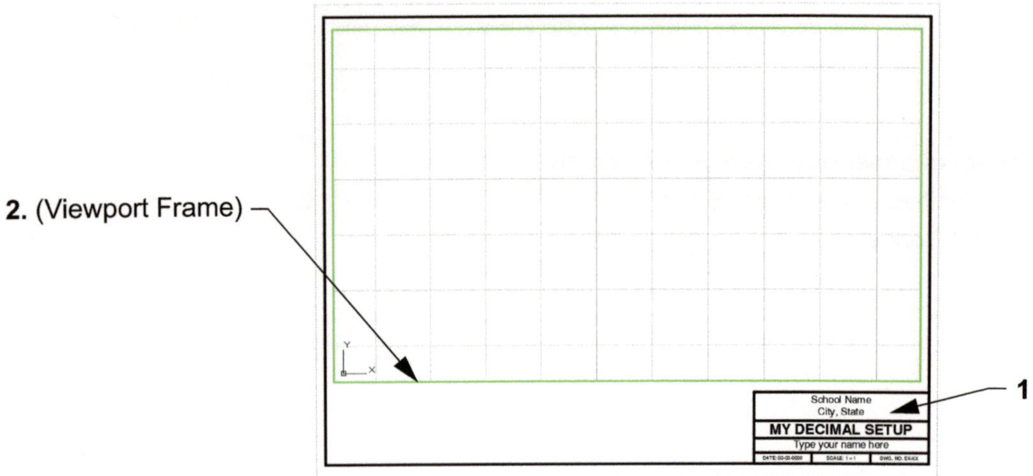

3. Adjust the scale of the viewport as follows:
   A. You must be in Paper Space.
   B. Click on the viewport frame.
   C. Unlock the viewport if it is locked.
   D. Select the **viewport scale** down arrow ▾.
   E. Select a scale.

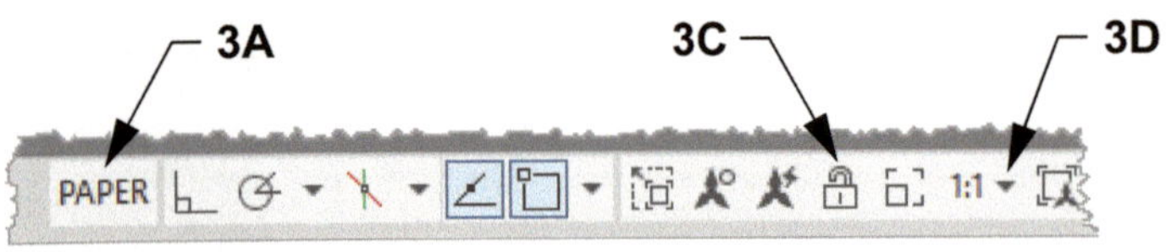

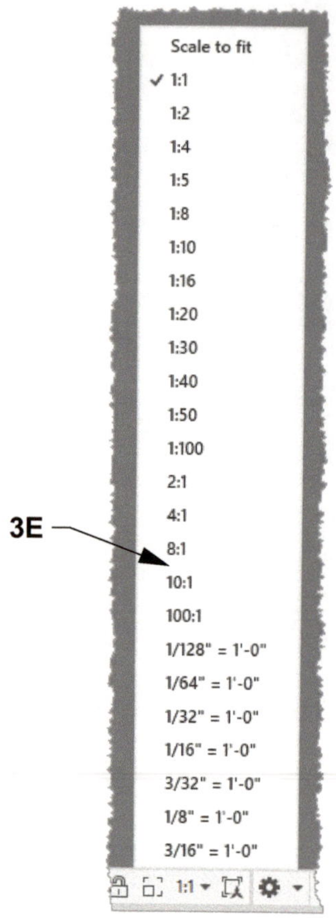

4. **Lock** the viewport.

Now you may zoom as much as you want and it will not affect the adjusted scale.

Note: Adjusting the scale of the viewport will be discussed more in Lesson 27.

Plotting from a Layout Tab

The previous Page Setup instructions were to select the printer and paper size. Now you need to specify how you want to plot the drawing. You will find the **Plot** dialog box almost identical to the Page Setup dialog box.

Note: You must create a **Page Setup** before Plotting from a Layout Tab. If you have not created a Page Setup, refer to page 26-9 before proceeding.

How to plot from a Layout Tab

1. Open the drawing you want to plot.

2. Select the Layout Tab you want to plot.

3. Select the **Plot** command using one of the following:

Quick Access Toolbar = 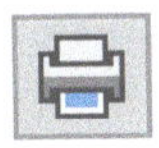
or
Ribbon = Output Tab / Plot Panel / 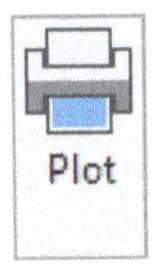

or
Application Menu = Print / Plot
or
Keyboard = *Plot <Enter>*

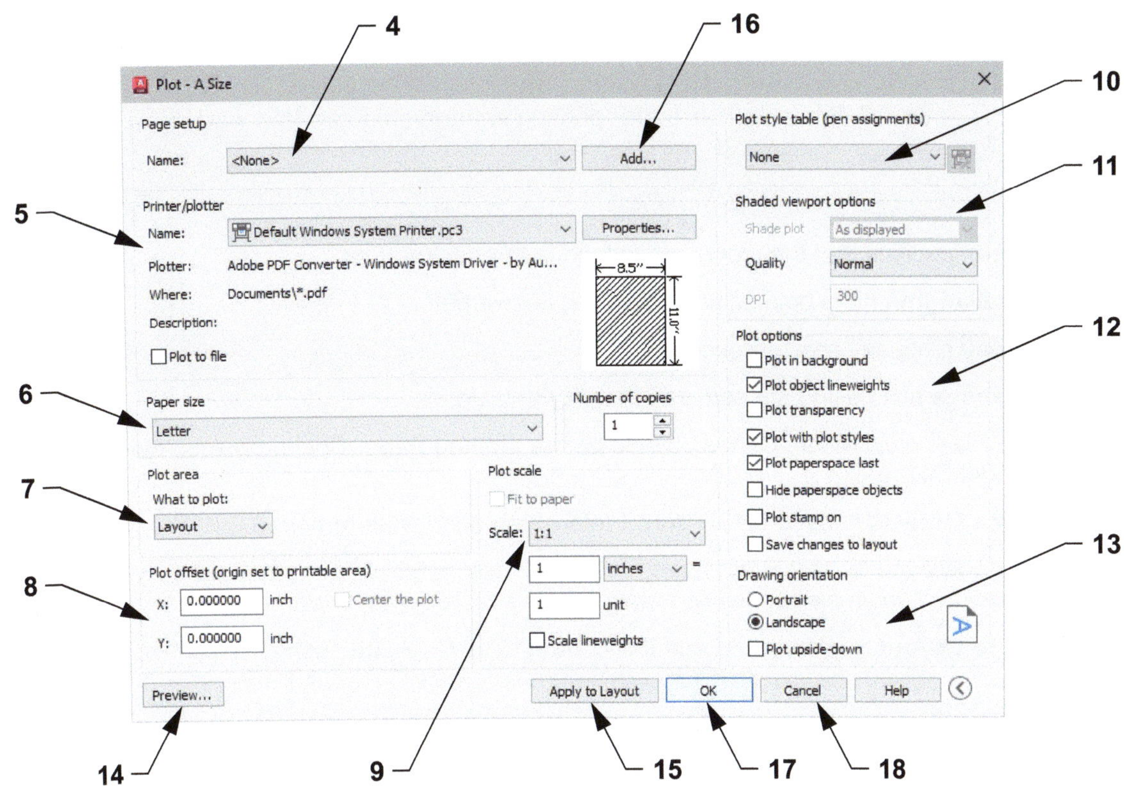

4. **Page Setup Name:** After you have selected the desired settings, you will save the new Page Setup and it will appear here. If you have previously created a Page Setup, you may select it from the drop-down list and all of the settings will change to reflect the previously saved Page Setup settings.

5. **Printer / Plotter:** Select the Printer that you want to use. All previously configured devices will be listed here. (If your printer / plotter is not listed, refer to "Add a Printer / Plotter" Appendix A.)

Continued on the next page...

6. **Paper size:** Select the paper size. The paper sizes shown in the drop-down list are the **available sizes for the printer that you selected**. If the size you require is not listed, the printer you selected may not be able to handle that size. For example, a letter size printer cannot handle a 24" X 18" size sheet. You must select a large format printer.

7. **Plot area:** Select the area to plot. Layout is the default.

   Limits plots the area inside the drawing limits. (This option is only available when plotting from Model Space.)

   Layout plots the paper size. (Select this option when plotting from a Layout.)

   Extents plots all objects in the drawing file even if out of view. (This option is only available if you have a viewport.)

   Display plots the drawing exactly as displayed on the screen.

   Window plots objects inside a Window. To specify the Window, choose Window and specify the first and opposite (diagonal) corner of the area you choose to plot. (Similar to the Zoom / Window command.)

8. **Plot offset:** The plot can be moved away from the lower left plot limit corner by changing the X and / or Y offset. If you have selected Plot area "Display" or "Extents", select "Center the plot".

9. **Scale:** Select a scale from the drop-down list or enter a custom scale.

Note: This scale is the Paper Space scale. The Model Space scale will be adjusted within the viewport. If you are plotting from a **Layout** Tab, normally you will use **plot scale 1:1**. (This may seem a little confusing right now. Scaling will be discussed more in Lesson 27.)

10. **Plot style table:** Select the Plot style table from the list. The Plot Styles determine if the plot is in color, black ink, or screened. You may also create your own. If you want to print in black Ink only, select Monochrome.ctb. If you want to print in color, select Acad.ctb

11. **Shaded viewport options:** This area is used for printing shaded objects when working in the 3D environment.

12. **Plot options:**

    Plot background specifies that the plot is processed in the background.

    Plot Object Lineweights plots objects with assigned lineweights.

    Plot transparency plots any transparencies.

    Plot with Plot Styles plots using the selected Plot Style Table.

    Plot paperspace last plots Model Space objects before plotting Paper Space objects. Not available when plotting from Model Space.

    Hide Paperspace Objects is used for 3D only. Plots with hidden lines removed.

    Plot Stamp on allows you to print information around the perimeter of the border such as: drawing name, layout name, date/time, login name, device name, paper size, and plot scale.

    Save Changes to Layout is what you check if you want to save all settings to the current Layout Tab.

13. **Drawing orientation:**

    Portrait means the short edge of the paper represents the top of the page.

    Landscape means the long edge of the paper represents the top of the page

    Plot Upside-down plots the drawing upside down.

Continued on the next page...

14. Select the **Preview...** button. Preview displays the drawing as it will plot on the sheet of paper.

Note: If you cannot see through to Model Space, you have not cut your viewport yet.

If the drawing appears as you would like it, press the ***<Esc>*** key and continue.

If the drawing does not look correct, press the ***<Esc>*** key and re-check your settings, then preview again.

Note: If you have any of the layers set to **"no plot"**, they will not appear in the preview display. The Preview Display only displays what will be printed.

15. **Apply to Layout** button: This applies all of the settings to the Layout Tab. Whenever you select this Layout Tab, the settings will already be set.

16. **Save the Page Setup:** At this point, you have the option of saving these settings as another Page Setup for future use on other Layout Tabs. If you want to save this setup, select the **Add...** button, type a name, and select **OK**.

17. If your computer **is** connected to the plotter / printer selected, select the **OK** button to plot, then proceed to **Step 19**.

18. If your computer is **not** connected to the plotter / printer selected, select the **Cancel** button to close the Plot dialog box and proceed to **Step 19**.

    Selecting **Cancel** will cancel your selected setting if you did not save the Page Setup as specified in **Step 16** above.

19. **Save the drawing:** This will guarantee that the Page Setup you just created will be saved to this file for future use.

Again, this is the concept only. The step-by-step instructions are shown in Exercise 26E on page 26-27.

Annotative Property

In Lesson 16, you learned to create a dimension style. In Lesson 25, you learned to create a text style.

In this Lesson, you will create a new dimension style and text style, but this time you will include the **Annotative Property** in both.

The Annotative property automates the process of scaling text, dimensions, hatch, tolerances, leaders, and symbols. The height of these objects will automatically be determined by the annotation scale setting. This will be discussed more in Lesson 27.

For now, I just want you to know how to select it when creating your new styles.

Text Style

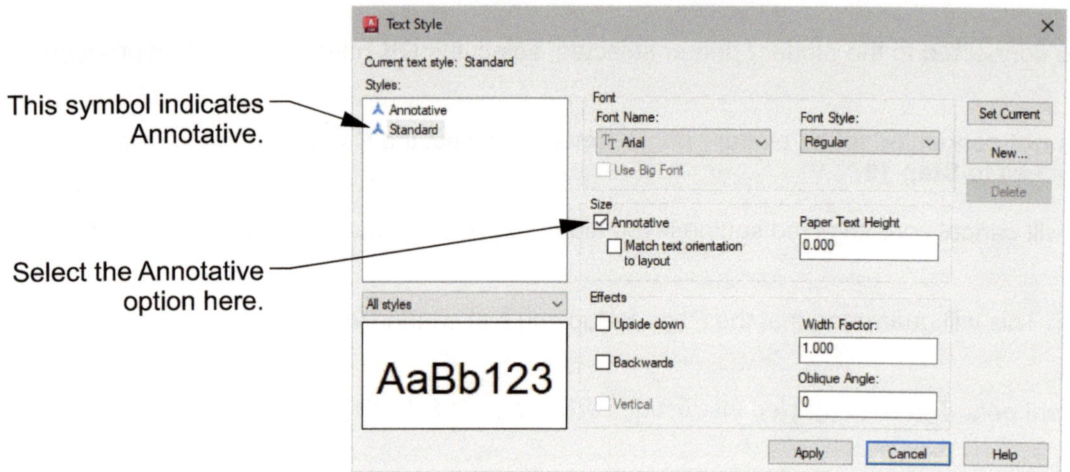

Dimension Style

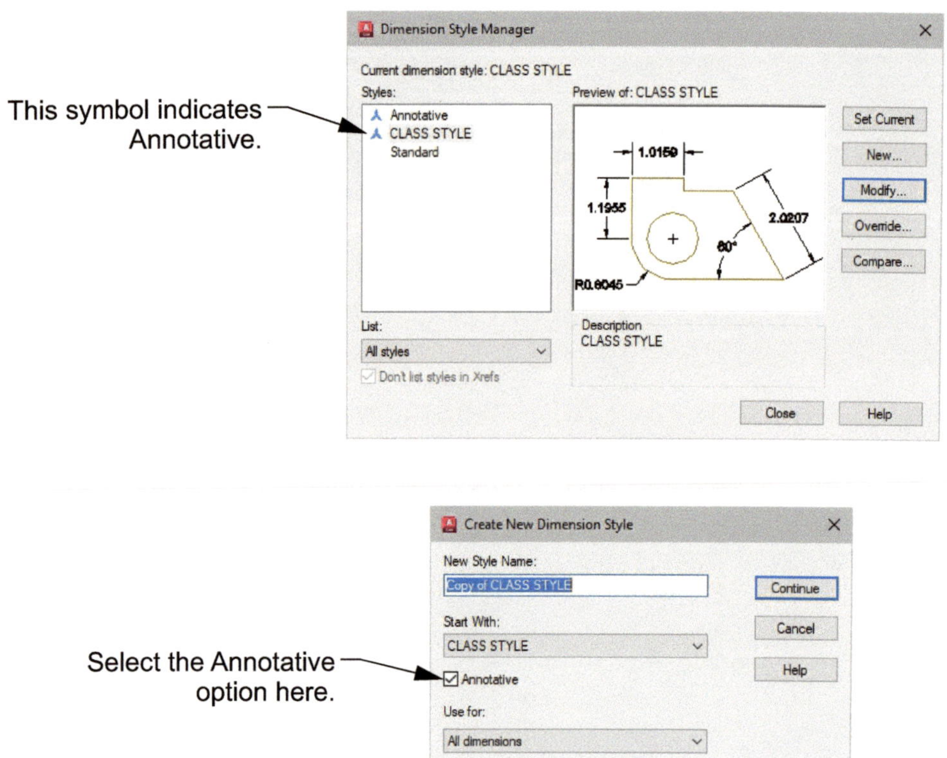

Exercise 26A: Create a Master Decimal Setup Template

The following instructions will guide you through creating a "master" decimal setup template. The "inch-helper" and "Border A-Metric" are examples of a master setup template. Once you have created this "master" template, you just open it and draw. No more repetitive inputting of settings.

New Settings

1. Begin your drawing with a different template as follows:
   A. Select the **New** command.
   B. Select template file **acad.dwt** from the list of templates for inch users.
   or
   Select template file **acadiso.dwt** from the list of templates for metric users.
   (**Note:** Do not select **"acad3D.dwt"** or **"acadiso3D.dwt"** by mistake.)

2. Set the **Drawing Units** as follows:
   A. Type ***units*** and then press ***<Enter>***.
   B. Change the **Length Type: Decimal** and **Precision: 0.000** or **0.00** as shown below.
   C. Select the **OK** button.

Inch users

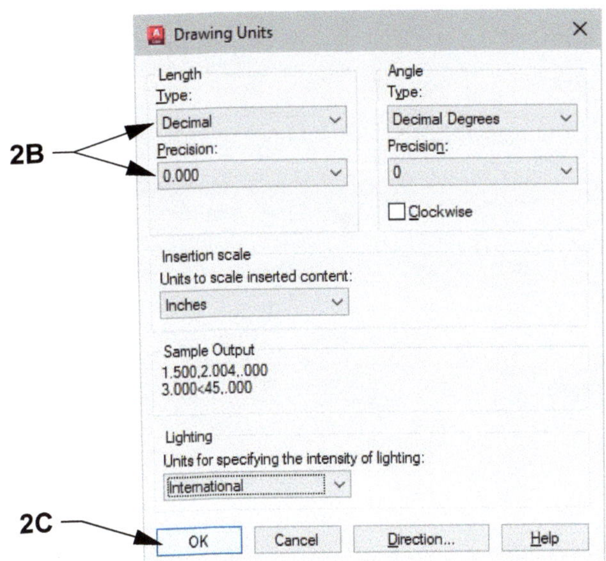

Metric users

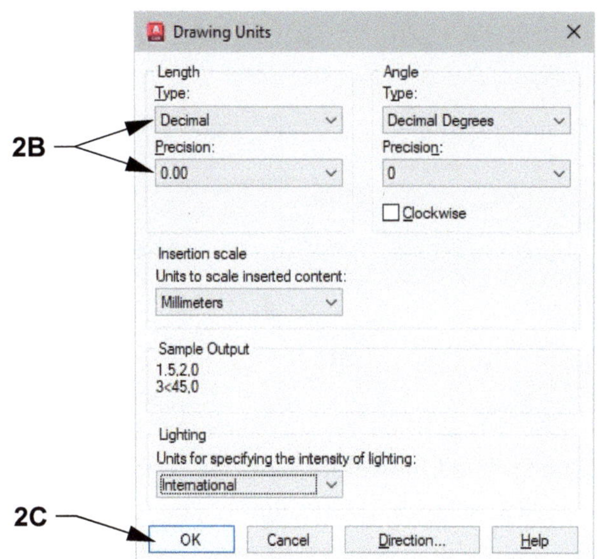

3. Set the **Drawing Limits**:
   A. Type ***Limits*** and then press ***<Enter>***.
   B. Enter lower left corner = ***0,0***
   C. Enter upper right corner = ***11,8.5*** inch users.
   or ***297,210*** for metric users.
   D. Use **"Zoom / All"** to view the new limits.
   E. Set your **Grids** to **on** to display the limits.

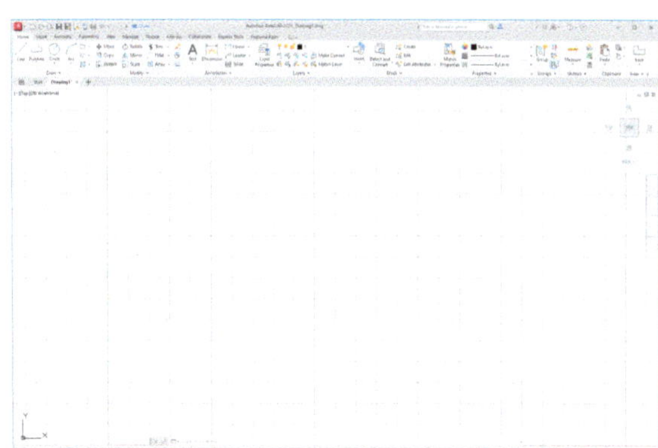

Continued on the next page...

Exercise 26A...continued

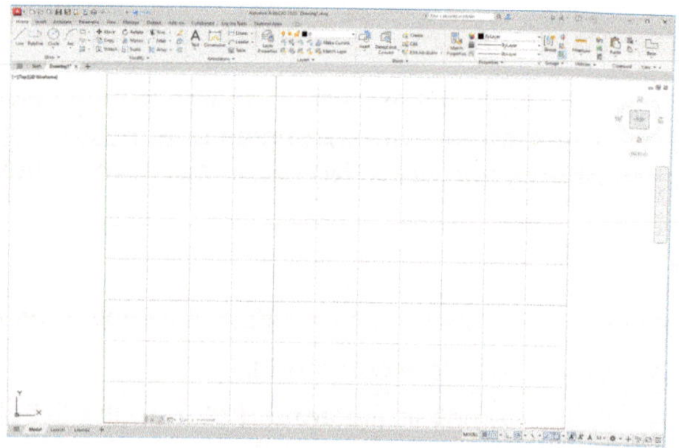

4. Set the **Grids** and **Snap** as follows:
   A. Type ***DS*** and then press ***<Enter>***.
   B. Select the **Snap and Grid** Tab.
   C. Change the settings as shown below.
   D. Select the **OK** button.

Inch users

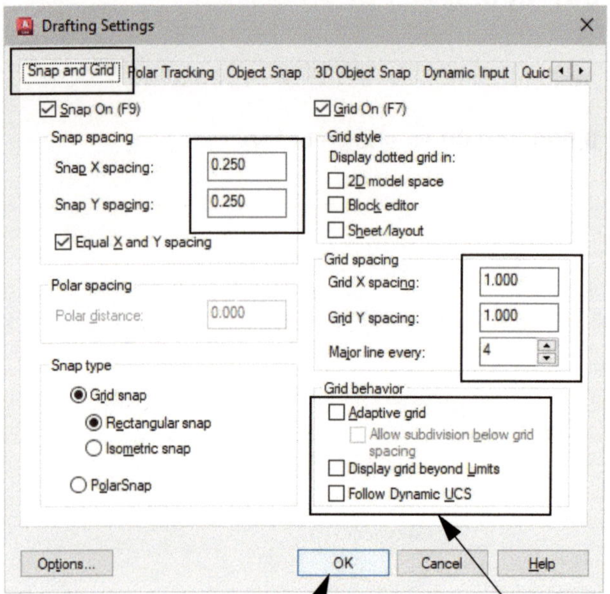

Metric users

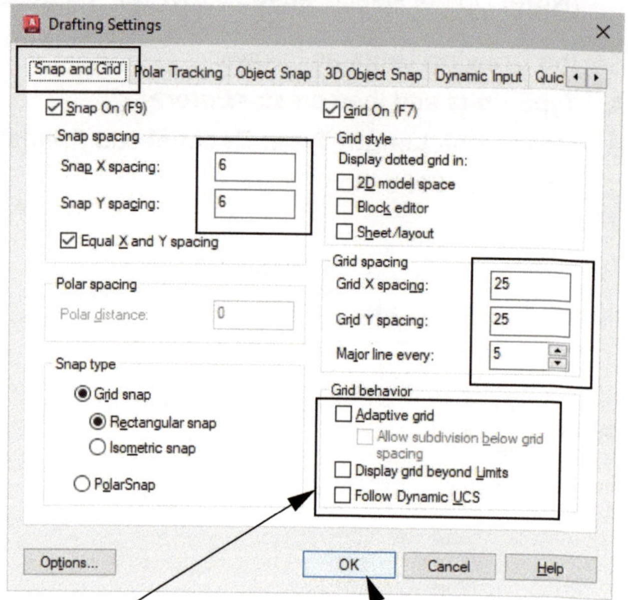

4D

Important:
Uncheck these.

4D

5. Change the **Lineweight Settings** as follows:
   A. Right click on **Lineweight** button located on the Status Bar. 
   B. Select **Lineweight Settings**. (Refer to page 3-10 for instructions.)
   C. Change to **Inches** or **Millimeters** and **Adjust Display Scale** as shown below.
   D. Select the **OK** button.

Inch users

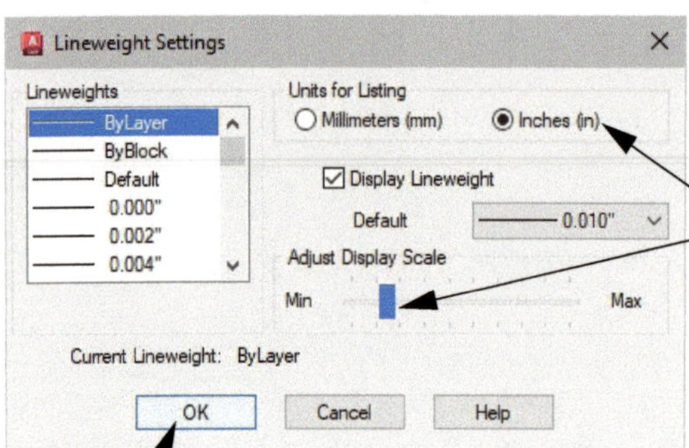

5C

5D

Metric users

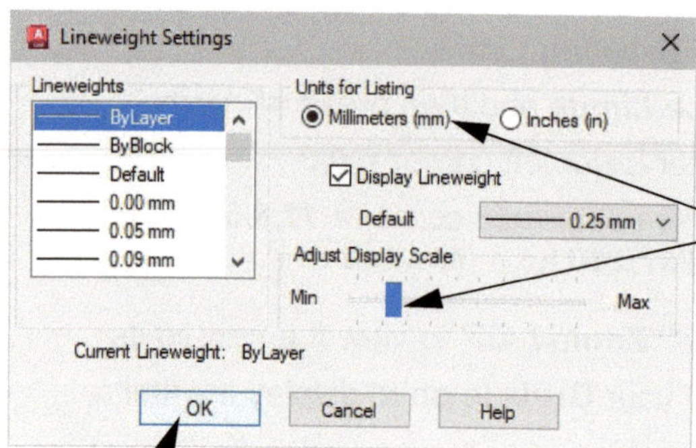

5C

5D

Continued on the next page...

Exercise 26A...continued

6. Create new **layers** as follows:
   A. Load the linetypes listed below. (Refer to page 3-13 for instructions.)
      CENTER2, HIDDEN, and PHANTOM2.
   B. Assign names, colors, linetypes, lineweights, and plotability as shown below.
      (Refer to page 3-12 for instructions.)
   C. Make the Object Line layer current and change the Viewport layer to **Not Plot** by clicking on the printer icon.

Inch users

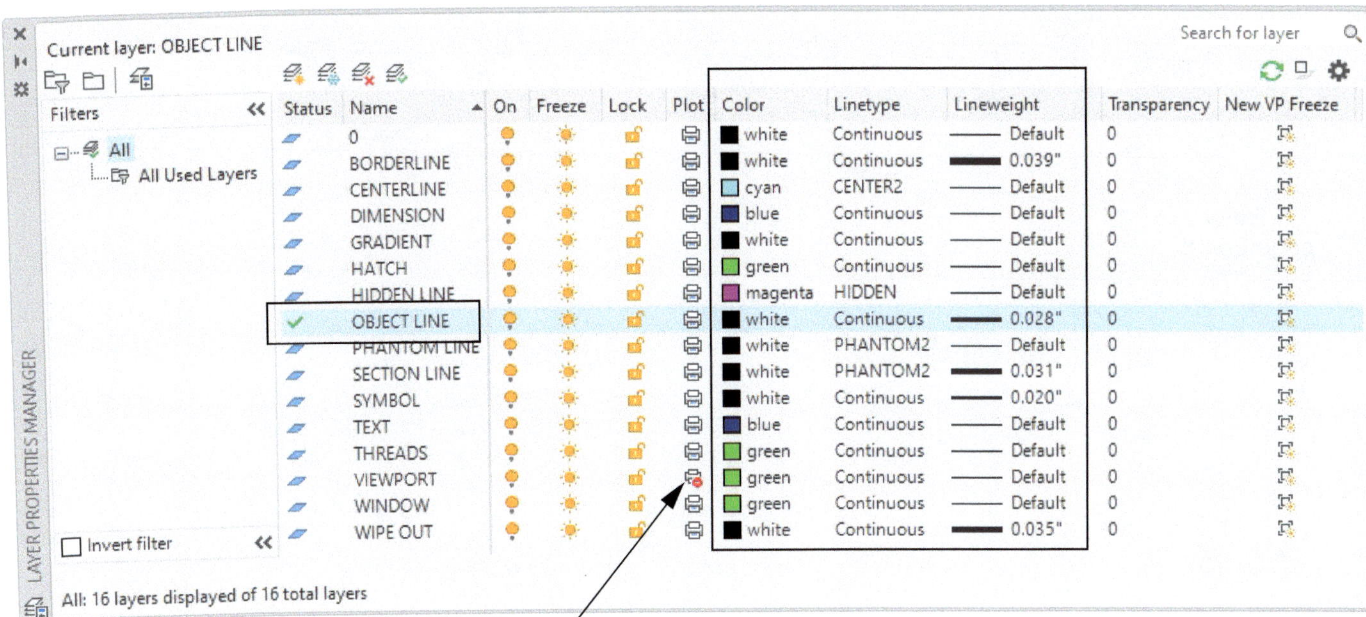

Name	Color	Linetype	Lineweight	Transparency
0	white	Continuous	Default	0
BORDERLINE	white	Continuous	0.039"	0
CENTERLINE	cyan	CENTER2	Default	0
DIMENSION	blue	Continuous	Default	0
GRADIENT	white	Continuous	Default	0
HATCH	green	Continuous	Default	0
HIDDEN LINE	magenta	HIDDEN	Default	0
OBJECT LINE	white	Continuous	0.028"	0
PHANTOM LINE	white	PHANTOM2	Default	0
SECTION LINE	white	PHANTOM2	0.031"	0
SYMBOL	white	Continuous	0.020"	0
TEXT	blue	Continuous	Default	0
THREADS	green	Continuous	Default	0
VIEWPORT	green	Continuous	Default	0
WINDOW	green	Continuous	Default	0
WIPE OUT	white	Continuous	0.035"	0

Not Plot

Metric users

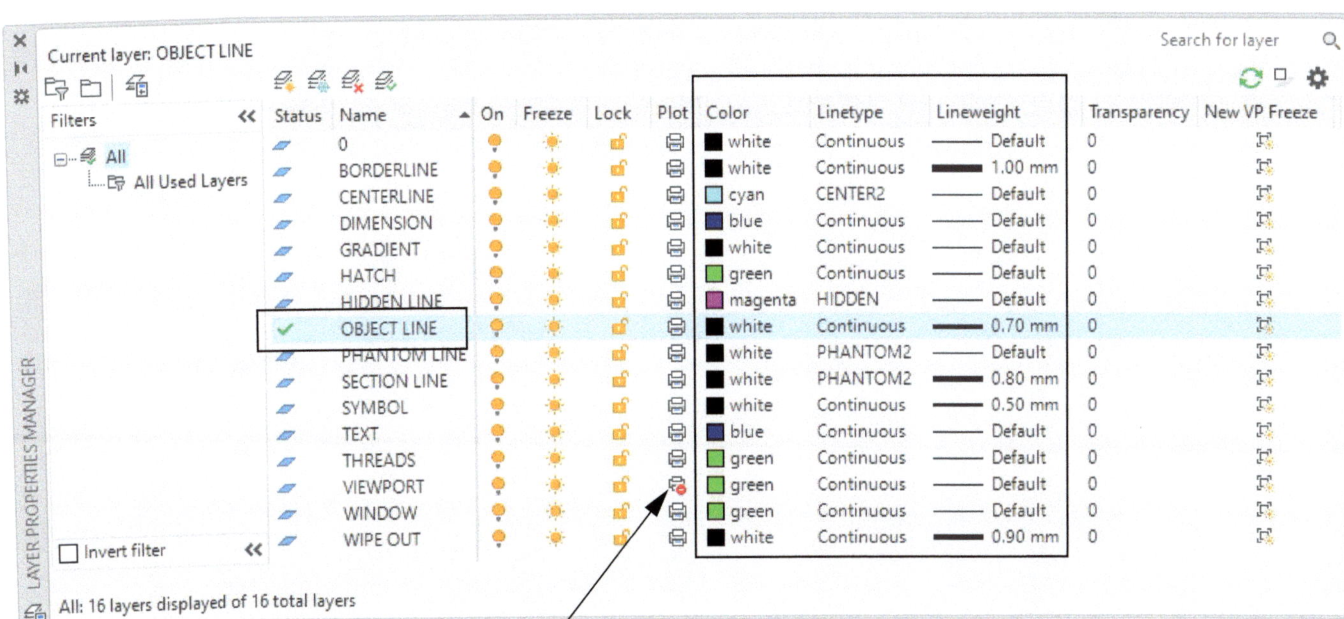

Name	Color	Linetype	Lineweight	Transparency
0	white	Continuous	Default	0
BORDERLINE	white	Continuous	1.00 mm	0
CENTERLINE	cyan	CENTER2	Default	0
DIMENSION	blue	Continuous	Default	0
GRADIENT	white	Continuous	Default	0
HATCH	green	Continuous	Default	0
HIDDEN LINE	magenta	HIDDEN	Default	0
OBJECT LINE	white	Continuous	0.70 mm	0
PHANTOM LINE	white	PHANTOM2	Default	0
SECTION LINE	white	PHANTOM2	0.80 mm	0
SYMBOL	white	Continuous	0.50 mm	0
TEXT	blue	Continuous	Default	0
THREADS	green	Continuous	Default	0
VIEWPORT	green	Continuous	Default	0
WINDOW	green	Continuous	Default	0
WIPE OUT	white	Continuous	0.90 mm	0

Not Plot

Continued on the next page...

Exercise 26A...continued

7. Create a new **Text Style** as follows:
   A. Select Text Style. (Refer to page 25-2.)
   B. Create the text style **TEXT-CLASSIC** using the settings shown below.
   C. When complete, select **Set Current**.
   D. Select **Close**.

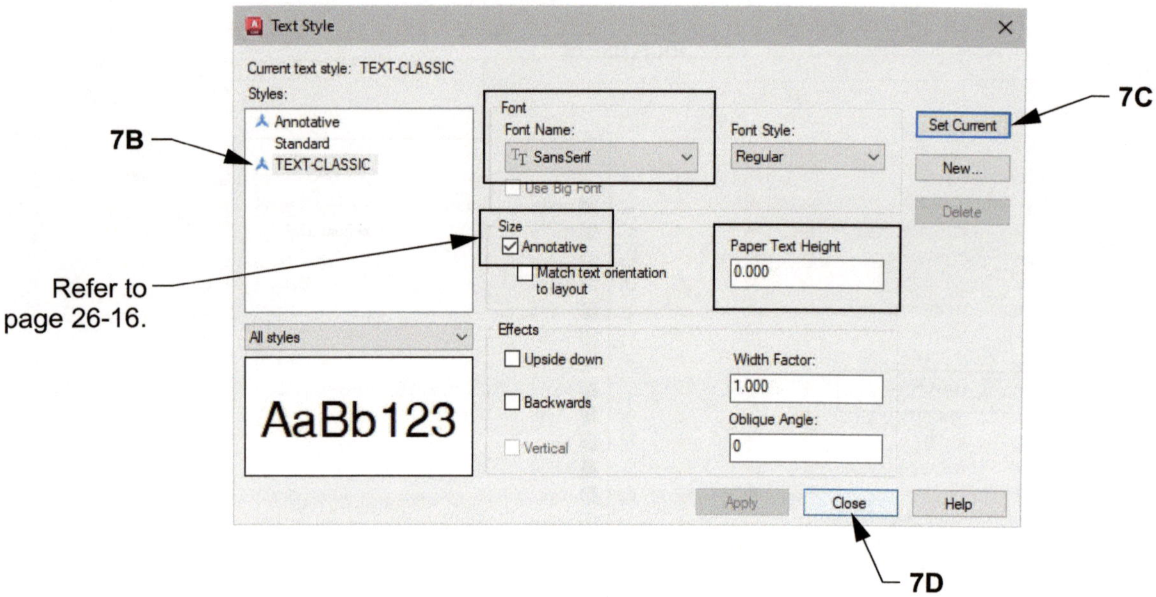

8. Create a new **Dimension Style** as follows:
   A. **Important:** Follow the directions on pages 16-7 through 16-10.
   B. All settings will be the same as pages 16-7 through 16-10 except the following:
   Name = DIM-DECIMAL
   Annotative = This new dimension style will be Annotative. (Refer to page 26-16.)
   Text Style = Select the newly created TEXT-CLASSIC.

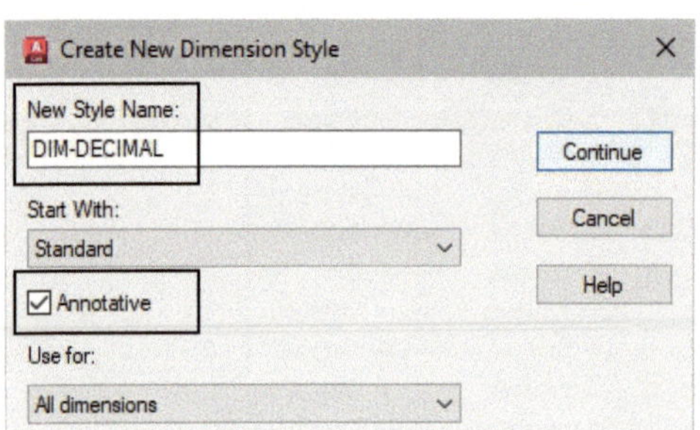

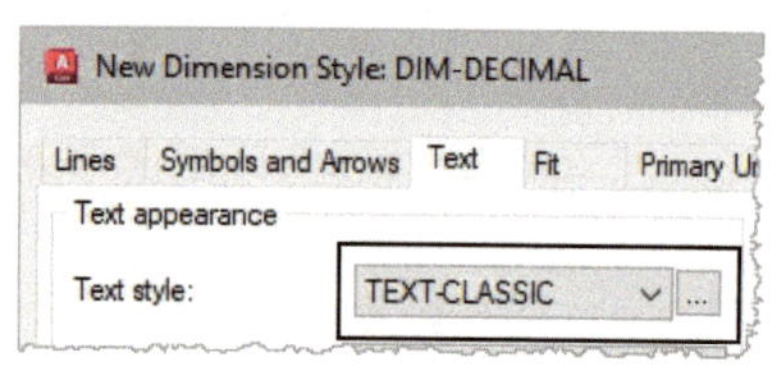

Continued on the next page...

Exercise 26A...continued

9. Create a new **Dimension Sub-Style** as follows:
   A. Follow the directions on page 18-9 and 18-10.
   B. All settings will be the same as pages 16-7 through 16-10 except the following:
   Annotative = This new sub-style will be Annotative. (Refer to page 26-16.)

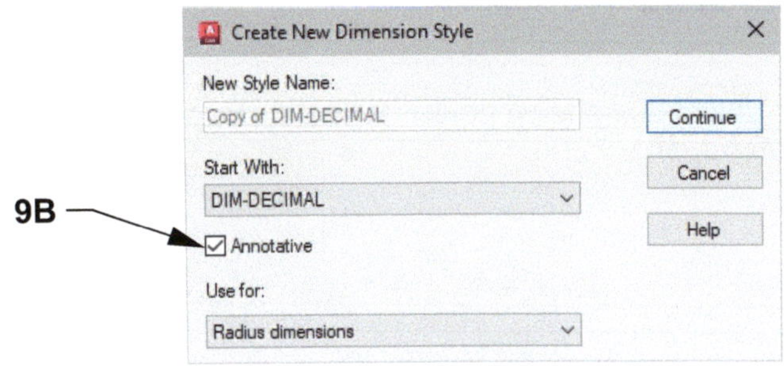

Your Dimension Style Manager should now appear as shown below:

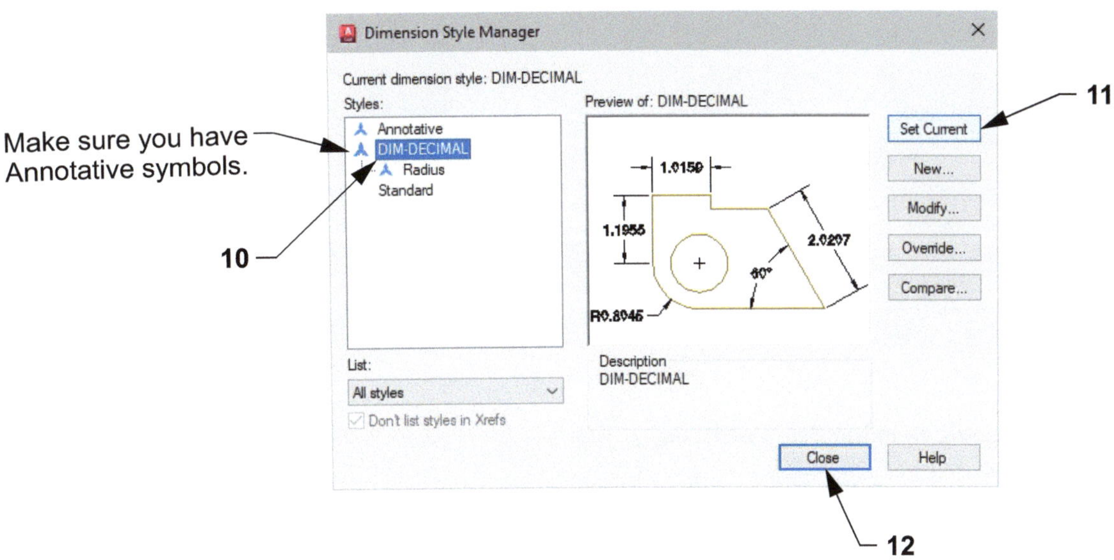

10. Select **DIM-DECIMAL**.

11. Select **Set Current**.

12. Select the **Close** button.

This next step is very important.

13. **Save** all the settings you just created as follows:
    Save as / AutoCAD Drawing Template / My Decimal Setup

*Now continue on to **Exercise 26B**...you are not done yet.*

Exercise 26B: Page Setup

Now you will select the printer and the paper size to use for printing. You will use the Layout1 Tab (Paper Space).

1. Open **My Decimal Setup** (if not already open).

2. Select the **Layout1** Tab. 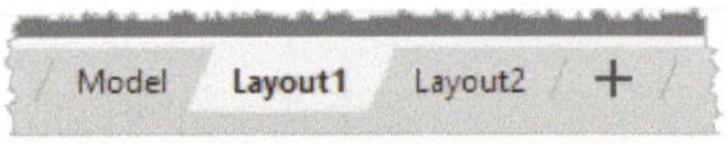

(Refer to page 26-3 if you do not have these Tabs.)

Note: If the **"Page Setup Manager"** dialog box shown below does not appear automatically, right click on the Layout Tab and select **Page Setup Manager**.

3. Select the **New...** button.

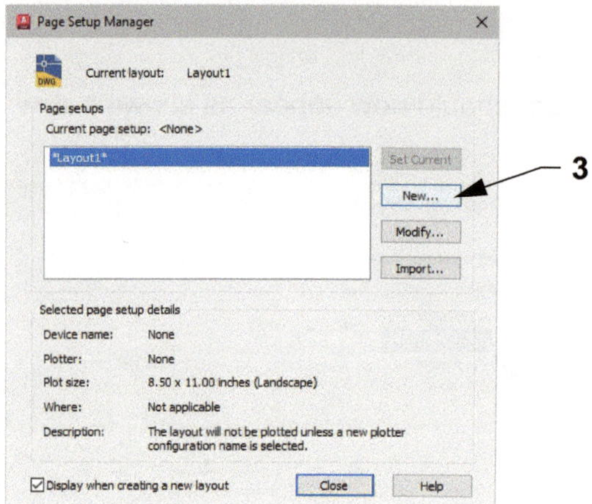

4. Select **<Default output device>** in the **Start with:** list.

5. Enter the New Page Setup name: **Setup A**

6. Select the **OK** button.

5

4

(I am assuming that your computer is attached to a printer. If not, select **Layout1**.)

6

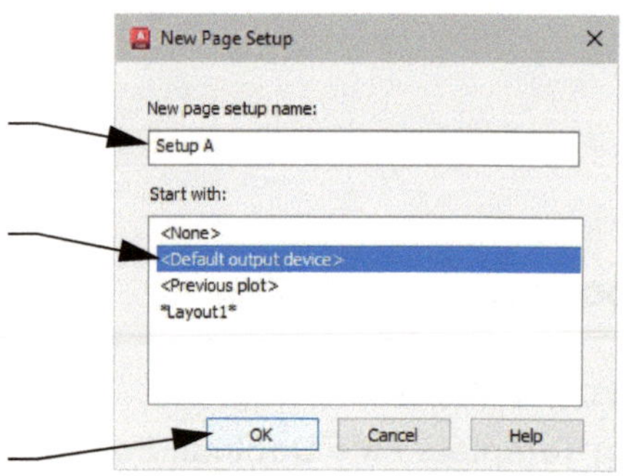

Continued on the next page...

Exercise 26B...continued

This is where you will select the **printer / plotter**, **paper size**, and the **plot offset**.

7. Select the **Printer / Plotter**
   Note: Your current system printer should already be displayed here. If you prefer another, select the down arrow and select from the list. If the preferred printer is not in the list, you must configure the printer. (Refer to **Appendix-A** for instructions.)

8. Select the **Paper size**.

9. Select **Plot offset**.

10. Set to **Inches** for inch users, or **Millimeters** for metric users.

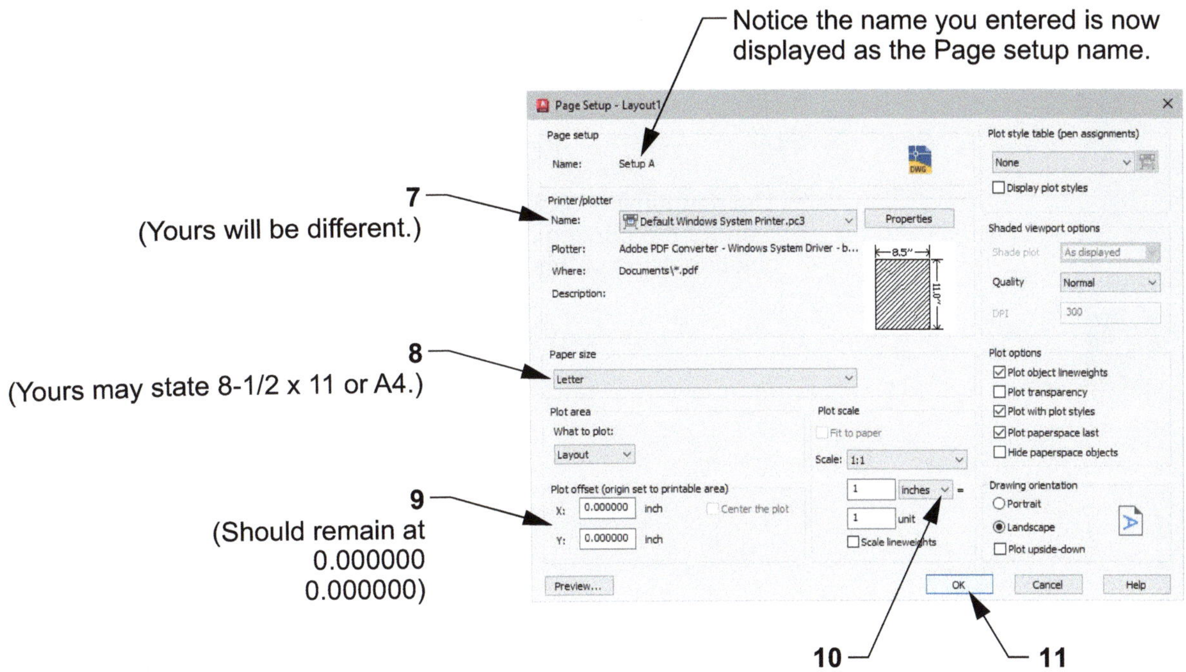

11. Select the **OK** button.

12. Select **Setup A**.

13. Select the **Set Current** button.

14. Select the **Close** button.

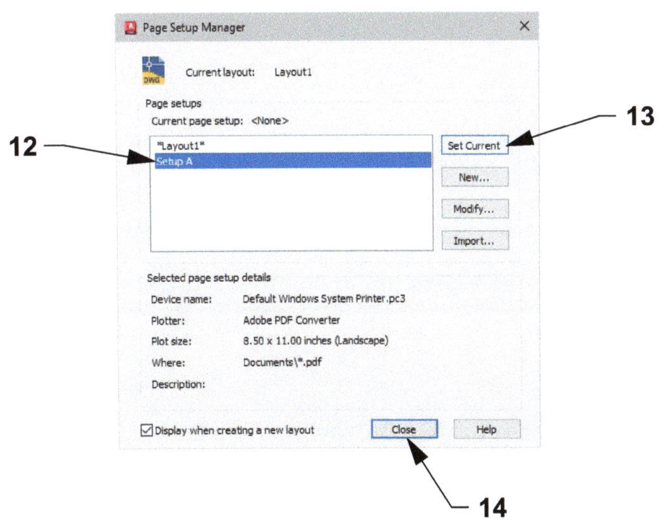

Continued on the next page...

Exercise 26B...continued

You should now have a sheet of paper displayed on the screen.
This sheet is the size you specified in the “Page Setup”.
This sheet is in front of Model Space.

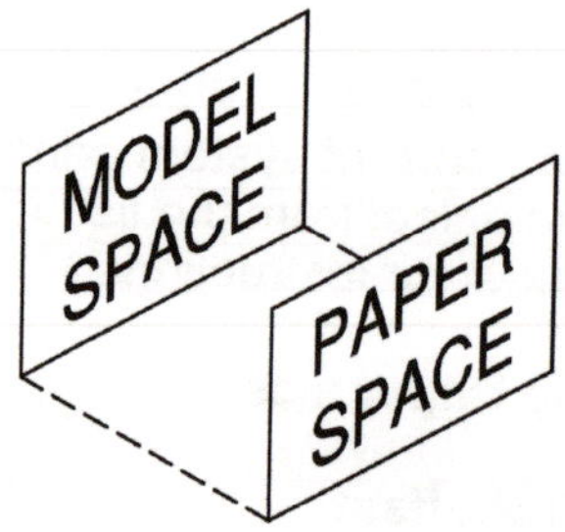

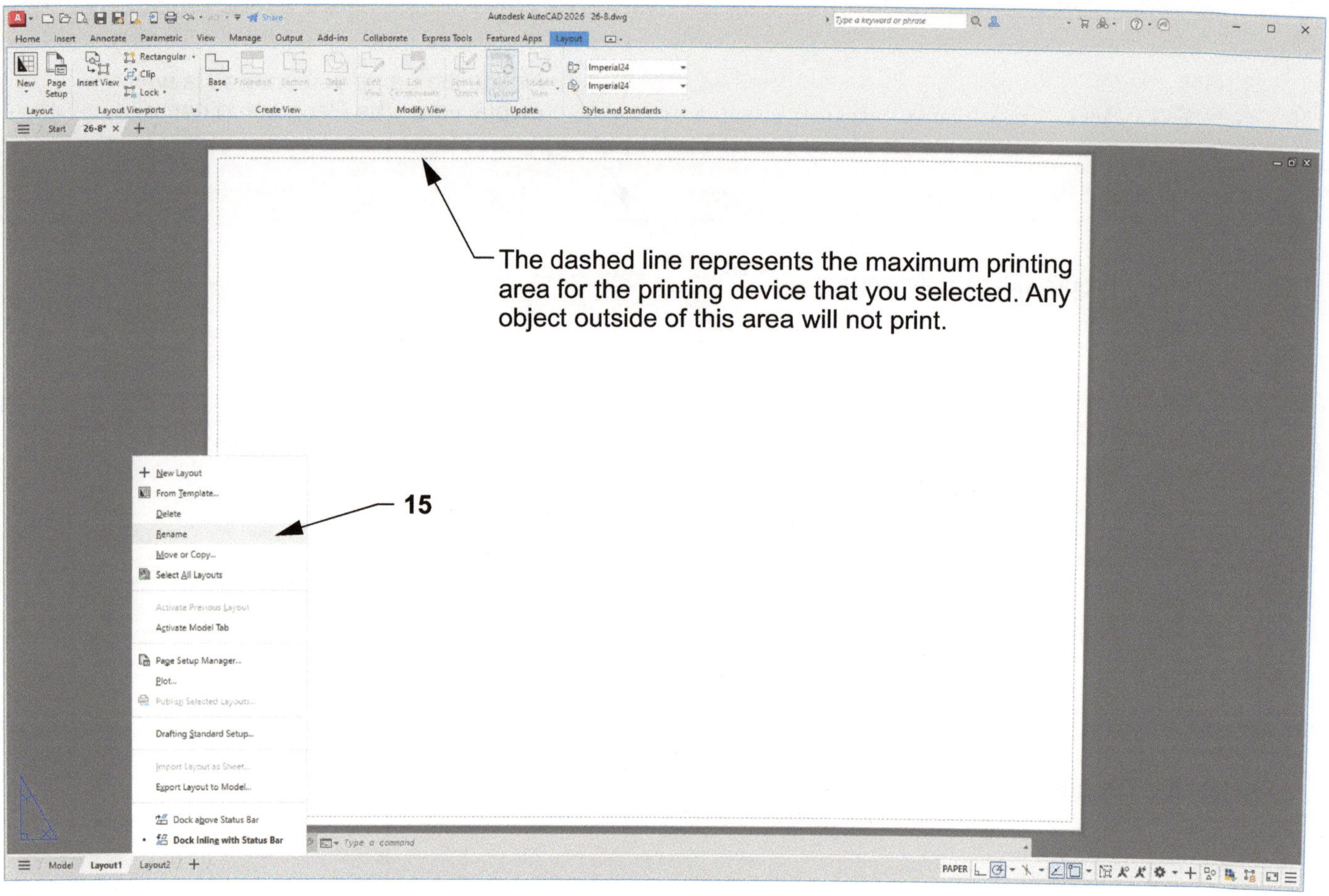

The next step is to rename the Layout Tab.

15. Right click on the active Layout Tab and select **Rename** from the list.

16. Enter the new Layout name **A Size** and then press ***<Enter>***.

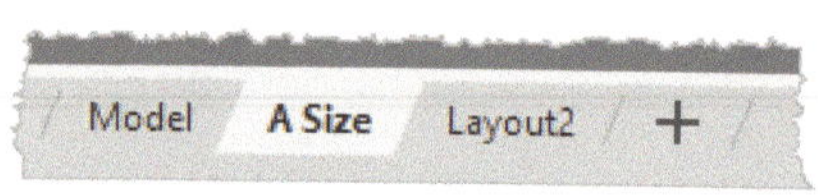

17. **Very important: Save as / AutoCAD Drawing Template / My Decimal Setup** again.

*Now continue on to **Exercise 26C**...you are not done yet.*

Exercise 26C: Create a Border and Title Block

1. Draw the border rectangle as large as you can within the dashed lines approximately as shown below using Layer Borderline.

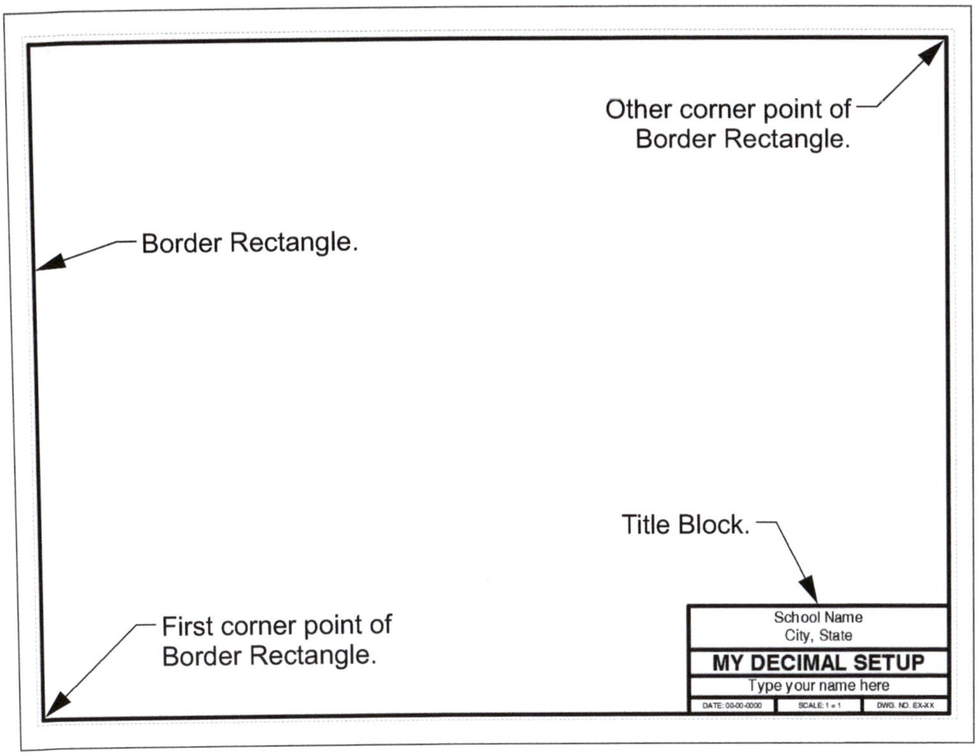

2. Draw the Title Block as shown using:
   A. Layers Borderline and Text.
   B. Multiline Text: Justify **Middle Center** in each rectangular area.
   C. Text Style = **TEXT CLASSIC**. Text Height: Varies.

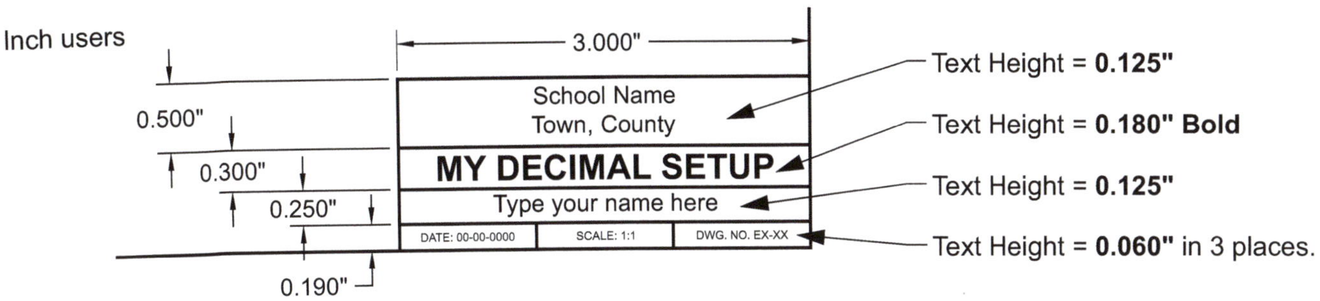

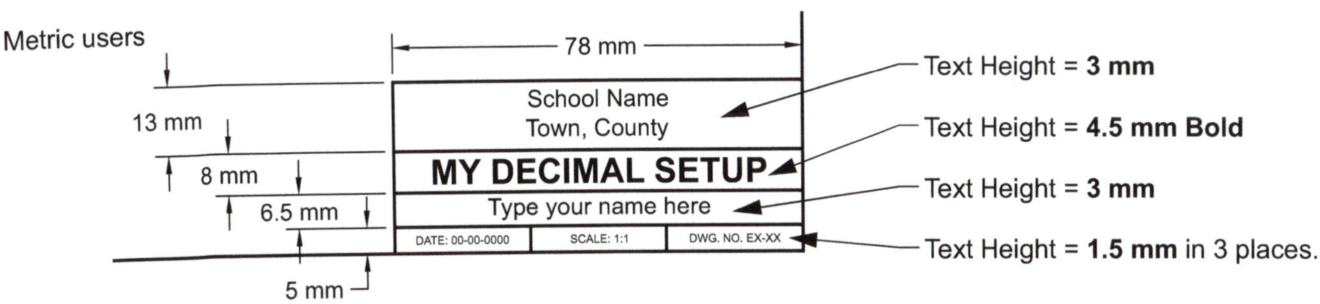

3. **Very important: Save as / AutoCAD Drawing Template / My Decimal Setup** again.

*Now continue on to **Exercise 26D**…you are not done yet.*

Exercise 26D: Create a Viewport

The following instructions will guide you through creating a **viewport** in the Border Layout sheet. Creating a viewport has the same effect as cutting a hole in the sheet of paper. You will be able to see through the viewport frame (hole) to Model Space.

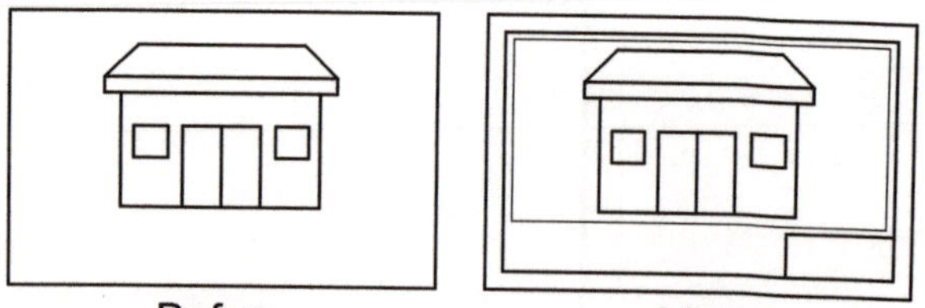

Before After

1. Open **My Decimal Setup** (if not already open).
2. Select the **A Size** Layout Tab.
3. Select Layer Viewport.
4. Type ***MV*** and then press ***<Enter>***. (Refer to page 26-6.)
5. Draw a single viewport approximately as shown. (Turn off **Osnap**.)

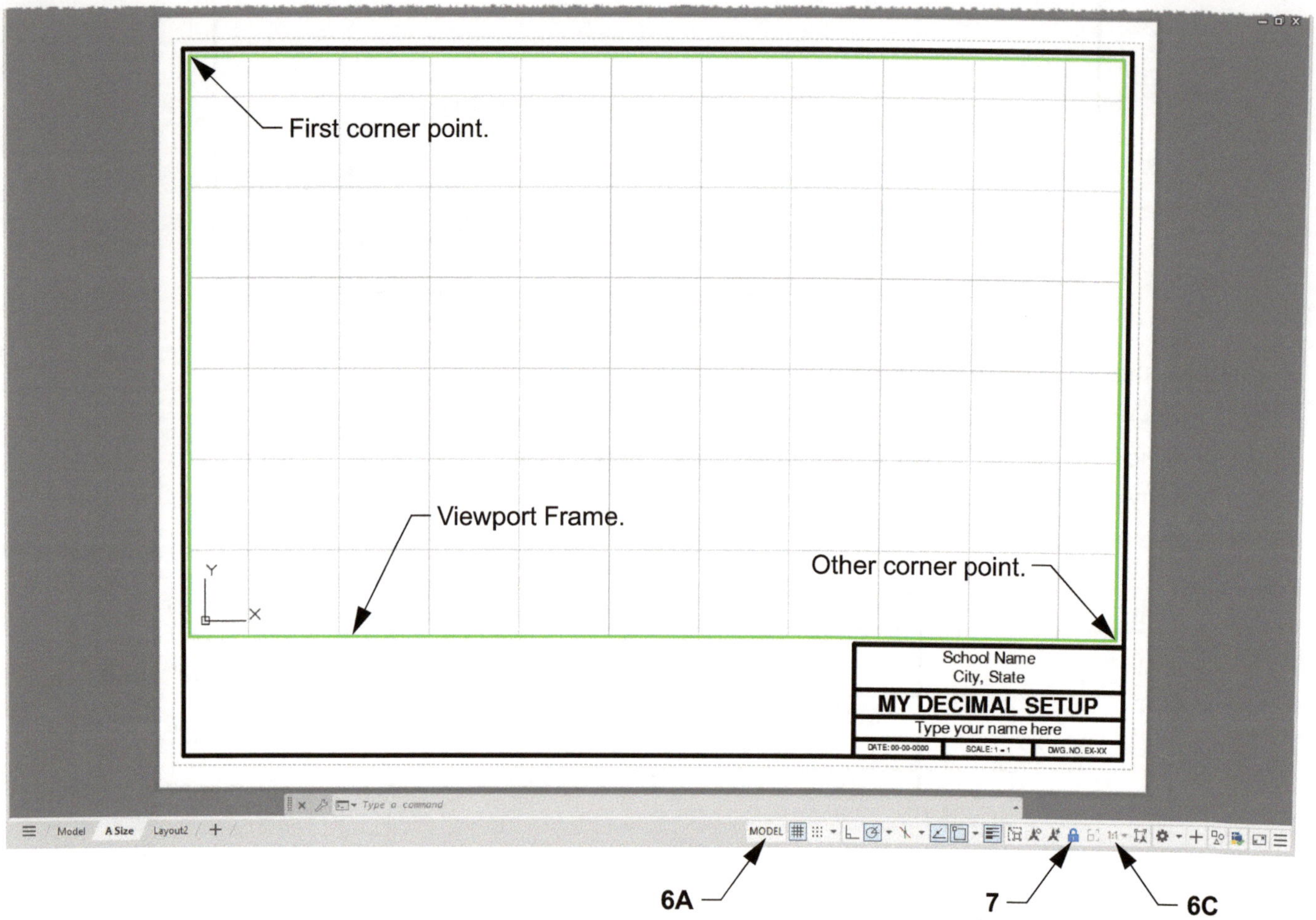

6. Adjust the viewport scale as follows:
   A. Confirm the **"MODEL"** button is showing.
   B. Select **"Zoom / All"**.
   C. Select the viewport scale of **1:1** (Refer to page 26-12).
7. Lock the viewport. (Refer to page 26-9.)
8. **Very important: Save as / AutoCAD Drawing Template / My Decimal Setup** again.

*Now continue on to **Exercise 26E**...you are not done yet.*

Exercise 26E: Plotting from the Layout

The following instructions will guide you through the final steps for setting up the master template for plotting. These settings will stay with **My Decimal Setup** and you will be able to use it over and over again.

1. Open **My Decimal Setup** (if not already open).
2. Select the **A Size** Layout Tab.
3. Select the **Plot** command.

Note: Make sure your settings match the settings shown below.

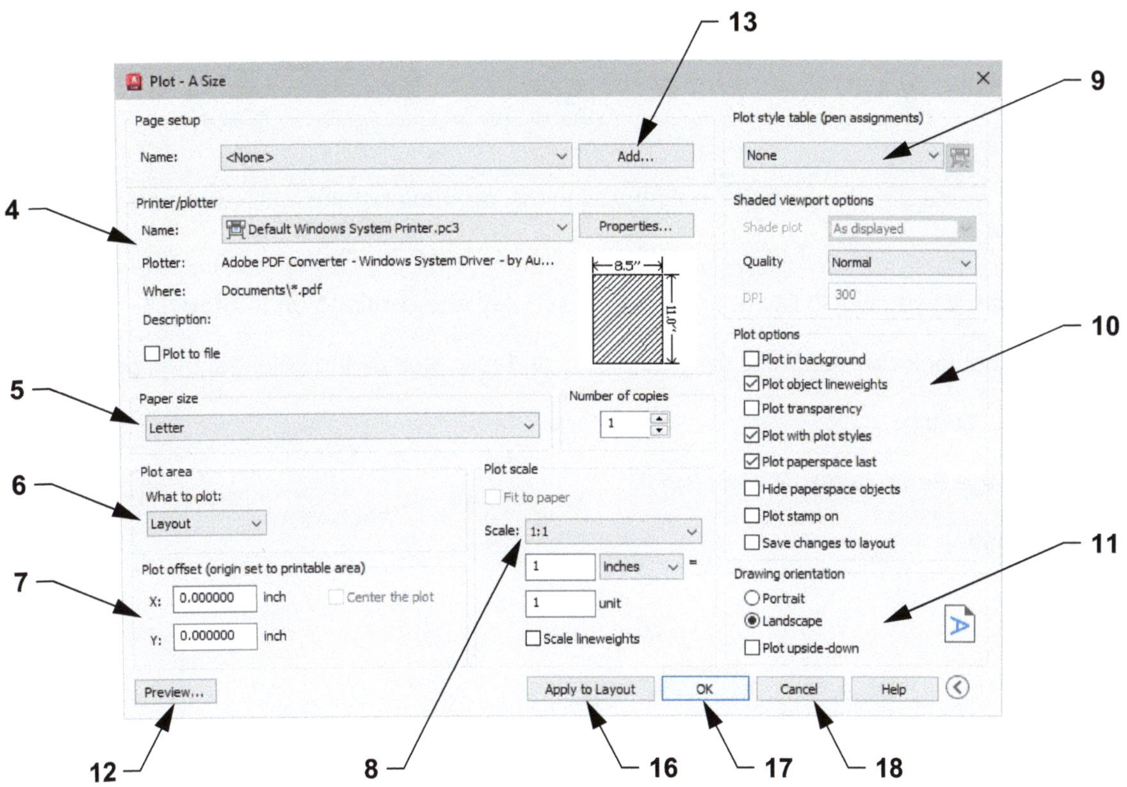

4. Select your printer.
5. Select the Paper Size.
6. Select the Plot Area.
7. Plot offset should be **0.000000** for **X** and **Y**.
8. Select scale **1:1** (**Note:** Make sure you select **Inches** for inch users, or **Millimeters** for metric users.
9. Select the Plot Style Table **Monochrome.ctb**
10. Select the Plot options shown.
11. Select Drawing Orientation: **Landscape**.
12. Select the **Preview...** button.

Continued on the next page...

Exercise 26E...continued

Note: The viewport frame and the grids will not appear in the **preview** because the Viewport layer is set to **No Plot** and grids never plot.

If the drawing appears as you would like it, press the ***<Esc>*** key and continue on to **Step 13**.

If the drawing does not look correct, press the ***<Esc>*** key and re-check all the settings, then preview again.

13. Select the **Add...** button.

14. Type the new Page Setup name: **Plot Setup A**

15. Select the **OK** button.

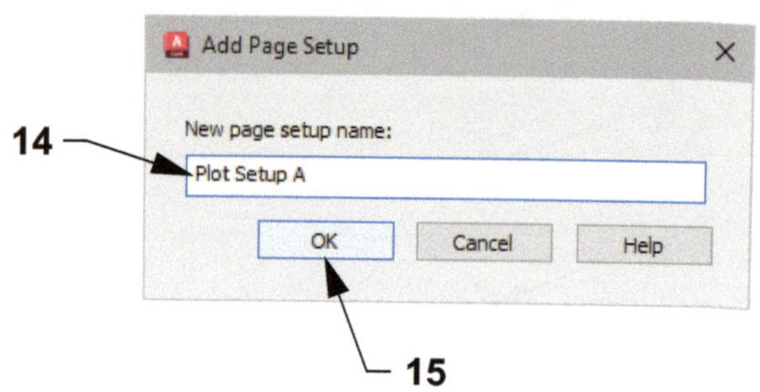

16. Select the **"Apply to Layout"** button.
The settings are now saved to the Layout Tab for future use.

17. If your computer __is__ connected to the plotter / printer selected, select the **OK** button to plot, then proceed to **Step 20**.

18. If your computer is __not__ connected to the plotter / printer selected, select the **Cancel** button to close the Plot dialog box and proceed to **Step 20**. **Note:** *Selecting **Cancel** will cancel your selected setting if you did not **Add** the Page Setup as specified in **Step 13**.*

You are almost done.

19. Select Layer Object Line. (You don't want Layer Viewport to be the current layer.)

20. Now Save all of this work as a Template one last time: **Application Menu / Save as / AutoCAD Drawing Template / My Decimal Setup**.

*That probably seemed like a lot of work, but you have now completed the **Plotting Page Setup** for the **My Decimal Setup.dwt**.*

Now you are ready to use this master template to create and plot many drawings in the future. In fact, you have one on the very next page.

Exercise 26F: Model Space and Paper Space

This exercise is in two parts. First you will draw the drawing in the **Model** Tab. Then you will create a layout to display parts of the drawing in three viewports. *Hopefully this will help you understand the differences between Model Space and Paper Space.*

Part 1.

1. Select **New** and select **My Decimal Setup** template (if not already open).
2. Select the **Model** Tab. 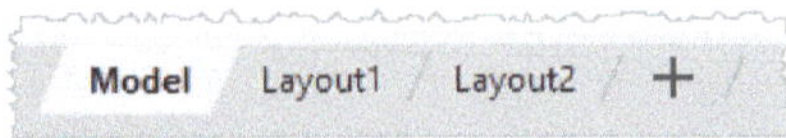

3. Draw the object shown below.

   Use Layer Object Line, do not use Layer Viewport and do not dimension.
4. Save the drawing as: **Ex-26F**

   (Make sure that you save it as a **drawing (.dwg)** and not as a template.)

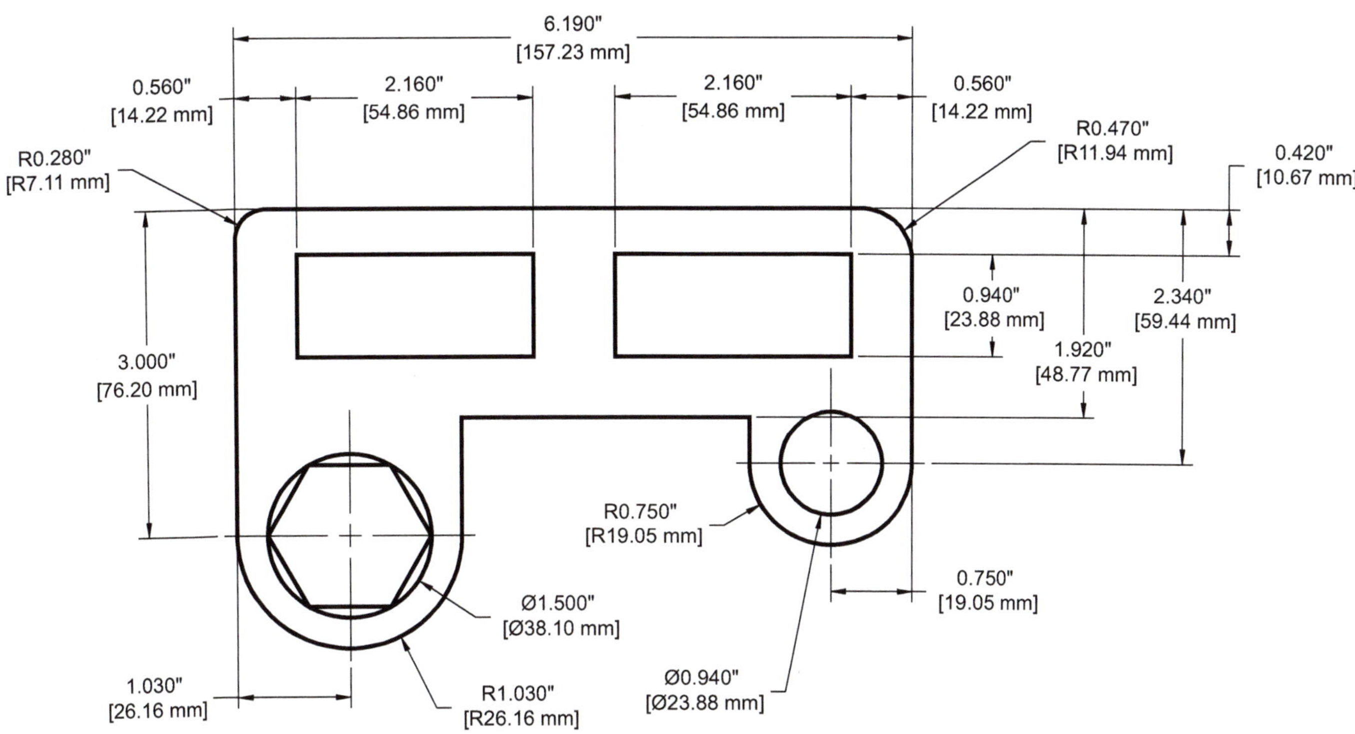

Continue to Part 2 on the next page.

Exercise 26F...continued

Part 2.

1. Select the **A Size** Tab. 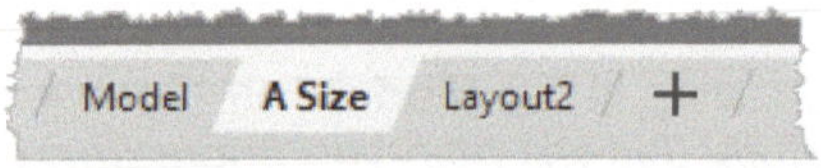

2. Confirm the MODEL / PAPER button says PAPER. 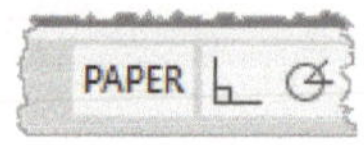

3. Erase the one existing viewport frame. (Click on the frame and select Erase.)

4. Select the Viewport layer and create three new viewports approximately as shown below.

5. Using Zoom and Pan, inside each viewport, try to display each viewport content as shown below. (**Note:** Scale is not critical at this time, but it will be in Lesson 27.)

6. Lock the viewport.

7. This time, I would like you to **print the viewport frames**, so you need to make the Viewport layer **plottable**. (Remove the **No Plot** sign 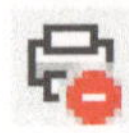 on the printer symbol in the **Layer Properties Manager**.)

8. Change the Title block: **Title: VIEWPORTS** **Scale:** NONE **Dwg no:** Ex-26F

9. Save the drawing again as: **EX-26F**

10. Plot using Plot Page Setup name: **Plot Setup A**. (Remember to view the preview.)

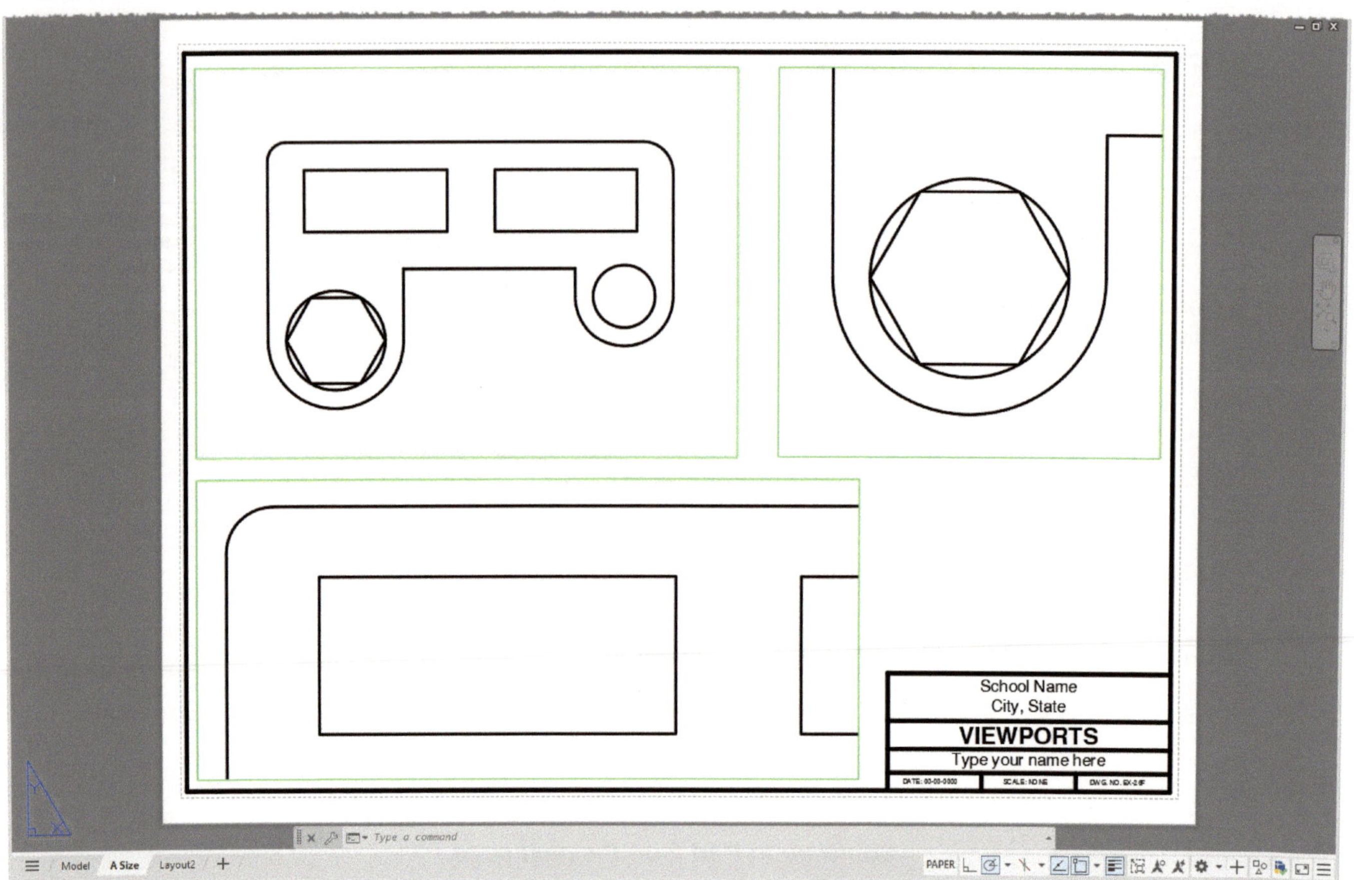

LESSON 27

LEARNING OBJECTIVES

After completing this lesson, you will be able to:

- Understand Scaled Drawings
- Adjust the Scale within a Viewport
- Understand Annotative Objects
- Understand Paper Space versus Model Space dimensioning
- Create a Master Feet-Inches Setup Template

Creating Scaled Drawings

In the lessons previous to Lesson 26, you worked only in Model Space. Then in Lesson 26, you learned that AutoCAD actually has another environment called Paper Space, or Layout. In this lesson, we need to learn more about why we need two environments and how they make it easier to display and plot your drawings.

Note: A very important rule in CAD you must understand: *"All objects are drawn full size".*

In other words, if you want to draw a line 20 feet long, you actually draw it 20 feet long. If the line is 1/8" long, you actually draw it 1/8" long.

Drawing and Plotting objects that are very large or very small

In the previous lessons, you created medium-sized drawings. Not too big, not too small. But what if you wanted to draw a house? Could you print it to scale on an 11" X 8-1/2" or 297 mm X 210 mm piece of paper? How about a small paper clip? Could you make it big enough to dimension? Let's start with the house.

Printing an entire house on an 11" X 8-1/2" or 297 mm X 210 mm sheet of paper

Remember the photo and picture frame example I suggested in Lesson 26? (Refer to 26-4.) This time, try to picture yourself standing at the front door of your house with an empty picture frame in your hands. Look at your house through the picture frame. Of course the house is way too big to fit in the frame. Or is it because you are standing too close to the house?

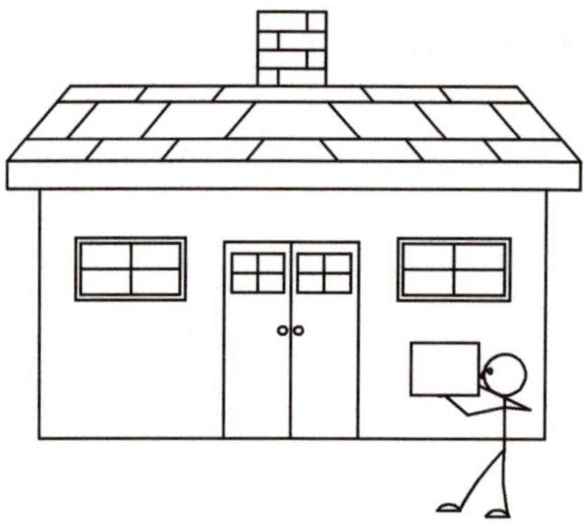

Now walk across the street and look through the picture frame in your hands again. Does the house appear smaller? Can you see all of it in the frame? If you could walk far enough away from the house, it would eventually appear small enough to fit in the picture frame in your hands. But the house did not actually change size, did it? It only appears smaller because you and the picture frame are farther away from it.

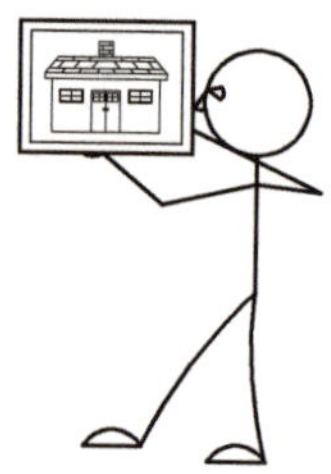

Adjusting the Viewport Scale

When using AutoCAD, walking across the street with the frame in your hands is called **adjusting the viewport scale**. You are increasing the distance between Model Space (your drawing) and Paper Space (Layout) and that makes the drawing appear smaller.

Example: A viewport scale of 1/4" = 1' would make Model Space appear 48 times smaller. But when you dimension the house, the dimension values will be the actual measurement of the house. In other words, a 30 foot line will have a dimension of 30'-0".

When plotting something smaller, like a paperclip, you have to move the picture frame closer to Model Space to make it appear larger. For example, you might use a viewport scale of 8 = 1.

Continued on the next page...

The following will take you through the process of adjusting the scale within a viewport.

1. **Open** a drawing.
2. Select a **Layout** Tab. (Paper Space.)
3. Create a new **Page Setup**.
4. Cut a new viewport or unlock an existing viewport.
5. Adjust the scale:
   A. You must be in Paper Space. (See below.)
   B. Select the viewport frame.
   C. Unlock the viewport, if locked.
   D. Select the viewport scale down arrow.
   E. Select the scale from the list of scales.

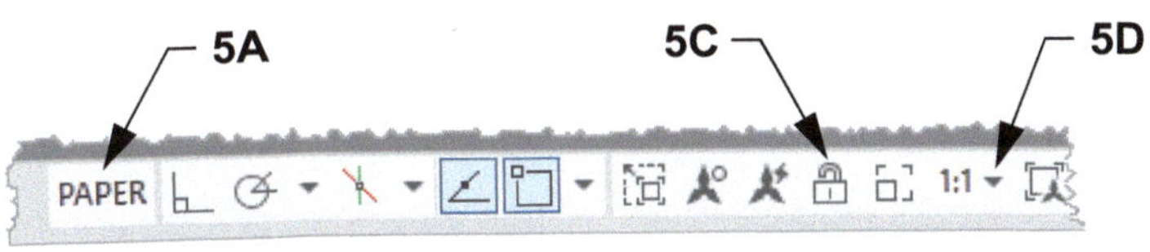

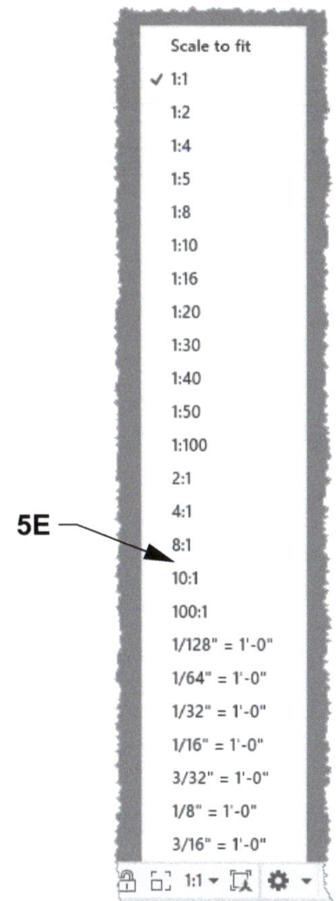

6. Lock the viewport.

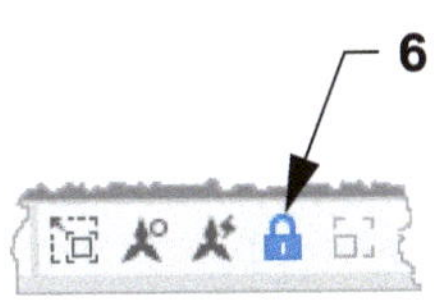

Note: If you would like to add a scale that is not on the list:

1. Type ***scalelistedit*** and then press ***<Enter>***.
2. Select the **Add...** button.
3. Enter the **Scale name** to display in Scale List.
4. Enter **Paper** and **Drawing units**.
5. Select the **OK** button.

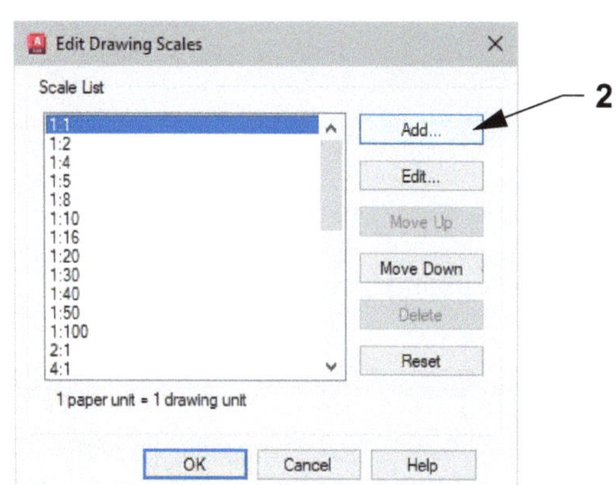

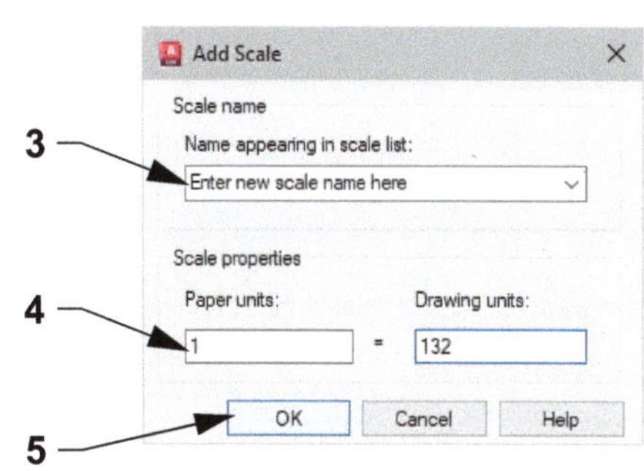

Annotative Objects

In Lesson 26, you created a new text style and dimension style. You also added the **"Annotative"** property to both simply by placing a check mark in the Annotative box.

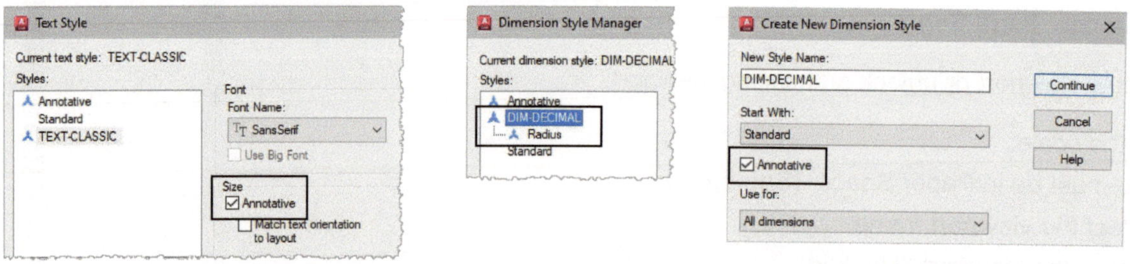

The Text Style and Dimension Style shown above are now **Annotative**. Notice the Annotative symbol  beside the style name. (No symbol by "Standard".)

"Annotative objects" are scaled automatically to match the scale of the viewport. For example, if you want the text, inside the viewport, to print 0.200" or 5 mm in height and the viewport scale is 1:2, AutoCAD will automatically scale the text height to 0.400" or 10 mm. The text height needs to be scaled by a factor of 2 to compensate for the Model Space contents appearing smaller.

The easiest way to understand how **Annotative property** works is to do it. So try this example. (When complete, you will save this file for use in Lesson 28.)

1. Start a New file using **My Decimal Setup.dwt**

2. Select the **Model** Tab (Model Space).

3. Draw a rectangle **3" [76.2 mm]** Long X **1.5" [38.1 mm]** Wide. Use Layer Object Line. Do not dimension.

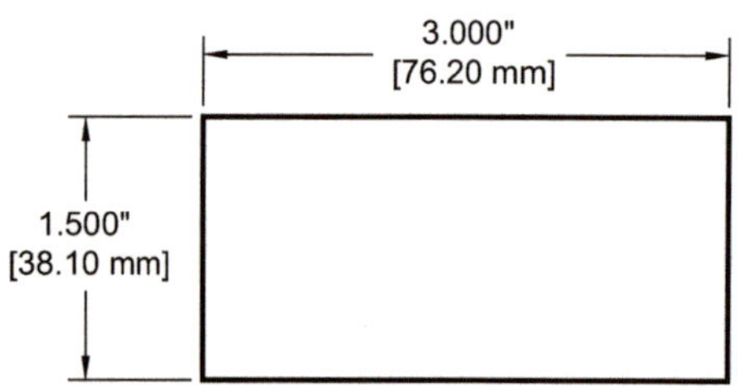

4. Select the **A Size** Tab (Paper Space).

5. Erase the existing single viewport frame.

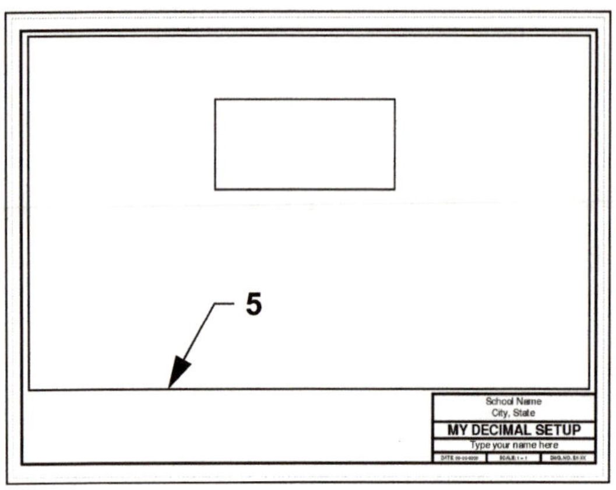

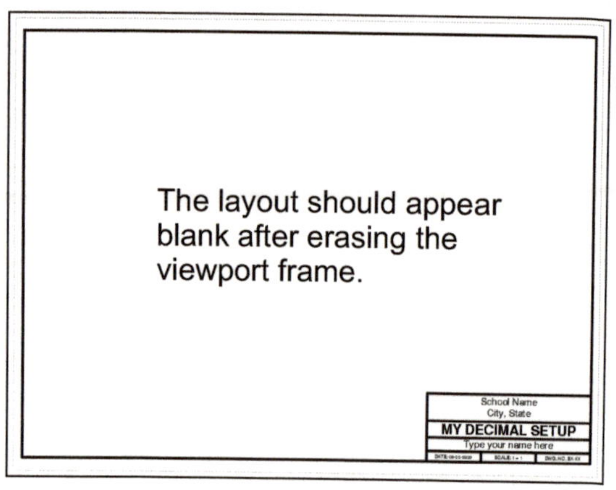

Continued on the next page...

6. Change current layer to Viewport and cut two new viewports as shown below. (See 26-6.)

7. Activate the left viewport (double click inside the viewport frame) and do the following:
   A. Adjust the scale of the viewport to **1:1**. (Refer to page 27-3.)
   B. **Pan** to place the rectangle in the center of the viewport. (Refer to page 26-8.)
   C. **Lock** the viewport. (Refer to page 27-3.)

8. Activate the right viewport (click inside the viewport frame) and do the following:
   A. Adjust the scale of the viewport to **1:4**.
   B. **Pan** to place the rectangle in the center of the viewport.
   C. **Lock** the viewport.

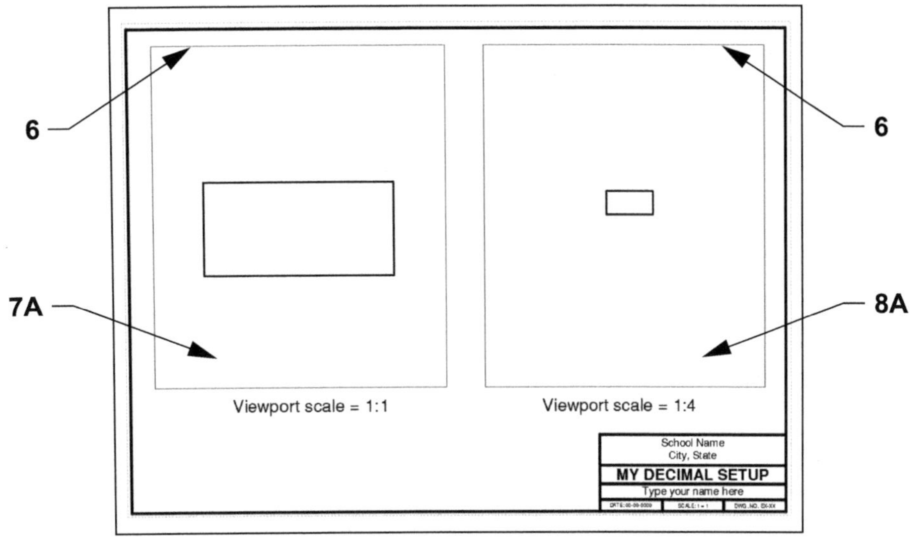

Your screen should appear approximately like this.
Note: My grids are off.

9. Activate the **left-hand** viewport. (Double click inside left-hand viewport.)

10. Change current layer to Text and add **0.200" [5 mm]** height text, Scale **1:1** as shown using text style TEXT-CLASSIC. (**Note:** Text style TEXT-CLASSIC is annotative.)

11. Change current layer to Dimension and add the dimension shown using dimension style DIM-DECIMAL. (**Note:** Dimension style DIM-DECIMAL is annotative. The dimension shown is in inches, metric users dimensions will appear differently.)

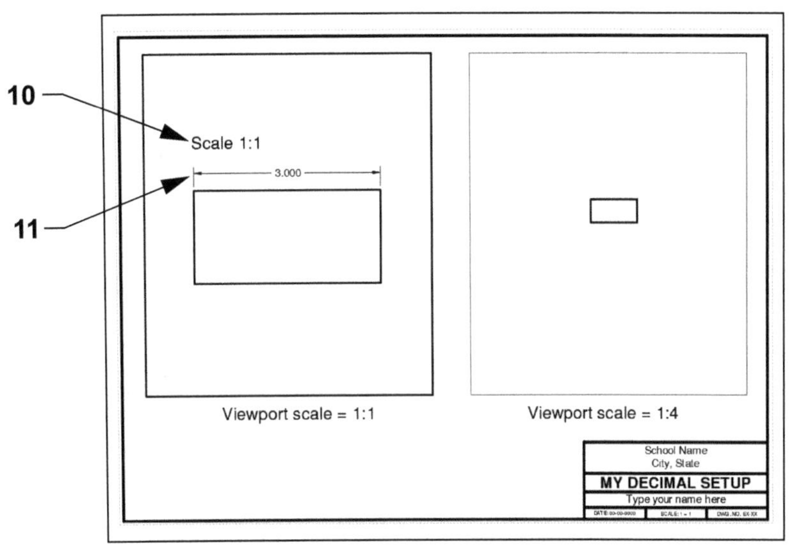

Continued on the next page...

12. Activate the **right-hand** viewport. (Click inside right-hand viewport.)

13. Change current layer to Text and add **0.200" [5 mm]** height text, Scale **1:4** as shown using text style TEXT-CLASSIC. (**Note:** Text style TEXT-CLASSIC is annotative.)

14. Change current layer to Dimension and add the dimension shown using dimension style DIM-DECIMAL. (**Note:** Dimension style DIM-DECIMAL is annotative. The dimension shown is in inches, metric users dimensions will appear differently.)

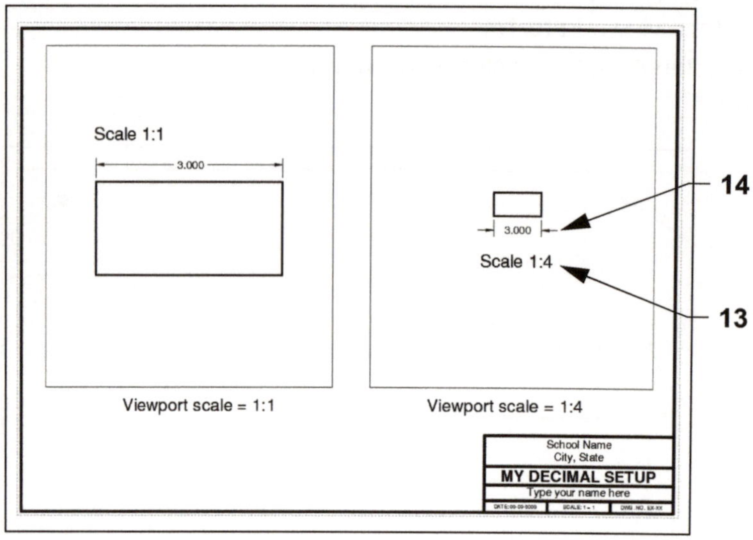

Notice that the text and dimensions appear the same size in both viewports.

15. Now select the **Model** Tab. (Model Space.)

Notice there are two sets of text and dimensions.

One set has the annotative scale of 1:1 and will be visible only in a 1:1 viewport. One set has the annotative scale of 1:4 and will be visible only in a 1:4 viewport. But you see both sets when you select the Model Tab.

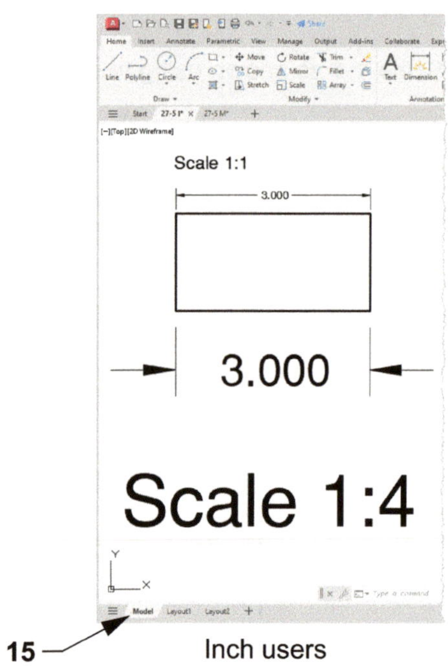

Inch users

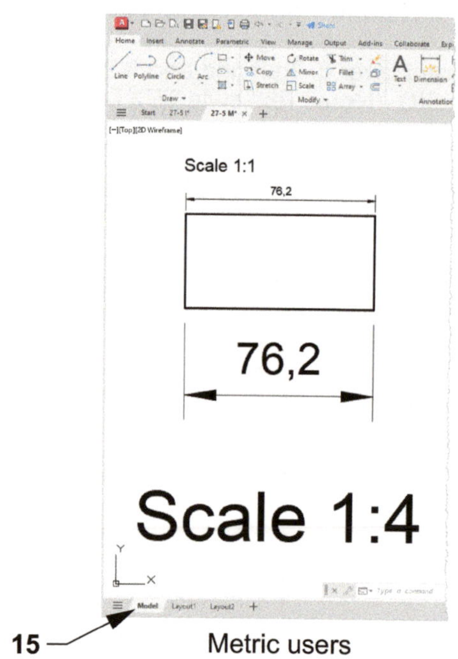

Metric users

(In Lesson 28, you will learn how to assign multiple annotative scales to one set of text or dimensions so you will not have duplicate sets in the Model Tab.)

Continued on the next page...

16. Place your cursor on any of the text or dimensions. An "Annotative symbol" will appear. This indicates this object is annotative and it has only one annotative scale. (In Lesson 28, you will learn how to assign multiple annotative scales to a single annotative object so it will be visible in multiple viewports.

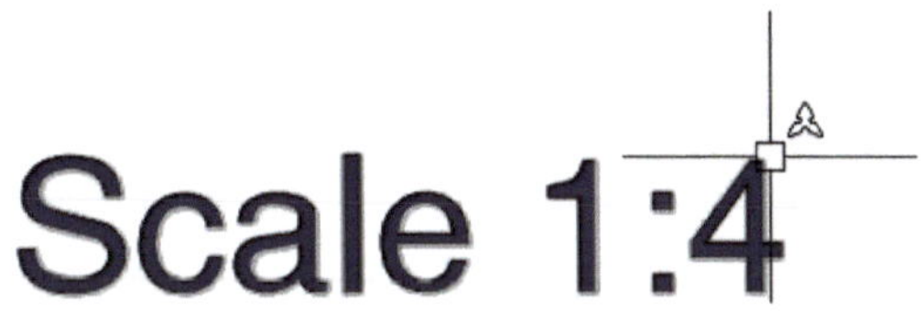

17. Click once on the **Scale: 1:4** text. The Quick Properties box should appear. (**Quick Properties** button on the Status Bar must be **on**.) *Notice the text height is listed twice.*

Paper text height (inch users) **= .200** This is the height that you selected when placing the text. When the drawing is printed, the text will print at **0.200"**.

Model text height (inch users) **= .800** This is the desired height of the text (**.200**) factored by the viewport scale (**1:4**). The viewport scale is a factor of **4**. (**4 X .200 = .800**)

Paper text height (metric users) **= 5** This is the height that you selected when placing the text. When the drawing is printed, the text will print at **5 mm**.

Model text height (metric users) **= 20** This is the desired height of the text (**5**) factored by the viewport scale (**1:4**). The viewport scale is a factor of **4**. (**4 X 5 = 20**)

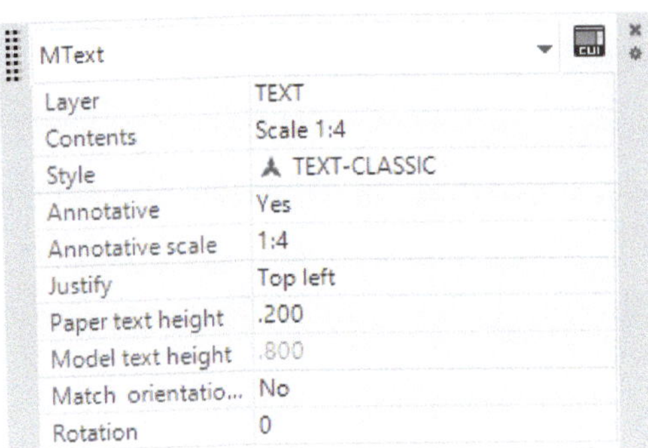

MText	
Layer	TEXT
Contents	Scale 1:4
Style	TEXT-CLASSIC
Annotative	Yes
Annotative scale	1:4
Justify	Top left
Paper text height	.200
Model text height	.800
Match orientatio...	No
Rotation	0

Quick Properties
Inch users

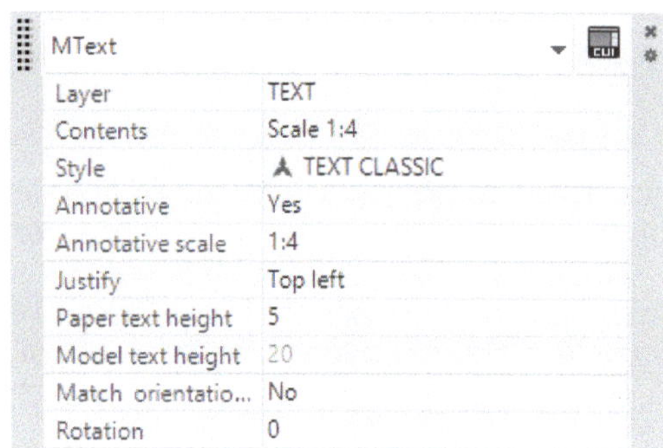

MText	
Layer	TEXT
Contents	Scale 1:4
Style	TEXT CLASSIC
Annotative	Yes
Annotative scale	1:4
Justify	Top left
Paper text height	5
Model text height	20
Match orientatio...	No
Rotation	0

Quick Properties
Metric users

18. **Important:** Save this drawing as: **Annotative objects**. (You will need it for Lesson 28.)

Note: If an object is annotative, AutoCAD automatically adjusts the scale of the object to the viewport scale. The most commonly used Annotative objects are: Dimensions, Text, Hatch, and Multileaders. Refer to the Autodesk Assistant or AutoCAD Help Menu for more information. (In Lesson 28, you will learn how to assign multiple annotative scales to a single annotative object.)

Paper Space Dimensioning

Some people prefer to dimension in Paper Space. Paper Space dimensions are Trans-spatial. Trans-spatial means that you may place the dimension in Paper Space while the object you are dimensioning is in Model Space. Even though the dimension is in Paper Space, it is actually attached to the object in Model Space.

Example:

1. You draw a house in Model Space.
2. Now select the Layout Tab.
3. Cut a viewport so you can see the drawing of the house.
4. Go to Model Space, adjust the scale of the viewport (Model Space), and lock it.
5. Now go to Paper Space and dimension the house.

Should you dimension in Paper Space or Model Space? This is your choice. Personally, I dimension in either space depending on the situation. Here are some things to consider.

Paper Space dimensioning

Pros

1. You never have to worry about the viewport scale.
2. All dimensions will have the same appearance in all viewports.
3. If you Xref the drawing, the dimensions do not come with it. (Xref is not discussed in this Workbook.)

Cons

1. If you move the objects in Model Space, or adjust the viewport scale, sometimes the dimensions do not move with them. (If this happens, type: ***dimregen*** and then press ***<Enter>***.)
2. Dimensions will not appear in the Model Tab. They appear only in the Layout Tab in which they were placed. If you go to another Layout Tab, they will not appear.

Annotative dimensioning in Model Space

Pros

1. Dimensions always appear and move with the objects.
2. Dimensions will appear in all Layout Tabs within viewports.

Cons

1. If you External Reference (Xref the drawing), the dimensions also are included.

Exercise 27A: Create a Master Feet-Inches Setup Template

The following instructions will guide you through creating a "master" Feet-Inches Setup Template. The "inch-helper" is an example of a master setup template. Once you have created this "master" template, you just open it and draw. No more repetitive inputting of settings.

Note to Metric users: This template is for **"Architectural Inch"** units of measurement, which are no longer used in countries where **"Metric"** units of measurement are used. It is still good practice for metric users, so have a go and see how you do. There may be occasions when you have to use architectural drawings that use the inch system.

New Settings

1. Begin your drawing with a different template as follows:
   - A. Select the **New** command.
   - B. Select template file **acad.dwt** from the list of templates. (**Not** acad3D.dwt.)
   - C. Save the drawing as: **My Feet-Inches Setup** but keep the drawing open.
2. Set the **Drawing Units** as follows:
   - A. Type ***units*** and then press ***<Enter>***.
   - B. Change the **Type** and **Precision** as shown.
   - C. Select the **OK** button.

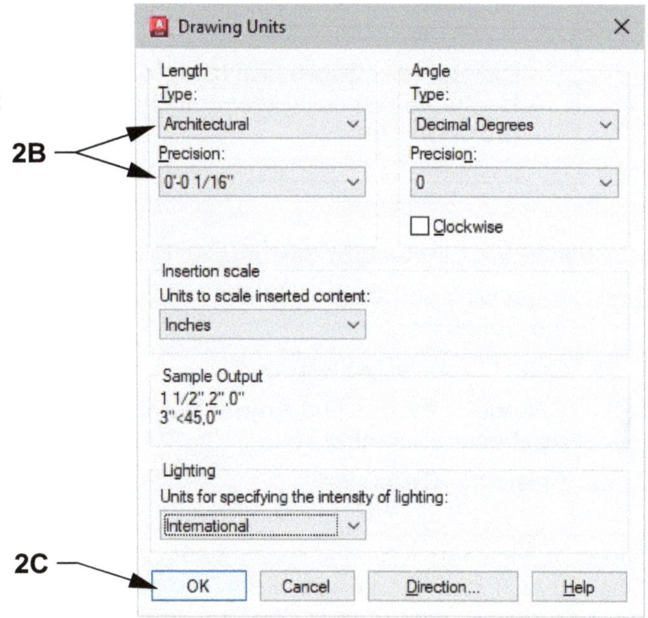

3. Set the **Grids** and **Snap** as follows:
   - A. Type ***DS*** and then press ***<Enter>***.
   - B. Select the **Snap and Grid** Tab.
   - C. Change the settings as shown below.
   - D. Select the **OK** button.

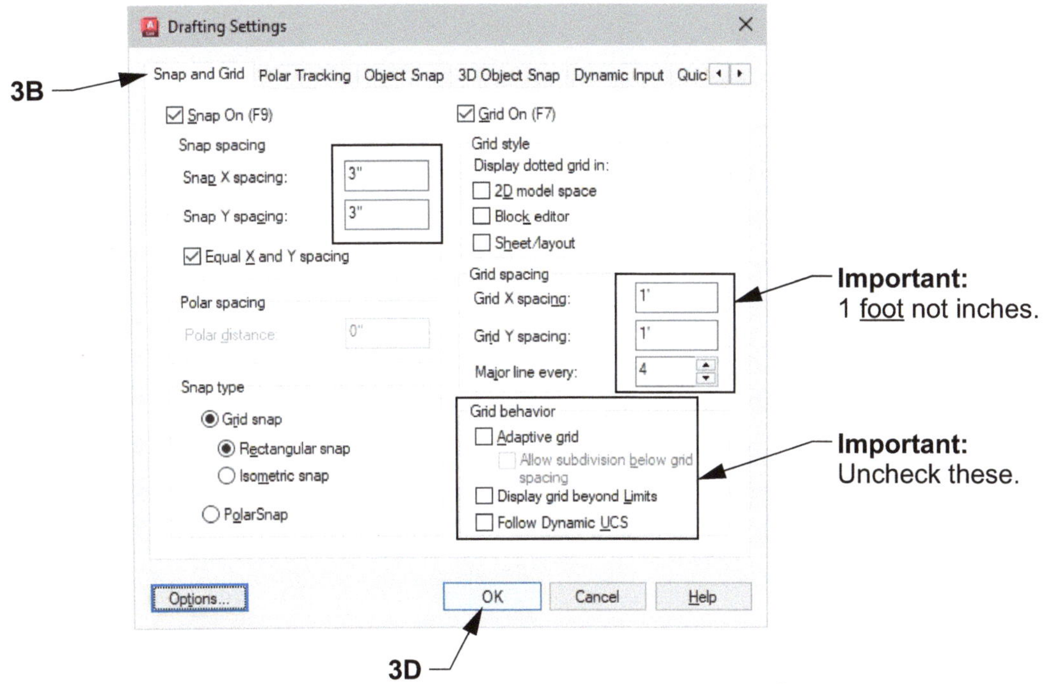

Continued on the next page...

Exercise 27A...continued

4. Set the **Drawing Limits** (size of the drawing area) as follows:
   A. Type ***Limits*** and then press ***<Enter>***.
   B. Enter lower left corner = ***0'-0",0'-0"***
   C. Enter upper right corner = ***44',34'***
   (Notice this is feet, not inches.)
   D. Use **"Zoom / All"** to view the new limits.
   E. Set your **Grids** to **on** to display the limits.

If your screen does not appear as shown here, check these:
- Did you use, **Zoom / All** (**Step 4D**)?
- Is your Grid spacing feet or inches?
- Is the Grid **on**?

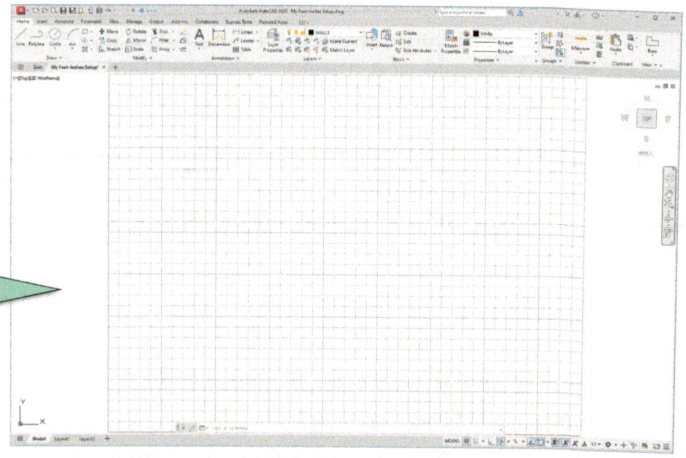

5. Change the **Lineweight Settings** as follows:
   A. Right click on **Lineweight** button located on the Status Bar.
   B. Select **Lineweight Settings**.
   C. Change to **Inches** and **Adjust Display Scale** as shown. (See note.)
   D. Select the **OK** button.

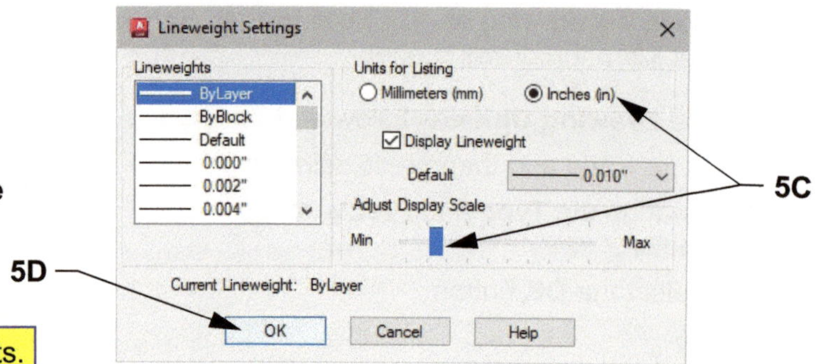

Note: Metric users will be showing metric lineweights. Change to inches, then **re-save** the drawing. Close down AutoCAD, then re-open **My Feet-Inches Setup**.

6. Create new **layers** as follows:
   A. Load the linetype DASHED. (Refer to page 3-13 for instructions.)
   B. Assign names, colors, linetypes, lineweights, and plotability as shown below.
   Make the Viewport layer **Not Plot** and the Walls layer **current**. (Refer to page 3-12 for instructions.)

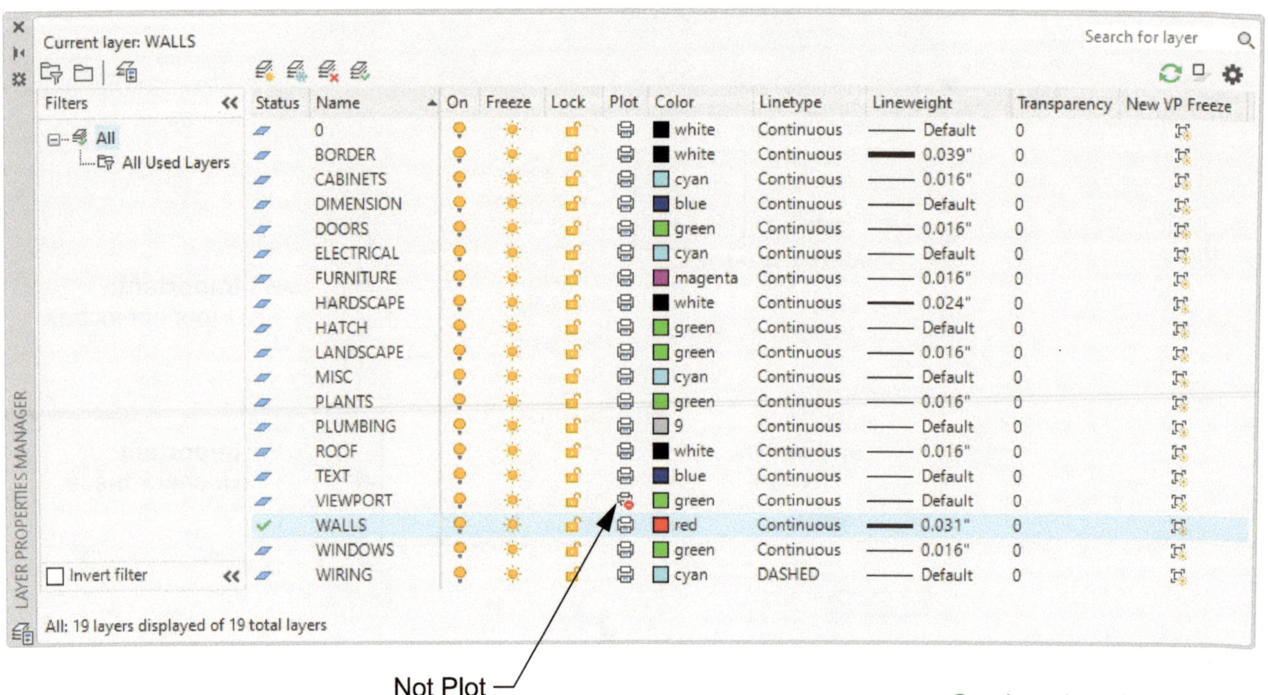

Status	Name	Color	Linetype	Lineweight	Transparency
	0	white	Continuous	Default	0
	BORDER	white	Continuous	0.039"	0
	CABINETS	cyan	Continuous	0.016"	0
	DIMENSION	blue	Continuous	Default	0
	DOORS	green	Continuous	0.016"	0
	ELECTRICAL	cyan	Continuous	Default	0
	FURNITURE	magenta	Continuous	0.016"	0
	HARDSCAPE	white	Continuous	0.024"	0
	HATCH	green	Continuous	Default	0
	LANDSCAPE	green	Continuous	0.016"	0
	MISC	cyan	Continuous	Default	0
	PLANTS	green	Continuous	0.016"	0
	PLUMBING	9	Continuous	Default	0
	ROOF	white	Continuous	0.016"	0
	TEXT	blue	Continuous	Default	0
	VIEWPORT	green	Continuous	Default	0
✓	WALLS	red	Continuous	0.031"	0
	WINDOWS	green	Continuous	0.016"	0
	WIRING	cyan	DASHED	Default	0

Continued on the next page...

Exercise 27A...continued

7. Create two new **Text Styles** as follows:
   A. Select **Text Style**. (Refer to page 25-2.)
   B. Create the text style TEXT-CLASSIC and TEXT-ARCH using the settings shown below.

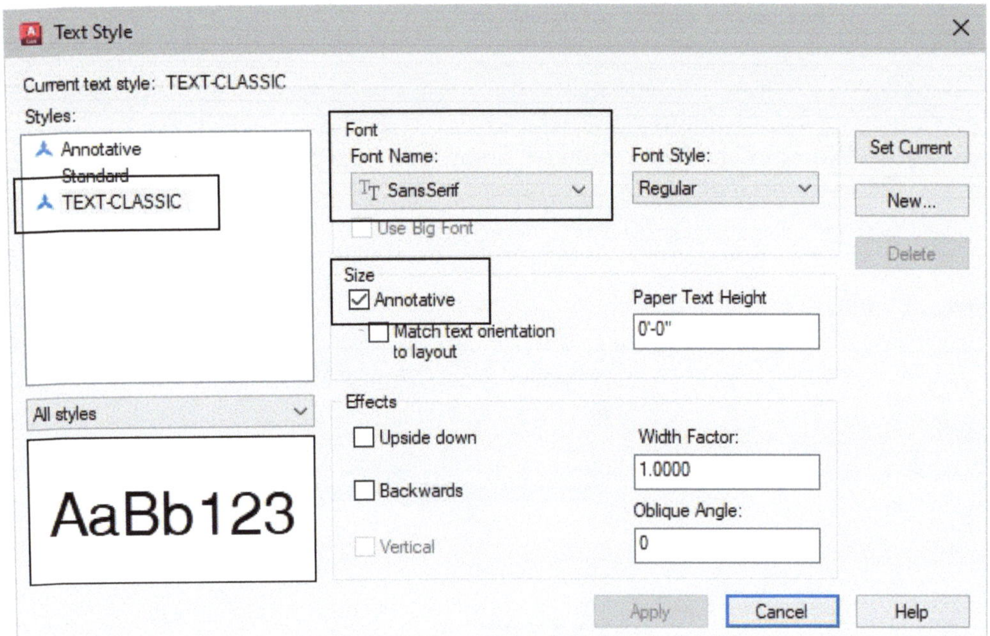

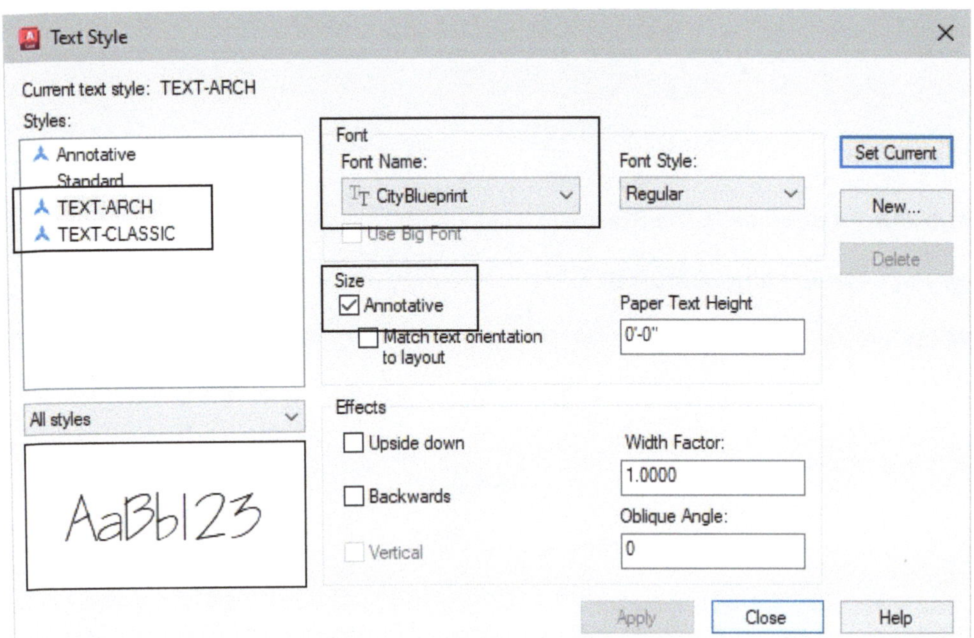

This next step is very important.

8. **Save** all the settings you just created as follows: Save your drawing again as: **My Feet-Inches Setup**

*Now continue on to **Exercise 27B**…you are not done yet.*

Exercise 27B: Page Setup

Now you will select the printer and the paper size to use for printing. You will use the Layout1 Tab (Paper Space).

1. Open **My Feet-Inches Setup** (if not already open).
2. Select the **Layout1** Tab. 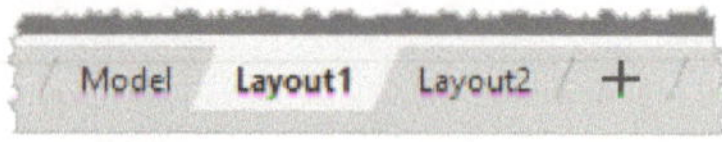

(Refer to page 26-3 if you do not have these Tabs.)

Note: If the **"Page Setup Manager"** dialog box shown below does not appear automatically, right click on the Layout Tab and select **Page Setup Manager**.

3. Select the **New...** button.

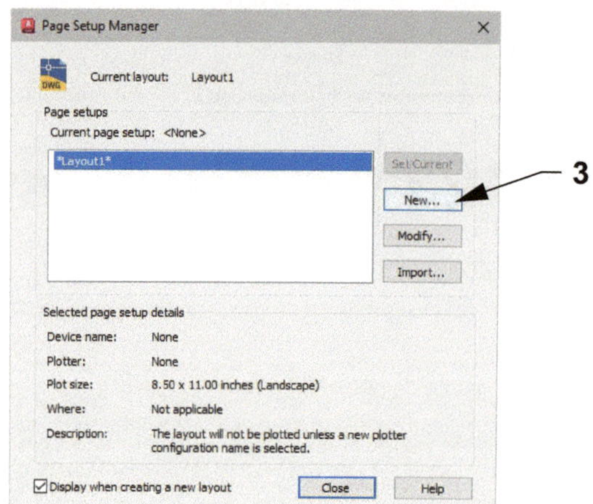

4. Select **<Default output device>** in the **Start with:** list.
5. Enter the New Page Setup name: **Setup B**
6. Select the **OK** button.

5

4

(I am assuming that your computer is attached to a printer. If not, select **Layout1**.)

6

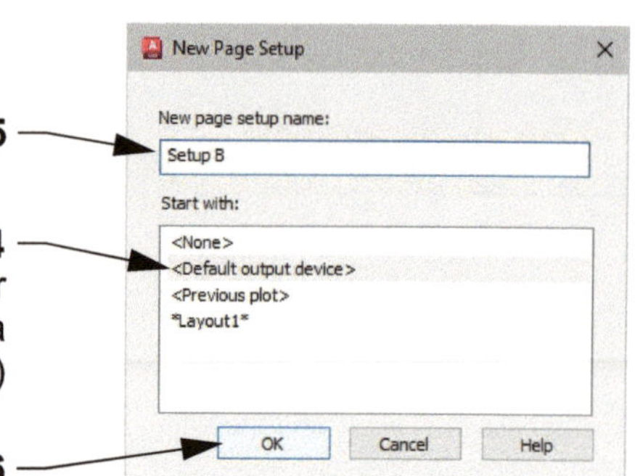

Continued on the next page...

Exercise 27B...continued

This is where you will select the **printer / plotter**, **paper size**, and the **plot offset**.

7. Select the **Printer / Plotter**
   Note: Your current system printer should already be displayed here. If you prefer another, select the down arrow and select from the list. If the preferred printer is not in the list, you must configure the printer. (Refer to **Appendix-A** for instructions.)

8. Select the **Paper size**.

9. Select **Plot offset**.

Notice the name you entered is now displayed as the Page setup name.

7 (Yours will be different.)

8 (Yours may state 8-1/2 x 11 for inch users. Or A4 for metric users.)

9 (Should remain at 0.000000 0.000000)

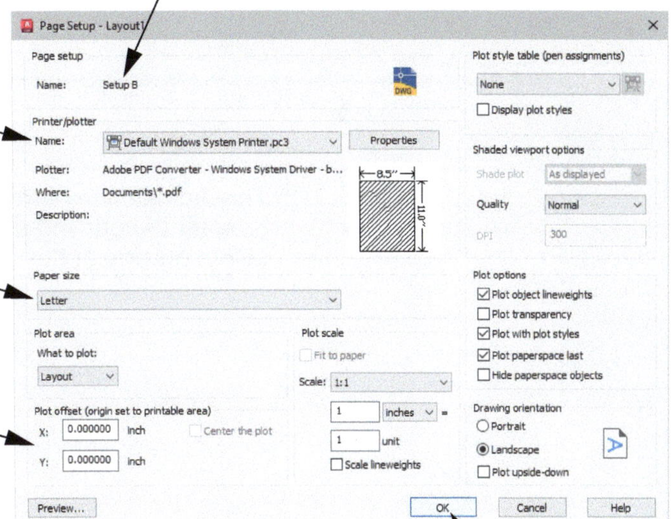

10

10. Select the **OK** button.

11. Select **Setup B**.

12. Select the **Set Current** button.

13. Select the **Close** button.

11 12

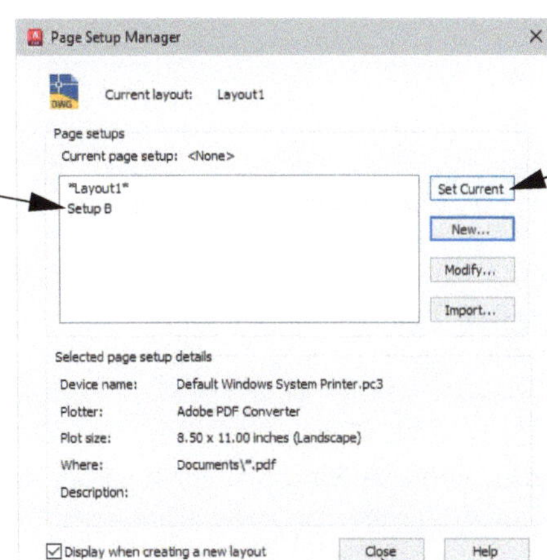

13

Continued on the next page...

Exercise 27B...continued

You should now have a sheet of paper displayed on the screen.
This sheet is the size you specified in the "Page Setup".
This sheet is in front of Model Space.

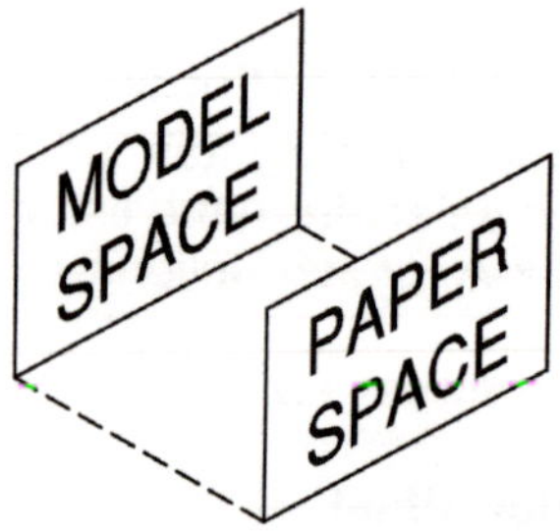

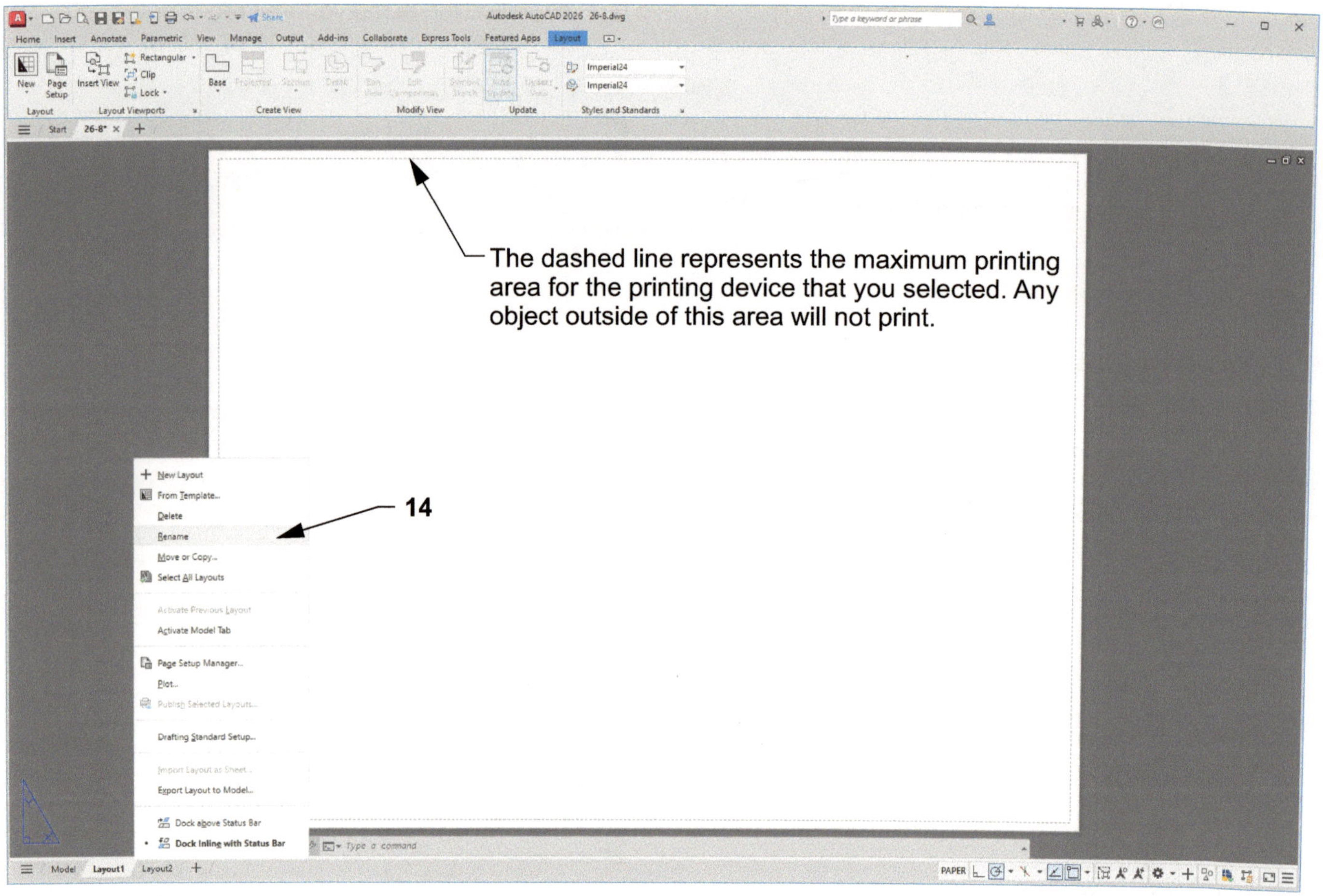

Rename the Layout Tab

14. Right click on the active Layout Tab and select **Rename** from the list.

15. Enter the new Layout name **A Size** and then press ***<Enter>***.

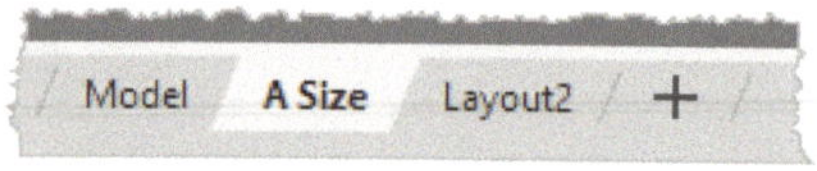

16. **Very important: Save as / AutoCAD Drawing Template / My Feet-Inches Setup** again.

Now continue on to ***Exercise 27C****...you are not done yet.*

Exercise 27C: Create a Border and Title Block

1. Draw the border rectangle as large as you can within the dashed lines approximately as shown below using Layer Border.

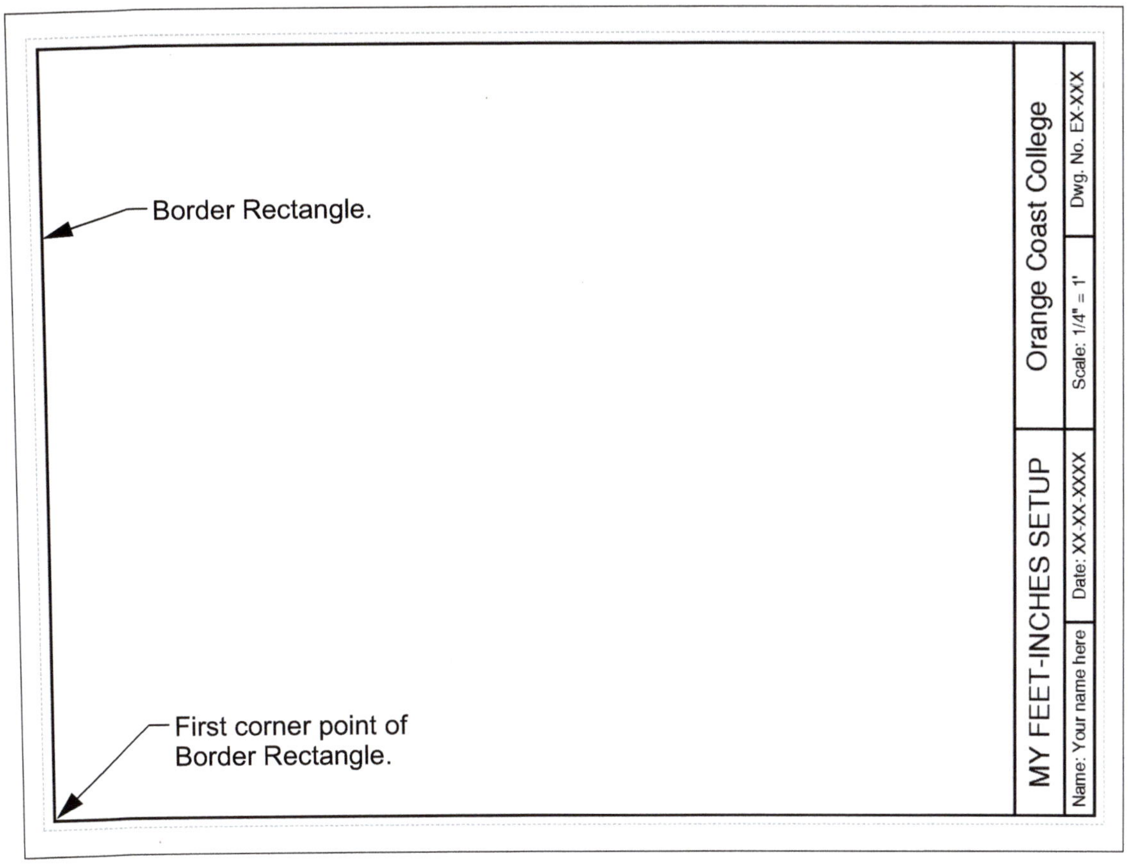

2. Draw the Title Block as shown using:
   A. Layers Border and Text.
   B. Single Line Text: Justify **Middle** in each rectangular area. (**Note:** The diagonal lines are an aid to find the center of each rectangle and can be erased after the text has been placed into position.)
   C. Text Style = **TEXT CLASSIC**.
   Text Height: Varies.

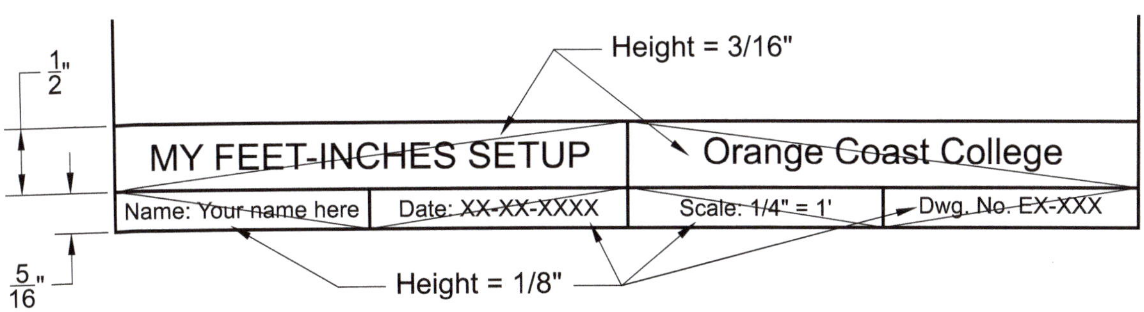

3. **Very important: Save as / AutoCAD Drawing Template / My Feet-Inches Setup** again.

*Now continue on to **Exercise 27D**...you are not done yet.*

Exercise 27D: Create a Viewport

The following instructions will guide you through creating a **viewport** in the Border Layout sheet. Creating a viewport has the same effect as cutting a hole in the sheet of paper. You will be able to see through the viewport Frame (hole) to Model Space.

1. Open **My Feet-Inches Setup** (if not already open).
2. Select the **A Size** Layout Tab.
3. Select Layer Viewport.
4. Type ***MV*** and then press ***<Enter>***. (Refer to page 26-6.)
5. Draw a Single viewport approximately as shown. (Turn off **Osnap** temporarily.)

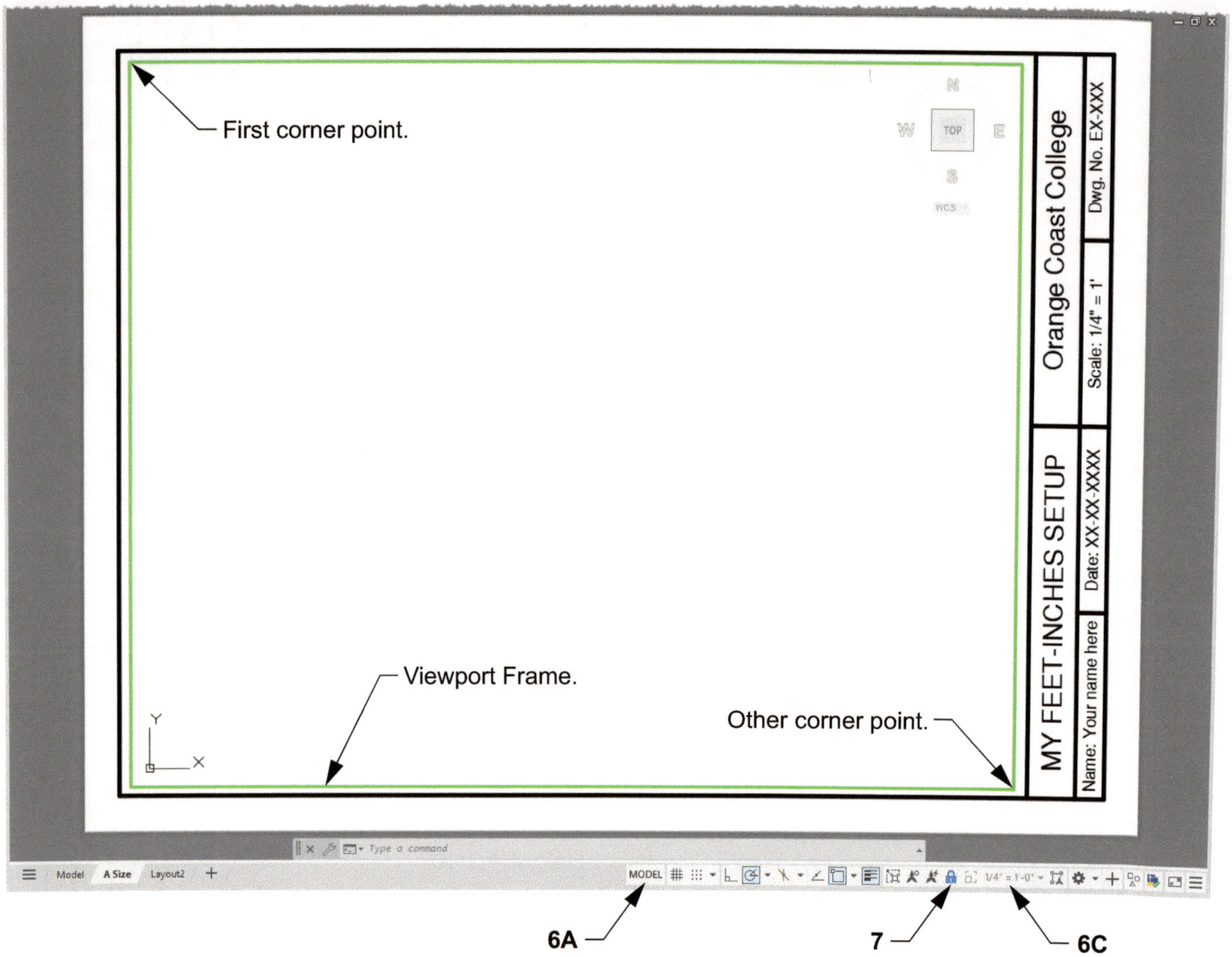

6. Adjust the viewport scale as follows:
   A. Confirm the **"MODEL"** button is showing.
   B. Select **"Zoom / All"**.
   C. Select the **viewport scale** of **1:4" = 1'-0"** (**Note:** inches =feet.)
7. **Lock** the viewport. (Refer to page 27-3.)
8. **Very important: Save as / AutoCAD Drawing Template / My Decimal Setup** again.

*Now continue on to **Exercise 27E**...you are not done yet.*

Exercise 27E: Plotting from the Layout

The following instructions will guide you through the final steps for setting up the master template for plotting. These settings will stay with **My Feet-Inches Setup** and you will be able to use it over and over again.

In this exercise, you will take a short cut by importing the **Plot Setup A** from **My Decimal Setup**. (**Note:** If the **Printer**, **Paper size**, and **Plot Scale** is the same, you can use the same Page Setup.)

Note: If you prefer not to use Import, you may go to page 26-27 and follow the instructions for creating the **Plot- Page Setup**.

1. Open **My Feet-Inches Setup** (if not already open).
2. Select the **A Size** Layout Tab.
   You should be seeing your border and Title Block now.
3. Select the **Plot** command.
4. Select **Import…** from the **Page setup** drop-down menu

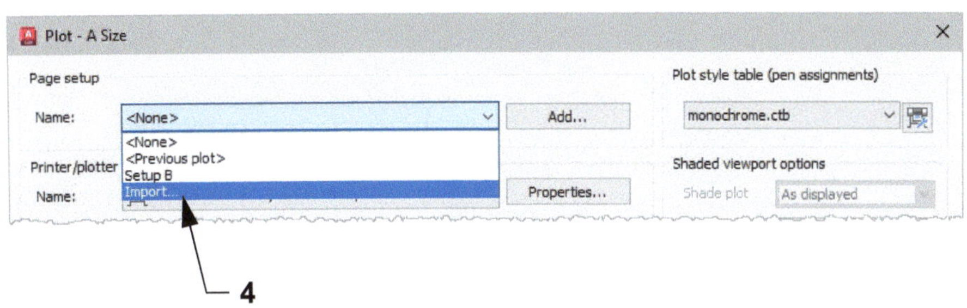

5. Find **My Decimal Setup.dwt** as follows:
   A. Select Files of type: **Template [.dwt]**
   B. Select **My Decimal Setup.dwt**
   C. Select the **Open** button.

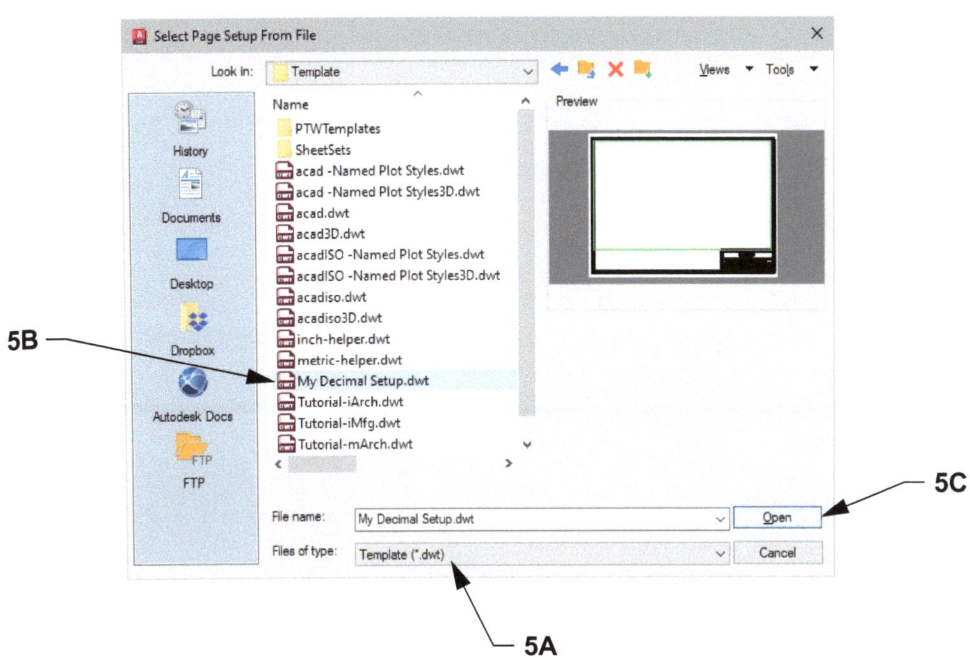

Continued on the next page...

Exercise 27E...continued

6. Select **Plot Setup A** from the Page Setups list. (**Note:** If you do not have a **"Plot Setup A"** in My Decimal Setup, refer to page 26-27.)

7. Select the **OK** button.

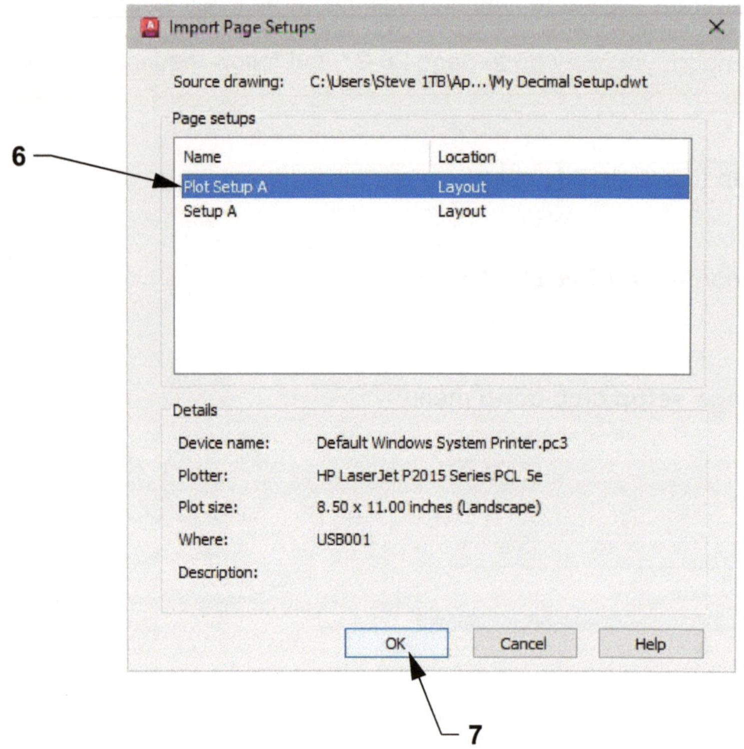

8. Select **Plot Setup A** from the Page list. (**Note:** It was not there before. You just imported it in from My Decimal Setup.)

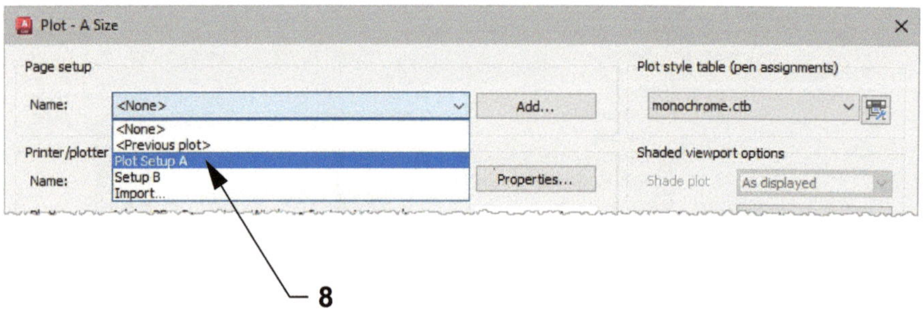

Continued on the next page...

Exercise 27E...continued

9. Check all settings.

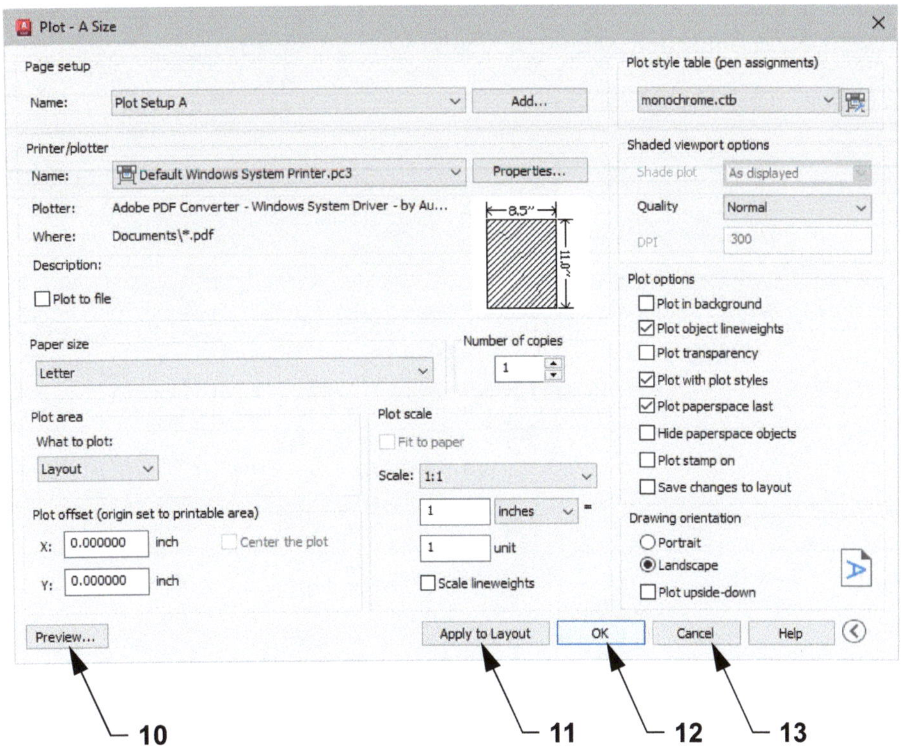

10. Select the **Preview...** button.

If the drawing appears as you would like it, press the ***<Esc>*** key and continue.

If the drawing does not look correct, press the ***<Esc>*** key and check all your settings, then preview again.

Note: The viewport frame and the grids will not appear in the preview because the viewport is on a **no plot** layer and grids never plot.

11. Select the **Apply to Layout** button.
    (*The settings are now saved to the Layout Tab for future use.*)

12. If your computer **is** connected to the plotter / printer selected, select the **OK** button to plot, then proceed to **Step 14**.

13. If your computer is **not** connected to the plotter/ printer selected, select the **Cancel** button to close the Plot dialog box and proceed to **Step 14**.

14. **Very important...** Save as **My Feet-Inches Setup** again.

*Now continue on to **Exercise 27F**....you are not done yet.*

Exercise 27F: Create a Dimension Style

1. Open **My Feet-Inches Setup** (if not already open).

2. Select the **Dimension Style Manager** command. (Refer to page 16-6.)

3. Select the **New...** button.

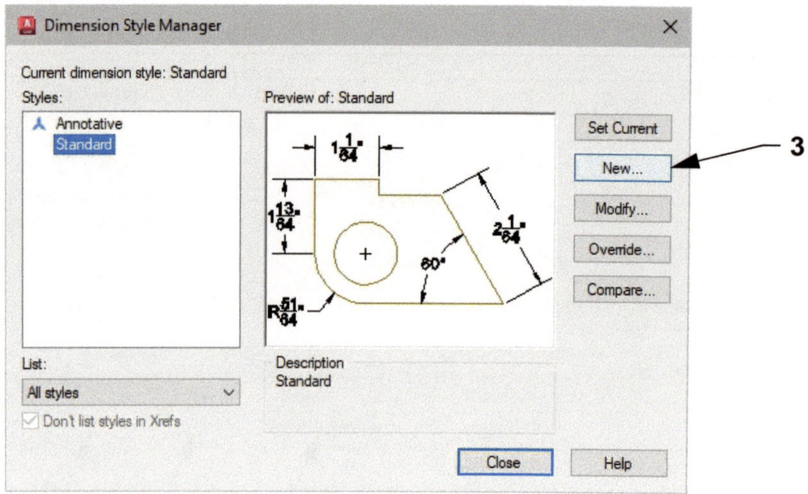

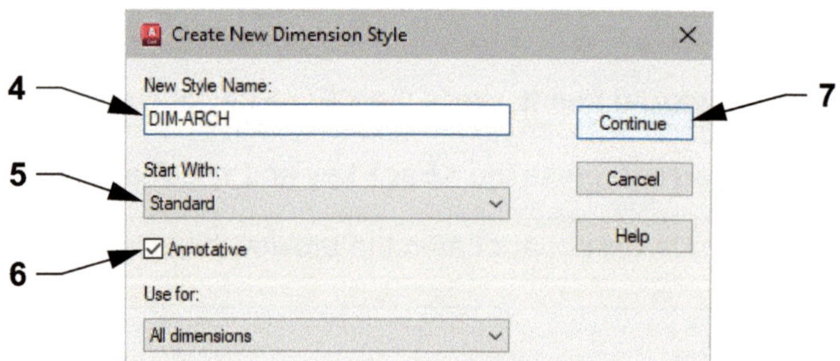

4. Enter **DIM-ARCH** in the “New Style Name:” box.

5. Select **Standard** in the “Start With:” box.

6. Select the **Annotative** box.

7. Select the **Continue** button.

Continued on the next page...

8. Select the **Primary Units** Tab and change your settings to match the settings shown below.

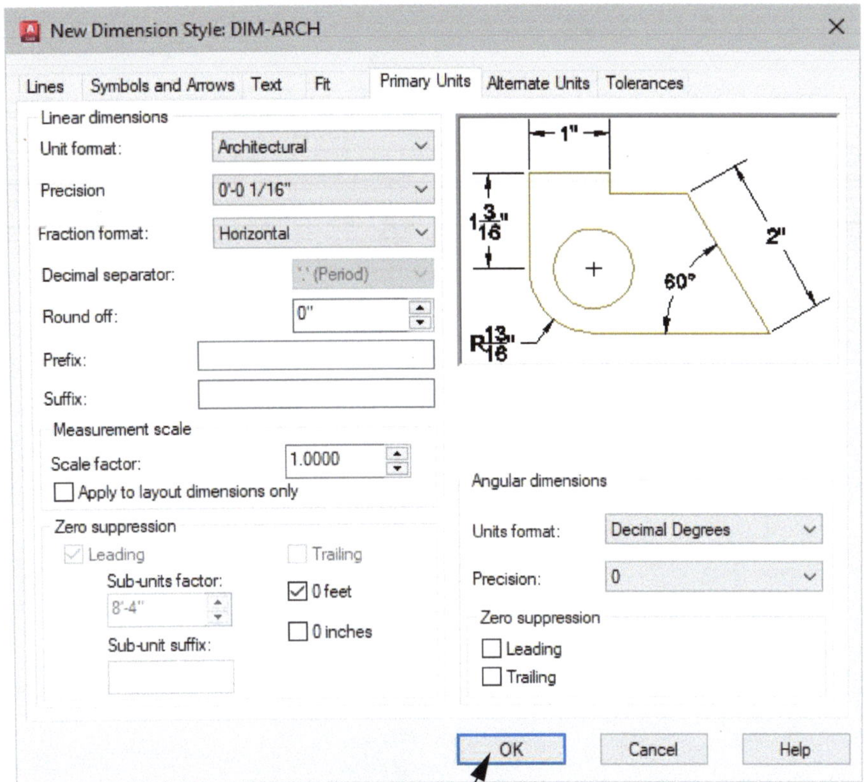

Do not select the **OK** button yet.

9. Select the **Lines** Tab and change your settings to match the settings shown below.

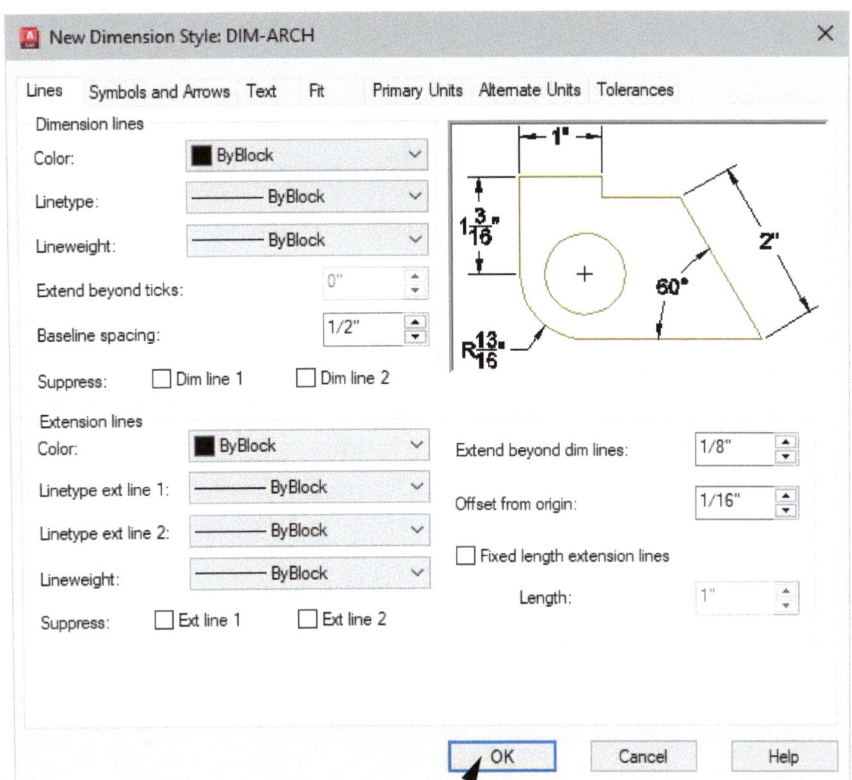

Do not select the **OK** button yet.

Continued on the next page...

Exercise 27F...continued

10. Select the **Symbols and Arrows** Tab and change your settings to match the settings shown below.

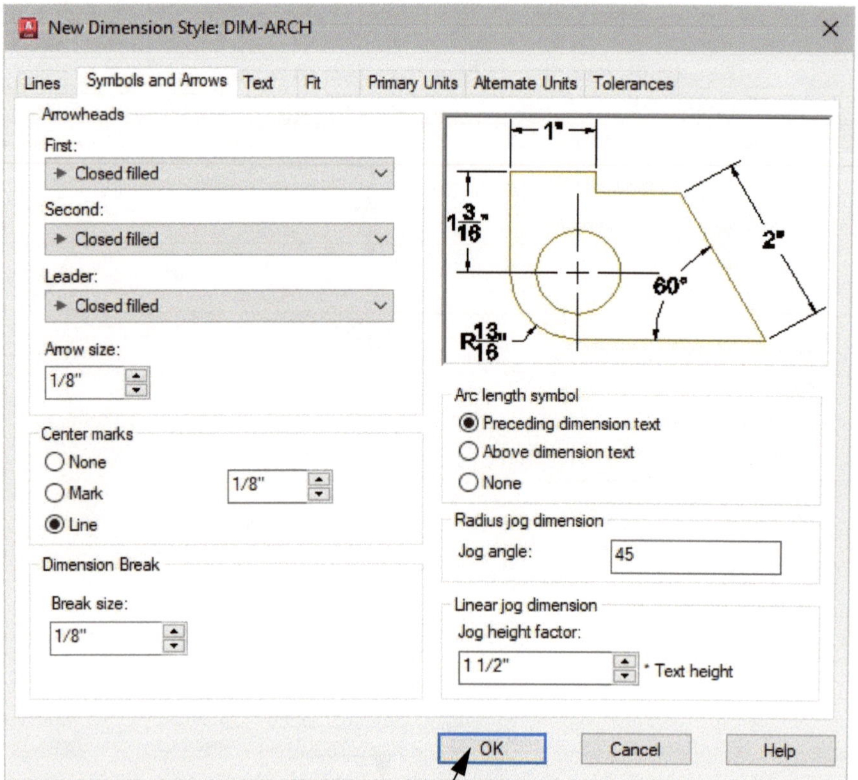

Do not select the **OK** button yet.

11. Select the **Text** Tab and change your settings to match the settings shown below.

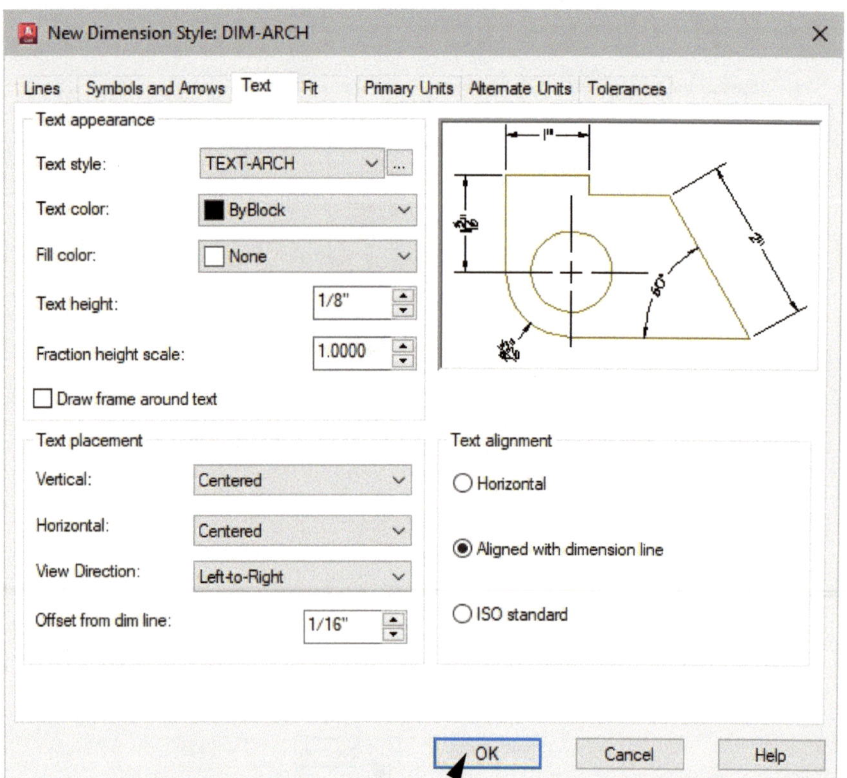

Do not select the **OK** button yet.

Continued on the next page...

12. Select the **Fit** Tab and change your settings to match the settings shown below.

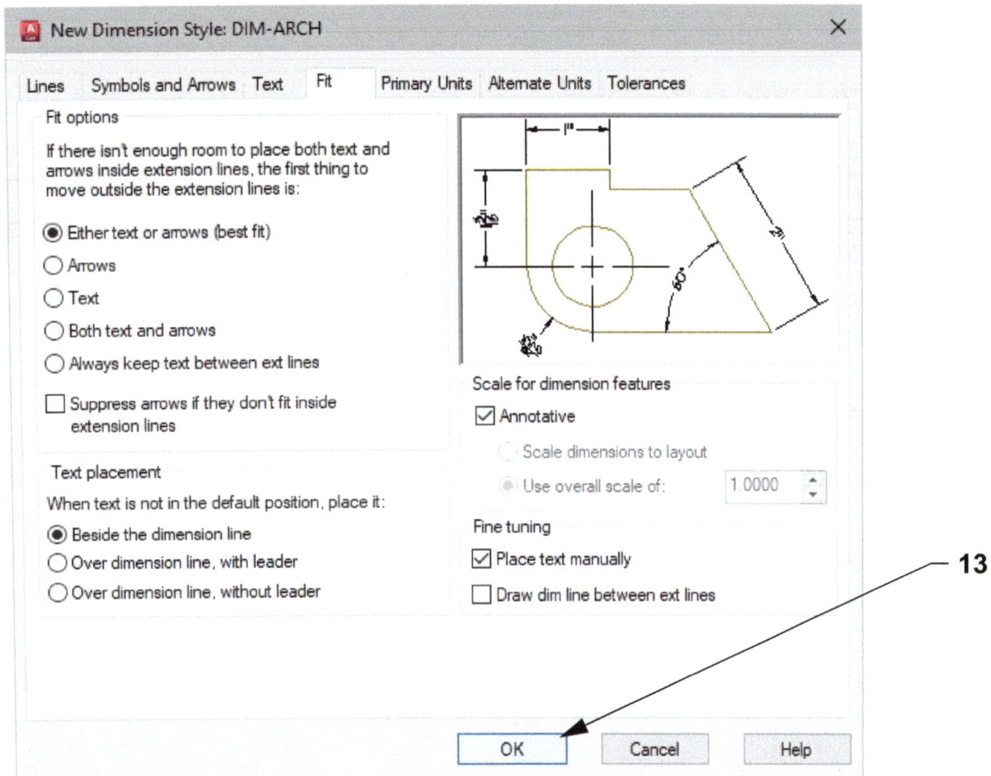

13. Now select the **OK** button.

Your new **DIM-ARCH** dimension style should now be in the list.

14. Select the **Set Current** button to make your new style **DIM-ARCH** the style that will be used.

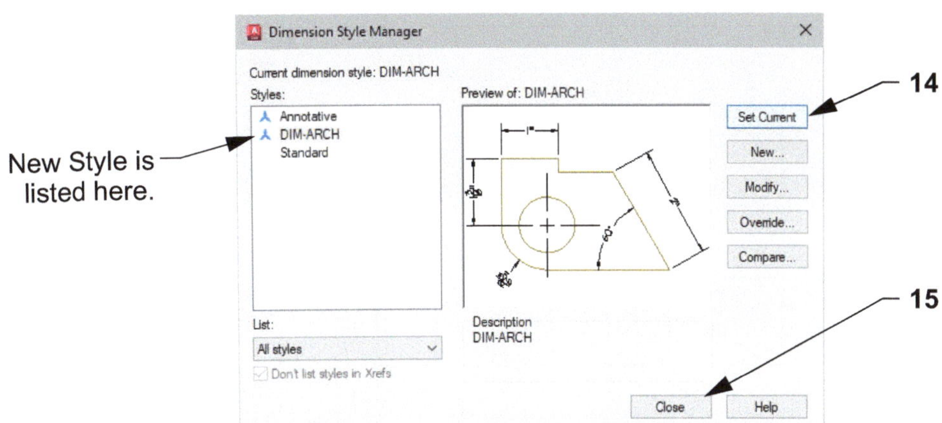

15. Select the **Close** button.

16. **Important:** Save your drawing as **My Feet-Inches Setup** again.

17. Change the current layer to Walls. (So Layer Walls will always be the current layer when you start a New file.)

18. Now Save all of this work as a **Template**: **Application Menu / Save As / AutoCAD Drawing Template / My Feet-Inches Setup**. (Refer to page 2-3.)

Again, that probably seemed like a lot of work, but you are now ready to use this master setup template to create and plot many drawings in the future.

Exercise 27G

1. Select **New** and select **My Feet-Inches Setup.dwt** (template).
2. Draw the classroom shown below.
3. Use Layers Walls, Furniture, Doors, and Windows.
4. Dimension as shown inside the viewport.
5. Use Dimension Style DIM-ARCH and Layer Dimension.
6. Edit the **Title** and **Ex-XX** by double clicking on the text. Do not erase and replace.
7. **Save** as **Ex-27G**
8. Plot using Page Setup **Plot Setup A**. (Refer to page 27-19.)

Drawing hints:

1. Consider using **Array** (Lesson 13) and **Offset** (Lesson 12).
2. Remember you have to Unlock the viewport to use Pan inside the viewport.
3. If you change the scale of the viewport, change the viewport scale back to **1/4"=1'** and lock the viewport.
4. If your dimensions are gigantic, you are not inside the viewport.

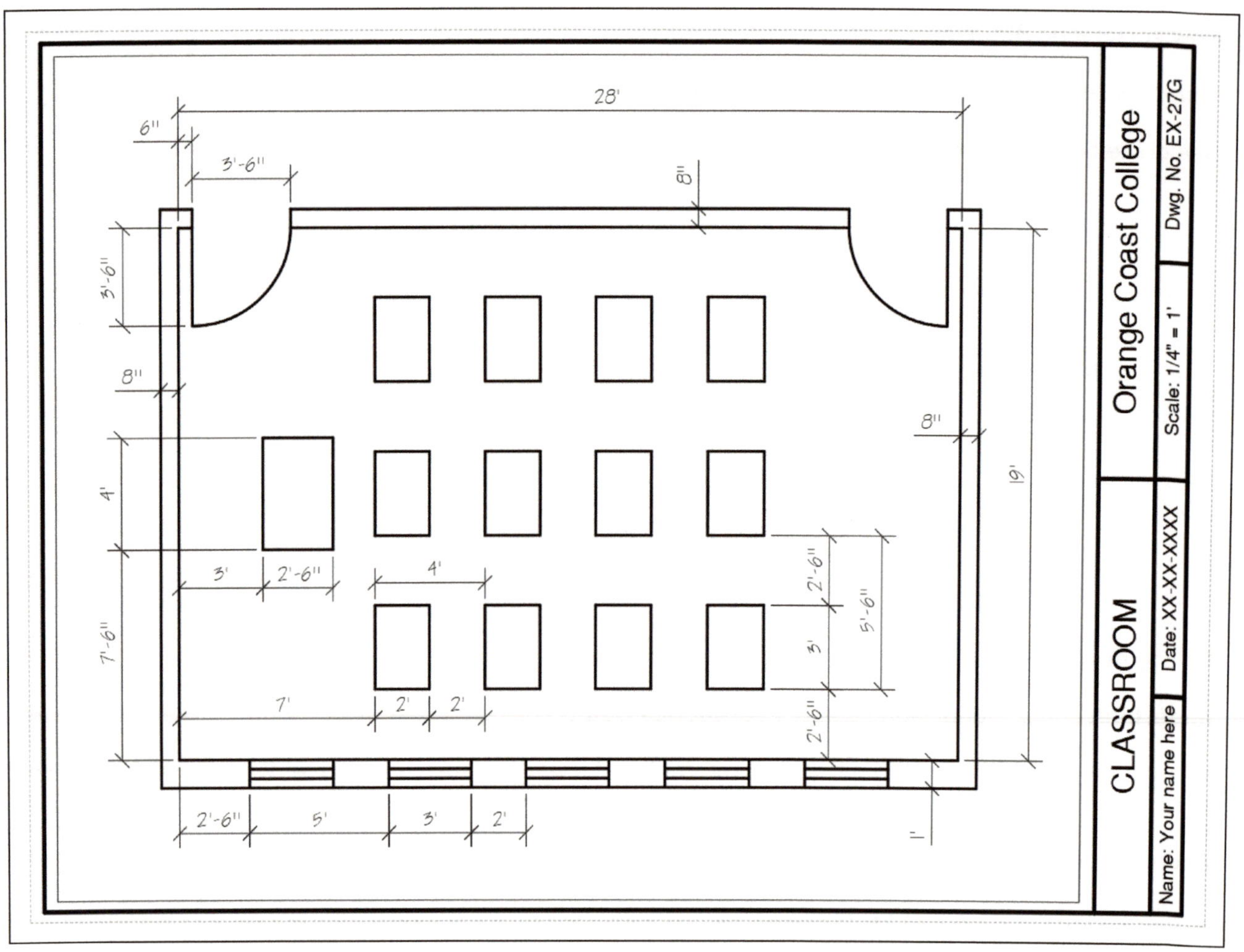

Exercise 27H-Inch

Step 1.

1. Select **New** and select **My Decimal Setup.dwt** (template).
2. **Unlock** the viewport.
3. Change the viewport scale to **8:1** (important: not 1:8).
4. Lock the viewport.
5. Turn **Snap** and **Grids** off or change them relative to the new viewport scale.
6. Draw the Paperclip on Layer Object Line.
7. Dimension as shown using Dimension Style DIM-DECIMAL. (Dimensions should be inside the viewport, in Model Space. Confirm the "MODEL" button 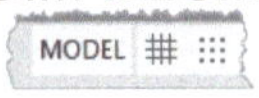

is displayed.)

Drawing hints:

The entire Paper clip can easily be drawn using Offset (Lesson 12) and Fillet (Lesson 7) or Circles and trim.

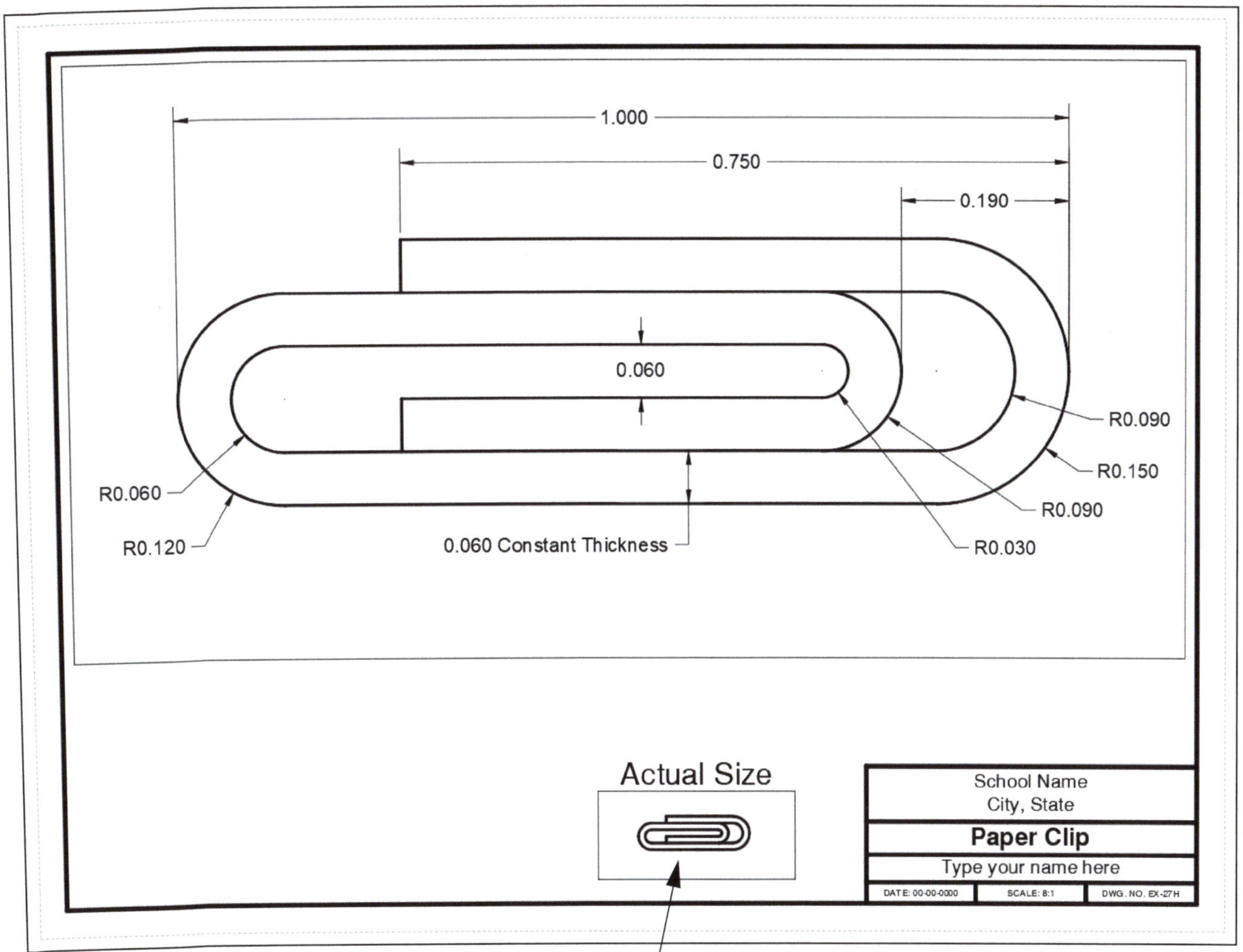

Step 2.

1. Select Paper Space. 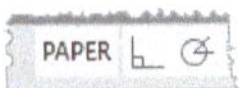

2. Select the Viewport layer and cut a second small viewport next to the Title Block.
3. Double click inside the small viewport to make it active. Use **Zoom / All** (You should now see the paper clip.)
4. Change the viewport scale of the **small** viewport to **1:1** and Lock the viewport.
5. Add **"Actual Size"** text **0.200"** height in Paper Space. Use: Layer Text.
6. Edit the **Title** and **Ex-XX** by double clicking on the text. Do not erase and replace.
7. **Save** as **Ex-27H-I**
8. Plot using Page Setup **Plot Setup A**. (Refer to page 26-27.)

Exercise 27H-Metric

Step 1.

1. Select **New** and select **My Decimal Setup.dwt** (template).
2. **Unlock** the viewport.
3. Change the viewport scale to **8:1** (important: not 1:8).
4. Lock the viewport.
5. Turn **Snap** and **Grids** off or change them relative to the new viewport scale.
6. Draw the Paperclip on Layer Object Line.
7. Dimension as shown using Dimension Style DIM-DECIMAL. (Dimensions should be inside the viewport, in Model Space. Confirm the "MODEL" button 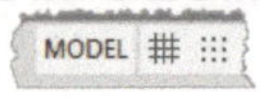

is displayed.)

Drawing hints:

The entire Paper clip can easily be drawn using Offset (Lesson 12) and Fillet (Lesson 7) or Circles and trim.

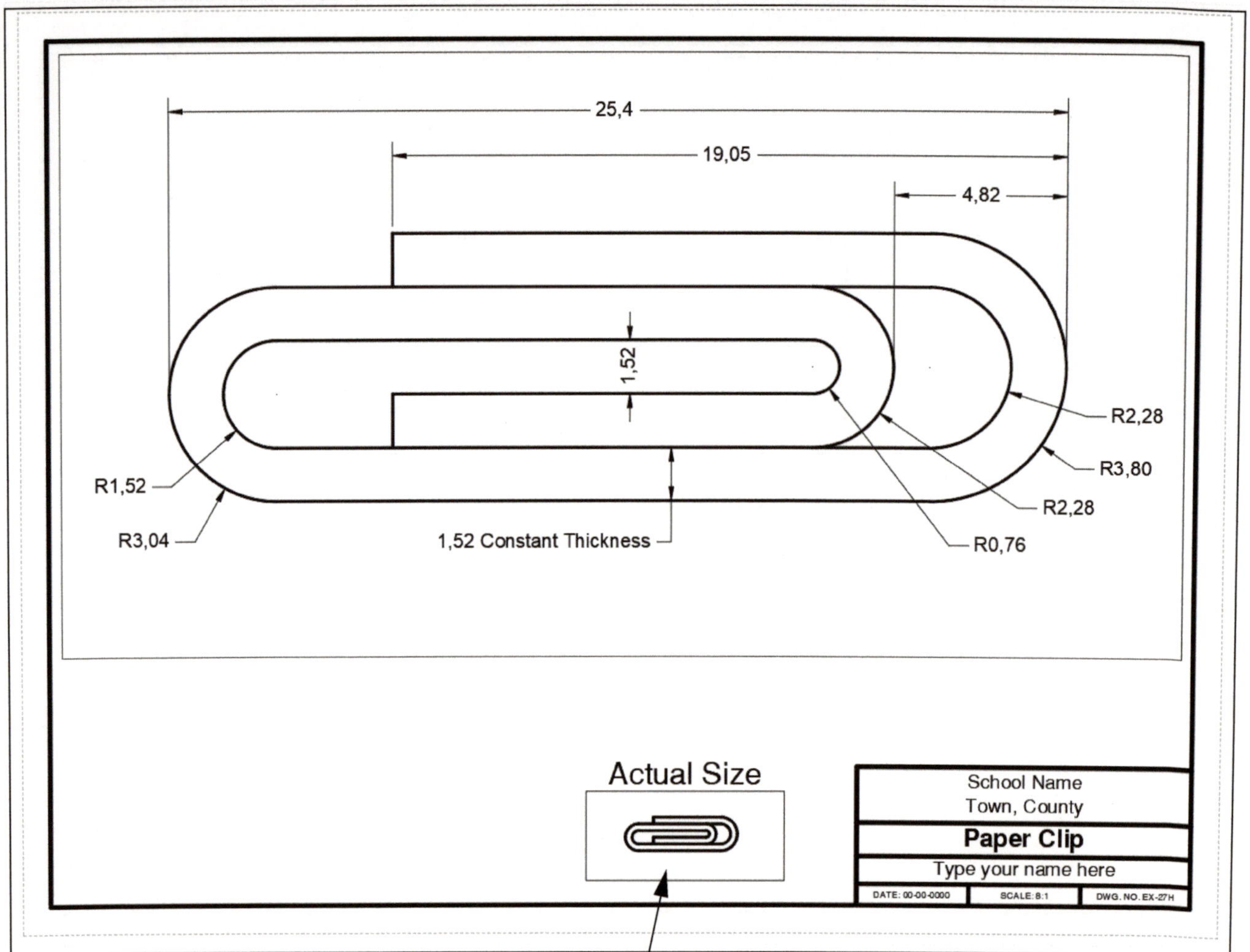

Step 2.

1. Select Paper Space. 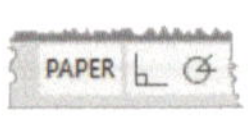

2. Select the Viewport layer and cut a second small viewport next to the Title Block.
3. Double click inside the small viewport to make it active. Use **Zoom / All** (You should now see the paper clip.)
4. Change the viewport scale of the **small** viewport to **1:1** and Lock the viewport.
5. Add **"Actual Size"** text **5 mm** height in Paper Space. Use: Layer Text.
6. Edit the **Title** and **Ex-XX** by double clicking on the text. Do not erase and replace.
7. **Save** as **Ex-27H-M**
8. Plot using Page Setup **Plot Setup A**. (Refer to page 26-27.)

LESSON 28

LEARNING OBJECTIVES

After completing this lesson, you will be able to:

- Assign multiple annotative scales to a single annotative object
- Remove an Annotative Scale from a Viewport
- Assign annotative scales to Hatch sets

Assigning Multiple Annotative Scales

In the previous lesson, you learned how annotative text and dimensions are automatically scaled to the viewport scale. But in order to have an annotative object appear in both viewports, you placed two sets of text and dimensions. In this lesson, you will learn how to easily assign multiple annotative scales to a single text string or dimension so that you will not need to duplicate them each time you create a new viewport. You will just assign an additional annotative scale to the annotative object.

Again, the easiest way to understand this process is to do it. The following is a step-by-step example.

1. Open **Annotative Objects.dwg** from the previous lesson and select the **A size** Layout Tab. *If you did not complete the example from the previous lesson, go back and do it now. (Refer to page 27-4.)*

2. Make the right-hand viewport active. (Double click inside the viewport.)

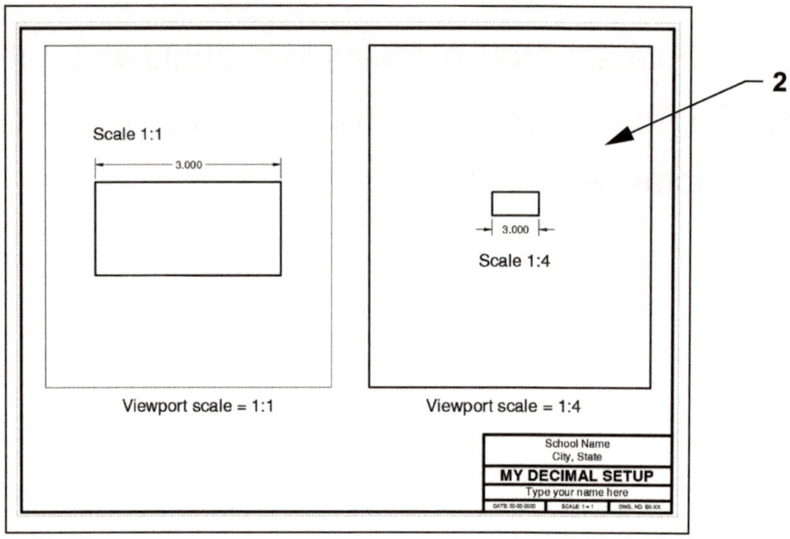

3. Erase the text and the dimension in the right-hand viewport only.

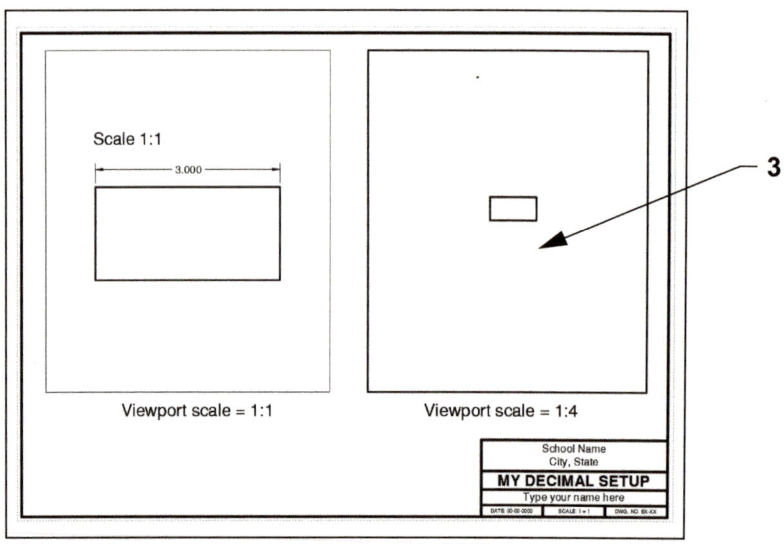

4. Display all annotative objects in all viewports as follows:

   A. Select the **Annotation Visibility** button located in the lower right corner of the Drawing Status Bar.

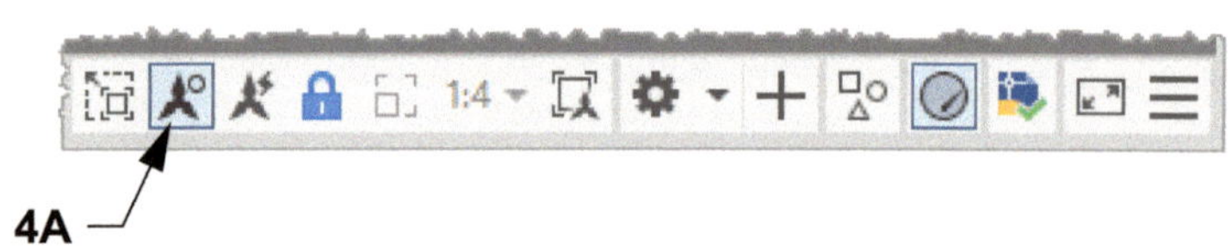

Continued on the next page...

On (Blue icon) Displays **all** Annotative objects in all viewports (example below).

Off (Gray icon) Displays **only** Annotative objects that have an annotative scale that matches the viewport scale.

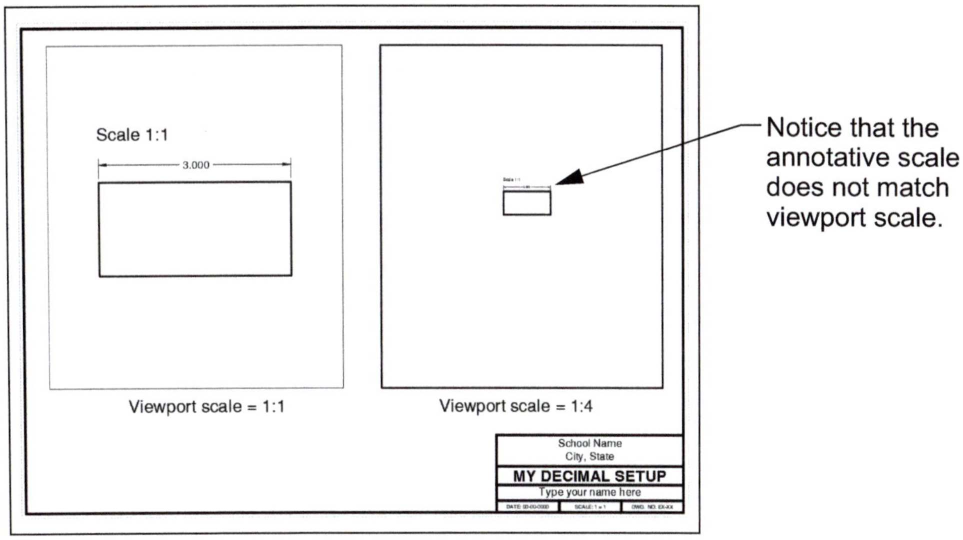

The dimensions and text are now displayed in both viewports. But the annotative scale of the dimensions and text in the right-hand viewport do not match the scale of the viewport. (Notice they are smaller.) **The scale of the annotative objects must match the scale of the viewport**. *Follow the next steps to assign multiple annotative scales to an annotative object.*

5. Place your cursor near the dimension in the right-hand viewport. Notice the **single annotative symbol**. This single symbol indicates the annotative dimension has only one annotative scale assigned to it.

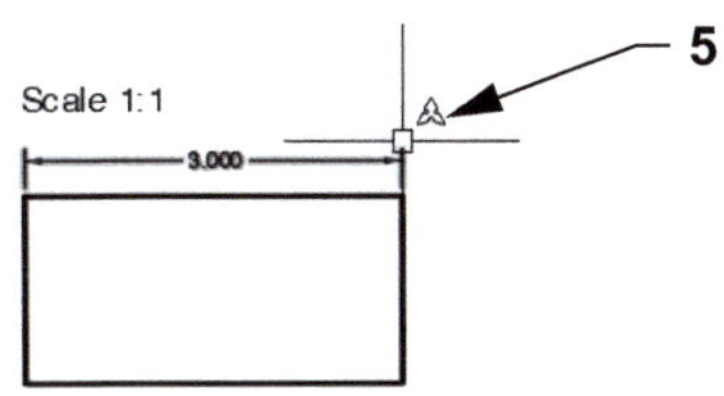

6. Select the **Annotate** Tab.

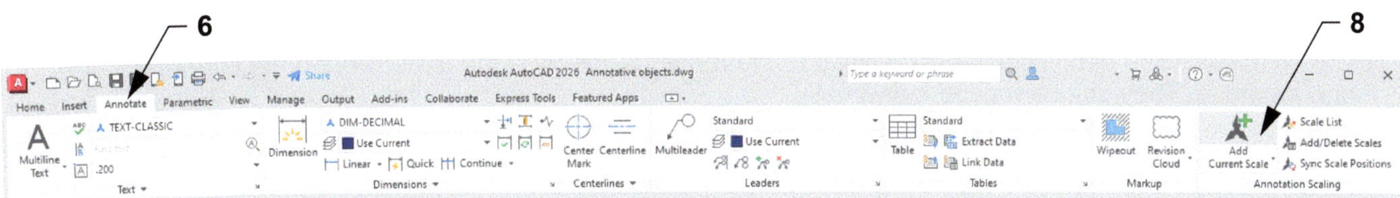

7. Select **only** the dimension in the right-hand viewport.

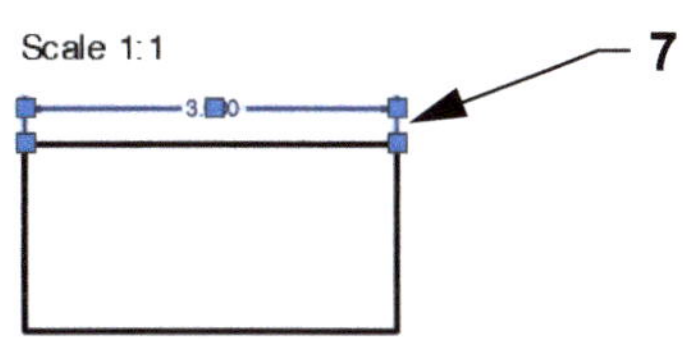

8. Select the **Add Current Scale** tool on the **Annotation Scaling** Panel.

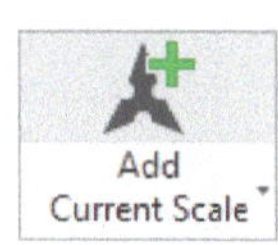

Continued on the next page...

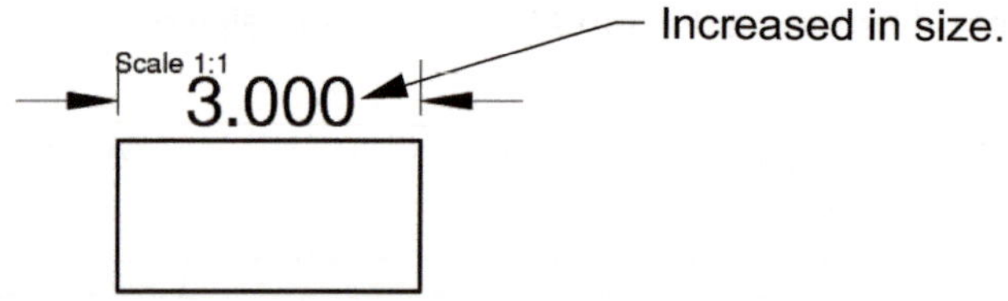

The annotative dimension should have increased in size as shown above.

9. Turn **off** the **Annotation Visibility**. (Click on button. The icon should turn Gray.)

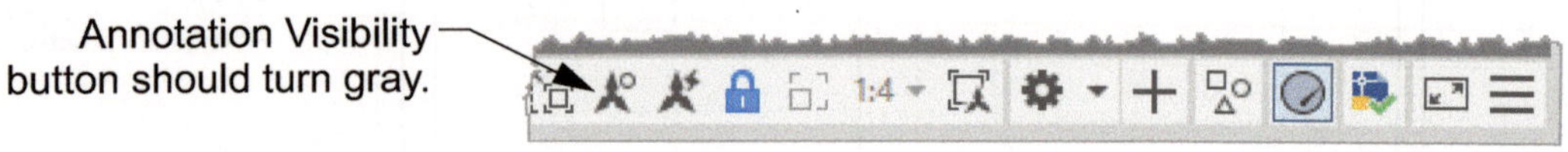

Notice the text in the right-hand viewport is no longer visible. When the ***Annotation Visibility*** *is* ***off****, only the annotative objects that match the viewport scale will remain visible. The dimension is the only annotative object that has the 1:4 annotative scale assigned to it.*

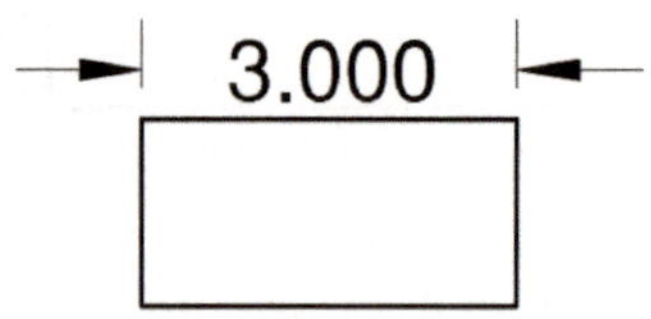

Annotation Visibility off

10. Place your cursor near the dimension. Notice two annotative symbols appear now. This indicates two annotative scales have been assigned to the annotative object.

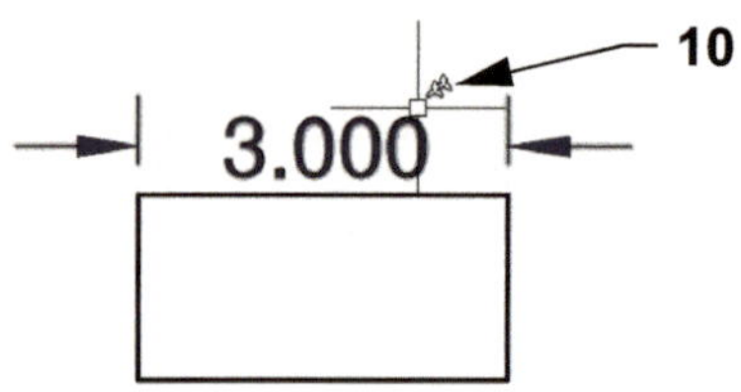

11. Click on the dimension to display the grips and drag the dimension away from the rectangle approximately as shown below. (The dimension was too close to the rectangle.) Notice the dimension in the left-hand viewport did not move. They can be moved individually.

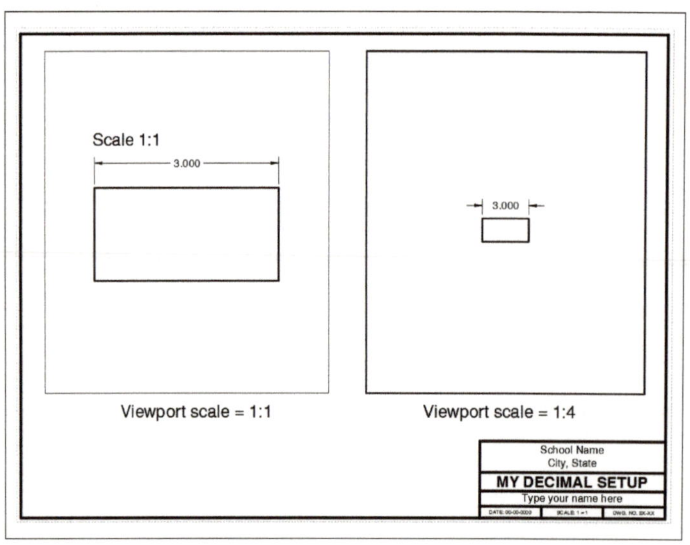

Removing an Annotative Scale

If you have an annotative object such as a dimension that you would like to remove from a viewport, you must remove the annotative scale that matches the viewport scale. **Do not delete** the dimension because it will also be deleted from all of the other viewports. This sounds complicated but is very easy to accomplish.

Problem: I would like to remove **dimension A** from the right-hand viewport. But I do not want **dimension B** in the left-hand viewport to disappear.

Solution: I must remove the **1:4** Annotative scale from **dimension A**.

(Refer to the step-by-step instructions below.)

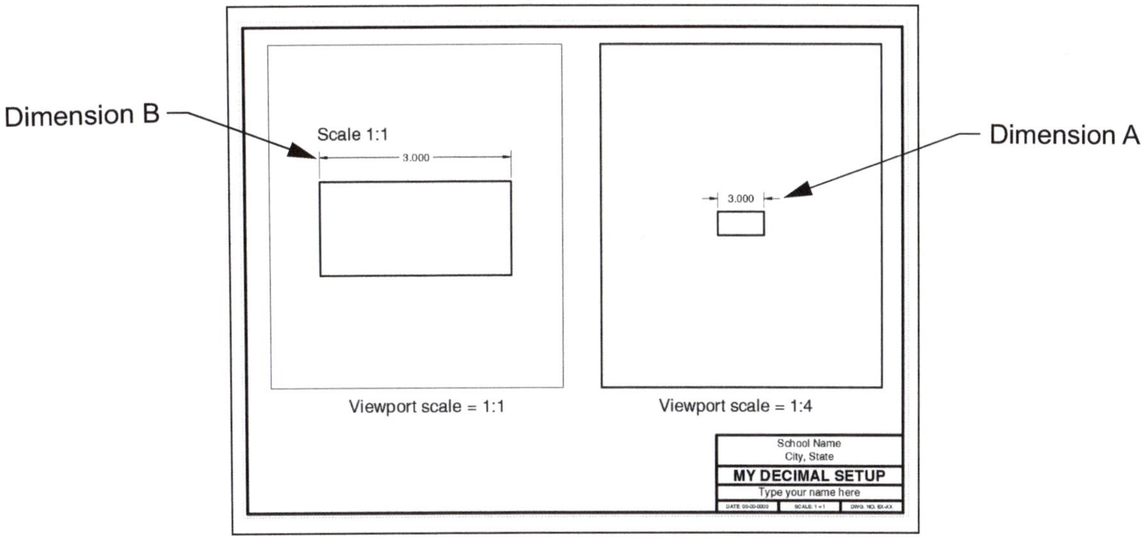

1. Select the **Annotate** Tab on the Ribbon.
2. Select **dimension A** shown above. (You must be inside the viewport.)
3. Select the **Add / Delete Scales** tool located on the **Annotation Scaling** Panel.

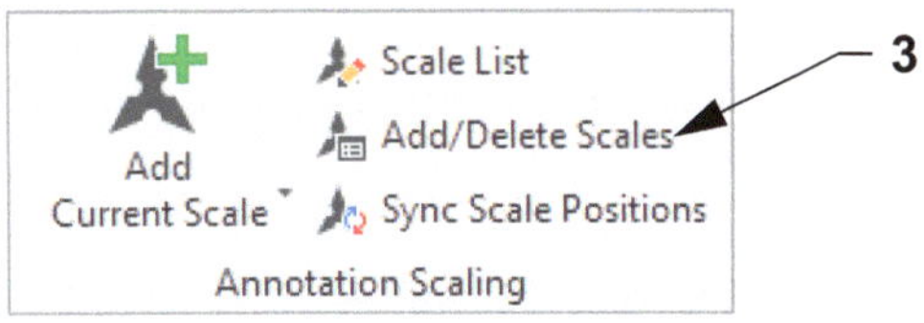

4. Select the Annotative scale to remove. **1:4**
5. Select the **Delete** button.
   (Remember, you are deleting the annotative scale from the dimension. You are not deleting the dimension. The dimension still exists, but it will not have an annotative scale of 1:4 assigned to it. As a result, it will not be visible within any viewport that has been scaled to 1:4.)
6. Select the **OK** button.

Continued on the next page...

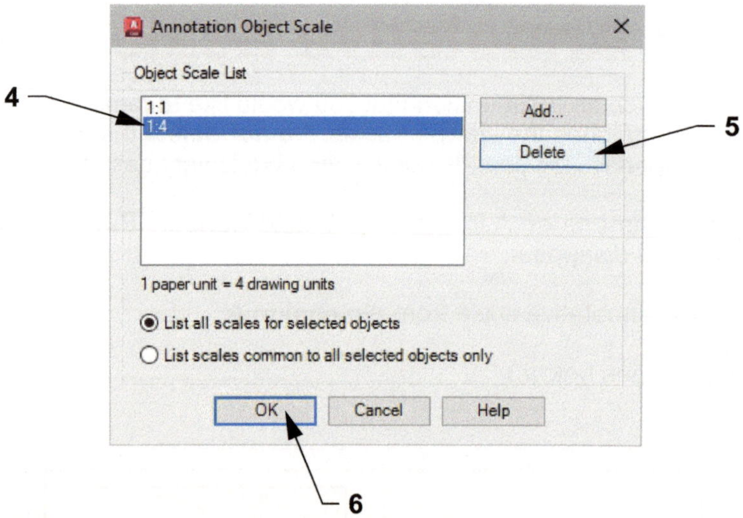

Note: Dimension A has been removed and **dimension B** remains.

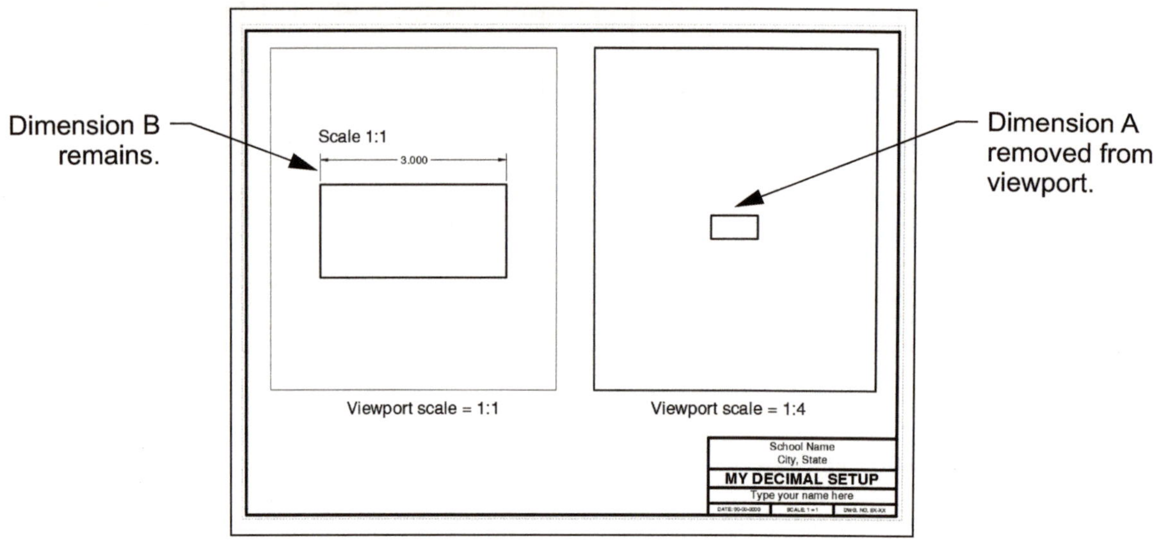

Annotative Hatch

A Hatch may be Annotative also. You may select the Annotative setting as you create the Hatch set or you may add the annotative setting to an existing Hatch set.

How to select the Annotative setting as you create the Hatch set

1. Select the **Hatch** command (refer to Lesson 15) and select the desired settings including **Annotative**. The Hatch set created will be **Annotative**. (Blue is **on**.)

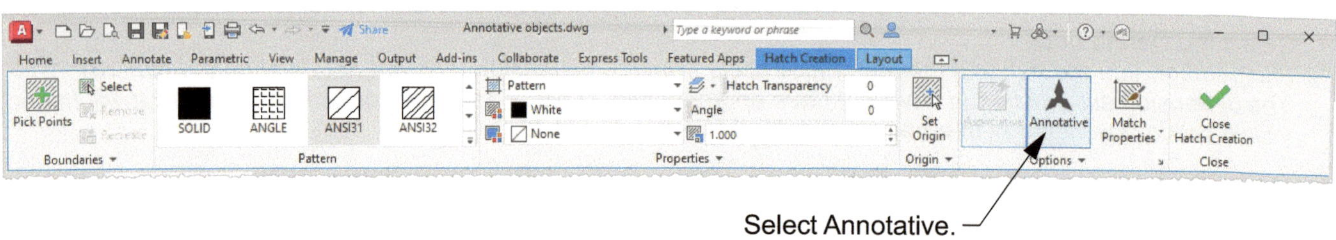

Continued on the next page...

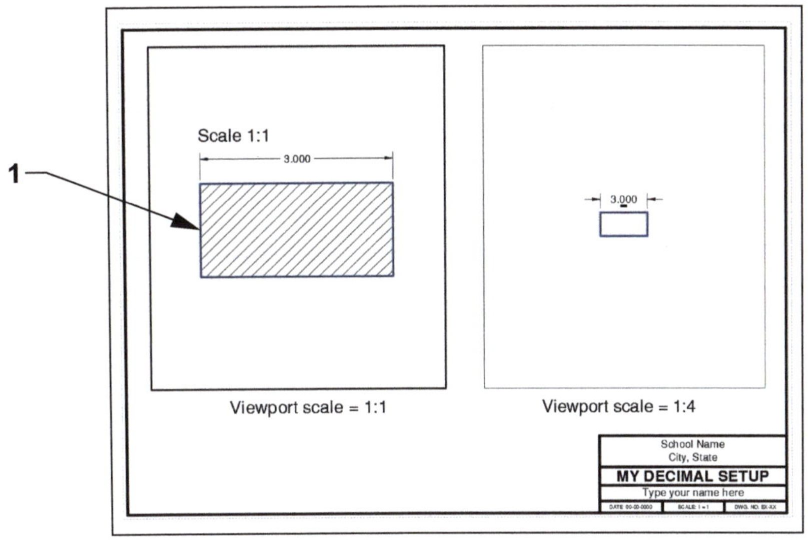

How to change an existing Non-Annotative Hatch set to Annotative

1. Click on the existing Hatch set.

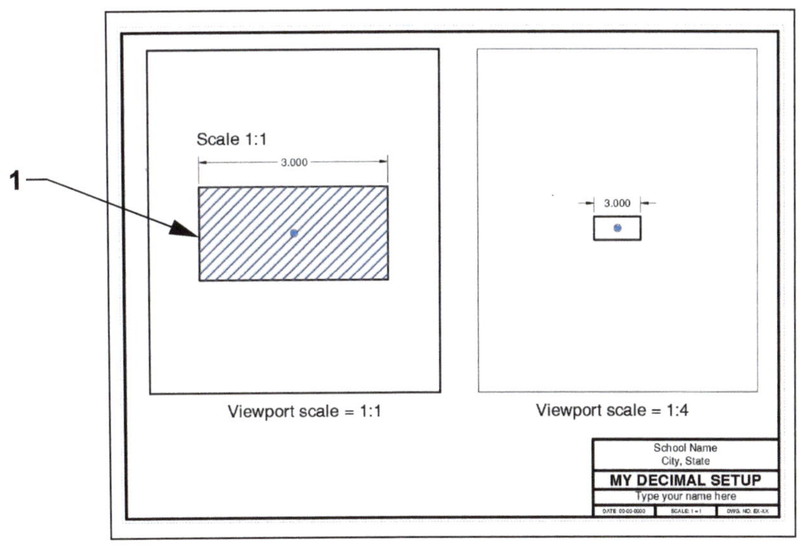

The Hatch Editor will appear.

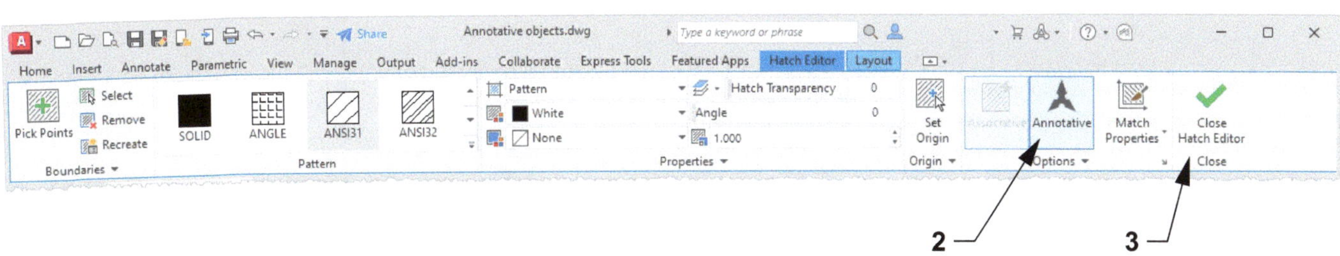

2. Select the **Annotative** option.

3. Select **Close Hatch Editor**.

Note: The Non-Annotative Hatch is now **Annotative**.

How to assign multiple Annotative scales to a Hatch set

1. Create the hatch in one of the viewports using **Annotative Hatch**. (Refer to page 28-6.)

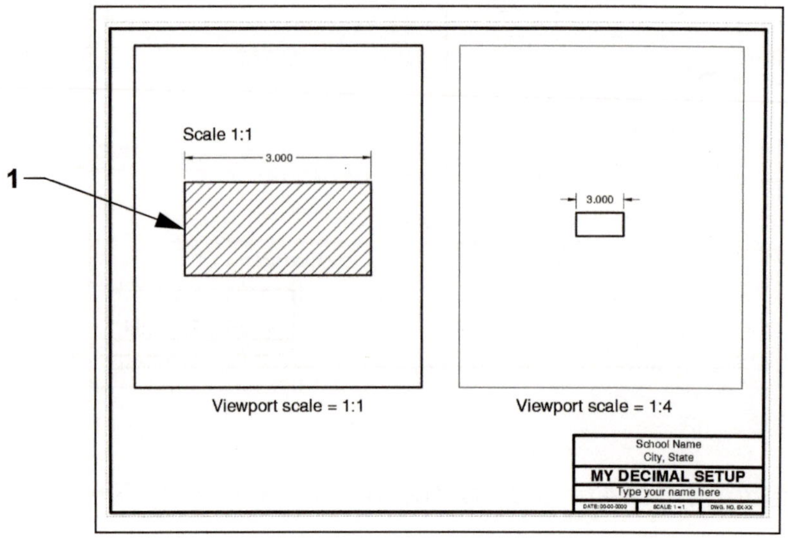

2. Turn on **Annotation Visibility**. 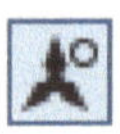 (Blue.) (Refer to page 28-2.)
3. Select the **Hatch set** to change.

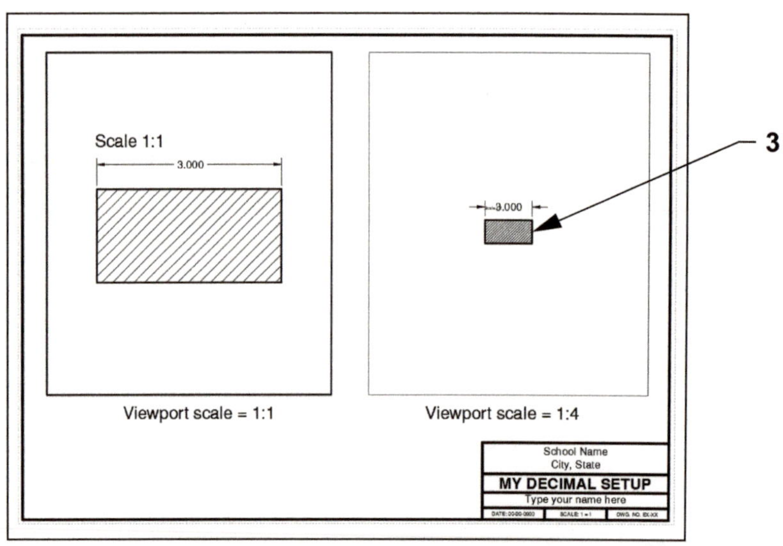

Continued on the next page...

4. Select the **Annotate** Tab.

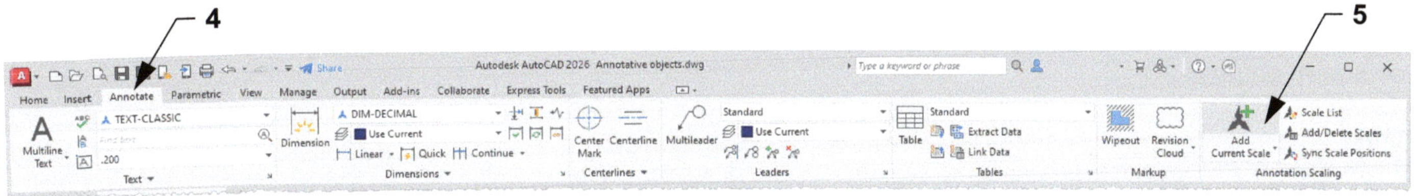

5. Select the **Add Current Scale** tool.
6. Turn **off** Annotation Visibility.  (Gray.)

Now the appearance of the Hatch sets should be identical in both viewports.

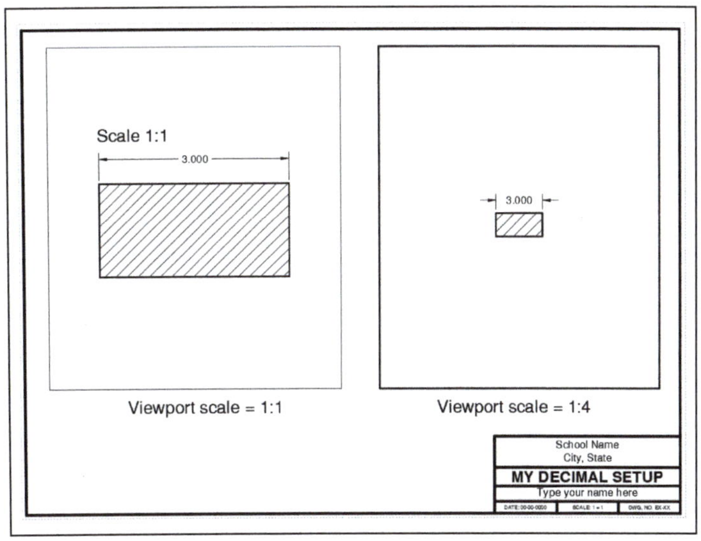

Exercise 28A

The following exercise will give you practice adding annotative scales to existing dimensions in order to display them within viewports with different scales. Follow the instructions below as you manipulate the drawing to appear as shown on page 28-12.

Step 1.

1. Open **Ex-27G**
2. Select the **A Size** Tab.
3. Erase the one existing viewport (click on it and select delete).
   (The classroom will disappear from Paper Space because the hole is gone.)
4. Select the Viewport layer and **create** four new viewports approximately as shown below.

Notice each viewport is a hole in Paper Space and you are looking through to Model Space. Also notice the dimensions do not appear.

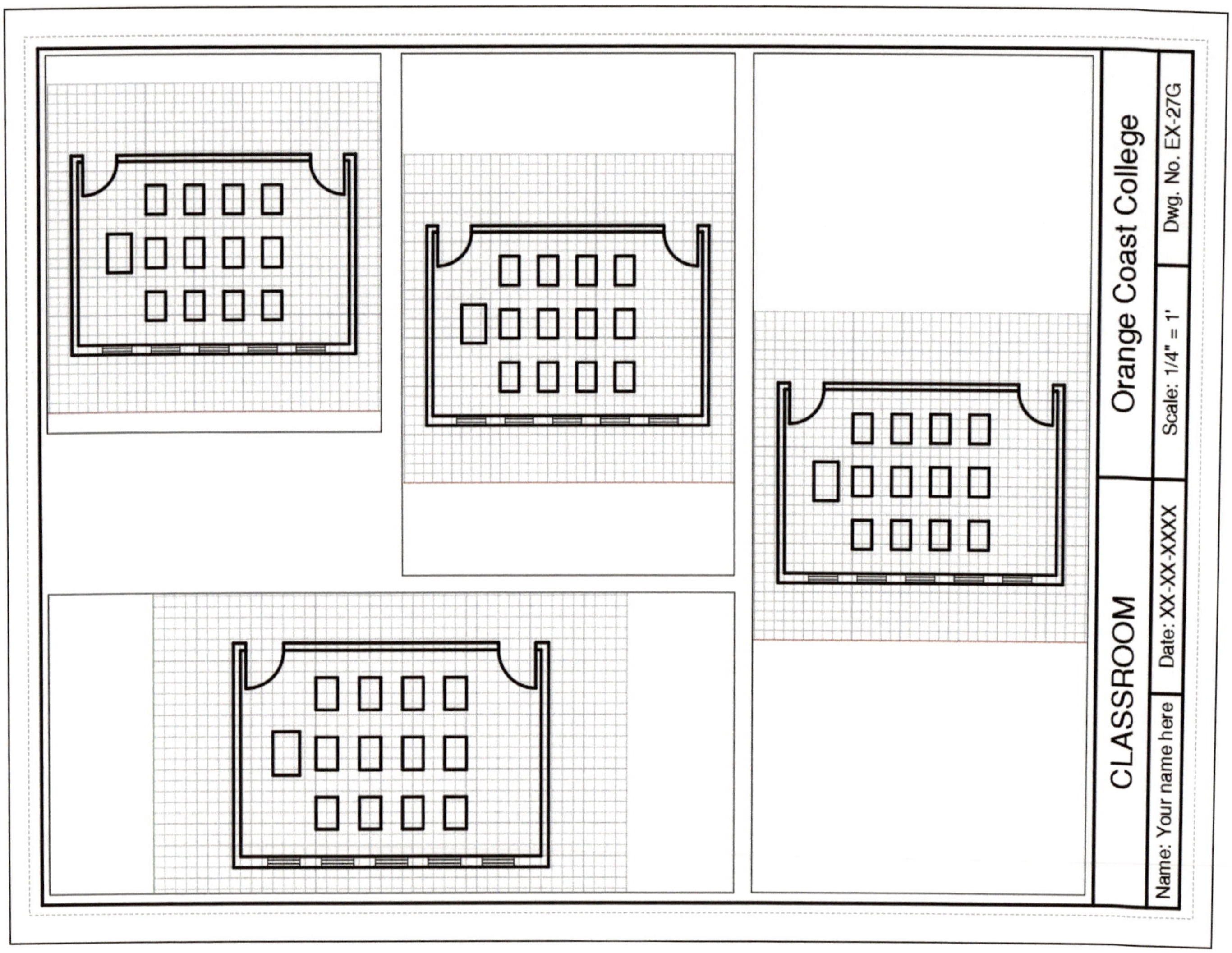

Continued on the next page...

Exercise 28A...continued

Step 2.

1. **Adjust the scale** of each viewport as indicated below.
2. Use **Pan** to move the image within the viewport without changing the scale.
3. Add the **VP SCALE** labels.
   Place them in **Paper Space**.
   Text style: TEXT-CLASSIC
   Text Height: 1/8"
4. **Lock** each viewport.

Notice the annotative dimensions appear only in the 1/4" = 1' viewport.
The dimensions currently have only 1/4" = 1' annotation scale.

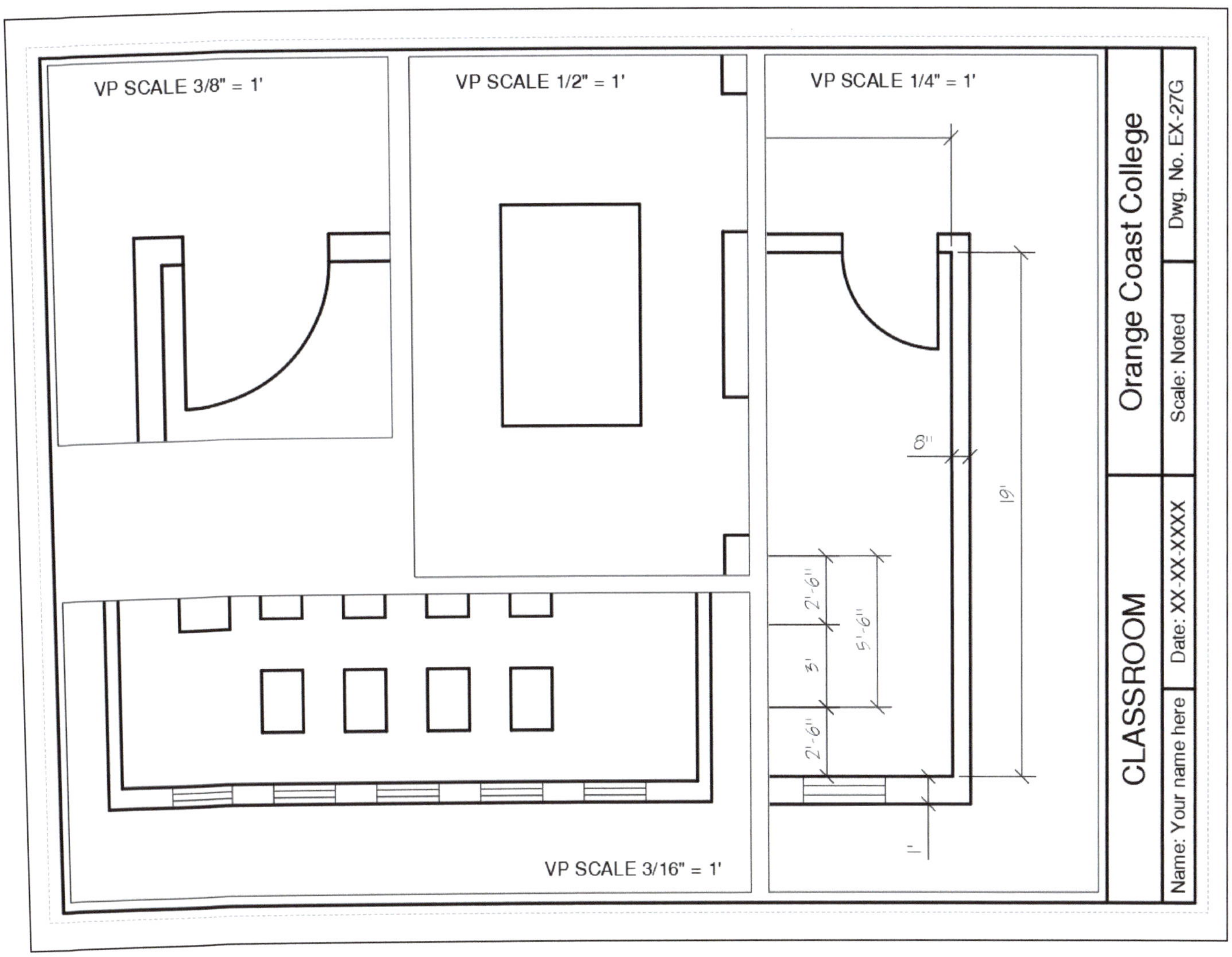

Note: I have turned the grids off for clarity.

Continued on the next page...

Exercise 28A...continued

Step 3.

1. Turn on **Annotation Visibility**. 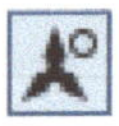 (Blue.) (Refer to page 28-2.)

You should see dimensions in all viewports now.
And they are all different sizes.

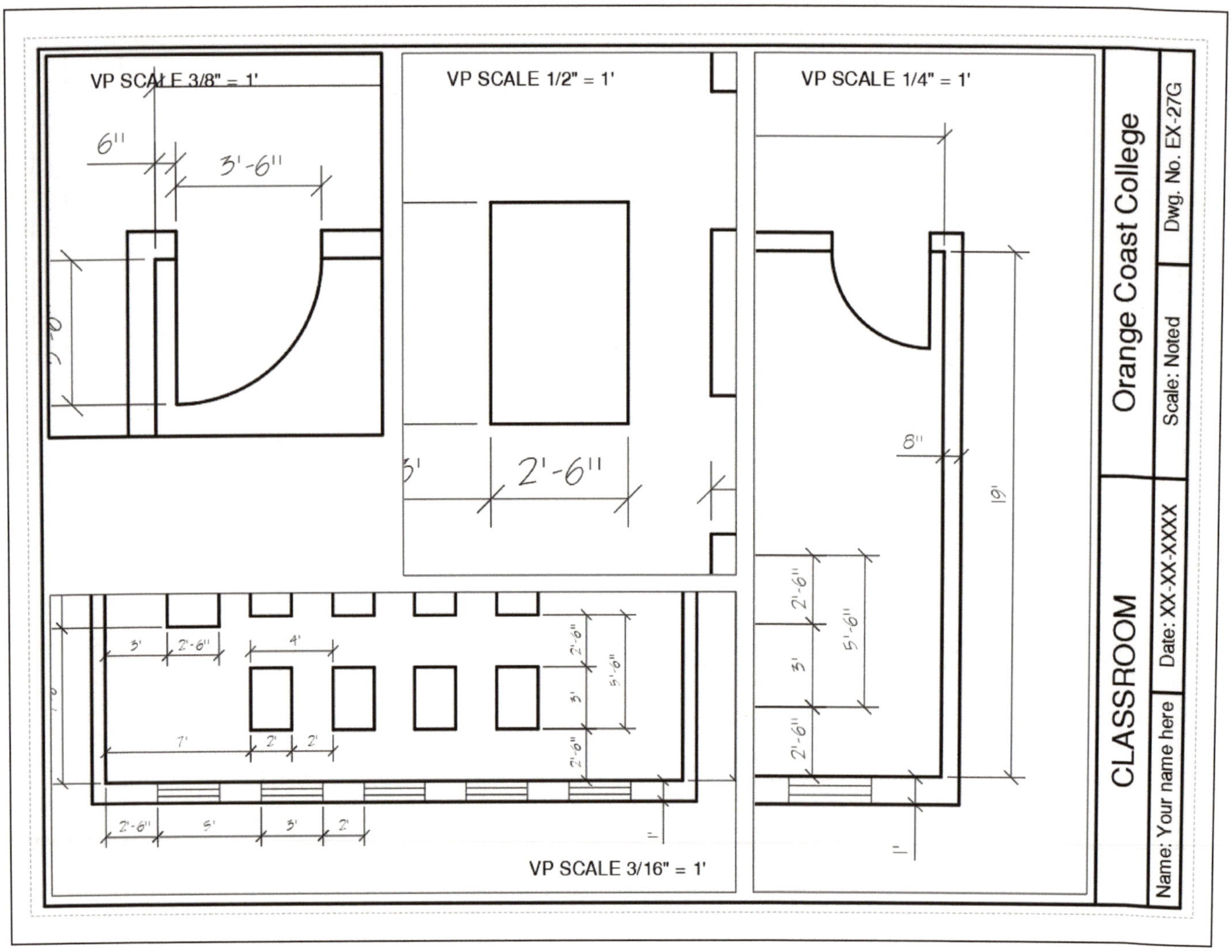

Continued on the next page...

Exercise 28A...continued

Step 4.

1. Add **annotative scales** to the dimensions shown in the viewports below:
   A. Select a viewport (double click inside viewport).
   B. Select the dimensions that you want to stay visible in that viewport.
   C. Select the **Annotate** Tab. (Refer to page 28-3.)
   D. Select Add **Current Scale** button. This will automatically **add** the **current viewport scale** to the **selected objects**.
   E. Go on to the next viewport and repeat **A** through **D** above until they are all done.

Step 5.

1. Turn off **Annotation Visibility**.  (Gray.) The unwanted dimensions should have disappeared.
2. **Save** as **Ex-28A**

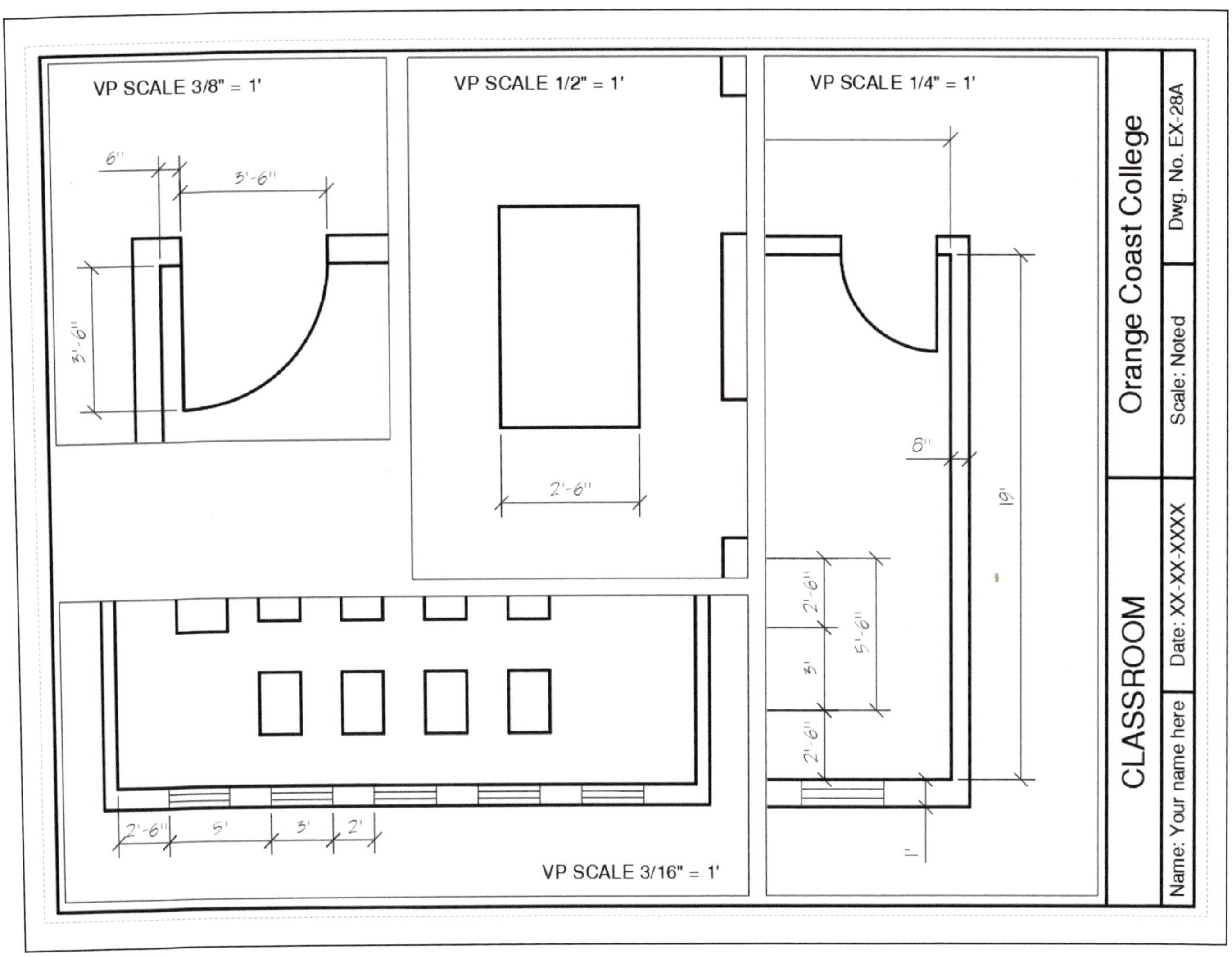

Exercise 28B

The following exercise will give you practice with Hatch within scaled viewports using the **annotative scale** option. Follow the instructions below while you manipulate the drawing to appear as shown.

Step 1.

1. Start a **New** file using **My Decimal Setup.dwt**
2. Select the **A size** Layout Tab.
3. Erase the one existing viewport.
4. Select the Viewport layer and create three new viewports approximately as shown.
5. **Adjust the scale** of each viewport as indicated.
6. **Lock** each viewport.
7. Activate the **1:1** viewport. (Double click inside the viewport.)
8. Draw a **2" X 2" [50.8 mm X 50.8 mm]** rectangle inside the **1:1** viewport as shown using Layer Object Line. (The rectangle will appear in the other two viewports because all viewports display Model Space.)

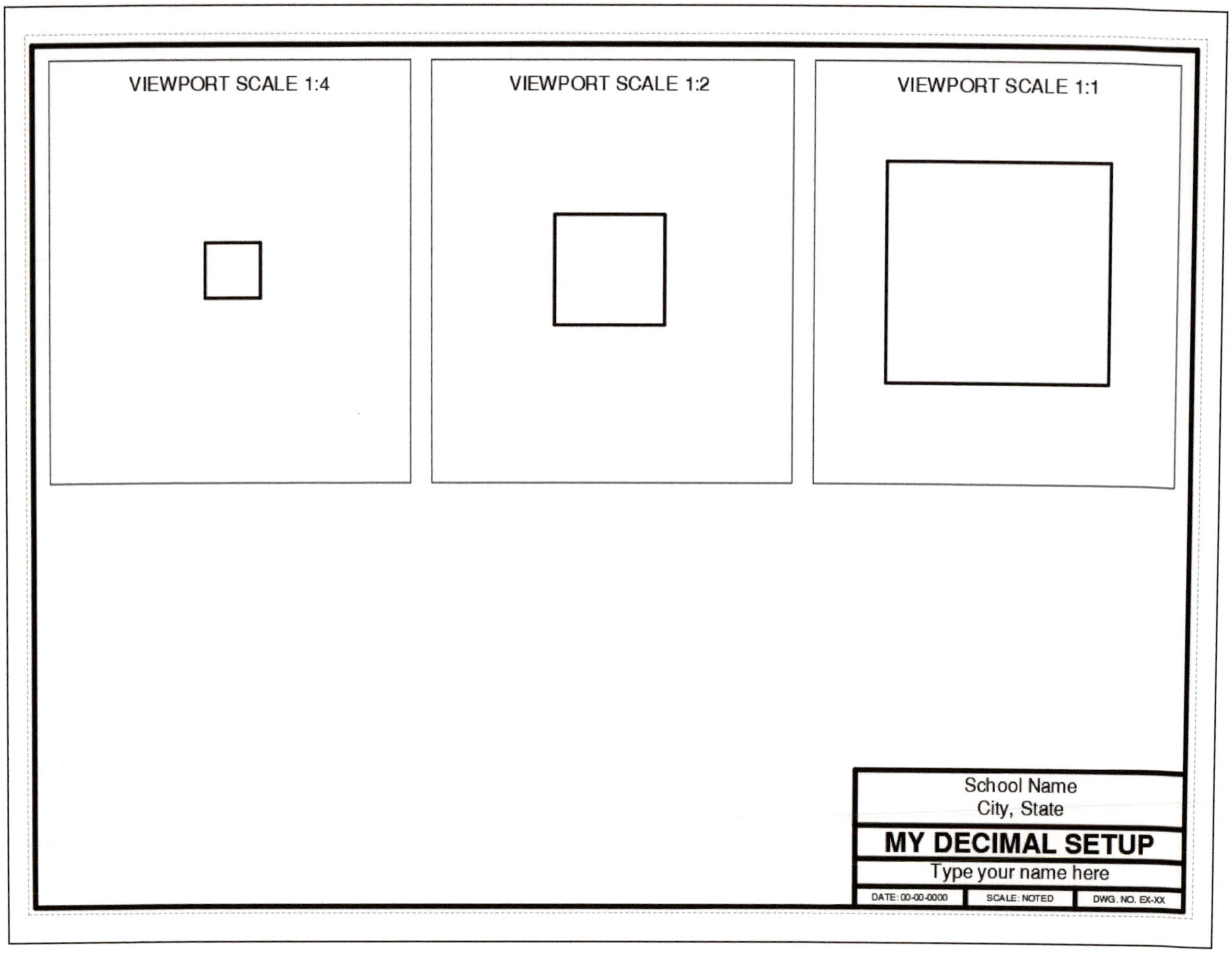

Continued on the next page...

Exercise 28B...continued

Step 2.

1. Select the Layer Hatch.
2. Hatch the **1:1** rectangle using:
   A. Pattern: **ANSI31**
   B. Angle: **0**
   C. Scale: **1.000**
   D. Select the **Annotative** option.

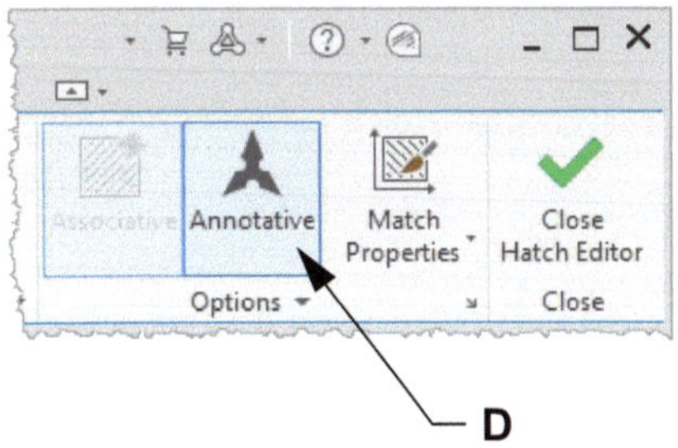

3. Place the Hatch inside the rectangle as shown and press ***<Enter>***.

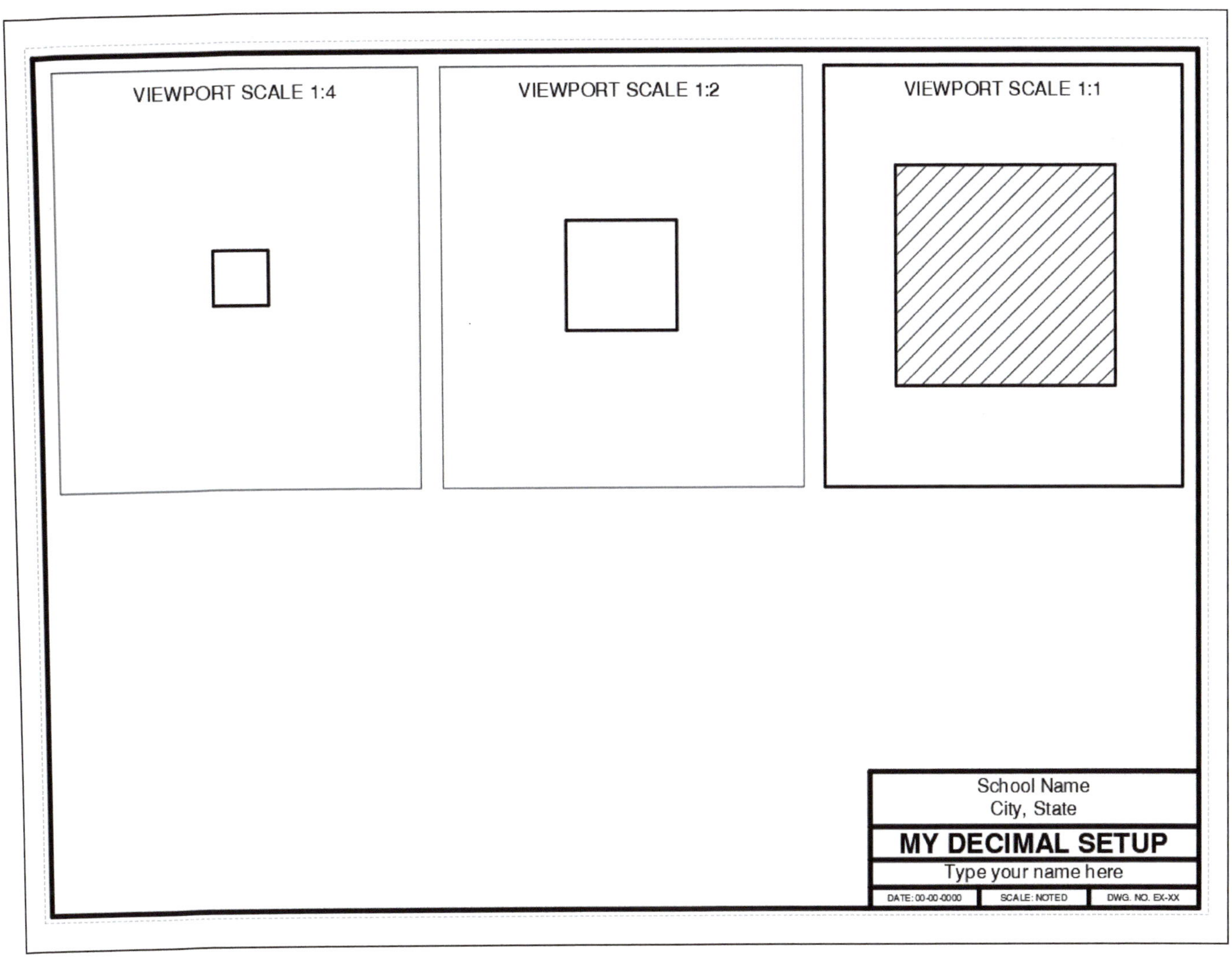

Note: If a Hatch set appears in all rectangles, you probably forgot to select the **Annotative box** or you have **Annotation Visibility on**.

Continued on the next page...

Exercise 28B...continued

Step 3.

1. Turn on **Annotation Visibility**.  (Blue.) (Refer to page 28-2.)

The Hatch set appears in all of the viewports.
But the sizes are all different.

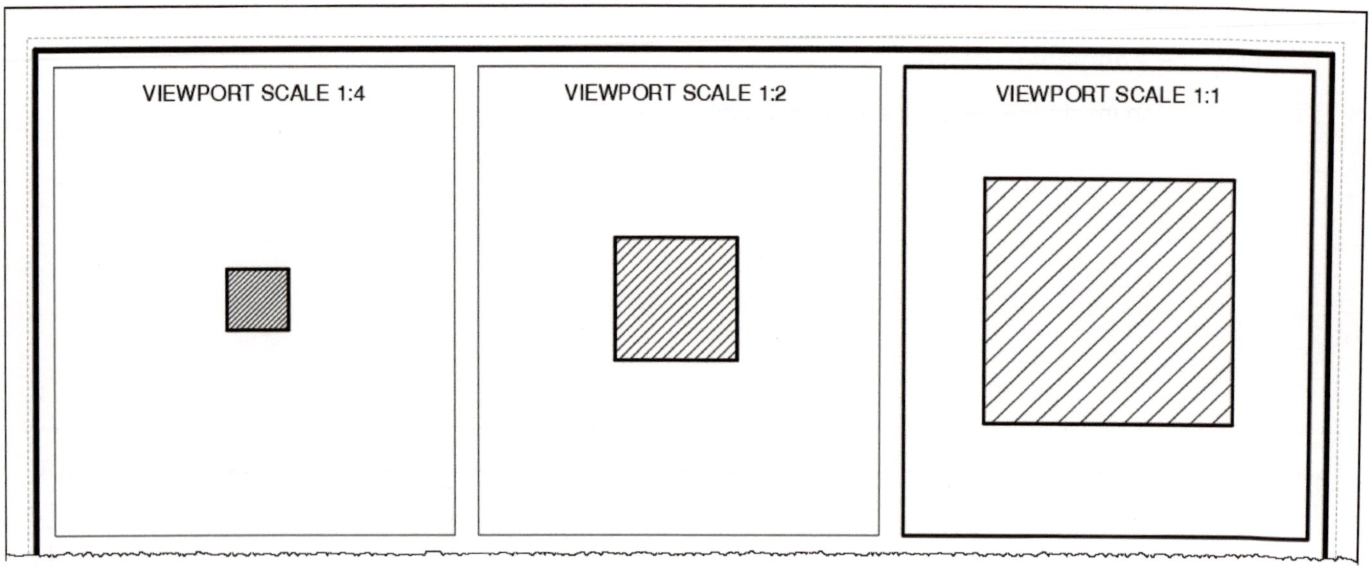

2. Assign the correct Annotative scale to the Hatch sets in viewport **1:4** and **1:2**.

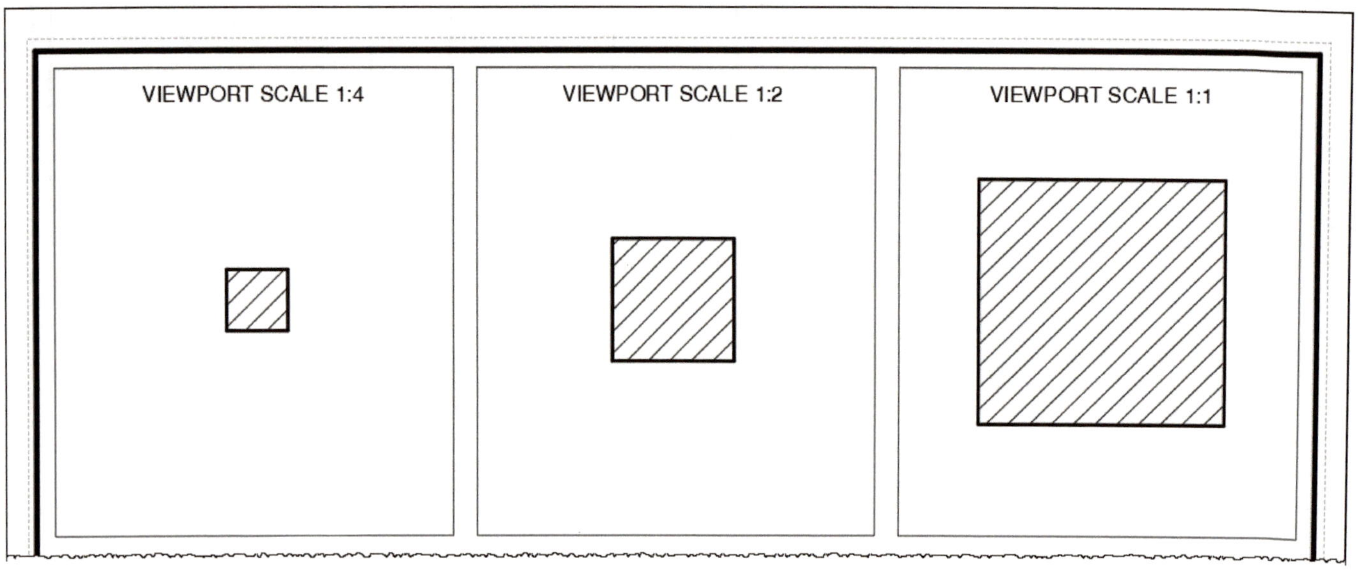

3. Turn off **Annotation Visibility**.  (Gray.) (Refer to page 28-2.)

4. **Save** as **Ex-28B**

LESSON 29

LEARNING OBJECTIVES

After completing this lesson, you will be able to:

- Understand Blocks
- Create a Block
- Insert a Block
- Insert a Block using Automatic Block Placement
- Re-define a Block
- Count the number of Blocks in a Drawing
- Purge unused Blocks
- Create Multileaders with Blocks attached
- Use the Collect Multileader tool

Blocks

A **Block** is a group of objects that have been converted into **one** object. A Symbol, such as a transistor, bathroom fixture, window, screw or tree, is a typical application for the block command. First a Block must be **created**. Then it can be inserted into the drawing. An inserted Block uses less file space than a set of objects copied.

Creating a Block

1. First draw the objects that will be converted into a Block.

   For this example, a circle and two lines are drawn.

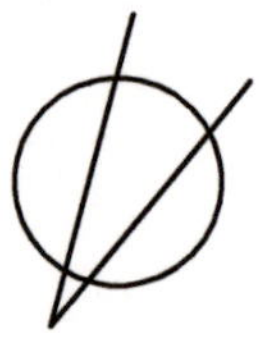

2. Select the **Create Block** command using one of the following:

   Ribbon = Insert Tab / Block Definition Panel /
   or
   Keyboard = *B <Enter>*

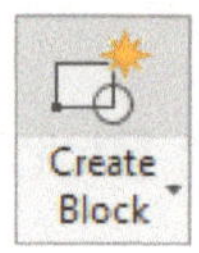

3. In the **Block Definition** dialog box, enter the new block name in the **Name** box.

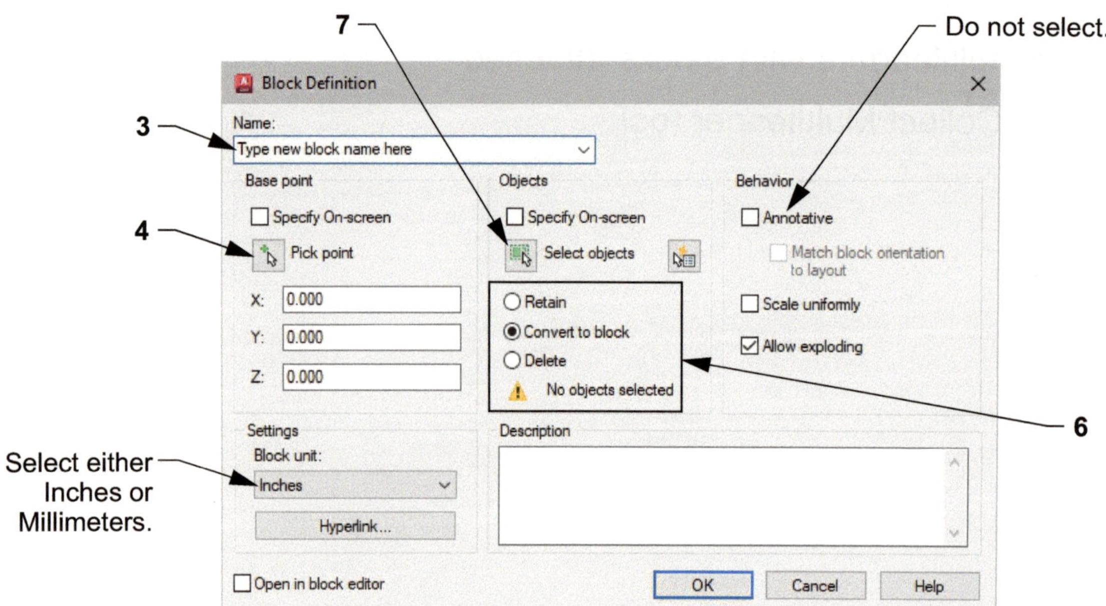

4. Select the **Pick point** button. (Or you may type the X, Y, and Z coordinates.) *The **Block Definition** box will disappear and you will return temporarily to the drawing.*

5. Select the location where you would like the insertion point for the Block. *Later when you insert this block, the block will appear on the screen attached to the cursor at this insertion point. Usually this point is the **Center**, **Midpoint**, or **Endpoint** of an object.*

Continued on the next page...

5

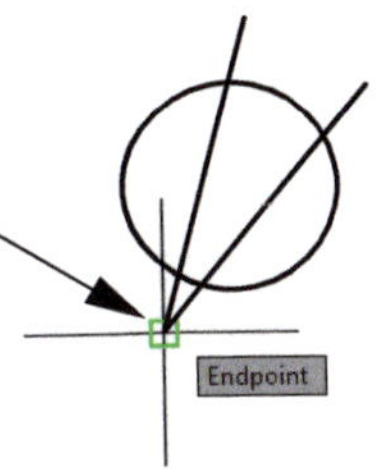

Notice the coordinates for the basepoint are now displayed. (Don't worry about this. Use **Pick point**, and you will be fine.)

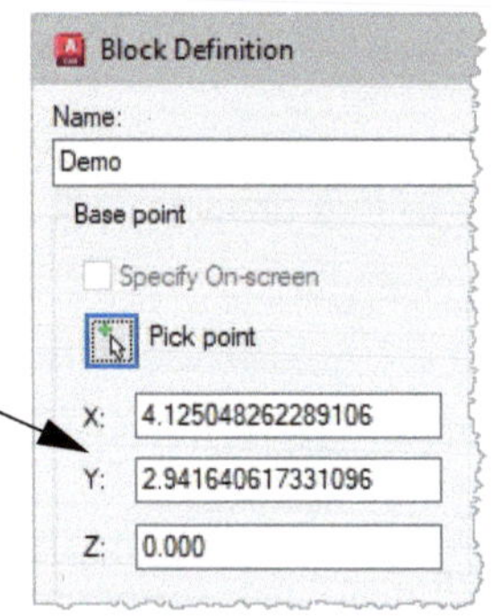

6. Select one of the options described below. It is important that you select one and understand the options below.

   Retain: If this option is selected, the original objects will stay visible on the screen after the block has been created.

   Convert to block: If this option is selected, the original objects will disappear after the block has been created, but will immediately reappear as a block. It happens so fast you won't even notice the original objects disappeared.

   Delete: If this option is selected, the original objects will disappear from the screen after the block has been created. (This is the one I use most of the time.)

7. Select the **Select objects** button.

   The Block Definition box will disappear and you will return temporarily to the drawing.

8. Select the objects you want in the block, and then press ***<Enter>***.

Selection Window.

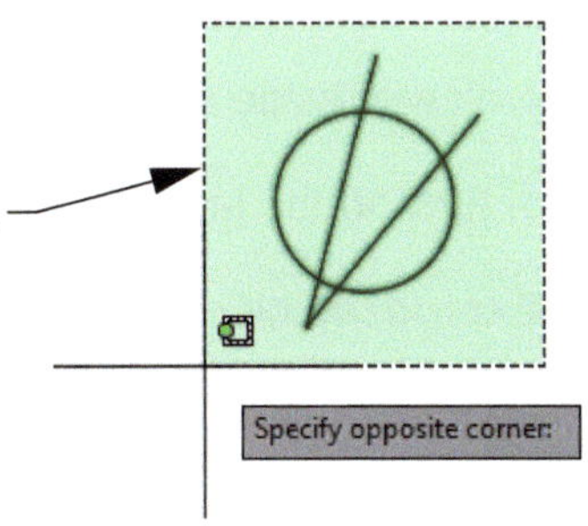

The Block Definition box will reappear and the objects you selected should be displayed in the Preview icon area.

10 Preview.

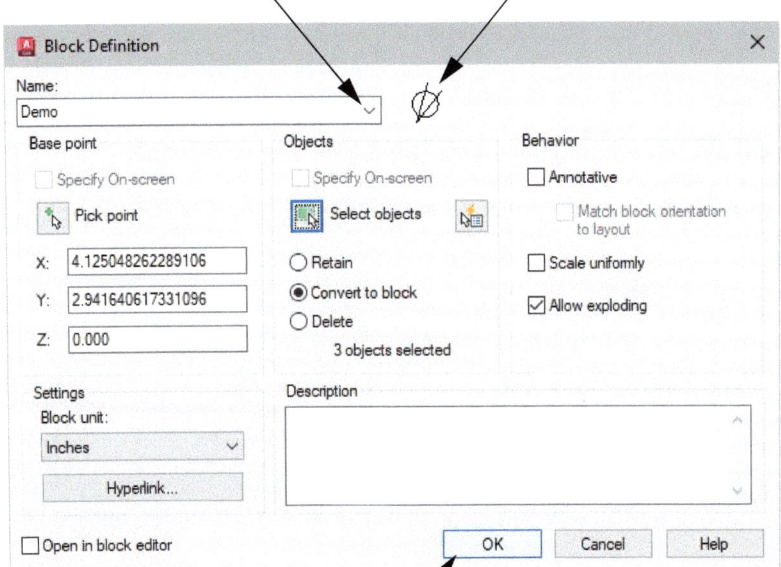

9

Continued on the next page...

9. Select the **OK** button. *The new block is now stored in the drawing's block definition table.*

10. To verify the creation of this Block, select **Create Block** again, and select the Name (✓). A list of all the blocks in this drawing will appear. (Refer to page 29-5 for inserting instructions.)

Additional definitions of Options

Block unit: You may define the units of measurement for the block. This option is used with the "Design Center" to drag and drop with Autoscaling. The Design Center is an advanced option and is not discussed in this book.

Hyperlink: Opens the **Insert Hyperlink** dialog box, which you can use to associate a hyperlink with the block.

Description: You may enter a text description of the block.

Scale uniformly: Specifies whether or not the block is prevented from being scaled non-uniformly during insertion.

Allow exploding: Specifies whether or not the block can be exploded after insertion.

How Layers affect Blocks

If a block is created on Layer 0:

1. When the block is inserted, it will take on the properties of the current layer.
2. The inserted block will reside on the layer that was current at the time of insertion.
3. If you **Freeze** or turn **off** the layer, the block was inserted onto, the block will disappear.
4. If the Block is **Exploded**, the objects included in the block will revert to their original properties of Layer 0.

If a block is created on Specific layers:

1. When the block is inserted, it will retain its own properties. It **will not** take on the properties of the current layer.
2. The inserted block **will reside** on the current layer at the time of insertion.
3. If you **Freeze** the layer that was current at the time of insertion, the block **will** disappear.
4. If you turn **off** the layer that was current at the time of insertion, the block **will not** disappear.
5. If you **Freeze** or turn **off** the block's original layers, the block **will** disappear.
6. If the Block is **Exploded**, the objects included in the block will go back to their original layer.

Inserting Blocks

A Block can be inserted at any location within the drawing. When inserting a Block, you can Scale, Rotate or Explode it.

1. Select the **Insert** command using one of the following:

   Ribbon = Insert Tab / Block Panel / 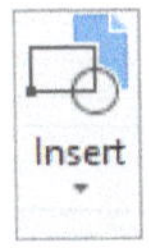

   or
   Keyboard = *insert* <*Enter*>

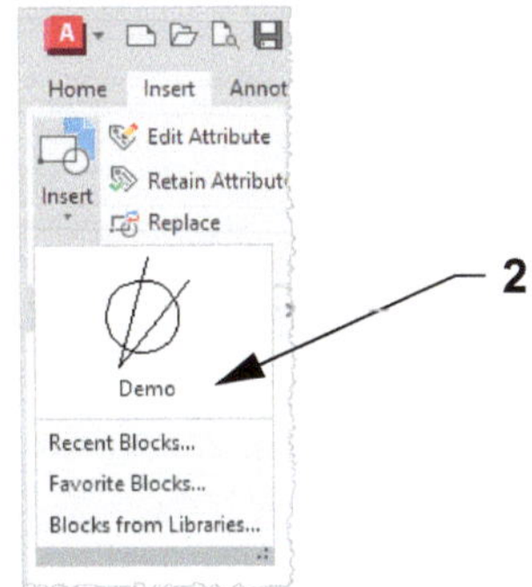

2. Select the **Block** you want to insert, from the Insert Window.

3. Select the insertion location for the Block by moving the cursor and pressing the left mouse button.

CAD TIP! If you create a new drawing file with no recent or current Blocks, the **Insert** drop-down window will be blank. Select **Recent Blocks...** to access the **Blocks Palette**. You may also type ***insert*** and then press ***<Enter>*** to access the Blocks Palette, which is discussed below.

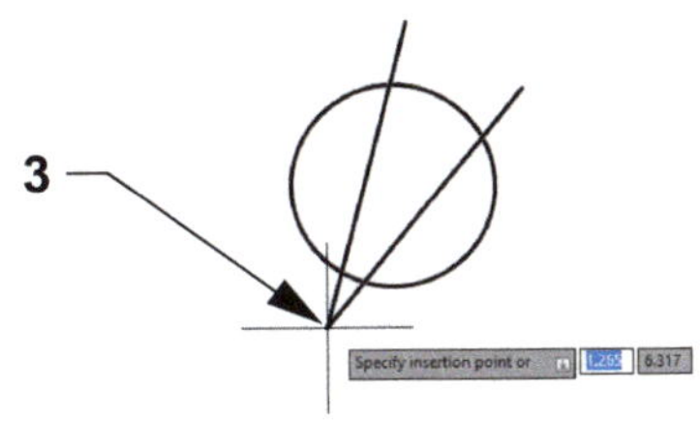

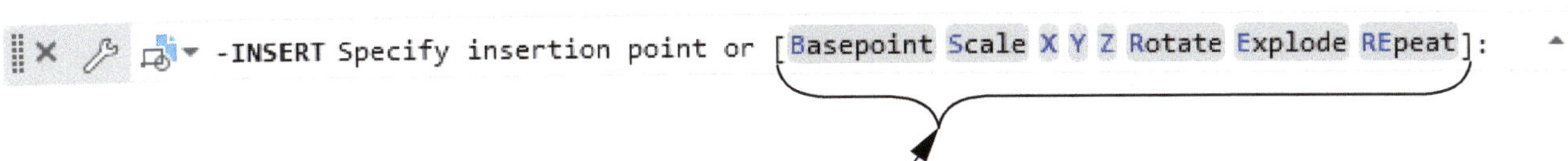

Note: If you want to change the **Basepoint** or **Scale** or to **Rotate** the Block before you actually place the Block, press the right-hand mouse button and you may select an option from the menu, or select an option from the Command Line, shown above.

You may also **"preset"** the insertion point, scale, rotation, or insert recent Blocks or Blocks from Libraries. This is discussed next.

Presetting the Insertion Point, Scale, or Rotation

You may preset the **insertion point**, **scale**, or **rotation**, using the **Blocks Palette** instead of at the Command Line.

1. Select the **Insert** command.

2. Select **Recent Blocks...** from the drop-down menu.

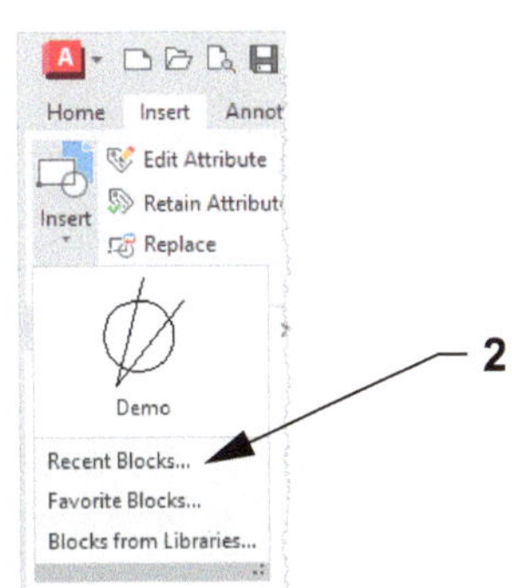

Continued on the next page...

3. The **Blocks Palette** will open. Any recent Blocks used in the current drawing or previously opened drawings, will appear in the **Recent Blocks** Window.

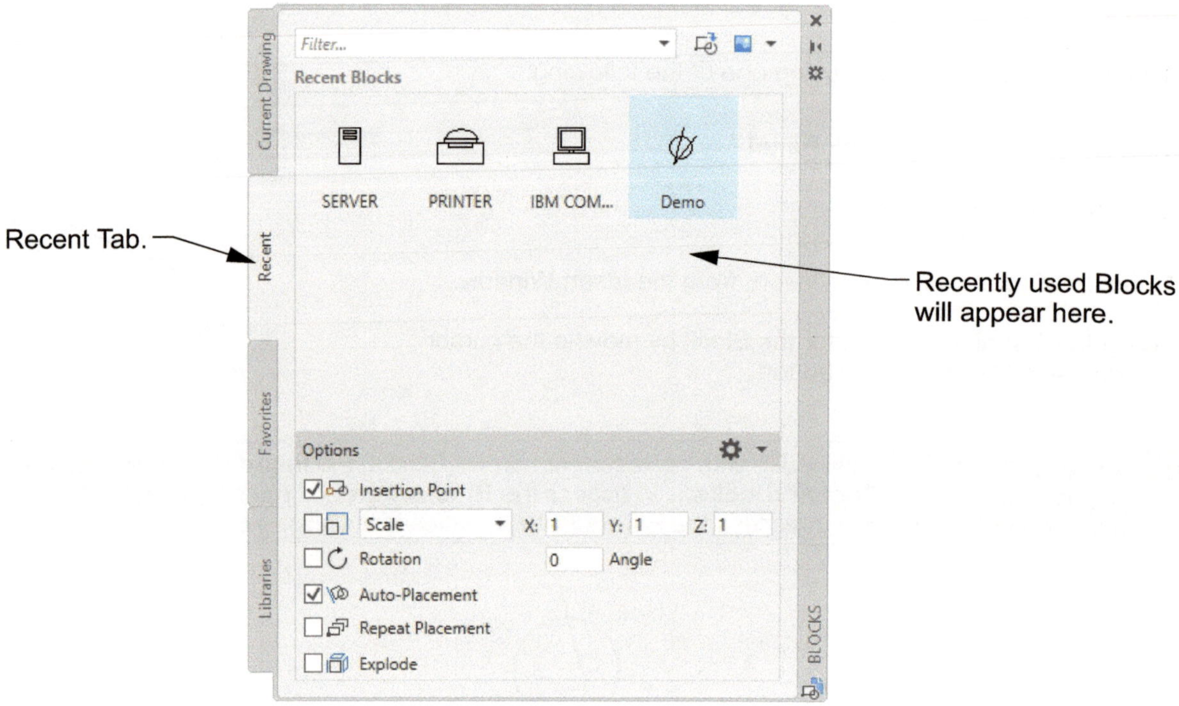

4. If there are no recent Blocks used, select the **Current Drawing** Tab.

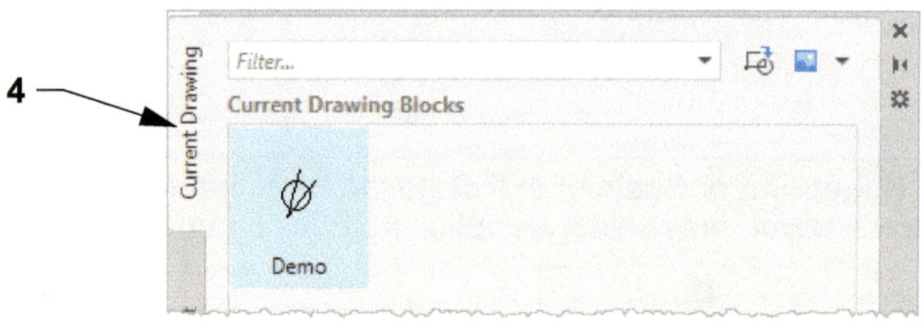

5. If you want to insert the Block at a specific location, uncheck the **Insertion Point** box.

6. Enter the location coordinates required in the **"X"** and **"Y"** fields. (**Note:** The **"Z"** field is used for 3D, so leave it as it is for 2D drawing.)

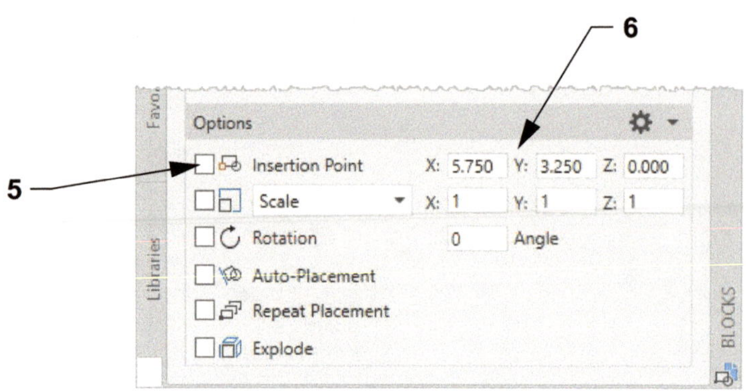

Continued on the next page...

7. If you want to insert the Block at a specific scale, uncheck the **Scale** box.

8. Enter the scale factors in the **"X"** and **"Y"** fields. You can set the **"X"** and **"Y"** fields at different scale factors if you require the Block to be scaled non-uniformly.

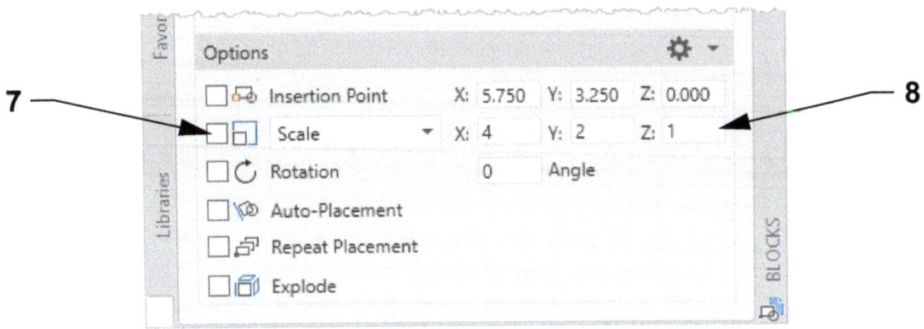

9. If you want to insert the Block at a specific angle, uncheck the **Rotation** box.

10. Enter the rotation angle you require for the Block, in the **Angle** field.

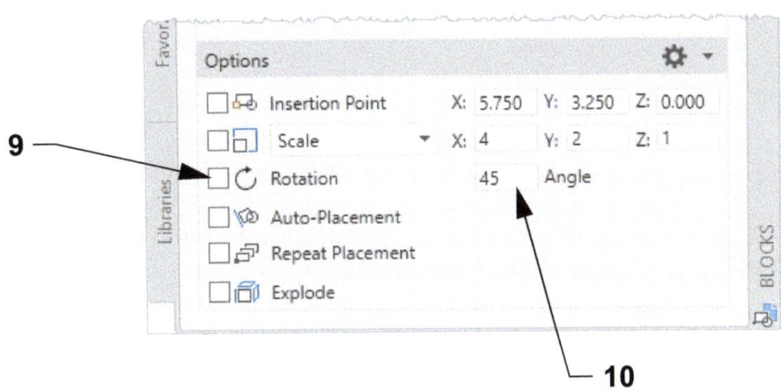

11. If you want to insert more than one instance of the Block, check the **Repeat Placement** box.

12. If you want the Block to be exploded into its constituent parts, check the **Explode** box.

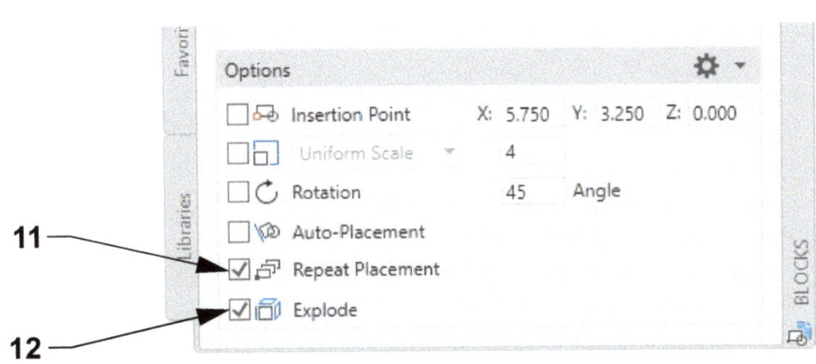

Note: If you check the **Explode** box, the non-uniformed scale factor will not work. Instead it will just scale to the "X" factor that you have set, and in a uniformed scale. So in the example above, the Block will scale at four times its original size.

Continued on the next page...

13. Once you have completed selecting any of the options in the **Insertion Options** fields, click on the Block you want to insert. Then left click in the drawing area for the Block to be placed.

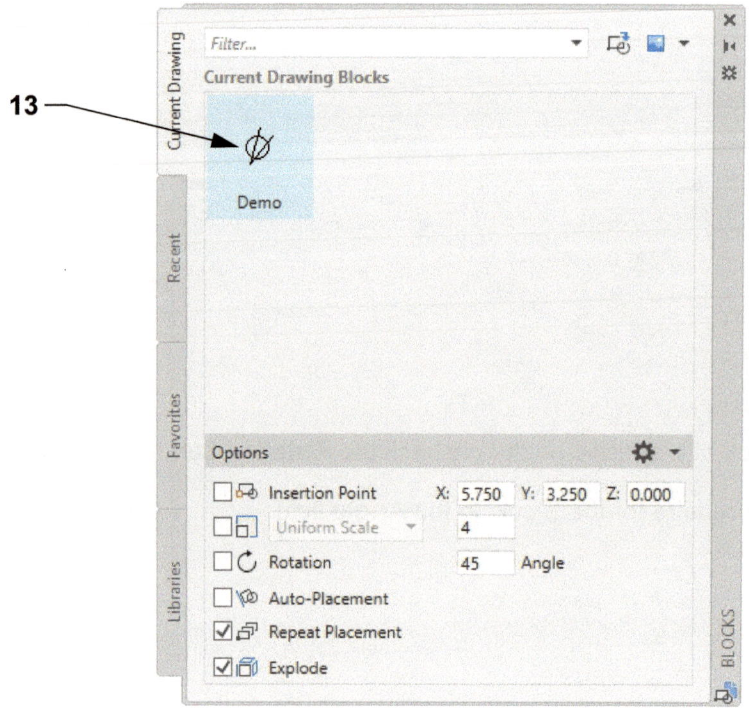

If you want to insert a Block from a drawing stored on your system that is not currently available in the **Blocks Palette**, select the **Libraries** Tab, then select the **Browse** button to search for a Block on your system.

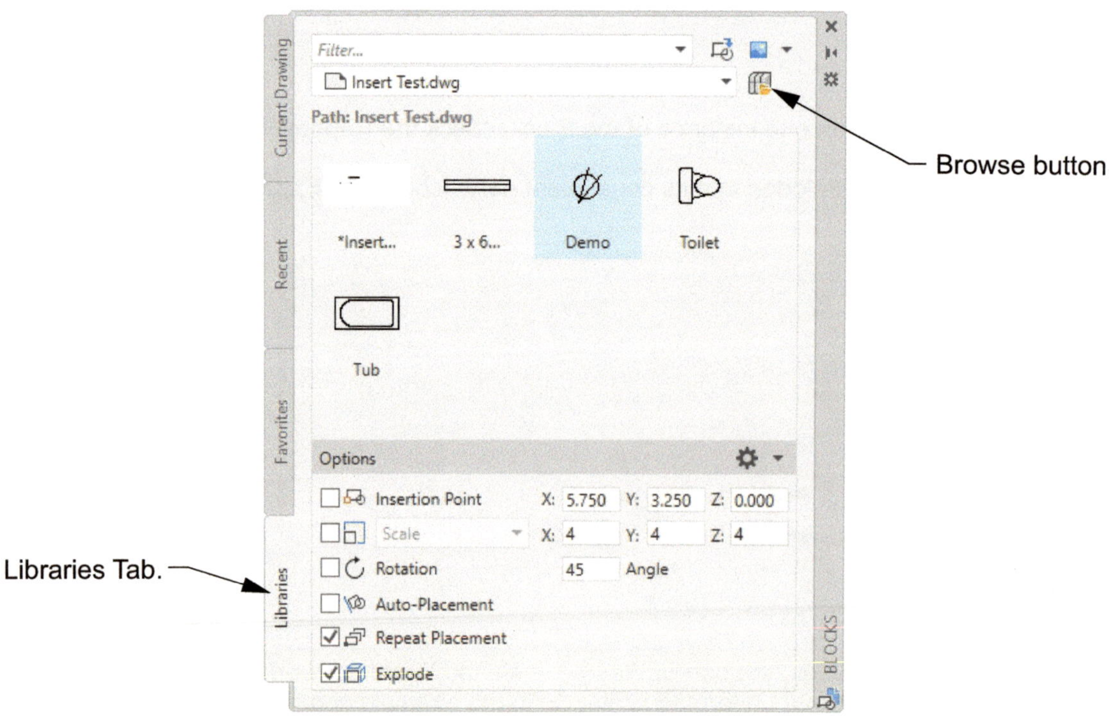

Automatic Block Placement

Automatic Block Placement allows you to quickly position and insert a new block into a drawing by providing placement suggestions, based on existing blocks of the same type that have already been inserted into the drawing. For example, if you have placed a block of an office chair by a desk in a room, and want to place the same chair in an adjacent room with the same type of desk, this command makes similar positioning easier.

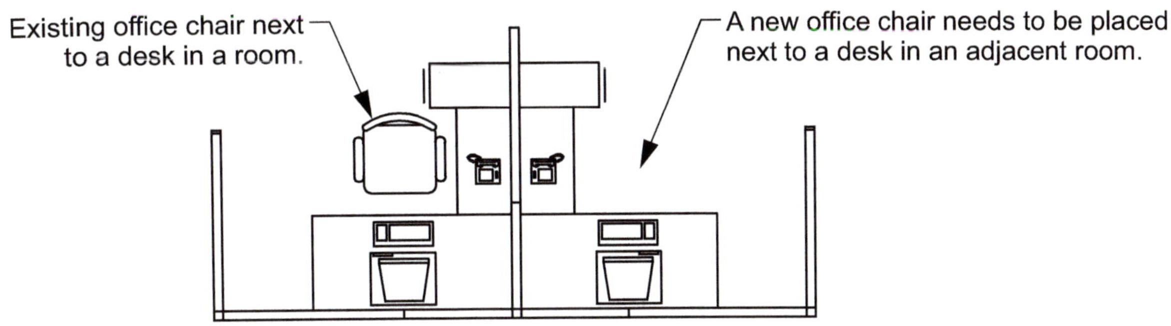

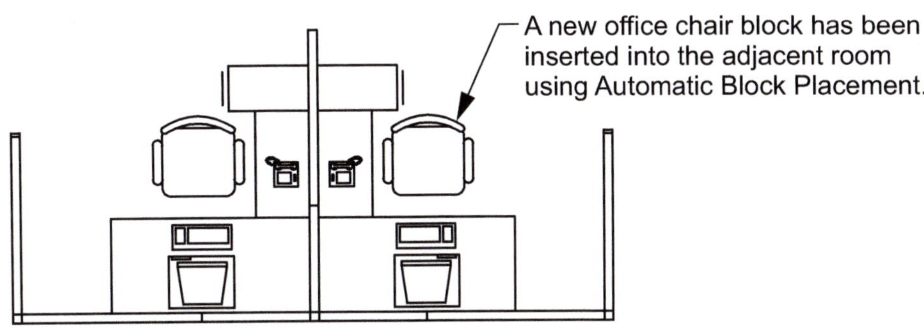

How to insert a block using Automatic Block Placement

1. Open the **Blocks Palette.**
2. Check the **Auto-Placement** box.
3. Select the block you want to insert into the drawing.

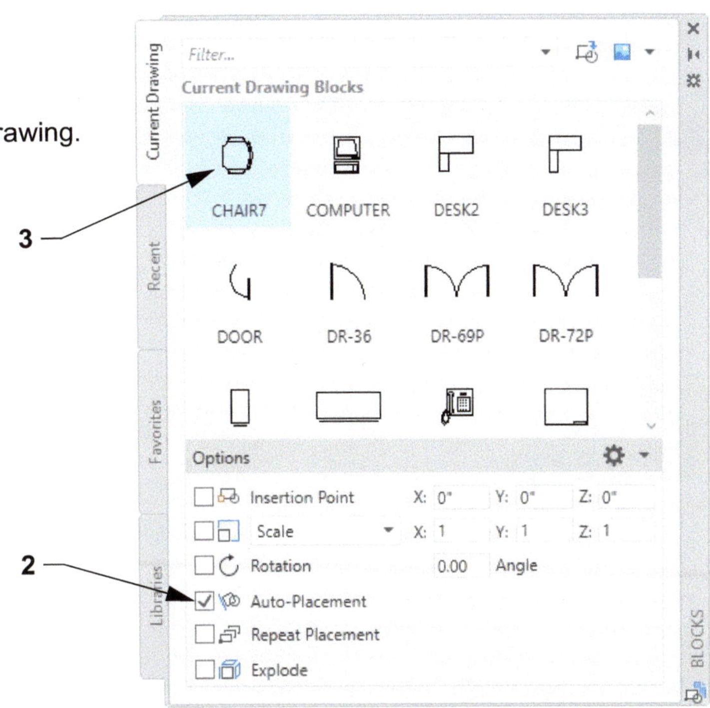

Continued on the next page...

4. Move the cursor to approximately the position where you want to place the block.

5. As you move the block to the position, the desk edges will highlight in orange.

6. Left click to place the block in the suggested position.

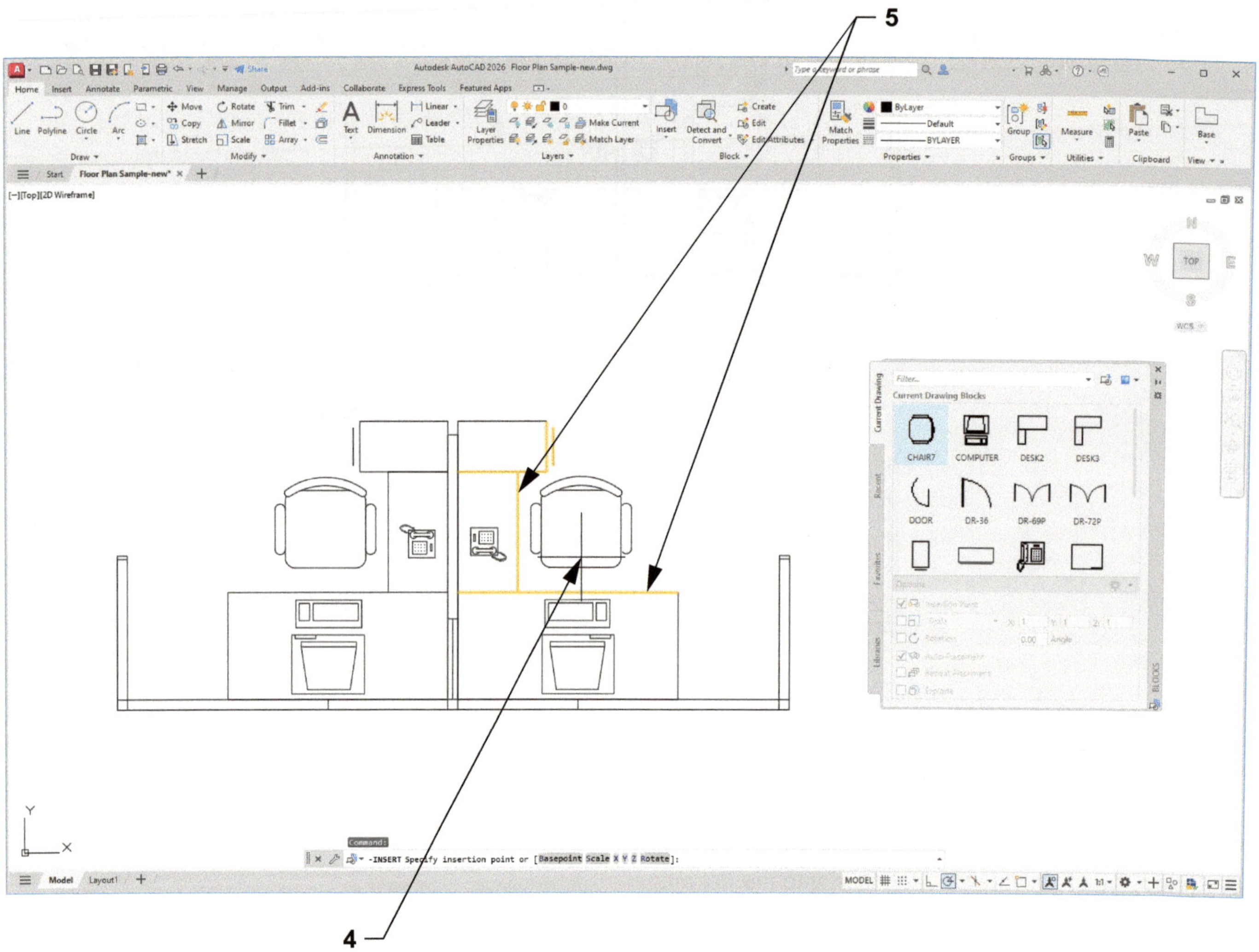

Note: If you do not want to place the block in the suggested position, press the ***<Tab>*** key for alternative placement suggestions. If you want to temporarily hide the suggestions for the block placement, hold down the ***<Shift>*** + ***W*** keys.

Re-Defining a Block

You can change the design of a block previously inserted by re-defining it in the **Block Editor**.

1. Select the **Block Editor** command using one of the following:

Ribbon = Insert Tab / Block Definition Panel /
or
Keyboard = *bedit <Enter>*

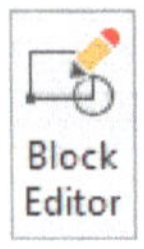

*The **Edit Block Definition** dialog box will appear.*

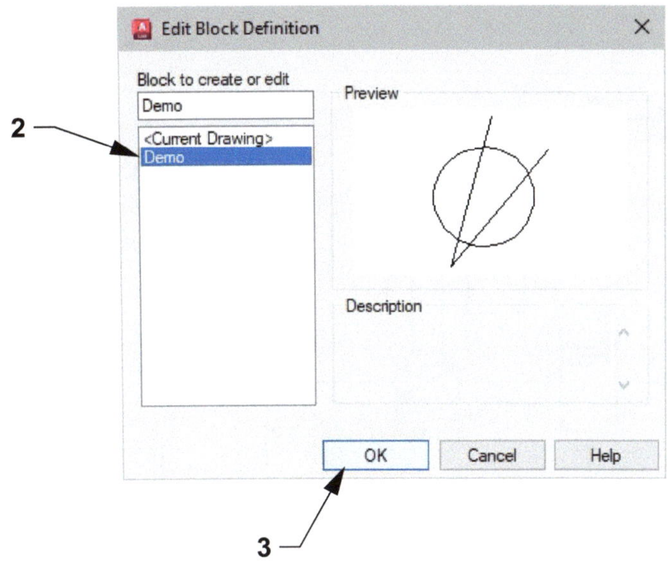

2. Select the name of the Block that you want to change.
3. Select the **OK** button.
4. The Block that you selected should appear large on the screen. You may now make any additions or changes to the block. You can change Tabs and use other Panels such as Draw and Modify. But you must return to the **Block Editor** Tab to complete the process.
5. Return to the **Block Editor** button if you selected any other Tab while editing.
6. Select the **Save Block** tool from the **Open/Save** Panel.
7. Select the **Close Block Editor** tool.

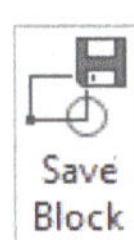

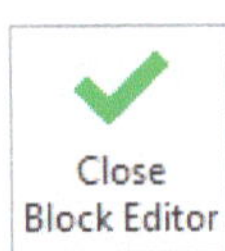

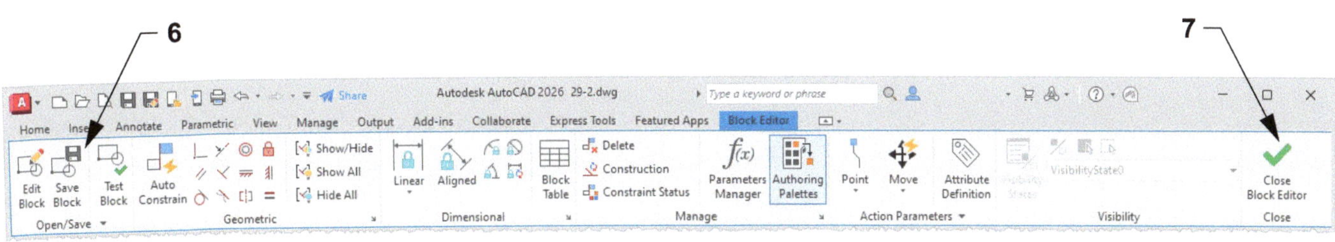

You will be returned to the drawing, and **all previously inserted** blocks with the **same name** will be updated with the changes that you made.

Count the Number of Blocks in a Drawing

You can quickly count the number of blocks in a drawing by using the **Count** command. After selecting the Count command, the **Count Palette** will open, showing you how many blocks there are in your drawing, along with the names of those blocks.

You can also create a table with the blocks you select, and then insert the table into your drawing. The table will display the name of the blocks, as well as how many blocks there are of each type.

The example below shows a network diagram with various types of blocks, along with the Count Palette that shows the name of and total count for each type of block.

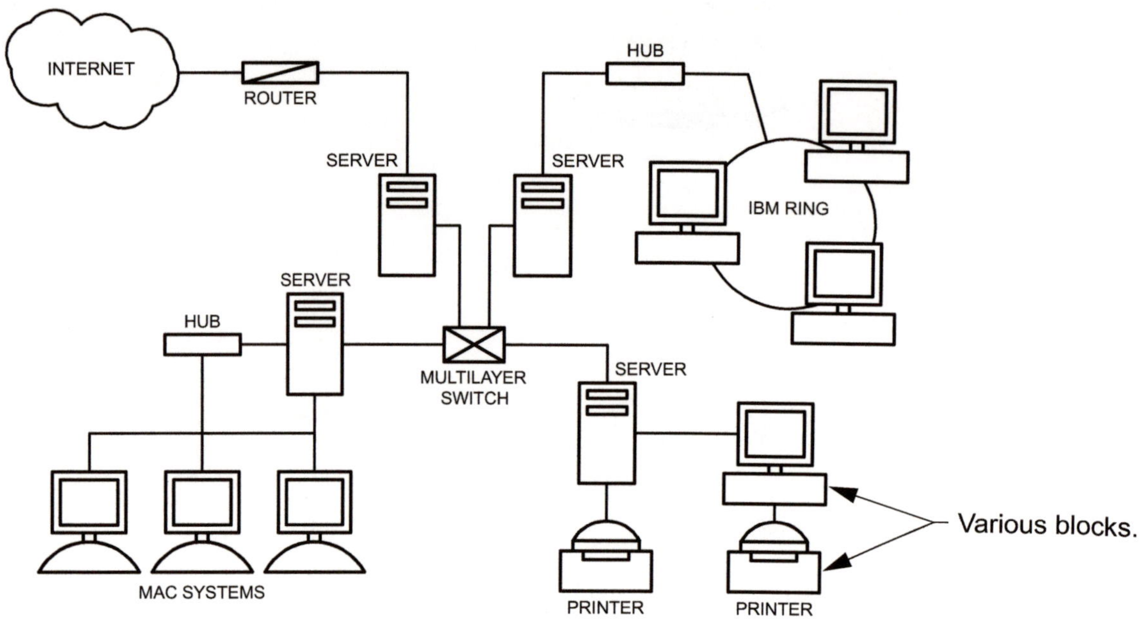

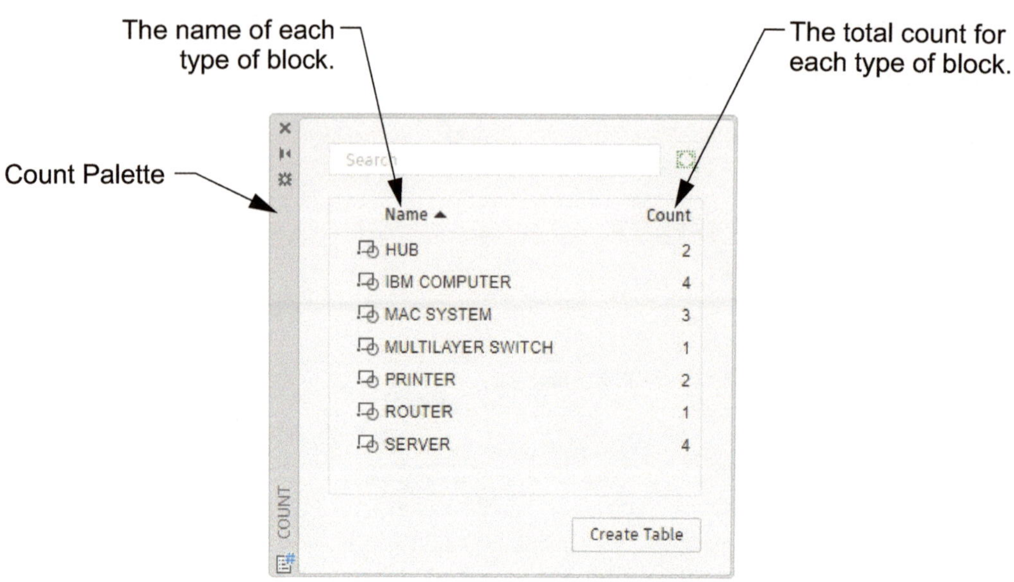

How to count the number of blocks in a drawing

1. Select the **Count** command using one of the following:

Ribbon = View Tab / Palettes Panel / 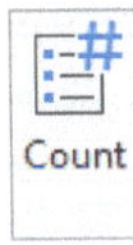

or
Keyboard = ***countlist <Enter>***

2. The Count Palette will open.

How to create and insert a table into the drawing

1. Open the **Count Palette**.
2. Select the **Create Table** button.
3. Place a checkmark in the top field if you want the table to show all the blocks, or select fields individually.
4. Select **Insert**.

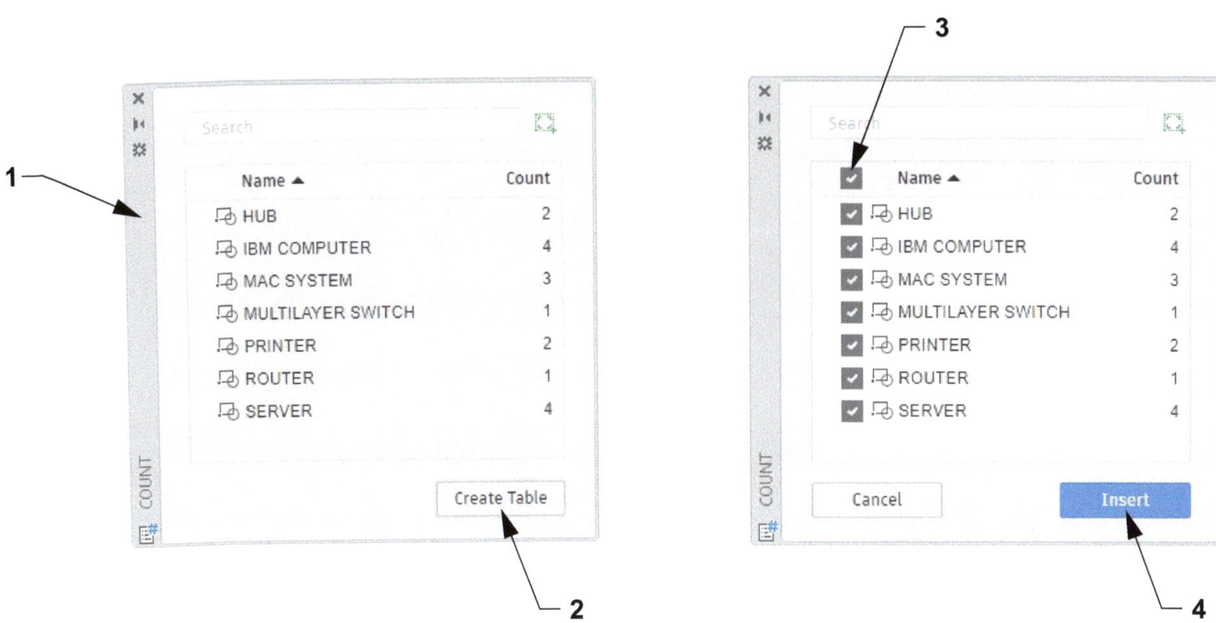

5. Move the table to the location where you want it to appear in the drawing.
6. Left click to place the table in the drawing.

Item	Count
HUB	2
IBM COMPUTER	4
MAC SYSTEM	3
MULTILAYER SWITCH	1
PRINTER	2
ROUTER	1
SERVER	4

Note: The style and height of the text in the table is determined by your current Text Style. If you want a different text style and/or height, you must change it before selecting the table insert button.

Purge Unwanted and Unused Blocks

You can remove a block reference from your drawing by erasing it; however, the block definition remains in the drawing's block definition table. To remove any **unused** blocks, dimension styles, text styles, layers, and linetypes, you may use the **Purge** command.

How to delete unwanted and unused blocks from the current drawing

1. Select the **Purge** command using one of the following:

   Ribbon = Manage Tab / Cleanup Panel / Purge
   or
   Keyboard = ***purge <Enter>***

2. Select the **+** sign beside Blocks. (It will change to a **–** sign.)

3. Select the block that you want to purge by clicking in the box to insert a check mark.

Note: Only **unused** blocks will be listed.

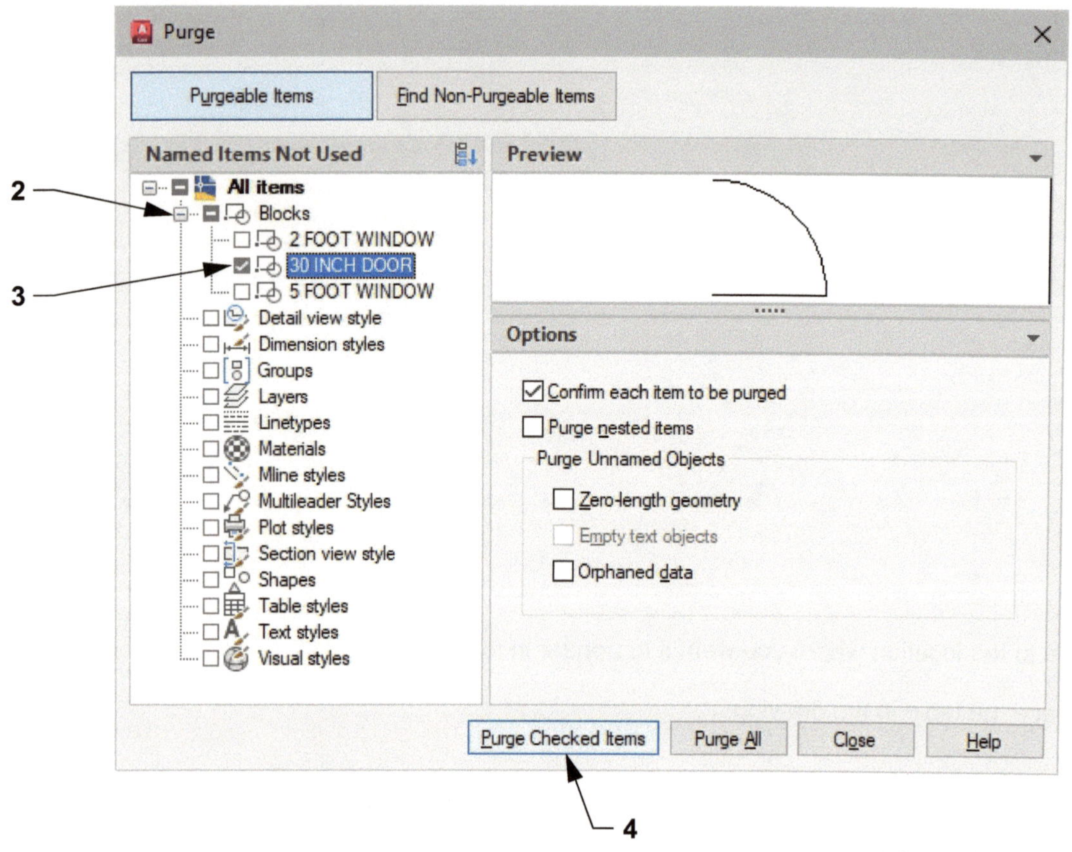

4. Select the **Purge Checked Items** button.

Continued on the next page...

5. Select **Purge this item** in the **Purge - Confirm Purge** dialog box.

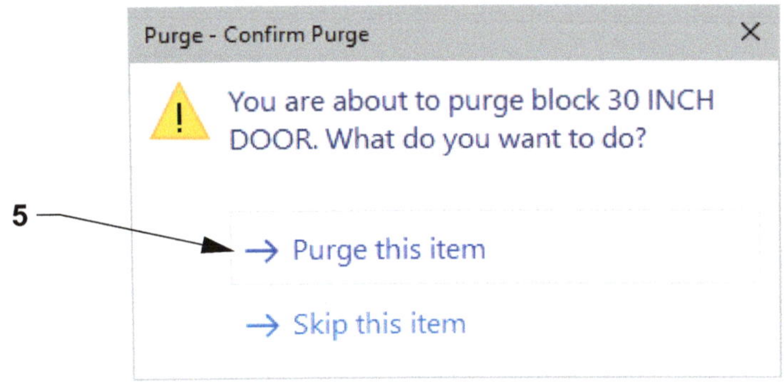

6. The block has now been removed from the drawing and no longer shows in the **Named Items Not Used** list.

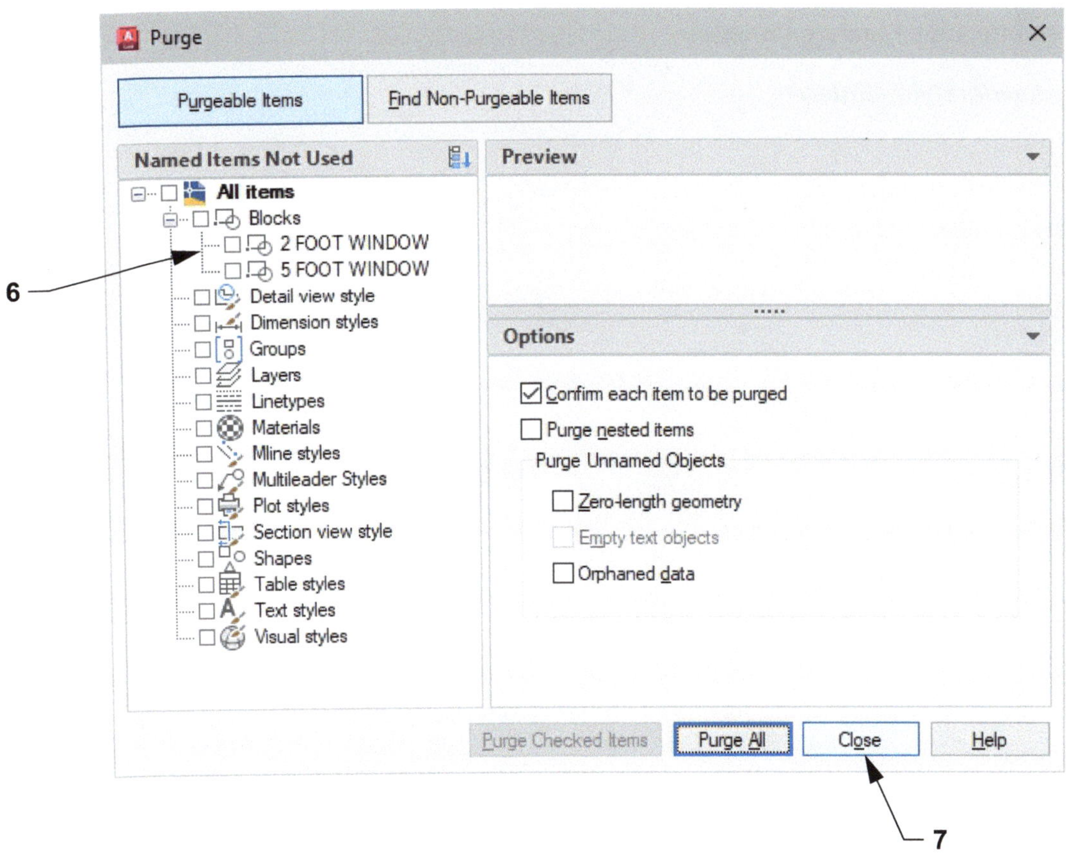

7. Select **Close**.

Note: When you create a block, it is saved within the drawing you created it in. If you open another drawing, you will not find that block.

Multileader and Blocks

In Lesson 19, you learned about multileaders and how easy and helpful they are to use. (*I saved this command for this lesson because it works best with Blocks attached to the leader.*) In this lesson, you will learn another user option within the Multileader Style Manager that allows you to **attach a pre-designed block** to the landing end of the leader. To accomplish this, you must first create the style and then you may use it.

Here are a few examples of multileaders with pre-designed blocks attached:

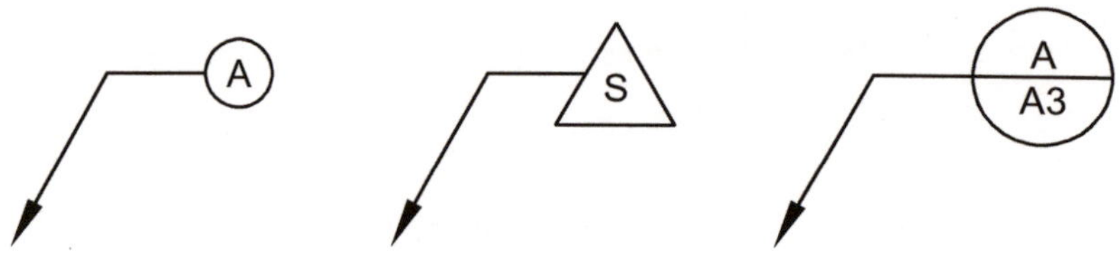

Part 1. Create a new Multileader Style

1. Select the **Multileader Style Manager** command using one of the following:

Ribbon = Annotate Tab / Leaders Panel / ↘
or
Keyboard = *mleaderstyle <Enter>*

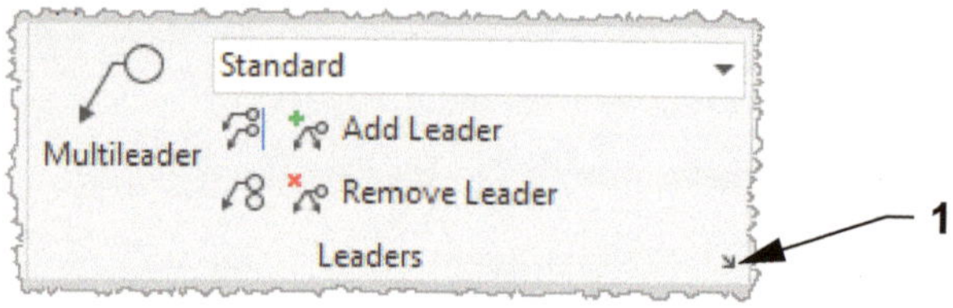

2. Select the **New...** button.

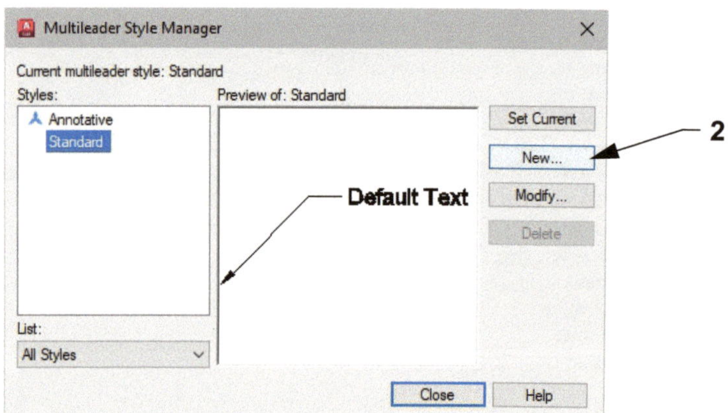

3. Enter **New style name**
4. Select a style to **Start with:**
5. Select the **Annotative** box.
6. Select the **Continue** button.

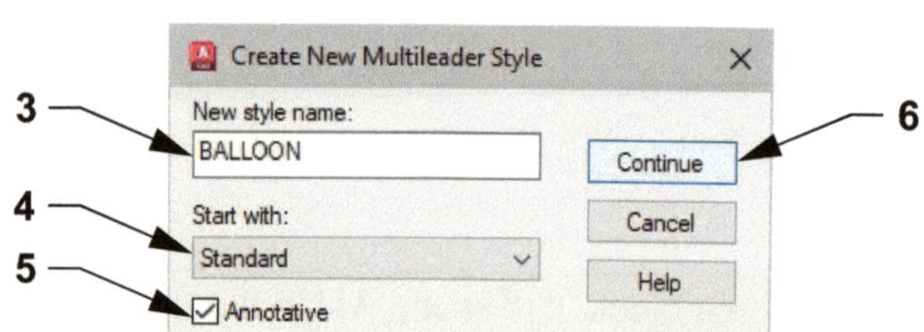

Continued on the next page...

7. Select the **Content** Tab.

8. **Multileader type:** Select **Block** from drop-down menu.

9. **Source block:** Select **Circle** from the drop-down menu.
   Note: You have many choices here. These are AutoCAD pre-designed blocks with attributes.

10. **Attachment:** Select **Center Extents**.
    Note: This selection works best with Circle, but you will be given different choices depending on which source block you select.

11. **Color:** Select **ByLayer**.
    ByLayer works best as it acquires the color of the current layer setting.

12. **Scale:** Select **1.000**

13. Select **OK** button.

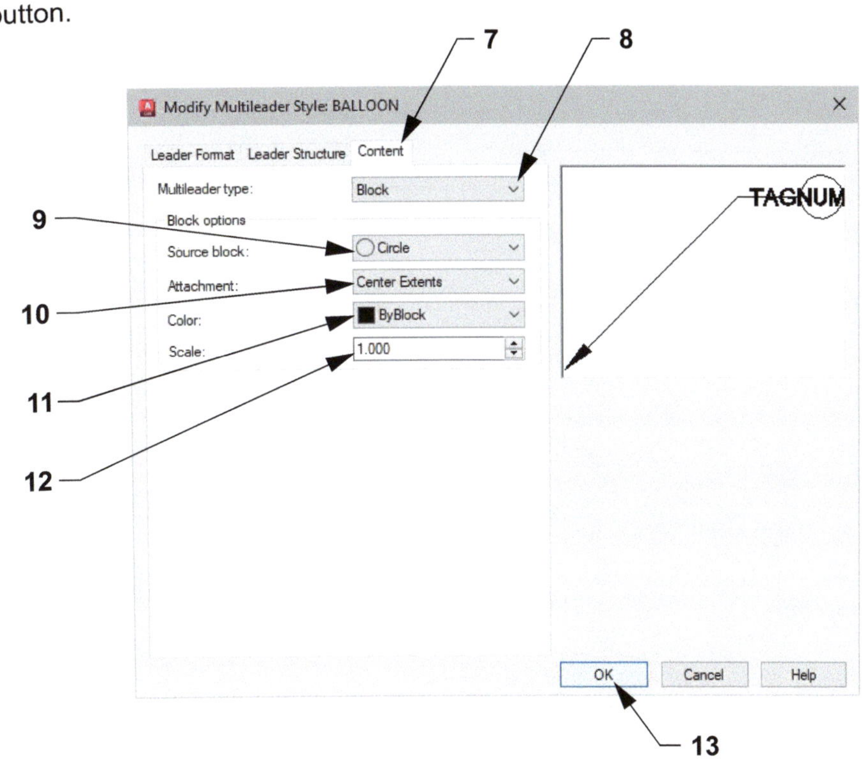

Your new multileader style should be displayed in the Styles list.

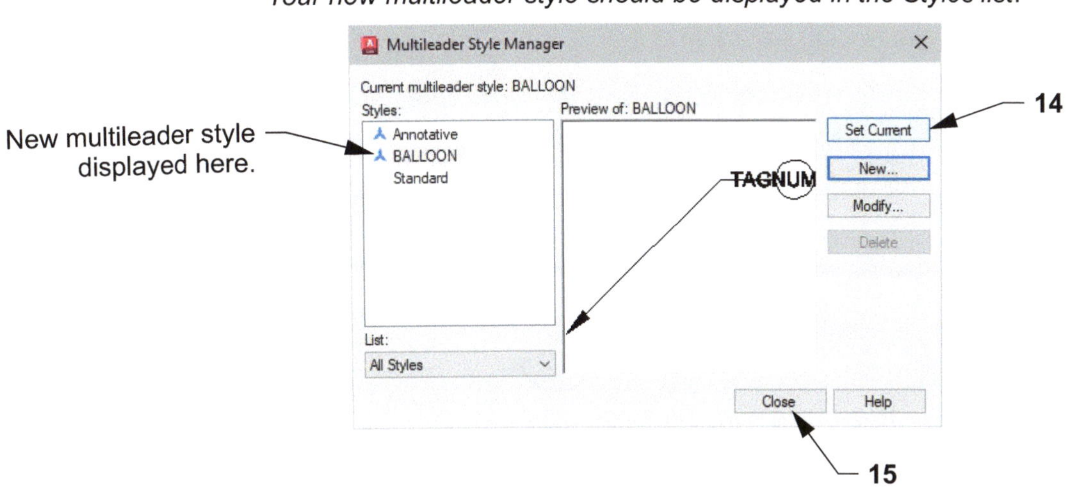

14. Select **Set Current**.

15. Select the **Close** button.

Part 2. Using the Multileader with a Block Style

1. Select the **Annotate Tab / Leaders Panel**.
2. Select the **Style** from the Multileader drop-down list.

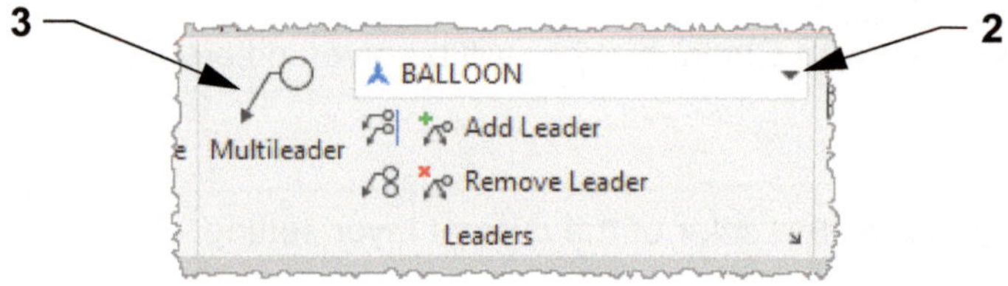

3. Select the **Multileader** tool.
4. The **Select Annotation Scale** box may appear. Select **OK** for now.

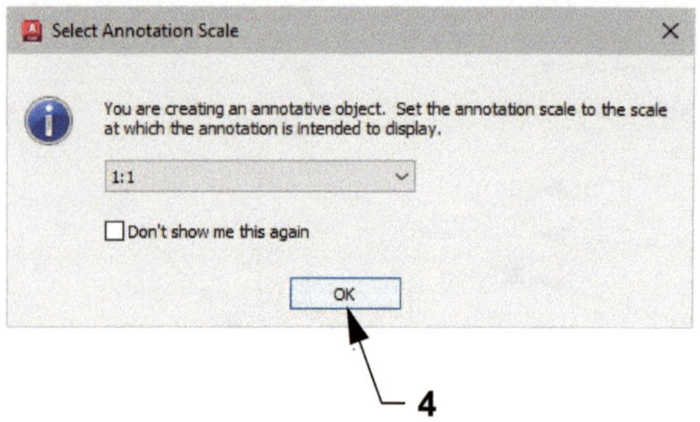

The following will appear on the Command Line:

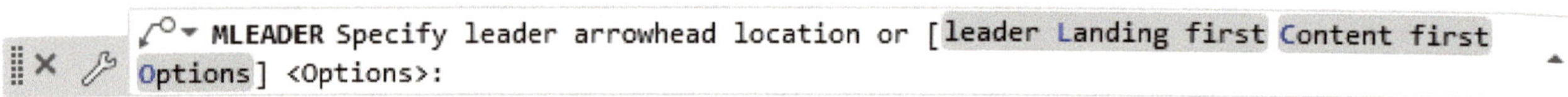

5. Place the desired location of the arrowhead (left click).

The following will appear on the Command Line:

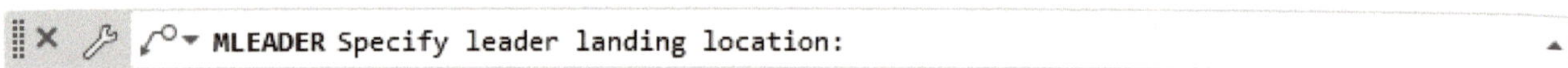

6. Place the desired location of the landing (left click).

The next step is where the pre-assigned **Attributes** activate.
Refer to the *Advanced AutoCAD® Exercise Workbook* for Attributes.

7. The Edit Attributes dialog box will appear. Type the **number** or **letter** of the multileader tag and then select **OK**.

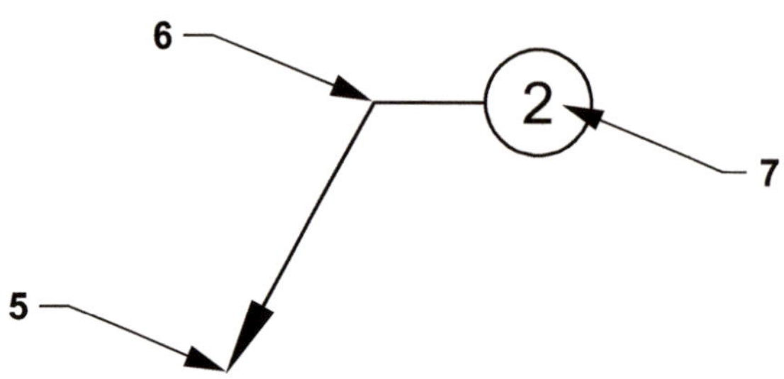

Collect Multileader

In Lesson 19, you learned how to **Add**, **Remove**, and **Align** multileaders. Now you will learn how to use the **Collect Multileader** tool.

If you have multiple leaders pointing to the same location or object, you may wish to **Collect** them into one Leader.

Example:

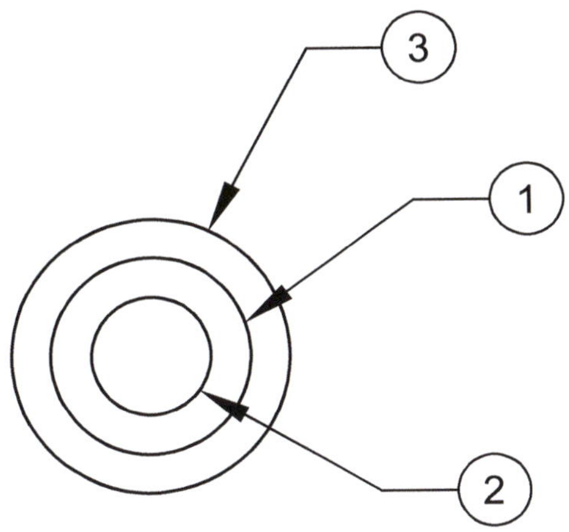

Before Collect

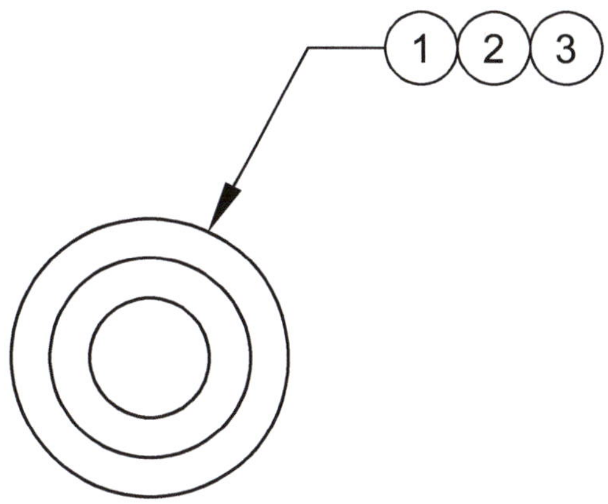

After Collect

How to use Collect Multileader

1. Select the **Collect Mutileader** tool.

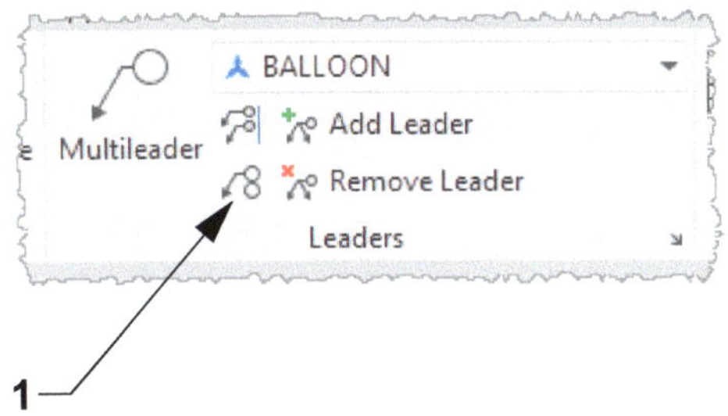

2. Select the multileaders that you want to combine and then press ***<Enter>***.

   Note: Select them one at a time in the order you want them to display:

   1, 2, 3, A, B, C, etc.

3. Place the combined leader location. (**Ortho Mode** should be **off**.)

Exercise 29A

Step 1.

1. Start a New file using **My Decimal Setup.dwt**.
2. Select the **Model** Tab.
3. Draw the objects shown below approximately as shown. Use Layer Symbol.
4. Do not dimension.
5. Create a block of each.
   A. **Important:** Don't forget to select a **basepoint** on each object.
   B. Select **"Delete"** so they disappear as they are made.
   C. **Do not include the name** when selecting the objects for each block.
6. **Save** as **Ex-29A**

5B

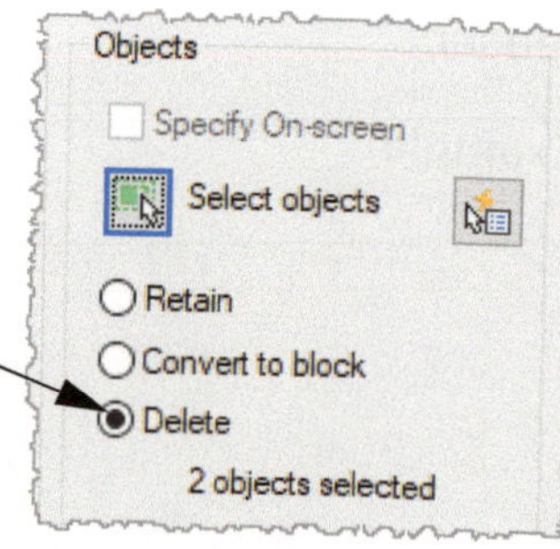

HUB

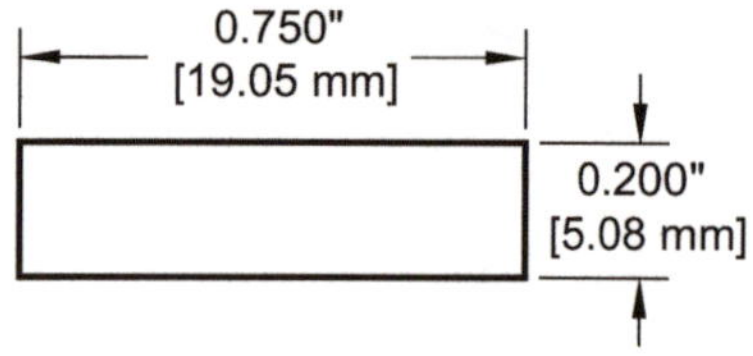

MULTILAYER SWITCH

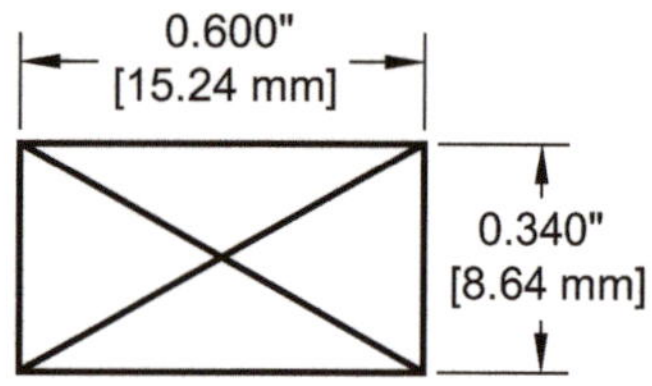

ROUTER

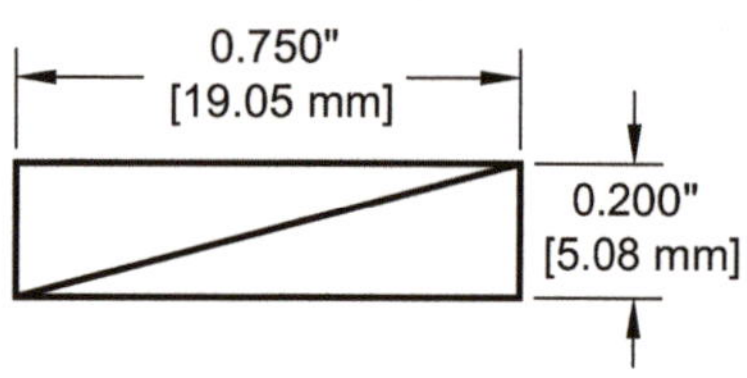

SERVER

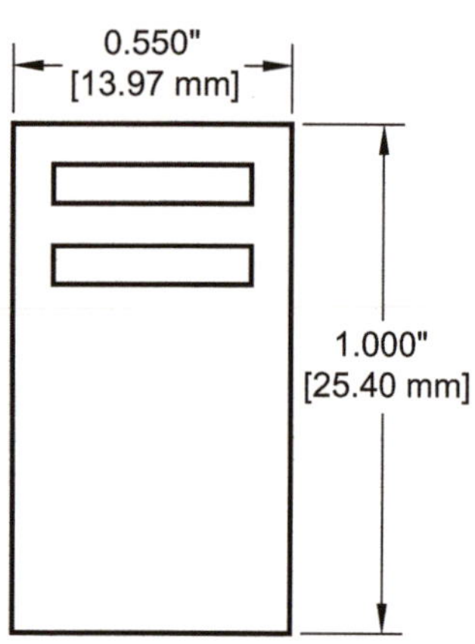

IBM COMPUTER

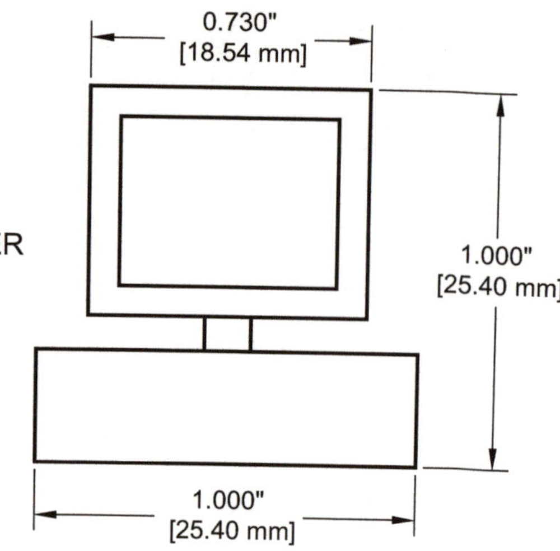

PRINTER

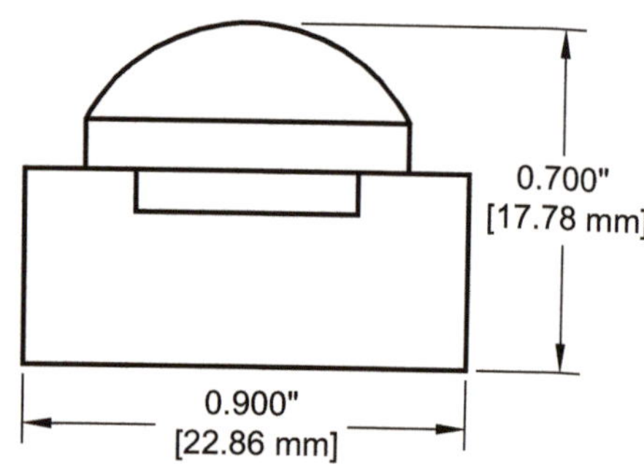

MAC SYSTEM

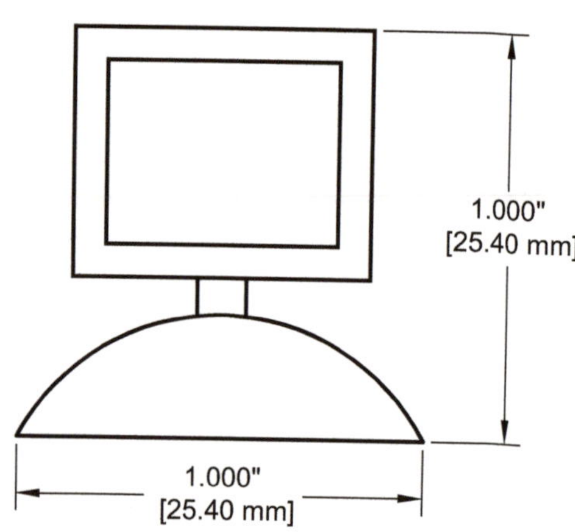

Continued on the next page...

Exercise 29A...continued

Step 2.

1. Open **Ex-29A** (if not already open).
2. Select the **A Size** Tab and confirm the **Model** Tab is displayed.
3. Draw the Network Diagram approximately as shown within Model Space.
4. Insert the blocks you previously created in **Step 1**.
5. Add the labels. Use Text Style TEXT-CLASSIC. Text height **0.125" [3.2 mm]**.
6. Edit the Title Block: SCALE: NONE
7. **Save** the drawing as **Ex-29A** again.
8. **Plot** using plot page setup **Plot Setup A**

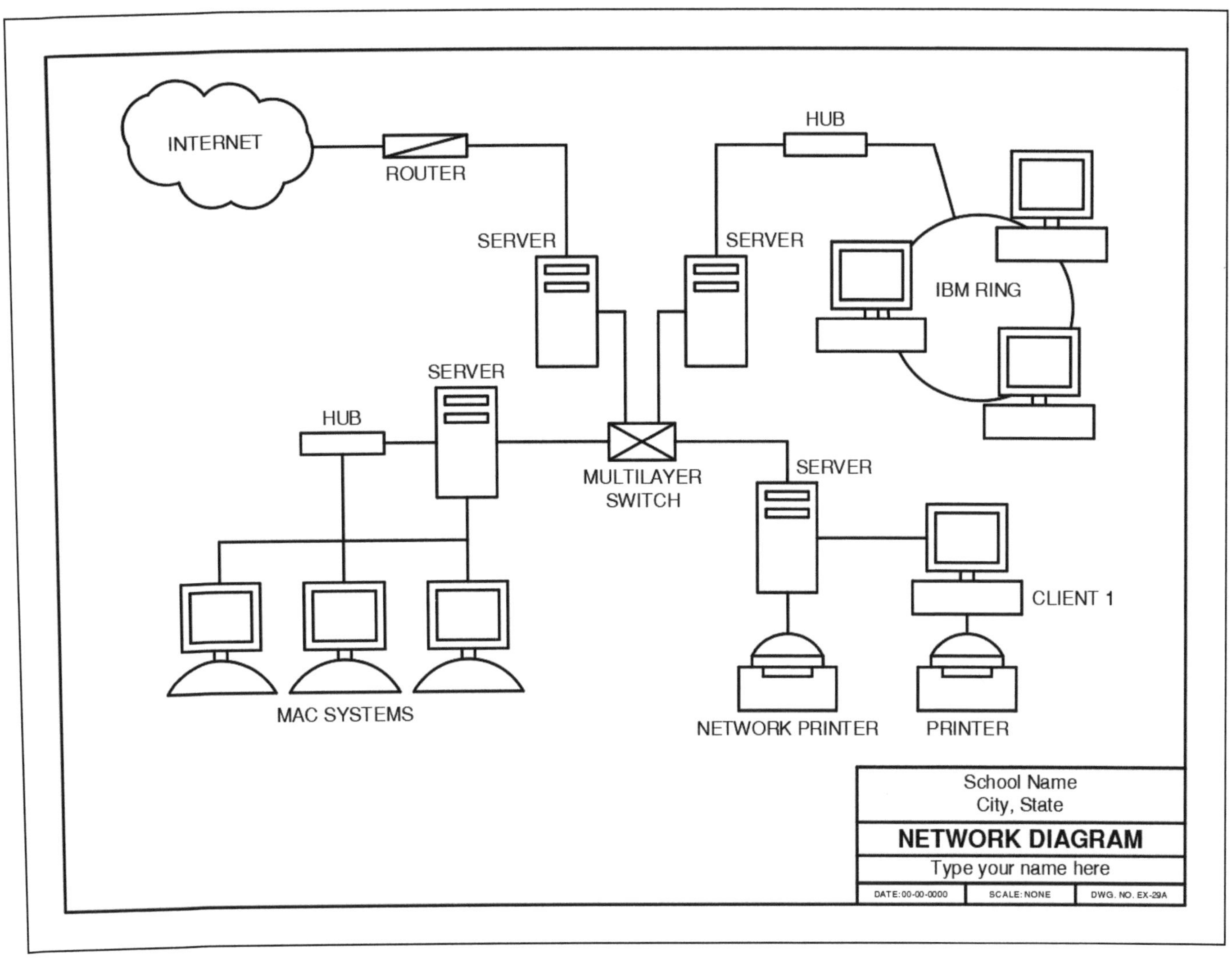

Continued on the next page...

Exercise 29B

Step 1.

1. Start a New file using **My Feet-Inches Setup.dwt**.
2. Select the **Model** Tab.
3. Draw the objects shown below approximately as shown. Use layers specified.
4. Do not dimension.
5. Create a block of each.
   A. **Important:** Don't forget to select a **basepoint** on each object.
   B. Select **"Delete"** so they disappear as they are made.
   C. Use the numbers for name. **Do not include the number** when selecting the objects for the block.
6. **Save** as **Ex-29B**

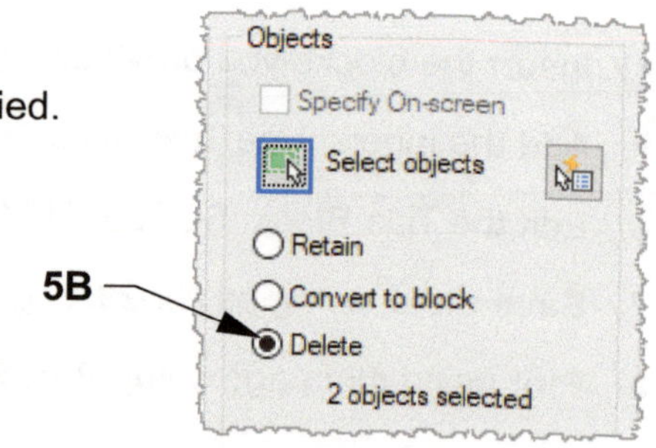

Note: Your blocks will appear much smaller and thinner than the blocks shown below. These have been enlarged for clarity.

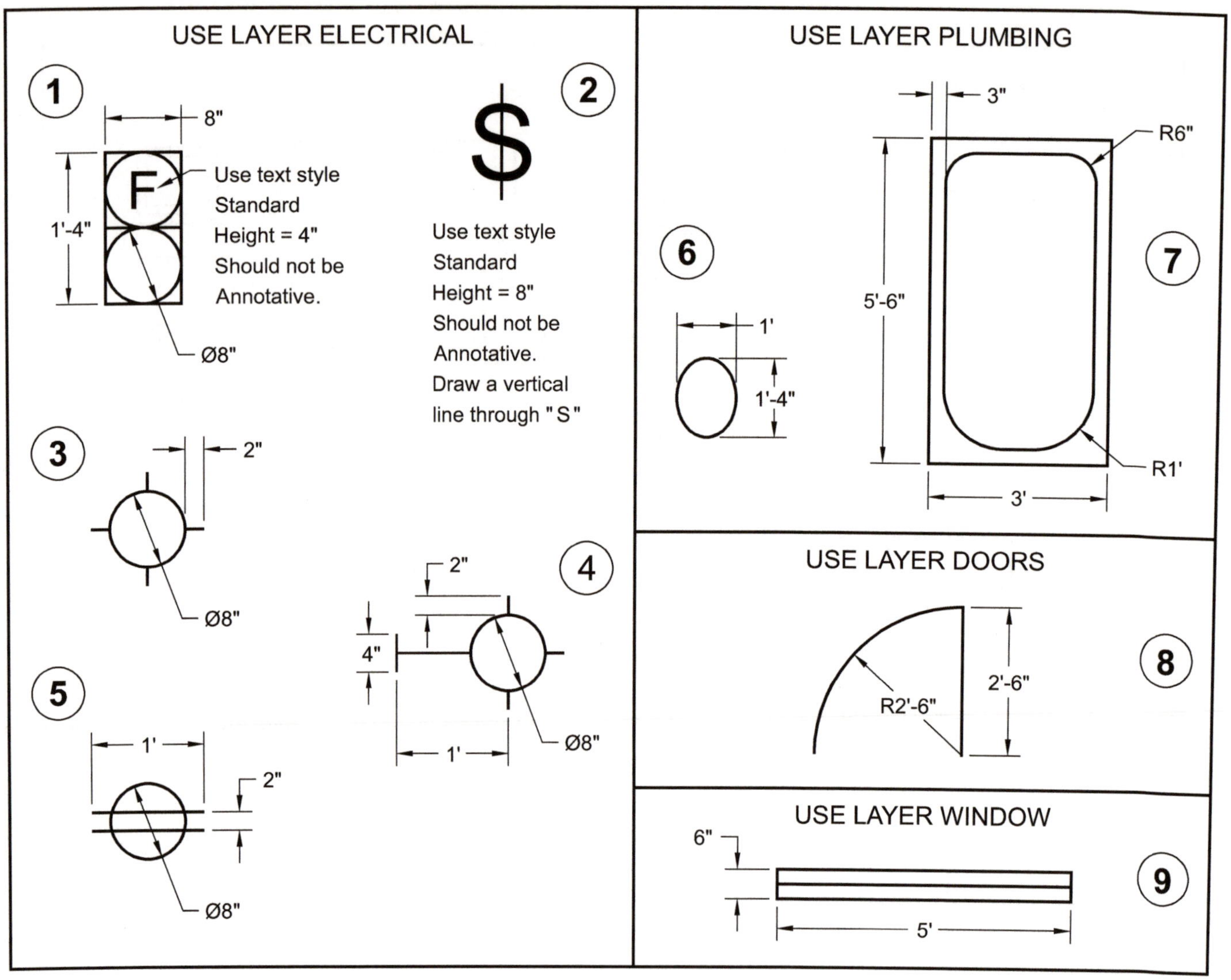

Continued on the next page...

Exercise 29B...continued

Step 2.

1. Open **Ex-29B** (if not already open).
2. Select the **A Size** Tab.
3. Draw the floor plan approximately as shown below inside the viewport.
   (**Note:** The viewport scale should already be adjusted to **1/4" = 1'**.)
   A. The walls are **6"** wide.
   B. The space behind the door is **4"** and counter depth is **24"**.
4. Insert the (8) DOOR Blocks on Layer Doors.
5. Insert the (9) WINDOW Blocks on Layer Windows.
6. Dimension as shown using Dimension Style DIM-ARCH. (Your dimension text will appear different. I used text style TEXT-CLASSIC because it is easier for you to read.)
7. **Save** the drawing as **Ex-29B** again.

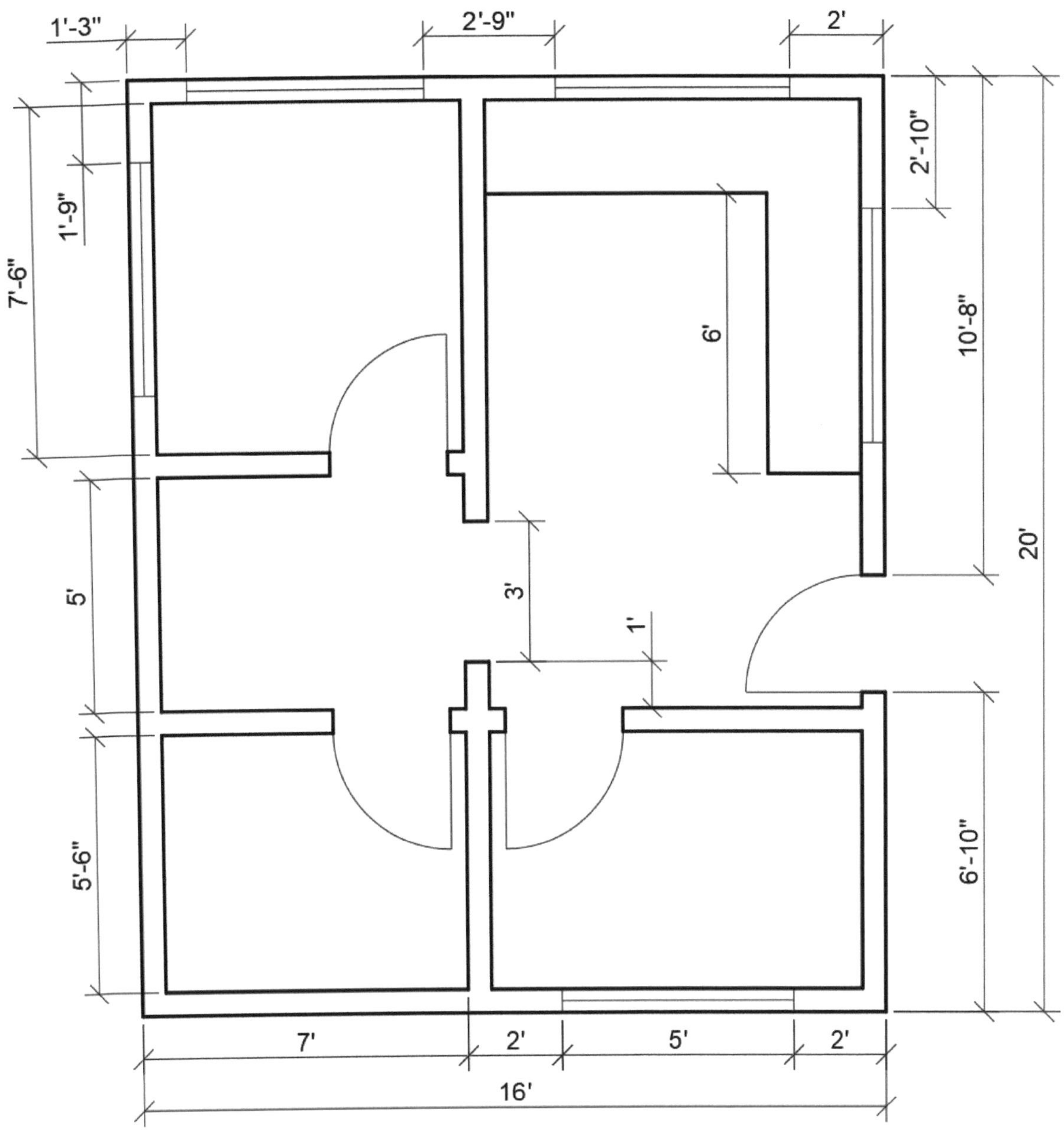

Continued on the next page...

Exercise 29B...continued

Step 3.

1. Open **Ex-29B** (if not already open).
2. Add the Electrical Blocks (1, 2, 3, 4 and 5) on Layer Electrical.
3. Add the Plumbing Blocks (6 and 7) on Layer Plumbing.
4. **Save** the drawing as **Ex-29B** again.

Note: The Dimension layer has been temporarily turned off so you can see the electrical and plumbing blocks easier.

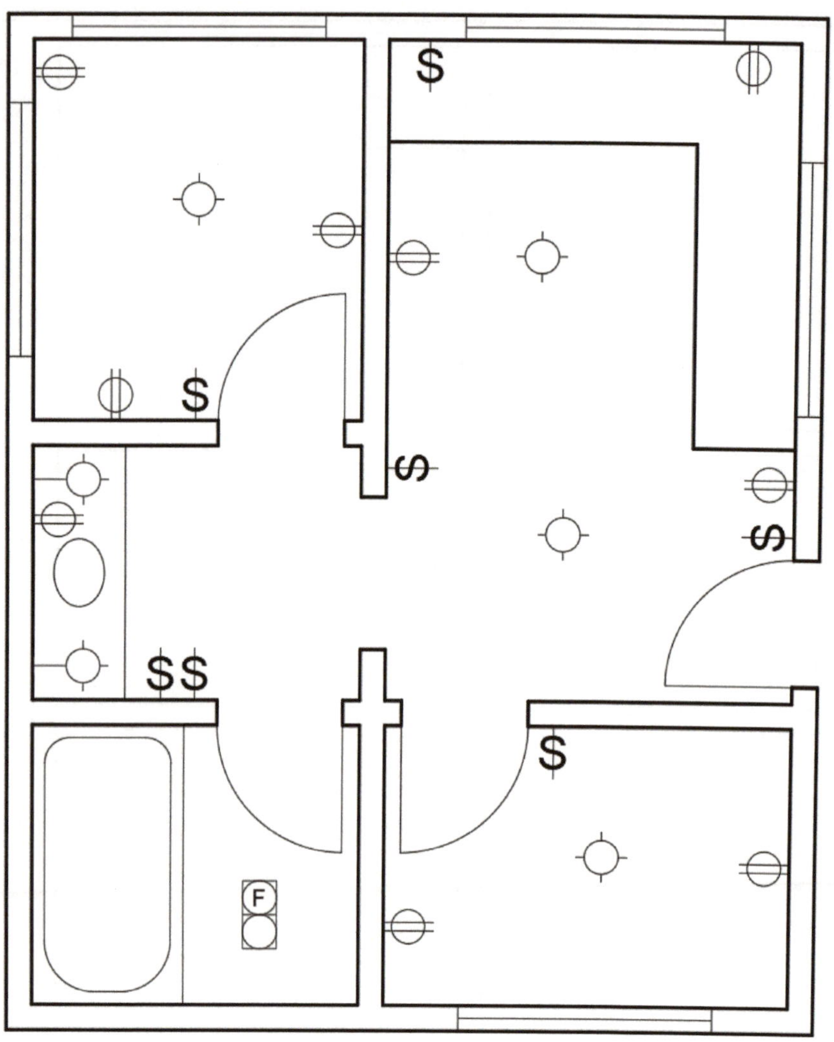

Continued on the next page...

Exercise 29B...continued

Step 4.

1. Open **Ex-29B** (if not already open).
2. Add the Wiring.
   A. Use Layer Wiring.
   B. Use Arc, (Start, End, Direction).
3. Notice that your dashed lines do not appear like the example below.
   A. Type ***LTS*** and then press ***<Enter>***. (This is the **Linetype Scale** setting.)
   B. Type ***0.25*** and then press ***<Enter>***. (This scales all linetypes).
4. **Save** the drawing as **Ex-29B** again.

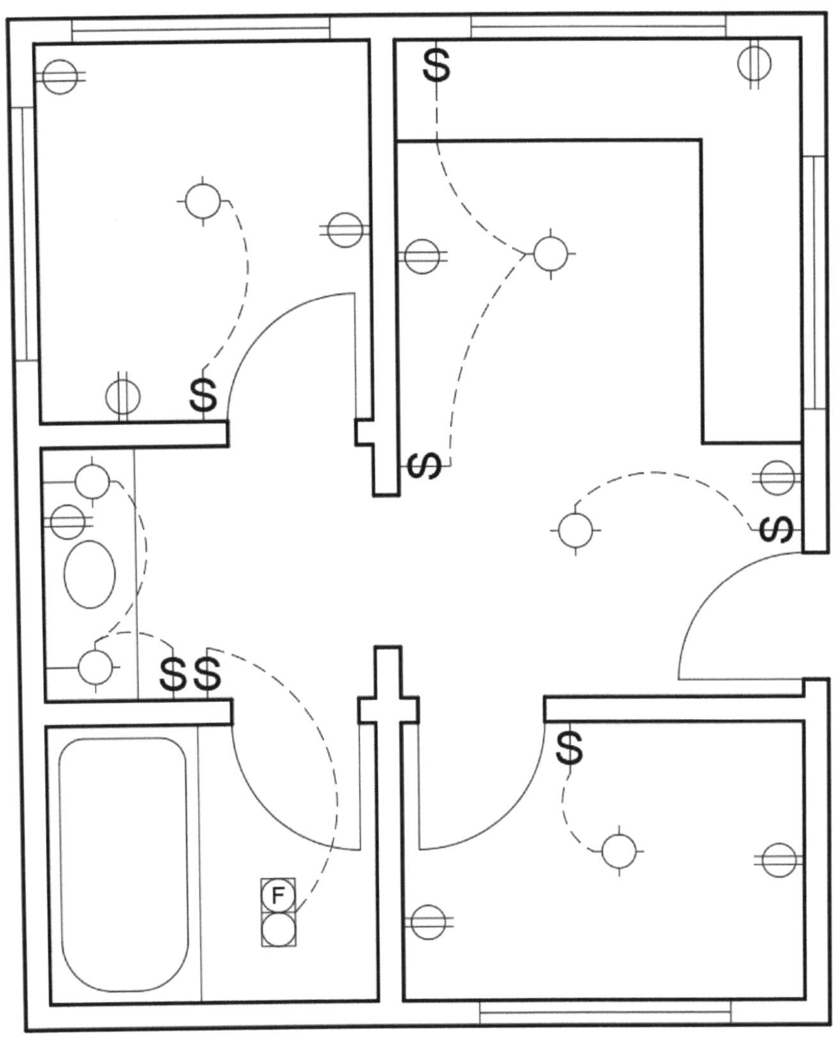

Continued on the next page...

Exercise 29B...continued

Step 5.

1. Open **Ex-29B** (if not already open).
2. Erase the existing viewport Frame. (The floor plan will disappear.)
3. Select Viewport as the current layer and create three new viewports approximately as shown.
4. Adjust the scale of each viewport.
5. Use **Pan** to find the area to view as shown below.
6. **Lock** each viewport.
7. Add the dimensions in the two viewports on the right by adding the current object scale. (Refer to pages 28-3 and 28-4.)
8. **Add** the viewport scale labels (**3/16"** height) in Paper Space (not Model Space).
9. Make viewport Frames **plottable**. (Refer to page 3-8.)
10. Edit the Title Block.
11. **Plot** using plot page setup **Plot Setup A**
12. **Save** as **Ex-29B**

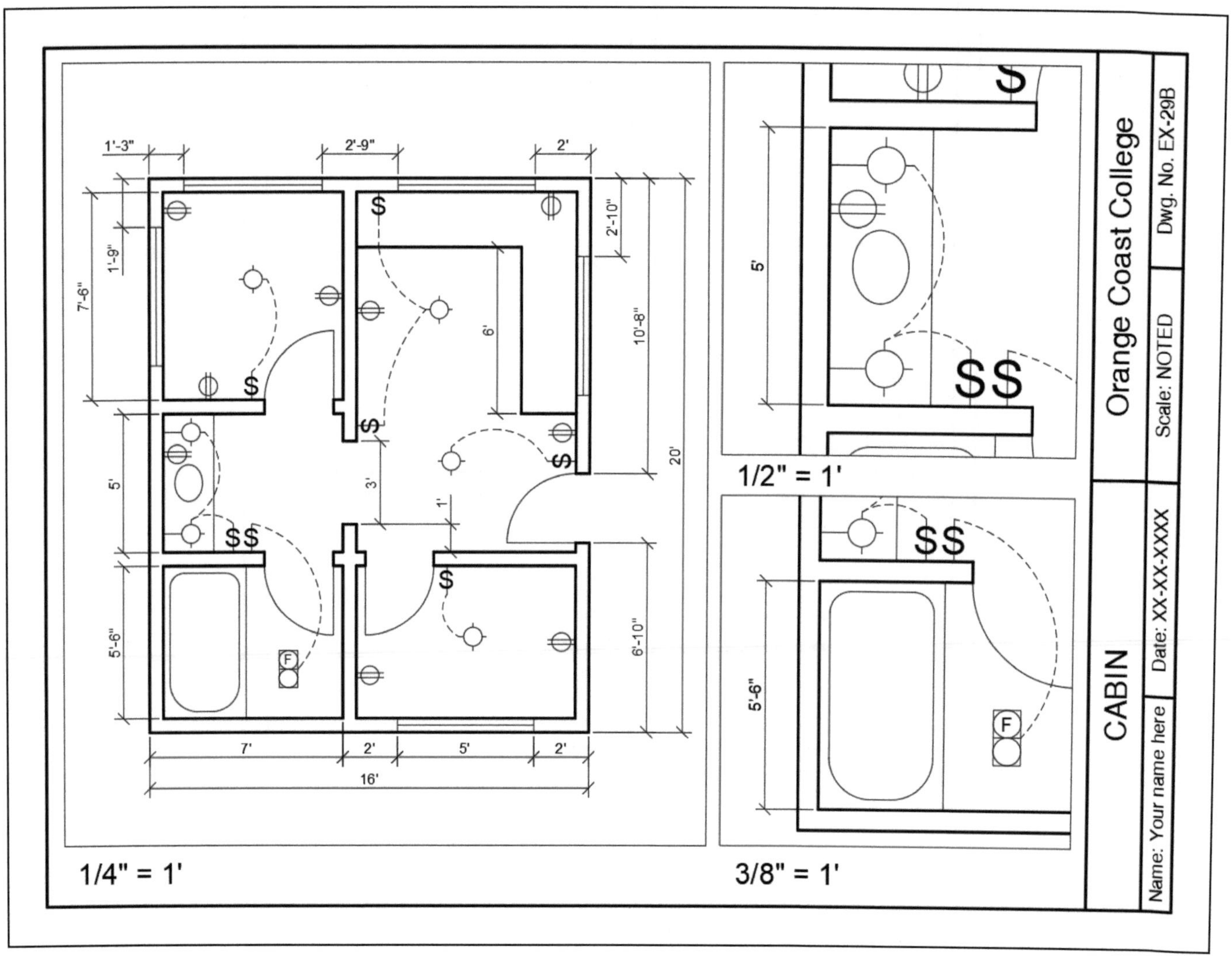

Exercise 29C-Inch

Step 1.

1. Start a **New** file using **My Decimal Setup.dwt**.
2. The viewport should already be scaled to 1:1 and locked.
3. Draw the objects shown below.
4. Draw the **Break** line using the Polyline command with the Spline option. (Refer to page 24-3.)
5. Draw the Hatch using Pattern ANSI31 and Layer Hatch.
6. Draw the Threads.
   A. Use Layer Threads.
   B. Thread lines are **0.0625"** distance apart.
7. Dimension as shown using Layer Dimension and dimension style = DIM-DECIMAL.
8. **Save** the drawing as **Ex-29C-I - STEP 1**.

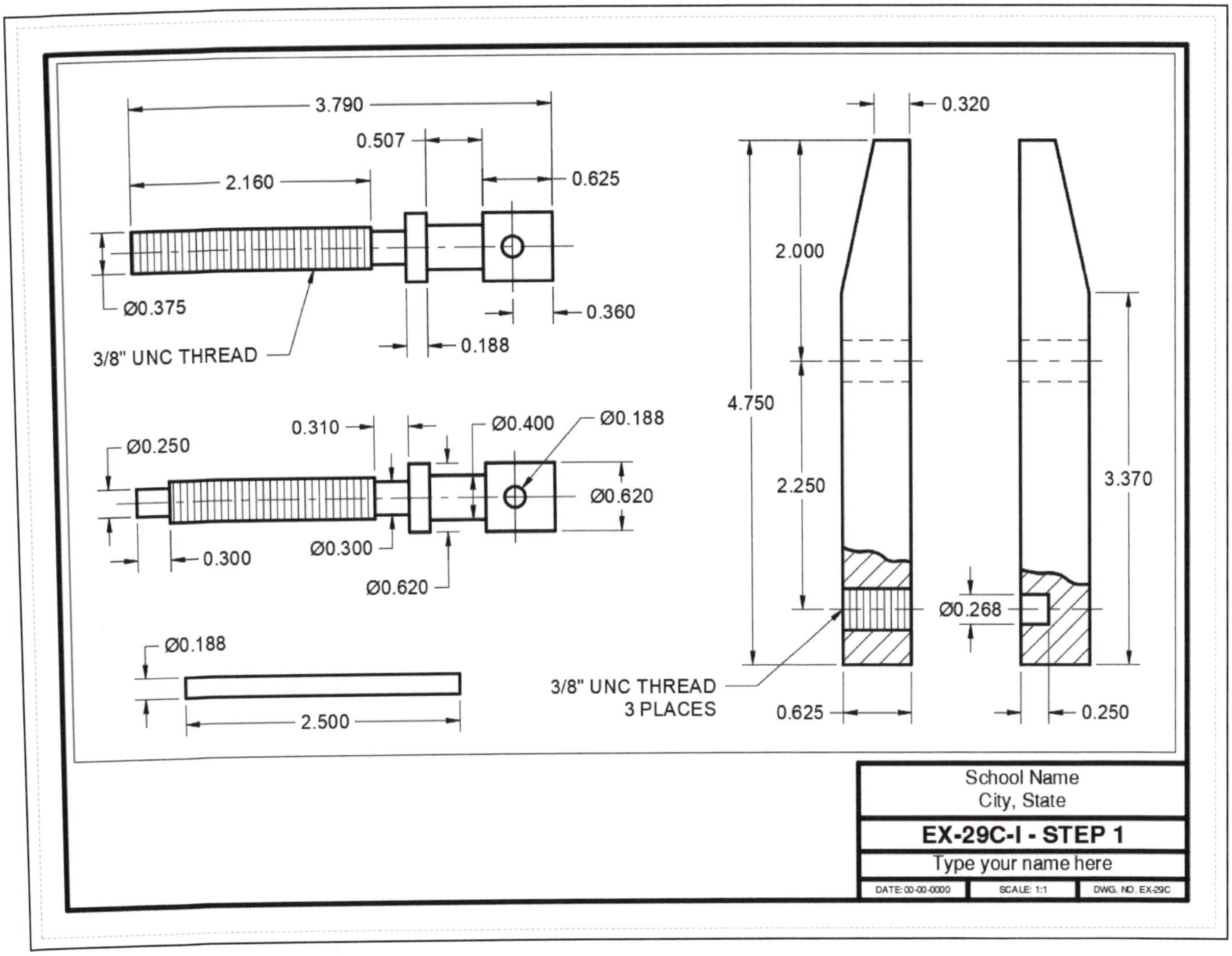

Continued on the next page...

Exercise 29C-Inch...continued

Step 2.

1. Open **Ex-29C-I - STEP 1** (if not already open).
2. **Freeze** the Dimension layer.
3. **Assemble** the parts as shown below.
4. Use **Move**, **Mirror**, **Rotate**, **Trim**, **Erase**, and any other command that you need.
5. The distance between the Jaws should be **1.000"**.
7. **Save** the drawing as **Ex-29C-I - STEP 2**.

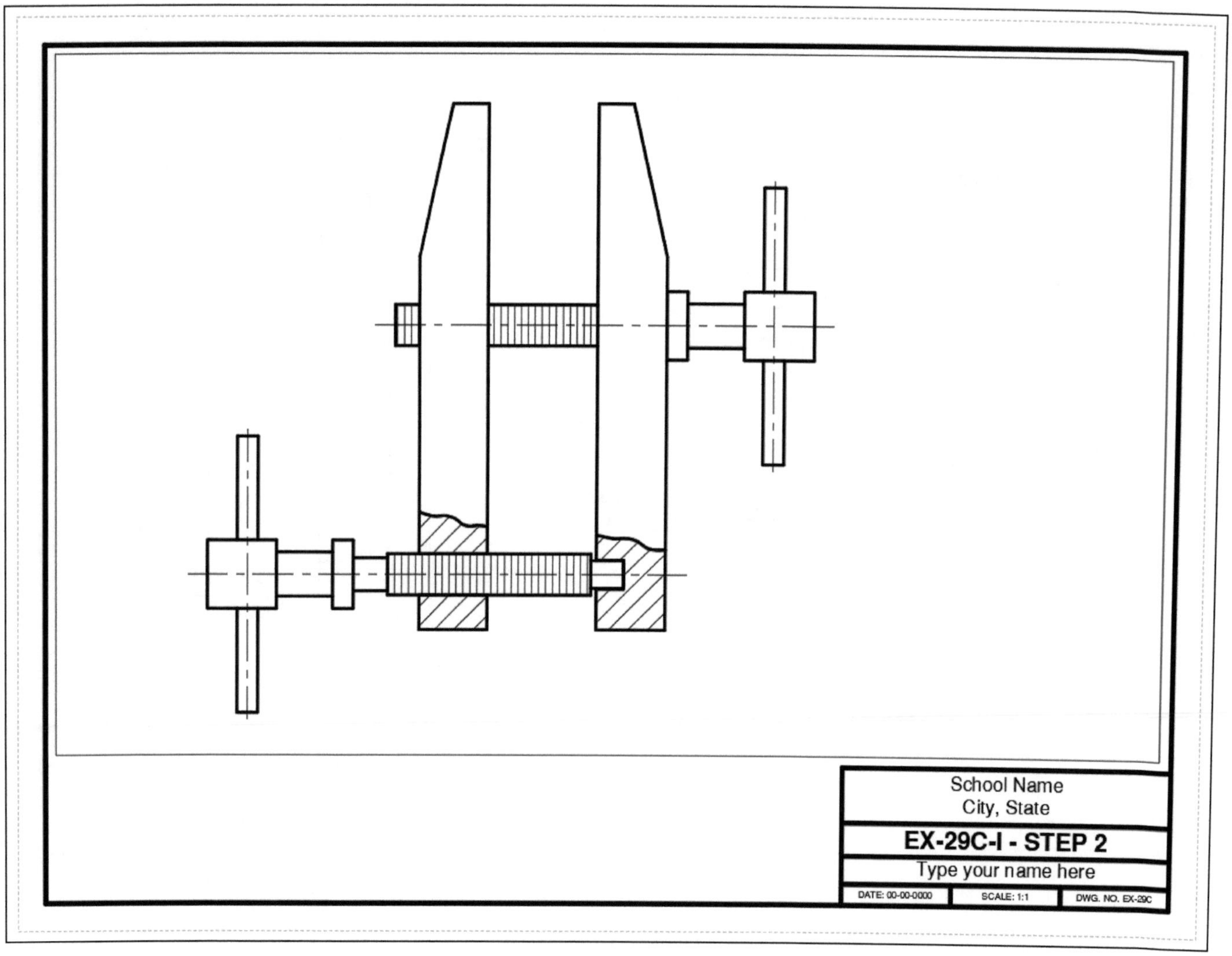

Continued on the next page...

Exercise 29C-Inch...continued

Step 3.

1. Open **Ex-29C-I - STEP 2** (if not already open).
2. Draw the list of parts in Paper Space above the Title Block as shown.
3. Use Multiline Text to enter text. Justify = Middle Center. Text height = **0.125"** and **0.062"**.
4. Add Ballooned Leaders as shown in Model Space. Scale **1.5**. Use Collect and Align.
5. Save as **CLAMP ASSEMBLY** and plot using plot Page Setup **Plot Setup A**

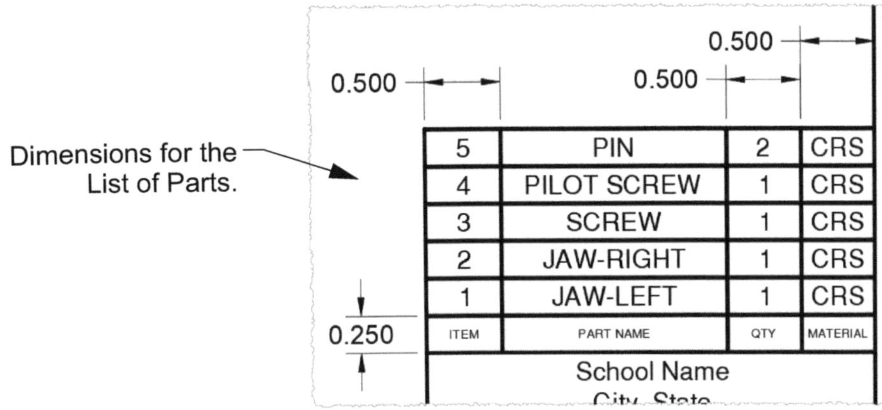

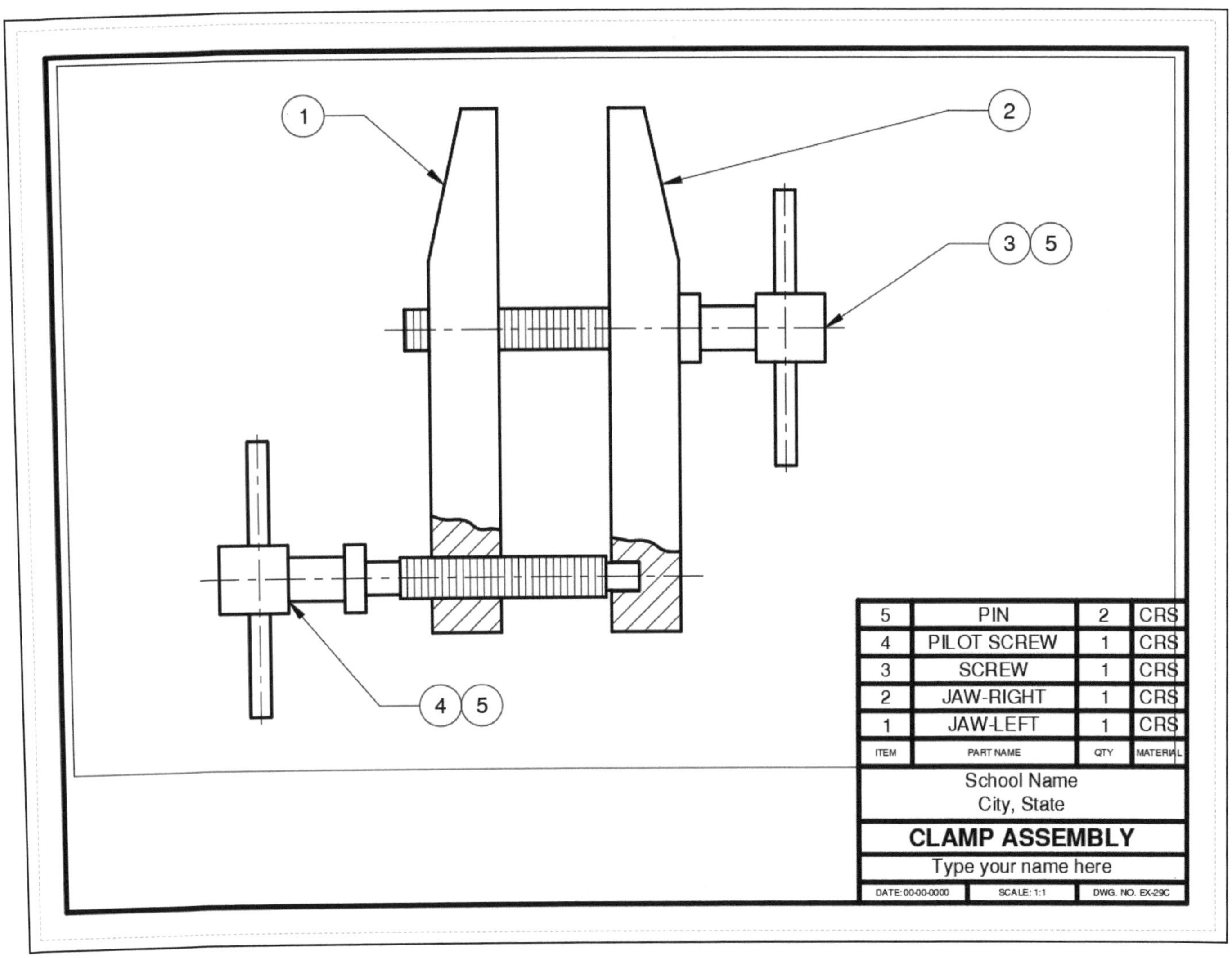

Exercise 29C-Metric

Step 1.

1. Start a **New** file using **My Decimal Setup.dwt**.
2. The viewport should already be scaled to 1:1 and locked.
3. Draw the objects shown below.
4. Draw the **Break** line using the Polyline command with the Spline option. (Refer to page 24-3.)
5. Draw the Hatch using Pattern ANSI31 and Layer Hatch.
6. Draw the Threads.
   A. Use Layer Threads.
   B. Thread lines are **1.5 mm** distance apart.
7. Dimension as shown using Layer Dimension and dimension style = DIM-DECIMAL.
8. **Save** the drawing as **Ex-29C-M - STEP 1**.

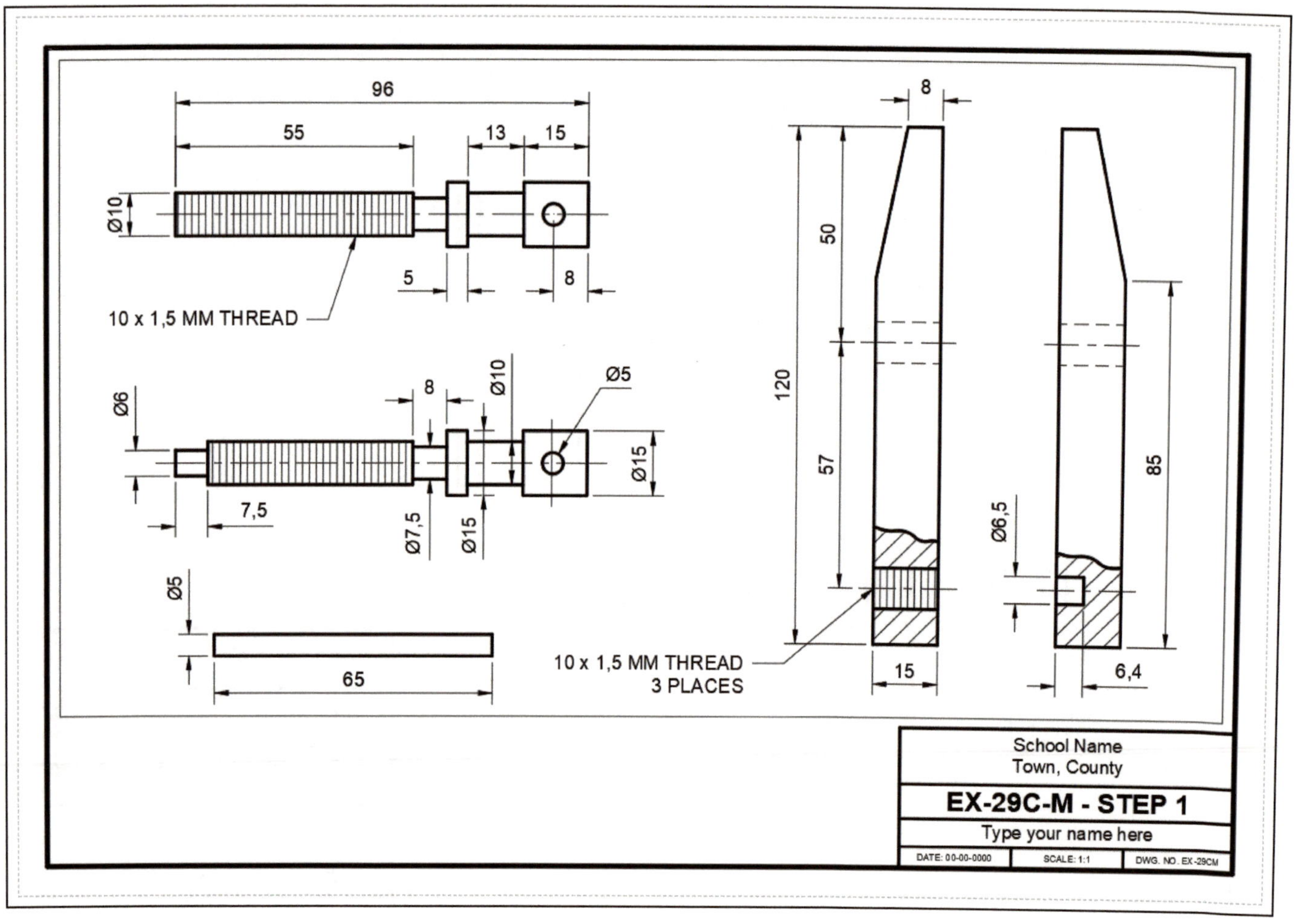

Continued on the next page...

Exercise 29C-Metric...continued

Step 2.

1. Open **Ex-29C-M - STEP 1** (if not already open).
2. **Freeze** the Dimension layer.
3. **Assemble** the parts as shown below.
4. Use **Move**, **Mirror**, **Rotate**, **Trim**, **Erase**, and any other command that you need.
5. The distance between the Jaws should be **25 mm**.
7. **Save** the drawing as **Ex-29C-M - STEP 2**.

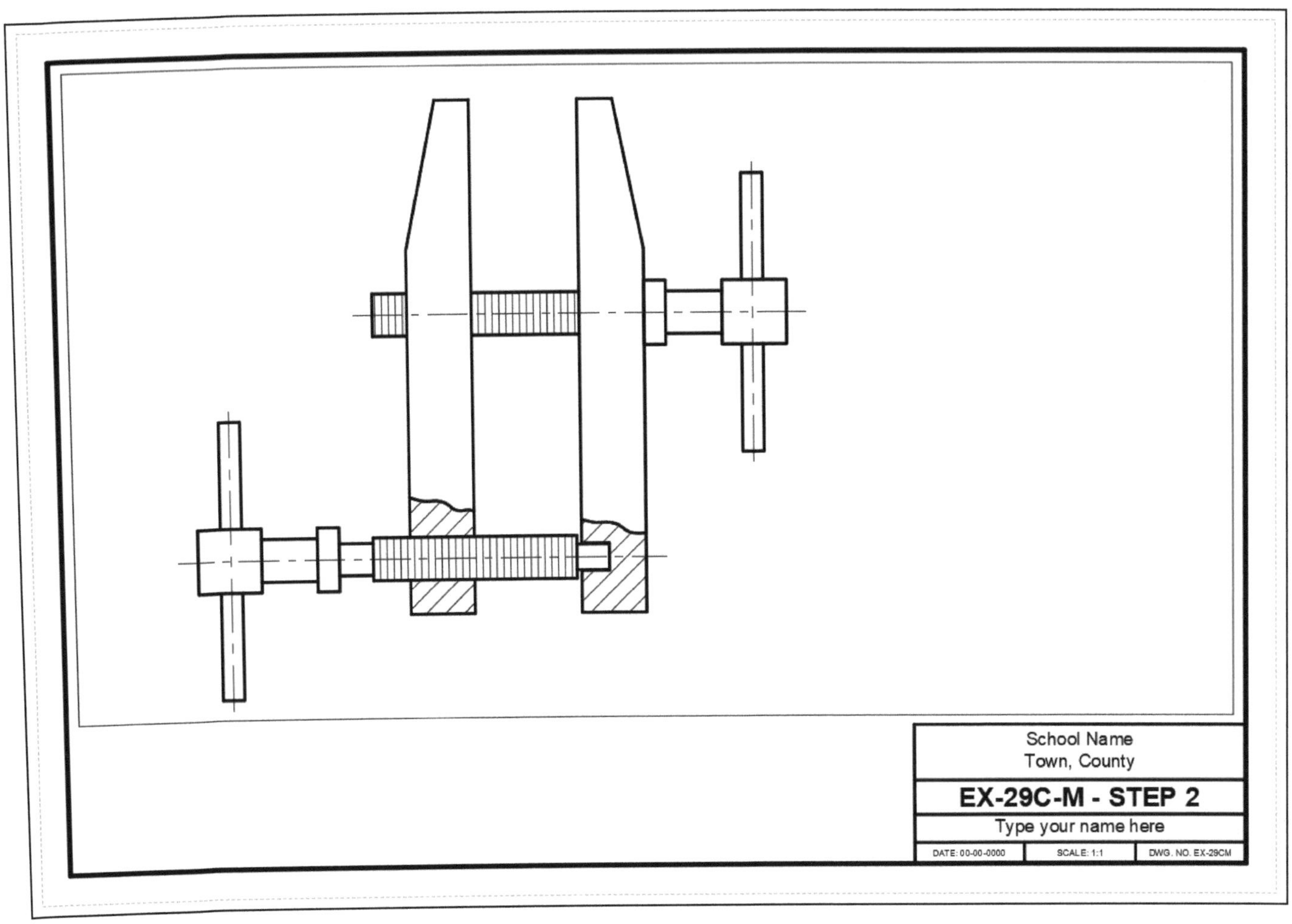

Continued on the next page...

Exercise 29C-Metric...continued

Step 3.

1. Open **Ex-29C-M - STEP 2** (if not already open).
2. Draw the list of parts in Paper Space above the Title Block as shown.
3. Use Multiline Text to enter text. Justify = Middle Center. Text height = **3.2 mm** and **1.5 mm**.
4. Add Ballooned Leaders as shown in Model Space. Scale **1.5**. Use Collect and Align.
5. Save as **CLAMP ASSEMBLY** and plot using plot Page Setup **Plot Setup A**

Dimensions for the List of Parts.

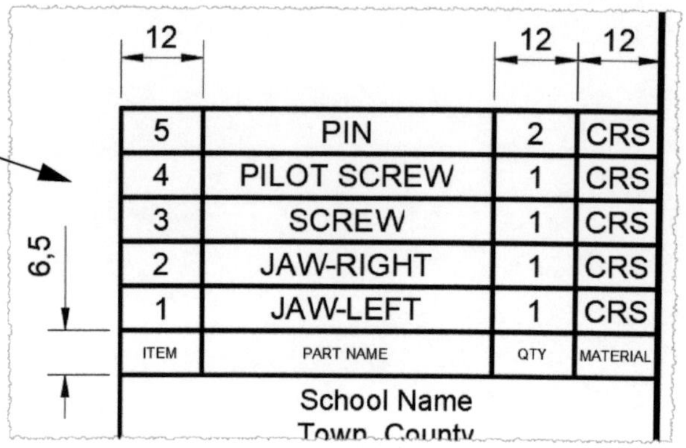

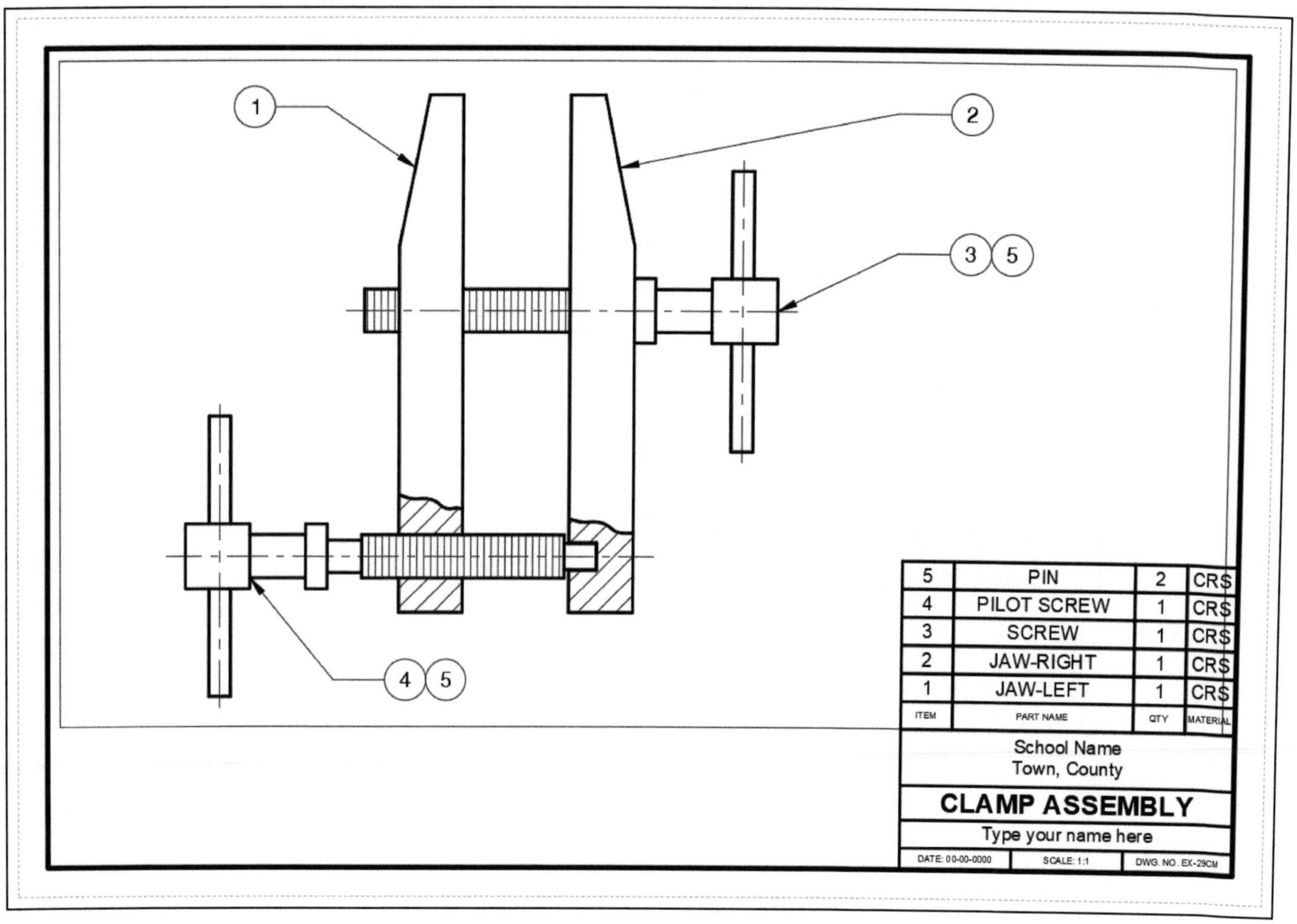

LESSON 30

LEARNING OBJECTIVES

After completing this lesson, you will be able to:

- Use six useful "Express Tools"
- Import a PDF file into AutoCAD as usable geometry
- Import a PDF file with SHX fonts
- Compare the differences between two versions of any drawing
- Store your designs in the Cloud for other users to view

Text – Arc Aligned

ArcText allows you to place text along the arc's curve. You may control the appearance of the text easily using the ArcAlignedText Workshop Dialog box shown below.

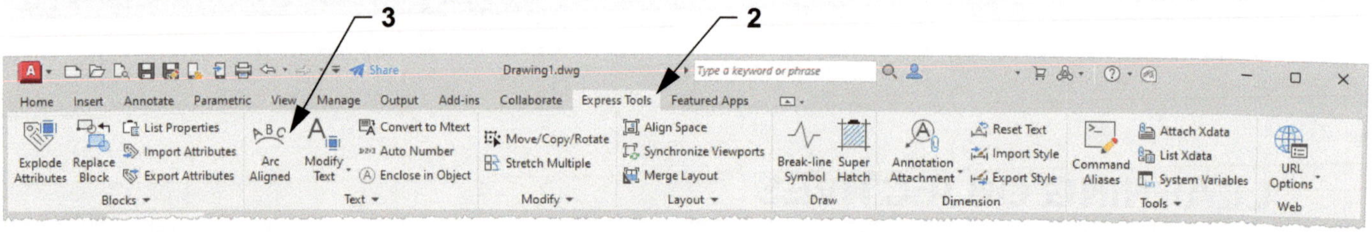

How to create ArcText

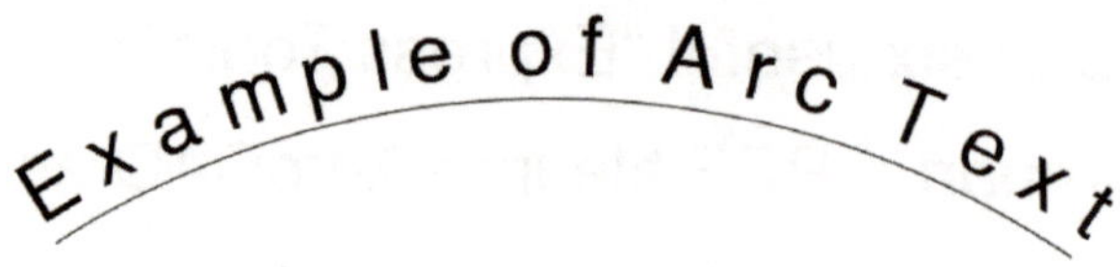

1. Draw an arc.
2. Select the **Express Tools** Tab.
3. Select the **Arc Aligned** tool.
4. Select the arc.

The **ArcAlignedText Workshop - Create** dialog box will appear.

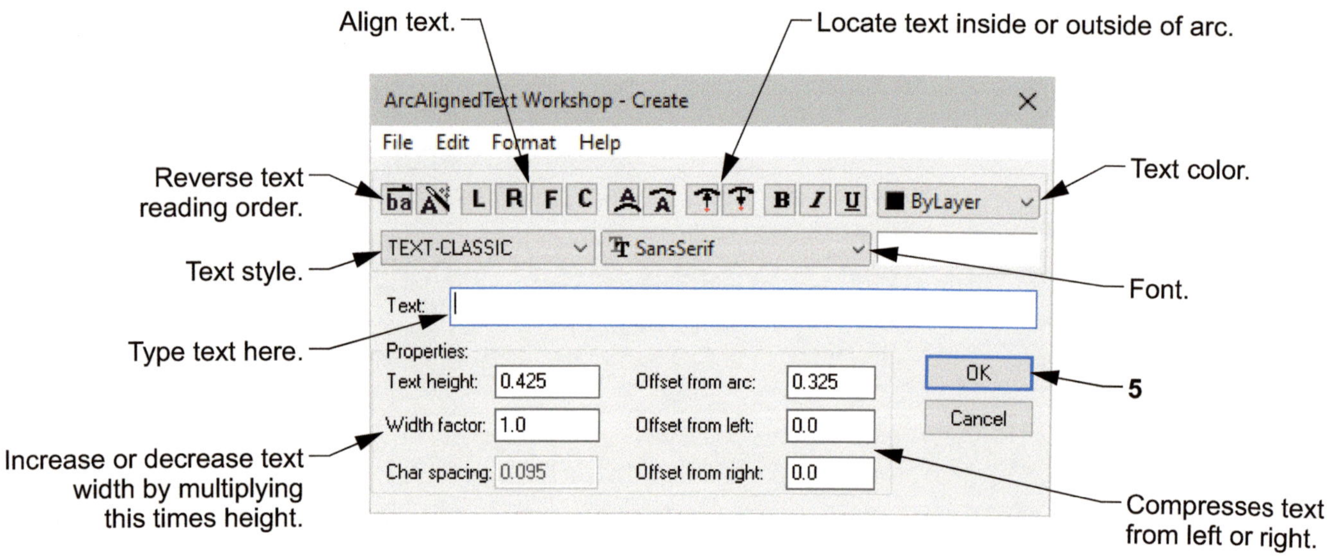

5. Select the options you want and then select the **OK** button.

How to edit the ArcText

1. Select the **Arc Aligned** tool.
2. Select the existing text. The **ArcAlignedText Workshop - Modify** dialog box will appear.
3. Make changes and select the **OK** button.

Text – Modify Text

Modify Text has five options, **Explode**, **Change Case**, **Rotate**, **Fit**, and **Justify**, to increase your manipulation of existing text.

1. Select the **Express Tools** Tab.
2. Select the **Modify Text** ▾ down arrow.
3. Select one of the options listed below.
4. Select the text to be modified.
5. Select ***<Enter>*** to stop.

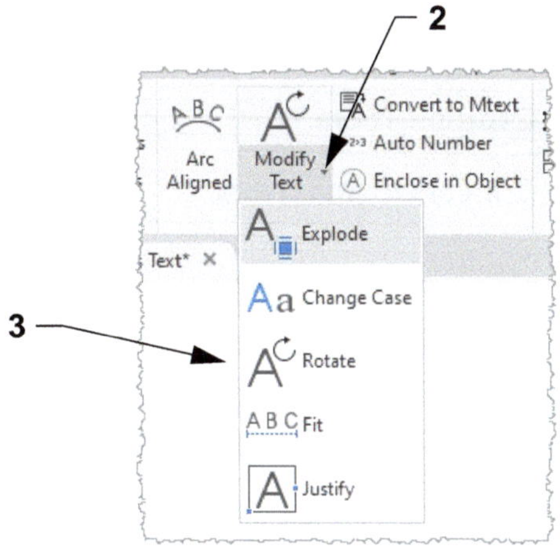

Explode: Text explodes into lines and arcs.

Change Case: Changes text case to one of the following:

Sentence case: The first letter of the sentence is uppercase. All others lowercase.

lowercase: All letters lowercase.

UPPERCASE: All letters uppercase.

Title: The first letter of each word uppercase.

tOGGLE cASE: Changes uppercase to lowercase and vice versa.

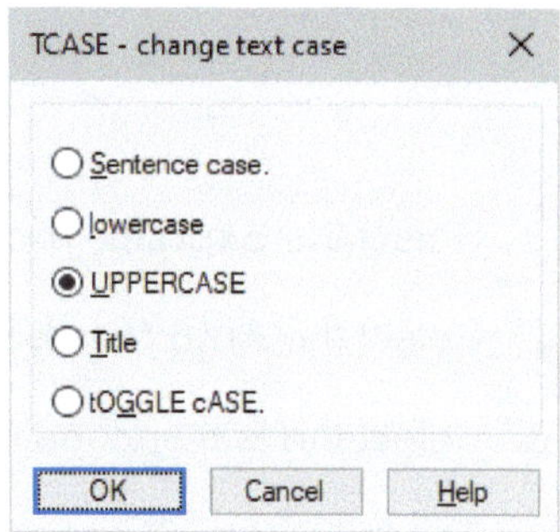

Rotate: Allows you to rotate the existing text to the "most readable" orientation or to a specific angle.

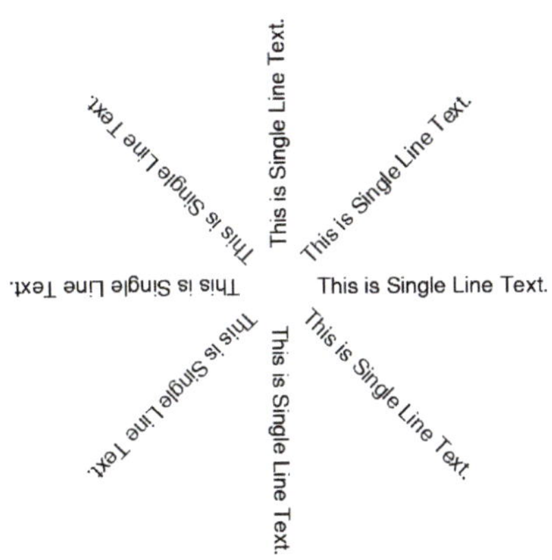

Before Rotate

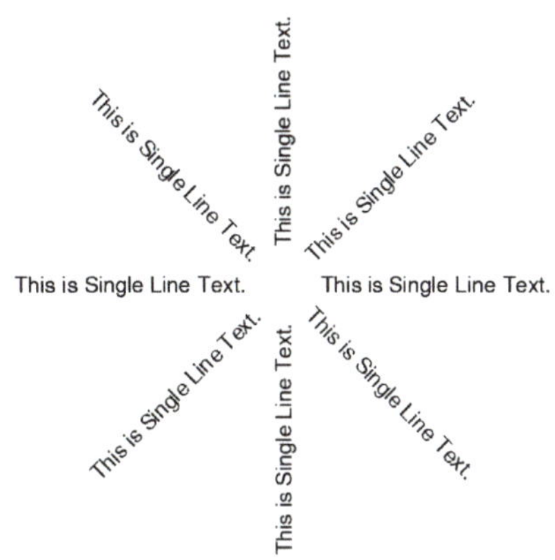

After Rotate

Fit: Allows you to compress or stretch text between two points.

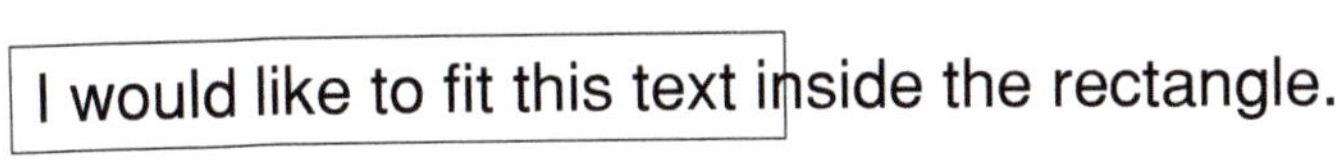

Before Fit

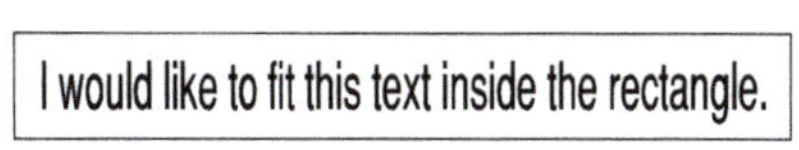

After Fit

Justify: Allows you to change the justification. You also can do this with **Properties**.

Text – Convert to Mtext and Auto Number

Convert to Mtext allows you to convert text created with Single Line text command to Multiline text. This is very helpful if you would like to modify the single line text because multiline text editor includes many additional editing options.

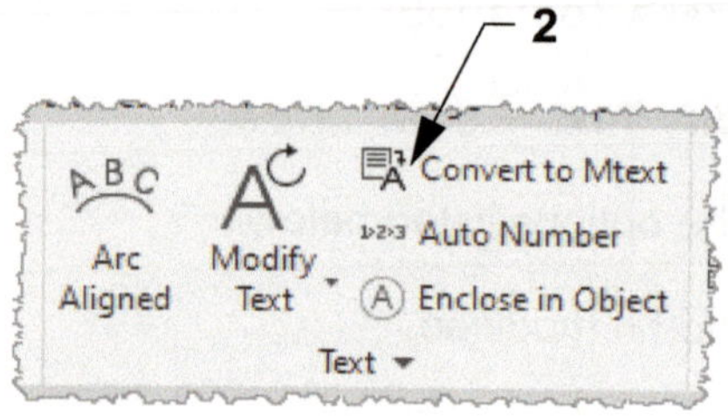

1. Select the **Express Tools** Tab.
2. Select the **Convert to Mtext** tool.
3. Select the text to be converted.
4. Press ***<Enter>***.

Auto Number allows you to add sequential numbers to existing text.

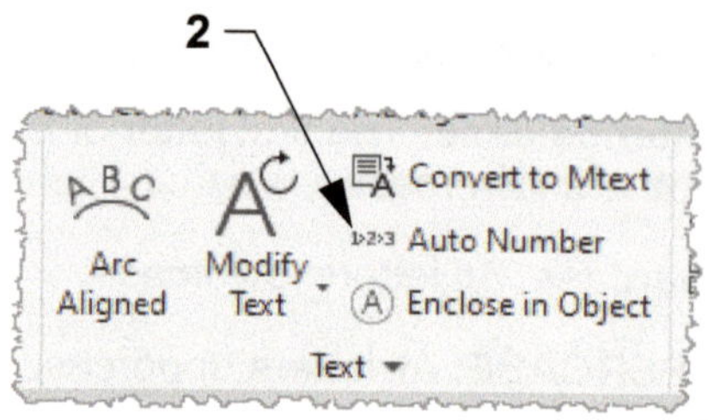

1. Select the **Express Tools** Tab.
2. Select the **Auto Number** tool.
3. Select the text and then press ***<Enter>***.

   The following will appear on the Command Line:

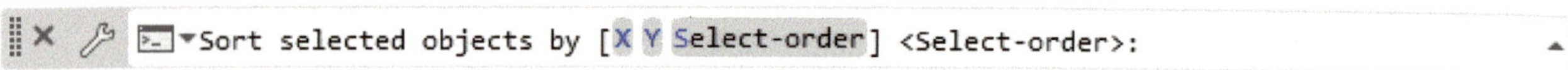

4. Press ***<Enter>***.

   The following will appear on the Command Line:

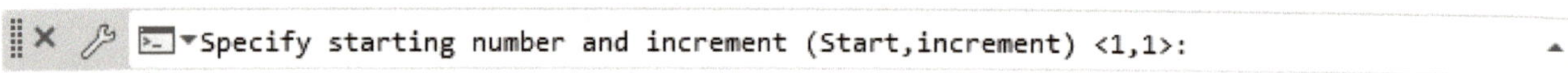

5. Press ***<Enter>***.

   The following will appear on the Command Line:

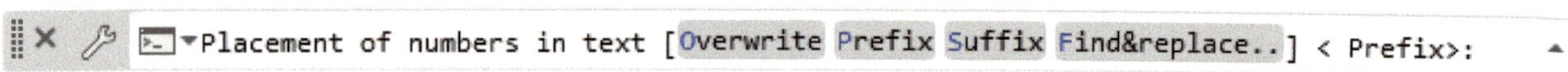

6. Press ***<Enter>***.

This is line one. This is line two. This is line three.	1 This is line one. 2 This is line two. 3 This is line three.
Before Auto Number	After Auto Number

Text – Enclose in Object

Enclose in Object allows you to place a circle, slot, or rectangle around text.

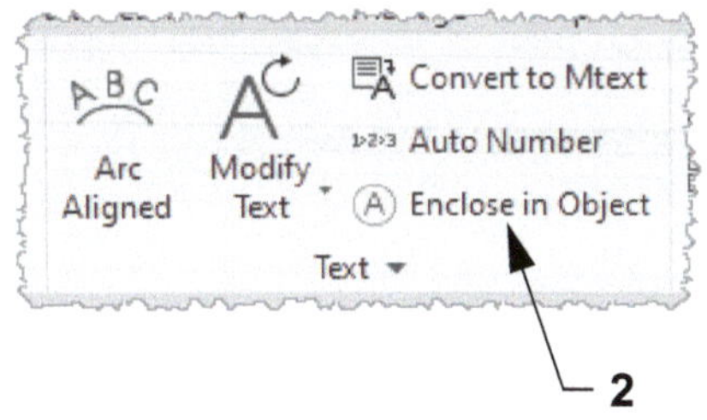

1. Select the **Express Tools** Tab.
2. Select the **Enclose in Object** tool.
3. Select the text to be enclosed.
4. Press ***<Enter>***.

   The following will appear on the Command Line:

   ```
   Enter distance offset factor <0.3500>:
   ```

5. This determines the offset from the text. It multiplies this number times the text height. Press ***<Enter>***.

   The following will appear on the Command Line:

   ```
   Enclose text with [Circles Slots Rectangles] <Circles>:
   ```

6. Select the enclosure you require.

   The following will appear on the Command Line:

   ```
   Create circles of constant or variable size [Constant Variable] <Variable>:
   ```

7. If you select **Constant**, AutoCAD will select the largest text height and multiply it times the offset distance to determine the size for all. If you select **Variable**, each offset distance will be calculated based on the individual text height. Press ***<Enter>***.

Example:

	Variable			Constant		
Circle	A	A	AAA	A	A	AAA
Slot	A	A	AAA	A	A	AAA
Rectangle	A	A	AAA	A	A	AAA

Draw – Break-Line Symbol

The **Break-line Symbol** command creates a polyline and inserts the break-line symbol.

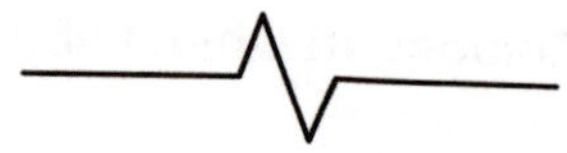

How to create a Break-line Symbol

1. Draw three Lines as shown below.

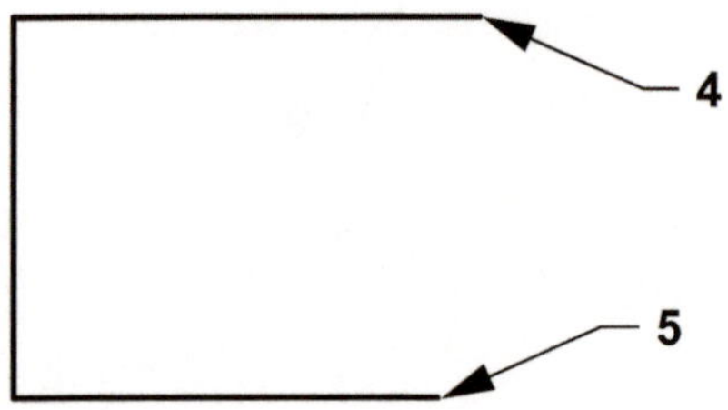

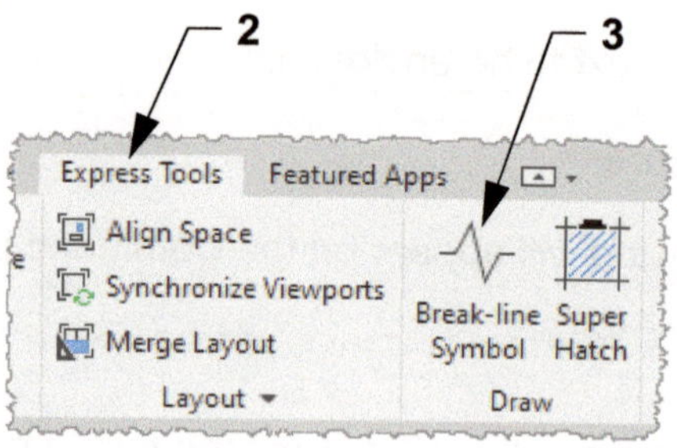

2. Select the **Express Tools** Tab.
3. Select the **Break-line Symbol** tool.

   The following will appear on the Command Line:

   `Specify first point for breakline or [Block Size Extension]:`

4. Snap to the first point.

   The following will appear on the Command Line:

   `Specify second point for breakline:`

5. Snap to the second point.

   The following will appear on the Command Line:

   `Specify location for break symbol <Midpoint>:`

6. Press ***<Enter>*** to locate the break at the Midpoint.

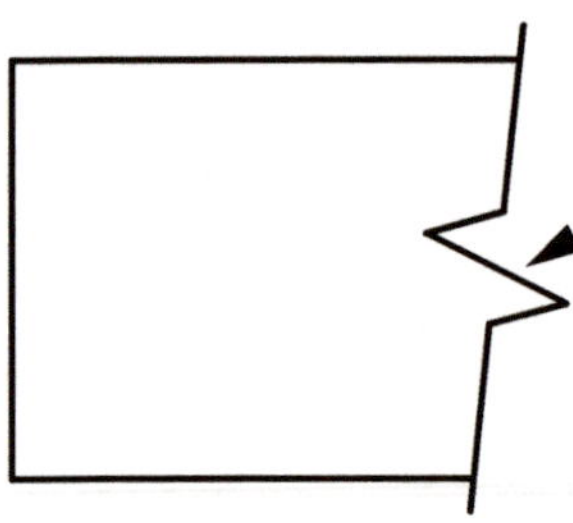

Options:

Block: You can select another block to be used instead of the Break-line Symbol.

Size: You can designate the size of the Break-line Symbol.

Extension: You can designate the amount the Break-line overlaps the existing line.

Tools – Command Aliases

AutoCAD has many preset aliases for common commands. For example, **C** for **Circle**. These aliases are stored in the **acad.pgp** file. Using the **Command Aliases** tool, you can **Add**, **Remove**, or **Edit** an Alias.

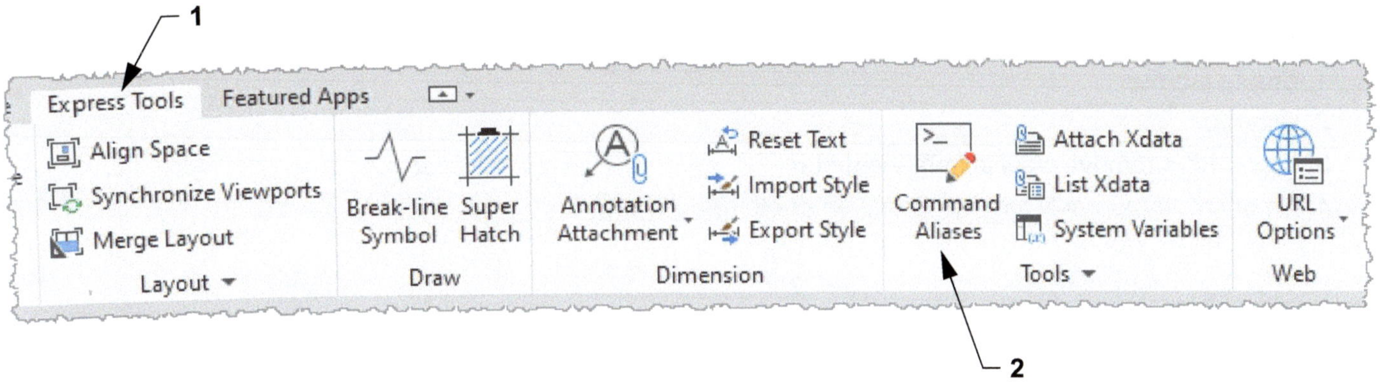

How to use the Command Aliases tool

1. Select the **Express Tools** Tab.
2. Select the **Command Aliases** tool.

The **acad.pgp - AutoCAD Alias Editor** appears.

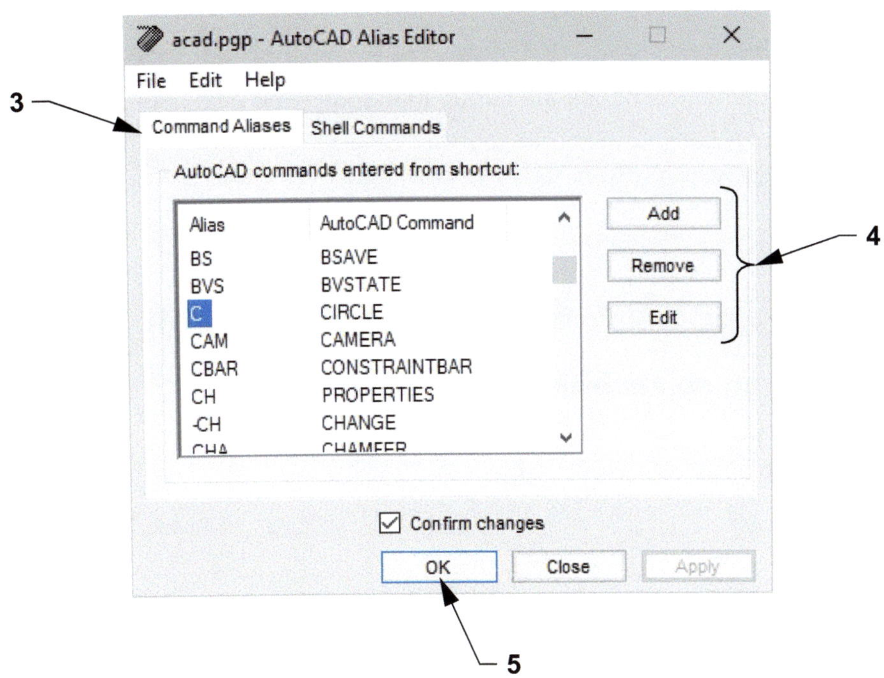

3. Select the **Command Aliases** Tab.
4. Select the **Add**, **Remove**, or **Edit** button.

   Add: You will specify an **Alias** and select the corresponding command. Then select **OK**.

   Remove: Click on the Alias and select the **Remove** button.

   Edit: Enter the new Alias to replace the existing Alias. Then select **OK**.

5. Select the **OK** button.

Creating and Managing a Group

The **Group** tools allow you to group a set of objects and give them a name.

Select: **Home Tab / Groups Panel**

You may:

1. Create a group.
2. Ungroup.
3. Edit: Add or remove objects from the group.
4. On or Off: Temporarily turn the Group on or off. (Blue is **on**.)

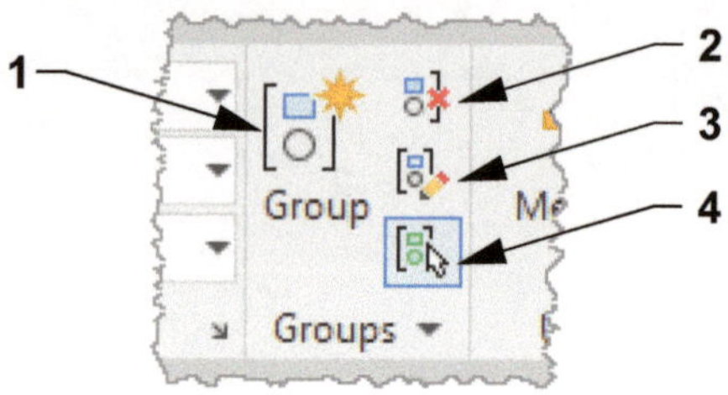

How to create a Group

1. Select the Create Group tool (1).
2. Select the objects to Group.
3. You may give the group a name now or use the Group Manager. (See next page.)

How to Ungroup

1. Select the Ungroup tool (2).
2. Select the group to Ungroup.

How to Add or Remove an object from the group

1. Select the Edit tool (3).
2. Select the group to edit.
3. Select Add or Remove.
4. Select the object/s to Add or Remove and then press ***<Enter>***.

How to turn a Group On or Off

1. Select the On or Off tool (4).
2. Select the group to turn off or on.

Note: You may temporarily turn off a group, move an object within the group, and then turn the group on again. The moved object will remain in the group but now will be in a different position.

Purge a Group

A Group, no longer needed, can be purged. (Refer to page 29-14.)

Continued on the next page...

Managing the Group

1. Select the **Groups** down arrow ▾

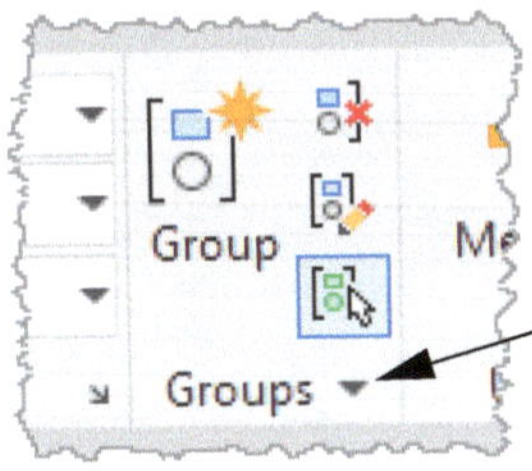

2. Select **Group Manager** or **Group Bounding Box**.

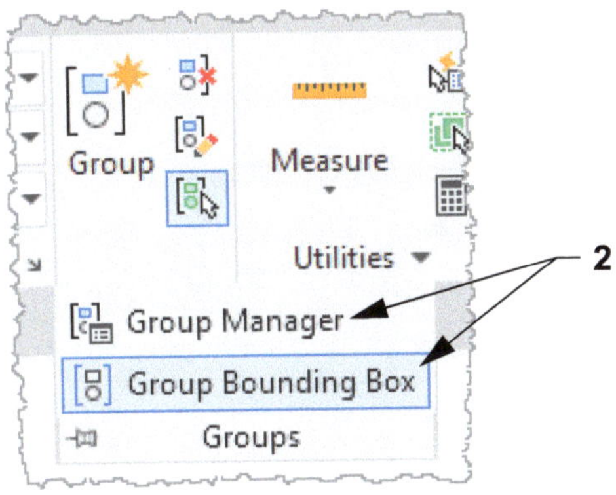

Group Manager

The Group Manager allows you to Identify, Create, and Change a group.

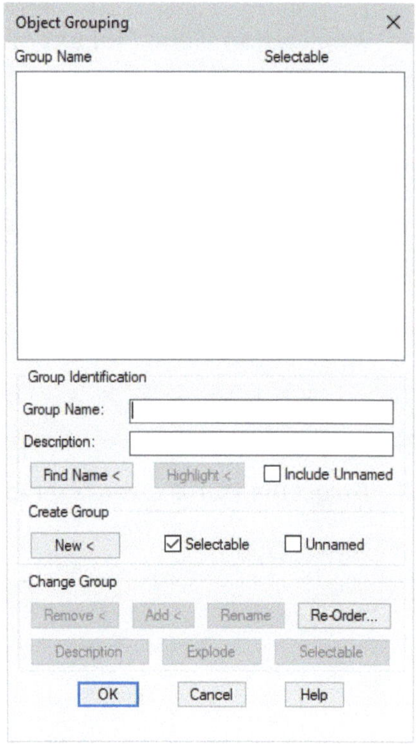

Group Bounding Box

When you select a previously created group, AutoCAD identifies it using a Group Bounding Box or grips. The default is a Group Bounding Box. If you want grips, unselect this option.

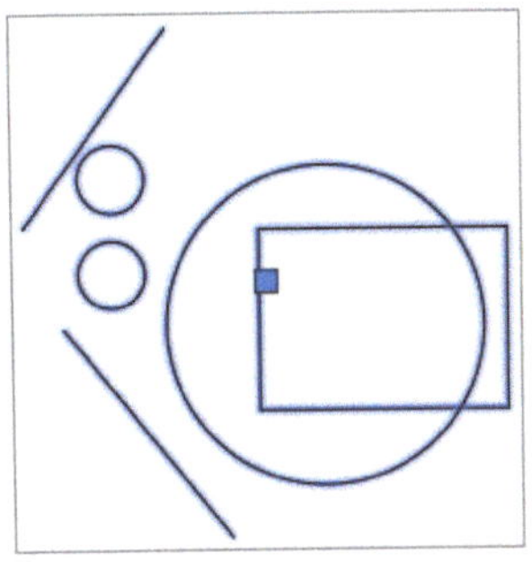

Group Bounding Box on

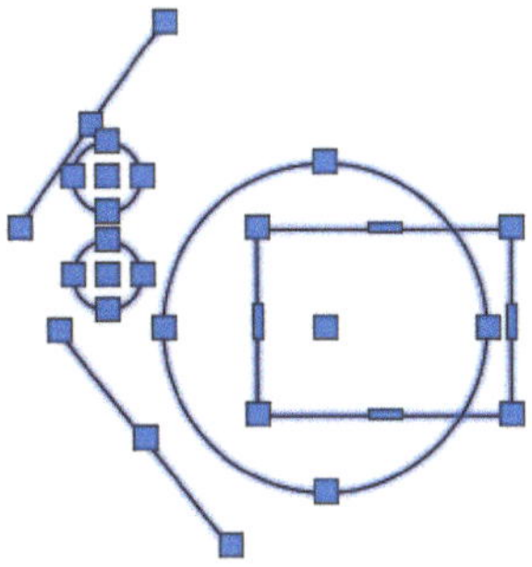

Group Bounding Box off

Import a PDF File into AutoCAD

PDF files are now a common way of sharing your designs and for receiving designs from others. AutoCAD now allows you to import the geometry and TrueType text of a PDF file of a design. There are various ways to import a PDF file.

How to import a PDF file

1. Select the **PDF Import** command using one of the following:

   Ribbon = Insert Tab / Import Panel / PDF Import ▾ / PDF Import
   or
   Keyboard = *pdfimport <Enter>*

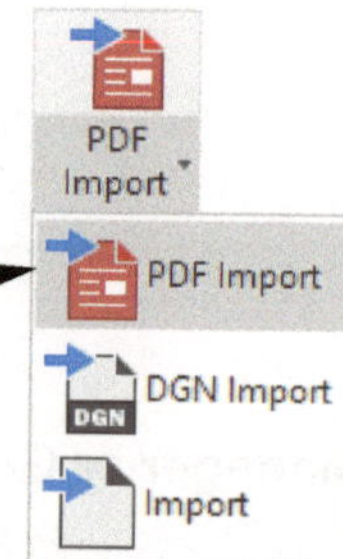

   or
   Application Menu / Import / PDF

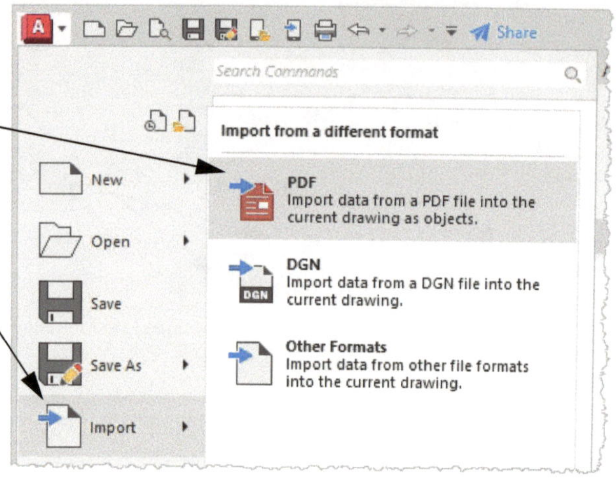

2. Select the PDF file you want to import from the **Select PDF File** dialog box.

3. Select **Open**.

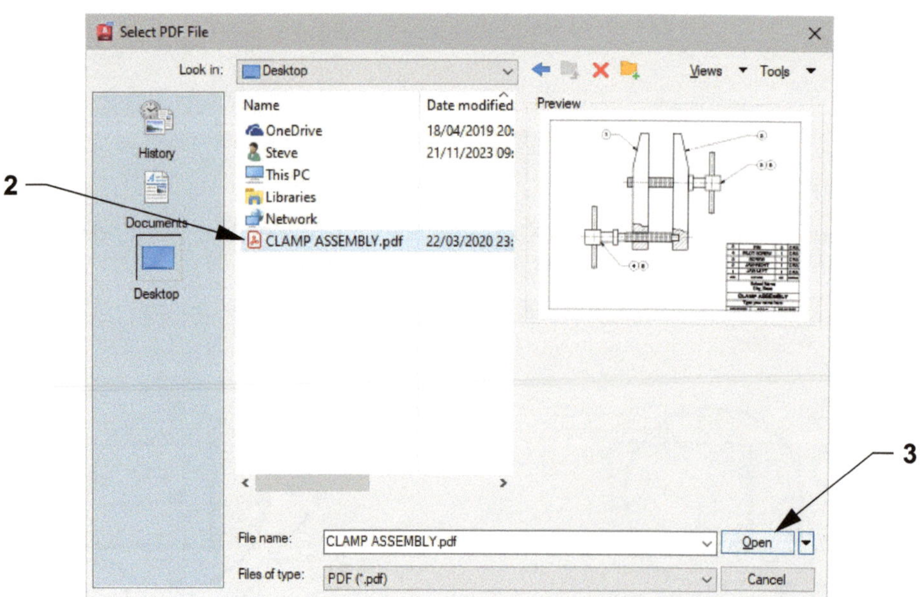

Continued on the next page...

4. The **Import PDF** dialog box will appear with various options. Most of the options in the dialog box are self-explanatory, but there are a few that need explaining. The dialog box defaults are shown in the image below.

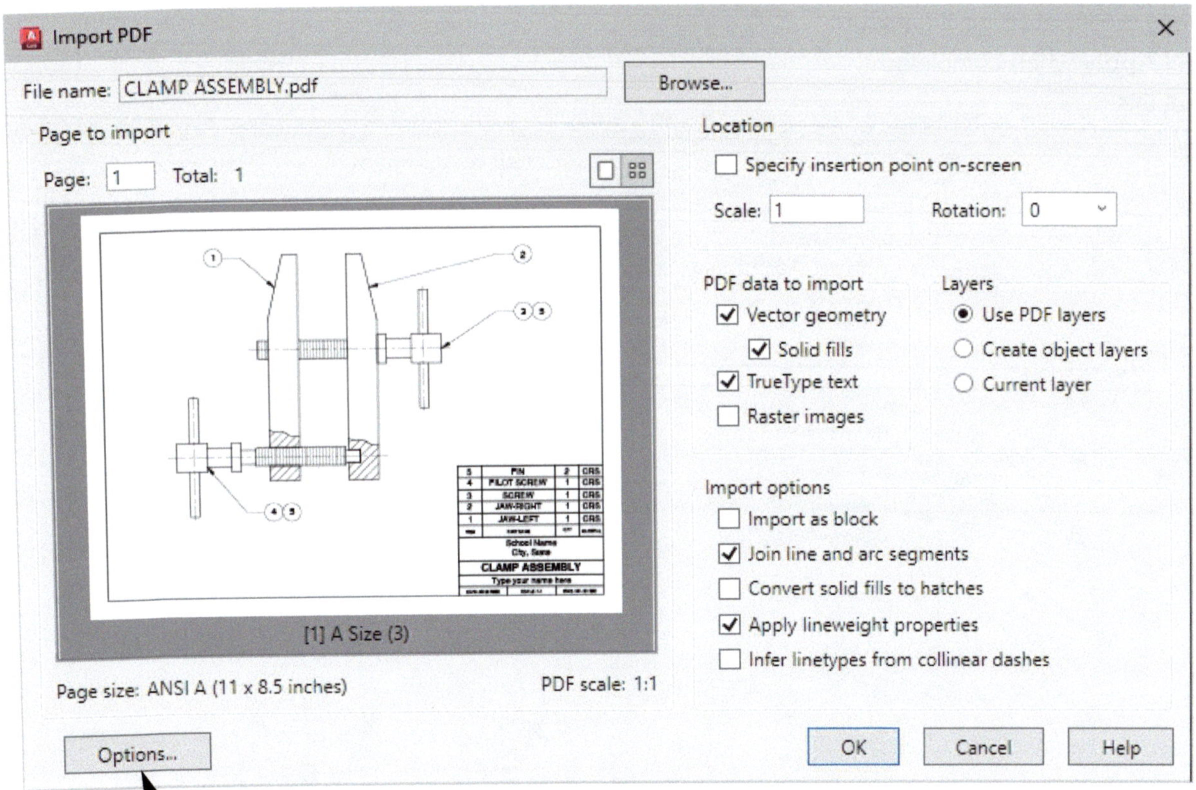

Options... button. (See below.)

Import options

Convert solid fills to Hatches: If there are any solid filled objects in the PDF you want to import, they will be converted into Hatches when this **option is checked**, but it does add extra to the processing time.

Infer linetypes from collinear dashes: If you have any lines that are created using a linetype such as a Centerline or a Dash-Dot line, they will be converted to a single polyline when this **option is checked**. If the option is left **unchecked** and you have a Centerline or a Dash-Dot line in the PDF file to be imported, these lines will be converted to single dashes and dots. So it is best to check this option if you have any linetypes other than a single continuous line in the PDF file you want to import.

PDF Data to Import

Raster images: If there are any Raster images in the PDF file to be imported, these images will be attached to the drawing when this **option is checked**. You can specify a location for the attached images by selecting the **Options...** button, which will open up the **Options** dialog box, where you can specify a location for the attached images. It is good practice to save the attached raster images in the same location as you save the drawing in.

Options... button.

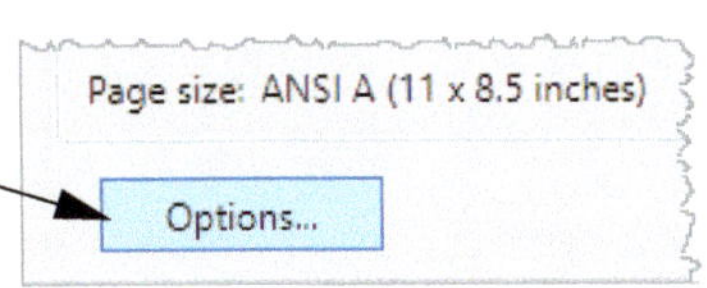

Continued on the next page...

A. Select the **+** sign next to **PDF Import Image Location**.

B. Select **PDF Images**.

C. Select **Browse**.

D. Choose the location for the image.

E. Select **Apply** when completed.

F. Select **OK**.

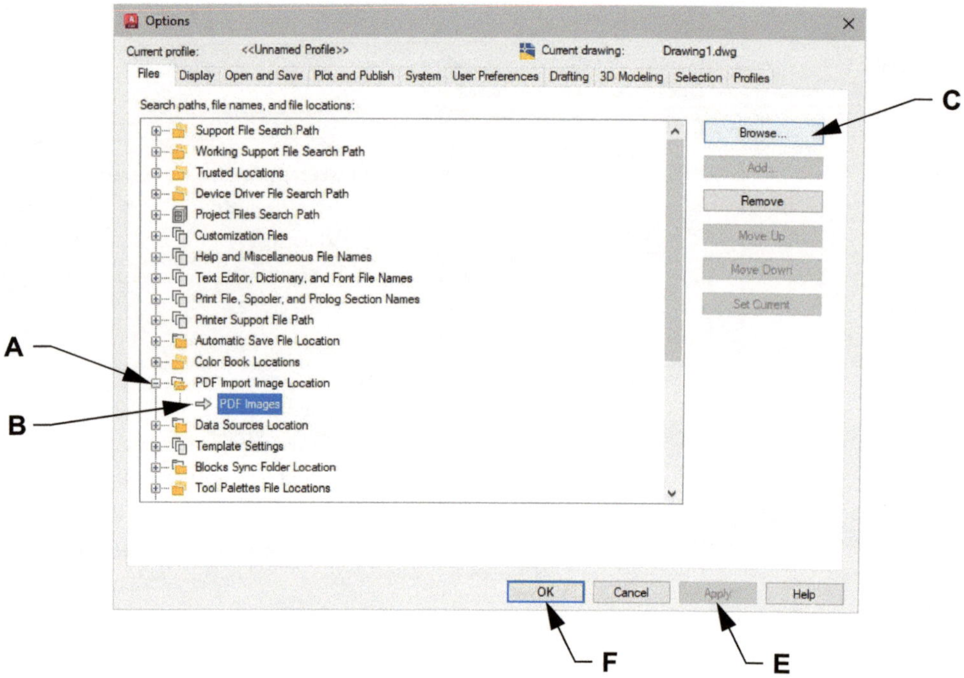

5. After you have made all your selections in the **Import PDF** dialog box, select **OK** to insert the PDF file.

The image below is an example of Exercise 29C - I - Clamp Assembly that was plotted as a PDF file and then imported into AutoCAD using the **PDF Import** command.

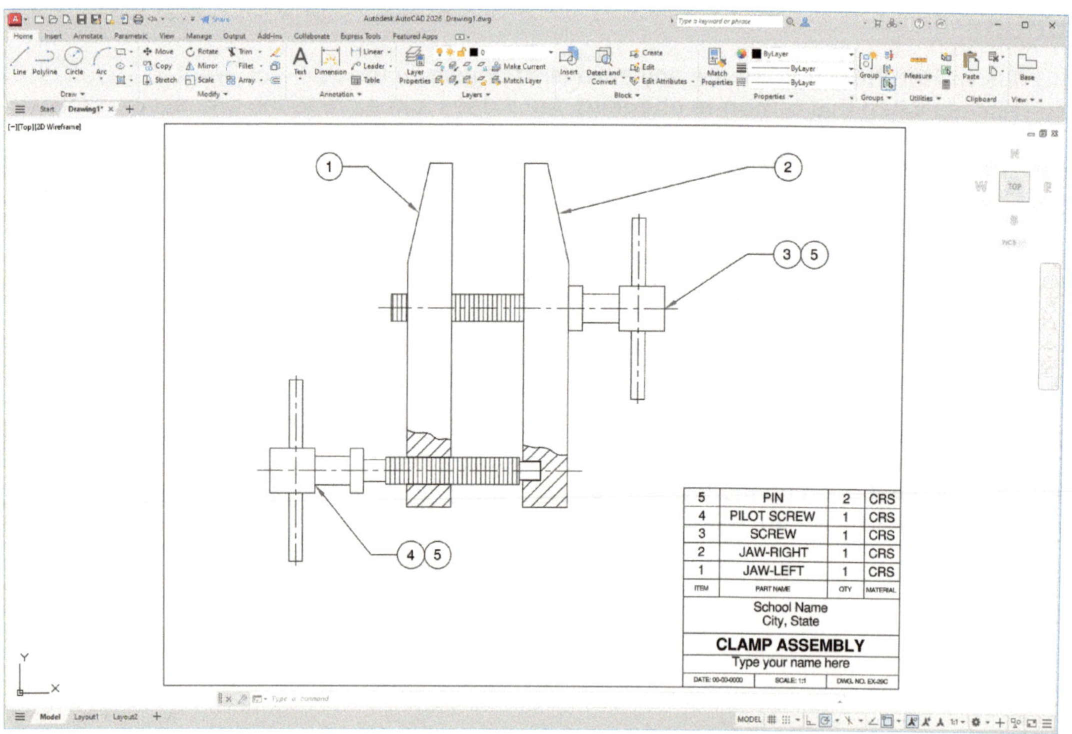

Note: Text with **TrueType fonts** are imported as text objects if they exist in the PDF file you want to import. This means you will be able to edit any text that is TrueType font text.

Import a PDF File with SHX Fonts

If you import a PDF file into AutoCAD that contains **SHX fonts**, they will be imported as separate polyline objects. AutoCAD can now convert those polyline objects to original SHX fonts by using the SHX font recognition tool.

How to convert the polyline SHX text imported in a PDF file

1. Import a PDF file that contains SHX fonts. For this example, I have imported a PDF file that contains three different SHX fonts.

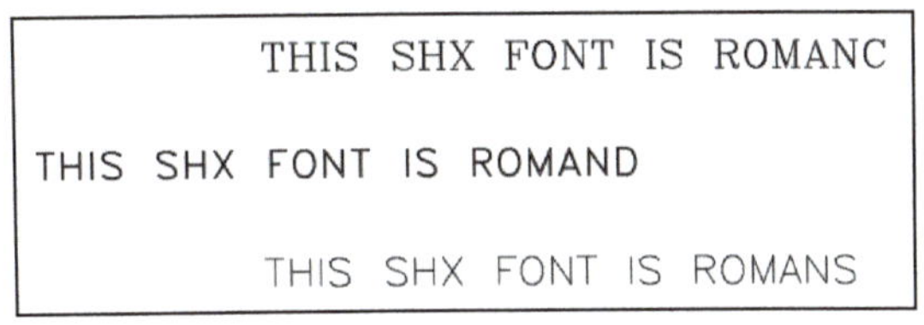

2. Select the SHX font **Recognition Settings** command:

Ribbon = Insert Tab / Import Panel / Recognition Settings 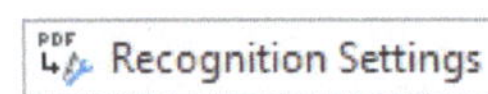

The **PDF Text Recognition Settings** dialog box will open.

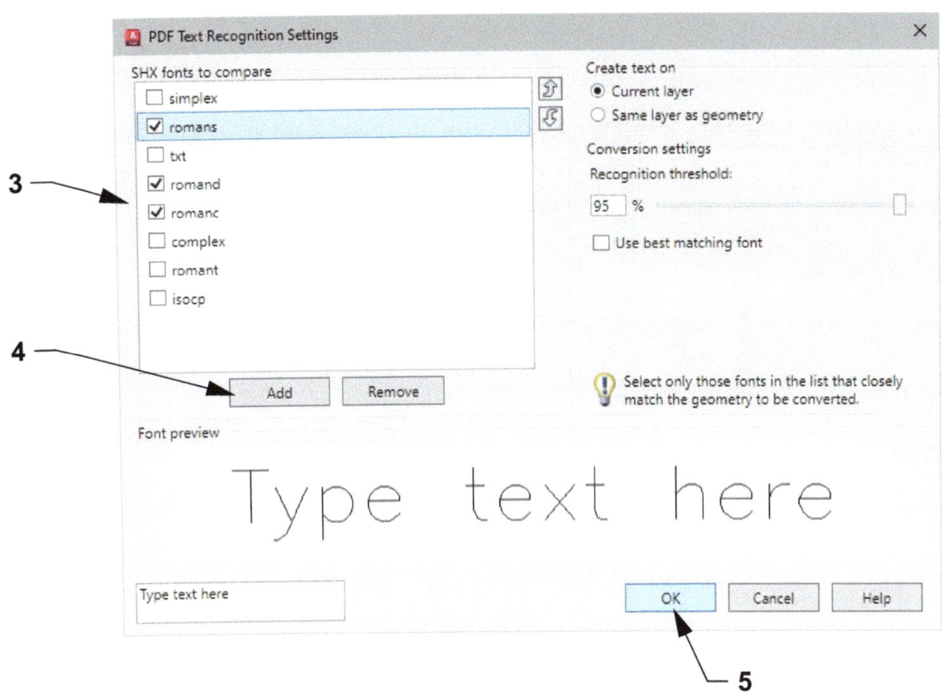

3. Select the fonts that closely match the polyline SHX fonts that you imported.

4. If your SHX font is not listed in the **SHX fonts to compare** Window, select the **Add** button and then select a suitable font from the list of available SHX fonts included with AutoCAD. Or you may wish to add your own SHX font.

5. Select the **OK** button.

Continued on the next page...

The following will appear on the Command Line:

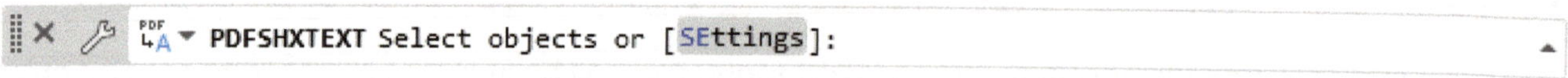

6. Select the polyline SHX fonts using a Crossing Window selection and then press ***<Enter>***.

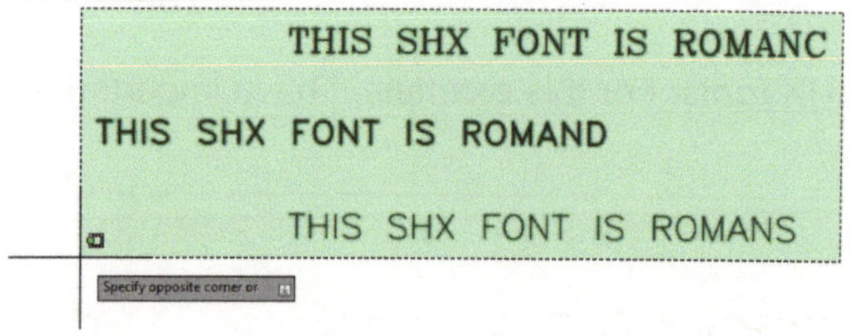

The **Recognize SHX Text** dialog box will appear informing you that the selected polyline SHX fonts were converted correctly.

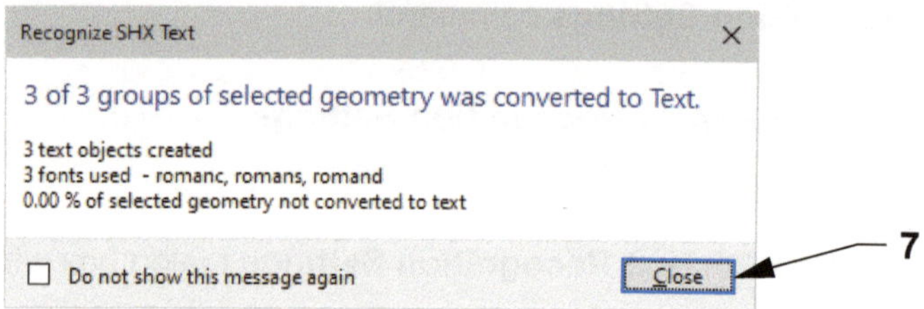

7. Select **Close**.

If you double left click on any of the recently imported SHX fonts, the **Text Editor** Ribbon Tab will open up where you can now fully edit the SHX fonts.

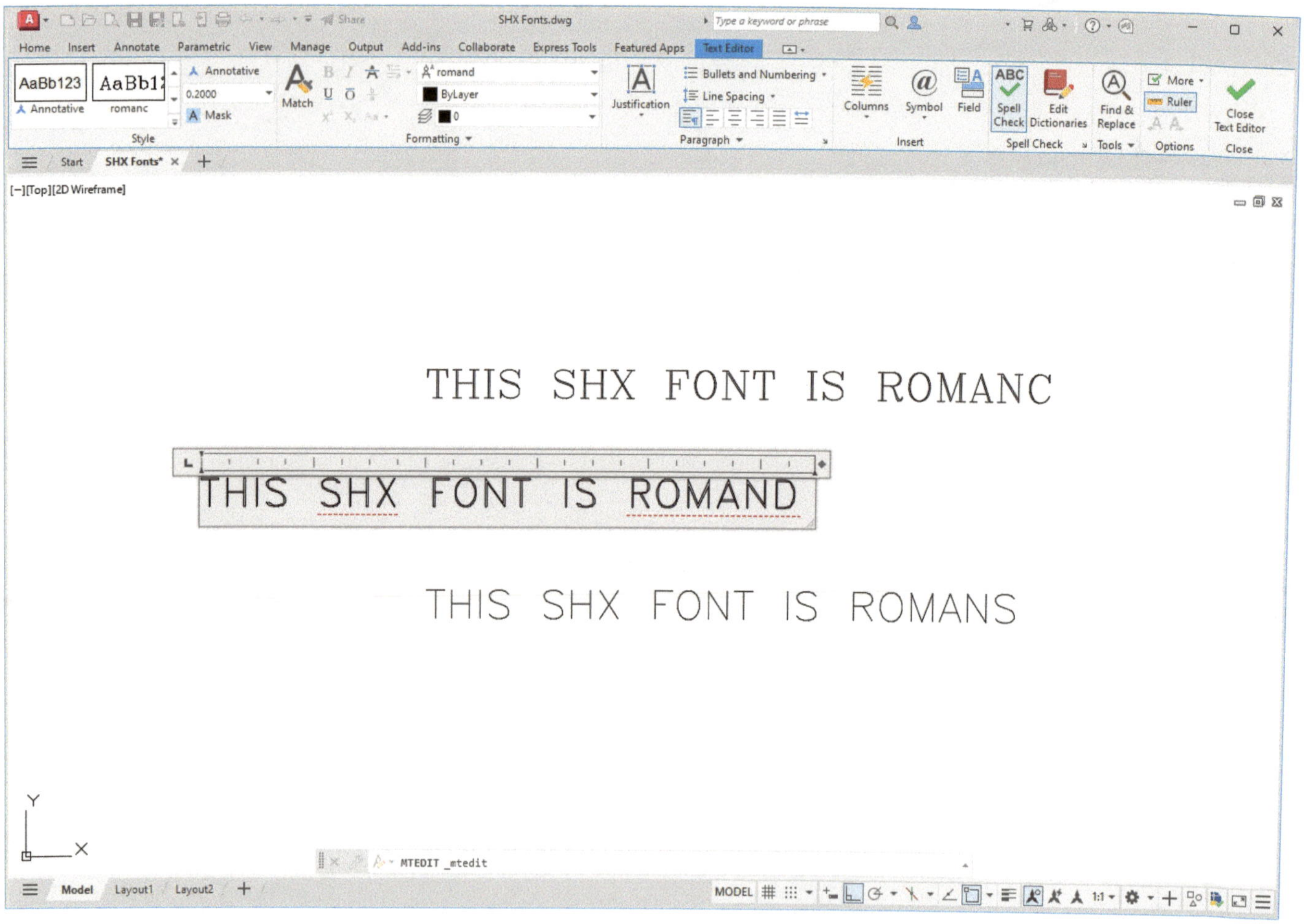

Drawing Compare

The **Drawing Compare** tool allows you to compare the differences between two versions of a drawing. You can edit the Base drawing (open drawing) with a Reference drawing while in the compare state. If there are any differences identified between the two versions of the drawing, the differences will be enclosed within Revision Clouds.

The example below shows two versions of a floor plan. There is a small yet critical change between the two versions, but it may be difficult to identify this small change. There is also a large change that can easily be identified.

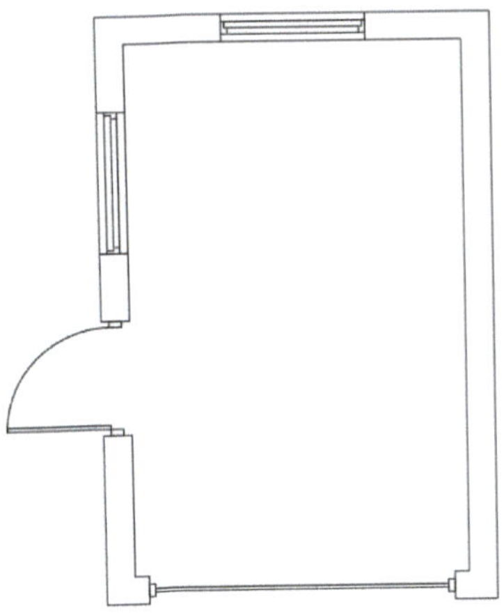

Floor Plan-Rev 1

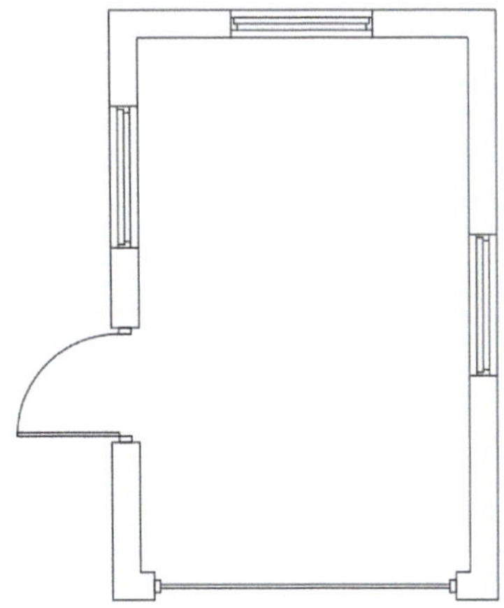

Floor Plan-Rev 2

After loading the two versions of the floor plan using the Drawing Compare tool, you can clearly see that the left-hand door opening has been moved to a different location. There is also an extra window added in the Floor Plan-Rev 2 Reference drawing that is not in the Floor Plan-Rev 1 Base drawing. Seeing the differences between the drawings, you can decide whether you need to make further changes in your current drawing.

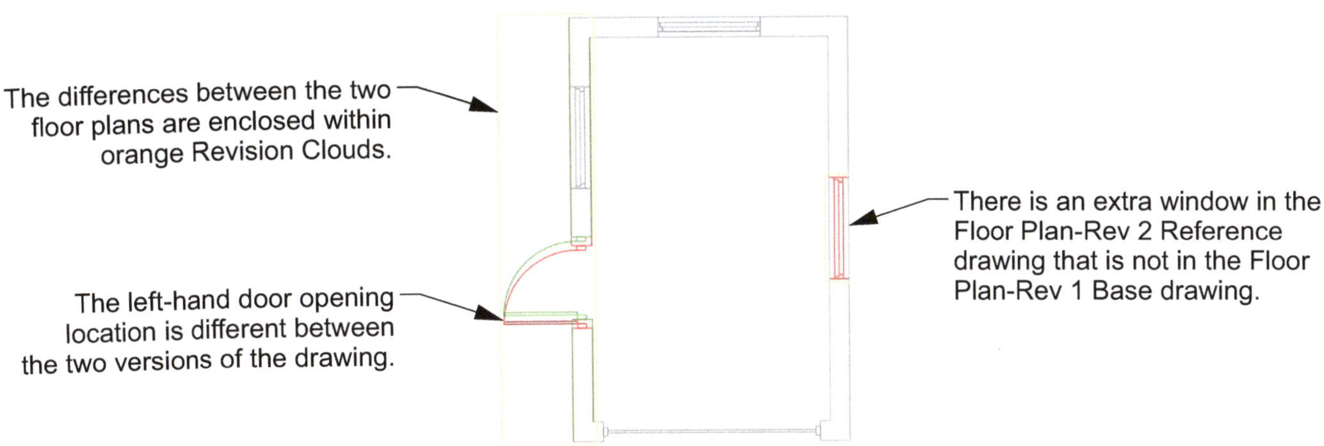

Selecting the Drawing Compare tool

The Drawing Compare tool can be accessed from the main Application Menu if there are no drawings active. If there is a drawing active, select the **Drawing Utilities** Menu in the **Application** Menu.

How to select Drawing Compare with no drawings active

1. Select the **Application** Menu.
2. Select the **DWG Compare** command.

Continued on the next page...

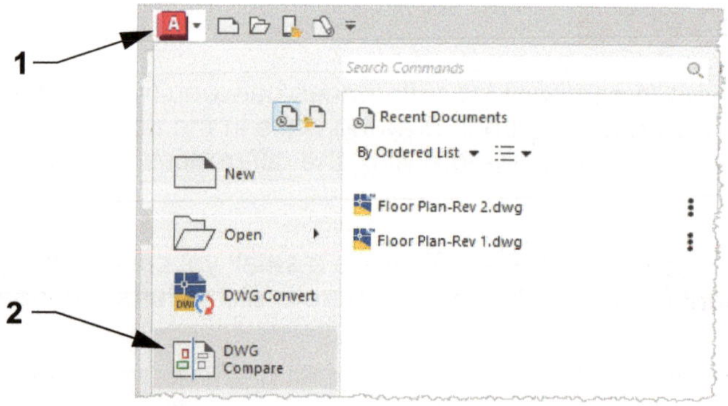

How to select Drawing Compare with drawings active

1. Select the **Application** Menu.
2. Select the **Drawing Utilities** Menu.
3. Select the **DWG Compare** command.

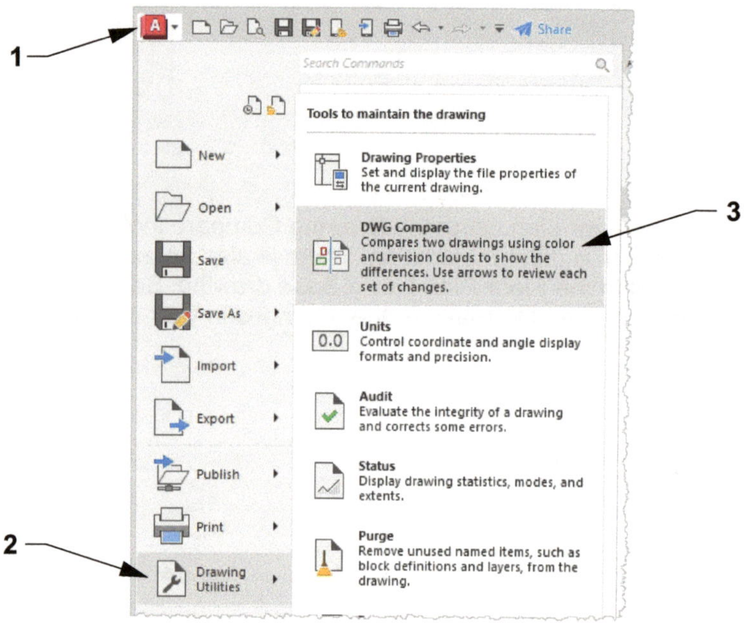

If a drawing is active, you may also select the Drawing Compare tool on the **Collaborate** Ribbon Tab.

1. Select the **Collaborate** Tab.
2. Select the **DWG Compare** tool.

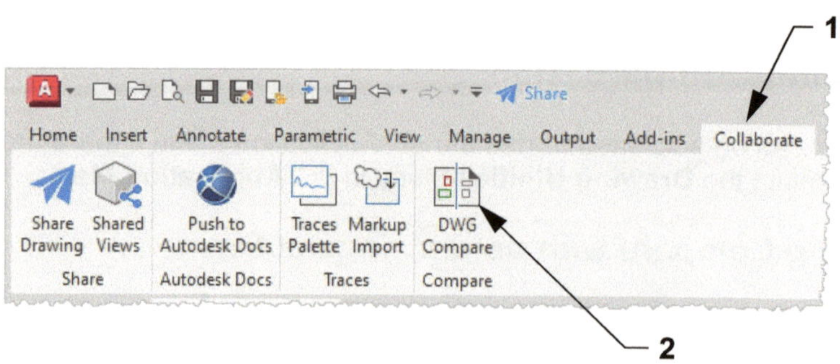

Continued on the next page...

Example of Using the Drawing Compare tool

For this example, let's use the Floor Plan shown earlier.

1. Select the **Drawing Compare** tool.
2. In the **DWG Compare** dialog box, select the first drawing (Base drawing) you want to compare by selecting the browse button and then locating the file in the **Select a drawing to compare** dialog box.
3. Repeat Step 2 to select the second drawing (Reference drawing) to compare.
4. Select the **Compare** button.

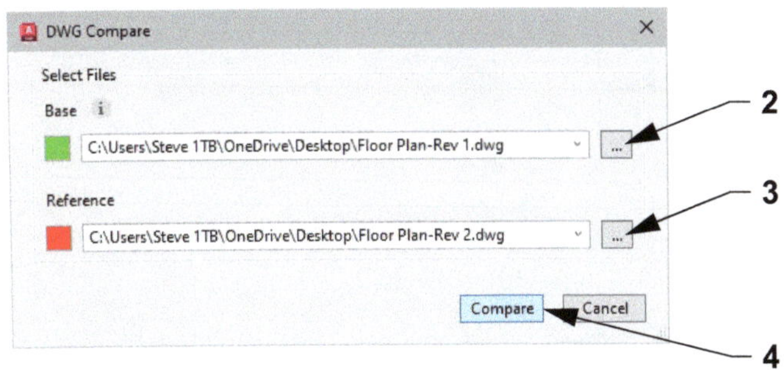

The first drawing selected is called the **Base** drawing (open drawing). The second drawing selected is called the **Reference** drawing.

The Base drawing will open in its own drawing Tab and will be fully editable. The Reference drawing also will open in the same drawing Tab but will be overlaid and locked (non-editable). The **DWG Compare** Toolbar will automatically display at the top of the drawing area, with various tools that allow you to control the appearance of the compared drawing results. The available tools include the **Import Objects** tool, which allows you to import objects from the Reference drawing to the Base drawing.

If there are any differences between the Base drawing and the Reference drawing, the differences will be displayed in green on the Base drawing, and displayed in red on the Reference drawing. Any elements that remain unchanged between the Base drawing and the Reference drawing will be displayed in gray. You can change the colors if you want, by selecting the **Settings** button ⚙ on the **DWG Compare** Toolbar.

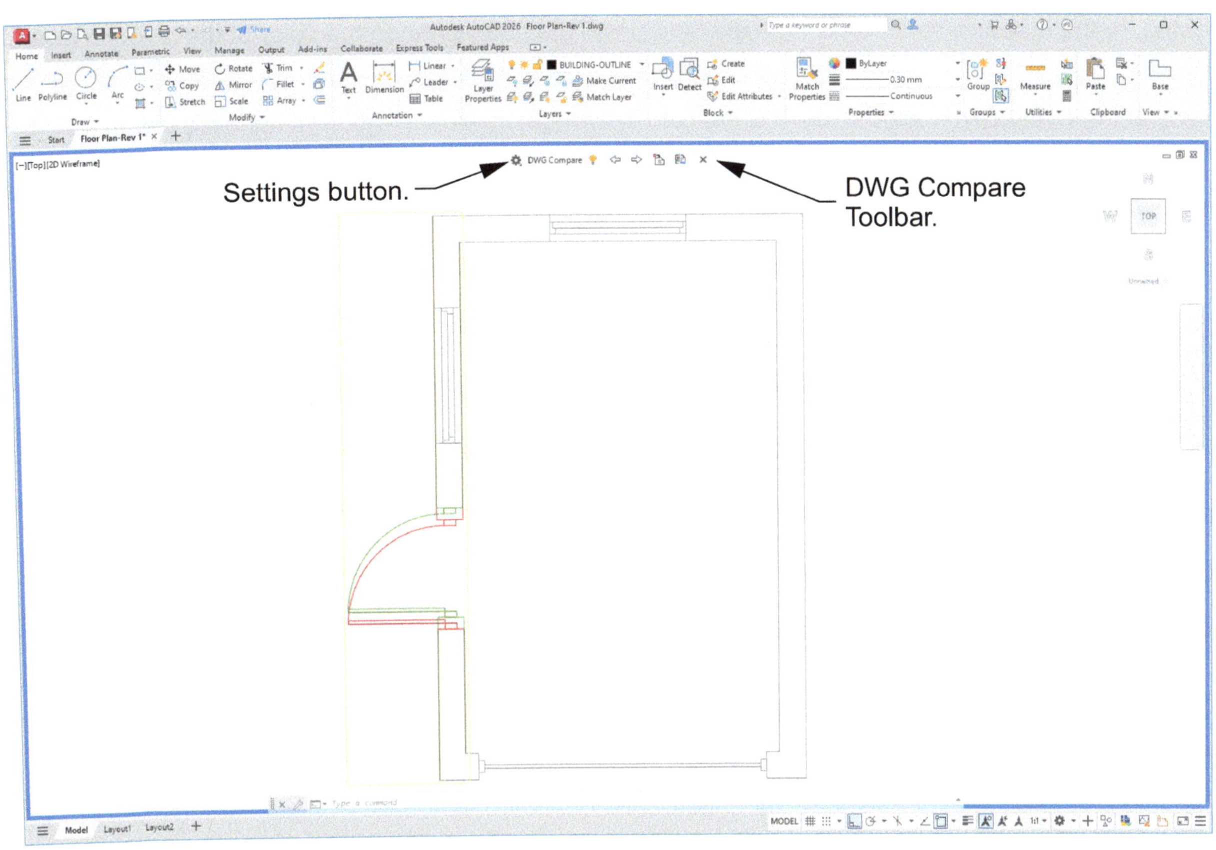

The DWG Compare Toolbar tools and their functions

Selecting the **Settings** button allows you to change the appearance of the Base and Reference drawings, and also choose whether to display Hatches or Text objects.

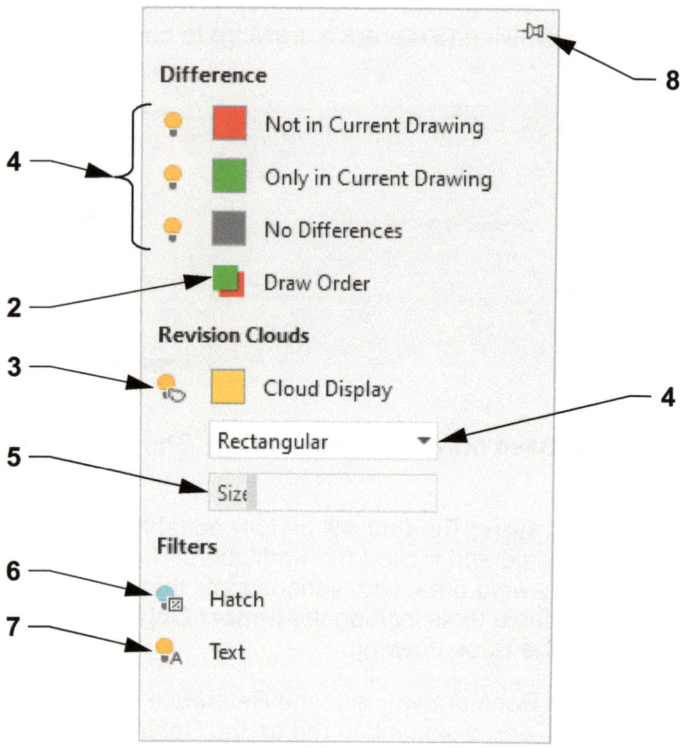

1. By selecting the appropriate **lightbulb**, you can turn off the graphics of the Base drawing, the Reference drawing, or the graphics common to both drawings. You may also change the color of the drawing graphics by selecting the appropriate color swatch.
2. Selecting the **Draw Order** allows you to bring the Reference drawing to the front and vice versa.
3. Selecting the **Cloud Display** lightbulb allows you turn the Revision Cloud graphics on or off. You may also change the color of the Revision Cloud graphics by selecting the color swatch.
4. By selecting the **Cloud Display** down arrow, you can change the shape of the Revision Cloud from Rectangular to Polygonal.

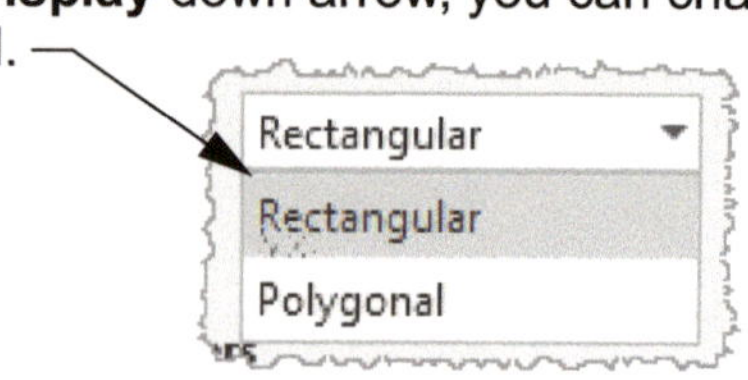

5. Selecting the slide bar allows you to change the size of the margins between the Revision Cloud and the compared objects. The bigger the number, the bigger the margin.
6. Selecting the **Hatch** lightbulb allows you turn on or off any Hatch objects that may be in the Base or Reference drawing. The Hatch Lightbulb is **off** by default.
7. Selecting the **Text** lightbulb allows you to turn on or off any Text objects that may be in the Base or Reference drawing. The Text lightbulb is **on** by default.

Continued on the next page...

8. You may pin the **Settings Panel** so it remains open during the drawing compare session. You can also move the Settings Panel to any location on the screen.

Additional DWG Compare Toolbar tools

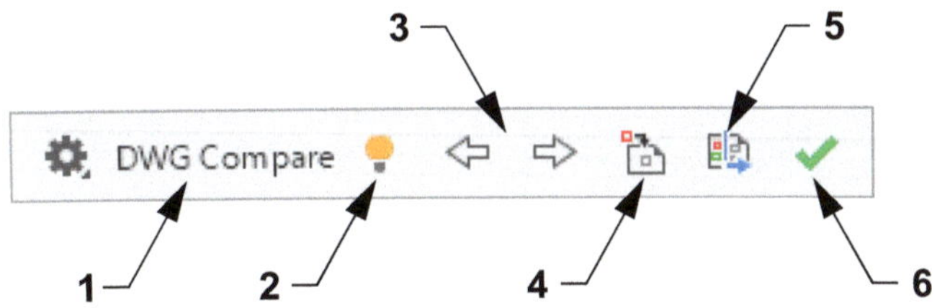

1. Hover the cursor over the **DWG Compare** text to see a **Color Key** for the Base and Reference drawings.
2. Selecting the **lightbulb** allows you turn on or off the compare graphics.
3. Depending on how many Revision Cloud sets there are in the compare session, you can select the left and right arrows to automatically zoom into each individual set.
4. Selecting the **Import Objects** tool allows you to import objects from the Reference drawing to the Base drawing. The following pages explain the Import Objects tool in more detail.
5. The **Export Snapshot** tool allows you to save a new drawing file with the comparison results of the Base and Reference drawings, including the compare graphics.
6. Select the **Exit Compare** tool once you have made any changes to the Base drawing, or when you are otherwise ready to close the compare session.

Example of Using the Import Objects tool

In the previous Floor Plan example, the Reference drawing has an additional window and also the door location has moved. This example will show you how to use the **Import Objects** tool to add the additional window to the Base drawing and how to move the location of the door to match the Reference drawing using the **Stretch** command.

Adding the additional window to the Base drawing

1. Select the **Import Objects** tool on the **DWG Compare** Toolbar.
2. Select objects: Use a Window selection from **P1** to **P2** to select the additional window and then press ***<Enter>***.

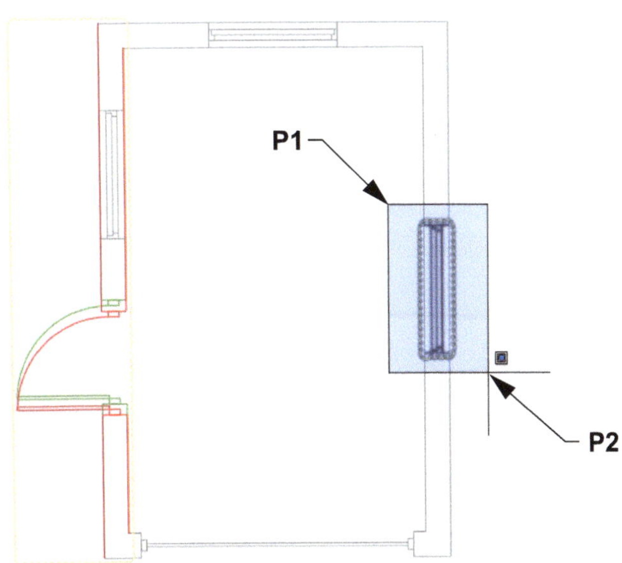

Continued on the next page...

3. The **Import Objects** tool will end automatically. The Revision Cloud that encased the additional window will disappear and the additional window color will turn gray. The gray color of the window means that the additional window is now in the Base drawing as well as the Reference drawing.

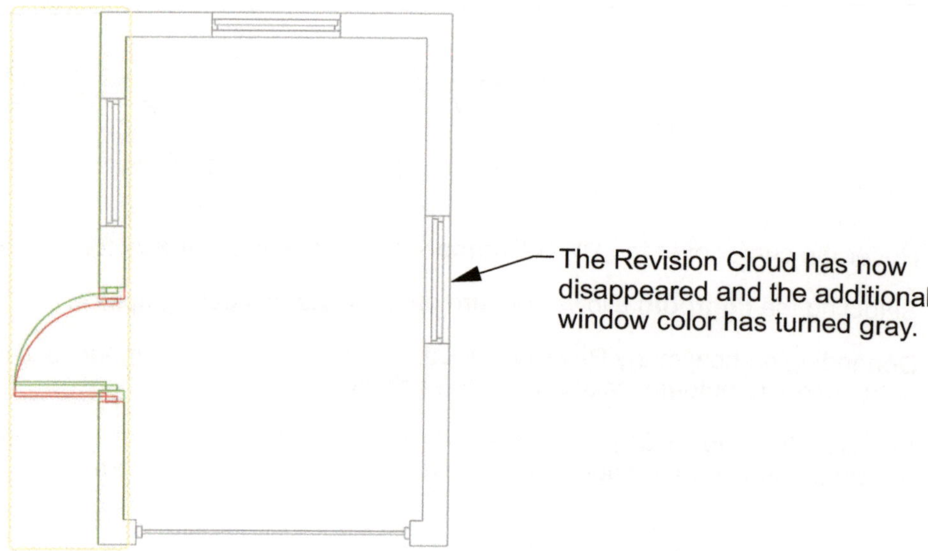

Moving the door location of the Base drawing to match the Reference drawing

1. Select the **Stretch** command on the **Modify** Panel of the **Home** Tab.
2. Select objects: Use a Crossing Window selection from **P1** to **P2** to select the elements of the door and wall that need to be moved and then press ***<Enter>***.
3. The selected objects will highlight blue. These are the Base drawing objects.

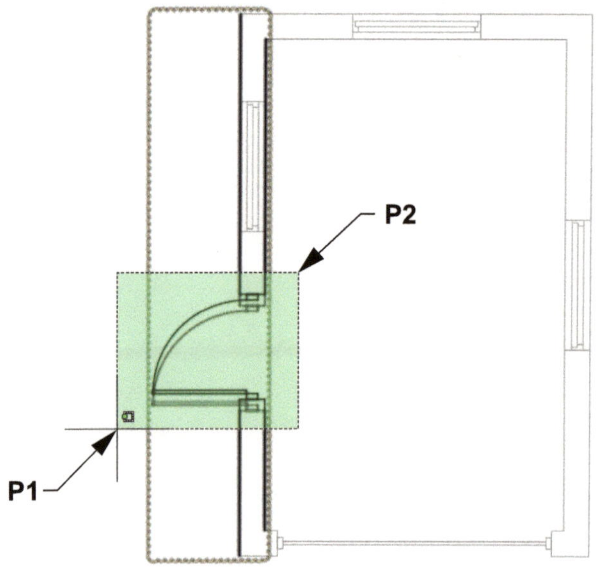

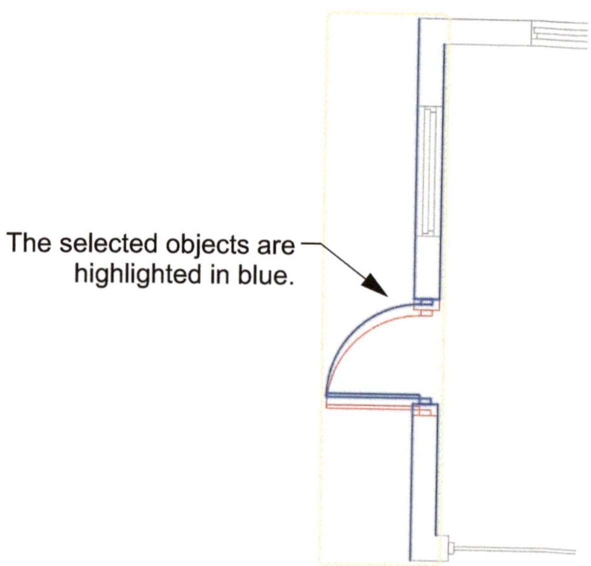

Continued on the next page...

4. Using the Endpoint Object Snap, select a point on the door or wall (in blue), drag the objects to the new location on the Reference drawing (in red), and then left click to end the **Stretch** command.

5. The Revision Cloud that encased the door and wall will disappear, and the color of the door and wall will turn gray. The Base drawing is now identical to the Reference drawing.

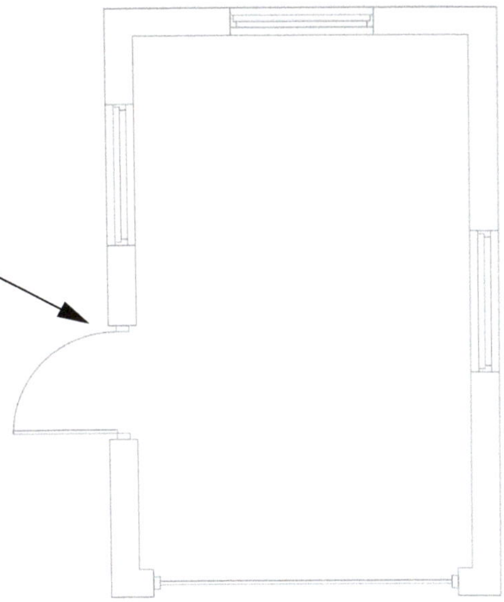

6. Select the **Exit Compare** tool to end the compare session.

Shared Views

The **Shared Views** tool allows you to store your designs in the Cloud for other users to view. Those you share your designs with can view, measure, and comment on the designs and share feedback with you. To use the Shared Views tool, you will need to sign in to your Autodesk® account and be a subscription user of AutoCAD. You must also be connected to the internet.

Selecting the Shared Views tool

The Shared Views tool can be accessed from the main **Application** Menu or from the **Collaborate** Tab. You will need an active drawing to share the view.

Selecting Shared Views from the Application Menu

1. Select the **Application** Menu.

2. Select **Publish**.

3. Select **Share View**.

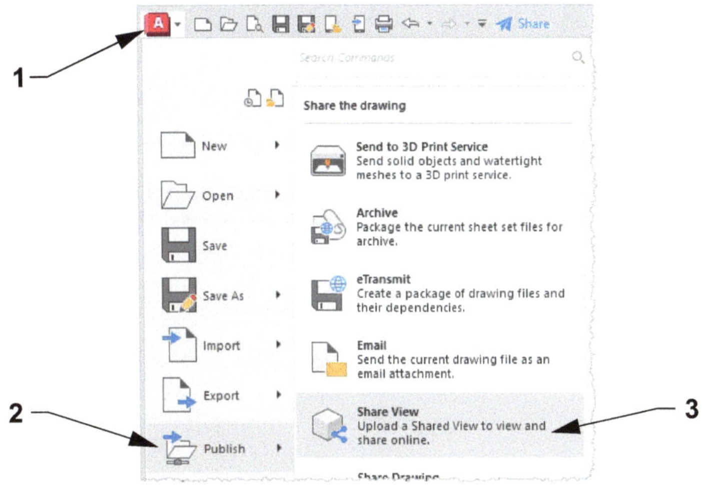

Continued on the next page...

Selecting Shared Views from the Collaborate Tab

1. Select the **Collaborate** Tab.

2. Select **Shared Views**.

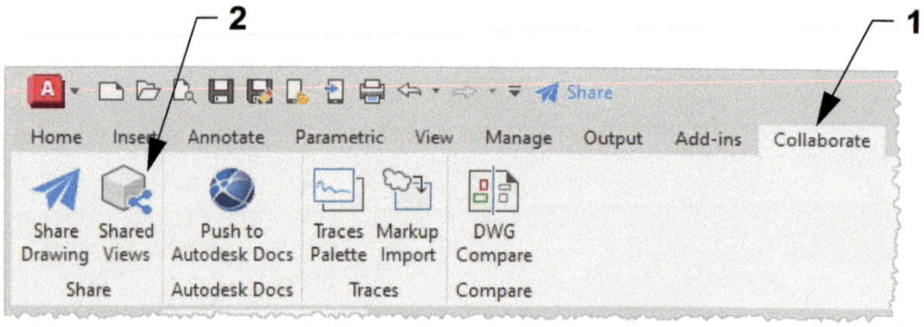

After selecting the **Shared Views** tool, the **Shared Views Palette** will open.

Create a new Shared View to share on the Cloud

1. Select the **New Shared View** button.

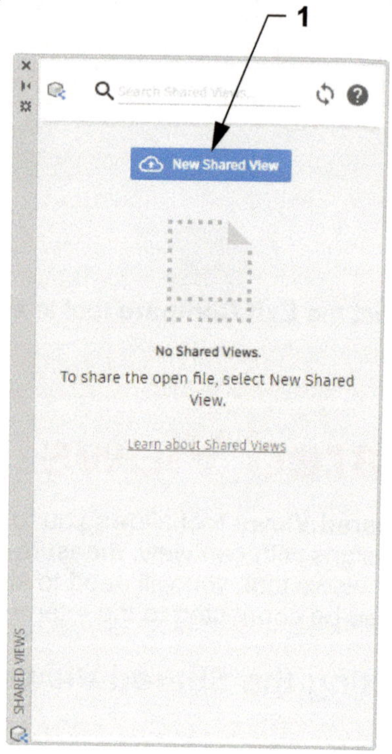

2. In the **Share View** dialog box, enter the name you want to call the shared view and check or uncheck any elements of the view you want to share.

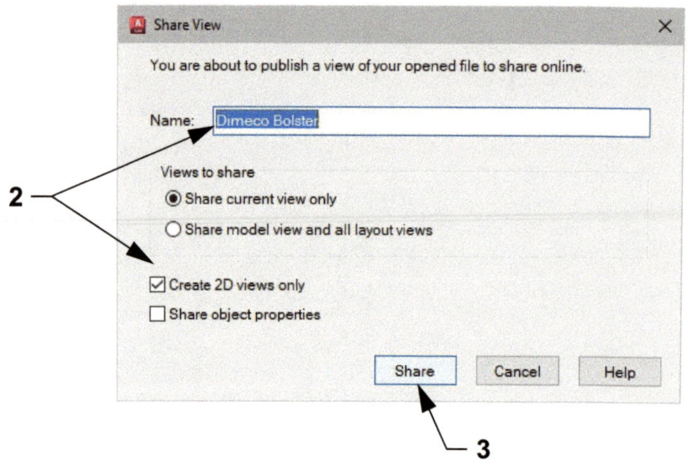

3. Select the **Share** button.

Continued on the next page...

4. The **Share View - Processing Ready to Start** dialog box will appear. Select **Proceed**.

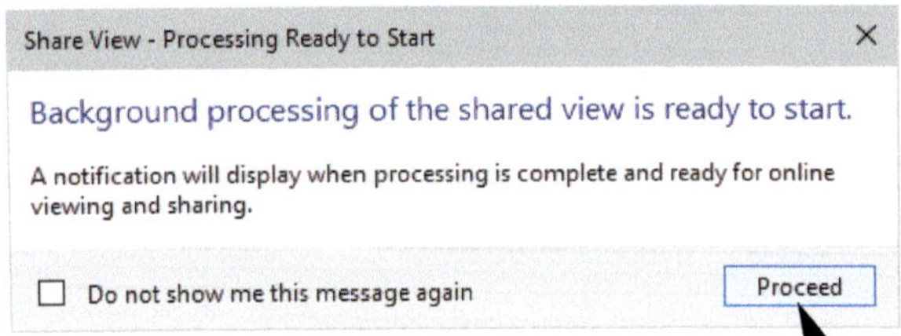

5. The Shared View tool will continue working in the background. When the view is uploaded to the Cloud, a bubble notification will appear at the bottom right-hand corner of the screen informing you that the share view upload is complete.

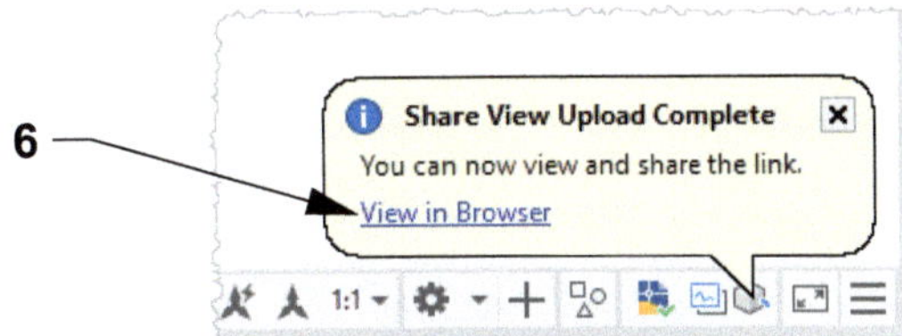

6. To view the shared view in the Cloud, select **View in Browser**.

7. Your internet browser will open showing you the shared view in the Cloud.

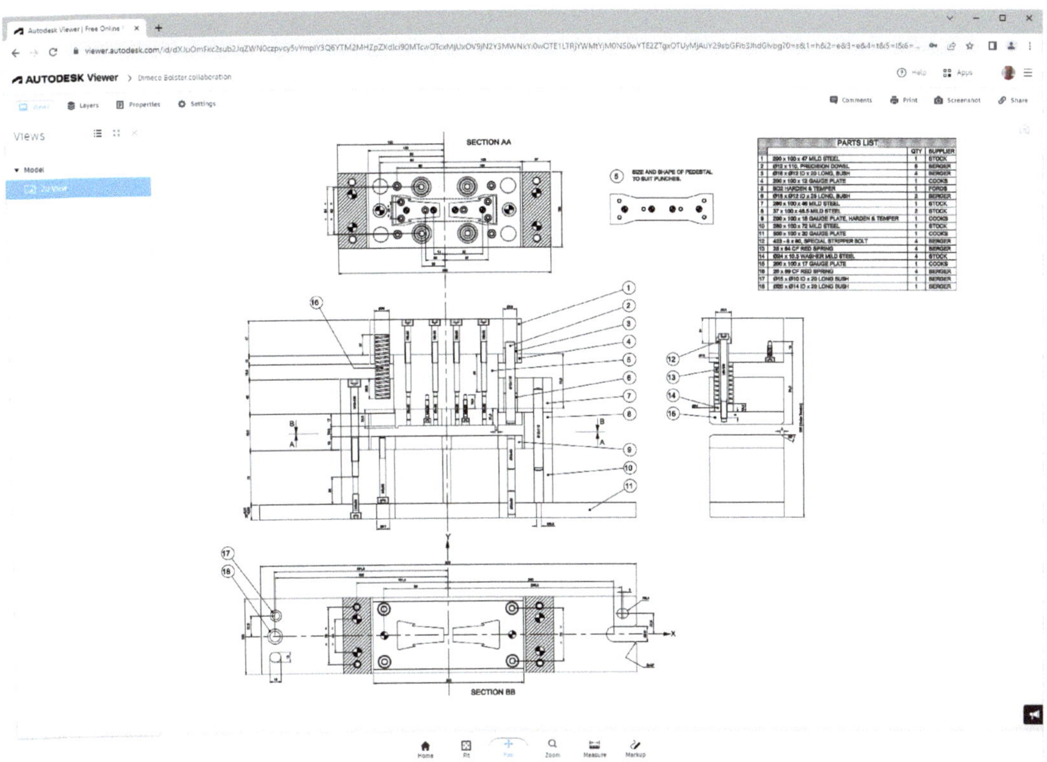

8. You will also receive an email notifying you that your file is ready to view. Click on **View File** to open your internet browser, which will automatically take you to your shared view in the Cloud.

Continued on the next page...

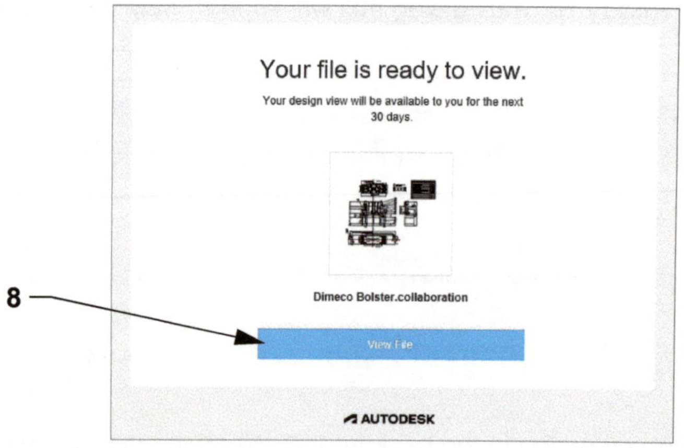

9. Select the **Refresh** button on the Shared Views Palette and your shared view will appear in the Palette Window.

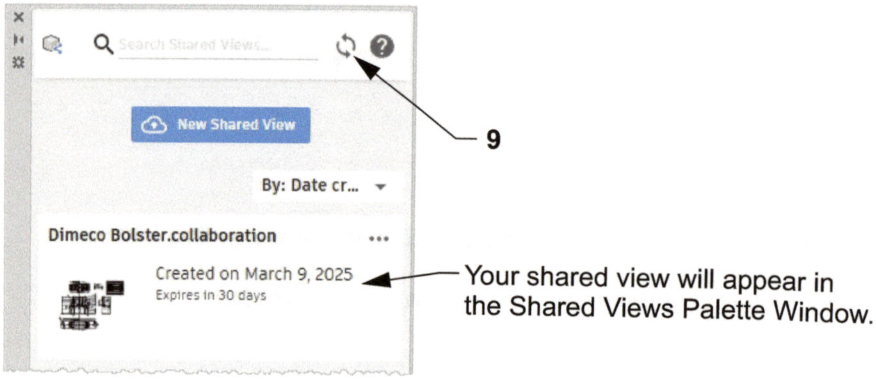

Adding comments to the Shared View and sharing your view

1. Select the shared view drawing in the Shared Views Palette Window.

2. Select **Add a comment**.

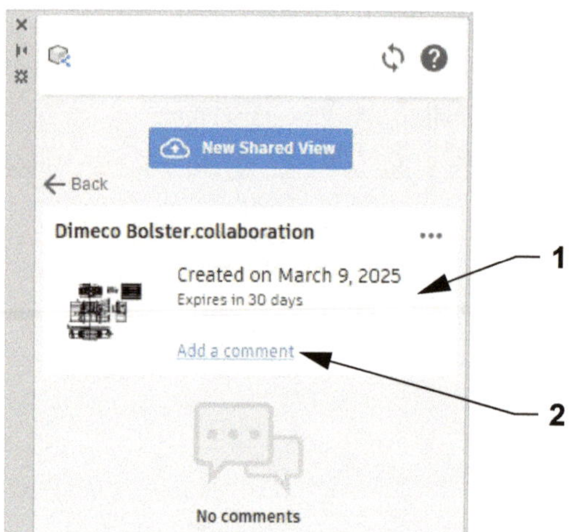

Continued on the next page...

3. Your internet browser will open showing your shared view in the Cloud. Select **Comments**.

4. Type a comment in the Comment Window and then click **Post**.

5. Select **Share**.

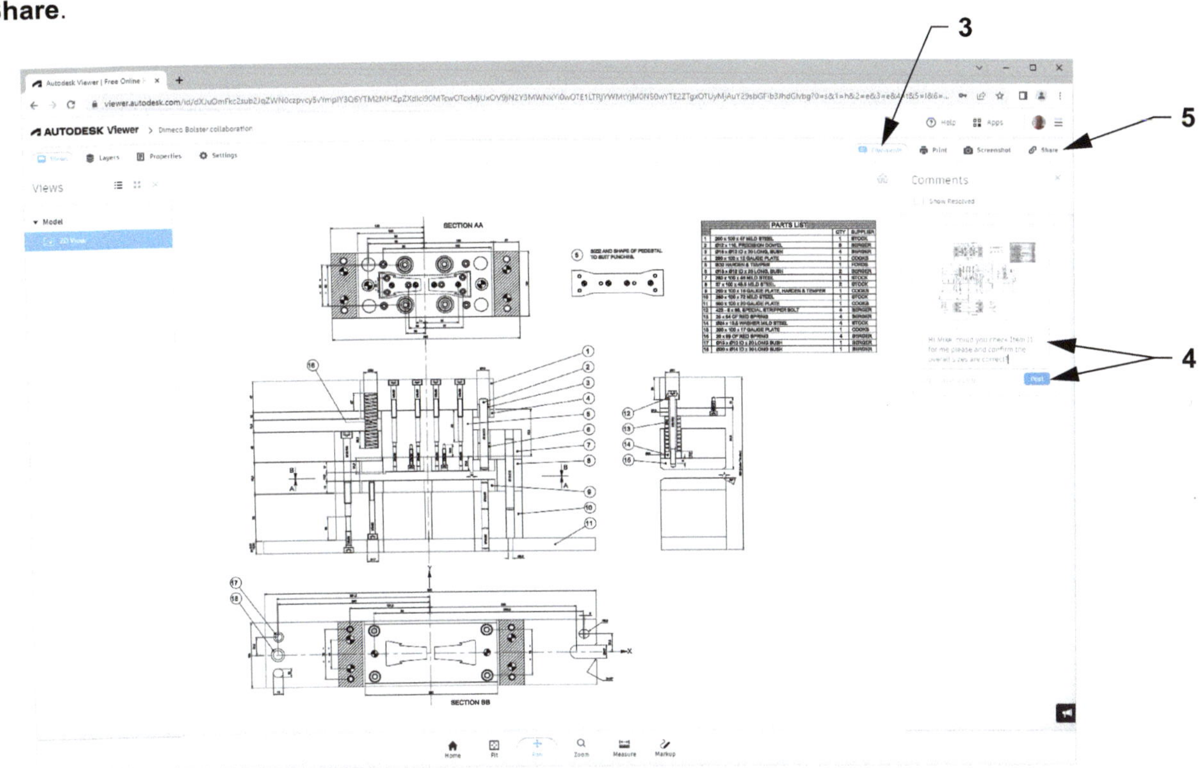

6. Select **Copy**, which copies the link to your shared view in the Cloud.

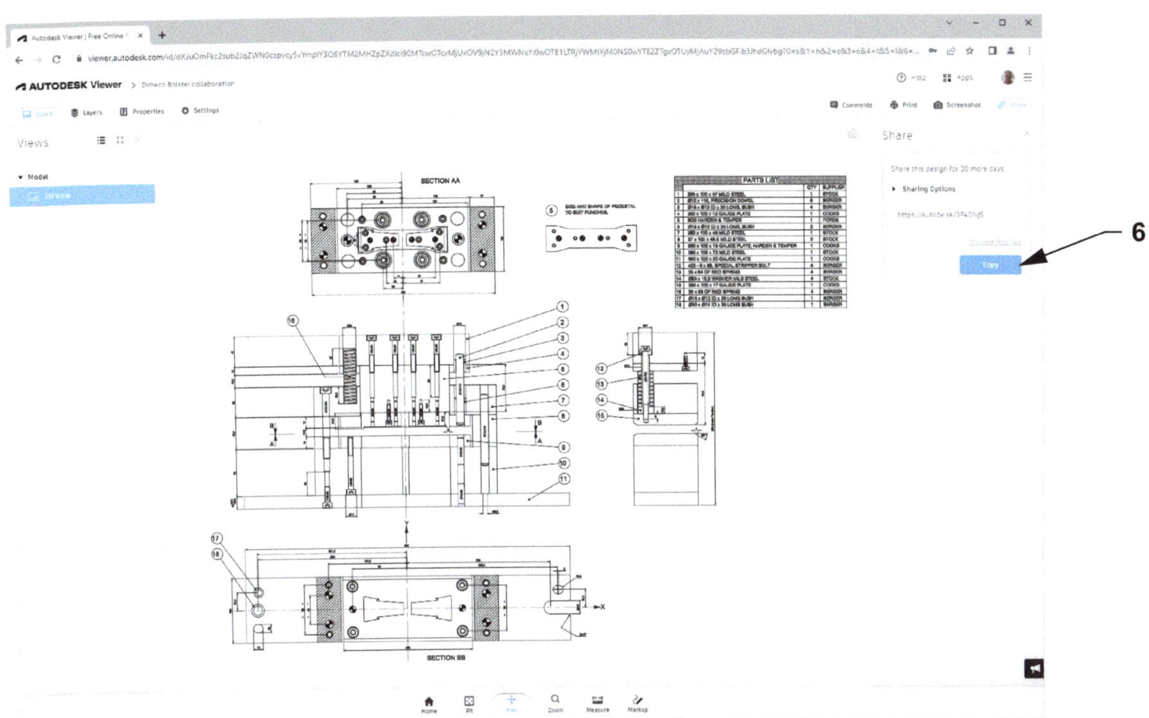

7. You can now send the copied link to another user so they can view your shared view and make any comments to share with you if necessary.

Continued on the next page...

8. Select the **Refresh** button to see your comment in the Shared Views Palette.

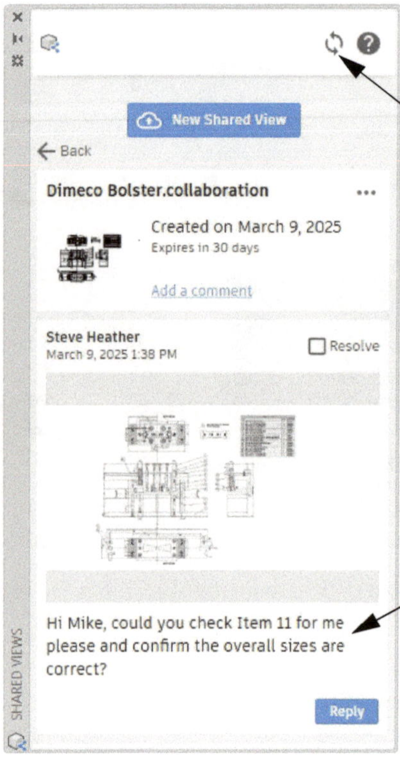

8

Your comment will now appear in the Window.

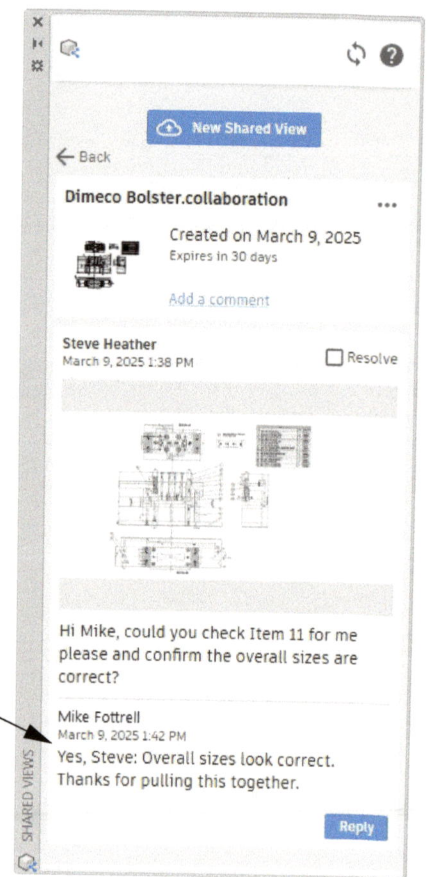

If the user you shared the view with makes a comment, it will appear in the Shared Views Palette Window. You will also receive an email notification that another user has made a comment on your shared view.

CAD TIP! In the future, if you want to add a comment on the shared view, just click **Reply** to enter your comment in the Shared Views Palette Window.

Exercise 30A

Instructions:

1. Start a **New** file using the **My Decimal Setup.dwt**.
2. Draw the arc shown using **"Center, Start, End"**.

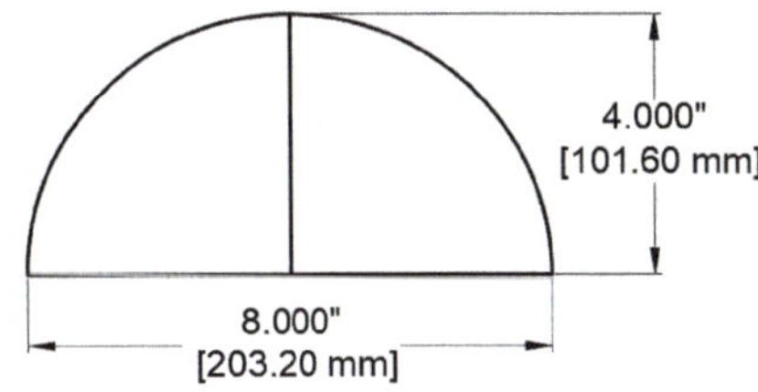

3. Create the **Arctext** shown.

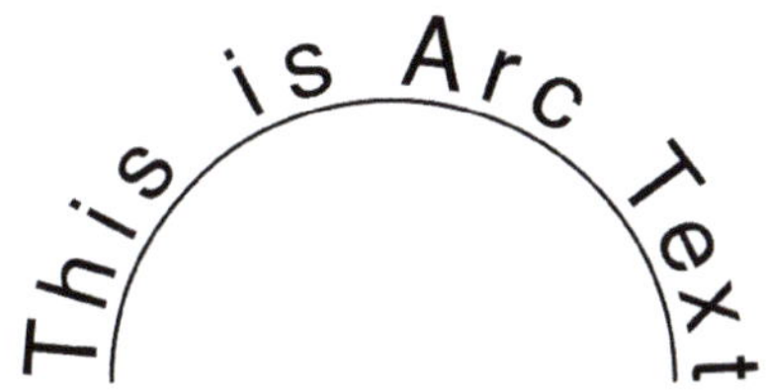

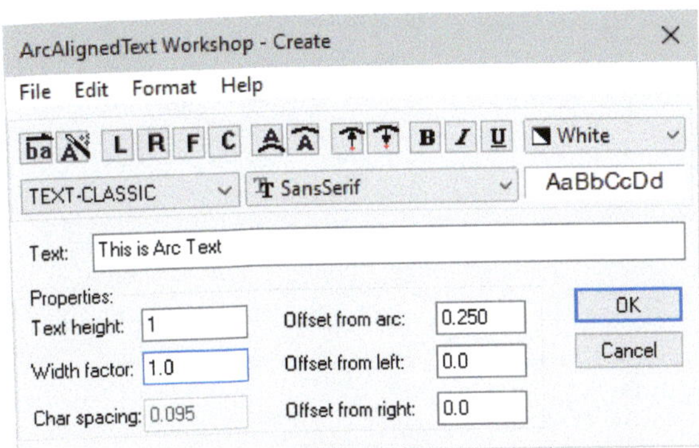

Inch users

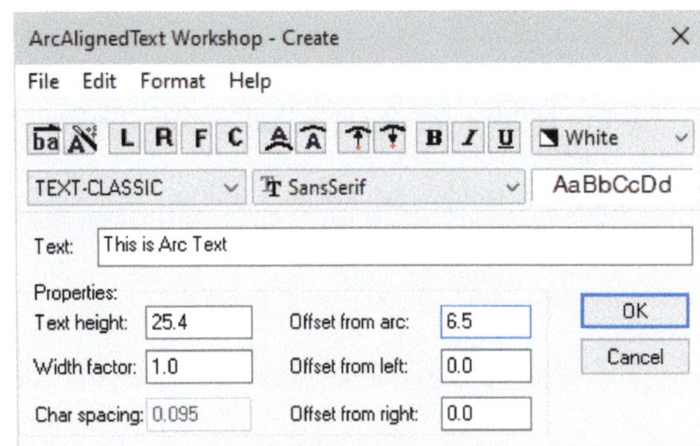

Metric users

4. **Save** as **Ex-30A**

Exercise 30B

Instructions:

1. Open **Ex-30A** (if not already open).
2. Add the text on the inside of the arc as shown.

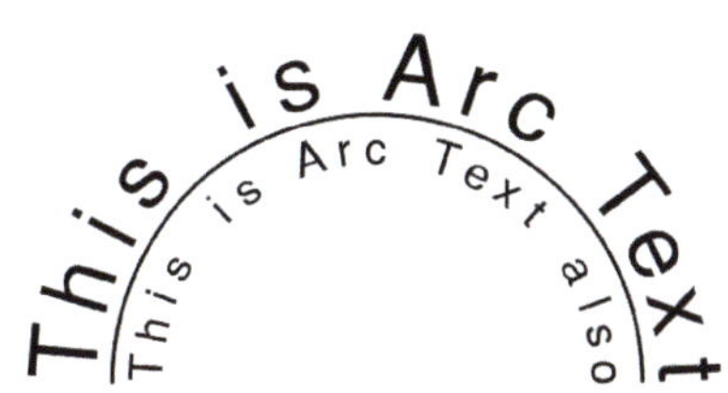

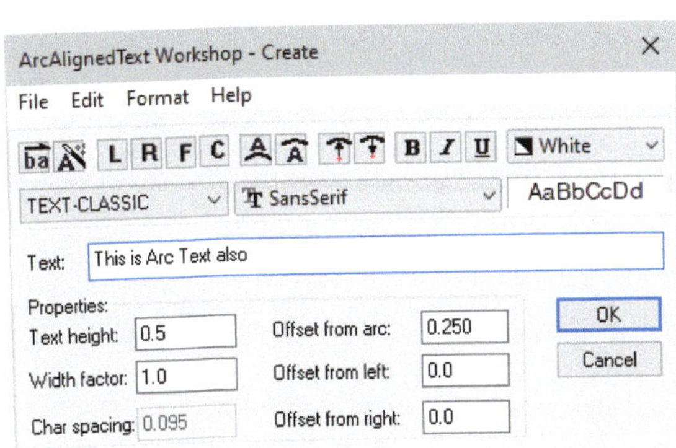

Inch users

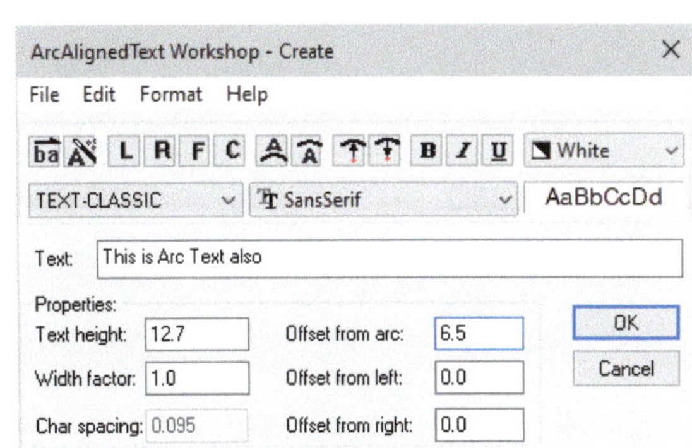

Metric users

3. **Save** as **Ex-30B**

Exercise 30C

Instructions:

1. Start a **New** file using the **My Decimal Setup.dwt**.
2. Draw the letters as shown below.
   A. Use text style TEXT-CLASSIC.
   B. Text height shown.

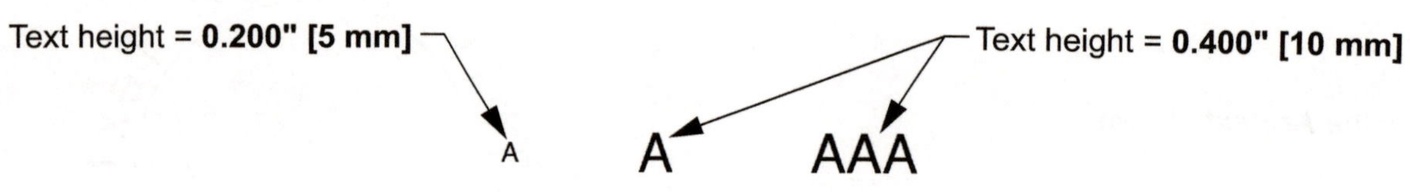

3. Copy the letters three times as shown below.
4. **Enclose** the letters as shown below.

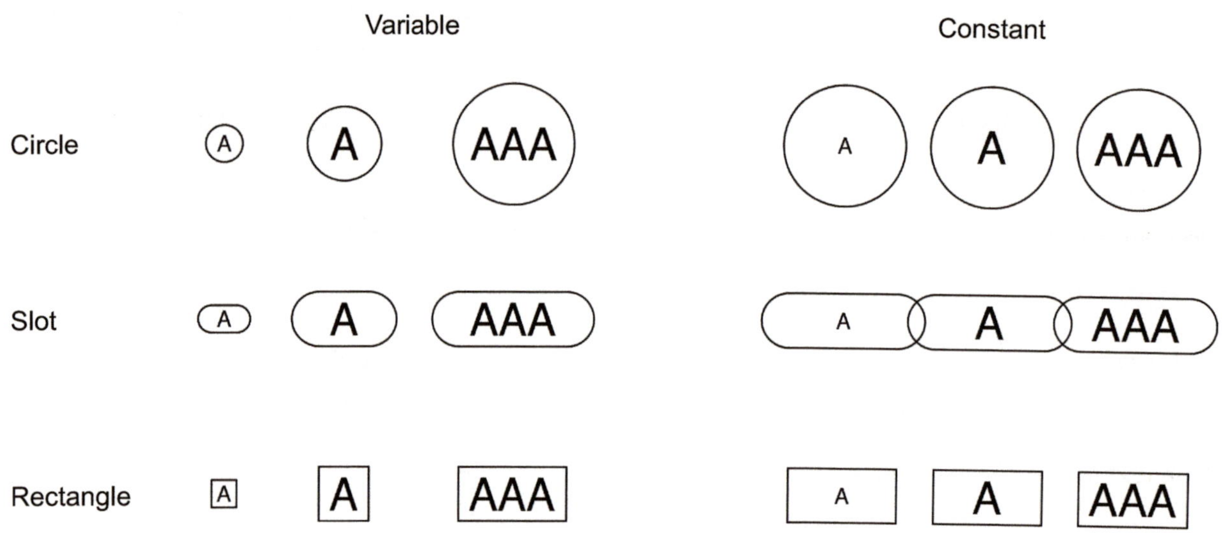

5. **Save** as **Ex-30C**

Exercise 30D

Instructions:

1. Start a **New** file using the **My Decimal Setup.dwt**.
2. Draw the objects shown below approximately as shown.

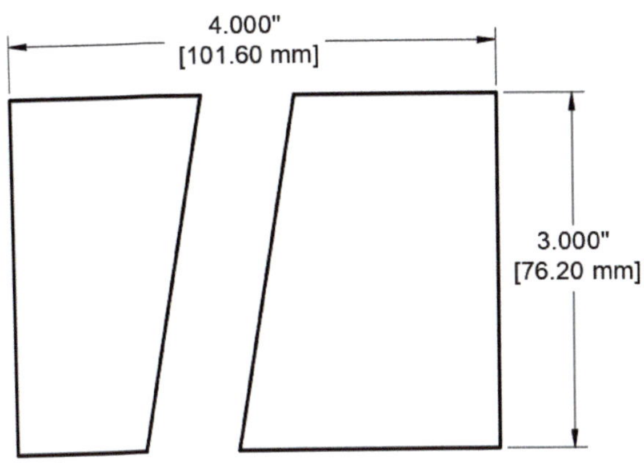

3. Add the **Break-line Symbol** to both lines as shown below.

   Note: Use Object Snap **Nearest** to place the Break-line Symbol accurately on the line.

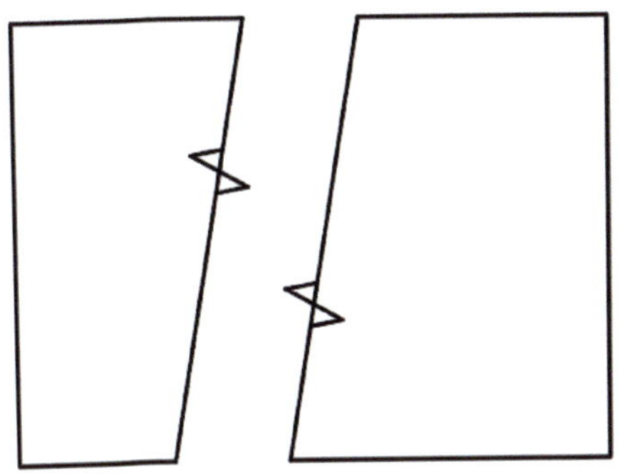

Size: **0.300" [7.6 mm]**

Extension: **0.180" [4.6 mm]**

4. Delete the Line under the Break-line Symbol and add the Hatch as shown below.

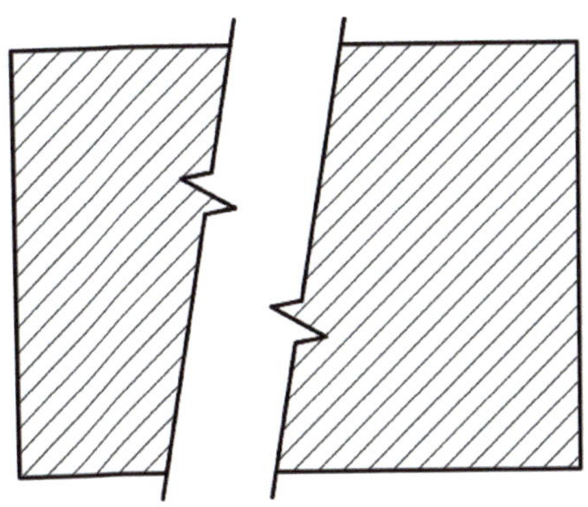

Hatch Type: **ANSI31**

Scale: **1.000**

Delete the line before Hatch.

5. **Save** as **Ex-30D**

Exercise 30E

Instructions:

1. Open **Ex-10B** and then immediately save it as **Ex-30E**.
2. Erase the large center circle as shown below.

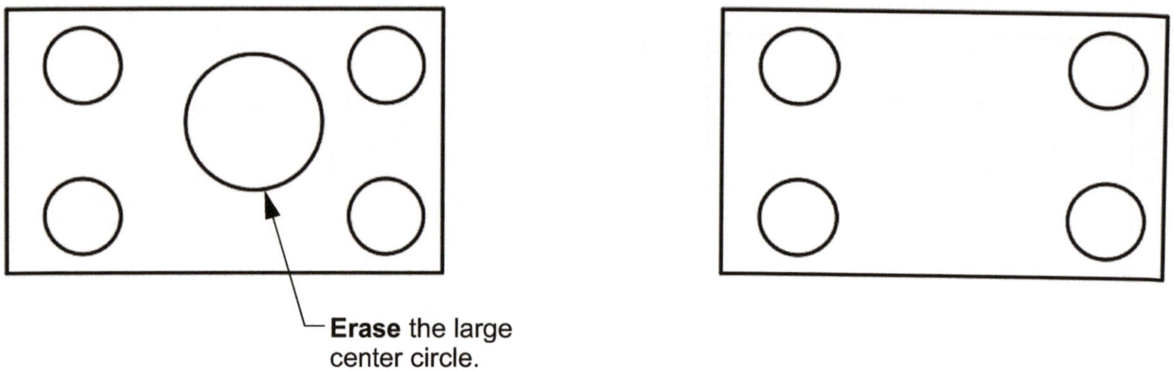

3. Select the **Drawing Compare** tool on the **Collaborate** Tab.
4. In the **Select a drawing to compare** dialog box, locate and select **Ex-10B**.
5. Select the **Import Objects** tool on the **DWG Compare** Toolbar.
6. Left click on the large center circle enclosed in a Revision Cloud and press ***<Enter>***.

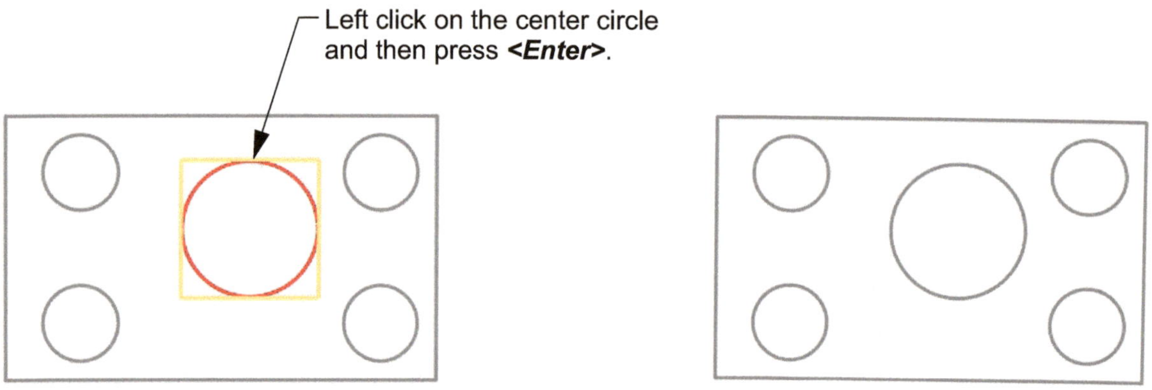

7. Select the **Exit Compare** tool on the **DWG Compare** Toolbar, to end the compare session.
8. **Re-save** as **Ex-30E**

Appendix A: Add a Printer / Plotter

The following are step-by-step instructions on how to configure AutoCAD for your printer or plotter. These instructions assume you are a single system user. If you are networked or need more detailed information, please refer to the Autodesk Assistant or AutoCAD Help Menu.

Note: You can configure AutoCAD for multiple printers. Configuring a printer makes it possible for AutoCAD to display the printing parameters for that printer.

1. Type in ***plottermanager*** and then press ***<Enter>***.
2. Select **Add-A-Plotter Wizard**.

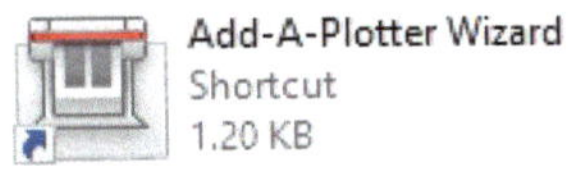

3. Select the **Next** button.

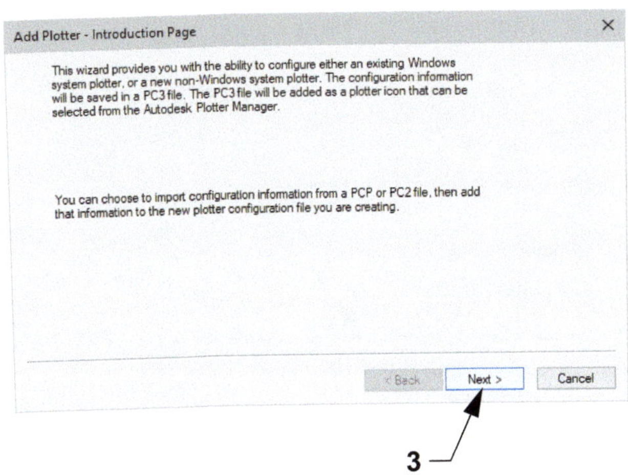

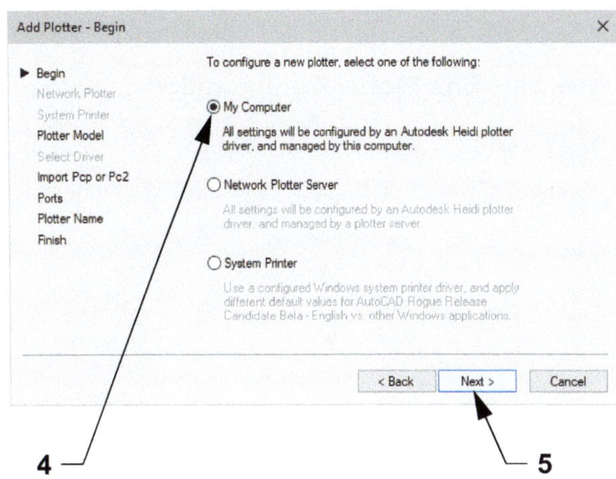

4. Select **My Computer**.
5. Select the **Next** button.

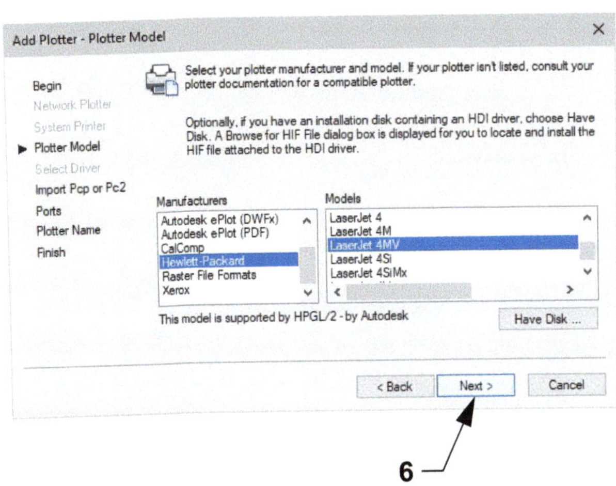

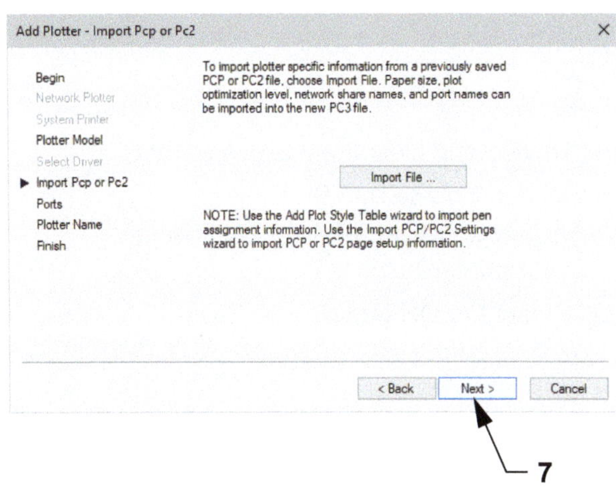

6. Select the **Manufacturer** and the specific **Model** desired and then select the **Next** button.

   (If you have a disk with the specific driver information, put the disk in the disk drive and select **"Have Disk ..."** button, then follow instructions.)

7. Select the **Next** button.

Continued on the next page...

8. Select:
   A. **Plot to a port**.
   B. Then select **Next**.

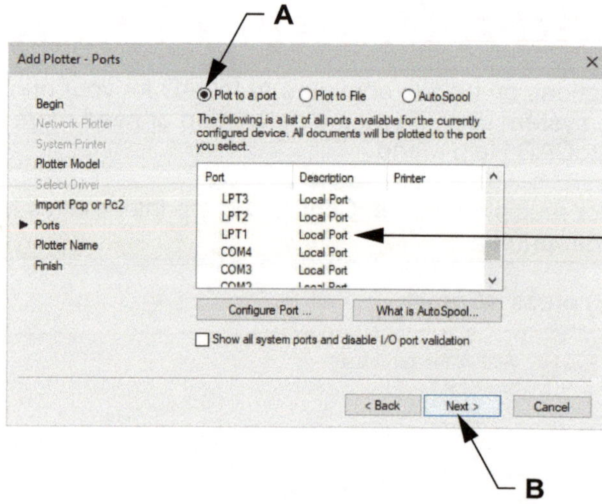

Select **LPT1** only if you are configuring your printer and it is attached to this computer.

Not necessary to select any if the selected printer is **not attached** to this computer.

9. The printer name that you previously selected should appear. Then select the **Next** button.

10. Select the **Edit Plotter Configuration ...** button.

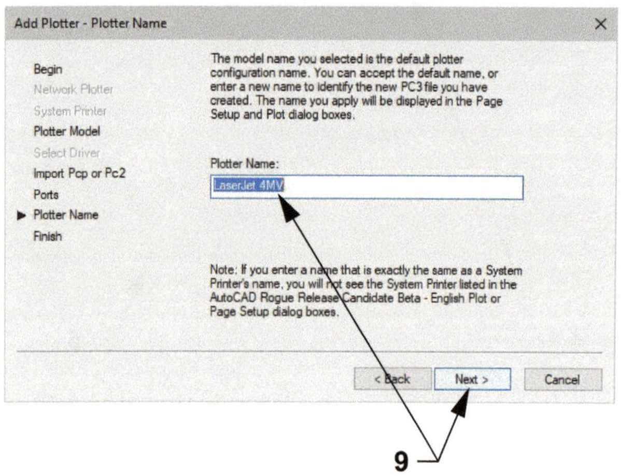

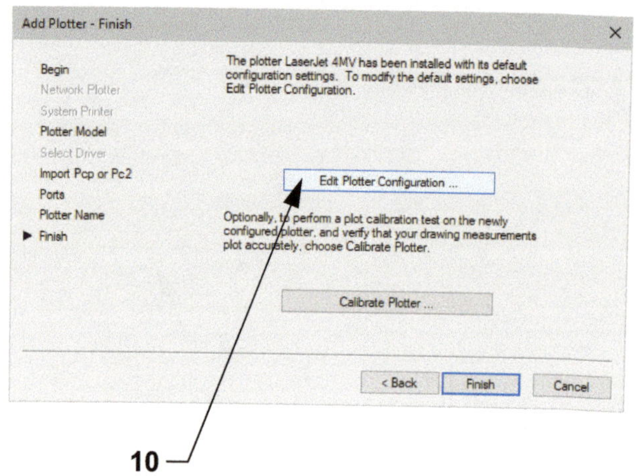

11. Select:
    A. Device and Document Settings Tab.
    B. Media: Source and Size.
    C. Size: (Select the appropriate size for your printer / plotter.)
    D. Select the **OK** button.

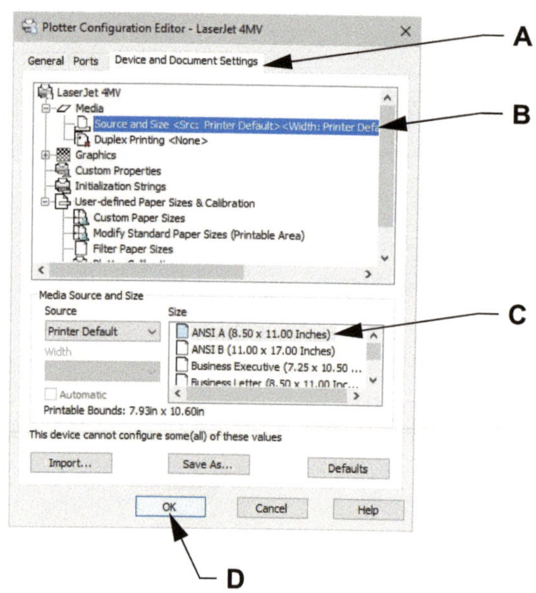

12. Select the **Finish** button.

13. Type in ***plottermanager*** and then press ***<Enter>***.

*Is the **printer / plotter** there in the list?*

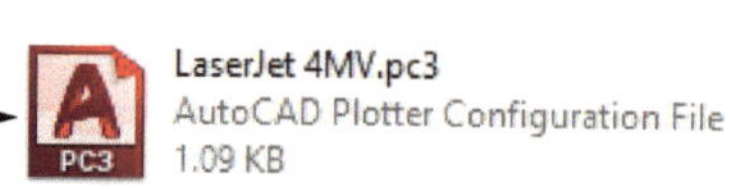

Appendix B: Dynamic Input

To help you keep your focus in the "drawing area", AutoCAD has provided a command interface called **Dynamic Input**. You may input information within the Dynamic Input instead of on the Command Line.

The first point displays the (Absolute: X, Y) distance from the Origin.

Enter the **X** dimension, press the ***<Tab>*** key, enter the **Y** dimension and then press ***<Enter>***.

The second and next points display (relative: distance and angle) from the last point entered. Enter the **distance**, press the ***<Tab>*** key, enter the **angle**, and press ***<Enter>***.

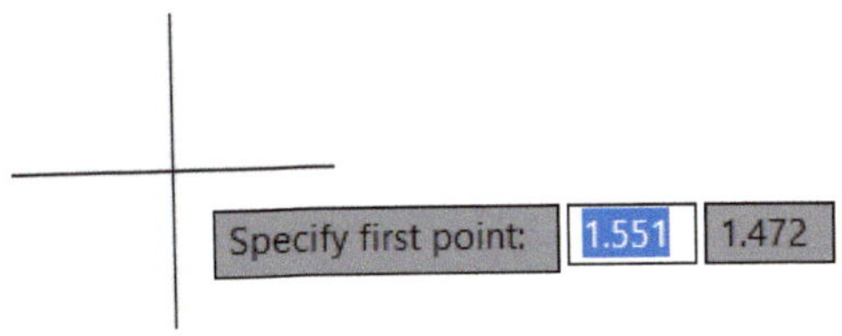

First point.

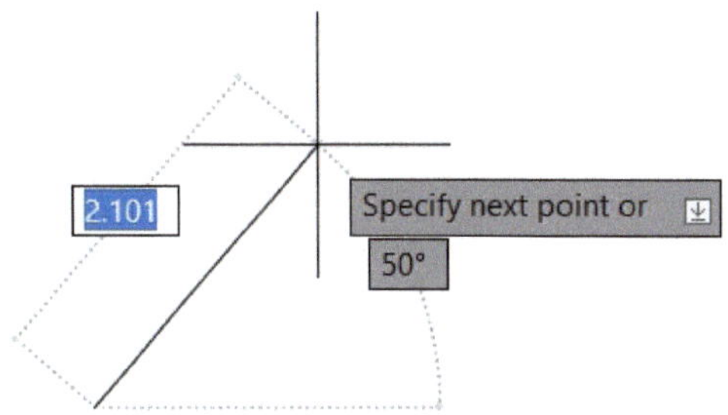

Second and next points.

Note: Some users find Dynamic Input useful, some find it distracting. After completing the steps on the following pages, you can decide whether you want to use it. It is your choice.

How to turn Dynamic Input on or off

Select the **Dynamic Input** button on the Status Bar or use the ***<F12>*** key.

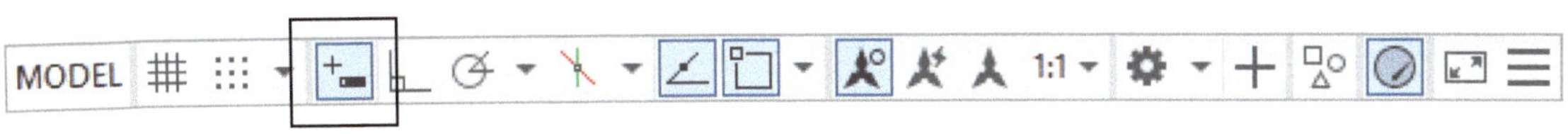

Note: Refer to page 1-12 for information on adding tools to the Status Bar.

Dynamic Input has three components

1. Pointer Input.
2. Dimensional Input.
3. Dynamic Prompts.

You may control what is displayed by each component and turn each on or off.

Refer to the following pages for more detailed descriptions.

Pointer Input

Pointer Input is only displayed for the **first** point.

When Pointer Input is enabled (**on**) and a command has been selected, the location of the crosshairs is displayed as coordinates in a tooltip near the cursor.

Example:

1. Select the **Line** command.

2. Move the cursor.

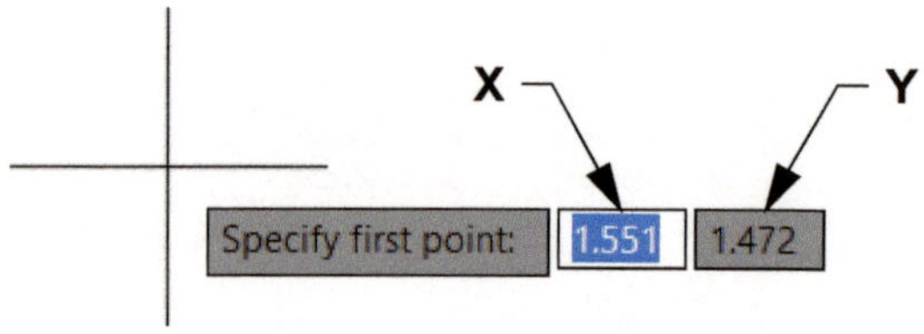

The two boxes display the cursor location as **X** and **Y** coordinates **from the Origin**. (Absolute coordinates.)

3. You may move the cursor until these coordinates display the desired location and press the left mouse button or enter the desired coordinate values in the tooltip boxes instead of on the Command Line.

How to change Pointer Input settings

1. Right click the **Dynamic Input** button on the Status Bar and select **Dynamic Input Settings**.

2. In the **Drafting Settings** dialog box select the **Dynamic Input** Tab.

3. Under **Enable Pointer Input**, select the **Settings** button.

4. In the **Pointer Input Settings** dialog box select:

   Format: (Select one as the display format for **second** or **next points**.)

   Polar or **Cartesian** format.

   Relative or **Absolute** coordinates.

   Visibility: (Select one for tooltip display.)

   As soon as I type coordinate data. When pointer input is turned on, displays tooltips only when you start to enter coordinate data.

   When a command asks for a point. When pointer input is turned on, displays tooltips whenever a command prompts for a point.

   Always - even when not in a command. Always displays tooltips when pointer input is turned on.

5. Select **OK** to close each dialog box.

Dimensional Input

When Dimensional Input is enabled (**on**), the tooltips display the distance and angle values for the **second** and **next points**.

The default display is **Relative Polar** coordinates.

Dimensional Input is available for: **Arc**, **Circle**, **Ellipse**, **Line**, and **Polyline**.

Example:

1. Select the **Line** command.
2. Enter the first point.

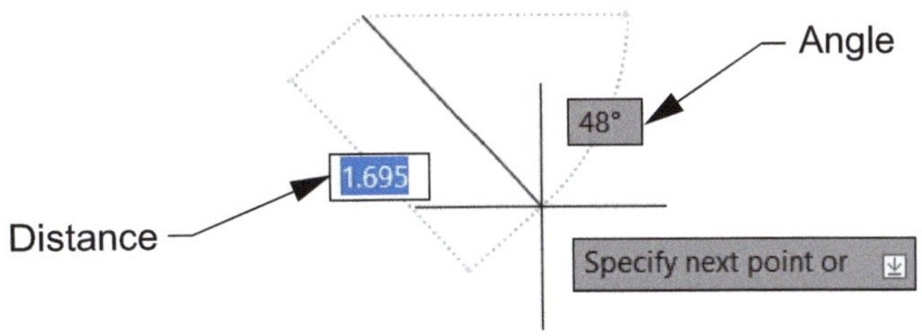

3. Move the cursor in the direction you desire and enter the **distance**, ***<Tab>*** key, and the **angle** ***<Enter>***.

How to change Dimensional Input settings

1. Right click the **Dynamic Input** button on the Status Bar and select **Dynamic Input Settings**.
2. In the **Drafting Settings** dialog box select the **Dynamic Input** Tab.
3. Under **Dimension Input**, select the **Settings** button.
4. In the **Dimension Input Settings** dialog box select:

   Visibility: (Select one for tooltip display.)

   Show only one dimension input field at a time. Displays only the distance dimensional input tooltip when you are using grip editing to stretch an object.

   Show two dimension input fields at a time. Displays the distance and angle dimensional input tooltips when you are using grip editing to stretch an object.

   Show the following dimension input fields simultaneously. Displays the selected dimensional input tooltips when you are using grip editing to stretch an object. Select one or more of the check boxes.

5. Select **OK** to close each dialog box.

Dynamic Prompts

When Dynamic Prompts are enabled (**on**), prompts are displayed.

You may enter a response in the tooltip box instead of on the Command Line.

Example:

1. Select the **Circle** command.
2. Press the right mouse button for command options or you may also press the down arrow to view command options. Select an option by clicking on it or press the up arrow for the option menu to disappear.

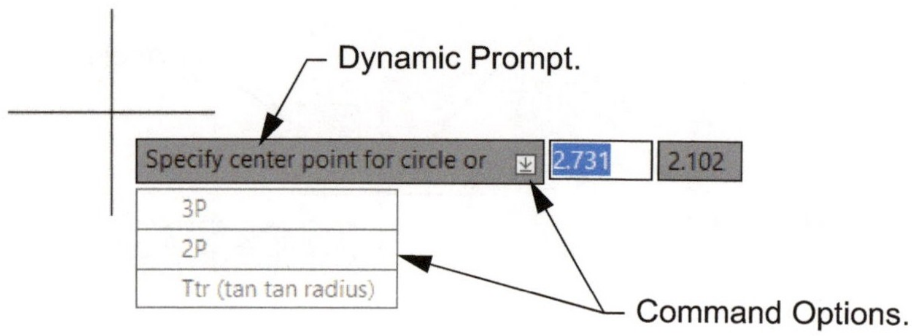

How to change the Color, Size or Transparency of tooltips

1. Right click the **Dynamic Input** button on the Status Bar, then select **Dynamic Input Settings**.
2. In the **Drafting Settings** dialog box, select the **Dynamic Input** Tab.
3. At the bottom of the dialog box, select the **Drafting Tooltip Appearance** button.
4. Select the **Colors** button. Under **Context:** select **2D Model Space**. Under **Interface element:** select **Drafting tool tip** or **Drafting tool tip background**. Select a color from the Color drop-down list. Then select **Apply** and then the **Close** button.
5. Under **Size**, move the slider to the right to make tooltips larger, or to the left to make them smaller. The default value, **0**, is in the middle.

   (I like 2, a little larger than the default.)
6. Under Transparency, move the slider. The higher the setting, the more transparent the tooltip.
7. Under **Apply to**, choose an option:

   Visibility: (Select one for tooltip display.)

   Override OS settings for all drafting tooltips. Applies the settings to all tooltips, overriding the settings in the operating system.

   Use settings only for Dynamic Input tooltips. Applies the settings only to the drafting tooltips used in Dynamic Input.
8. Select **OK** to close each dialog box.

Appendix C: Frequently Asked Questions

Q. My Command Line has disappeared, how do I get it back?

A. Press ***<Ctrl+9>*** on your keyboard. You can also use the same method to turn off the Command Line.

Q. How do I change the Model Space background color?

A. Right click with your mouse and select **Options / Display** Tab **/ Colors** button, in the **Context** Window select **2D Model Space**, then select the **Color** drop-down list and change the color. Select **Apply & Close**. You can restore the original colors by selecting **Restore classic colors**.

Q. How do I change the AutoCAD theme color?

A. Right click with your mouse and select **Options / Display** Tab, in the **Window Elements** Window select the **Color scheme** drop-down list, then change the theme color to either **Dark** or **Light**. Select **Apply**, then **OK**.

Q. How do I reset AutoCAD to default factory settings?

WARNING!: Resetting AutoCAD to default factory settings will remove any custom templates, Toolbars, Ribbon Tabs, and any modifications you have made. If you have any custom templates, it is advisable to save them in a different location before resetting.

A. To reset AutoCAD to the default factory settings using Windows 10:

1. Select the Start Menu.
2. Scroll down and select the AutoCAD 2026 folder.
3. Select **"Reset Settings to Default"**.

A. To reset AutoCAD to the default factory settings using Windows 11:

1. Select the Start Menu.
2. Select the **All apps** button.
3. Scroll down and select the AutoCAD 2026 folder.
4. Select **"Reset Settings to Default"**.

Note: While previous versions of AutoCAD could be used with earlier operating systems, AutoCAD 2026 is not compatible with anything before Windows 10.

Notes:

Appendix D: AutoCAD Quick Key Guide

Toggle (⇔) Functions

F1 ⇔ **Display Help**
F2 ⇔ **Extended Command History List**
F3 ⇔ **2D Object Snap Mode**
F4 ⇔ **3D Object Snap Mode**
F5 ⇔ **Isoplanes**
F6 ⇔ **Dynamic UCS**
F7 ⇔ **Grid Mode**
F8 ⇔ **Ortho Mode**
F9 ⇔ **Snap Mode**
F10 ⇔ **Polar Mode**
F11 ⇔ **Object Snap Tracking**
F12 ⇔ **Dynamic Input**

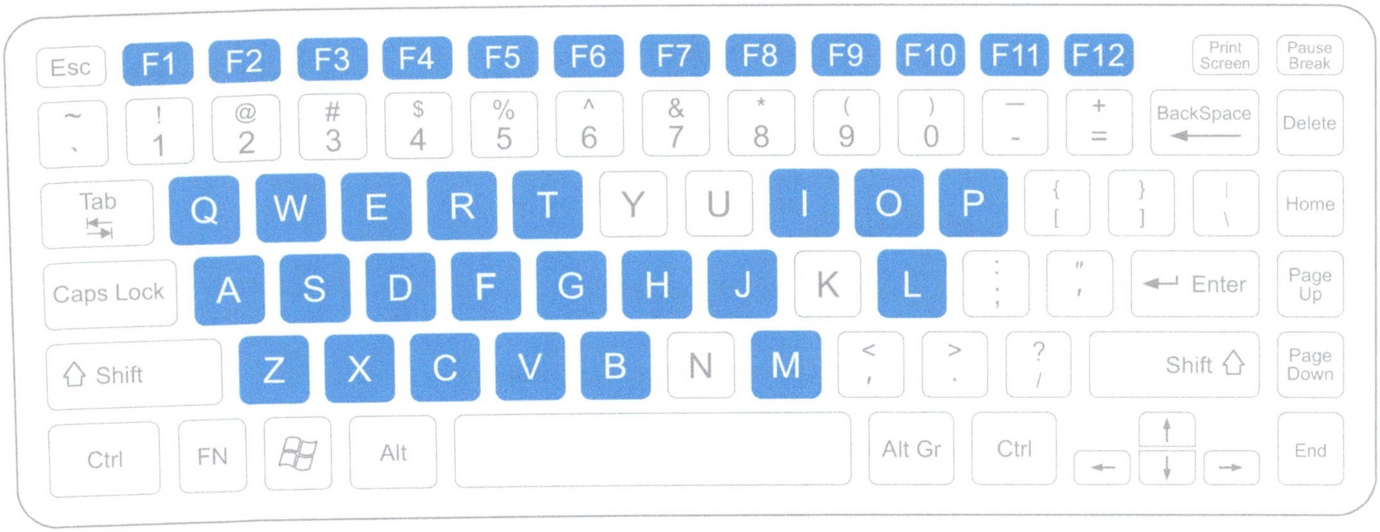

One-Key Shortcuts (After entering the one-key shortcut, press *<Enter>*.)

A	**ARC**	Arc created.
B	**BLOCK**	Block definition created from selected objects.
C	**CIRCLE**	Circle created.
D	**DIMSTYLE**	Dimension style created and modified.
E	**ERASE**	Erases objects(s) from drawing.
F	**FILLET**	Fillets (rounds) edges.
G	**GROUP**	Groups selected objects and saves as group.
H	**HATCH**	Hatch pattern or solid or gradient shade fills area or object.
I	**INSERT**	Inserts a block or drawing.
J	**JOIN**	Joins objects to form one object.
L	**LINE**	Line segment(s) created.
M	**MOVE**	Moves one or more object(s) a specific distance in a specified direction.
O	**OFFSET**	Offset circles and parallel lines and curves created.
P	**PAN**	For panning around a drawing.
Q	**QSAVE**	Quick saves current drawing.
R	**REDRAW**	For redrawing, clears display in current viewport.
S	**STRETCH**	Stretches object(s) selected (crossed with selection window or polygon).
T	**MTEXT**	Multiline text object(s) created.
V	**VIEW**	Views (camera, layout, named, and present) saved and restored.
W	**WBLOCK**	Writes block (collected objects) to new drawing file.
X	**EXPLODE**	Separates object into its constituent parts.
Z	**ZOOM**	Zooms in or out of current viewport.

Drawing Management Keyboard Combinations

<Ctrl> + A	Select all.
<Ctrl> + N	Create new.
<Ctrl> + O	Open existing.
<Ctrl> + P	Plot dialog box.
<Ctrl> + Q	Quit (close drawing, but not save).
<Ctrl> + S	Save.
<Ctrl> + *<Page Down>*	Page to next Tab.
<Ctrl> + *<Page Up>*	Page to previous Tab.
<Ctrl> + *<Tab>*	Go to next file.
<Ctrl> + *<Shift>* + *<Tab>*	Go to previous file.

Workflow Keyboard Combinations

<Esc>	Cancel current command.
<Ctrl> + C	Copy.
<Ctrl> + *<Shift>* + C	Copy to clipboard (using basepoint).
<Ctrl> + V	Paste object(s).
<Ctrl> + *<Shift>* + V	Paste block(s).
<Ctrl> + X	Cut object(s).
<Ctrl> + Y	Redo.
<Ctrl> + Z	Undo.

Other General Keyboard Combinations

<Ctrl> + 0	Clears the screen of all Ribbons and Toolbars.
<Ctrl> + 1	Property Palette on/off.
<Ctrl> + 3	Tool Palette on/off.
<Ctrl> + 9	Command Line on/off.
<Ctrl> + F	Running Object Snaps on/off.
<Ctrl> + G	Grid on/off.

INDEX

D

E

F

G

H

I

U

V

W

X, Y, Z

Final Notes about AutoCAD®

Whether a student, teacher, professional, or home user, you can make efficiency a daily part of the task at hand with AutoCAD® software. Meticulously refined with the drafter in mind, AutoCAD facilitates efficient day-to-day drafting with features that increase speed and accuracy while saving time. No matter what you are drawing, you will find that AutoCAD will help you achieve an unmatched level of aesthetic precision and professionalism in delivering your final product.

What's New in AutoCAD 2026 and this *Exercise Workbook*?

As ever, the latest version of AutoCAD builds and improves on the versatile features of previous program releases. And this *AutoCAD® 2026 Exercise Workbook* includes updated explanations, illustrations, and instructions to explain each feature.

The most significant addition to this latest AutoCAD release is Activity Insights. A new palette gives individual users and collaborators access to a helpful summary of what has been done to a drawing in each session. Plus, the results can be filtered by time, users, and activities. (See Lesson 2.)

On top of upgrading this edition of the *Exercise Workbook* to vibrant, full color, numerous improvements have been made throughout the book. For instance, based on feedback from users of the extensive *AutoCAD® 2D Resource Kit*, we have added an AutoCAD Quick Key Guide, an invaluable list of function keys, time-saving, one-key shortcuts, and helpful keyboard combinations for the most efficient use of AutoCAD. (See Appendix D.)

Preset Drawings Accompanying this Book

To get the supplied drawings that go with this book, download the file:
workbook-helper.zip
from our website:
https://books.industrialpress.com/autocad/workbook-helper.zip

Enter the address into your web browser and the download will start automatically. Once the file has been downloaded, you can unzip it to extract both drawing files.

AutoCAD Versions

Note that newer editions of AutoCAD can exist with earlier versions of AutoCAD and AutoCAD LT. But you may be limited to using certain versions based on your operating system. AutoCAD 2026 cannot be used with operating systems earlier than Windows 10.

For information on available trial versions and to purchase AutoCAD software, please visit Autodesk's web page at **www.autodesk.com**.

If you are attending a school, check with your instructor regarding potential educational discounts.

For More Information

For a list of other titles in this bestselling series of *AutoCAD® Exercise Workbooks* published by Industrial Press, as well as related digital offerings, see page iv at the front of this book.

For further information about these and other Industrial Press products, please visit us at **books.industrialpress.com** and **ebooks.industrialpress.com**.